A RESOURCE-BASED HABITAT VIEW FOR CONSERVATION

Butterflies in the British Landscape

To the memory of my uncle Ian Scott Hawke Dennis

When attending his funeral in 1983 my attention was drawn to the images of three butterflies in the stained glass window in Bickerton Church, near Chester, constructed in thanksgiving for his recovery from losing both his legs at Tours, France in 1938. Some artistic licence is evident in the butterfly images, but one is unmistakably a red admiral whilst the others depict a fritillary and a small pierid. The coincidence of butterflies in this church memorial, a rare sight, and my, by then, ingrained interest in butterflies made a lasting impression on me as did his fortitude, tenacity and humour in dealing with life's nastier surprises.

A RESOURCE-BASED HABITAT VIEW FOR CONSERVATION

BUTTERFLIES IN THE BRITISH LANDSCAPE

Roger L. H. Dennis

NERC Centre for Ecology and Hydrology, Maclean Building, Benson Lane, Crowmarsh Gifford, Wallingford, Oxon OX10 8BB, UK

and

Institute for Environment, Sustainability and Regeneration, Room s122, Mellor Building, Staffordshire University, College Road, Stoke on Trent ST4 2DE, UK

WILEY-BLACKWELL

A John Wiley & Sons, Ltd., Publication

Blackwell Publishing was acquired by John Wiley & Sons in February 2007. Blackwell's publishing program has been merged with Wiley's global Scientific, Technical and Medical business to form Wiley-Blackwell.

Registered office: John Wiley & Sons Ltd, The Atrium, Southern Gate, Chichester, West Sussex, PO19 8SQ, UK

Editorial offices: 9600 Garsington Road, Oxford, OX4 2DQ, UK
The Atrium, Southern Gate, Chichester, West Sussex, PO19 8SQ, UK
111 River Street, Hoboken, NJ 07030-5774, USA

For details of our global editorial offices, for customer services and for information about how to apply for permission to reuse the copyright material in this book please see our website at www.wiley.com/wiley-blackwell

Library of Congress Cataloguing-in-Publication Data
Dennis, Roger L. H.
A resource-based habitat view for conservation : butterflies in the British landscape / Roger L. H. Dennis.
p. cm.
Includes bibliographical references and index.
ISBN 978-1-4051-9945-2 (hardback : alk. paper) 1. Butterflies – Habitat – Conservation – Great Britain. 2. Butterflies – Ecology – Great Britain. 3. Butterflies – Monitoring – Great Britain. 4. Wildlife conservation – Great Britain. I. Title.
QL555.G7D464 2010
639.9′75789–dc22

2009028752

ISBN 978-1-4051-9945-2 (hb)

A catalogue record for this book is available from the British Library.

Set in 9/11pt Photina by Graphicraft Limited, Hong Kong
Printed and bound in Malaysia by Vivar Printing Sdn Bhd

1 2010

CONTENTS

FOREWORD

Butterflies are arguably the most well-known group of insects and in Britain have been studied by naturalists for over 300 years. Our knowledge has grown exponentially in recent decades as researchers have used them to examine pressing issues ranging from population ecology and evolution to biodiversity loss, habitat fragmentation and climate change. Yet, such is the complexity of the natural world that, despite this immense knowledge, huge gaps still remain in our understanding. This book aims to fill one of those gaps, on the vital topic of butterflies and their habitats.

The book is both timely and important. Butterflies are in steep decline and are among our most threatened wildlife groups. Losses have been far greater than in other well-recorded groups such as plants and birds. Butterflies are also valuable environmental indicators and their declines reflect a huge loss of biodiversity. If we can understand the causes and reverse the decline of butterflies, we can help a wide range of other species.

It has been my pleasure to have known the author, Roger Dennis, for all my working life. We are exceptionally fortunate that he has brought his encyclopaedic knowledge of butterflies to bear on such a pressing conservation issue. The book brings together a wealth of information and blows away some woolly thinking on the many interacting factors that make up a butterfly habitat. It is the culmination of a lifetime's work and painstaking observations from one of our leading researchers.

With characteristic generosity, Roger has offered to donate all the royalties from the book towards the work of Butterfly Conservation. The book will thus help conservation in two ways: firstly by providing an incredibly valuable synthesis that can underpin our work; and secondly by providing funds for us to take practical action through our landscape-scale projects. I am extremely grateful on both counts and am sure the book will become a benchmark for researchers for many generations to come.

Dr Martin Warren
Chief Executive
Butterfly Conservation

PREFACE

This book has two clearly defined and linked objectives. First, to outline concepts and issues associated with a resource-based approach to animal habitats; second, to present an intricate view of butterfly habitats, specifically those in Britain which have been studied to a greater depth than anywhere else on Earth. These beautiful creatures, indicators for the health of many other organisms, are rapidly declining and under substantial threat from increasing human pressures. It is hoped, then, that this book will encourage a deeper appreciation of butterfly habitats, the resources they need, and advance their conservation. Butterfly enthusiasts who have read E. B. Ford's exquisite *Butterflies* must have pondered about his two beautifully-worded sentences on the enigma of the butterfly habitat (see Chapter 1 leader). The mystery perhaps is that it has taken so long to conjure up a formula for identifying a butterfly's habitat. The solution provided here is one that everyone should be able to appreciate without much difficulty. Each of us can enumerate cherished resources that ensure our own well-being and survival, and that enhance our lives. For butterflies, a habitat is no different: places for food and sites for roosting, meeting mates and basking, all within easy reach of one another. What we have to remember about butterflies is that as holometabolous insects, their development, their lives, are characterized by a distinct metamorphosis, with well-defined larval and adult stages triggered by environmental cues. Eggs, larvae, pupae and adults tend to occur at distinct times of the year. Thus, any understanding of habitat has to absorb stage-defined activities and requirements; larvae primarily form the vehicle for growth, developing in phases (called instars), and adults are the active stage, focused on mating and seeking out plant resources for their young. Larvae and adults have very different mobility, thus the space–time scales of resource packages for them differ. The concept of habitat herein engages this view of resource units; an analogy is the shift in physics from the atom to subatomic particles – once understood to exist, they cannot thereafter be ignored.

The resource-based or functional view of habitats occurred to me in 1996 when writing the text for a booklet, produced with Tim Shreeve, on the butterflies of the smaller British islands. It arose out of a need to be clear about what we meant by migration. In the process of considering Robin Baker's succinct and functionally apt definition for migration (movement of individuals between habitat units), I became aware that it was necessary to determine exactly what was meant by a habitat unit. Casual doodling using Venn diagrams resulted in a figure in the island booklet (Dennis and Shreeve, 1996:45) that forms the basis for the idea; the concept and its implications were expanded on in later papers with three colleagues, Tim Shreeve, Hans Van Dyck and David Sheppard (Dennis *et al.*, 2003, 2006b, 2007; Shreeve *et al.*, 2004).

The format for this book is an unusual one and very different from those I have previously produced. The key feature of each chapter is a set of statements, referred to perhaps over-optimistically as principles, but which can be interpreted simply as points or ideas for discussion. These principles are highlighted, bulleted and bold, and this gives them a formal appearance. But, they are not laws – nothing so grand – and there is no excuse for treating them as such. Some principles may essentially define rules, which of course have exceptions. Each principle is expanded on in subsequent paragraphs but briefly; to my immense frustration much more could have been added. It is hoped that the points made provide a platform for discussion and argument (essay topics), for tests and experiments (dissertations, projects). The reason this format was chosen is that an attempt is made to simplify what is understandably a complex subject. Butterflies are but one component of ecosystems, a term coined for environmental systems by Tansley in 1935 to describe the intricate relationships of organisms with each other and the abiotic environment.

Following four introductory chapters, the book investigates butterfly habitats in a variety of contexts, particularly against a background of environmental variation in space and over time. The aim here is to provide a context for explanation. Yet, explanation in ecology is no simple matter. It involves many variables, some proximal in causation, others embedded more deeply in the causal nexus; any may cause a system failure, and the explanatory variables vary in significance with changes in space–time scales. Explanation is thus plucked precariously from intricate webs of cause and effect, and one should try to visualize problems in a general systems framework (von Bertalanffy, 1951), which we may crudely envisage as boxes linked by flows. Therefore, against my nature, I have broken up this web into statements so that bits of the subject can be chewed over one at a time. In doing so, enthymemes (missing premises) become a real hazard, and the reader should always consider what other factors may affect a given relationship or statement.

I should here point out what this book tries to do and does not do. It is an introduction to terms (shown in bold), concepts and ideas and is not a manual of techniques or resources for different species; nor does it entertain axiology or political statements. That is surely a next step. The basic objective of the book is to strip the epidermis off habitat understood as vegetation so that the layers of complexity beneath can be plainly inspected. How does our knowledge on butterfly resources stand up to this examination? A scan through the appendices, kindly collated by my colleagues Peter Hardy and Tim Shreeve, provides a depressing answer to this question: there are substantial gaps in our knowledge, at least in print. I should point out that new references for the book, including some in press, were adopted up to 1 January 2009. I suspect, from making enquiries for the appendices, that much of this information does actually exist amongst the many knowledgeable amateur entomologists we have in the UK. It is essential that this is drawn into a single database for use by Butterfly Conservation. I have omitted concepts that could not be followed by a UK 'A' level geography or biology student. It has to be said that they would have a familiarity with numerical techniques. Reference is made to different statistical methods, and the content assumes that readers understand the significance of the concept of variation in organisms – for plants as well as animals – and their resources (range of values, quartiles, variance) for evolution, and of statistical significance itself. The adage that 'nothing remains constant for long' should become apparent.

In trying to explain what butterflies need and do, there is a danger of spoiling the magic for many readers of observing butterflies in the countryside. In defence, I would refer to the distinction made by Hospers in 1946 (see Lockwood and Latchininsky, 2008) between a thin and thick sense of beauty. A thin sense of beauty pertains to our sensual enjoyment of a place or object by virtue of its 'appearance'. A thick sense of beauty involves an engagement with the qualities and values that this 'appearance' conveys to an observer. In current vulgar usage these terms are perhaps unfortunate; there may be a suggestion of superficial in 'thin' and of insensitive in 'thick', but there is no reason why a capacity to enjoy should not be enhanced through understanding the fine detail. It is hoped that the approach here will give pleasure to those who enjoy inquiring deeper into the natural world and form the basis for conserving not just butterflies but all organisms. There is a precedent in that there is still much to learn and to secure for future generations.

RLHD
March 2009

Please go to www.wiley.com/go/dennis/butterflies for further resources

ACKNOWLEDGEMENTS

No scientific book is entirely one's own work. With great pleasure I have a number of colleagues and friends to thank for their help. First, there are those with whom I have worked closely over the years, Dr Tim G. Shreeve and Peter B. Hardy. Tim and I have been researching butterfly biology together since 1987; as a Reader in Biology at Oxford Brookes University, and from a lifetime of field experience, he always brings scientific rigour and realism to research ideas. We have enjoyed an immensely fruitful collaboration; there are very few with his extensive understanding of the broad sweep of butterfly biology in Europe. He has valuably scanned the entire book and has updated Table 3.1 and Appendices 5 and 6 with data from his own research programmes and those of his PhD students. Peter has been responsible for many of the appendices to the book; those on hostplants and nectar sources are a mine of information and have taken him years to accumulate. Peter is the ultimate conservationist – to nature he thinks and does no harm but an immense amount of good; he is one of Butterfly Conservation's dedicated atlas recording servants. Peter has been indulgent beyond the call of duty; I have concocted a survey or a project and he has willingly undertaken it, occasionally not knowing what I was after, but trusting that it was going to be useful and spending unquantifiable amounts of time amassing data!

Other friends have taken on extensive tasks. Eddie John, kindly taking on this task at a very difficult time for him, read through the entire text and attachments picking up, inevitably, the many errors I had made. Three other friends have also scrutinized parts of the text, Professor John W. Dover (Chapters 6 and 7), Dr John G. Hodgson (Chapter 8 and Appendix 2c) and Professor Tim R. New (much of the book, but Chapter 9 in detail); in the book readers will find the many contributions that John Dover has made over the years as a pioneer in landscape and agrarian butterfly ecology. Several people gamely contributed to the data in Appendix 6 (Dr Tim Shreeve, Peter Hardy, Dr Peter J. Russell, James (Jim) E. Pateman and Eddie John). My own daughter, Pamela, with help from my wife, Margaret, took on the unenviable task of formatting and checking the references and saved me, but not herself I suspect, many a headache. To these gallant friends I convey my grateful thanks as I do also to those below who have contributed in various ways to the book. The text has many illustrations. The vast majority of butterfly images, quite superb, are kindly provided by Peter Eeles; my thanks for his patience in dealing with my repeated requests. The reader will find the photographs and many more at his website: www.ukbutterflies.co.uk. Dr Barry Fox also provided a number of remarkable photographs from his PhD thesis on habitat and microhabitat aspects of the white admiral butterfly; readers will find his discoveries on this butterfly a revelation and my only regret is that there was not space to add more of them. Other illustrations and figures were contributed by Dr Jim Asher, Professor Michel Baguette, Dr Caroline R. Bulman (Butterfly Conservation: BC), Robin Curtis, Dr Brian N. K. Davis, H. T. Eales, Dr Ian S. Evans, Shane Farrell (from his site at http://www.flickr.com/photos/shane58), Professor Marie-Josée Fortin, Dr Adrian Fowles (Countryside Council for Wales: CCW), Nick Greatorex-Davies (who, with his typical kindness, undertook a dedicated trip to obtain the photograph for Fig. 7.4), Dr Jenny Hodgson, Crispin Holloway, Dr Nick J. B. Isaac (Centre for Ecology and Hydrology: CEH), Dr Jochen Jaeger (measures of landscape fragmentation), Dr Keith Porter, Professor Jeremy A. Thomas and Professor Hans Van Dyck. Ken J. Willmott, very trustingly, sent a copy of his confidential, fascinating report on the purple emperor for me to read. Other colleagues have patiently answered my enquiries or undertaken other work: Dr Keith Bland (hibernating small tortoiseshells in Scotland), Dr Tom M. Brereton

(BC), Dr Leonardo Dapporto (illustrations), Harry Eales (Northumberland sites), Dr Simone Fattorini (illustrations), Richard Fox (BC; corrections to Appendix 1), William Gill (Brereton Heath, Cheshire survey of alder buckthorn), Loren Hardy, Dr Stephen and Gail Jeffcoate (Surrey), Dr Jenny Joy, Phillip M. Kinder (Appendix 7), Dr Otakar Kudrna (taxonomy), Chris Laithwaite (geography texts), Dr Ashley C. G. Morton (populations), Dr David Nash and Ian Rippey (Irish sites and islands), Dr Adrian Spalding (Cornish sites), Professor Tim H. Sparks (Box 8.3), Nigel Stone (Shropshire) and Dr W. Gerry Tremewan (taxonomy). To these kind people, and others who have helped and that I may inadvertently have failed to mention, my grateful thanks. All have saved me making many a blunder. Dr Martin Warren (Chief Executive of Butterfly Conservation) has kindly provided a foreword for the book and has answered many enquiries on butterfly conservation as well as correcting Appendix 1. Much of the work since 2002 has been produced on one machine; this was constructed by Peter Mather (Knutsford) and his intervention at times when it did not behave is, to put it mildly, much appreciated! Finally, none of this work would have been possible but for the brave and skilful intervention of Dr E. Raymond S. Ross; to him my ever grateful thanks for facilitating sufficient pain-free thinking time and mobility to do any useful work at all these last 6 years.

The book has been produced whilst I have enjoyed research fellowships at Staffordshire University (Institute for Environment, Sustainability and Regeneration), working with Professor John Dover, and the Centre for Ecology and Hydrology initially at Monks Wood, Huntingdon and latterly at Wallingford, Oxon. The support of these institutions for this work and other research is most gratefully acknowledged.

In producing the book every effort has been made to contact holders of copyright to obtain permission to reproduce copyright material. However, if any have been inadvertently overlooked, the publishers will be pleased to make the necessary arrangements at the first opportunity. I would like to take the opportunity here of thanking all those publishing companies that gave permission to publish material in this book. Copyright of illustrations is retained by authors contributing them to the book.

Finally, I have been most fortunate – and am most grateful – to have had a highly professional production team at Wiley-Blackwell to look after the processing of the manuscript, in the form of Rosie Hayden and Jane Andrew. The way Jane Andrew has sifted out and dealt with the numerous issues arising in the manuscript has my entire admiration and gratitude. My thanks also to Ward Cooper for taking on the project.

Chapter 1

WHAT IS A HABITAT? AN AWKWARD QUESTION

A collector who is a careful observer is often able to examine a terrain and to decide, intuitively as it were, whether a given butterfly will be found there, and that rare being the really accomplished naturalist will nearly always be right. Of course he reaches his conclusion by a synthesis, subconscious as well as conscious, of the varied characteristics of the spot weighed up with great experience; but this is a work of art rather than science, and we would gladly know the components which make such predictions possible.
(Ford, 1945, p. 123, courtesy of HarperCollins Publishers)

It cannot be over-emphasized just how important the concept of a habitat is in butterfly ecology and for biology generally. The habitat has long been regarded as the basic unit around which both theoretical advances and practical applications have been generated in animal ecology and population biology (Watt, 1947, Elton, 1966; Southwood, 1977; Knight and Morris, 1996). More important today is its role in conservation; to conserve organisms we need to know what we mean when we talk about protecting or conserving a species' habitat (Dennis *et al.*, 2007). It is no good vaguely pointing to a wood, as say 'a silver-washed fritillary *Argynnis paphia* habitat', and declaring that 'it needs to be conserved'. It is usually obvious that something does need to be done, but just what about the wood needs to be managed or conserved: the wood as it is, the wood as it was, elements and sections of the wood, or specific resources used by the butterfly in parts of the wood? We have long understood that simply to preserve the bounds of vegetation units will almost certainly not adequately conserve a specific organism in them. The vegetation unit may often be a distinct land unit and therefore the legal or ownership unit for management; that does not mean that some management practice should be applied broadcast across the entire unit in one sweep. As the memorable quotation above from Ford (1945) illustrates, the concept of habitat is elusive when transferred to the ground. Difficulties in producing a functional definition of habitat are not surprising; different organisms, including butterflies, vary enormously in their environmental associations and in their responses to environmental gradients (Thomas, 1994). Furthermore, among butterflies the different sexes and stages require different conditions for existence and are therefore not necessarily found in identical locations (Wiklund, 1977; Dennis *et al.*, 2003).

DEFINITIONS OF HABITAT

Developments in understanding habitats have been frustrated by inconsistencies in the definition and treatment of an organism's habitat and in the lack of precision over terms (Table 1.1; Hall *et al.*, 1997). Typically, but variably, a habitat refers to an identifiable locality (viz., site, place, situation, residence) or to the environment (viz., topography, soils, vegetation types, environmental conditions) and its subdivisions (i.e., microhabitats) occupied by an organism; but practical guidance is noticeably lacking. In *The Concise Oxford Dictionary* (Thompson, 1995), habitat is: (i) the natural home of an organism, and (ii) a habitation (Latin = it dwells). The emphasis is on a place where an organism lives, and it lives there because it has the requirements for existence. In this book, we do not move away from this basic understanding of habitat. But, there is a need to know how to identify the space and what is essential to it.

A Resource-Based Habitat View for Conservation, 1st edition. By Roger L. H. Dennis. Published 2010 by Blackwell Publishing.

Table 1.1 Definitions of habitat.

- Place, living space, where an organism lives (Odum, 1963)
- The place where an organism lives, characterized by its biotic or physical characteristics (Whittow, 1984)
- Type of environment in which an organism lives (Collin, 1988).
- The locality, site and particular type of local environment occupied by an organism; *ece*, local environment; *oike*, oikos (Lincoln *et al.*, 1982)
- Place where a species normally lives, often described in terms of physical factors such as topography and soil moisture and by associated dominant forms (e.g., intertidal rock pools or mesquite woodland). Definitions in ecological literature vary widely but consensus for the following: key environment features related to a species; habitat and vegetation classifications may be concordant, but not always so; subdivisions occur – microhabitats (Calow, 1999)
- Habitat is a suite of resources and environmental conditions that determine the presence, survival and reproduction of a population (Caughley and Sinclair, 1994; Weddell, 2002)
- Habitat is a zone (area) comprising a set of resources, consumables and utilities, for the maintenance of an organism. The resources occur in union and/or intersect and may also be equivalent; links between resource outlets are established by individual searching movements of the organism (Dennis and Shreeve, 1996)
- Habitat [as] the resources and conditions present in an area that produce occupancy – including survival and reproduction – by a given organism. Habitat is organism-specific; . . . it is the sum of the specific resources that are needed by organisms (Hall *et al.*, 1997)
- Habitat is 'the sum of the abiotic and biotic factors essential to the life and reproduction of the species within its natural geographic range' (Haslett, 2007)

DISTINGUISHING HABITAT FROM BIOTOPE AND VEGETATION UNITS

Habitat is most frequently, and it is emphasized *wrongly*, treated as being synonymous with a particular vegetation category or alternatively with a **biotope** (Webb, 1993). Biotope, as in the case of a habitat, has also been described in different ways; some definitions of biotope relate it to a habitat (e.g., Lincoln *et al.*, 1982) but, to avoid confusion, it is sensible to define it in stand-alone terms. Collin (1988) defines biotope as a small area with uniform biological conditions (climate, soil, altitude, etc.). This is sound if biological conditions are understood by the reader to have variance. For those who adopt a literal interpretation of 'uniform', but then no space is entirely uniform, **biotope is understood to mean a region (area, space; e.g., woodland, heath, cliff, dune complex) that is distinguished by particular environmental conditions; therefore, it will tend to contain a characteristic assemblage of organisms** (Calow, 1999). The treatment of habitat as a vegetation unit or a biotope belies reality, as Ford well understood, and this is most unfortunate if only because vegetation associations, and biotopes, are capable of being described at a hierarchy of levels, the bounds between which are often arbitrary. For example, woods and meadows are distinct biotopes but share many plant species, and meadows may well have emerged as a minority integral part of the ancient Holocene forests, grassland that was initially maintained by large herbivores (Vera, 2000; Peterken, 2009). Moreover, and more importantly, vegetation and biotope both typically describe a space that is either insufficient or surplus to that occupied by a specific organism but rarely neatly demarcating it. A vegetation hierarchy is clearly illustrated in the five-volume classification of British vegetation types (Rodwell, 1991–2000). A monumental work, it describes the natural vegetation units as 'communities', and their descending tiers in the hierarchy as 'sub-communities' and exceptional cases as 'variants', within the following major vegetation types: aquatic vegetation, swamps and tall-herb fens; grasslands and montane vegetation; heaths and mires; woodlands and scrub; salt-marsh, sand dune and sea-cliff communities; and weed vegetation.

The key to understanding the relationship between an organism's habitat and vegetation is to appreciate that vegetation is infinitely variable, even to the extent, as Rodwell explains, of having to allocate a field sample of vegetation to a community even though one or more of the marker plant species for the unit is missing. Variability inevitably suggests hierarchy in inclusiveness of plant species; thus, by way of example, in the general introduction to Rodwell's volumes is tabulated details for one category of mesotrophic grasslands, *Centaurea-Cynosurus* grassland (knapweed/crested dogs-tail grass, coded MG5) and the table illustrates the sub-communities, one of which is the *Galium verum* (lady's bedstraw) sub-community. This approach is not unique; the classification technique has close affinities with the Braun-Blanquet method (see Box 5.2) and Braun-Blanquet and Tuxen (1952) identified a similar unit on the European mainland called *Centaureo-Cynosuretum cristati*. Butterflies are faced, then, with a variety of major vegetation structures and substrates and infinitely varied vegetation and therefore plant associations. From this five principles become immediately apparent about butterfly–vegetation associations to those studying butterflies in the field.

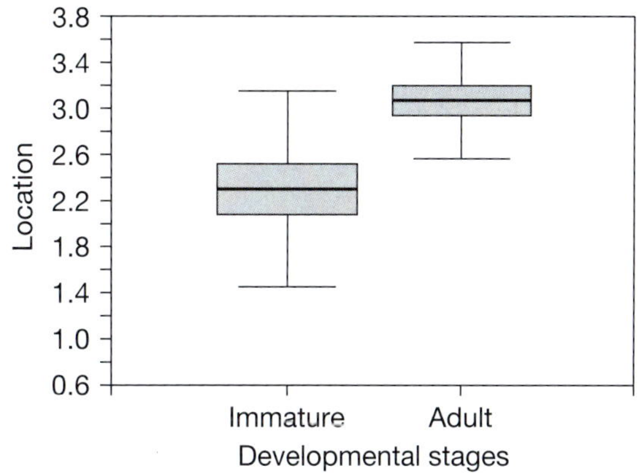

Fig. 1.1 Vegetation types and substrates occupied by early developmental stages and adults of British butterflies. Immature and adult stages occupy distinct vegetation. Means (lines), standard errors (2 standard errors, grey boxes) and 1 standard deviation (whiskers) are shown for 60 British butterfly species. Distinct resource uses, $N > 5$–19 for each stage of each species ($t_{(59)} = -10.04$, $P < 0.0001$). Locations for resource uses are scored as: 0, underground; 1, soil surface; 2, short turf; 3, tall herbs; 4, shrubs and shrub edge; 5, climbers, trees and wood edge. (From Dennis *et al.*, 2003, courtesy of Blackwell Publishing.)

Principles relating to butterfly–vegetation associations

- **P1.1: Different butterfly stages often occupy different vegetation types and substrates.**
- **P1.2: Different individuals of the same species and developmental stage often occupy different vegetation types.**
- **P1.3: Individuals (a population) of a butterfly species rarely occupy the entire zone of a vegetation unit.**
- **P1.4: A key resource for a butterfly (larval hostplant, nectar flower) is rarely restricted to a single vegetation unit or community.**
- **P1.5: Resources for a butterfly will tend to be distributed differently over a vegetation unit(s).**

Different butterfly stages are typically observed to occupy distinctly different vegetation types and substrates (e.g., surfaces such as rock types, walls, soils); the statistical pattern is a significant one over all British butterflies (Fig. 1.1; **P1.1**). Though undoubted examples of species exist where all stages are found in one vegetation type (e.g., purple hairstreak *Favonius quercus* on or under an oak tree), inevitably, because of the contrasting needs of eggs, larvae, pupae and adults, distinctions in vegetation associations among butterfly stages are invariably found in a butterfly community. Larvae, but particularly adults, often occupy very different vegetation types in relation to their behaviour (e.g., searching for mates and nectar feeding), and eggs and pupae of some species can be distributed over different vegetation types and on different substrates (**P1.2**). Thus, silver-studded blue *Plebejus argus* adults actively use areas of scrub as well as different categories of calcareous heath on the Great Ormes Head, North Wales (Dennis, 2004b; Dennis and Sparks, 2006) and the peacock *Inachis io* seeks out mates on a variety of vegetation types including bare ground but lays eggs on tall herbs (nettles) (Baker, 1972; Dennis and Sparks, 2005).

A butterfly is rarely found to occupy the entire area of an apparently homogeneous vegetation unit, nor the entire area of hostplant zone in a locality (**P1.3**). This fact is well known from Professor Jeremy Thomas's detailed research on the Adonis blue *Polyommatus bellargus*, the eggs being distributed in relationship to turf height and microfeatures (thus microclimate), the surfaces chosen varying between the two broods (Thomas, 1983a; Roy and Thomas, 2003). Thus, the speckled wood *Pararge aegeria*, which is regarded as a typical woodland butterfly, is not expected to occur evenly throughout woodland and can be abundant or absent from different but similar-looking parts of the same wood. This is a distinctly noticeable feature of densely packed conifer woods, such as the Macclesfield Forest in east Cheshire, UK; here, the butterfly is found along the forest edge and tracks, but is absent a few metres into the deep, dark, litter-strewn interior which is deficient of ground vegetation (R. L. H. Dennis, personal observation); it is also a feature of deciduous woods, though to a lesser extent (P. B. Hardy, unpublished data). When it is realized that key butterfly resources can occur, and are used, in very different vegetation types (**P1.4**), even nettle patches for common nymphalids (see Fig. 4.1) (Dennis, 2008a), the mismatch between butterfly species and vegetation units is something that has to be expected and not treated as an exception. Not even a single resource, such as a hostplant for a locally monophagous butterfly – nor then, potentially, butterfly eggs and larvae – is invariably restricted to a single vegetation unit. For example, cow wheat *Melampyrum pratense* for the heath fritillary *Melitaea athalia* is found in woodland clearings in Kent and on heathland associated with bilberry *Vaccinium myrtillus* in Exmoor combes (Kemp *et al.*, 2008). It then follows that different resources supporting a butterfly (e.g., larval hostplant and nectar flower species) will tend to occupy different parts of the same vegetation unit or different vegetation units altogether (**P1.5**). The same *apparent* resource will also vary significantly in quality throughout the same biotope. Thus honeysuckle (*Lonicera* sp.) in full sunlight may be used by the white admiral *Limenitis camilla* for nectar feeding but is unsuitable for oviposition and larval development (Fig. 1.2) (Fox, 1996). A simple exercise that the reader can undertake to appreciate this point would be to map the local distribution of cuckoo flower *Cardamine pratensis* and hedge garlic *Alliaria petiolata*, both larval hostplants for orange-tip *Anthocharis cardamines* and green-veined white *Pieris napi*, and the occurrence of eggs on these plants. The distributions of the plants will be found to overlap but differ significantly for landform and vegetation types (Dennis, 1982a).

The habit, therefore, of referring to vegetation units (e.g., grassland, heath, woodland) as habitats is at best unhelpful and at worse misleading and ultimately results in poor ecological science and bad conservation practice (Dennis *et al.*, 2006b) – they are either vegetation or biotope. To make the point clear, some schematic examples are illustrated in Boxes 1.1 and 1.2. In Box 1.1, an assumption is made that we can identify the bounds of a butterfly's habitat, at this stage simply defined as the place where the butterfly lives out its existence. Three situations are illustrated: where the habitat is something less than the vegetation unit within which it occurs, where it almost exactly equals the bounds of the vegetation unit, and finally where it crosses different vegetation units. Of course, the habitat comprises different needs (resources) for different stages of development and the sexes; the most obvious are the hostplants that larvae eat and nectar flowers used by adults. The plants (nectar plants, larval hostplants) that form part of the habitats in these cases are components of the vegetation units. Without mapping these particular resources we do not know their relationship to the area used as a habitat within which butterflies are largely restricted. But, different situations may be envisaged. For example, in Fig. B1.1b the hostplants and nectar plants are perhaps expected to occur throughout the triangular field, in Fig. B1.1a they probably have a distribution limited to the zone marked as a habitat; whereas in Fig. B1.1c they are more likely to occur in different vegetation zones, if only because in this last case the habitat does so too, and indeed outside the zone marked as habitat where conditions may not be suitable for egg laying. The point is: we do *not* know any of this until the resources are mapped or at least inspected. The implication, discussed at length in the following chapter, is that hostplant presence does not necessarily mean existence of a habitat for a butterfly; it has to be in suitable condition, as do other items making up the habitat. To drive the point home, in Box 1.2 two other schematic diagrams illustrate the break up of the habitat zone, which we accept here as the place a butterfly lives and is found, into distinct resources. Just two possible combinations out of many are illustrated. Just how many combinations exist when more resources are considered can be appreciated by glancing briefly at Box 6.3.

A start along these lines of identifying habitat was made many years ago by Wiklund (1981), to whom the credit for this approach should go, but butterfly biologists were deflected from following the path taken in this book by a shift in focus from single to multiple populations and the need to have simple mechanisms to delimit habitat patches rapidly during field surveys. The pattern of study being followed by Christer Wiklund (1977, 1981) and Jeremy Thomas (1984) was effectively abandoned with the burgeoning of multiple

Fig. 1.2 White admiral resources in woodland biotopes. (a) Shade woodland biotope for honeysuckle *Lonicera periclymenum* drapes, at Bentley Wood (Hampshire/Wiltshire border, UK), required by female white admiral *Limenitis camilla* for egg laying and for larval development. Inset is *L. camilla* feeding on honeysuckle. (b) Contrasting biotope of woodland clearing at Bentley Wood that is inappropriate for white admiral *L. camilla* egg laying and larval development, but suitable for adult feeding and for broad-bordered bee hawk-moth *Hemaris fuciformis* L. (Sphingidae) development. (a, b, courtesy of Barry Fox.)

Box 1.1 **Relationship between habitat, vegetation units and biotope**

The habitat of an organism does not have to equate with the area of vegetation or land units (i.e., fields). It may occupy an area less than the land or vegetation unit, match it almost exactly or extend over several vegetation and land units. The schematic diagrams illustrate these alternatives. It is assumed that the habitat can be identified based on the locations of individuals belonging to different stages.

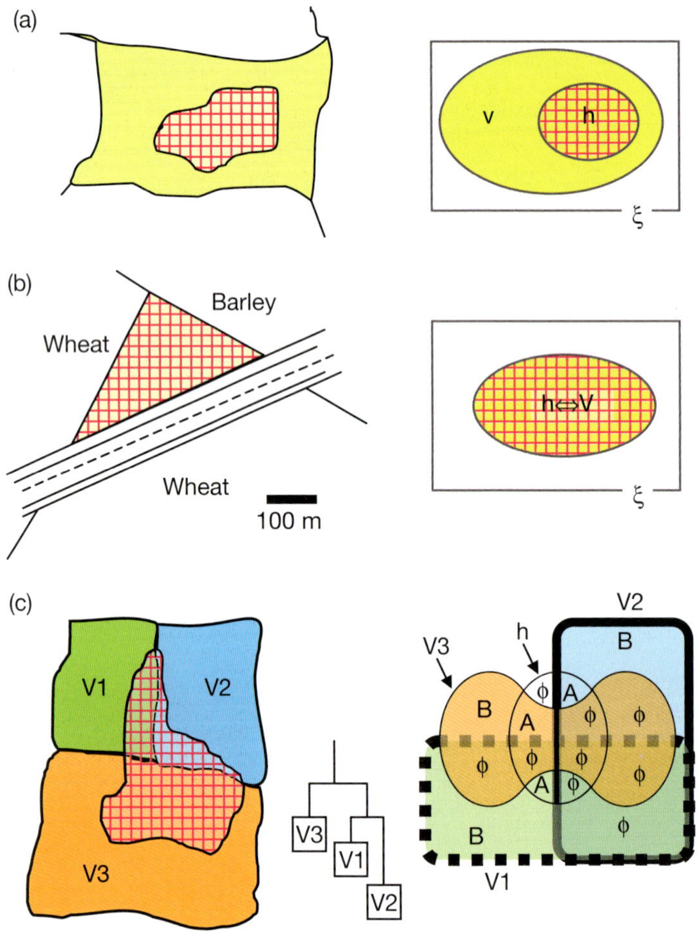

Fig. B1.1 (a) Habitat area is smaller than a vegetation unit and this is a subset of the vegetation unit (e.g., marsh fritillary *Euphydryas aurinia* habitat based on an area of devil's-bit scabious *Succisa pratensis* in lowland purple moor-grass *Molinia caerulea* pasture, called 'rhos' in Wales). (b) Habitat area is the same as the vegetation unit; there is one to one correspondence between them (e.g., remnant tall-herb grass and scrub in field corner abandoned with dual carriageway development and occupied by meadow brown *Maniola jurtina*). (c) Habitat area overlaps a single biotope made up of the vegetation units in which V1 and V2 have more similarities in plants than have either V1 and V2 with V3. In the Venn diagram, 'A' indicates habitat as a subset of one of the vegetation types, 'B' is the complementary set (a vegetation unit but not habitat) and ϕ the empty set (the combination does not exist in this example) (e.g., wet (V1) and dry (V2) heathland and wooded (birch) (V3) heath all with *Molinia caerulea* and used by large skipper *Ochlodes sylvanus* on Lindow Common, Cheshire (Dennis and Williams, 1987)).

Yellow, gold, lime and pale blue are vegetation types (v); pink cross hatching, supposed habitat area within vegetation zones (h); pink cross hatching over yellow, overlap of vegetation and habitat in Venn diagrams for (a) and (b).

Box 1.2 **Schematic maps to illustrate the variable relationships among some resources, habitat and vegetation patchworks**

Habitat is not a uniform homogeneous entity or spatial unit. It comprises various resources used by a species and there is no reason why these should overlap let alone exactly coincide. Instead of illustrating what is understood to be a habitat in vegetation or land units, the schematic diagrams 'map' three resources that necessarily form a part of habitats for butterflies within field boundaries to illustrate two situations. First, where all these occur within a single field and, second, where these variably cross field boundaries. The diagrams also qualify a key resource used by butterflies, the larval hostplant, identifying a zone within the hostplant patch where oviposition is restricted, to impress that a resource may not be all that it seems to be. Immediately, it becomes obvious that the bounds of a habitat are not that clear.

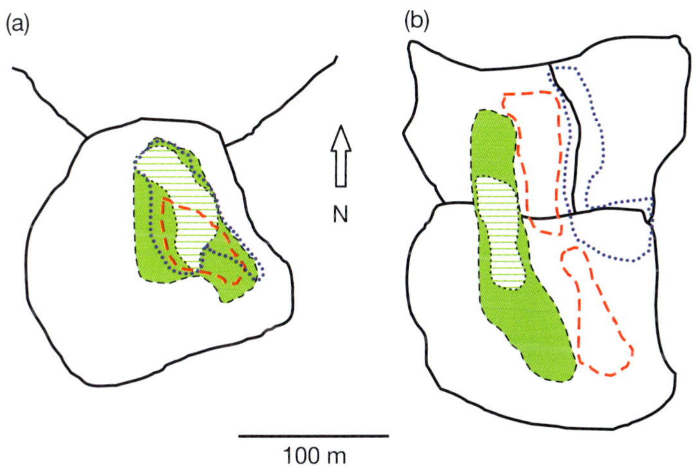

Fig. B1.2 (a) Complementary resources (e.g., mate location, nectar) coincide with the hostplant area (but not entirely with the hostplant area that is suitable for egg laying) and within a single vegetation unit. (b) Complementary resources do not coincide with the hostplant area and both hostplants and complementary resources cross vegetation units. Typically, the area over which a hostplant is exploited by a butterfly will generally be a fraction of the total hostplant area and resource zones are not expected to coincide with vegetation units. Thick continuous lines, vegetation unit boundaries; green areas, hostplants; shaded areas, hostplants suitable for oviposition and breeding; red dashed lines, nectar zones; blue dotted lines, mate location zones.

population (metapopulation) studies, though Thomas in his continued study of *Maculinea* ecology has constantly impressed on us the crucial part played by intrahabitat quality (Thomas *et al.*, 1998a, 1998b). What did Wiklund envisage? He understood that the 'Essential requirements for the survival of most butterfly populations are that (1) males and females can find each other and mate, (2) adult females can deposit the eggs in such a way that the larvae can find their hostplants and (3) adults can find food in order to live long enough to perform both activities mentioned above'. Thus, the area occupied by any given butterfly population should include a 'mating habitat', a 'breeding habitat' and a 'foraging habitat'. 'This being a functional classification of habitats, means that these habitat types can be spatially separate or identical' (Wiklund, 1981:164). Wiklund distinguished primary from secondary, and from novel and unsuitable 'breeding habitats' based on the larval hostplants used (Wiklund, 1981). What in this book are called 'resources' (Dennis *et al.*, 2003), Wiklund referred to as a specific habitat type. The question is: how do they integrate to form a complete butterfly habitat? Is there anything missing? Is it possible to map this habitat identified by Wiklund?

The crux of the matter is as follows. Because reference is habitually made to butterfly habitats, a clear definition of habitat becomes axiomatic. Just what definition of habitat is applied can have serious consequences for developments in butterfly biology and particularly for butterfly conservation (Dennis *et al.*, 2003, 2006b, 2007; Shreeve *et al.*, 2004); for this reason alone the topic deserves careful thought and detailed consideration. In all empirical and theoretical population work, habitat is implicitly or explicitly a bounded space (e.g., Southwood, 1977; Baker, 1978; den Boer and Reddingius, 1996; Hanski and Gilpin, 1997; Ehrlich and Hanski, 2004); butterfly biologists draw boundaries on maps purported to enclose habitat patches. The fundamental problem is that it is rarely evident what, in detail, this space comprises and there is typically implicit, even explicit, treatment of an organism's habitat as a particulate, homogeneous and invariant unit – and inevitably this is a vegetation type, biotope or even the area occupied by a single plant species. But it is none of these things and examples in this book attempt to illustrate just what is involved in the notion of a butterfly habitat. Habitat is necessarily the location where an organism lives out its life cycle. As such, it follows that it should be possible to map the bounds of a habitat in terms of life history requirements. The next chapter begins to outline a suitable procedure to this end.

Chapter 2

A SIMPLE MODEL FOR BUTTERFLY HABITATS

Engaging finer resource units within a biotope is the ecological equivalent of subatomic particles to the atom in physics. Much as physicists wouldn't think now of truncating their lower limit of attention to an atom, no longer can ecologists restrict their thinking to a vegetation unit as habitat.

HABITAT MODEL

How best then to define a habitat? A useful way of conceptualizing the problem is to regard species as requiring a set of resources and conditions in order to function (e.g., reproduce, develop and survive) (Wiklund, 1977; Dennis and Shreeve, 1996; Hall *et al.*, 1997; Caughley and Sinclair in Weddell, 2002). In effect, conditions are resources: **a resource is any component of the environment that can be used by an organism and varies in quality and quantity** (Lincoln *et al.*, 1982). In local population terms, **a habitat is that collection of resources and conditions that ensures the persistence of a population at a site** (Dennis *et al.*, 2006b); it is well to understand at the outset that there is no expectation in this definition that habitat quality and bounds remain constant over time. A convenient way of categorizing such resources for insects, or more widely for arthropods, is to consider separately each life history stage and each sex. In this way, at a minimum, an adult butterfly would require resources for egg laying, mate location, thermoregulating, resting, roosting, feeding and predator escape, much as do birds and other vertebrates (Southwood, 1977). Each stage can be treated similarly and the resources mapped. The habitat is simply the logical extension of this reasoning, and is defined by the intersection and combination of these resources, the links made by flights of adults and movement of larvae (Dennis and Shreeve, 1996). A formal definition is provided in Box 2.1. It is important to realize that resource use and tolerance of conditions are found to vary from one location to another; this variability across a species' range can be expressed by the variance (or other measures of variability) in resource traits. The definition of habitat can thus be extended to the species level as: the combination of resources and conditions occurring at locations and required for populations of the species to persist within its realized geographical range. A similar definition of habitat for the entire species, if factors are read as resources, is suggested by Haslett (2007): 'the sum of the abiotic and biotic factors essential to the life and reproduction of the species within its natural geographic range.'

The resources required by each life history stage, which comprise the habitat, may usefully be divided further into two groups: **consumables** and **utilities**. Consumables are materials taken in directly by larval and adult individuals. These include: hostplant parts, oxygen, water, minerals, salts, nectar, pollen, sap excretions and organic soups from rotting corpses. It would also include materials transferred between conspecifics at mating (Karlsson, 1995, 1996). Utilities describe the conditions for existence and persistence, such as physical sites for various activities (e.g., mate location) and stages (e.g., pupation sites), suitable conditions for somatic maintenance, development, activity (e.g., maintaining body temperatures, flight) and reproduction (i.e., suitable local climates and microclimates), and enemy-free space. So viewed, a functional definition of habitat becomes neatly linked to Hutchinson's concept of an organism's hyperdimensional niche (Vandermeer, 1972; Whittaker *et al.*, 1973). What is the difference? Habitat describes real ground conditions, resources and physical objects used by butterflies

A Resource-Based Habitat View for Conservation, 1st edition. By Roger L. H. Dennis. Published 2010 by Blackwell Publishing.

Box 2.1 **A resource-based definition of habitat**

The habitat model is based on resource distributions and individual movements. Habitat is the intersection and union of necessary complementary resources and conditions for an organism, linked by stage to stage movements of most individuals (>95%) in the population ensuring population maintenance (Dennis and Shreeve, 1996). Population maintenance is typically described as the intrinsic rate of population increase $r \geq 1$, where $r = b - d$; b is the instantaneous birth rate and d is the instantaneous death rate (Lincoln *et al.*, 1982).

For simplicity, resources are shown schematically as sets or envelopes; the elements of sets are arbitrary units of ground space based on fine-scale responses of individual butterflies (e.g., 1 m units) as illustrated by the grid in Fig. B2.1c. Resources are connected by daily search flights (so-called trivial flights) to give habitats by combination and overlap (Fig. B2.1a); by substitution (Fig. B2.1b) (e.g., cuckoo flower *Cardamine pratensis* is a hostplant, nectar source and roost substrate for orange-tip *Anthocharis cardamines*); by combination of neighbouring resources (Fig. B2.1d); and by combination of isolated resources linked by daily trivial flights (Fig. B2.1e) or seasonal movements (Fig. B2.1f). Figure B2.1c illustrates a small part of Fig. B2.1b to show that at a finer resolution there may be non-resource zones (areas passed over but not necessarily used) within resource zones and habitats.

Distinctions are typically made between explorative (so-called trivial) flights within habitats and direct linear movements between habitats (Van Dyck and Baguette, 2005), but see Fig. 6.3.

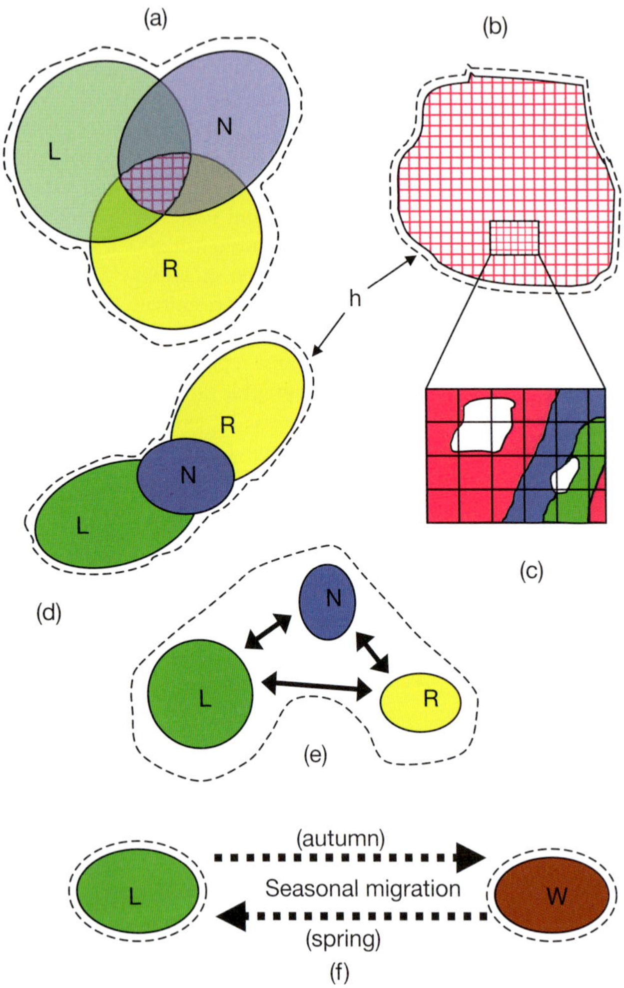

Fig. B2.1 (a) Intersection and union of resources comprising a habitat. (b) Equivalence and equality of resources comprising a habitat. (c) A small part of (b) in which green is the hostplant, blue is nectar, pink is jointly nectar and hostplant, and white is neither (possibly trivial space) but may or may not be used for other activities such as pupation, adult resting, etc. (d) Contiguous union of resources making up a habitat. (e) Disjointed union of resources linked by trivial flights making up a habitat. (f) Disjointed non-union of resources linked by seasonal migration, each part comprising a seasonal habitat (e.g. brimstone *Gonepteryx rhamni* (inset)).

A maximum of three resources are illustrated in each diagram: N, nectar resource (blue); L, larval resource (green); R, roost sites (yellow); W, overwintering sites (brown). The habitat boundary (h) is indicated by a dashed line; the habitat core is illustrated by pink cross shading.

with geographical co-ordinates that allow the spatial bounds occupied to be mapped. **Niche**, on the other hand, describes the biological (attribute) space occupied by a species in a graph using vectors or axes of influential agents. A formal definition is provided in Box 2.2. A helpful guide in making the distinction here is that elements describing a habitat are recognized by a species as having proximal (immediate) significance for them, whereas a niche includes elements that occur more distally along the causal chain in incidence, ones that underlie the resources being used. Niche dimensions also include biological interactions beyond the definition of a habitat, including predator–prey interactions. Accurate recognition of the habitat – measurements taken from numerous locations or habitats – is a prerequisite for part determination of the niche, which otherwise can only be notional.

Defined in this way, habitats for many species will be readily discernible from the following:

1 The overlap or contiguity of resources (see Fig. B2.1a–c).

2 The movements of adults searching for, visiting and returning to distinct resource zones (Fig. B2.1d).

In other cases (e.g., painted lady *Vanessa cardui*), the bounds of the habitat are far from obvious. In these situations, where resources are not compact but diffuse (Fig. B2.1e), there is no reason why a habitat should be an obvious spatial structure, neither to a butterfly ecologist nor for that matter to a butterfly (Delibes *et al.*, 2001). The issue is: can we adequately recognize a habitat as an individual working system? Kluge (1990) refers to four criteria that taken together define a working system. The first two (spatial and temporal boundaries) refer to pattern and the second two (cohesion and integration) refer to process. Whether or not a collection of resources makes a habitat – the working system – depends entirely on the relative scale in the distribution of resources to that of the movement of mobile life history stages. For butterflies, a habitat is clearly discernible where movements of adults and larvae achieve integration and cohesion of resource outlets, as for instance by a population of the meadow brown *Maniola jurtina* (Brakefield, 1982a, 1982b). However, it is less obvious in situations where movements occur over wide geographical areas as in the case of the brimstone *Gonepteryx rhamni* (Pollard and Hall, 1980). This

Box 2.2 The use of habitat and niche in butterfly biology

The terms 'habitat' and 'niche' have a long history and their meaning has changed over time. Thus, no single definition can be plucked from the past that satisfactorily distinguishes these terms; for instance, niche has frequently been used as a synonym for habitat. The term niche evolved through Grinnell (1914 to 1928), Elton (1927), Kostitzin (1935), Levin (1966, 1968), MacArthur (1968), Pianka (1969), May (1974), Hutchinson (1944, 1957, 1978) and other authors. The history and these references are summarized in Schoener (1989). Throughout, the emphasis has been distinct from spatial notions of habitat. For example, Elton (1927:63–64) in *Animal Ecology* specified the concept:

> It is . . . convenient to have some term to describe the status of an animal in its community, to indicate what it is *doing* and not merely what it looks like, and the term used is 'niche'.

The focus has also been on (the competition for) food. But Elton did not limit niche to food, giving clear examples of competition for nesting sites in birds.

A major advance was made with Hutchinson (1957, 1978). Hutchinson (1978:158) referred to a footnote of his in a 1944 paper in *Ecology* on phytoplankton periodicity:

> The term niche . . . is here defined as the sum of all the environmental factors acting on the organism; the niche thus defined is a region of an *n*-dimensional hyper-space, comparable to phase-space of statistical mechanics.

Later, Hutchinson (1957:416) expanded on this concept (Fig. B2.2a, b). He envisaged a species (S_1) occupying a space (area) within axes (two variables X_1, X_2) for two niche variables. When the variables are independent (not correlated) the axes will be at right angles to one another and the areas occupied by the species will be rectangular (Fig. B2.2a). When correlated the angle between the axes is <90° (Pearson $r = \cos\theta$, the angle subtended by the axes). Hutchinson continued:

> We may now introduce another variable, x_3 and obtain a volume, and then further variables $x_4 \ldots x_n$ until all the ecological factors relative to S_1 have been considered. In this way an *n*-dimensional hypervolume is defined, every point in which corresponds to a state of the environment [physical and biological] which would permit the species S_1 to exist indefinitely.

He referred to this space as the **fundamental niche** for species S_1. For a second species S_2, a similar

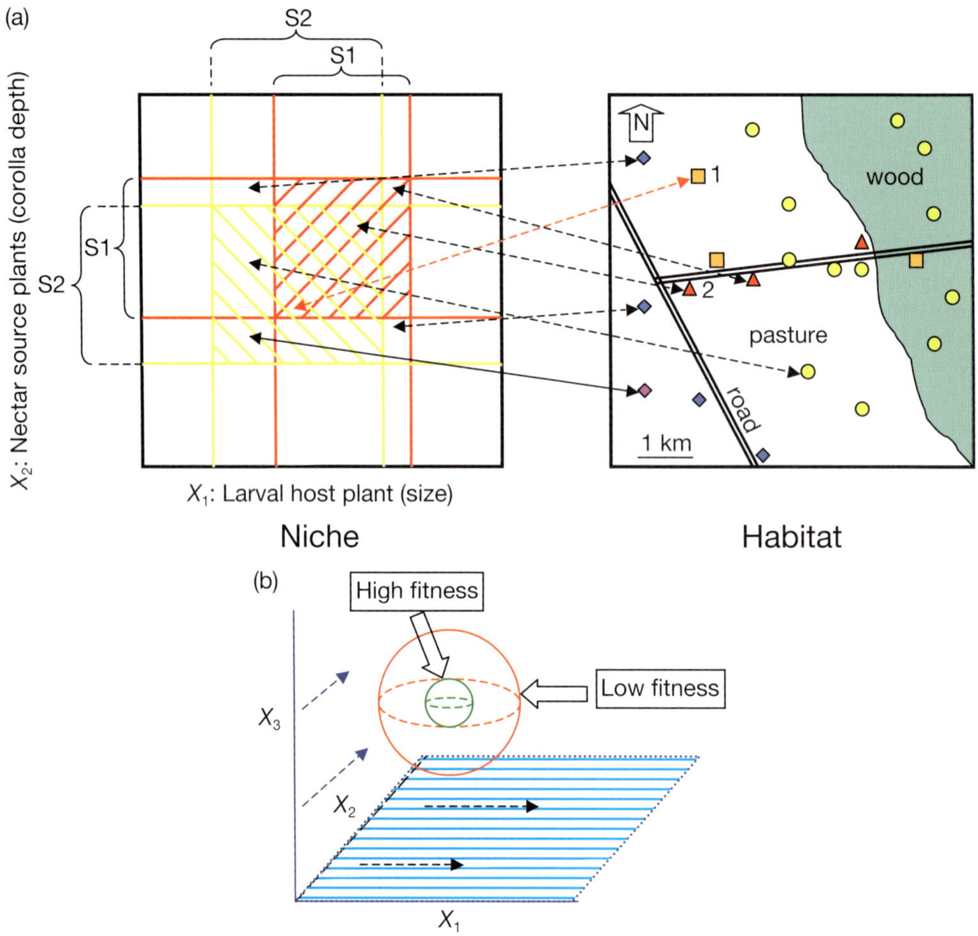

Fig. B2.2 (a) Schematic diagrams to illustrate the difference between niche and habitat. Fundamental niches for two species (S_1, S_2) are defined here by just two variables (X_1, X_2, examples given for ease of interpretation), abstracted from many, in a two-dimensional space (niches) and their link to the physical space they occupy in the landscape (habitats). On theoretical grounds, owing to competition, only one species is expected to persist where their potential for survival overlaps. Diagonal shading indicates distinct zones for the two species and where the diagonal lines cross, they would overlap, determined from samples taken in the surrounding region. They may actually be found together for a number of reasons (competition not activated). Symbols indicate populations of the species: a triangle for S_1, a circle for S_2 and a square for both species together; a diamond indicates that neither species occurs (pink when this may be owing to chance; blue where either hostplant or nectar sources are not in the suitability range of either species). In the example, two sites (1 and 2) are indicated from the niche overlap: 1, where both species occur, 2, where only one species occurs. It should be understood that for simplicity just two variables are illustrated (here labelled as single measures of hostplants and nectar sources to help understanding), but that niche separation envisages many such variables. Such niche separation, and potential competition, typically involves species using the same consumer resources (larval hostplants, nectar flowers). However, two butterfly species may use the same hostplant individuals and yet not come into any contact (e.g., *Anthocharis cardamines* on flower pods and *Pieris napi* on leaves). (b) Niche shown as an *n*-dimensional hypervolume for three variables. The placement of species within a niche axis is unlikely to be absolute (unlikely to have the same fitness through the range of the niche variable occupied) but will vary and have a frequency distribution with a mean and variance. The larger sphere may indicate the zone of the fundamental (potential) niche whereas the smaller sphere indicates a zone of higher fitness (realized niche). (Redrawn from Vincent, 1990, Fig. 4.5, courtesy of Routledge and Taylor and Francis.)

hypervolume can be described. The fundamental niche defined in this way is supposed to completely describe a species' ecological properties. Owing to competition, predation and, as we now fully appreciate with metapopulation dynamics, chance local population extinction, a species usually does not occupy its entire fundamental niche in these niche axes. This lesser space occupied is referred to as its **realized niche**. It becomes obvious that if there is a fundamental and realized niche, then there must also be a **fundamental habitat** and a **realized habitat**; this follows simply because measurements of a species' niche must have been made in the real world, the ground the species occupies. Niche is thus defined for an 'abstract space', but this has been transferred from measurements in real 'biotope space' from locations where species are actually found (the habitat space) compared with where they are not found (matrix space).

Owing to problems in applying Hutchinson's concept, it has largely been overtaken by a utilization concept (Schoener, 1989). The modern theory of niche is closely tied to understanding competition amongst organisms (cf., Gause's competitive exclusion principle in Chapter 5). Thus, Schoener develops a decision tree to relate niche overlap to competition, according to niche theory. The issue here is that potential niche is reduced by competition (predation) to the realized niche. Inevitably, to understand space competed over, it is necessary to have measurements of all resources and conditions that make the space inhabitable or uninhabitable. Niche now has a more practical meaning than envisaged by Hutchinson and encompasses 'utilization distribution' and is defined for a particular population; it identifies the fractional use of resources arranged along one or more dimensions called **niche axes**. It is important to understand that it is an abstraction of measurements taken from habitat and non-habitat bounds. In acquiring a species' fundamental niche and fundamental habitat, it is necessary that surveys are made at many points throughout a species' range over time. Not only does this obtain a measure of variance for habitat and niche axes but it also determines the degree of change, mutability over time, in line with changes in selection pressures, biodiversity and chance. The habitat and niche are both variable and mutable. Thus, it is essential to distinguish between *the* fundamental and realized niche and habitat for a particular region occupied by one or more species, and *a* realized niche and habitat for a species, a specific site example encompassing a population or many, integrated local populations (a metapopulation, see Chapter 6).

Why is this important? The reason is that there is an essential difference between biologists whose primary interest is in this variability coupled with species dynamics and interactions, and that of managers of sites who want clear answers to questions – not about something abstract called niches – but about the habitat they have to manage. Where is it? What is it? What does it comprise at site *X*? What can be done to improve the site? The key thing for those concerned with conserving butterflies, is that study of a species at many locations (in a region) will identify its fundamental niche and fundamental habitat, which are both abstractions (since not all resources and conditions that a species could use occur at any one location) but, owing to variability in conditions, resources and chance factors, a number of specific sites will display variable examples of the realized habitat during any one season. Even so, at any one site, the fundamental habitat can be ascertained by study of the site over a number of years. Only then, will it become clear just what resources and conditions are used, and why or why not, from year to year. It many cases it may simply be a matter of population size or changes in population density over the site, or the result of stochastic events (e.g., weather events).

The separation of habitat and niche in this book accords with observations by Whittaker *et al.* (1973). They urged the separation of 'niche' and 'habitat', a plea that has largely been ignored. Their justification is that they identified three interpretations of 'niche': (i) the 'functional niche', that is, the position or role of a species in the community with an emphasis on comparative behaviour and interactions with other species; (ii) the 'habitat niche' or 'place niche concept' which is interpreted here specifically to imply a zone in physical space; and (iii) the combined niche, as (i) + (ii). In this book (ii) is the habitat and niche is understood to be an abstraction of habitat, often a fractional abstraction (e.g., limited axes investigated) of (i) + (ii).

A number of measures exist for describing niche breadth and niche overlap between species (see Krebs, 1989, for following references). A simple measure of niche breadth is Levins (1968), *B*:

$$B = 1/\sum p_j^2 \text{ or } B = Y^2/\sum N_j^2,$$

where p_j is the proportion of individuals found on or using resource state j, N_j is the number of individuals found in or using resource state j, and $Y = \sum N_j$ = total number of individuals sampled. A measure of overlap is given by Renkonen's (1938) percentage similarity measure (also Schoener's index (1970), *O*):

$$P_{lm} = [\sum^n (\text{minimum } p_{il}, p_{im}]100,$$

where P_{lm} is the percentage overlap between species *l* and species *m*, p_{il} and p_{im} is the total proportions of resource *i* used by species *l* and *m*, and *n* is the number of resource states.

may be simply because of logistical problems of tracing such movements (e.g., the monarch *Danaus plexippus*; Garland and Davis, 2002). Alternatively, it may be because individuals involved in such movements do not return to the *same* areas (e.g., red admiral *Vanessa atalanta* and *V. cardui*).

KEY ISSUES IN THE HABITAT MODEL

There are three aspects of the habitat model that deserve some additional explanation: the status of resource-free or 'empty' spaces, the nature of movements and the open versus closed designation of habitats.

The matrix or so-called empty space

First, habitats will always include ground (substrates, vegetation surfaces, plants, etc.) that is never, or is very rarely, used by individuals belonging to a single population and occupying a single site. Examples are particular flowers that because of corolla structure are inaccessible for nectar feeding, ground between hostplant patches, as well as hostplants occurring in unsuitable surroundings, for instance crucifers such as hedge garlic *Alliaria petiolata* which are hostplants for the orange-tip *Anthocharis cardamines* but which are overtopped and shaded out by shrubs. At a trivial, but nevertheless physical, level, it includes the 'empty space' between individual flowers used as nectar sources, here called **trivial space**. Such 'unused' spaces will always form part of butterfly habitats. The area largely unused outside the habitat is frequently referred to as the **matrix**; it becomes obvious then from what has just been said that the matrix is everywhere, inside and outside habitats. Later, reference is made to original work by Vanreusel and Van Dyck (2007) on the green hairstreak *Callophrys rubi* in Belgium, which demonstrates how habitats can actually be identified in the field and how this inevitably includes non-resource spaces over which individuals pass.

Movement in and between habitats

The second aspect of the habitat model is how butterflies move between resources. A distinction is frequently made between the routine movements (the **trivial flights** of Johnson (1969) or the **routine movements** of Van Dyck and Baguette (2005)) typically associated with foraging, and displacement or **dispersal flights**, which some consider are used by individuals in transfers between habitats. The distinction between the two is that the former tends to be slower and weaving, involving returns, and typically including searching behaviour and inspections; the latter conversely is faster, linear and directed. It remains a moot question whether these types of movement, respectively, are entirely restricted to habitats and non-habitats (see Fig. 6.3) (Dennis and Hardy, 2007). There is no reason why movements should not be directed, for instance, when individuals move between two distinct resource zones within a habitat (e.g., between a hostplant patch and a detached nectar source as in the ridge checkerspot *Euphydryas editha* at Del Puerto Canyon, California; Gilbert and Singer, 1973) or why movements in the matrix should not involve searching flights as has been as demonstrated by disorientating *Maniola jurtina* on releasing them in unfamiliar surroundings (Conradt *et al.*, 2000; Dennis, 2004a). Even so, routine movements strongly suggest foraging, resource seeking and resource use and are characteristic of habitat occupancy.

One individual's behaviour does not determine a habitat. Habitats must necessarily be defined on the basis of data for an adequate sample of individuals – the habitat bounds will appear to be fussy, the core surrounded by a dense web of tracks (T. G. Shreeve, personal communication) – and the emphasis must necessarily be on this being determined in natural circumstances. However, it is always difficult to interpret what artifice may be present when applying artificial releases of individuals. The reader is directed to Singer (2004) for cautionary notes on, and suitably adjusting their perception to, the impact of human influences in field and laboratory experiments.

Open versus closed populations and species

The distinction between 'open' versus 'closed' populations is closely tied into a description of habitats as diffuse and compact (Warren, 1992a; Pollard and Yates, 1993a). Despite variations in common usage, in systems theory terminology (Von Bertalanffy, 1951, 1962) the former terms effectively describe a system's dynamics as opposed to the latter which describe a system's structure; that is, population behaviour or mobility compared to habitat structure. Inevitably

Table 2.1 Relationships between population dynamics and habitat resource structure.

		Population dynamics	
		Open	Closed
Habitat resource structure	Compact	✓b	✓✓a
	Diffuse	✓✓a	✗c

a, expected relationship; b, occurs but is subordinate to open/diffuse for any species affected (e.g., orange-tip *Anthocharis cardamines*); c, does not occur.

describing populations as 'open' or 'closed' has connotations for habitat structure and it is worth considering the implications of using these terms.

How a population behaves, in terms of mobility, does not logically have to reflect on resource distribution (Table 2.1). One would expect that an 'open' population would have diffuse resource distributions and a 'closed' population would have compact resources, and on this basis a system's dynamics would match the system's structure. A conundrum appears inasmuch as one of the other two possible relationships breaks down. A population of a species having diffuse complementary resources cannot be closed; this would imply that a partial set of its resources are sufficient for persistence. But then some species have an open population structure despite all their resources apparently occurring in one place (e.g., *Anthocharis cardamines*; Wiklund and Åhrberg, 1978; Dennis, 1982a, 1982b). Consequently, the 'open' versus 'closed' designation describes something other than just apparent resource aggregation and habitat structure, particularly when conferred on the entire species. As all species disperse from habitat patchworks, to a variable degree over time and space, the 'open' versus 'closed' designation is not particularly useful. Much of the confusion would be avoided if descriptions of mobility and structure were kept separate, and this is the usage applied here. Of course, none of this implies that there is mutual exclusion between movement and resource distribution; on the contrary, mobility is primarily driven by resource distributions – resource stability, co-occurrence and size. Key variables are hostplant phenology, hostplant–habitat lifespan, and resource aggregation, frequency and abundance (Dennis *et al.*, 2003). For instance, it is to be expected that a designation of 'open' or 'closed' will correlate with hostplant generalism and specialism, respectively (Ries and Debinski, 2001). Agents influencing movement are explored in Chapter 6.

QUALIFYING RESOURCE OUTLETS

The model for habitats described in Box 2.1 is limited, for the sake of illustration, to a few resources. It is well to remember that butterflies depend on a wide array of resources for their maintenance and survival. Currently, a database on butterfly resources for British and European butterflies is being constructed at the Institute for Environment, Sustainability and Regeneration at Staffordshire University and at the School of Life Sciences, Oxford Brookes University (Dennis *et al.*, 2008). The key variables and attribute states listed under developmental stage and activity for this database are outlined in Table 2.2. Below – because butterflies often have choice of surfaces, substrates and consumables – reference is often made to resource outlets, implying multiple opportunities in cases where the resource is not singular. Even when it is, resource elements are rarely identical or equivalent as they vary in quality, and butterflies exact choice within the parameters of energy supply, local and micro climatic contrasts and search time (Dennis and Sparks, 2005).

Two basic concepts emerge in discussions of resource use:

1 Specialization. A key attribute in describing butterflies is their degree of specialization. Butterflies are generally described as being **specialists** (non-varied) or **generalists** (varied) in resource use. This is most often used in relation to the range of larval hostplants exploited. It is important to realize that this is not the only usage of these terms; for instance, there is no reason why this concept should not be extended to other resource use such as nectar or utilities, or more widely to occupancy of vegetation units or biotopes. Generalist and specialist butterfly species are often defined in different ways (e.g., Dennis *et al.*, 2004; Fox *et al.*, 2006). The important thing is to make it clear just to what these terms refer (Box 2.3).

Table 2.2 Binary state resource attributes used for an ecological classification of butterflies in Britain. Each state is coded as present (1) or absent (0). (From Shreeve *et al.*, 2001, with kind permission of Springer Science and Business Media.)

Variable/stage	Variable	State
Life history	Overwintering stage	Egg
		Larvae
		Pupae
		Adult
	Hibernation site	Buried
		Surface
		Short sward
		Tall sward
		Shrub
		Tree
		Liana
		Rock\cliff face
	Generations	Biennial
		Univoltine
		Univoltine + partial second generation
		Bivoltine
		Multivoltine
	Symbionts (ants)	Monophagous
		Oligophagous
		Polyphagous
	Egg-laying type	Single
		Small batch
		Large batch
Eggs	Egg-laying substrate	Bare earth or ground artefact
		Short turf/herbs
		Tall/mature herbs
		Shrub
		Tree trunk
		Canopy
		Liana
	Egg-laying locality	Light
		Partial shade
		Shade
		Recesses
		Hostplant edges
Larvae	Hostplant family	Apiaceae
		Aquifoliaceae
		Araliaceae
		Betulaceae
		Brassicaceae
		Cannabaceae
		Caprifoliaceae
		Caprifoliaceae
		Celastraceae
		Cistaceae
		Cornaceae
		Cyperaceae
		Dipsacaceae

Table 2.2 (*continued*)

Variable/stage	Variable	State
		Ericaceae
		Fabaceae
		Fagaceae
		Geraniaceae
		Gramineae
		Grossulariaceae
		Lamiaceae
		Plantaginaceae
		Polygonaceae
		Primulaceae
		Resedaceae
		Rosaceae
		Salicaceae
		Scrophulariaceae
		Tropaeolaceae
		Urticaceae
		Violaceae
	Hostplant range	Monophagous
		Oligophagous I (1 hostplant per location)
		Oligophagous II (>1 species per location)
		Polyphagous
	Hostplant phenology	Biennial
		Annual/biannual
		Short-lived perennial
		Long-lived perennial
	Hostplant growth form	Short herb/grass
		Tall herb
		Shrub
		Tree
		Non-plant
		Liana
	Hostplant part used	Flowers/pod
		Leaf
		Bud
		Stem
	Larval environment	Buried
		Ground layer
		Field layer
		Shrub layer
		Canopy layer
		Attended
	Hostplant patchiness	Large patch
		Small patch/single plant
Pupae	Pupal location	Buried
		Ground layer
		Field layer
		Shrub layer
		Trunk layer
		Canopy layer
		Attended

Table 2.2 (*continued*)

Variable/stage	Variable	State
Adults	Adult feeding	Herb flower
		Grass/sedge/ergot
		Shrub/tree flower
		Honeydew
		Sap
		Decaying plants
		Animal
		Mineral
		Hostplant coincident
	Adult roosting location	Rock/soil
		Trunk/fence
		Short turf
		Grasses
		Tall herb
		Shrub
		Tree canopy
		Liana
		Exposed
		Communal
		On hostplant
		Off hostplant
	Mate-locating location	Rock/cliff
		Bare earth
		Short herbs
		Tall herbs
		Shrubs
		Tree canopy
		Nectar site
		Hostplant site
		Physical edge site
		Light edge site
		Hilltop
	Basking sites	Rock/bare earth
		Short herb
		Grasses
		Tall herbs
		Shrub
		Tree canopy

2 Substitutability. Resource types may also be referred to as substitutable (e.g., different nectar plants) or non-substitutable (e.g., a single hostplant or adult nectar source) giving rise to the concepts of resource **supplementation** and resource **complementation**, respectively (Tilman, 1982). Two nectar sources supplement adult feeding whereas a single nectar source and larval hostplant complement adult and larval feeding. Substitutability is a key component determining the importance of resource variables. A monophagous butterfly clearly has no capacity for substitutability; without the one hostplant, this butterfly cannot persist. On the other hand, a number of different surfaces may be chosen by males for territorial activity when seeking mates; some may be better than others, but the absence of the most appropriate surfaces may not

Box 2.3 **The generalist–specialist spectrum in British butterflies**

Generalists and specialists are typically distinguished by their breadth of habitat or resource use, typically that of hostplants (Lincoln *et al.*, 1982), but there are no set boundaries for classifying species. Thus the classification of herbivory in butterflies by Wiklund and Åhrberg (1978) identifies four categories of phagy: **monophagy**, feeding on single plant species; **oligophagous I** (monophagous strategy), feeding on plants in one family but typically on one in each location; **oligophagous II** (polyphagous strategy), feeding on several plants in one family in each location; and **polyphagy**, feeding on more than one plant family. This differs from the classification of Steffan-Dewenter and Tscharntke (1997) (monophagous, species with only one main hostplant; oligophagous I, hostplants within only one plant genus; oligophagous II, hostplants in more than one genus but within a single family; polyphagous, several hostplants in several plant families) A different classification from these is used below (Table B2.3b). From a resource-based habitat viewpoint, it is important to bear in mind that generalist–specialist distinctions can be made for different resource types, consumables and utilities, as well as for biotopes. A butterfly may be a larval hostplant specialist but a nectar generalist. If there are only three types of resources specialism (hostplant, nectar feeding, utilities) – and in detail there are obviously more – then there are seven combinations of specialism. So, although there is a significant tendency for different types of resource specialism to be significantly correlated (Fig. B2.3), not all resource variables are significantly correlated (Table B2.3a). The distinction also depends on spatial scale (single site, region, geographical range), the unit of interest (species, among populations, among individuals in a single population, developmental stage of an individual) (Singer, 1971; Chew, 1977; Wiklund, 1981) and the status of resources (Table B2.3a).

Table B2.3a Correlations between habitat resource variables.

Variable	Number of hostplants (all)	Nectar specialization	Utilities	Number of biotopes	Niche breadth
Number of hostplants (all)	1.00				
Nectar specialization	0.18†	1.00			
Utilities	0.25	0.10	1.00		
Number of biotopes	0.54†	0.10	0.50†	1.00	
Niche breadth	0.78†	0.35†	0.20	0.64†	1.00

Correlations (Pearson *r*) between resource variables († $P < 0.05$). Nectar specialization is the number of nectar sources standardized on records. Utilities include number of outlets for pupae and adults excluding nectar.

Table B2.3b Phagy (hostplant range) use in British butterflies.

Hostplant status	Monophagous	Oligophagous I	Oligophagous II	Polyphagous
Britain (main)	23 (38.3%)	10 (16.7%)	21 (35.0%)	6 (10.0%)
Britain (all)	2 (3.3%)	5 (8.3%)	35 (58.3%)	18 (30.0%)
Northwest Europe	3 (5.0%)	1 (1.7%)	34 (56.7%)	22 (36.7%)

Percentages are for 60 British species. Modified table $\chi^2_{(6)} = 50.98$, $P < 0.0001$. There is a surplus of monophages and deficiency of oligophages and polyphages for British species when main plants alone are considered, whereas the classification of the full British hostplant list is not that different from the northwest Europe list (Bink, 1992). Monophagous, one plant; oligophagous I (narrow), one plant per region/locality and from one family only; oligophagous II (broad), several plants from one family; polyphagous, plants from >1 family. In all, 92 plants are used as main hostplants and 244 plants as main and subsidiary hostplants.

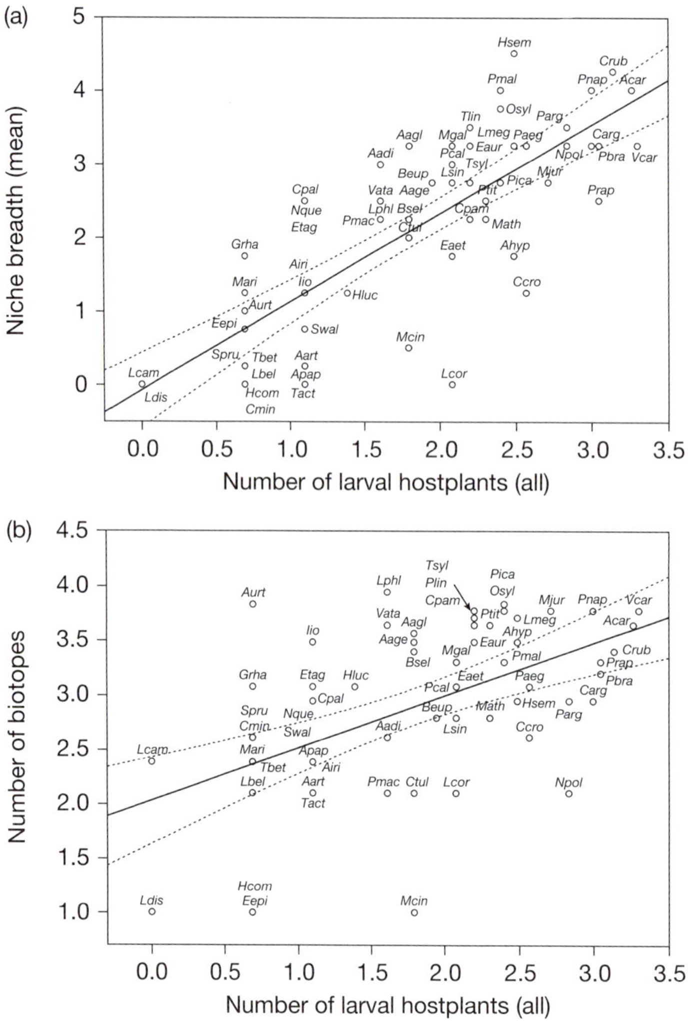

Fig. B2.3 (a) Niche breadth related to log number of all hostplants (all: main, secondary and novel). Niche breadth is range in Ellenberg values (see Chapter 5) (mean of light, soil moisture, pH and nitrogen scores for hostplants; Pearson $r = 0.78$, $R^2 = 61.0\%$, $P < 0.0001$). (b) Number of biotopes related to number of all hostplants (both log) ($r = 0.54$, $R^2 = 28.7\%$, $P < 0.0001$).

The degree of specialism clearly depends on just what is being referred to (see Chapter 5 for a wider explanation). Resources will be wider for a whole species taken over its geographical range than for a local population of the same species. Individuals also vary in the breadth of resources they are willing to use. Generalism or specialism involves choices made by individuals of different developmental stages (larvae and adults). As such, conflict in selectivity can arise, as for hostplants, between adults (egg laying) which more narrowly seek out suitable hosts, and larvae (herbivory) which will tend to be selected for diet breadth to maximize probability of survival on hostplants chosen by females (Wiklund, 1975). Usually, females ovipositing on suitable plants, and host choice, is largely genetically determined. However, mistakes are essential for evolutionary shifts in host choice and other factors may lead them to selecting unusual hosts:

- A low probability of finding alternative hosts.
- The insect has a high rate of egg production.
- Alternate hosts fluctuate in density (Jaenike, 1978).
- Variation in the threshold of acceptance of alternative plants.

Wiklund (1981) found individual thresholds to vary enormously in Swedish swallowtail *Papilio machaon*, in effect, for there to be generalist and specialist females. The different choices can be maintained in a community if the lower fecundity and higher offspring survival

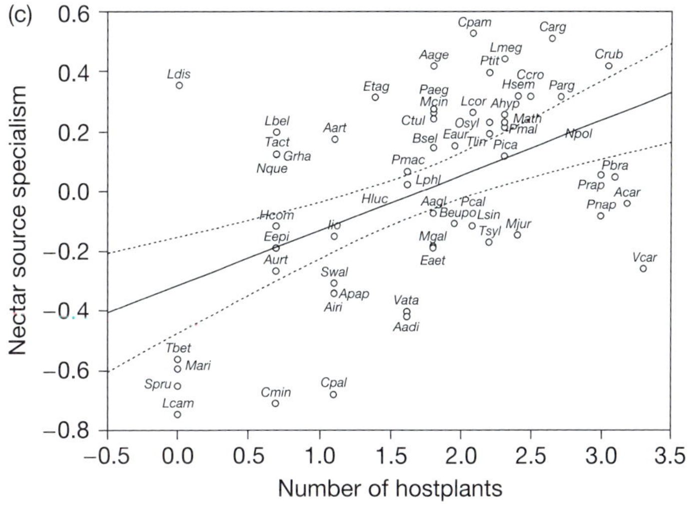

Fig. B2.3 *(continued)*
(c) Relationship between nectar source specialism (nectar sources standardized on feeding records) and larval hostplants (log, main and secondary) used by British butterflies ($r = 0.48$, $R^2 = 23.0\%$, $P < 0.0001$). Codes for species include the first generic letter and first three letters of the species name (see Appendix 1). Regression line with 95% confidence limits are shown. (From Hardy *et al.*, 2007, courtesy of Elsevier.)

associated with a specialist strategy is balanced by the higher fecundity and lower survival associated with a generalist strategy (Wiklund, 1981). Although survival will usually be highest on one plant this may not be consistent but be affected by extraneous conditions such as climate which affects the suitability of different hostplants for *Anthocharis cardamines* in dry and wet substrates in Sweden from year to year (Wiklund and Åhrberg, 1978).

The driving force behind specialism and generalism in resource use is predictability in resource availability and survival; a predictable outcome encourages specialization, and an unpredictable outcome generalism (Slobodkin and Sanders, 1969; Scriber, 1973; Futuyama, 1976; Wiklund, 1977). As such, generalism and specialism are closely linked to life history strategies. As Wiklund explains, when there exists a clear ranking in suitability of hostplant, this will inevitably lead to butterfly populations making a number of adjustments as to phenology, physiology and morphology to strengthen the bond (Ehrlich and Raven, 1965; Feeny, 1975). A number of factors affect predictability. It is enhanced by resource abundance, equivalence and longevity (stability) (Slobodkin and Sanders, 1969), and by the suitability of a resource, thus consistency in fecundity and survival of offspring to adulthood (Dethier, 1954; Feeny, 1975; Wiklund, 1977). Other factors considered to lead to specialism are selection for enemy-free space whether from predators (Ohsaki and Sato, 1999; but see Keese, 1997), parasites (Ohsaki and Sato, 1999) or competitors (MacArthur, 1958; Diamond, 1973) as these indirectly affect the suitability of the host medium. Field studies are urgently needed on the effect of hostplants and their milieu on insect fitness – at present these are rare (Mayhew, 1997). Fastidiousness affects predictability if only because of variation in host context, but a fraction of an apparent resource may actually be a suitable medium for oviposition; a resource needs to be very large or highly apparent and easily accessible to be a stable, consistent reserve for long-term exploitation and a predictable outcome. Against this, resource predictability may not be obvious to a human observer; the key is that it is predictable to the target organism. Specialization will also be greatly affected by migration (Wiklund and Åhrberg, 1978). In a two host model in which performance, a genetically determined quantitative character, is viewed against changes in population density, specialization is shown to depend on the amount of migration – intermediate migration rates resulting in specialization and small and large migrations leading to generalism (Ronce, 2001).

As regards resource status, the classification differs significantly if a resource type (viz., hostplants) is grouped into those that are 'main' (key or core ones) or 'all'. In the case of larval hostplants, all hostplants include those which are secondary for development and population persistence (see Appendix 2a). When tabulated, the distribution of herbivory types is similar to that over the European continent. The figures for classes in this table are not constant, but shift with additions, as new resources are found (e.g., Scotch argus *Erebia aethiops*), and deletions, owing to regional extinctions (e.g., English chequered skipper *Carterocephalus palaemon*).

necessarily be critical for the occupancy of a site or for mating success.

As an introduction to the resources listed for British butterflies in the appendices, it is useful to draw up some basic principles on resources and habitats. These are expanded on in the sections immediately following on consumable and utility resources, but are given here as a list:

- **P2.1: There are strong taxonomic relationships between butterfly and plant (hostplant taxa, nectar plants) that limit use of plants available in a biotope or vegetation unit.**
- **P2.2: Not all resource use differences among species are hard-wired to phylogeny; differences are also driven by distinctions in life history and environment (i.e., phenology, overwintering stage, voltinism, size, wing loading, pattern, etc.).**
- **P2.3: Some resource types are more limiting to habitat occupancy than others owing to the variability in requirements by species for different resources.**
- **P2.4: Climate (regional climate, local climate, microclimate, weather conditions) is a key factor in the exploitation of all resources, consumables as well as utilities, especially in high latitude regions. In effect, climatic requirements (light, heat, etc.) are resources.**
- **P2.5: A key to understanding resource use is the link between behaviour and substrate use.**
- **P2.6: Habitat dimensions for a butterfly species are crucially tied into butterfly mobility and the capacity for flight.**

Phylogeny, the evolutionary history of a group or lineage, has led to distinctions among species in resource use; there are strong links between specific butterfly and larval hostplant families that reflect on evolutionary associations dating back to the origins of flowering plants and butterflies (Ehrlich and Raven, 1965; Dennis, 1993a). It is important that special techniques are applied in comparative studies to account for relationships when differences between taxa (i.e., species) are sought for aspects of biology, ecology and life history (Harvey and Pagel, 1991; Purvis and Rambaut, 1995). A number of resources used by butterflies are closely tied to phylogeny (**P2.1**); in the case of larval hostplants this relationship is substantially obligatory. It may also be evident in other resource use such as adult feeding (nectar flower species; Hardy *et al.*, 2007) and mate location sites (Dennis and Shreeve, 1988). But not all resource use is explained by phylogeny (**P2.2**); instead relationships may be more closely related to differences in life history (e.g., emergence time is linked to overwintering stage; nectar plant use relates to proboscis length) that cut across lineages. Resource types (e.g., nectar, hostplants, mate location sites) in a habitat space assume importance in terms of the variability of outlets tolerated by a species (**P2.3**) and on their abundance and spatial arrangement (see Chapter 3). Absence of a suitable hostplant is obviously decisive. But it should not simply be assumed that no resources other than hostplants are critical for a species at a site. Just how critical a utility resource(s), such as mate location sites or roosts, is depends on its impact on productivity, development and survival. One of the most important resources, if not the most important resource, is the existence of light and warmth (**P2.4**). This includes suitable local climates, microclimates and weather conditions for each stage and phase of development. Climate is just one of a number of environmental conditions affecting butterflies which has been repeatedly shown at a variety of scales to be critical for butterfly populations, especially at higher latitudes where warmth and sunlight are at a premium (Shreeve, 1992a; Dennis, 1993a; Dover *et al.*, 1997; Karlsson and Wiklund, 2005). Warm, sunny conditions increase the number of potential resource elements (e.g., hostplants, mate location sites) that can be exploited by a butterfly species (Dennis, 1993a; Dennis and Sparks, 2005, 2006). Determining the importance of different resource types and outlets can only be achieved through careful autecological (species-specific) survey. This involves following individual productivity and survival through a life cycle and undertaking behavioural studies; to understand just what is important in a habitat space, it is necessary that behaviour is tied into the use of surface and substrate (**P2.5**) (Dennis, 2004a). Finally, just how large a habitat is likely to be for a species, whether it is possible for the assortment of resources a species requires to comprise a habitat, depends on how far resources are apart from one another in relation to the routine movements individual butterflies undertake throughout a day (**P2.6**). On the basis of these parameters a habitat can be described and it requires that the resources and movements be mapped as demonstrated by Vanreusel and Van Dyck (2007).

CONSUMABLES

Larval hostplants and herbivory

The key consumable resource for butterflies is their larval hostplants, which reflects on the tight phylogenetic relationship between butterfly and plant taxa (Ehrlich and Raven, 1965). The link is an ancient one involving co-evolution in an arms race between herbivores (butterflies) and plant chemical and physical defences; theoretical treatments accounting for plant–herbivore associations are given in Ehrlich and Raven (1965), Feeny (1975), Futuyama (1983) and Ackery (1988) (see Porter *et al.*, 1992, for a summary). This relationship between butterfly and hostplant is manifest even at the scale of the British Isles (see Appendix 2b) (Porter *et al.*, 1992). A list of hostplants is presented in Appendix 2a and statistics on host use in Appendix 4. It would have been thought that this list would be complete and definitive; unfortunately, this is far from the case. Plants have been given hostplant status from records of egg laying and laboratory rearing as well as larval incidence and repeated, clear evidence for successful breeding in the wild. Thus, we distinguish, as best we can, the status of plants from the records. Many of the records have simply been repeated from one author to the next (Hardy *et al.*, 2007; Dennis *et al.*, 2008). The most reliable will be those based on careful field observations over a number of years. A recurrent problem then – deserving emphasis – is the uncertainty concerning the status of hostplants (e.g., Scotch argus *Erebia aethiops*; see Franco *et al.*, 2006). Much survey work needs to be carried out to determine the main as opposed to secondary or subsidiary use by larvae (see Kemp *et al.*, 2008) and their success in using the hostplants listed in Appendix 2a. An original approach is being made at Oxford Brookes University by determining **phytoliths** (opaline silica found in plant cell walls) specific to grass species in the frass of grass-feeding larvae (T. G. Shreeve, personal communication).

From an insect/plant co-evolutionary viewpoint, hostplants must fulfil all the following criteria in order to be suitable:

1 Larvae must be stimulated to feed on the plant.
2 The plant must not contain substances toxic to larvae.
3 The plant must not have physical obstacles that cannot be overcome.
4 The plant must be nutritionally adequate for larval growth.
5 The microhabitat of the plant must be suitable for development.
6 The phenology of the plant and larvae must be synchronized.
7 The plant must be sufficiently abundant.
8 The milieu of the plant must provide suitable surroundings (for rest, anti-predator escape, etc.).

British butterfly species differ not only in their hostplant species but in the number of hostplants used – the degree of hostplant specialism or generalism (see Box 2.3; Appendix 4). The number of hostplants used by a butterfly will, of course, depend on how large an area is considered. A butterfly may be a regional or island-wide polyphage but a local or site monophage. On a national scale, most British butterflies are oligophagous and very few are truly polyphagous (e.g., holly blue *Celastrina argiolus*). Another, much simpler and perhaps equally useful, way of classifying hostplant use is just to record the number of hostplants used successfully by larvae (see Appendix 4). Compared with nectar sources, the current list of larval hostplants for the 60 native British butterflies is restricted numerically and taxonomically (Hardy and Dennis, 2008); the current list (1 July 2008) includes 269 larval hostplants (with species clearly distinguished), 89 of which are key to persistence in habitats and 58 are alien, comprising 21.6% of the total. The correlation across species for the three different classes of hostplant (total, main, main/secondary) is high and significant (Table 2.3). Some species, even some widely different butterfly taxa, share hostplant families and host species (e.g., Heteropterinae, Hesperiinae and Satyrinae use grasses; Parker, 2003). An important group (based on main hostplants, $N = 23$; all hostplants $N = 2$) are monophagous and therefore ultimately hostplant specialists (see Appendix 2a), but many more butterflies may rely on a single hostplant within a single site.

Some principles relating to larval hostplants in habitats are as follows:

- **P2.7: Larval hostplants are the key resource for populations; given a suitable climate, the quality and quantity of the hostplant primarily determines population size.**
- **P2.8: For butterflies that can exploit more plants there is a greater likelihood that they will have supplementary hostplants at a single site.**
- **P2.9: Supplementary hostplant species will generally broaden the habitat base for a species.**

Table 2.3 Correlations between different classes of butterfly larval hostplants and nectar plants.

	Hostplants (main)	Hostplants (native)	Hostplants (total)	Non-native hostplants (%)	Non-native nectar plants	Native nectar plants	Nectar plants (all)	Nectar specialism
Hostplants (main)	1.00							
Hostplants (native)[1]	0.63*	1.00						
Hostplants (total)[2]	0.61*	0.97*	1.00					
Non-native hostplants (%)[3]	−0.06	−0.00	0.20	1.00				
Non-native nectar plants[4]	0.32†	0.47*	0.51*	0.24	1.00			
Native nectar plants	0.40*	0.56*	0.60*	0.13	0.84*	1.00		
Nectar plants (all)	0.39*	0.56*	0.60*	0.18	0.91*	0.98*	1.00	
Nectar specialism[5]	0.45*	0.49*	0.46*	−0.18	0.10	0.17	0.16	1.00
Non-native nectar plants (%)[6]	0.23	0.40*	0.45*	0.26†	0.92*	0.64*	0.74*	0.05

Spearman correlations: † $P < 0.05$, * $P < 0.01$. All variables are number of plants except for percentages and nectar specialism. 1, Main and secondary native plants; 2, all hostplants including non-native (alien) plants; 3, percentage of total number of hostplants including non-native ones; 4, non-native (alien) plants; 5, total number of nectar plants used standardized by number of records (Hardy *et al.*, 2007); 6, non-native plants as percentage of total number of nectar plants.

- **P2.10: The extent to which a hostplant(s) is used within sites depends very much on context, weather conditions and the seasonal climate.**
- **P2.11: Hostplant chemistry and physical defences are key to the life history of butterfly species, underlying number of broods and chemical defences.**
- **P2.12: Many butterfly species share hostplants, especially grass feeders, but usually do not exploit the same area or part of the hostplant.**
- **P2.13: Butterfly species vary widely in larval host use, from generalists (polyphages) to specialists (monophages). Specialism is a distinctive feature of occupation of single sites.**
- **P2.14: Annual plants rarely form the primary larval hostplant.**

Larval hostplants are the key habitat resources for any species; without the presence of a hostplant, there can be no reproduction and any 'population' (e.g., a migrant swarm of *Vanessa cardui*) is temporary (**P2.7**). However, it must be emphasized that the presence of a hostplant, even in suitable condition, is not sufficient to determine the existence of a habitat. Although numbers of butterfly species are limited in hostplant use (see above), many butterflies often use more than one hostplant over a region (e.g., *Celastrina argiolus*). If they do, the chances are that the number of hostplants for a species occurring and used at any site will mirror, but be a smaller fraction of, the number of hostplants used over the wider region (**P2.8**). As hostplants are key in population production and maintenance, the more hostplants (any resource, in fact) a species has, the greater the potential for habitat creation and expansion. The spatial aspects of habitat are considered in the next chapter; here, it is sufficient to provide a foretaste of later sections on habitat structure. Supplementary resources are likely to broaden the potential base of the habitat as different plants are likely to occur in different conditions and in different parts of a site (**P2.9**). It is important to understand that butterfly larvae are mobile and can make choices about food items (see Chapter 4). Individual larvae also vary in their choice of and 'performance' on different plants and plant parts (Singer *et al.*, 1988). At the same time, it is worth emphasizing that the existence of a habitat is governed by climate; a woodland butterfly, like the speckled wood *Pararge aegeria*, cannot persist long if woodlands occur in areas too cold for the butterfly to be active. In the same way, access for a butterfly and local climate (e.g., light, warmth) determines which part of a site can be used, or is more suitable (see section on thermoregulation below), despite the presence of what looks to be suitable patches of hostplant (**P2.10**).

It becomes quickly obvious to those observing different butterfly species in one area that hostplants do not provide the same opportunities to these different species; if all hostplants are abundant, this may appear rather puzzling. The answer lies in what is invisible to the human eye – food chemistry and physical defences (e.g., thickened cell walls, canal networks, **laticifers** (elongated secretory cells found in leaves and/or stems that produce latex as secondary metabolites)) (see Fox, 1996). A change in one chemical in a plant (e.g., nitrogen) is often accompanied by changes in other chemicals (e.g., sugars, defence chemicals). A study by Cizek *et al.* (2006) nicely demonstrates what has been suspected for some time, that foliage feeders on plants with quantitative defences (e.g., woody plants and grasses protected by macromolecules, leaf toughness, low water and/or nitrogen content) have fewer broods each year than those feeding on plants with qualitative defences (e.g., specific allelochemicals) (**P2.11**).

A further observation is that although many species share the same plant species (e.g., bird's foot trefoil *Lotus corniculatus* feeders such as the dingy skipper *Erynnis tages* and common blue *Polyommatus icarus* (Gutiérrez *et al.*, 2001); grass feeders such as the meadow brown *Maniola jurtina* and gatekeeper *Pyronia tithonus* (Dennis, 2004a)), they are not equally abundant nor necessarily found concentrated in the same parts of the same site (**P2.12**). For grass feeders, using the same grass species, this may relate to exploiting those hostplants in different areas of a site; plants that are subject to different levels of herbivory thus accumulate different levels of silica (phytolith) defences (Massey and Hartley, 2007). They may also not share the same plant part (see Appendix 4); the green-veined white *Pieris napi* and *Anthocharis cardamines* exploit the same crucifer hostplants, but *P. napi* eats the leaves and *A. cardamines* the developing pods. An extension of this observation is that species vary substantially in degrees of specialism and generalism in host use (**P2.13**). These terms are most satisfactorily applied to the scale of single sites, the habitat, as it is at this scale where it matters for conservation purposes and can be unequivocally demonstrated. Thus part of the reason that *P. icarus* and *E. tages*, which both share *L. corniculatus* as larval food, do not occur in identical areas is that *E. tages* is more specialized in host use and for other resources. Hostplants range over a variety of life forms, including trees, climbers, grasses and forbs (see Appendix 2). One interesting observation is that annual plants figure among larval hostplants less frequently than expected (**P2.14**) (Kemp *et al.*, 2008). This may relate to their spatial and temporal unpredictability; they figure prominently in disturbed sites and are quickly lost with vegetation succession. Only one British butterfly species persists on an annual plant, the heath fritillary *Melitaea athalia* on cow wheat *Melampyrum pratense*, and this likely reflects on cow wheat's greater predictability than most annuals with its semi-parasitic life history and ant dispersal. Even so, *M. athalia* is declining more rapidly than most species in Britain (Fox *et al.*, 2006).

Nectar sources and adult food

Adults feed on more than just nectar flowers (e.g., purple emperor *Apatura iris*) and these observations are recorded in Appendix 3. Beck and Fiedler (2009) found that feeding habit links strongly with adult lifespans of butterflies, with pollen feeders living longer than fruit feeders, which in turn live longer than nectar feeders. Adult feeding, including nectar consumption, may well compensate for inadequate resources in larval feeding (e.g., European map *Araschnia levana*) (Mevi-Schütz and Erhardt, 2005). Interestingly, some adults feed on the flowers of plants that are consumed by their larvae (Hardy *et al.*, 2007). The intriguing evolutionary implications, including potential mutual benefits of food for pollination, are discussed by Erhardt and Mevi-Schütz (2009). Both *Pieris napi* and *Anthocharis cardamines* do this; in the case of the latter species, the flowers form the larval food. Nectar and other adult food resources can form a vital part in the life history of butterflies (Hardy *et al.*, 2007; Erhardt and Mevi-Schütz, 2009). Nectar, and dissolved pollen, includes a wide variety of important chemicals (e.g., alkaloids, amino acids, lipids, sugars, vitamins) as do other adult food resources which include ash, earth, water, fruit, honeydew, sap, animal excretions, cuckoo spit and corpses (e.g., nutrients, trace elements, nitrogenous compounds, sugars) (Porter *et al.*, 1992; Shreeve, 1992a; Dennis, 2005a). **Puddling** is a commonly observed activity in congregations of male butterflies in tropical countries and, in the UK, butterflies have been observed to puddle singly (e.g., purple hairstreak *Favonius quercus* on a Dorset beach; R. L. H. Dennis, personal observation). Nectar availability has been found to increase butterfly longevity and fecundity (Stern and Smith, 1960; Murphy *et al.*, 1983; Murphy, 1984; Wiklund and Karlsson, 1984; Boggs, 1987; Shreeve, 1992a; Boggs and Ross, 1993; but see also Erhardt and Mevi-Schütz, 2009), particularly for income breeders as opposed to capital breeders relying more on larval investments (Boggs, 1990; Bergstrom and Wiklund, 2002), and yet nectar plants are often disregarded in the determination of habitat patchworks for conservation purposes. Nectar sources can clearly influence the microdistribution of butterflies at sites (within habitats) (e.g., *Maniola jurtina*; Brakefield, 1982a) and limit populations (western checkerspot *Euphydryas chalcedona* in USA (Murphy *et al.*, 1984); chequered skipper *Carterocephalus palaemon* in Scotland (Ravenscroft, 1994b)). Even so, use of nectar plants by adults is catholic compared with their limited hostplant dependency and owing to limitations of space it is only possible to tabulate the most prominent nectar sources for British butterflies in Appendix 3 (a web site there gives a full listing). A previous list of butterfly nectar sources is given in Porter *et al.* (1992).

Although perhaps not to the same degree, much as there are hostplant specialists, there are nectar use specialists (see Appendix 4) (Tudor *et al.*, 2004). The reasons for variable nectar flower species use among British butterflies have been explored by Porter *et al.* (1992) and Corbet (2000). They found that profitability of butterfly foraging depends both on flower and butterfly physical structures, specifically on the corolla depth and clustering of flowers and the proboscis (tongue) length and **wing loading** (wing area to body mass) of butterflies. Small, short-tongued butterflies do not visit deep flowers (but see Tiple *et al.*, 2009) and butterflies with high wing loading generally confine visits to flowers that exist in clusters or that are very nectar-rich. Thus lineage relationships emerge: among British butterflies nymphalids, which have long tongues and high wing loading, tend to visit deep massed flowers, whereas lycaenids with short tongues and low wing loading do not. Scent and colour – butterflies have a very wide visual spectral sensitivity – very likely act as cues for appropriate nectar flower sources (Boggs, 1987; Omura *et al.*, 1999; Anderson, 2003; Omura and Honda, 2005) and butterfly species are found to have preferences for different flower colours (Tiple *et al.*, 2009). Butterflies also have the capacity for rapidly associating (learning) resource quality with cues such as colour (Weiss, 2001; Weiss and Papaj, 2003).

For conservation purposes, it is important to know which food sources are most valuable for adults. Key plant families and species, and non-plant food sources for adults, are listed in Appendix 3. A major feature, differing from larval hostplants, is that the list for nectar plant species used is much longer (N = 435 species), with non-native nectar plant species forming a larger proportion of the total (N = 218 alien versus 217 native; 50.1%) than do non-native larval hostplant species (Fig. 2.1). If wider plant groups are taken (including plants not distinguished to species level), then the figures are 531 plant groups: 273 non-native and 251 native plants (Hardy and Dennis, 2008). Divisions are found in plant life history that, influencing such physical attributes as corolla depth and flowering period, affect their value as nectar sources for butterflies (Hardy and Dennis, 2008).

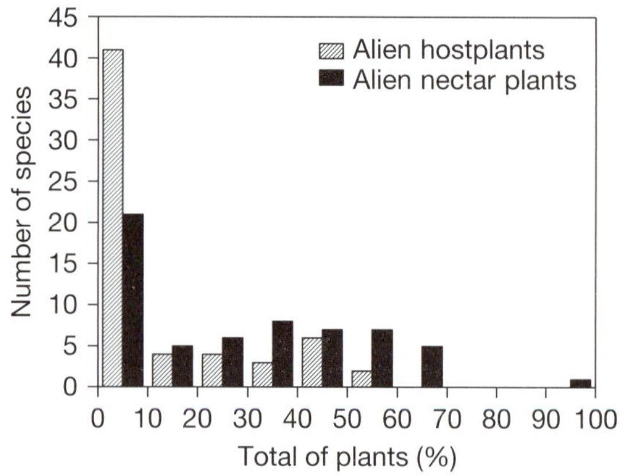

Fig. 2.1 Use of non-native (alien) plants as nectar sources and larval hostplants by British butterflies (Wilcoxon matched pairs test: $T = 193.0$, $Z = 3.80$, $P = 0.0002$, $N = 60$; medians: alien hostplants 0%, alien nectar plants 27.9%).

Some principles can be drawn up relating to adult feeding:

- **P2.15: Butterflies use many more nectar flowering plant species as adults than they do hostplant species as larvae.**
- **P2.16: Some species use a wide range of non-nectar sources for adult food.**
- **P2.17: Trees and annual plants do not figure prominently in adult nectar food.**
- **P2.18: Butterflies share many nectar flowers; some plants are prominent nectar flower species whereas others are under-utilized or non-producers of nectar.**
- **P2.19: Some butterfly species are nectar generalists, while others are nectar specialists exploiting few nectar flowering plant species.**

The prime observation in adult feeding is that it is generally less restricting on species generally, and on individuals at sites, than larval host exploitation (**P2.15**), though it would be unwise to confer to this observation the status of an invariable rule; the large heath *Coenonympha tullia* has been found on sites that, temporarily at least, lacked any nectar sources (R. L. H. Dennis, personal observation). But then many butterflies use, and may depend on, a range of non-flowering sources as adult food (**P2.16**); an important example is *Apatura iris* (see Appendix 3e; Hardy *et al.*, 2007). Trees often provide poor nectar sources (Tiple *et al.*, 2005), as do annual plants (**P2.17**) (Corbet, 2000; Kitahara *et al.*, 2008). In the case of annual plants, with single flowers, this may well relate to their providing limited **aliquots** (an integral portion) of food (Corbet, 2000) and for foraging insects this may mean that energy expended in foraging exceeds energy uptake. Further investigation is needed to understand why tree flower nectar (e.g., hawthorn *Crataegus* sp., rowan *Sorbus aucuparia* L., elder *Sambucus nigra* L.) is not so widely used; perhaps a factor is nectar quality, but part of the answer probably lies with marginal biotope (ex-habitat) locations and vulnerability to predation in wood edge situations. In nectar feeding, some flowering plants are prominent in being used by many species (thistles *Cirsium* sp., bramble *Rubus fruticosus*), whereas others are less often used than expected and others produce no nectar (e.g., wood anemone *Anemone nemorosa* L.) (**P2.18**). This can be owing to a long tubular corolla, too long for many species (e.g., three failed attempts were observed by a male *Maniola jurtina* to feed on foxglove *Digitalis purpurea* on Alderley Edge (Cheshire, UK) in July 2007; R. L. H. Dennis, personal observation), but it is often because the plants flower when most butterflies have not yet emerged in spring (e.g., cuckoo flower *Cardamine pratensis*) or are in the early stages of hibernating in the autumn (e.g., ivy *Hedera helix*). These plants can nevertheless be vitally important for the few species using them, in the case of the two examples, *Anthocharis cardamines* and *Vanessa atalanta*, respectively. In the same way, some butterfly species are found to be generalists when feeding on flowering plants whereas others are specialists (**P2.19**). This observation is often linked to varying availability of suitable nectar plants in different biotopes; woodland butterflies, especially fritillaries, tend to be specialists (Tudor *et al.*, 2004). Corbet (2000) has isolated an important relationship between tongue length and wing loading in butterflies; butterflies heavy for their wing areas need larger energy reserves. Butterflies with high wing loading tend to have long tongues, thus can access most flowers, and concentrate feeding on plants with masses of flowers (e.g., nymphalids on *Buddleja davidii*). Butterflies with low wing loading may be restricted by competition to plants with fewer flowers, but owing to their smaller size and shorter proboscises, thus more rapid nectar uptake, may be able to consume smaller quantities but sufficient nectar to better advantage (Tiple *et al.*, 2009).

UTILITIES

Butterflies require utility resources at every stage in development. Just what they use for different activities is complicated by the range of choice at various scales, as for nectar and hostplant selection, from the whole site (selection among resource outlets, areas or patches), to part of a single resource patch or area, to single resource items or neighbouring items and, in turn, to their parts and surfaces (Fig. 2.2). Choice at these various scales is often linked to distinctive butterfly behaviour (Appendix 6) but much has yet to be learnt how choice is linked to biology, learning, environmental variation and opportunity. Some behaviours are not fully understood (e.g., grooming; Dockery, 2005). For instance, how is adult roost site selection affected by roosting behaviour (whether single or gregarious), availability of different features or weather conditions? How are adult basking sites affected by basking mode – whether a species basks side on (lateral basker) or with fully exposed wings, and whether the wings absorb energy (absorbance basker) or reflect it onto the body (reflectance basker) (Shreeve, 1992a; Dennis, 1993a)? Clearly territorial perches depend on a butterfly species being a territorial percher as opposed to a patroller, but do they differ in their associations with physical structures for acquiring mates (Dennis and Shreeve, 1988)? Some utility resources are intimately associated with consumable resources. For example, sites for egg laying and substrates used for egg laying are obviously proximate to the hostplant; similarly, so are structures used by larvae when not feeding, particularly night feeders. But, how capable are species in 'sensing' resources they seek in different situations? Many questions remain for the next generation of butterfly ecologists; do not expect them all to be conveniently answered here!

Below, a brief account is given of utility resources, and accompanying behaviour, demonstrated to be of importance for individual development and survival, together with relevant references that attempt to explain the range in behaviour. An introduction to butterfly ecology and behaviour is provided in Dennis (1992a), the contributing authors are referenced at the outset of each topic.

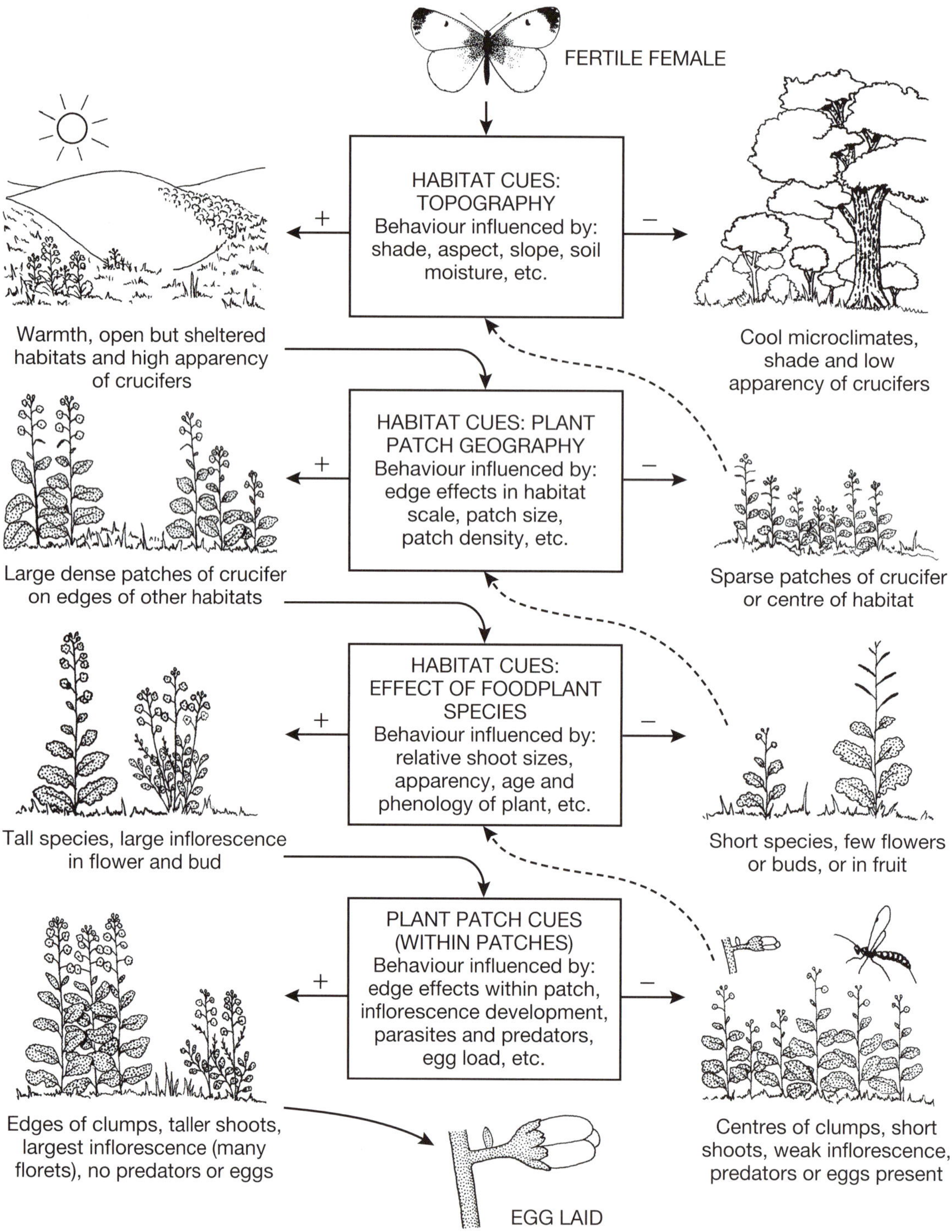

Fig. 2.2 Hierarchy in decision making for a fertile female orange-tip butterfly *Anthocharis cardamines*. The butterfly responds to cues at different scales: landscape, habitat, within hostplant patches and for individual host plants. (From Porter, 1992, courtesy of Oxford University Press.)

Adult basking sites and behaviour

(Shreeve, 1992a:22–30)

Whatever the stage of development, maintaining body heat (**thermoregulation**) is central to every activity during a butterfly's existence (Clench, 1966). Butterflies are largely **ectotherms**, their body temperatures dependent largely on ambient conditions of temperature and radiation, and they have developed mechanisms for exploiting these conditions in all life history stages (Dennis, 1993a). Whether an adult butterfly is feeding, mate locating, evading predators, roosting or hibernating, it is also maintaining body warmth and water balance; choice of sites and substrates for whatever reason will be undertaken with these objectives. Body temperature is influenced not just by surrounding ambient temperatures (microclimate, not screen shade temperatures) but also by solar radiation and air movement. Butterflies in all stages have developed a variety of behavioural ploys, along with physical and physiological mechanisms, to maintain warmth (Figs 2.3 and 2.4; see also Fig. 4.5). Thermoreceptors located in the wings and antennae of some darkly pigmented butterflies (e.g., *Pachliopta aristolochiae* (Fabricius, 1775)) protect them from heat damage while sun basking (Campbell *et al.*, 2002). The behaviour adopted depends on the need for activity and the relation of body temperature to ambient conditions (Table 2.4); each butterfly species has optimal body heat for different activities

Fig. 2.3 Male grayling *Hipparchia semele* lateral or side basking whilst territorial perching on sand dunes in sunny but cool conditions at Pembrey Country Park, Dyfed, South Wales. (Courtesy of Tim Shreeve.)

Fig. 2.4 Red admiral *Vanessa atalanta* dorsal basking, head down and hindwings appressed, on an apple tree trunk in a Great Orme garden, North Wales. (See Dennis and Asher, 2009.)

Table 2.4 Basic patterns of activity in temperate butterflies in relation to body temperature and solar radiation. The behavioural responses of butterflies vary in relation to initial body temperatures depending on ambient conditions, previous thermoregulatory behaviour and previous activities. A, active; I, inactive.

	Sun in	Sun out
Body cold	A: Roost, settle; use warm surfaces & surroundings; appression & flight / I: Night: roost. Daytime: roost, settle & select warm surroundings	A: Basking behaviour; shelter, sun aspect, warm surroundings & micro-landforms / I: Basking behaviour, feeding & resting
Body hot	A: Reduced flight activity; use of cooler surroundings / I: Night: roost. Daytime: rest, roost, in/on surroundings	A: Selection of cooler surroundings; evasion of direct sunlight when settled / I: Settled (rest), feeding or daytime roosts in shade

(Kingsolver, 1985). When it is cold and cloudy, butterflies select surfaces that maintain body heat; when the sun is not obscured, the hotter the butterflies become. If body temperature goes above optimal they will reduce activity, becoming inactive or switching their activity to cooler surroundings. When the sun is out and the butterflies are cold but still active, they will use shelter and sunny positions, and absorb energy by basking, by selecting warmer surfaces, and by undergoing body and wing **appression** – making contact with the underlying surface (Fig. 2.4). When overheated and the sun is out, they will actively evade direct radiation, and adopt cooler environments and reduce activity; when settled, if in the sun, they will close their wings and align 0° to the sun's rays. This is a brief and necessarily oversimplified version of responses to different conditions; a more detailed account of thermoregulation is given in Dennis (1993a).

Most species are **dorsal baskers**; energy is transferred to the body by absorbance or reflectance. Many fewer are obligate lateral absorbance baskers (e.g., *Callophrys rubi*, *Gonepteryx rhamni*), though all butterflies can absorb energy when their wings are closed, even when feeding. But, even apparently obligate **lateral baskers** are sometimes observed to bask with their wings open (e.g., grayling *Hipparchia semele*; P. B. Hardy in Dennis, 1992b) and normally dorsal basking species may adopt full lateral basking with appropriate postural changes (e.g., meadow brown *Maniola jurtina*; Maier, 1998).

There are some crucial principles concerning thermoregulation and habitats:

- **P2.20: Individual butterflies respond to changing conditions by adjusting to their surroundings; cooler and warmer sites are selected within a habitat space as appropriate to maintaining an adequate body temperature for activity.**

• **P2.21: Different surroundings and surfaces are needed at different times (for different time scales) depending on conditions of radiation, ambient air temperature and air movement, as well as individual activity.**
• **P2.22: As resource use depends on body heat and water balance, supplementary resources cater for different circumstances and surroundings accessible to most individual adults and larvae in the population.**
• **P2.23: A vital part of thermoregulation by individuals is the selection of surroundings (landscape components) and surfaces at a range of spatial scales.**
• **P2.24: Use of landscape components at one scale, for one activity, can dominate their use at another scale.**
• **P2.25: Species differ in resource use and biology, thus behaviour within the same sites, and therefore differ in the surroundings used for retaining body heat and water.**

A fundamental part of thermoregulation is the active selection of different surroundings and surfaces by individuals subject to varying conditions of radiation, heat and wind movement (**P2.20**). Individuals in a single population require different surroundings and surfaces within the habitat space for thermoregulation from season to season (broods), on separate days, at different times of day; and even from minute to minute, in response to sun angle, radiation load, ambient air temperature and wind (**P2.21**). During a day, male *Pararge aegeria* follow the emergence of new sun spots in a wood (Davies, 1978) and male *Anthocharis cardamines* continually shift the path of flyways along which they patrol to remain in the sun (Dennis, 1982b). Weather changes can be very sudden and responses by individuals are often immediate. An essential prerequisite of resource use within habitat space is suitable conditions of light, warmth and moisture for exploitation. Supplementary resources in different circumstances form key habitat components for species (**P2.22**); exposed, sheltered and shaded hostplant and nectar patches will be used in different conditions, for instance, calm, windy and hot bright weather, respectively. Perching or settling heights vary for many species throughout a day and throughout each season; higher perches are usually used earlier and later in the day, very likely relating to roosting, whereas, lower perches are used earlier (especially) and later in the season, ground temperatures providing the warmest conditions unless wet (e.g., small tortoiseshell *Aglais urticae*) (Hardy and Dennis, 2007).

One aspect of behaviour to have been isolated recently is that butterflies are responding to the thermal environment at different spatial scales (**P2.23**). Nymphalids in the Bollin Valley, Cheshire, UK were found to be selecting perch sites in spring at the scale of the local topography (south-facing valley sides), at the scale of mini-landforms (molehills, bare earth), and at the scale of micro-features on the mini-landforms (south-facing sheltered slopes, hollows on molehills, protection of objects such as sticks, and grass margins of bare ground patches) (Dennis, 2004c; Dennis and Sparks, 2005). In doing this, it was evident that the selection of landforms at one scale dominated the use of others at a different scale (**P2.24**). In the case of perching nymphalids in the Bollin Valley, side of the valley (slope aspect) dominated the use of mini-landforms; molehills were not used on the shaded valley side or out of direct sunlight in the cool spring conditions. Clearly, just how individuals react to weather conditions, how they thermoregulate, and what objects they use to do so, depends on their resource use within the same site (**P2.25**). A butterfly with its key resource, a hostplant, in an exposed part of a site (e.g., *A. urticae*) will use very different landform components and elements for thermoregulation, than one in a more sheltered part of the same site.

Mate location sites, substrates and behaviour (Shreeve, 1992a:34–45)

For populations to persist, the sexes need to be able find one another; in effect, they are a resource for one another and the reader is recommended to a detailed examination of mate behaviour by Wickman (2009). Small animals like butterflies do this by using visual and olfactory (chemical) signals (Boppré, 1984; Silberglied, 1984); much still needs to be known about the contribution of scent and whether butterflies engage in scent marking. The time the sexes spend in mate location and mating varies; pairs can remain in copula for several hours and in this situation are vulnerable to predators (Wiklund, 1985). Typically, male butterflies mate repeatedly during their lifetime whereas there are limits to the number of times female butterflies mate (Shreeve, 1992a) – some females are deterred from further mating by receiving a mating plug or **sphragis** from a male, a seal over the bursa copulatrix (e.g., marsh fritillary *Euphydryas aurinia*, *Hipparchia semele*) (Porter, 1992). The majority of females are **monandrous** (usually mate once), though females mate several times (**polyandry**) in many butterfly species (e.g., swallowtail *Papilio machaon*, *Pieris* sp.) (Karlsson, 1995, 1996). Mated females typically evade males by using distinctive **mate-rejection behaviour** and occupying areas away from concentrations of males, the rest of their active time being spent feeding and egg laying (Shreeve, 1992a; Shreeve *et al.*, 2006). An interesting finding by Karlsson (1995) is that in polyandrous butterflies, males deliver heavier and more nutrient-rich ejaculates (**spermatorphores** or sperm packages; see Porter, 1992:56) than males in species where females mate only once. This suggests that males in polyandrous species are selected for having high-quality reserves, and indeed males in polyandrous species have higher amounts of nitrogen in their abdomens, probably a form of sexual selection (Karlsson, 1996; Wickman, 2009). Thus, high-quality males can also be a limited resource for females (Boggs and Gilbert, 1979). Polyandry may be indicative of variation in larval food or other environmental factors that affect adult size and a suite of morphological variables is correlated with increasing polyandry (Wickman, 2009).

Receptive females are a scarcer resource for males, a rapidly diminishing one as the season progresses, and there is intense competition among males for females, the latter benefiting from **nuptial gifts** (Boggs, 1990). Thus, males invest most of their energy in acquiring mates and fighting off or evading competition from other males. In doing so, they have developed two basic ploys; one is to search for mates whilst in flight (**patrol**), the other is to settle and wait (**perch**) (Scott, 1974; Shreeve, 1992a). In both activities, landscape cues are used and patrolling is not a random exploration of the habitat (Wickman, 2009). In some species, such as the wall brown *Lasiommata megera* and large skipper *Ochlodes sylvanus*, these activities are associated with time of day or weather conditions (Fig. 2.5) (Dennis, 1982–1983; Dennis and Williams, 1987). Species that perch differ in morphology (larger thorax : body ratio, higher wing loading, higher **aspect ratios** (wing length to breadth ratio); Wickman, 1992), even individuals of the same species (e.g., *Pararge aegeria*; Van Dyck *et al.*, 1997) do so in relation to individual male strategy. In both these activities they can to all appearances ignore competition from other males (e.g., *Pieris napi*; Dennis, 1982b) or engage in defence by being territorial (occupying and defending a specific area). Such defence is most aggressive when linked to distinct landmarks and their surrounds (a territory); territory holders are those with some physical advantage in fighting capability such as size and aerial manoeuvrability (Kemp and Wiklund,

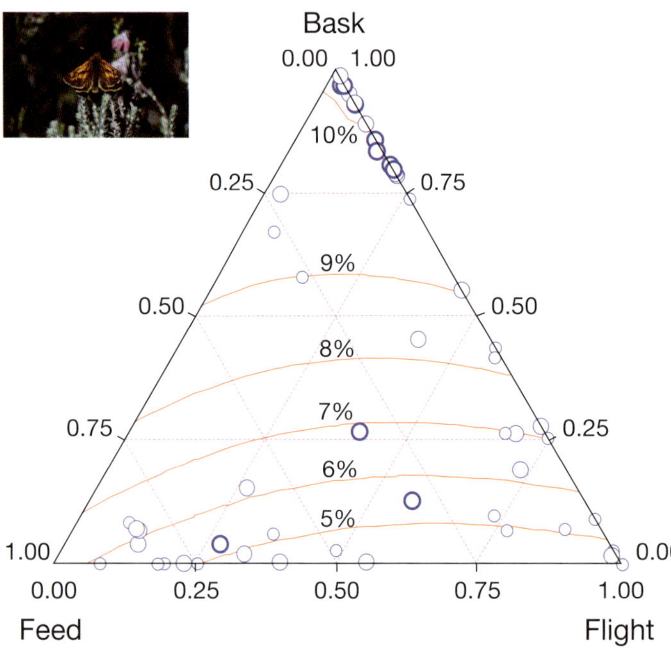

Fig. 2.5 Territorial behaviour and opportunism in mate locating in male large skippers *Ochlodes sylvanus* (inset). Interspecific interactions occupy the entire ternary graph although territoriality peaks when butterfly males are settled and basking typically during afternoons. *O. sylvanus* males defend territories to acquire mates (basking), but also patrol habitats for them (flight), particularly wooded or shrubby edges, and even attempt courtship and skirmish with other males when feeding. Each symbol is one of 54 individual cases monitored on Lindow Common, Wilmslow, Cheshire, UK in 1983 for <300 s (small), 300–600 s (medium) and >600 s (large); isolines indicate interactions (% total time), skirmishes with other males and attempted courtship with females (From Dennis and Williams, 1987, courtesy of the Lepidopterists' Society, USA.)

2001). Territories are typically established at sites where females are most likely to pass or are likely to solicit mating: resource zones, particularly larval hostplant patches (e.g., *Aglais urticae*; Baker, 1972) or nectar flower patches (e.g., purple-edged copper *Lycaena hippothoe*; Fischer and Fiedler, 2001), but especially at physical landmarks forming flyways in sunshine (e.g., vegetation and substrate edges, cross-roads, tracks, rock outcrops, peaks, sunspots) (Baker, 1972; Dennis and Shreeve, 1988). In some species there are clearly key landmarks on which numerous males aggregate to form what in effect is a **lek** in which territorial defence is reduced (e.g., *Plebejus argus*, *Ochlodes sylvanus*; Dennis and Williams, 1987; Dennis, 2004b). The ploy used at any time or situation is likely to be determined by opportunities for mating and energy expenditure (Baker, 1972; Dennis and Shreeve, 1988; Rutowski, 1991). The key habitat issue is that not all of any one site occupied by a butterfly population presents equivalent opportunities for locating mates; zones of patrolling and perching males generally occupy small fractions of habitat areas.

The following principles concern mate location relating to habitats:

- **P2.26: The key to successful mate location is apparency of the sexes to one another, but many species seem to lack physical (bright colours; ultraviolet reflectance) or chemical (distinctive pheromones for distant communication) apparency.**
- **P2.27: There is exacting competition for mates and males tend invariably to be aware of conspecifics, especially females, when engaged in activities other than mate location (e.g., feeding).**
- **P2.28: Landscape components (landforms, landscape features) enhance apparency for mate location.**
- **P2.29: A key feature of site selection for activities, and therefore for mate location, is that they ensure optimal flight conditions (optimal body temperatures).**
- **P2.30: Larger landscape features (areas) are used more by territorial patrolling butterflies than by territorial perching butterflies, which tend to return to perches after sorties with passing insects.**
- **P2.31: As a consequence of competition for territories, individual (male) butterflies spread out over sites, some adopting suboptimal positions. Where sites are limited in number, this can break down or evolve into a situation in which males can become concentrated at sites to form leks.**
- **P2.32: A number of landforms and features are used in common by different species, although commonality of landscape components for mate location can have a taxonomic bias.**
- **P2.33: The distribution of females over a habitat is not the same as the distribution of virgin females, or females prepared to mate again, that are available to males.**

The key to mate location is apparency, either in vision, scent or both (**P2.26**). An increasing amount is being discovered about visual apparency in sexual communication, and the importance of vision and patterns in the ultraviolet range (Silberglied, 1984) but, although use of pheromones is widely appreciated in moth communication, less is known about communication using scent and scent marking among butterflies (Boppré, 1984; Wickman, 2009). Several observations are relevant. First, individuals of numerous butterfly species do not seem to be aware of one another until they are very close and even then territorial males will often rise to meet and chase non-conspecifics (R. L. H. Dennis, personal observation), even different insect taxa; this not only takes them away from perch sites but also wastes energy. Second, not all butterfly species immediately recognize their hostplants using scent cues and will be seen 'tasting' non-hostplants with their tarsi (e.g., *Vanessa atalanta*, large white *Pieris brassicae*; R. L. H. Dennis, personal observation). This suggests that scent communication may not figure prominently in distant sexual communication. Third, landscape elements are regularly used by species for mate location and are clearly of great importance in creating contacts for the sexes; again, this suggests that scent or vision alone is not enough to tag conspecifics. Reasons were outlined above for severe competition amongst males for females; an indication of this is that every opportunity is used for courtship. Thus, males will attempt courtship at nectaring sites and will investigate all other insects that resemble females of their own species during daily routines (**P2.27**). This is a cautionary note not to interpret resource use simply as sequential resource exploitation; males are always capable of multitasking when impressing females is one of the objectives!

The key principle in this section relating to habitats is that landscape elements and features enhance mate location (**P2.28**). They do this by simply being more evident, apparent, than the butterflies themselves. These features vary in scale from small piles of gravel to mountain peaks (Shields, 1967; Dennis, 1982–1983; Dennis and Shreeve, 1988; Dennis and Dennis, 2006, 2007). But not just any landscape feature will do for mate location; males need to be in optimal condition to intercept passing females and competitors, and bearing in mind that such situations are likely to draw the attention of enemies, to be in condition also to evade predators. Thus, perching and patrolling zones, in mid-latitudinal regions such as Britain, tend to be warm, in shelter and sunshine (**P2.29**). In using landscape elements, patrolling butterflies adopt large swathes of landscape in the sunshine, such as wood edges, hedgerows, lanes and river banks (e.g., *Anthocharis cardamines*, *P. brassicae*), whereas perching butterflies pick smaller zones, 'peaks' in the landscape including hedge intersections, lane junctions, a gateway, a sunspot, a cluster of molehills or a hostplant patch (e.g., *Inachis io*, *Aglais urticae*, comma *Polygonia c-album*) (**P2.30**). This distinction is most evident when the same butterfly switches between perching and patrolling ploys (e.g., *Lasiommata megera* (Dennis, 1982–1983), *Ochlodes sylvanus* (Dennis and Williams, 1987)). The severe competition for perch sites typically results in an approximately even distribution (referred to as **over-distribution**) of males throughout sites. This does not mean that there is a super-abundance of sites or that individual site values for mate location are the same. The contrary situation occurs, in fact; dominant males pick the best sites and subordinate ones poorer sites where contacts are likely to be less productive. However, when the male population is large and competing for the same mate location perch, then competition can break down with males 'sharing' the site and forming leks (**P2.31**) (Baker, 1972); females then have the pick of males or yield to the most persistent male. It is inevitable that different species use similar landforms and landscape features for mate location. This more typically occurs if they belong to the same taxon; that is, there is a phylogenetic bias for mate location sites (**P2.32**). Thus nymphalids use similar territorial perches (field corners, bare ground on slope summits) and pierids use similar flyways for patrolling (e.g., south-facing wood edges). An important observation is that the distribution of available females within habitats is not the same thing as the distribution of females (**P2.33**) (Wickman, 2009). The former will be in much shorter supply and the behaviour of recently mated and unmated females is typically very different, including their orientation to landscape features such as bushes (e.g., small heath *Coenonympha pamphilus*; Wickman, 2009).

Egg-laying sites and substrates

(Porter, 1992:46–72)

Butterfly eggs are not simply dropped anywhere while the female is in flight; egg structure, the shape and composition of each species' egg, is intimately associated with placement in its microenvironment (Porter, 1992). The interior of the egg of *Polyommatus icarus* is silvery, akin to a thermos flask, and it is interesting to speculate on whether this contributes to their having such an extensive geographical range. Butterfly larvae depend largely on female behaviour for their ultimate destinations on hostplants. Although later instar larvae can be highly mobile and crawl many metres (e.g., *Inachis io*), female butterflies are responsible for choice of biotope, local climate, including light and shade environment, hostplant species and patch, position on the patch and ultimately the precise destination of young eggs and larvae, whether on the plant, on a nearby plant species or on a different substrate. There is then the decision of how many eggs to lay, behaviour that is a species-specific attribute and linked to hostplant abundance and defences against predators and parasites (see Brakefield *et al.*, 1992:119). Many deposit a single egg per visit to a hostplant (e.g., *Anthocharis cardamines*), whilst others may lay a small clutch (e.g., *Thymelicus* sp., Duke of Burgundy *Hamearis lucina*) or a large batch (e.g., *I. io*, *Aglais urticae*) (see Appendix 5). Porter (1992) neatly illustrates the selection of egg sites by a female as a descending series of choices (see Fig. 2.2). This topic is explored fully in Chapter 4 and here only a few basic issues are outlined, including some principles of egg laying relating to habitats:

- **P2.34: Distinct cues (visual, scent) are applied by females in selecting destinations for eggs over several spatial scales from biotope to egg site.**
- **P2.35: Only a fraction of locations (habitat) populated by a butterfly species are generally**

suitable for the hostplant(s) and a fraction of the hostplant resource is suitable for the eggs and young larvae.
• P2.36: Local climate and microclimate contrasts are key variables in female site selection for oviposition.
• P2.37: Eggs are not invariably placed on the hostplant, but are necessarily always located near the hostplant.

An extensive discussion is provided by Porter (1992) of the visual and scent cues used by female butterflies in the process of egg laying. It is clear that females do use sensory cues for selecting oviposition sites at several scales from the biotope to the position on the hostplant or nearby substrate, but it is not clear what these cues are for different species (**P2.34**). Butterflies that lay on apparent (large or bright) plants seem to have little difficulty in locating a host (e.g., *A. cardamines*) but, for those using unapparent plants, it is clear that visual cues are insufficient for the recognition of a suitable host and cues involving direct contact with a plant (tactile) are required (e.g., *Pieris brassicae*, *Vanessa atalanta*; R. L. H. Dennis, personal observation). Part of the selection process is the detection of potential predators as well as determining the suitability of the plant as a platform for the eggs and as food for the young larvae. The hostplant area usually forms a fraction of a butterfly's habitat, and in turn only a fraction of the area of hostplant is suitable for placement of a species' eggs and for the early development of young larvae (**P2.35**). Some of this fractional access is structural; for instance, *Thymelicus lineola* may only select leaf sheaths of grasses such as cock's foot *Dactylis glomerata* and creeping soft-grass *Holcus mollis* that provide tight enough leaf sheath 'envelopes' within which the small batches of eggs and their unhatched larvae can overwinter (Porter, 1992).

A key part of this fractional suitability of hostplant patches and hostplant individuals relates to local climate and microclimate (**P2.36**). Eggs laid in unsuitable heat, light and humidity conditions have a substantially reduced probability of the larvae hatching and surviving. Some eggs are laid to be in full light (e.g., northern brown argus *Aricia artaxerxes*, *Polyommatus icarus*, *Hipparchia semele*) whereas others are laid in partial or full shade (e.g., *Pararge aegeria*, white admiral *Limenitis camilla*, wood white *Leptidea sinapis*). There is necessarily variation in site selection on the basis of microclimate depending on latitude and altitude in Britain; south-facing sites becoming increasingly important northwards and at higher elevations (Dennis, 1993a). Placement of eggs is not invariably on hostplants (**P2.37**). Substrates other than the hostplant are likely to be used by butterflies that overwinter in the egg stage when the plant part being used dies down (e.g., high brown fritillary *Argynnis adippe*); the exceptions are those exploiting woody perennials (e.g., *Favonius quercus*) (Wiklund, 1984). However, a number of species developing directly also lay on non-hostplant substrates (see Appendix 5b). A classic example is silver-washed fritillary *Argynnis paphia* which places eggs in crevices in bark on the mossy side of trees (Frohawk, 1934), but *Polyommatus icarus* will also place eggs on plants that are non-hostplants (Dennis, 1984a). These non-hostplant oviposition sites are typically used where hostplants are abundant and may well be a ploy to evade enemies, particularly parasites, but Wiklund (1984) also describes situations where the hostplant may simply not be a sufficiently supportive structure to carry the weight of an egg-laying female (e.g., Baltic grayling *Oeneis jutta* uses dwarf birch *Betula nana* stems).

Adult rests and roosts (Brakefield *et al.*, 1992:98–99)

Butterflies may well spend >50% of their time inert roosting, when their enemies, insectivorous birds, rodents and arthropods such as spiders, are very much active and hunting. Yet, so little is apparently known about this vitally important behaviour that no more than a paragraph could be allocated to the subject in *Ecology of Butterflies in Britain* (Brakefield *et al.*, 1992). What we do know is that butterfly species adopt distinctive roosting behaviour, and are selective of substrate and locations, when roosting (see Appendix 5a). Some species are found in more exposed sites, others are concealed; some roost communally, others alone. This is one area where a great deal of work is to be encouraged; however, some principles are suggested for consideration:

• P2.38: Site structures are key to roost sites, form a fraction of the site occupied by a butterfly population and differ from those used when butterflies are stationary, but potentially active (i.e., thermoregulation).

Fig. 2.6 Ranch fencing at Brereton Heath, Cheshire, UK, used by wall brown *Lasiommata megera* for roosting, mate location, egg laying and larval development. Inset is a marked *L. megera* male roosting under the top-most fence spar.

- **P2.39: Daytime roosts ('rests') may well differ from night-time roosts as the predator community changes from night to day.**
- **P2.40: Specialism and generalism among species in roosting very likely relates to degree of crypsis in butterflies.**
- **P2.41: Behaviour during roosting provides clues as to the choice of roost sites; behaviour is likely to be geared to thermoregulation and facility for escaping enemies.**
- **P2.42: Roosts for any one species will vary among sites in relation to available substrates and locations; use of a particular roost is not proof of optimality of the chosen site.**
- **P2.43: Roost sites vary in location and substrate with weather conditions and local climate contrasts.**

There is clearly active selection by individual butterflies of locations to roost at night as well as during daylight hours (**rests**) when conditions are adverse, either cold, wet, subject to gales, or unusually hot. The structures used – vegetation types, rock or soil substrates, human landscape features, landforms – differ substantially from those used by butterflies when potentially active, when they are thermoregulating or resting between activities such as nectar feeding or egg laying. Roost structures used typically represent some fraction of the entire site occupied by a population (**P2.38**). For example, in a study of *Lasiommata megera* on Brereton Heath, Cheshire, roosting was mainly confined to ranch fencing and trees fringing the site (Fig. 2.6) (Dennis, 1986a). Similarly, on Great Ormes Head in North Wales, roosting by *Plebejus argus* is concentrated on scrub and bracken fringing the open calcareous heath occupied by the larval hostplants (Dennis, 2004b). Daytime rests may well differ from night-time roosts as the predator community changes (**P2.39**). This does not seem to be the case in *P. argus* on Great Ormes Head, where butterflies spend cold, wet, windy days as they would spend cold, wet, windy nights. However, the Neotropical butterfly *Manataria maculata* (Hopffer) (Satyrinae) moves between communal roosts on the ground and individual night roosts in trees (Hedelin and Rydell, 2007); it seems to avoid rodents by night and birds by day. Roosting and resting is closely tied up with the importance of enemy-free space. Endler (1984) found that specialist species that rest on a single background are more cryptic than generalist species that use different backgrounds and species that rest under leaves and are not visible from above and are not very cryptic (**P2.40**). This observation, which needs to be tested among European butterflies, warns us against assuming that specialism and generalism are simply restricted to consumer resources.

Choice of roost (rest) sites is undoubtedly closely tied into roosting behaviour (**P2.41**). The position adopted during roosting, head up or down, or aligned sideways, and the height above the ground suggest that the butterflies position themselves for active escape. At Brereton Heath, Cheshire, it was interesting that *L. megera* used the middle spar of the ranch fencing significantly more often than the lower or uppermost spars and that they typically hang upside down (Dennis, 1986a), in warm locations and high enough for a rapid exit on being disturbed. In grassland, butterflies (e.g., *Polyommatus icarus*, *Maniola jurtina*) often roost relatively high up on stems; lower, damper positions would possibly make them more vulnerable to arthropods. In the case of Glanville fritillary *Melitaea cinxia* on the Isle of Wight, tall herbs for roosting occur in areas outside the main hostplant areas (Willmott, 1985). Locations chosen for roosts in populated sites are unlikely always to be optimal for roosting (**P2.42**); choice is necessarily made from what is available. *P. argus* on the Great Orme, caught out by sudden changes of weather on exposed, higher slopes are forced to use tufts of grass bent back by prevailing winds (R. L. H. Dennis, personal observation) and even human debris, one individual roosting inside an empty plastic bag caught in a hollow. Roost sites undoubtedly vary with weather and local climate (**P2.43**). On warm, calm nights, more *P. argus* will roost on shorter vegetation on the calcareous heath; during cool, windy weather they retreat to the marginal scrub. Daytime rests on cold, windy days are typically on the leeward sides of low scrub, more individuals selecting the larger, prostrate, ellipsoidal-shaped cotoneaster bushes (see Fig. 4.6) (Dennis and Sparks, 2006).

Larval sites for resting and moulting

(Brakefield *et al.*, 1992; Warren, 1992a)

Hostplants not only have to provide a source of food for larvae, they also have to supply a suitable environment for resting when larvae are not eating but when digestion is still in progress, sites for larvae to moult (**ecdysis**, shedding skins) between instars, for evading enemies – predators and parasites – and, for many species, hibernation. Butterflies generally remain in the larval stage for the longest period since the larval stage is the commonest stage for hibernation (see Appendix 5e). Thus, the hostplant surroundings are critical for long-term survival as well as for growth and development into the next stage.

Most principles associated with larvae relate to hostplants as consumer resources. But, there are key principles affecting the hostplant environment for larvae as utilities:

- **P2.44: The hostplant environment can influence the capacity of butterfly larvae to access and exploit the host.**
- **P2.45: The condition of the hostplant can affect survival of larvae other than by providing suitable food.**
- **P2.46: Local climate and microclimate of the hostplant environment affect development when not feeding, and survival during the entire larval stage.**
- **P2.47: Both the composition and structure of the contextual vegetation influence the capacity of larvae for evading enemies, obtaining rest and hibernating.**

Very little is known of the impact of the larval environment on larval survival. Although doctoral theses have been undertaken on a number of species (see Dennis, 1992a), of their nature these tend to have provided a limited view of how variation in the larval environment can influence mortality levels (Warren, 1992a). Study of this stage is labour intensive and it is difficult to follow cohorts of larvae from eggs to pre-pupae at one location. Nevertheless, from an intensive and intricate study of *Limenitis camilla* by Barry Fox (1996, 2005), we have been given some astonishing insights into how a plant in the wrong location can affect the capacity of a larva to access food (**P2.44**). This butterfly builds a platform for feeding, isolated from the stem, which includes a latrine. In the shade it is possible to build a suitable structure; in full sun it is not owing to an inability to cope with copious amounts of sticky secretion produced when the larvae incise sunlit honeysuckle *Lonicrea periclymenum* L. (see Fig. 4.10c). This work also illustrates well how mechanical defences in plants can deter host use other then consumption of the plant. For instance, **trichomes** (outgrowths, hair or spine; see Fig 4.10c) can frustrate the construction of hibernacula and tents; in this way they present an attachment hurdle (Southwood, 1973), though some butterfly species can remove them by mowing or **swiping**.

At a much simpler level, premature or late flowering in *Cardamine pratensis*, whose flowering parts are consumed by *Anthocharis cardamines* seeking unopened buds for herbivory, can force the butterfly onto unsuitable plants. When early and already in-pod, feeding is retarded and individuals are exposed for longer periods to enemies; when late, flowers may be few, in short supply and already occupied by eggs, forcing individuals to compete with larvae that are cannibalistic (Dennis and Hardy, 2006). The hostplant is additionally a platform for many other larval activities, as well as one used by competitors and enemies (i.e., ants, spiders, beetles). Inevitably, there is variation in the presence of insectivorous arthropods and, as larvae increase in size, variation too for ground and aerial predators. Such variation in host associations brings with it variation in survival, one that may conflict with the host's suitability as food (**P2.45**).

In the British climate, which is marginal for so many butterfly species, the local climate and microclimate of hosts has an important part to play in survival and development both during feeding and during intervening periods of rest (**P2.46**). Digestion continues following feeding and food assimilation is affected by the heat environment – whether plants are in sun or partial shade, exposed or sheltered, and have structural features such as embayments that create their own shelter and heat traps. Direct sunlight enables the gregarious marsh fritillary *Euphydryas aurinia* larvae to reach suitable temperatures to feed outside their protective webbing and to feed more rapidly (see Fig. 4.9) (Porter, 1982), thus evading parasites more easily, in which case local conditions enhancing exposure would be a premium over those that reduce it. No doubt, similar behaviour may be found in the gregarious nettle-feeding nymphalids *Aglais urticae* and *Inachis io*. A different example is provided by hare's-tail cottongrass *Eriophorum vaginatum*, a core hostplant of *Coenonympha tullia*; when growing in tussocks it is less liable to immersion as the winter watertable rises in upland mires (Dennis and Eales, 1997). The submergence of overwintering *C. tullia* larvae is known to have an adverse effect on their survival (Joy and Pullin, 1997). An important area for research is the influence of the vegetation context of hostplants for larval survival. A common indication of the significance of host context and structure on survival is that larvae of many butterflies feed at night (e.g., *Maniola jurtina*, *Hamearis lucina*) and are not evident during the day; the layers of ground vegetation into which they disappear, including the hostplants, forms a vital structure for rest and evading predators (**P2.47**). It is likely that butterfly larvae are particularly vulnerable to predators when moulting between instars; appropriate cover is then essential for these transformations. It will also be of immense importance for the construction of hibernacula for those many species overwintering as larvae.

Pupation sites (Brakefield *et al.*, 1992)

Pupation sites are selected by pre-pupal larvae and therefore have to be within crawling distance from the hostplant; but the actual distances full grown larvae are able to travel is not known and for conservation purposes alone is in urgent need of study. From the discovery of *Inachis io* larvae on roads (R. L. H. Dennis, personal observation), it would seem that 10 m, or even 20 m as in the case of *Aglais urticae* (John, 2004) or *Pieris brassicae* (Kinder, 2009), is not an impossible distance to cover. Pupae rely on concealment, crypsis and warning patterns (**aposematism**) for surviving searches by enemies, and on sufficiently robust structures to withstand the rigours of weather and seasonal climate changes during what can be long periods of being static. Pupae may be attached firmly to objects, be suspended from them, occur in grass tents, silk webs or cocoons or even be underground (see Appendix 5d).

Some principles relating to pupation sites are suggested here:

- **P2.48: Suitable structures for pupation can be a limiting resource for pupation sites within access of larval hostplants.**
- **P2.49: Structures selected by pre-pupae for pupation have the capacity of affecting apparency of pupae to potential enemies.**
- **P2.50: Structures vary in their capacity for deterring predators from gaining access to pupae.**
- **P2.51: Sufficiently robust structures are required with suitable microclimates for the survival of pupae over long periods in harsh conditions, including hibernation.**

Clearly, the greater the distance for pre-pupae to travel from hostplant patches to suitable sites for pupation, the more energy reserves are consumed in this task and the less remains for pupation, surviving conditions during pupation and allocation to adult resources on emergence (**eclosion**). Moreover, exposure to enemies (e.g., birds, rodents) increases during transit to pupation sites. A premium is then placed on there being neighbouring suitable sites for pupation with access for larvae (**P2.48**) and these sites vary considerably in form and structure for different species. Pupal sites may be in soil (e.g., *Hipparchia semele*) or ant nests under stones (*Plebejus argus*), in live vegetation on green shoots (e.g., *Lasiommata megera*), in taller structures on dead shoots, stalks of plants or sticks (e.g., *Anthocharis cardamines*) or higher up still in crevices on walls (e.g., *Pieris brassicae*) (P. Kinder, personal communication). The two British butterflies (i.e., white-letter hairstreak *Satyrium w-album* and black hairstreak *S. pruni*) that pupate on trees, do so on their arboreal hosts, on leaves as well as twigs, and eclose that same summer. These structures vary in the extent to which they conceal pupae (**P2.49**). Those exposed above ground require further defences in the form of crypsis (e.g., *Aglais urticae*; Brakefield *et al.*, 1992) or warning colours (e.g., the bright green overwintering pupae of *P. brassicae*; Rothschild *et al.*, 1977); *S. pruni* is most extraordinary among British butterflies in that the pupae resemble bird droppings. Some species have the capacity for matching up with the colour of background substrates (e.g., swallowtail *Papilio machaon*, *Leptidea sinapis*, *A. cardamines*) (Brakefield *et al.*, 1992). Similarly, different structures vary in the degree to which they deter access to enemies; some species (*Pieris* sp.) even regularly use corners and crevices in human-made structures (Fig. 2.7; **P2.50**). Those in ant nests have particularly formidable defences against generalist intruders but may still be exposed to specialist parasitoids (e.g., *Neotypus* sp., as in the case of *Neotypus melanocephatus* (Gmelin) (Hymenoptera: Ichneumonidae) which attacks caterpillars of the dusky large blue *Maculinea nausithous* on the continent; Anton *et al.*, 2007) (see section on symbionts as resources below). Butterfly pupae cannot move 'house' when conditions become more severe and those hibernating in this stage necessarily have to adopt strong enough structures to survive decay, grazing, browsing and 'erosion' during the winter months (**P2.51**) (e.g., *P. machaon*; Dempster *et al.*, 1976).

Fig. 2.7 Small white *Pieris rapae* pupa occupying a crevice deep inside a gap between fence posts at Meols Common, Wirral, Cheshire, UK. (Courtesy of Shane Farrell; inset courtesy of Peter Hardy).

Parasitoids and predators in the resource zones (Warren, 1992a; Shaw, 2006; Shaw *et al.*, 2009)

Predators and **parasitoids** (an organism living within the body of another organism eventually causing its death; Lincoln *et al.*, 1982) inflict substantial casualties on butterfly populations, often the highest mortality among the range of factors including losses from weather-related effects (e.g., the shortfall in eggs compared to potential fecundity). This is most evident from detailed key factor analysis which determines mortality of individuals from samples of all stages through butterfly development in a population. Warren (1992a) gives examples in his chapter on butterfly populations in the *Ecology of Butterflies in Britain*. Predators include vertebrates (e.g., birds, rodents, lizards) as well as many invertebrates (e.g., beetles, ants, spiders); the former tend to reduce numbers of larger stages (late instar larvae, pupae, adults), whereas the latter are preponderant in the deaths of earlier stages (eggs, small larvae). Parasitoids largely fall into two invertebrate orders (i.e., Diptera (e.g., Tachinidae, Bombyliidae) and Hymenoptera (e.g., Braconidae, Ichneumonidae, Eulophidae)) (Shaw and Askew, 1976; Shaw, 1997); the larvae develop in or on the butterflies and the adults are free-living. There is a tendency to class particular parasitoid species as egg, larval or pupal parasitoids, but many overlap their attack on different stages; egg parasitoids include tiny Chalcidoidae (e.g., Trichogrammatidae), Proctotrupoidea and Scelionidae, which manage to complete their development within the butterfly egg.

It cannot be emphasized sufficiently how little we know of the impact of these agents, particularly parasitoids, on populations of different butterfly species. What we do know has been ably summarized by Shaw (2006) and Shaw *et al.* (2009) who provide a key of insect parasitoids of European butterflies and a list of parasitoid species together with their host butterflies. Detailed case studies in this labour-intensive field are few but notable as this is an area of vital concern for butterfly conservation as well as for the specialist parasitoids (e.g., *Ichneumon eumerus* in mountain Alcon blue *Maculinea rebeli*; Thomas and Elmes, 1993) that depend on particular butterfly species hosts. There is little space to enter much detail on all these groups here, but some terms and concepts on parasitoids will increasingly confront readers on butterflies. Parasitoids are broadly divided into various groups. **Primary parasitoids** are those that attack a Lepidoptera host, whereas **secondary parasitoids** (**hyperparasitoids**) in turn parasitize the primary parasitoid. Further division of the latter distinguishes those that are **true hyperparasitoids**, which attack the primary parasitoid while it is still active inside the butterfly host, and **pseudohyperparasitoids**, which attack once the primary parasitoid is free of the resource provided by the butterfly host. Two other terms in increasingly common usage are **idiobiont** and **koinobiont**; the former describes parasitoids that develop in a dead host, whilst the former develop in a host that is still alive and developing. Parasitoids may also feed externally (**ectoparasitoids**) or internally (**endoparasitoids**), be **solitary** or **gregarious**, and involve **multiparasitism** (more than one primary parasitoid involved) and/or **superparasitism** (multiple attacks by the same species on one individual). This gruesome but fascinating area of butterfly biology is ably described by Shaw *et al.* (2009) who expand lucidly on the sheer complexity of the subject in such a way that the reader is left in no doubt as to the importance of this subject.

The following principles can be drawn up on parasites and predators relating to habitats:

- **P2.52: Predators and parasites can exact very high mortalities on butterflies at all stages in their development, though clear bias will be found for the stage attacked in any butterfly population during any brood by a particular parasitoid species.**
- **P2.53: Contrasting levels of predation and parasitization of any butterfly species are to be expected within sites (habitats) as well as between sites, if only because both butterfly predators and parasites are also subject to enemies.**
- **P2.54: Contrasting levels of predation and parasitization of a butterfly species occur over time, within and between broods, affecting periods even of single developmental stages of the butterfly at single sites.**
- **P2.55: Parasites have distinct cues, visual and scent, to track down appropriate butterfly prey within the habitat and may well be locked onto chemical signals in larval host-plants and emanating from larval feeding damage by butterflies.**

Both predators and parasitoids can cause very high mortality of individuals belonging to any stage of development (**P2.52**). Clear examples exist for high mortality caused by parasitoids (e.g., small white *Pieris rapae* (Richards, 1940; Dempster, 1967, 1968); scarce swallowtail *Iphiclides podalirius* (Stefanescu *et al.*, 2003); *Pieris brassicae* (Kristensen, 1994), *Aglais urticae* (Pyörnilä, 1976–1977; John, 2004); *Melitaea cinxia* (van Nouhuys and Hanski, 2004)). Mortality by parasitoids impacts on early stages. Predation by invertebrates and vertebrates can also inflict high mortality on early stages. And occasionally there is evidence that predation of adults by birds can also cause substantial losses (e.g., green-veined white *Pieris napi*; Dennis, 1993b); wing damage (e.g., in *Coenonympha tullia*; Dennis *et al.*, 1984, 1986) and direct observations of bird predation on butterflies (Collenette, 1935; Dennis, 1993b) suggest substantial mortality from bird predation, even during hibernation (e.g., losses of *A. urticae* to wrens; Leverton, 1998; Bland, 1999). A clear observation is that different predators and parasite species concentrate on different life stages of butterflies (Shaw *et al.*, 2009). These authors also emphasize what we do not know. It is clear from autecological surveys (e.g., on *Maculinea* species (J. A. Thomas *et al.*, 1998a); on *M. cinxia* (van Nouhuys and Hanski, 2004)) that large differences exist in mortality between and within sites (habitats) (**P2.53**) (Dennis, 2005b); the problem is that for the vast majority of butterfly species, this phenomenon has not been studied at all. Yet, it is a crucial aspect of butterfly habitats. Butterflies may concentrate eggs and thus larvae in parts of sites frequented or avoided by predators, perhaps relating to their own susceptibility to predation, or more or less evident to parasitoids. The consequences can obviously be profound for population persistence and gives full meaning to the importance of enemy-free space and conditions for avoiding enemies on sites. Just what these are for different butterfly species is not clear, but from mortality levels in populations it is manifestly of the utmost importance to understand.

It is clear that predation and parasitization vary over a season and between years (**P2.54**) (Dennis, 2005b; Shaw *et al.*, 2009). Parasitism tends to increase as a season progresses (Dempster, 1984; Askew and Shaw, 1986). A key issue in predation and parasitism is the cues used by enemies for searching out prey (**P2.55**). Birds clearly rely on vision for tracking down adult butterflies, but increasingly the importance of scent, particularly that of hostplants, is emerging as the cue that parasitoids use for seeking out early stages of butterflies (Isingrini *et al.*, 1985; Rietdorf and Steidle, 2002; Emden *et al.*, 2008).

This aspect of butterfly habitats is arguably the one we know least about and yet is crucial for long-term persistence of populations, as suggested by recent publications on male-killing inherited bacteria such as *Wollbachia* (see Creaser *et al.*, 2008 for references). There is clearly enormous potential for complex interactions among butterfly and parasitoid species' communities at sites, a research area ripe for investigation.

Symbionts and enemy-free space (Thomas in Brakefield *et al.*, 1992; Porter *et al.*, 1992)

One group of butterfly species (Lycaenidae) is unusual in its symbiotic association with ants, involving defence in exchange for food (nectar) (Hinton, 1951). This is no small matter in habitat components. Lycaenids represent about one-third of world butterflies (approximately 17.2–18 k butterfly species; Ackery 1984) a proportion much the same as found in British butterflies: 17 of 60 species (28.3%). Basic observations relating to these relationships are as follows:

- **P2.56: Butterflies engaged in symbiotic associations with ants have specialized organs for appeasing and rewarding ant attendants which, for a number of species, involve phylogenetic links between butterfly and ant taxa.**
- **P2.57: Ant defences make a sufficient difference to survival to be critical for survival and to ensure long-term population persistence in a number of lycaenid species.**
- **P2.58: Lycaenid butterfly populations may be constrained by local and microclimatic factors influencing the presence of ant populations.**
- **P2.59: For a number of highly specialized lycaenid species, ant association involves occupation of ant nests and structures created by ants.**
- **P2.60: Specialist parasitoids have evolved the means for adapting to and overcoming the defences of their lycaenid hosts.**

In the British Isles some 16 of the 17 lycaenid species are known to attract the attention of ants, and in the field ant association has been demonstrated for 14 species (see Appendix 5e). Lycaenid butterflies have evolved a range of structures and behaviours to ant attendance (e.g., honey gland or dorsal nectary organ, tentacle, cupola, sound production) which are ably described by Professor J. A. Thomas (in Brakefield *et al.*, 1992) (**P2.56**) (see Appendix 6). Thomas classes lycaenid feeding and adaptations into two broad groups: (i) those that are primarily phytophagous and remain on the larval hostplant; and (ii) those that are aphytophagous and build up the larger part of their body mass on ant brood, regurgitants and detritus as well as other symbiont associates such as coccids and aphids. In effect, the latter have become predators on their protectors. Again, this is no small matter, with large blue *Maculinea arion* acquiring 99% of their bulk in ant nests on ant brood (Fig. 2.8) (Thomas and Wardlaw, 1992). In lycaenid species that have been studied in detail, it is clear that ant-mediated defence greatly increases butterfly survival (e.g., American silvery blue *Glaucopsyche lygdamus* (Doubleday): 9.9% survival from egg to adults with ant defence versus 0.7% survival in its absence; Pierce and Easteal, 1986; see also Pierce *et al.*, 2002) (**P2.57**). The situation is likely to be the same in British species such as *Polyommatus bellargus* which are vigorously attended by ants (J. A. Thomas, 1983a). This defence extends usually from the second larval instar to adult emergence. Even butterflies expanding their wings and emerging from ant nests (e.g., *Plebejus argus*) have been observed attended by ants (C. D. Thomas, 1983).

From the view of butterfly habitats, two things become clear. First, for butterflies dependent on ant association, incidence of the right ant species in sufficient numbers is a critical determinant of the butterfly habitat. In some cases, the species of ant does not seem to be crucial (e.g., *Polyommatus icarus*), but in other butterfly species (e.g., *Plebejus argus* on black ants *Lasius niger* (L.) and *L. alienus* (Foerster); *Maculinea arion* on *Myrmica subuleti* (Mienert)), the association is more hard-wired. In these cases, without the right ant there will be population failure regardless of abundance of hostplants on which eggs are laid and, at least early, feeding occurs. An essential component for ant species

(a)

(b)

Fig. 2.8 Large blue *Maculinea arion* larvae: (a) attended by ants in a *Myrmica sabuleti* ant nest, and (b) amongst the ant brood over which it periodically glides to feed on ant larvae. (Courtesy of Jeremy Thomas.).

is warm, dry local climates and microclimates (**P2.58**); in Britain, *M. sabuleti* ants require shorter turf in which to thrive. It is possible, as Thomas (1975b) suggests, that the microclimate of soil near oak trees may explain the presence or absence of *Favonius quercus*; the pupae are ant attended without which viability may be seriously compromised.

The second observation is that many lycaenid species associated with ants would not adopt subterranean structures in their absence. Ants convey larvae and pre-pupae of *P. argus* and *Maculinea arion* into their nests and for other lycaenids (e.g., *P. bellargus*) create earth cells; these provide insulated environments against weather and concealment against predators and parasites (**P2.59**). In effect, any study of a lycaenid butterfly, including hairstreaks, cannot ignore microhabitat resources that could be linked to ant attendance.

The crucial point about symbiont associations from the habitat vantage is that further layers of restrictions are placed on a butterfly's resource use, not just an additional resource in the Venn diagram of associations, but specific conditions for the butterfly linked to the presence of the ant host (e.g., large nests in the case of *M. arion*; J. A. Thomas in Emmet and Heath, 1990) as well as suitable conditions for the ant host itself. A lycaenid butterfly dependent on ants will be found on a fraction of the area of distribution of host ants, which in turn will be associated with a fraction of the larval hostplant cover. It should be noted that despite such defences, a number of highly specialized, monophagous parasitoids are able to bypass them (**P2.60**), as in the case of *Ichneumon eumerus* on the mountain Alcon blue *Maculinea rebeli* (Thomas and Elmes, 1993).

Hibernation and aestivation sites

(Brakefield *et al.*, 1992; Warren, 1992a)

The response of butterflies to adverse seasons for activity is to become dormant or to migrate to another region with suitable conditions for survival. Dormancy at low latitudes, including Mediterranean regions, during the hot, dry season is referred to as **aestivation**. Some butterfly species found in Britain aestivate in hotter climates on the continent; *Maniola jurtina*, mainly females, take to scrub and woodland during the hot, dry summers of Italy and Greece to become active again, to lay eggs, in late summer (Scali, 1971). It is not surprising then that the butterfly adopts more wooded biotopes in extremely hot weather in Britain (Dennis, 2004a). As some activity is possible for butterflies even in the hottest weather, adaptations accompany aestivation (e.g., seasonal polyphenism in adult wing morphology) matching butterflies to their surroundings in response to predation (Brakefield and Larsen, 1984). Conversely, conditions are usually insufficiently warm for activity in high latitude winters and butterflies then hibernate, becoming torpid with body temperatures reflecting ambient temperatures. **Hibernation** sites are probably underestimated as a resource for butterflies, simply because this phase is least apparent to observers. Yet, it is crucially important for survival. To summarize some significant points:

- **P2.61: High latitude butterflies spend more time in hibernation sites than engaged in other resource use.**

• **P2.62: Hibernation sites are locations of concealment and/or insulation; as adaptations for long periods of torpor, micro-sites are likely to differ for the same stage used for resting when more active.**
• **P2.63: All developmental stages are known to hibernate though there is phylogenetic bias for the stage used.**
• **P2.64: The importance of proximity to larval hostplants for different hibernating stages ranks: eggs > larvae > pupae > adults.**
• **P2.65: Associations of stage hibernation with different biotopes, biomes and climate regions are indicative of the pressures associated with hibernation in different conditions.**
• **P2.66: Selection pressures on species during hibernation are highlighted by the construction of hibernacula.**

Sites for hibernation are rarely considered in population studies for high latitude butterfly species, but this resource is required, by many species, for longer than other resources during an individual's lifespan (typically 5 months hibernating) (**P2.61**). Occupying the correct site is important for two reasons:
1 To survive the seasonal cold and damp conditions – without becoming diseased
2 To avoid being eaten.

Because of seasonal cold and the length of time that butterflies are torpid, sites selected for hibernation are likely to differ from those typically selected by the same stage during more active periods (**P2.62**) – either in location relative to vegetation structures and/or construction where larvae actively modify the immediate environment. Butterflies are particularly vulnerable when torpid and rely on concealment and crypsis (Brakefield *et al.*, 1992). High latitude butterflies typically hibernate in one developmental stage; in Britain, the only exception is *Pararge aegeria*, which can overwinter as a larva or as a pupa. Even so, across species, butterflies hibernate in different stages and there is phylogenetic bias for the stage that hibernates (**P2.63**) (Dennis, 1993a). As different stages hibernate, very different sites and conditions are needed for hibernation by an assemblage of different British species at any location (see Appendix 5e). Some points follow. Proximity of larval hostplants is more crucial for hibernating eggs and larvae than for pupae and adults (**P2.64**), limited as they are by larval mobility following loss of winter fat reserves and the need to feed up in the spring. By

Fig. 2.9 A cluster of peacocks *Inachis io* hibernating on the concrete roof of a air-raid shelter at Bradley Fold, Radcliffe, Manchester, UK. (Courtesy of Peter Hardy.)

comparison, pre-pupal larvae have a single journey to undertake which can therefore exceed any distance undertaken by larvae that have to return to the hostplant to continue feeding in spring. Hibernation sites are also more restricted spatially for pupae than adults, as they are limited by the range of movement of final instar larvae from their final hostplant location. Adults, as in the case of nymphalids (e.g., *Aglais urticae*, *Inachis io*) and the pierid *Gonepteryx rhamni*, can obviously select from potential hibernation sites over a wide region, including human structures (Fig. 2.9). An indication of pressures on hibernating stages is the change in frequencies of stages with different biomes and biotopes for latitude and altitude (**P2.65**) (Dennis, 1993a). With higher altitude and latitude there is an increase in the proportion of species overwintering as larvae and a decrease in overwintering eggs and adults. This may well relate to the shorter growing season and need then to feed up on new, nutrient-rich foliage in the spring to complete development. If this is the case, then the importance of hibernacula for larvae is underpinned (see below).

Little is known statistically about the range of hibernation sites used by individual butterfly species. One thing is clear: the conditions required across species vary in exposure to ambient conditions and complexity in construction. Hibernation sites differ in height above ground, from the use of trees and ivy-clad walls and crevices (e.g., adult hibernating nymphalids and *Gonepteryx rhamni*) to the soil surface (e.g., *Lycaena phlaeas*); surfaces differing in height above the ground contrast for minimum temperatures, the coldest sites being on the bare ground surface, especially in the absence of snow cover (see next section), and the warmest being deep underground as for *Maculinea arion* in nests of *Myrmica sabuleti*. Hibernation sites vary in the use of vegetation cover from none (e.g., *Satyrium w-album* and *Favonius quercus* eggs) to substantial cover (e.g., marbled white *Melanargia galathea*), and in complexity of hibernacula (e.g., webs of *Euphydryas aurinia* larvae). Choice and construction in hibernation sites can be very precise, as observed in the construction of **hibernacula** *sensu stricto* (i.e., a tent-like structure or sheath made from a leaf or other material, in which a larva hibernates; e.g., small blue *Cupido minimus* in a flower head of kidney vetch *Anthyllis vulneraria* or surrounding vegetation) (**P2.66**). Some hibernation sites are delicate, not withstanding they have to persist for months through harsh weather conditions. A case in point are the leaf sheaths of Yorkshire fog *Holcus lanatus* in which the small skipper *Thymelicus sylvestris* overwinter as young larvae within cocoons. Hibernacula (see Appendix 6b) provide both concealment and insulation from seasonal climate and weather; these too can be delicate structures and their appropriate construction crucial for survival (*Limenitis camilla*; Pollard, 1979).

CONDITIONS AND CONDITIONERS

Each butterfly species currently in northern Europe has great spatial extent (geographical range; Dennis *et al.*, 1991) and apparent longevity (>20 k to >5 m years; Dennis and Schmitt, 2009) and has been, as currently, subject to a wide range of environmental conditions (i.e., weather, climate, soil characteristics) at contrasting scales, local as well as regional. In this arena the environmental factors can be referred to as **conditioners** (New, 2007). Thus, habitat has to be considered in this wide context of variable and varying conditions; habitat will always have spatial-scale relevance in relation to human objectives.

Climatic agents as conditioners

Butterflies, as ectotherms, are dramatically affected by atmospheric (weather or climatic) agents (viz., insolation, temperatures, humidity, winds, precipitation), as noted above in the section on thermoregulation. Climatic variables are distinguished by their regional and local contrasts and gradients, changing constantly, while weather conditions are even more variable for single sites. These contrasts and changes can be dramatic for butterflies, even on scales as fine as seconds (e.g., the passage of a cloud or downpour), often exceeding their functional tolerances (Dennis, 1993a; Dennis and Sparks, 2006; Howe *et al.*, 2007). In particular, by influencing butterfly physiology, there is an immediate impact on behaviour and suitability of resource components and ability to exploit resources. In Box 2.4 the distinction between climate and weather is outlined, together with some basic features of climate gradients at scales from the region to the locality, landforms, vegetation and substrates. Two key messages are implicit, if not explicit, in these details on climate. First, regional weather and climate is 'moderated' by local topography, vegetation structure and substrates. Second, because of this, local climates and microclimates are predictable from data on regional climates and local landscape topography (i.e., altitude, slope angle, aspect), vegetation and soils. Extending this reasoning, armed with requirements for a butterfly species to successfully exploit specific resources, it becomes entirely feasible to determine what parts of the landscape are habitable, or can be made habitable, for it. It cannot be emphasized enough how important tolerance to weather and climate conditions is for butterfly habitat suitability – indeed, the reason for this dedicated section. Climate and weather may be regarded as conditioners of consumer and utility resource use (New, 2007).

A number of generalizations can be made:

- **P2.67: Fractional and biased resource use by butterfly species often (invariably) has an atmosphere, climate or weather aspect to it, evident in the distribution of individuals on the resource compared to that available at a site.**
- **P2.68: Weather changes are often responsible for rapid changes in fractional resource use in butterflies.**
- **P2.69: At sites with a greater range (greater extremes) in conditions, a greater range of outlets may be required of resources for population maintenance.**
- **P2.70: The effect of weather and climate on butterfly species' habitat (resource) use depends on the degree to which it is moderated by topography, vegetation structure, substrates and human landscape features.**
- **P2.71: As local topography and microenvironment modify regional weather and climate, local landscape heterogeneity allows species to change location within trivial distances for**

Box 2.4 **Weather and climate: contrasts at different space–time scales**

Weather is the moment by moment change in atmospheric conditions: sunlight, temperature, precipitation, cloud cover, humidity, wind speed and direction and so on. **Climate** is traditionally described as average weather over a period of time, typically taken to be 30 years. However, this is rather simplistic as it involves much more, i.e., the variation and trends in conditions over time and space (Musk, 1988). If weather is an instantaneous record, a time slice, of atmospheric conditions, then climate is everything else. As records can be accumulated for any period of time (hours, days, months, seasons, years, decades), climatic data become virtually infinite and changes across time periods reveal a kaleidoscope of varying patterns. Pertinent points for butterfly biology are:

- Some atmospheric variables are more important than others for different seasons and butterfly stages (e.g., sunshine for summer and adults; rainfall/humidity for winter and hibernation; Beirne, 1955; Dennis and Sparks, 2007).
- There are scales of variability in weather/climate that are relevant for different stages and activity (behaviour) (e.g., seconds to minutes for adult egg laying; days for pupation).
- Gradients for *different* weather/climatic attributes are often correlated. At a broad regional scale, wetter areas in Britain are generally more humid, cooler, windier, cloudier and less sunny; at a microscale, windier positions are cooler.

Gradients in climate variables occur at a variety of scales: over continents, regions, landforms, and for vegetation types and substrates (rock, soil). The reasons for these gradients, a staple of school geography, are explained fully elsewhere (Geiger, 1965; Barry and Chorley, 1982) and summarized from a butterfly standpoint in Dennis (1993a). Steep climatic gradients and distinctive regions characterize relatively small land masses such as Britain (White and Smith, 1982; White and Perry, 1989); e.g., annual rainfall declines from 2000 to 60 mm in less than 20 km northwards down the Conway Valley from Snowdonia to Great Ormes Head, in North Wales. The country is roughly divided into quarters. A north–south line divides the humid west with cool winters from the dry east with cold winters; a west–east line divides the north with mild summers from the south with warm summers (White and Smith, 1982). The east–west division largely relates to competing influences of oceanic versus continental air masses; the north–south divide is a function of latitude, sun angle and length of daylight. Regional climatic gradients are greatly modified at a variety of scales, by the height and shape of the land surface, water bodies such as lakes, vegetation (e.g., forests, woodland, hedges) and human constructions (e.g., urban areas, roads, walls). These influences are usually labelled **local climates** and are to be differentiated from **microclimates**, which deal with contrasts at very fine scales such as the differences in temperature between the upper and lower side of a leaf or between the sides of a molehill or tree trunk. The greatest impact is imposed by mountain masses (altitude); with increasing elevation in Britain there is a drop in temperature (lapse rate of 7.1°C km^{-1}; White and Smith, 1982) and sunshine and an increase in frost, cloud cover, rainfall, snow cover, humidity and wind speed (see Chapter 8 for details).

The diagrams illustrate some typical features of local climates and microclimates important for butterfly biology. Radiation and temperatures are primarily influenced by aspect and sun angle. These two variables – key for regional gradients – control energy receipt at scales from major landforms (e.g., cuestas, drumlins, hills, quarries; Fig. B2.4a) to micro-landforms (e.g., plough drills, molehills; Fig. B2.4c), vegetation (e.g., tree trunks; Fig. B2.4h), and features in the human landscape (e.g., walls, hedges). The distribution of daytime temperatures on slopes is very different from radiation receipt (Fig. B2.4b). Temperature maxima follow the sun's rays increasing from the southeast via the south to the southwest. Radiation is at a maximum intensity on whichever slope is perpendicular to the sun's rays; maximum radiation therefore shifts from a relatively flat slope in summer to a steep slope in winter. A southern slope inclined at 20° at 50° north receives twice the radiation in January compared with a horizontal surface. However, the slopes that receive radiation later are drier, having warmed up during the earlier part of the day, so temperature maxima increase as the day progresses in spite of fact that radiation is symmetrical about the noon line. Hot slopes also tend to be dry ones. Elevation, slope angle and curvature influence heat retention. With greater range in elevation there is increasing disassociation between radiation receipt and temperature; areas at higher altitude lose energy rapidly at night, colder air sinks downhill and collects at foot slopes and in concavities. The general rule is that, at night, concave surfaces are cold and convex surfaces are warm, a pattern that *may* be reversed during the day. These patterns are highly predictable owing to the sun's azimuth. Less predictable at a local scale are wind speed and wind direction. Although wind speed maxima occur at high elevation and around coasts, and there is an average direction for prevailing winds for any

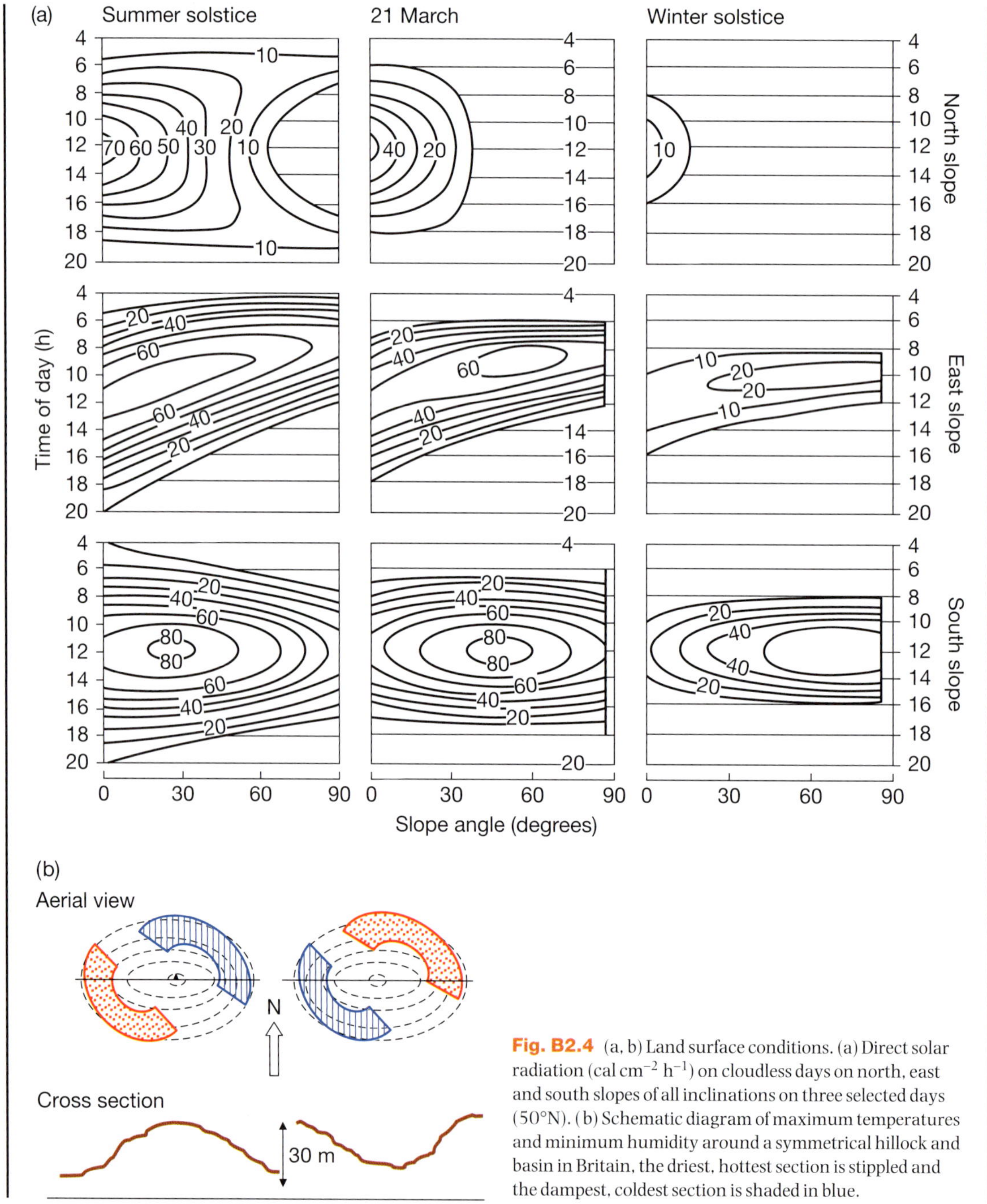

Fig. B2.4 (a, b) Land surface conditions. (a) Direct solar radiation (cal cm^{-2} h^{-1}) on cloudless days on north, east and south slopes of all inclinations on three selected days (50°N). (b) Schematic diagram of maximum temperatures and minimum humidity around a symmetrical hillock and basin in Britain, the driest, hottest section is stippled and the dampest, coldest section is shaded in blue.

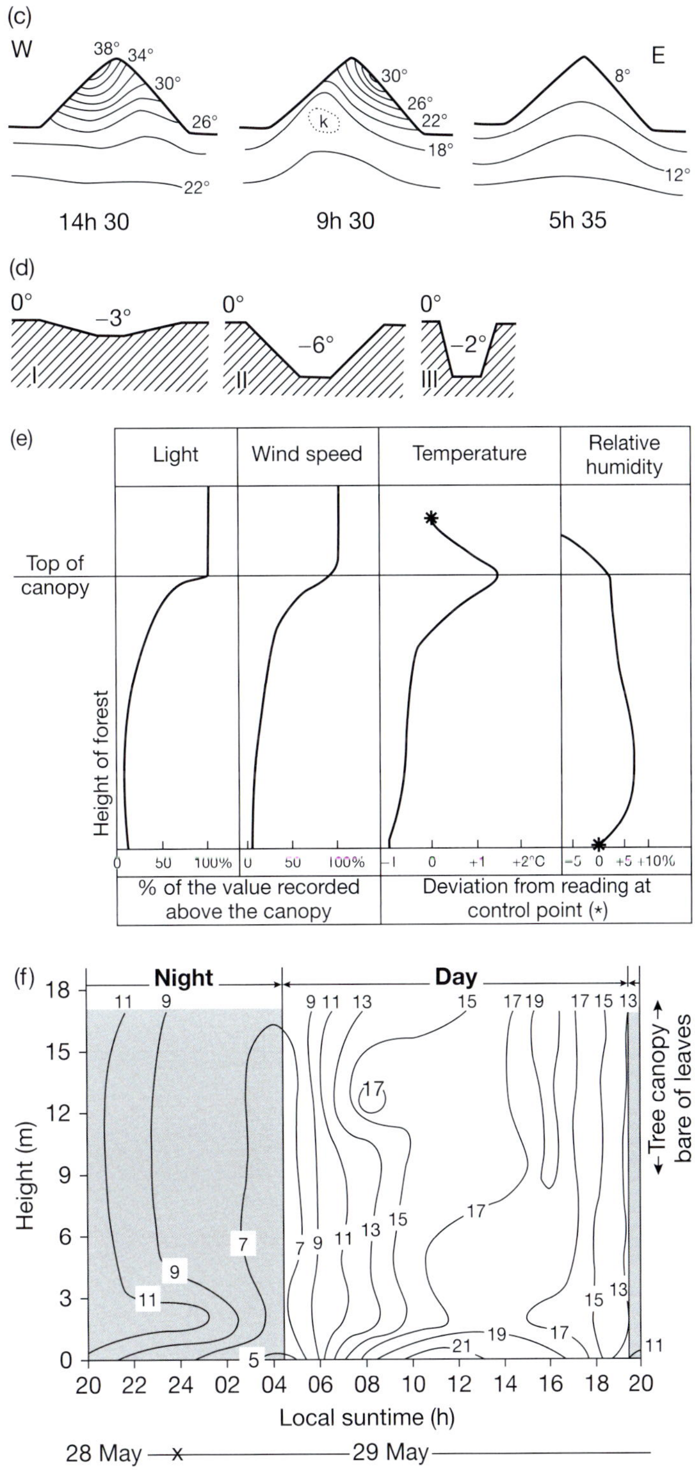

Fig. B2.4 *(continued)*
(c, d) Microtopography surface and subsoil temperatures. (c) Unplanted north–south crop drills with 45° sides, 13–15 cm deep and 26–30 cm apart before growth on a cloudless day, 31 August 1948, showing temperatures (°C) at three times; k = coldest site (after H. Lessmann). (d) Schematic diagram to illustrate night temperatures (°C) associated with cold air pockets in three types of small hollow (<10 cm dimensions) (as described by H. M. Bolz). (e–h) Forest climates. (e) Schematic diagram of conditions inside and above a mid-latitudinal deciduous forest in summer. (f) Diurnal variation in temperature within a deciduous forest near Montreal, Canada (46°N) in late May 1967; isopleths are in degrees C.

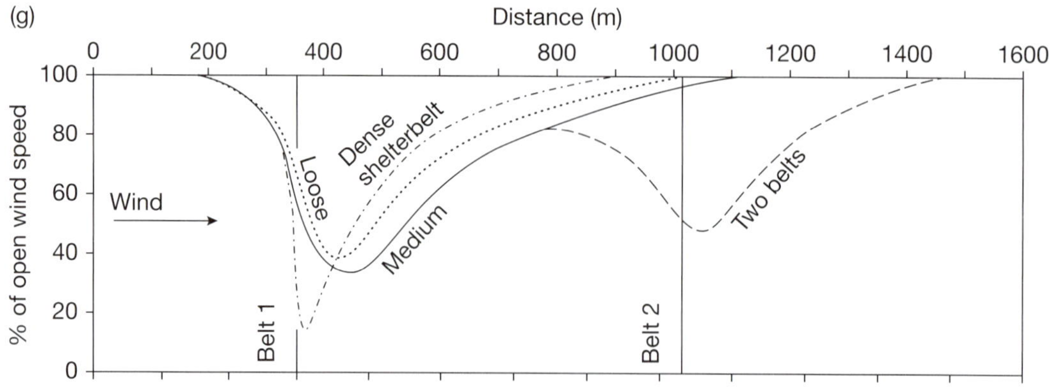

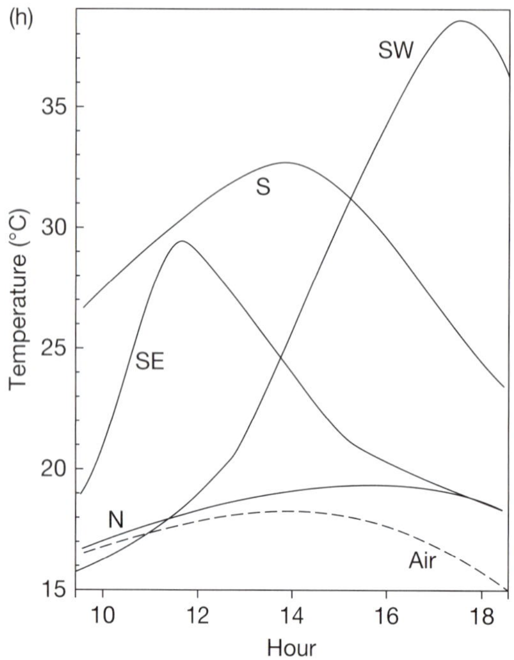

Fig. B2.4 (*continued*) (g) Influence of a shelterbelt that varies in density on wind velocity, and of two back-coupled, medium-dense shelterbelts. (h) Diurnal temperature variation for different aspects for the bark of a Sitka spruce tree (after N. Haarlov and B. B. Petersen).

location, wind direction is governed by rapid changes in air mass type (anticyclones, depressions, fronts) and deflected by topography and objects within sites. Physical features thus present windward and leeward sides that can vary greatly compared with sun and shade (see Box 7.6).

Moderation of weather and climate at local scales thus relates to a common set of influences, primarily: sun angle, aspect, elevation, slope curvature, exposure and insulation, and substrate composition and texture. Because of this, climate variables are predictable. Thus, White and Smith (1982) uses nine key landscape variables (i.e., grid east, grid north, distance to sea, elevation, aspect, slope east/west, slope north/south, change of slope, contour pattern declination) in multiple regression equations to calculate temperatures, rainfall, duration of bright sunshine, wind speed and snow depth for sites (<1 km resolution) for four seasons of the year. Many aspects of microfeatures have similar climatic patterns to those at local scales; thus north–south contrasts for plough ridges and cattle hoof divots are similar to those for hills and valleys. However, there are differences at very fine scales. For instance, in small depressions but a few centimetres deep, night-time temperatures can

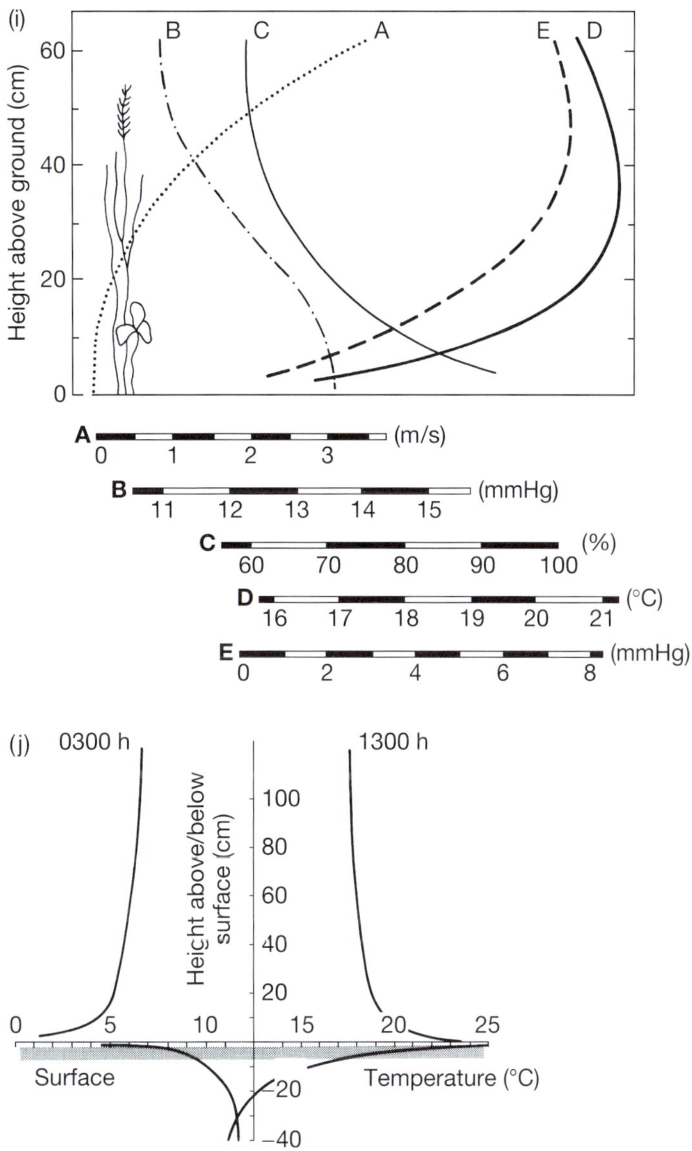

Fig. B2.4 (*continued*) (i, j) Ground temperatures. (i) Temperature, humidity and wind in a Scottish meadow on a sunny June day between 15.00 and 16.00 British Summer Time: (A) wind speed (m s^{-1}), (B) water vapour pressure (mmHg), (C) relative humidity (%), (D) temperature (°C), and (E) saturation deficit (mmHg) (after F. C. Waterhouse). (j) Air and soil temperatures associated with bare ground at Seabrook, New Jersey (39°34′N 75°13′E) on five bright days in spring; the diurnal range at screen level was 11°C, at 10 cm 15°C and at ground level >20°C. (a, c, d, h, i, redrawn from Geiger, 1965; e, f, based on material in JMB Advanced Geography Syllabus Paper 1 for 1972, Question 10; g, from Barry and Chorley, 1982, Fig. 7.11, p. 347, courtesy of Metheun and Co. Ltd and Taylor and Francis; j, after Crowe, 1971.)

be higher at the base of a steep slope than a gentle one, because steeper slopes are closer and heat from the soil warms the air at the base of the hollow (Fig. B2.4d). Even slope length and orientation can make for critical differences in humidity in small divots. Despite these differences for microclimates (the climate of vegetation and substrate surfaces at scales critical for performance in butterflies), such microclimates are eminently predictable using regression techniques and artificial neural networks (Bryant and Shreeve, 2002; Howe *et al.*, 2007). Vegetation and substrates have their own distinctive climate gradients (Fig. B2.4e–j). Thus, woodland and grassland have vertical profiles for diurnal temperature, humidity and wind speed, and reduce contrasts associated with bare ground.

mobility to continue resource exploitation and habitat maintenance.

- **P2.72: The impact of weather and climate on a butterfly's resource use depends on the developmental stage impacted and season of the year, thus type of resource use.**
- **P2.73: The response by butterflies at a locality (within the habitat space) to short-term marginal conditions is to change activity, in effect, resource use, or to change location within the habitat where they can continue the activity.**
- **P2.74: The response by butterflies to changing conditions within sites over the medium term, or to gradients imposed by changing conditions, that marginalize a subset of a resource, is to shift to a different subset of the resource(s) outlet. The alternatives are: population extinction, emigration or adaptation to conditions.**
- **P2.75: Habitat is a variable species-specific trait; populations of species may differ in resource use (use novel resources) or in the subset of a resource(s) exploited in relation to regional climate contrasts.**

Perhaps the key butterfly–climate principle is that involving fractional resource use (**P2.67**). Fractional resource use – the observation that butterflies use but a small fraction of an apparent available resource type (e.g., larval hostplant, nectar sources) – forms the basis of selection for resource quality. Fractional resource use emerges at all spatial scales from continental/regional down to micro-environment: the selection of the substrate surface. Climate and weather, if not the proximal cause of fractional resource use, may invariably be found to be an important distal (ultimate) cause of it. At a regional scale in the British Isles, the geographical range of very few butterflies extends as far as their larval hostplants (Dennis and Shreeve, 1991) and the reason for this 'gap' in exploitation is likely to be climate constraints on butterflies (Turner, 1986; Dennis, 1993a). At a local scale, numerous examples exist for the climate basis of fractional resource use. *Polyommatus bellargus* and silver-spotted skipper *Hesperia comma* are known to bias oviposition to shorter, hotter turf areas with bare ground (J. A. Thomas, 1983a; Thomas *et al.*, 1986), a pattern changing with climate warming (Roy and Thomas, 2003; Davies *et al.*, 2005, 2006). *Inachis io* and *Aglais urticae* use nettle patches in open, bright light environments and not under woodland shade (Dennis, 1984b). At even finer scales, *Argynnis paphia* lay eggs on the cooler north and west side of tree trunks (Asher *et al.*, 2001) not on the hottest surfaces facing south and southwest (Box 2.4); although this may relate in part to differences in bark surface character (e.g., moss) the ultimate reason is a climatic one. Similarly, the brown hairstreak *Thecla betulae* unusually concentrates territorial activity on the southeast side of trees (e.g., ash *Fraxinus excelsior*) in the early morning sun, having a capacity to be active at low ambient temperatures (11°C) (Middleton and Goodyear, 2008), whereas many other butterflies establish territories later in the day and exploit different aspects (Baker, 1972; Dennis and Shreeve, 1988). At the finest scale, there is a choice of the upper or lower surface of leaves for oviposition by *Aricia agestis* and *A. artaxerxes* (Ellis, 1999, 2003; Wilson *et al.*, 2002), very likely influenced by ambient temperatures.

There is a highly responsive dynamic to fractional resource use linked to weather and climate (**P2.68**); rapid changes in cloud, sunlight, wind speed, wind direction and temperatures are accompanied by equally rapid responses, changes in the use of exposed or sheltered slopes (e.g., *Plebejus argus*; Dennis and Sparks, 2006) and the use of microfeatures (e.g., *I. io* when perching on molehills; Dennis and Sparks 2005). Some landscapes experience much greater extremes in climate and weather (heat and cold, gales and calm, light, humidity, precipitation) than others. In such cases, it would be expected that periodic contrasts in fractional resource use will be observed ranging over different time periods from minutes to seasons, and that contrasting resource conditions are required for persistence at sites (**P2.69**). In Britain, greater extremes in wind speeds and surface temperatures are experienced in uplands and coastal margins. When integrated with seasonal temperatures, butterflies will be found to use more sheltered parts of the landscape in cool, windy summers than hot, calm ones (e.g., *P. argus*; Dennis and Bardell, 1996, Dennis and Sparks, 2006). Heterogeneity in topography, vegetation and substrate provide a template for variety in resource outlets that become more critical for butterflies experiencing thresholds for activity (e.g., feeding, flying). In fact, heterogeneity in landscape is vital for butterflies facing any kind of thermal or water budget threshold (**P2.70**). Local heterogeneity in landscape moderates climate and weather; cooler conditions are found on north-facing slopes and in shade, and warmer conditions on south-facing slopes and exposed to sunlight. With hummock and hollow terrain, and with open and shrubby vegetation, species can shift between resource outlets where they are found in cooler, warmer, drier, damper, windier or calmer conditions (**P2.71**) (Weiss *et al.*, 1988). How easily this is done depends on the stage involved – a finer grain of heterogeneity is clearly needed for larvae than adults, and more for species with less mobile adults than those that migrate over large distances. In hot, dry summers ovipositing butterflies can shift to northern and water-collecting concave slopes and shade; in cool, wet ones, hotter conditions may be found on south-facing, water-shedding convex slope segments and open but sheltered parts of sites. Where eggs have been broadcast over varied hostplant outlets in varied conditions over an entire site, subsequent changes in seasonal weather may nevertheless result in differential survival in relation to site heterogeneity, biased to areas that best moderate climate to suitable conditions.

Not all changes in climate impinge on butterfly biology. The nature of impact depends on three things:

1 The specific weather and climate changes.
2 Their timing as to season and time of day.
3 The developmental stage and activity (behaviour) of the butterfly (**P2.72**).

The passage of a depression may have limited impact on a hibernating pupa but will have profound repercussions for adults attempting to mate and lay eggs. Response to climate and weather inputs is tightly scale dependent. Adult butterflies, faced with sudden changes in weather, can respond in two ways depending, of course, on the nature of change in the weather (**P2.73**):

1 Conditions may not impinge on activity sufficiently to halt it, but enough to cause a shift in location within a site. Thus, a *Pyronia tithonus* may continue to feed on the same bramble bush but move to the lee side with a rise in wind speed, while a *Pararge aegeria* may take to woodland as daytime temperatures rise, abandoning the exposed fringe (Shreeve, 1984, 1985, 1986).
2 Where conditions cut across activity thresholds, then behaviour and activity change altogether. Thus, butterflies will bask with the passage of a cloud and take to daytime roosts with the passage of a rain storm. Such shifts in location or activity typically take place throughout a day.

In the medium term more profound responses occur (**P2.74**). With climate warming, there are likely to be shifts to using very different subsets of resources, especially larval hostplants. Already, such changes have been documented for *Hesperia comma* which now uses denser, taller sheep's fescue *Festuca ovina* than in the 1980s (cf., Davies *et al.*, 2005; Thomas *et al.*, 1986; Thomas and Jones, 1993). Such resource shifts can create changes in site geography for the butterfly and transform their regional geography over time. In the case of *P. aegeria*, this butterfly has been able to exploit resources previously unused and expand its geographical range in Britain (J. K. Hill *et al.*, 2002).

At the largest space–time scales, climate undoubtedly lies at the root of regional shifts in resource use, including substantial changes in behaviour and exploitation of novel resources (**P2.75**). Habitat is a species-specific attribute and varies as does any other species' trait. Towards the margin of geographical ranges, fewer resource outlets may provide suitable conditions for development, growth and survival than at the range core (Shreeve *et al.*, 1996a). In effect, the climate becomes marginal for resource use, a fact emphasized by cases where the types and structures of the resources have not necessarily changed (e.g., *Polyommatus icarus*; Howe, 2004). But marginal climates can create wholesale changes in behaviour and use of novel resources. In the Mediterranean region, *Hipparchia semele* and *Maniola jurtina* females aestivate in bushes during high summer, to reappear and continue reproduction in late summer; the males die out having already mated. This behaviour is absent in north Europe, but, as mentioned above, indications of potential for it in British *M. jurtina* are noted in their adoption of woodland during hot summers (Dennis, 2004a). The basic point here is that details of habitat cannot be safely derived from single sites or even single regions within the species' geographical range; an adequate sample of sites from throughout the range, representative of changing environments, local and micro-landscape heterogeneity, needs to be studied to understand a butterfly's habitat demands.

Edaphic agents as conditioners

It is clear to anyone observing changes in butterfly populations from one year to the next that they fluctuate in response to changing environmental conditions. The emphasis has been on seasonal climate and weather (Pollard, 1988; Roy *et al.*, 2001) rather than ground conditions and, as will become evident through the book, the two are linked. But soil (edaphic) conditions substantially affect resources for butterflies, in particular soil nutrients (e.g., nitrogen, calcium carbonate, sodium chloride), pH levels and soil water. Edaphic conditions are reflected in plants and butterfly species respond primarily to plant status. In Chapter 5 it will be shown how soil conditions can be assessed via plant species using Ellenberg scores.

Table 2.5 Basic recording fields for the UK Butterfly Biological Consumer Resources Database (BRD).

- Date of record
- UK county (post 1974) (Heath, 1976)
- Grid reference to 10 km square
- Butterfly species and Bradley and Fletcher (1986) code (see Emmet and Heath, 1991)
- Nectar plant species or substrate
- Larval hostplant species
- Larval host use status (E, egg laying, or eggs found in the field; L, larva(e) found in the field; N, successful breeding in natural conditions from egg to pupa; B, successful breeding in laboratory situation; F, failure to oviposit or develop successfully in laboratory situations; R, reference extracted from previous source)
- Nectar plant BRC code and family (Stace (1997) number)
- Number of individual butterflies observed for each entry in the file or record (counts of individuals)
- Source of record
- Reference source if not original, and sub-reference to earlier source, type of record and summary status of plant as a hostplant

Hostplant status, based on these records, is determined from the coding of records and designations in the literature into: M, main; S, subsidiary (auxiliary/secondary; generally regarded as insufficient without the presence of main hostplants); N, novel (alien plant to the British Isles); U, unsuitable/unlikely hostplant. X, denotes a likely mistake in oviposition or deliberate oviposition off the hostplant (Wiklund, 1984). The hostplant is also categorized as to any regional limitation: GL, geographical limitation (locality/area/region specified). See http://www.staffs.ac.uk/schools/sciences/gergraphy/links/IESR/staff_honfellows_rd_butterflydatabase.shtml.

RESOURCE DATABASE

This resource-based habitat approach to butterfly biology exposes two important issues. First, a butterfly habitat comprises a wide range of resources, other than larval hostplants, and the zone of each resource exploited by a species at a site is typically a fraction of each apparent resource. Second, despite more being known about British butterflies than perhaps for any other insect group, clearly little is known about many of the resources used by individual species. Yet, butterflies are argued to be an **indicator group** for environmental changes (J. A. Thomas *et al.*, 2004; Thomas, 2005) and focal (umbrella) organisms for sites. Indicator and focal species should be selected on precise criteria (Sanderson *et al.*, 2002; Coppolillo *et al.*, 2004) to be truly representative of a range of organisms and conditions. Thus, there is a strong argument for a resource database on butterflies (Dennis *et al.*, 2008). Two basic kinds are apparent: a biological resources database (BRD) and a site database (SD).

The UK BRD includes records on all aspects of resource use from observations over various sites and should be accessible to contribution from any observer. The data accumulating on adult (>49 000 observations) and larval (>5500 observations) feeding are growing rapidly and as understanding of other resource use develops (see Table 2.2) it will not be long before other resource types are included. Some of the current basic fields in this database are given in Table 2.5.

An SD is a very different construct and currently does not exist; it should provide the all-important conduit for spatially explicit site data on a range of resource variables and attributes. To understand properly how species persist in different landscapes it is essential to know what resources they access and use successfully *at a site scale*. It is one thing to have lists of resources that species use over an entire country; it is another to know just how they combine and are organized for distinct sites. At present, there is no inventory of site-based studies, which could comprise BSc and MSc dissertations as well as doctoral theses and reports from funded work by governmental and national, regional or local organizations (e.g., National Trust, Butterfly Conservation, County Trusts) as well as private unpublished, unfunded studies. An inventory is clearly a first step, followed by a meta-analysis of its contents, methods for which are now being advanced (Pullin and Knight, 2001).

The development of both databases should provide clear direction as to what data are missing on species and which species therefore require immediate attention in terms of data collection (Dennis *et al.*, 2008). The website for the BRD is noted in Table 2.5.

Chapter 3

BASIC PRINCIPLES FOR BUTTERFLY HABITATS

The hypothesis is the principal intellectual instrument in research. Its function is to indicate new experiments and observations and it therefore sometimes leads to discoveries even when not correct itself. (Beveridge, 1950, courtesy of Maryanne Kenny and The Blackburn Press, Caldwell, NJ, USA)

DESCRIBING VARIATION IN RESOURCES

Clearly, habitat is not always an easily identifiable space but it is not simply a vegetation 'type' – which is perhaps better referred to as biotope (Collin, 1988; Calow, 1999) whatever the scale selected to describe it, though a habitat may occupy one vegetation category or cut across several vegetation units (Wiklund, 1977, 1978; Dennis *et al.*, 2003, 2006b) (see Box 1.1). Now that guidelines have been provided to recognize habitats, we can move on to consider how habitats, in some detail, may impact on species. A straightforward way of approaching this is to consider possible generalizations, or principles, that can provide starting points for investigations into butterfly habitats. Habitats impact on their species' life history strategies and population dynamics through three distinct aspects of their resource distributions: **composition**, **physiognomy** and **connectivity**. This chapter concentrates on these basic components, and how they may affect the internal workings of the habitat. The details are derived from key autecological studies on British butterflies; the resources studied and references are listed in Table 3.1.

Resource composition

Composition refers to the occurrence (or absence) of a specific resource or resource component (e.g., one of several nectar sources or hostplants) and the variation in its make up and context. Context relates to the conditions in which a resource may occur, all of which affects its quality and therefore exploitation by individuals of a butterfly population. As an example of context, one vital resource, a hostplant, may occur in wetter or drier conditions, more basic or more acidic substrates, grow in the open or shade of trees, and be exposed or overtopped by other vegetation. Composition also includes density, frequency and abundance of a resource, though the latter two attributes are better dealt with under physiognomy (see below). The main resources for which variation in composition has been described within habitats are hostplants and nectar sources. These are listed for individual butterfly species in Appendices 2 and 3. However, there is also considerable variation in other resource outlets (e.g., roosts, mate location sites, thermoregulation sites, hibernation sites, pupal substrates, etc.) (see Appendix 5). For any life history stage or phase of activity, at a particular site, specific resources may be:

1 Single or multiple (e.g., as in the number of hostplants or nectar plants used) (see Appendix 4).

2 Entire or fractional (e.g., access to the whole of the hostplant or limited to parts of the hostplant).

3 Singular or transferable (i.e., a resource used by a single stage or several stages or phases; e.g., cuckoo flower *Cardamine pratensis* is used by orange-tip *Anthocharis cardamines* as a larval hostplant, nectar source and roost site).

4 Main or subsidiary. Wiklund (1981) describes host use as primary, secondary, unsuitable and novel. Obviously, distinctions in the status of resources may be categorized in different ways. An alternative division is between primary and auxiliary host use (e.g., Glanville fritillary *Melitaea cinxia* on the Isle of Wight

A Resource-Based Habitat View for Conservation, 1st edition. By Roger L. H. Dennis. Published 2010 by Blackwell Publishing.

Table 3.1 Detailed studies on butterflies within the British Isles.

Species	Resources studied*	References
Carterocephalus palaemon	AF, BA, HE, ML, OV	Collier, 1972, 1984, 1986; Ravenscroft, 1994a, 1994b, 1994c; Douglas, 2003
Thymelicus sylvestris	AF, FL	Hardy *et al.*, 1993; Warrington and Brayford, 1995; Goulson *et al.*, 1997a, 1997b
Thymelicus lineola	–	
Thymelicus acteon	AF, BA, FL, OV	J. A. Thomas, 1983c; Bourn and Thomas, 2002
Hesperia comma	BA, FL, HE, OV	Thomas *et al.*, 1986; Thomas and Jones, 1993; Hill *et al.*, 1996, 1999b; Warren *et al.*, 1999; Davies *et al.*, 2005, 2006
Ochlodes sylvanus	BA, HE, ML, OV, RO	Dennis and Williams, 1987
Erynnis tages	BA, FL, OV	Gutiérrez *et al.*, 1999; Gutiérrez, 2005; Wainwright, 2005b
Pyrgus malvae	HE, OV	Brereton, 1997
Papilio machaon	AF, FL, HE, ML, OV, PU	Dempster *et al.*, 1976; Dempster, 1995; Hoole *et al.*, 1999
Leptidea sinapis (reali)	AF, HE, ML, OV, RO	Warren, 1984b, 1985c; Warren *et al.*, 1986; Nelson *et al.*, 2001; Jeffcoate, 2006
Colias crocea	–	
Gonepteryx rhamni	FL, HB, HE, OV	Pollard and Hall, 1980; McKay, 1991; Gutiérrez and Thomas, 2000
Pieris brassicae		Feltwell, 1982
Pieris rapae	HE, OV	Richards, 1940
Pieris napi	AF, BA, HE, ML, OV, RO	Lees and Archer, 1974; Courtney, 1980, 1988; Courtney and Courtney, 1982; Dennis, 1982b, 1985c; Wilcockson, 2002
Anthocharis cardamines	FL, HE, OV	Courtney, 1980; Courtney and Duggan, 1983; Dennis, 1982a, 1982b, 1983a, 1983b, 1986b
Callophrys rubi	HE, OV	Shaw and Bland, 1994
Thecla betulae	AF, BA, HE, ML, OV	Thomas, 1974, 1975a; Smith and Clark, 2005; Middleton and Goodyear, 2008
Favonius quercus	AF, FL, HE, ML, OV	Thomas, 1975b; Crawford, 1996; Newland, 2009
Satyrium w-album	HE, ML, OV, PU	Davies, 1992
Satyrium pruni	AF, BA, HE, ML, OV	Thomas, 1974, 1975c
Lycaena phlaeas	AF, BA, FL, HE, OV	Dempster, 1971; Aspinall, 1987; Léon-Cortes *et al.*, 2000
Lycaena dispar	HE, ML, OV, RO	Duffey, 1968; Pullin *et al.* 1995; Pullin, 1997
Cupido minimus	BA, FL, HE, OV	Coulthard, 1982; Morton, 1985; Léon-Cortes *et al.*, 2003b
Plebejus argus	FL, HE, ML, OV, PU, RO, SY	Dennis, 1972b, 2004b; C. D. Thomas, 1983, 1985a, 1985b; Mendel and Parsons, 1987; Ravenscroft, 1990; Lewis *et al.*, 1997; C. D. Thomas *et al.*, 1998, 1999; Wilson, 1999; Dennis and Sparks, 2006; Hodgson, 2007
Aricia agestis	BA, HE, ML, OV	Bourn and Thomas, 1993; C. D. Thomas *et al.*, 2001; Burke *et al.*, 2005
Aricia artaxerxes	AF, BA, HE, ML, OV, RO	Clunas, 1986; Ellis, 1995, 1999, 2003; Wilson *et al.*, 2002
Polyommatus icarus	AF, BA, FL, HE, ML, OV, RO, SY	Dennis, 1984a, 1985a; Burghardt and Fiedler, 1996; Léon-Cortés *et al.*, 1999; Howe, 2004; Howe *et al.*, 2007
Polyommatus coridon	RO, SY	Brereton *et al.*, 2008
Polyommatus bellargus	FL, HE, OV, SY	Davies *et al.*, 1958; Thomas, 1983a; Bourn *et al.*, 1999; Whitfield, 1999; Bourn and Thomas, 2002; Roy and Thomas, 2003
Celastrina argiolus	AF, BA, HE, ML, OV, SY	Pollard and Yates, 1993b; Pollard and Moss, 1995; Willmott, 1999
Maculinea arion	AF, BA, HE, ML, OV, PU, RO, SY	Thomas, 1976, 1977a, 1977b, 1980a, 1995; Thomas *et al.*, 1989, 1998a, 1998b, 2009; Thomas and Wardlaw, 1990, 1992; Thomas and Elmes, 1998
Hamearis lucina	AF, BA, FL, HE, OV, PU	Oates, 1986, 2000; Oates *et al.*, 1986; Sparks *et al.*, 1994; Léon-Cortes *et al.*, 2003a; Turner *et al.*, 2009

Limenitis camilla	BA, FL, HE, OV	Pollard, 1979; Pollard and Cooke, 1994; Pollard and Moss, 1995; Joy *et al.*, 1999
Apatura iris	AF, BA, FL, HB, HE, ML, OV, PU	Willmott, 1987, 1990, 1994; Oates *et al.*, 2005; Oates, 2008
Vanessa atalanta	HB	Baker, 1969, 1984; Tucker, 1991; Pollard and Greatorex-Davies, 1998; Hardy and Kinder, 2000; Fox and Sleep, 2005; Dennis *et al.*, 2006c; Fox, 2007
Vanessa cardui	HB	Baker, 1969, 1984; Joy, 1996b; Pollard *et al.*, 1998; Wacher, 1998
Nymphalis (Aglais) urticae	BA, HE, ML, OV	Baker, 1972; Dennis, 1984b, 1985b, 1985d, 2004c; Pullin, 1988; Bryant *et al.*, 1997, 2000; Pollard *et al.*, 1997
Nymphalis polychloros		Pollard and Moss, 1995
Nymphalis (Inachis) io	BA, HE, ML, OV	Baker, 1972; Bryant *et al.*, 1997, 2000; Dennis, 2004c; Dennis and Sparks, 2005
Nymphalis (Polygonia) c-album	BA, HB, HE, ML, OV	Nylin, 1992; Nylin *et al.*, 1996; Wedell *et al.*, 1997; Bryant *et al.*, 2000; Dennis, 2004c; Braschler and Hill, 2007
Boloria selene		Ellis, 2000, 2001; Berney, 2002; Joy, 2002a, 2005; Stewart and Bourn, 2004; Bourn *et al.*, 2005; Jordan and Bourn, 2007; Andrews, 2008; Williams and Bourn, 2008
Boloria euphrosyne	BA, OV	Brereton and Warren, 1999; Joy, 2002b; Boardman, 2005
Argynnis adippe	BA, HE, ML, OV, PU	Warren, 1995; Warren and Oates, 1995; Clarke, 2005; Ford, 2008
Argynnis aglaja		Ford, 2008
Argynnis paphia	–	
Euphydryas aurinia	BA, FL, HE, ML, OV	Porter, 1982; Lavery, 1993; Warren, 1994; Bulman, 2001; Hobson *et al.*, 2002; Joyce and Pullin, 2003; Fowles and Smith, 2006; Bulman *et al.*, 2007
Melitaea cinxia	HE	Thomas and Simcox, 1982
Melitaea athalia	FL, HE, OV	Warren *et al.*, 1984; Warren, 1985a, 1985b, 1987a, 1987b, 1987c; Brereton *et al.*, 1998; Stewart *et al.*, 2001; Hodgson, 2007; Kelly, 2008; Vulliamy and Bulman, 2009
Pararge aegeria	BA, FL, HE, ML, OV	Davies, 1978; Shreeve, 1984, 1985, 1986, 1987; Pollard *et al.*, 1996; J. K. Hill *et al.*, 1999a, 2001
Pararge (Lasiommata) megera	BA, FL, HO, ML, OV, RO	Dennis, 1982–1983, 1983c, 1986a; Dennis and Bramley, 1985; Dennis and Dennis, 2006, 2007
Erebia epiphron	HE, OV	Boyd-Wallis, 1994; Shannon, 1995; J. K. Hill *et al.*, 2002
Erebia aethiops	BA, HE, ML, OV	Dennis, 1982c; Beaumont, 1995; Kirkland, 1996
Melanargia galathea	HE	Wilson, 1985
Hipparchia semele	BA, ML, OV, RO	Dennis, 1972a; Tinbergen, 1972; Findlay *et al.*, 1983; Shreeve, 1990; Joy, 1996a; Dennis *et al.*, 1998a, 1998b; Loram *et al.*, 2003; Robinson, 2008
Maniola (Pyronia) tithonus	FL, NE	Dover, 1996; Dennis, 2003, 2004a
Maniola jurtina	AF, BA, FL, HE, ML, NE, OV, PU, RO	Dowdeswell, 1961, 1981; Ford, 1964; Brakefield, 1979, 1982a, 1982b; Pollard, 1981; Shreeve, 1989; Goulson, 1991, 1993; Dover, 1996; Shreeve *et al.*, 1996b; Maier, 1998; Dennis, 2004a
Coenonympha pamphilus	HB	Pollard and Greatorex-Davies, 1997
Coenonympha tullia	FL, HB, HE, OV	Turner, 1963; Dennis *et al.*, 1984, 1986; Melling, 1987; Joy, 1991; Dennis and Eales, 1997, 1999; Joy and Pullin, 1997, 1999; Eales and Dennis, 1998; Wainwright and Ellis, 2004; Wainright, 2005a
Aphantopus hyperantus	FL	Dover, 1996; Sutcliffe and Thomas, 1996; Sutcliffe *et al.*, 1997a, 1997b

* Specific resources (consumables and utilities studied): hostplants, herbivory and larval sites (HE), oviposition and egg-laying sites (OV), adult feeding (AF), mate location and territoriality (ML), basking sites, thermoregulation and local/microclimatic conditions (BA), roosting and resting (RO), pupation sites (PU), symbionts (SY), hibernation sites (HB) and flyways (FL). Most sources also refer to conditions. Additional data can be obtained from the Butterfly Conservation's reports on sites and species; a few are mentioned here (e.g., for *Boloria selene*) for species lacking published material. References to studies in continental Europe are found in Settele *et al.* (2008).

uses ribwort plantain *Plantago lanceolata* as a main hostplant but final instar larvae may use buck's-horn plantain *Plantago coronopus*). Yet another is between primary (principal) and secondary hostplants, the latter occurring infrequently or sporadically, as seems to be the case in the holly blue *Celastrina argiolus* (see Appendix 2a).

Resources can also be described in many other ways that distinguishes their quality and that affects their exploitation by individual butterflies. An important point to bear in mind when a single resource is being considered is that it is rarely uniform in composition. To take a single hostplant, this can be described in terms of, for instance, growth form, nutrient status, water content, leaf size, age and longevity and height and so on (see Appendix 2). Other resources can be described in the same way (see Appendices 3 and 5), but have not yet been adequately studied across many species.

Resource physiognomy

Physiognomy refers to the geography of a resource patch. Each resource patch or the array of patches can be described in terms of:

- Location, both absolute in terms of coordinates, including elevation above sea level (x, y, z), and relative to other resource patches.
- Height (tallness).
- Size, for instance length and breadth (area).
- Shape, from circular to linear.
- Orientation.
- Slope.
- Fragmentation and comminution (the degree to which resource elements and individual resource patches are broken up by other resource or non-resource types; frequency of patches).
- Contagion, whether random in distribution, over-dispersed or clustered (aggregated) (Box 3.1).

Box 3.1 Pattern, shape and dimensionality in butterfly resources

Butterfly resources can be visualized as features belonging to different dimensions (0, point; 1, line; 2, patch; 3, volume). It is important to distinguish between what butterflies actually respond to and what can be adequately 'mapped'. Butterflies respond to the 'three-dimensioness' of structures of even the tiniest resource item. However, although in reality all resources have volume, and points have surfaces and volume as the resolution in human observation is increased, for mapping purposes resources tend to be allocated to one of the first three categories of dimensional features (e.g., a flowerhead is effectively a point, a hedgerow a line, a nettle clump a patch). Of course, just whether a feature is a point, line or surface depends on the scale of mapping; with finer scales, features gradually become promoted to higher dimensionality. In the case of multiple complementary resources (nectar flowers, larval hostplants, roost sites, mate location sites, etc.), one type of feature can occur within the mapped bounds (on, within or under) of others (e.g., single dispersed nectar flowers of ragwort within a patch of hostplant *Rumex acetosa* for small copper *Lycaena phlaeas*) but there is a logical order of association (Table B3.1a). Resource features may also variably intersect or be disjointed within

Table B3.1a Basic dimensions of landscape features and their natural associations.

		Placement on, within or under feature			
		Dot (point)	Line	Patch (surface)	3D space
Placement of feature	Dot (point)	✓	✓	✓	✓
	Line	✗	✓	✓	✓
	Patch (surface)	✗	✗	✓	✓
	3D space	✗	✗	✗	✓

✓, possible; ✗, do not usually occur but there may be exceptions.

a habitat space (see Box 2.1). Individual butterflies are transposed on these features, their geography shifting through each day as adults, and more dramatically in the form of different stages with seasons.

Objects have individual and collective size and shape; thus a flower used as a nectar source by a butterfly can occur singly or form part of a patch. The individuals can still often be counted, but a plant patch has a higher order of impact than the individual plant on an organism such as a butterfly in terms of patch size and shape. Often, individual resource items are not easily distinguished (e.g., the three-dimensional overlap of leaves on shrubs or of interlocked climbers, shrubs and trees making a hedge) and shape and size is then less meaningful for individual items making up such structures than the compound entity, as in the case of a hedge, scrub, grass patches, etc.

Together or singly, resources have pattern and shape. Pattern describes the distribution of individuals or patches made up of individuals within an area, usually a meaningful zone such as a habitat space. Shape can be described in many different ways, but shape usually refers to regularity and compactness. When resources exist as patches, or are designated as such by drawing an envelope about individual resource items, they can be described as regular or irregular in shape. Most patches of a resource in nature are irregularly shaped, though regular ones do exist and are important for some butterflies (e.g., a field of a crop such as cabbage for *Pieris brassicae*; Fig. B3.1a). Patterns of points (individual items) or patches (collections of items) can be usefully classified in two ways, first as regular or irregular, and second as **clustered**, random or **dispersed** (Fig. B3.1a). The two extremes are **absolutely dispersed**,

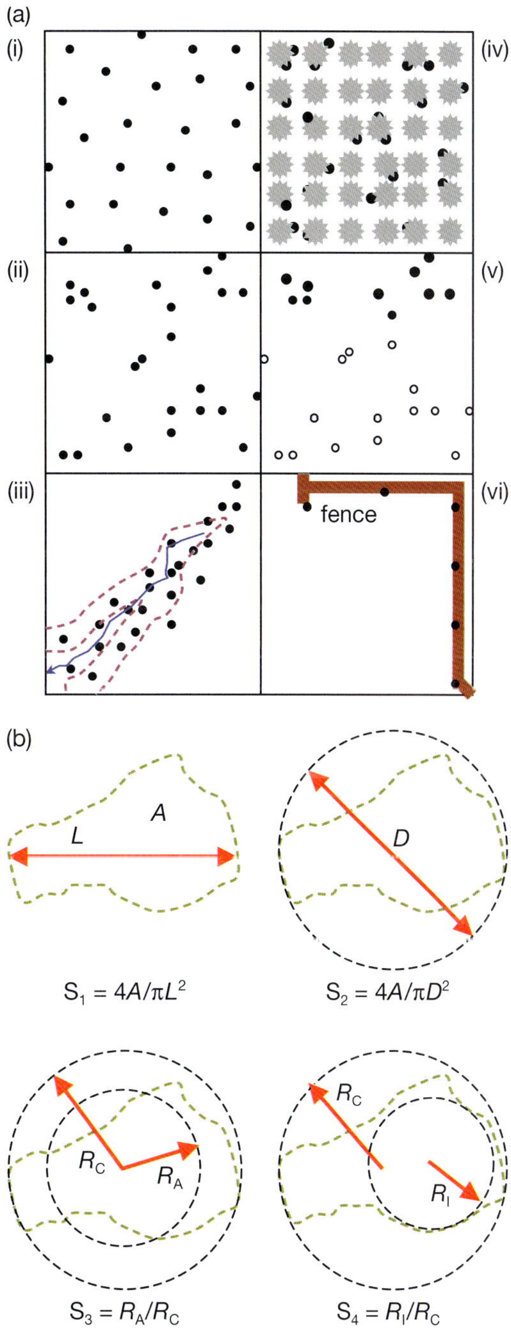

Fig. B3.1 (*right*) (a) Patterns of resource distribution within a habitat: (i) dispersed; (ii) random; (iii) clustered (e.g., crucifers in a small valley around a stream); (iv) dispersed and regular resource with a random distribution of the butterfly (cabbage patch with larval batches of large white *Pieris brassicae*); (v) a random distribution of resources with a clustered distribution of the butterfly or moth (e.g., ragwort of different sizes shown by area of circles with presence of cinnabar moth larvae *Tyria jacobaeae* (L.) shown by filled circle); and (vi) a clustered resource with dispersed butterflies (e.g., fence along field with territorial wall brown *Lasiommata megera*; Dennis and Bramley, 1985)). (b) Different methods for measuring compactness of resource items or patches. A, area; L, length of longest axis; D, diameter of smallest circumscribing circle; R_C, radius of smallest circumscribing circle; R_I, radius of largest inscribed circle; R_A, radius of circle with same area as shape ($=\sqrt{A/\pi}$).

where the points are distributed as if at the centres of hexagons, and **absolutely clustered**, where all the points are piled on top of one location on the surface. Just what type of distribution exists depends on the number of points and the size (bounds) to the area. To make sense of distribution patterns, the concept of a **random distribution** has been developed. It is possible using simple techniques to determine whether a distribution is **over-dispersed** (clustered or contagious), **under-dispersed** (evenly spaced) or random (Greig-Smith, 1964; Kershaw, 1973; Kellman, 1975; Gilbertson *et al.*, 1985; Ludwig and Reynolds, 1988). The essential features of a random distribution are:

1 The points have an equal chance of falling anywhere on the surface under investigation (in the squares of a grid if this is used in place of a prescribed surface).

2 The placing of a point is not affected by the placement of any other points.

The mere demonstration of non-randomness in butterfly individuals or their resources for that matter is of limited interest; what is of interest is the scale at which non-randomness occurs, as then it can be related to the environmental or other factors causing it. In a random distribution of points, the number of individuals per quadrat (for quadrats allocated randomly or systematically over the surface of the area investigated) follows a **Poisson series** e^{-m}, me^{-m}, $m2/2!e^{-m}$, $m3/3!e^{-m}$, $m4/4!e^{-m}$, etc., where m is the mean density of individuals and constant e is the base of natural logarithms. Successive figures of the series give the probability of quadrats containing 0, 1, 2, 3, 4 . . . *n* individuals, respectively, and the expected number of quadrats thus falling into each of these classes can be readily calculated. In a Poisson series the variance has the nice property of being equal to the mean and thus the ratio of these two values is equal to 1; various tests can be applied to determine whether a pattern of points is random or not (Ludwig and Reynolds, 1988). But, just whether a distribution of points is random or not depends very much on the scale of observations (**extent of area** studied or frame of map) and the scale of measurement or quadrats used (**mesh** or grain of mapped area), which together operate as a sieve with a mesh. Thus, on theoretical grounds, it can be demonstrated that for any contagious (clumped) population of points (mapped individual butterflies), the use of a Poisson series and tests of departure from it will reveal both random, contagious and regular distributions as the size of quadrats is steadily increased (from, say, 10 cm to 1 m). This relationship allows identification of the scale at which clumping occurs by re-sampling the population of points several times using quadrats of different sizes.

Other techniques are available for detecting pattern. The most frequently used is **nearest-neighbour analysis** developed by plant ecologists (Clark and Evans, 1954). This is specifically designed to measure patterns of points in two (or three) dimensions. It involves calculating the mean of distances between each point and its nearest neighbouring point. In some cases nearest neighbours form reflexive pairs, i.e., each point is the nearest neighbour of the other point. The number of nearest-neighbour distances is always the same as the number of points. The nearest-neighbour index is:

$$R = d_O/d_E,$$

where R is the index, d_O is the observed mean nearest-neighbour distance and d_E is the expected mean nearest-neighbour distance for a random arrangement of points:

$$d_E = 1/2\sqrt{p},$$

where p is the density of points per unit area (the number of points divided by area). The nearest-neighbour index ranges from 0 (absolutely clustered) through 1.0 (a random arrangement) to 2.15 (absolutely dispersed). A test of significance is provided by:

$$z = (d_O - d_E)/\mathrm{SE}d_O,$$

where z is a normal standard deviate (the sampling distribution is normal) and $\mathrm{SE}d_O$ is the standard error of the mean nearest-neighbour distance. As has been mentioned above, patterns can change over single sites and these measures of pattern therefore need to be used with caution. Nearest-neighbour analysis in any case assumes that the area the points are located within is infinite and points (plants, individual butterflies) are free to locate anywhere, neither of which is true. As nearest-neighbour index is influenced by the area selected; bounds are typically chosen for the test that brush against the sides of an enclosing space, usually a square (Fig. B3.1a). However, if one is interested in the pattern within a habitat space, the bounds of the habitat can be used (see p. 267). For other issues associated with the test, refer to Ebdon (1977).

Clearly the pattern can vary for resources and individual butterflies. The various combinations for a single resource and butterfly are shown in Table B3.1b. The most important aspect of shape from the vantage of resources for butterflies is their degree of **compactness**, as this fundamentally influences resource exploitation. This is basically a measure of how far a shape deviates from the most compact shape, a circle, the shape that has the smallest perimeter relative to area. Several measures of compactness are illustrated (Fig. B3.1b). An obvious simple measure of compactness is thus:

$$S_P = P/A,$$

where P is the length of the perimeter and A is the area (Pounds, 1963). But it is often difficult to obtain a measure of the length of perimeter and the index is not

Table B3.1b Combination of dispersion patterns for butterflies and their resources.

		Butterfly individuals		
		Dispersed (D)	Random (R)	Clustered (C)
Items or patches of a resource	Dispersed (D)	DD	DR (Fig. B3.1a (iv))	DC
	Random (R)	RD	RR	RC (Fig. B3.1a (v))
	Clustered (C)	CD (Fig. B3.1a (vi))	CR	CC

independent of units of measurement or absolute size of the shape being considered. Each of the indices (S_1 to S_4) is scaled in such a way that the most compact shape, a circle, has an index of 1.0 and as the shape becomes less compact the index falls in each case. Many other measures of shape exist (e.g., Boyce and Clark, 1964; Bunge, 1966) and others are being developed with advances in landscape studies (see Box 7.7).

As such, resources can be described as any other geographical feature (Evans, 1987) and at different scales.

Resource connectivity

Connectivity refers to the potential links between resource types and elements. Connectivity among resource patches, although distinct from connectivity among populations (habitat units) (see Hanski and Pöyry, 2007; Chapter 6), is similarly based on individual butterfly movements. Resource connectivity involves more than resource geography as it depends on adult and larval mobility and the degree to which paired resource outlets can be used consecutively by individuals. Measurements of connectivity for a resource-type patch (e.g., potential nectar flower patch) or patches (e.g., supplementary nectar flower patches) would require the measurement of numbers accessing the patch(es) and rate of access relative to isolation and barriers. The form that any resource connectivity algorithm takes would, of course, depend on what complementary or supplementary resource outlets are logically associated during a butterfly's life cycle and daily 'trivial' activity; it would involve very different parameters for adults and larvae. Resource connectivity can be described in terms of:

- Overlap, contiguity (contact, neighbourhoods) and isolation.
- Barriers and obstacles.

As these three aspects of resources vary in space and time, they provide useful headings for exploring some basic principles in resource availability. These principles or points provide useful templates against which to search for exceptions to expected relationships. In doing so, however, it is well to consider that several aspects of resource availability may be varying simultaneously over a habitat space.

RESOURCE VARIATION IN THE HABITAT SPACE

General principles of resource composition

- **P3.1: Only a fraction of potential resources is suitable for a butterfly species within any single biotope.**
- **P3.2: The suitability of a resource is affected by its context of non-resource items.**
- **P3.3: The greater the number of larval hostplants used by a butterfly species, the greater the number of vegetation units (biotopes) a habitat is expected to cover.**
- **P3.4: The greater the number of larval hostplants a butterfly species uses, the greater the variety of complementary resources a species is expected to exploit.**
- **P3.5: The more generalist in larval hostplant use a species is regionally, the more supplementary resources are expected to occur for it within a single site.**
- **P3.6: The wider the range in larval host use, the wider the parasitoid (and predatory) community exploiting the butterfly species within a single site.**

Some distinct relationships link resource composition to habitat status. Perhaps the most obvious is that in the absence of a resource the habitat is necessarily deficient (Leibig's 1840 **Law of the Minimum**; see p. 70). Where resources are missing from a breeding site a habitat can still function as a part habitat if, as in the case of the brimstone *Gonepteryx rhamni* where hibernation sites can so often be isolated from other resources (Pollard and Hall, 1980), the organism can disperse between the resource outlets. Two other well-known and expected relationships from butterfly autecology (Wiklund, 1974b, 1977; Thomas, 1984) are fractional resource suitability and resource context (**P3.1** and **P3.2**). An example of principle **P3.2** is found in the chequered skipper *Carterocephalus palaemon* which tends to occur on purple moor-grass *Molinia caerulea* hostplants on richer soils (i.e., flushes), the plants used by larvae being larger, persisting longer, with fewer flowers and higher nitrogen concentrations (Ravenscroft, 1994a, 1994b, 1994c) than those on nutrient-deficient soils. Butterflies are usually, one might say invariably, highly specific in resource use. This is well illustrated by the very different distributions of three butterflies (dingy skipper *Erynnis tages*, common blue *Polyommatus icarus* and silver-studded blue *Plebejus argus*) on one of their larval hostplants, bird's-foot trefoil *Lotus corniculatus* on the Creuddyn Peninsula in North Wales (Gutierrez *et al.*, 2001), and by nectar feeding among butterflies constrained by flower structure and butterfly morphology (Corbet, 2000). It is also well illustrated by the avoidance of small cuckoo flower plants by egg-laying *Anthocharis cardamines* (Dennis, 1982a, 1983a; Dennis and Hardy, 2006). A useful axiom follows: if any resource within a habitat is not 'pure', that is homogeneous in every respect, then the resource elements must have a variable context. This is invariably true since as resource patches are finite, and thus are described edge to centre context, parts of a patch will automatically differ for a butterfly approaching them. Context can make for substantial differences in population **fitness** (relative competitiveness of genotypes) and individual success. For instance, overtopping by other vegetation, such as tall herbs, shrubs or trees, can do the following:

- Affect the growth of key resources such as hostplants (e.g., hedge garlic *Alliaria petiolata* for *A. cardamines*; Dennis, 1982a).
- Reduce apparency of a resource, as in the case of hostplants for ovipositing females (Wiklund, 1984; Severns, 2008).
- Cause 'interference' by limiting access to preferred resources (e.g., oviposition in small tortoiseshell *Aglais urticae* (Dennis, 1984b); basking on molehills by peacock *Inachis io* (Dennis, 2004c)).
- Reduce larval development and increase mortality (e.g., *A. cardamines* (Courtney, 1986); predation in white admiral *Limenitis camilla* (Pollard, 1979; but see Fox, 1996, 2005)).

Four additional principles may also prevail based on the probability of occurrence (viz., stochastics) of varied resource outlets (**P3.3** to **P3.6**) linked to the key consumer resource, hostplants. As such, alongside increasing exploitation of an increasing number of hostplants locally is expected exploitation of a wider variety of other resource outlets, use of more vegetation types making up habitats, and the number of predators and parasites in turn preying on them (Shaw, 2006). These relationships are highly probable at a regional scale across populations or habitat units, where there is a good correlation between number of hostplants used and biotopes occupied, and between number of nectar sources used and hostplants exploited (Tudor *et al.*, 2004; Dennis *et al.*, 2005; Hardy *et al.*, 2007) (see Table 2.3, Chapter 5). It would be interesting to see if they also emerge within habitats and just how strong are the relationships. Exceptions occur in these relationships at a regional scale and must be expected at the within-habitat scale. Even so, butterflies that use a number of different hostplant species regionally would be expected to have access to more of the same kind of resource outlet (supplementary resources), in this case hostplants, within a single site.

General principles of resource physiognomy

- **P3.7: The more hostplants used by a butterfly within a site, the larger the area of the habitat.**
- **P3.8: The larger the area of a resource within a site, the more varied will its elements be physiognomically and the more opportunities there will be for exploitation.**
- **P3.9: The larger a resource within a site, the less critical are factors for exploitation associated with its spatial attributes.**
- **P3.10: Transferable resources (e.g., flowering plants used for larval hostplants and nectar) increase habitat compaction, and dispersion of resources within a habitat accompanies non-transferable resources.**

- **P3.11: Some resources are more likely to be linear (edge resources) than others and influence resource exploitation.**
- **P3.12: Resource types (and complements of resources) vary in physiognomy influencing resource status and habitat status for species.**
- **P3.13: Comminution (fragmentation on a fine scale; cf., compaction and context) affects the quality of a resource, generally adversely.**

The most obvious of physiognomic relationships links the size of a resource (area) to its influence on the variability of the resource; the larger the resource patch, the more variable are likely to be the resource elements or outlets (patches). In this respect, increasing resource dimensions works much like generalism for a resource type within sites. No two supplementary resources (e.g., two hostplants) will be ecologically equivalent and as such they are likely to dominate different parts of the same site where they occur, thereby increasing the area of the site (habitat) used (**P3.7**). Moreover, as resource elements are never entirely self-equivalent (homogeneous), however spatially limited their distribution, the larger the area covered by a resource patch(es) the more varied will be its elements for exploitation (**P3.8**). This occurs primarily because variation in substrate conditions (i.e., soils, geology, slope) increases with area. Naturally, there is not strict identity between several different (supplementary) resource outlets (i.e., hostplant species patches) and one abundant single resource outlet. Although for any hostplant species there is typically a hierarchy in oviposition preference and larval survival, this is usually more severe for a suite of different hostplant species than it is for a single hostplant species taken from the same site (e.g., swallowtail *Papilio machaon* in Sweden; Wiklund, 1975, 1981). As the dimensions of a resource at a site increase, the less critical are location parameters likely to be for a butterfly using that resource; with increase in site size there will generally be increased variation for slope, orientation, context and other spatial parameters (**P3.9**). This relationship will perhaps not be linear; the exact relationship will depend on environmental influences (i.e., geology, topography) (see Chapter 7).

Shapes and distributions (ranging from aggregated to over-dispersal as occurs with regular spacing) of complementary resources making up the habitat influence resource status and exploitation, much as single resource types do (see Chapter 4). Aggregated distributions are perhaps more commonly found than random or regular patterns (e.g., large copper *Lycaena dispar* and its hostplant water dock *Rumex hydrolapathum*; Webb and Pullin, 2000). When there is duplication (transferability) for resource types this increases habitat compaction (**P3.10**), but overall compaction depends on the distribution of the most diffuse resource type. Some resource types, typically utilities, tend to be more linear (less compact) than others and are often edge resources (**P3.11**). For instance, mate location sites (e.g., wood edges in *Anthocharis cardamines* (Dennis, 1982b); *Aglais urticae* and *Inachis io* (Baker, 1972; Dennis, 2004c; Dennis and Sparks, 2005) generally depend on local climate influences such as shelter and sun orientation. Often the whole constellation of resources is linear, as for instance habitats along hedgerows and lines of communication (Dover and Sparks, 2000). Along with differences in the spatial structure of linear and non-linear habitats are often differences in physiognomy and composition, usually owing to context, as well as differences in connectivity, influencing the suitability of the same resource types in these distinct settings for butterfly populations dependent on them (**P3.12**) (Clausen *et al.*, 2001). The comminution (fragmentation of elements) of a resource type has an effect related to whatever fills the gaps and influences individual fitness (predators, parasites, visual searching of females) (**P3.13**) (Summerville and Crist, 2001). It is rare for a context to fail completely as a resource – most surfaces provide at least temporary sites for settling adults – but this may be outweighed by negative aspects of the context elements (e.g., increased enemies).

General principles of resource connectivity

- **P3.14: The more abundant resources are within a site, the less likely are they to be isolated from one another and the more connected the habitat structure.**
- **P3.15: The greater the transferability of resources (equivalence), the more connected will be the resources making up a habitat.**
- **P3.16: The more diffuse (less compact, comminuted, fragmented) a resource or a set of resources, the less connected they are and the more mobile an organism needs to be to exploit them.**
- **P3.17: The degree of connectivity depends on the structural attributes of the matrix (context) separating resources.**

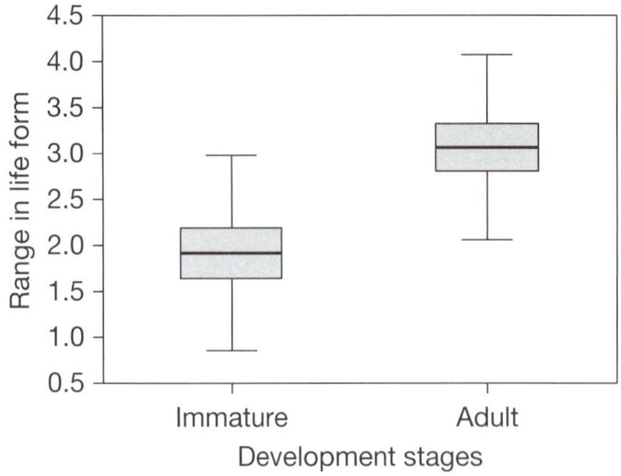

Fig. 3.1 Range in plant life form use by immature and adult stages of British butterflies ($n = 60$, $t_{(59)} = -6.66$, $P < 0.0001$). Life forms are coded as: 1, ground; 2, short herbs; 3, tall herbs; 4, shrubs and climbers; 5, trees. The range is the highest to the lowest value. Means for ranges (lines), standard errors (2 standard errors, grey boxes) and 1 standard deviation (whiskers) are shown for 60 British butterfly species.

Resource connectivity depends on the spatial structure of the resources and the mobility of two stages, larvae and adults. The range in resource types (life forms) are probably more varied for adults than for larvae as each adult sex has different requirements (Fig. 3.1) and they are more mobile than larvae, their morphology being adapted to different flight requirements (Berwaerts *et al.*, 2006). Consequently there are more adult resource types that are interconnected. Situations exist, such as the separation of breeding from overwintering sites in painted lady *Vanessa cardui*, where resources are not only isolated but differ in location every generation (Dennis, 1993a). Isolated resources can be critical for larvae, especially those that feed gregariously, rapidly consume their hostplants, and are forced to disperse to find alternative supplies (e.g., nymphalids such as marsh fritillary *Euphydryas aurinia*; Porter, 1982). The observation on distinctions between adults and larvae can be coined in clear probabilistic terms because differences in resource requirements are influenced by scale and mobility of these stages; but then little is known of the micro-niches exploited by larvae of most species, which could be highly varied.

Clearly, the greater the area occupied by resource types within any site (locality), the smaller the gaps in between them and therefore the more connected up the habitat will be for either mobile stage (**P3.14**). Connectivity is effectively maximized when resources become transferable, that is when hostplants double as nectar sources and roost sites and so on (**P3.15**). Any gap between resource types and resource elements decreases resource connectivity (**P3.16**), as the butterfly then has to expend energy and time in moving from one unit to another. More than this, the matrix (trivial space, amalgam or filling) between resource elements may act as a physical barrier, the more especially for larvae moving from depleted hosts to new hosts. Just what sort of a barrier the matrix (context) between resource outlets presents for organisms moving between them depends on its structural attributes (i.e., height, length, density, breadth, composition) (**P3.17**). Note that mobility is a plastic trait (Merckx *et al.*, 2003) and will tend to be related to abundance and variety (polyphagy) of hostplants and other resources exploited; individuals are likely to be increasingly successful in engaging movements, the more resources available to them within a region (Dennis *et al.*, 2003) though isolation will select for longer movements and greater migration capacity (Dennis, 1993a). The upside of this observation is that parasitoids may be less effective in colonizing butterfly hosts in more diffuse resource distributions (Shaw, 2006).

RESOURCE DYNAMICS WITHIN HABITATS

At any site, the composition, physiognomy and connectivity of all resources are forever changing; this process is termed **resource dynamics**. Resource dynamics occur at different temporal scales with different levels of predictability. Variation in the daily cycle is linked to predictable agents such as changes in daylight and temperature and associated variables such as humidity, as well as resource exploitation by organisms including butterflies; but it is also affected by unpredictable factors such as weather conditions. Variation over seasons is associated with plant phenology and seasonal climate. In the longer term (>1 year), variation in the resource base is linked to predictable vegetation cycles (i.e., vegetation successions; see Box 5.3) and the unpredictable intervention by external agents (e.g., humans, geomorphological events).

General principles of resource dynamics

- **P3.18: The resource base changes with time, influencing resource composition, physiognomy, connectivity and resource exploitation.**
- **P3.19: Potential resource types are available for specific time periods and for different lengths of time.**

Before looking in detail into the relationships generated by resource dynamics, two basic features of resources through time deserve a mention. First, all the spatial principles mentioned above are governed by time constraints; the composition of resources, their spatial structure and connectivity continually changes with time on scales from minutes (with the passage of a cloud) to millennia (involving glacial cycles) (**P3.18**). Changes on all these scales affect butterfly biology and resource selection by individuals within habitats, as for *Plebejus argus* on the Great Orme, North Wales (Box 3.2; Fig. 3.2) and *Inachis io* on the Carrs, Wilmslow, Cheshire (Dennis and Sparks, 2005, 2006) (see Box 4.1). Put simply, all resources are inconstant and although biotopes can be maintained for centuries (e.g., plagioseres) with biotope transformations (succession) habitats have a finite lifespan (e.g., dependence of heath fritillary *Melitaea athalia* on coppicing cycles and swaling; Warren, 1991; Bulman, 2004). Hostplant resource requirements of the summer and autumn/spring generations of the bivoltine butterfly, adonis blue *Polyommatus bellargus* (Roy and Thomas, 2003) provide a good example of seasonal shifts in resource requirements. The authors found a significant shift between the plants used for egg laying in each generation, with the hostplant composition occupied by summer-feeding larvae being broader and different to the autumn one. Measurements of soil temperature confirmed that the short, sheltered foodplants selected by ovipositing females in autumn placed the autumn/spring-feeding generation of larvae in the warmest available microclimates within sites. In late spring, egg-laying females avoided the hottest spots but extended egg laying into taller, less sheltered (relatively cool) turf where the microclimate was similar to that experienced by autumn/spring-feeding larvae. Similar brood shifts in hostplant use, and in fine-scale site exploitation, have been disclosed for wood white *Leptidea sinapis* (Jeffcoate, 2006) and were first shown for speckled wood Pararge aegeria (Shreeve, 1986).

Fig. 3.2 The Great Ormes Head, North Wales, illustrating the west side and biotope for both the grayling *Hipparchia semele* and silver-studded blue *Plebejus argus*. The right insets are a male (top) and female (bottom) *P. argus*, and the left inset is a male *H. semele*.

Box 3.2 Temporal shifts in habitat bounds of the silver-studded blue butterfly *Plebejus argus* in relation to environmental conditions

Habitat bounds are determined not just by the availability of resources accessible to butterflies, but the extent of the site used by them containing those resources. These bounds can change over different timescales from seconds, to days, seasons, years and decades; identifying habitat bounds necessarily has to factor in such variation. In many population studies it is tacitly assumed that habitat bounds are defined by larval hostplant areas or a single vegetation unit. The flux in habitat bounds over short periods in relation to weather conditions has been tested for *Plebejus argus* in two patches of a metapopulation for the butterfly on Great

(a)

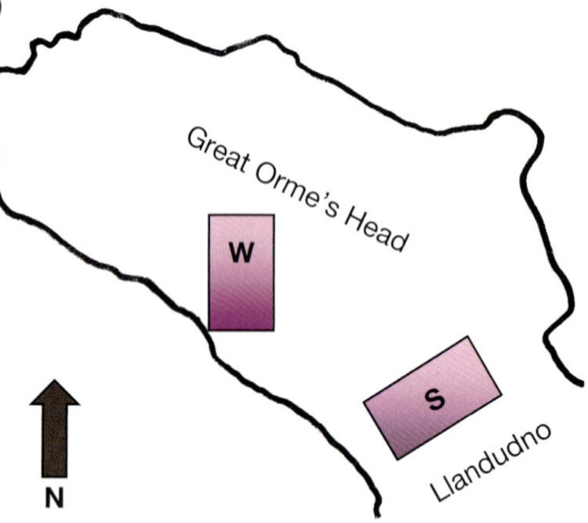

(b)

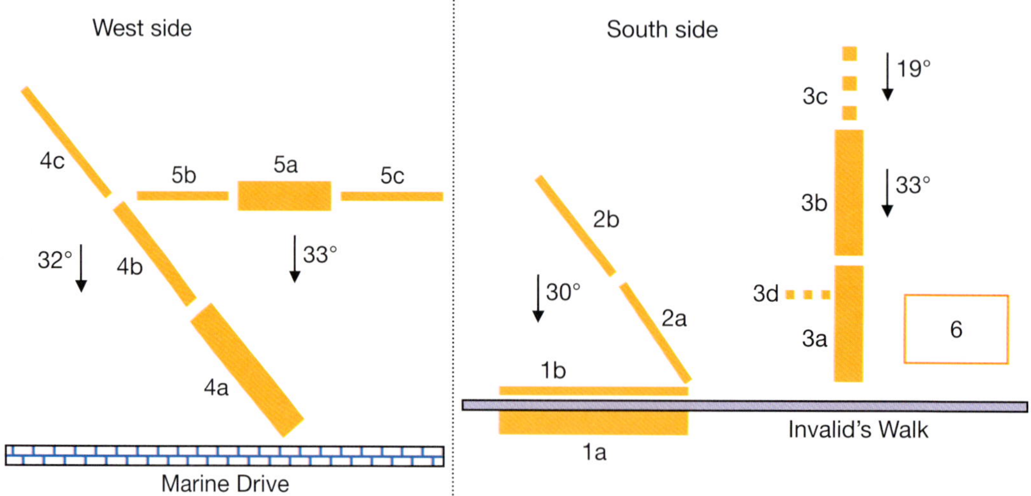

Fig. B3.2 (a) Great Ormes Head showing the location of two study areas. (b) Diagram illustrating the relative placement of transects and transect sections on the Great Ormes Head in the study areas. Transects are numbered 1 to 5, sections a to c. Line thickness is indicative of the amount of shrub cover with sections shown by thick lines having more scrub. The area marked as 6 is used for determining the relationship between the numbers of *Plebejus argus* resting and shrub (cotoneaster) size (see Fig. 4.6c). Arrows indicate the direction of downslope and their labels indicate mean slope angle. Regression parameters and weather variables associated with the proportional movement of individuals (arcsine transformed) between transect sections is recorded for 2004 and 2005 in Table B3.1. The influence of date is accounted. (After Dennis and Sparks, 2006, courtesy of Elsevier.)

Ormes Head, North Wales (see Fig. 3.2) (Dennis, 2004b; Dennis and Sparks, 2006). This species is not confined to larval hostplant areas or to a single vegetation unit. *P. argus* density is higher in the vicinity of shrubs that are used for roosting, resting, basking, mate location and shelter; a dominant proportion of the population adopts shrubby areas in cooler, cloudy and windy weather and larger shrubs are used by more butterflies (see Fig. 4.6c). Thus, shrubs are vital resource components and provide a refuge against inclement weather conditions. In warmer, sunnier and calmer conditions, the butterfly spends longer in flight and moves out onto calcareous heath dominated by hostplants (Tables B3.2a, b). In doing so, an increasing, even dominant, proportion of the population occupies exposed slopes adjacent to and above shrub-covered areas associated with the hostplant, areas that are entirely vacated during cool, windy, cloudy weather. At the same time, females become less noticeable to observers: they are apparent when they occupy bushes in cool weather, but are less visible than males when egg laying in sunshine and warmth. To human observers, the habitat bounds appear to change with conditions on scales of days to less than one day; in fact, on cool days a rapid transfer from shrubs to higher, open slopes can be witnessed within seconds when the sun comes out. In hot summers, the butterfly spreads out over the headland (Dennis and Bardell, 1996). Such seasonal differences in the occurrence of

Table B3.2a Significant stepwise regression models fitting arcsine-transformed proportion data of numbers on higher or more open sections for 2004, compared with lower and more shrubby adjacent sections for transects, to weather conditions.

Lower and/or shrubby section	Upper and/or open section	Proportion range (min.–max.)*	Predictor variable,† $B\pm SE$	F statistic	Size of sample	R^2	P
Males							
Up slope							
1a	1b	5.5–45.0	Ta 0.05±0.01	14.50	12	76.9%	0.0015
1a	2a, 2b	3.7–25.0	Ta 0.03±0.01	6.20	10	43.6%	0.019
3a	3b	21.8–56.7	Wd −0.48±0.25	3.81	11	29.7%	0.04
4a	4b	18.9–44.8	Ws −0.03±0.01	9.82	13	47.2%	0.005
4a, 4b	4c	4.8–28.2	Ta 0.71±0.21	10.86	13	49.7%	0.004
4b	4c	6.9–38.4	Ta 0.03±0.01	4.57	13	22.9%	0.028
Across slope							
5a	5b, 5c	0–46.9	Ws −0.15±0.03	18.55	11	67.3%	0.001
Females							
Up slope							
1a	1b	0–61.2	Wd −1.69±0.45 (in absence of Wd, Ta R^2 = 43.8%)	14.24	12	58.7%	0.002
1a	2a, 2b	0–27.2	Ta 0.04±0.01	9.84	10	55.1%	0.007
3a	3b	18.6–45.5	None				
4a	4b	14.6–50.1	None				
4a, 4b	4c	0–31.1	Ws −0.05±0.02 (sun 0.10±0.02)	12.05	13	23.2% (16.0%)	0.015 (both)
4b	4c	0–33.5	Ta 0.04±0.02	9.11	13	17.4%	0.023
Across slope							
5a	5b, 5c	0–53.3	Ws −0.13±0.04	8.83	8	59.5%	0.01

* Proportion of numbers in column 2 (upper and/or open section) to total numbers.
† Subsequent predictors and their estimates in parentheses. *B*, slope of regression line; Ta, mean air shade temperature; Wd, wind direction; Ws mean wind speed; sun, sunshine.

Table B3.2b Significant stepwise regression models fitting arcsine-transformed proportion data of numbers on higher or more open sections for 2005, compared with lower and more shrubby adjacent sections for transects, to weather conditions.

Lower and/or shrubby section	Upper and/or open section	Proportion range (min.–max.)*	Predictor variable,† *B*±SE	*F* statistic	Size of sample	R^2	*P*
Up slope							
1a	1b	2.4–45.5	Ta 0.05±0.01	13.68	14	53.3%	0.0015
1a	2a+b	14.7–47.2	Ta 0.03±0.008	10.31	14	46.2%	0.004
3a	3b	23.7–50.0	Ws −0.02±0.007	5.99	14	21.6%	0.024
3b	3c	10.0–42.1	Wd 0.09±0.07	1.49	14	11.0%	0.12
Across slope							
3a	3d	0–31.4	Sun 0.04±0.02	4.36	14	26.6%	0.03

* Proportion of numbers in column 2 (upper and/or open section) to total numbers.
† Subsequent predictors and their estimates in parentheses. *B*, slope of regression line; Ta, mean air shade temperature; Wd, wind direction; Ws mean wind speed; sun, sunshine.

adults in relation to seasonal weather conditions are likely to have a corresponding impact on the distribution of developmental stages and of adults for a subsequent season. An important message is contained in these observations for conservation: what part of a landscape may be defined as a habitat, and what part of it may appear to be most important for an organism, depends on just when and where surveys are carried out.

Secondly, resource types are generally only available for a fraction of the seasonal cycle and may be in a suitable state for exploitation for even shorter periods (**P3.19**). Resource requirements are generally needed consecutively, stage by stage, or associated with activity periods locked into seasonal and diurnal cycles. Some of these changes are predictable (e.g., daylight, seasons, vegetation succession); others are unpredictable (e.g., weather, catastrophes). Use of resources is linked closely with availability, and synchronicity among and within resource types drives life history strategies such as voltinism. A further twist to this issue is that physical resource requirements within sites can change, as in the case of silver-spotted skipper *Hesperia comma* which has taken to taller swards of sheep's fescue *Festuca ovina* during recent warming climate changes (C. D. Thomas *et al.*, 2001). In effect this butterfly has changed its resource type, because the context of appropriate resources has changed with changing climate.

General principles of resource composition

- **P3.20: Generalist species will tend to have access to a resource type for longer periods of time than specialist species.**
- **P3.21: Some resources have shorter lifespans than others.**
- **P3.22: Consumables will tend to be available for shorter periods of time than utilities.**

The key issues for resource composition over time involve comparisons between generalist and specialist species and between consumables and utilities. Generalist species would be expected to have access to a resource type for longer periods of time than specialist species, since different resource outlets usually occupy different conditions and their appearances are unlikely to be synchronized (**P3.20**). Thus, the greater the variety of supplementary resources of a resource type (for a stage or activity), the longer this resource type is likely to be available for an organism and the greater the likelihood of fitting in additional broods during any one season. Multiple broods of butterflies in Britain have usually been related to warmth and length of season; but how much do they also depend on resource (hostplant) variation?

Longevity of a resource for an organism to exploit depends on the resource type. This involves two timescales, the lifespan of the resource and seasonal and shorter periods of availability. All biotopes are subject to vegetation succession (see Box 5.3) and the lifespan of a resource typically correlates with the size of the life form, for herbs being short and for shrubs and trees much longer. As for seasonal availability, virtually all resource types reach a peak of suitability for specific butterfly stages and subsequently returning to a value of zero. Inevitably, some resources have shorter periods of exploitation (i.e., suitability for specific butterfly stages) than other resources that may bear little relation to their existences (lifespans) (**P3.21**). For example, bare ground and molehills are valuable thermoregulation sites and perches for territorial *Inachis io* during a brief period in spring and these substrates may persist all year but not be used in high summer in northern Britain when temperatures and perches are higher, and butterflies feed up for hibernation (Dennis, 2004c; Dennis and Sparks, 2005; Hardy and Dennis, 2007). Consumables will probably have a shorter lifespan than utilities since egg-laying females and developing larvae are extremely fastidious about the condition of hostplants and nectar quality is rapidly drained by competitors (**P3.22**) (Porter, 1992). The classic case among British butterflies is the limited use of annual plants for larval growth, the reason very likely being that the majority of butterflies have evolved to use perennial plants (Kemp *et al.*, 2008). Annuals may allow expansion of the area occupied but this can be short-lived, as in the case of brown argus *Aricia agestis* moving out from the chalk escarpment of the Chilterns, where it is dependent on the perennial rock rose *Helianthemum nummularium*, into the Vale of Aylesbury where it uses the annual cut-leaved cranesbill *Geranium dissectum* (Kemp, 1998).

General principles of resource physiognomy

- **P3.23: Spatial attributes of resources change continuously with time.**

- **P3.24: Larger resources of a type will remain suitable for a longer period than smaller resources.**
- **P3.25: Temporal resource availability depends on the location and spatial attributes of the resource type.**

Spatial attributes of resources change with time at a range of temporal scales from diurnal shifts in sunny spots with the sun angle and cloud cover (see Boxes 3.2 and 4.1) to the entire resource base with vegetation succession or catastrophe (e.g., flood) (**P3.23**). The period and timing of availability depends much on the spatial characteristics of a resource, i.e., its shape and orientation – a topic explored in the next chapter. The spatial characteristics of a resource can change if only because individuals' thresholds for resource use also change with time. This is evident through a day with the shift in a sun spot on a woodland floor used for perching by territorial *Pararge aegeria* (Davies, 1978; Shreeve, 1984) or over seasons as in the case of ovipositing *Hesperia comma* mentioned above (C. D. Thomas *et al.*, 2001). Larger resource outlets will tend to last longer as they are more likely to occupy a variety of different conditions and therefore the suitability of parts of the resource will peak at different times (**P3.24**). Just how long a resource is available will depend much on its location within topographic and substrate conditions of a site (e.g., aspect to sun angle, prevailing winds, substrate water), that is, on the spatial attributes of the resource type in relation to topography and substrate (**P3.25**). Varied topographies and substrates facilitate resource use for longer periods, but this in itself is insufficient to extend resource use. Disposition (spread, density, orientation, context, etc.) is crucial as will be discovered by anyone mapping crucifers for pierids over sites (Dennis, 1982a, 1983a, 1983b). The best resources may not be synchronized with demand for them, owing to their location within a site or just seasonal weather conditions (Dennis and Hardy, 2006).

General principles of resource connectivity

- **P3.26: As resources are never constant in time, connectivity also varies between resources and habitats.**

Neither resources nor non-resources (matrix) are constant in time (**P3.26**). Consequently, dimensions and relative locations of resources also change and this affects connectivity. Even in the absence of changes to the resource base, continual changes in the context of resources occur (i.e., other substrates, environmental conditions such as weather and sunlight) influencing links and barriers between resources and resource elements. This has important implications for understanding mobility in butterflies; variability in connectivity is the norm. Very small modifications to the hostplant context (i.e., dampness) may greatly affect barriers for larvae moving between hosts and similar shifts in the location of shade are sufficient to deter movements of adult butterflies between hostplant patches in adjacent meadows (e.g., *Anthocharis cardamines*; Dennis, 1982a, 1982b).

HABITATS, BUTTERFLY RESOURCES AND POPULATION STATUS

Local butterfly populations differ enormously in size (numbers, space occupied) and lifespan (persistence), and population size may be stable for extended periods (Box 3.3). However, populations do crash to extinction, and our 21st century British landscape which has no breeding butterflies over large parts of it bears witness to that well enough. Population dynamics has long been examined in relationship to **trophic** agents and relationships (i.e., nutrition, food chain links), such as herbivory, parasitism, predation and competition and **key factor analysis** has typically been applied to provide the answers (Warren, 1992a). This technique is labour intensive, as in auditing stage mortality it tracks individual survival, and consequently the findings tend to be restricted to few sites. Increasingly, a broader perspective is being taken of populations and their relationship to habitat structure, on how habitat components control butterfly population size and determine population lifespan. The broader perspective is being generated by the growing preoccupation with populations occupying multiple habitat patches (metapopulations; Hanski and Gilpin, 1997) (see Chapter 6) and the criticism that metapopulation dynamics models have failed to consider key components of habitats (Dennis *et al.*, 2003, 2006b). Although key factor analysis provided insights into some habitat components by following mortality and productivity through the life cycle of individuals, it failed to disclose resource use in all stages (e.g., pre-pupal larval selection of pupation sites, mortality during roosting). The impact of links between behaviour and substrate use in adults are not conventionally analysed by key factor analysis, and yet this could provide considerable insights into the influence of resource use on population dynamics. The data can be collected by extending simple transect recording and atlas observation techniques (Dennis, 2004a; Hardy and Dennis, 2007).

Here, attention is directed to some basic concepts associated with this broader approach, an illustration of how the three habitat components – resource composition, physiognomy and connectivity – and thus conditions within habitats determining habitat quality, may affect population status. The focus of attention is on identification of where different butterfly developmental stages engaged in specific behaviour are found, an

Box 3.3 Models of population stability and maintenance

Populations are part of complex natural systems involving many variables. As such, some stability is expected of them. Stability in systems requires negative feedback and is inherent in systems where mortality (e.g., enemies, disease, inclement weather) counteracts production and factors enhancing growth, development and reproduction (e.g., hostplants, nectar sources). Thus, *a priori*, periodic stability – population 'regulation' *sensu stricto* – is expected in butterfly populations; such negative feedback is largely driven by density dependence. Regulation, if only periodic, is accepted as part of a system's functioning; spatial variation in resource outlets and agents, the motility of individuals, increases the number of potential actions and reactions and the likelihood of negative feedback and stability.

Two basic models of population dynamics are the **equilibrium model** (Nicholson, 1933, 1954; May, 1976) and the **ceiling model** (Milne, 1957; Dempster and Pollard, 1981). The equilibrium model claims that population density is regulated in dynamic equilibrium, in which any deviation from the equilibrium is countered by density-dependent factors. Populations are considered to be maintained at equilibrium densities well below those their resources and environmental conditions can support. Regulation is envisaged as preventing overpopulation and extinction. The ceiling model makes no claims for an equilibrium population density. It considers that populations are limited purely by the availability of resources. **Density dependence** – negative feedback in population systems – only bites in when intraspecific competition comes into effect, that is, when the population approaches carrying capacity and population density starts to affect the supply of resources. The two models differ fundamentally in the way they portray population stability and persistence, particularly the frequency with which extinction is expected to occur among local populations (Dempster *et al.*, 1995). Distinctions have become increasingly polarized by controversy. For instance, the term 'regulation' is defined variously by different researchers. To some it is maintenance of population size around an equilibrium (Varley *et al.*, 1973). To many others, it means long-term population persistence and fluctuations within limits, with the lower limit >0 (Mountford, 1988; Murdoch, 1994). The debate is essentially a prolonged argument about how universally density-dependent factors operate in natural populations to maintain the sizes of these populations within finite, positive limits (Hanski *et al.*, 1993b).

Within the same populations, agents may generate feedback to population changes (density dependence and inverse density dependence) or have no discernible impact (density independence) (Fig. B3.3) (Odum, 1963). A number of agents are known that function in a density-dependent fashion. Prominent examples are enemies (disease, parasitoids, predators) and competitors of the same or different species. Other agents, even abiotic ones such as weather, can also *influence* populations in a density-dependent manner (Dennis, 1993a). However, density dependence may be countered by factors that operate in an inverse density-dependent way. For example, small populations may suffer from

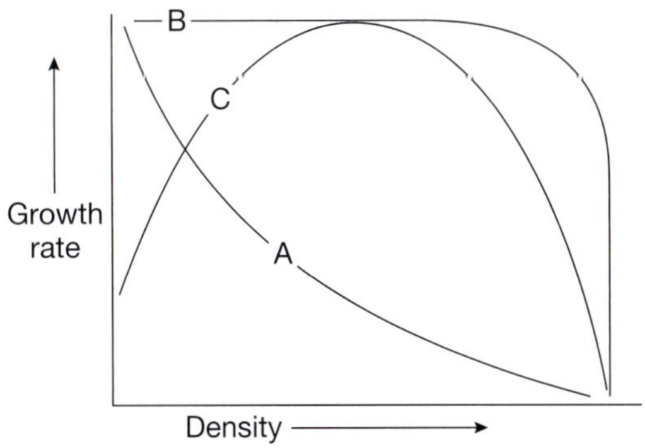

Fig. B3.3 Three patterns of population growth rate (per unit of population) in relation to population density: A, growth rate decreases as density increases (self-limiting or inverse density dependence); B, growth rate remains high until density becomes high and factors outside the population become limiting (density independent); C, growth rate is highest at intermediate densities (Allee type).

the Allee effect (Allee *et al.*, 1949; Kuussaari *et al.*, 1998) with depensation occurring at low density. A number of mechanisms are hypothesized to cause the Allee effect, including a decrease in mating frequency, reduced defences against predators, increased emigration rates and inbreeding depression (Fowler and Baker, 1991). The Allee effect has been observed in a large blue *Maculinea arion* population (Thomas, 1980a). Density-dependent reductions in mortality at lower densities may also be frustrated by density-independent factors. Small populations characteristically have greater variability in population size than large populations and are consequently particularly prone to demographic stochasticity and extinction (Gaston and McArdle, 1993). Moreover, agents that are expected to function in a typically density-dependent manner may not always do so. Usually, an increase in predators causes a decline in prey populations and an increase in prey triggers an increase in predators. However, predators may have positive effects on a prey population by influencing other components in the ecosystem (Abrams, 1992).

Much ambiguity occurs in discussions of density dependence. Problems have arisen in finding appropriate ways to measure or test it (Holyoak, 1993; Wolda and Dennis, 1993; Wolda *et al.*, 1994) and of interpreting it (Holyoak, 1993; Holyoak and Lawton, 1993; Dennis and Taper, 1994; Wolda *et al.*, 1994). Murray (1994) provides a lucid discussion of the most contentious issues. One of the problems has been to account why, if some density-dependent agents are ineffective, so many insect populations should be sparse when resources seem to be plentiful. The reason has long been evident from autecological studies on butterflies (Duffey, 1968; Thomas, 1984). The simple fact is the ceiling is much lower than expected. Dempster (1991) illustrates this for *Anthocharis cardamines*. Insect species generally have highly specialized resource requirements and select for resource quality, particularly for larval hostplants. The classic examples are presented by lycaenids of the genus *Maculinea* that have ant hosts (Thomas and Wardlaw, 1990, 1992). For example, in English *M. arion*, the basic reproduction rate depends on the number of ant nests of *Myrmica sabuleti*, not just on any *Myrmica* species occurring at the same sites, or on hostplant abundance (Hochberg *et al.*, 1992). As resources suitable for reproduction and survival may only be a fraction of those apparent to human observers, this suggests that competition is one density-dependent factor that will limit population growth (Murray, 1994). Lower ceilings are closer to the floor: small local populations are more likely to become extinct from demographic stochasticity than large populations. There is now much empirical evidence that most small populations persist simply because they are maintained by immigration.

approach that does not necessarily exclude a key factor type analysis. Four basic effects associated with these three resource components have been described that attempt to explain population status (numbers, and the size and shape of the site area occupied). They are referred to as **complementation**, **supplementation**, **neighbourhood** and **source-sink** (Box 3.4) (Dunning *et al.*, 1992). These concepts were engineered as much for multiple populations (metapopulations) as unitary populations on single sites but they more appropriately determine population status within single sites. Crucially, a scrutiny of resource components enables an understanding of occupied (i.e., sources, sinks and pseudosinks; see Chapter 6) and unoccupied habitats. Some fundamental principles emerge linking population incidence and size to the resource base. Each of the relationships outlined evokes one or more of the four characteristic effects noted above and works through resource composition, physiognomy and connectivity.

The following principles can be drawn up linking population incidence and size to resources:

- **P3.27: Populations are limited by whatever resource is in shortest supply or condition is exceeded; complementarity applies.**
- **P3.28: Habitat quality can be inferred from the location and abundance of butterflies on sites.**
- **P3.29: The larger the area of resources, the larger the population size.**
- **P3.30: Populations with supplementary consumable resources will tend to be more abundant and persist longer.**
- **P3.31: The denser the area of resources, the larger the population size.**
- **P3.32: The closer resource types and units are to one another and the fewer the barriers between them, the larger the population. Populations based on linear resources will tend to be smaller than populations based on more compact resources.**
- **P3.33: As resource geography and status is never even, neither will be population density. A reduction in population variability over time accompanies an increase in habitat heterogeneity.**
- **P3.34: Populations are smaller on sites with increased spatial variance (ruggedness) in the resource base.**

The first two principles describe local population status. The first principle (**P3.27**) follows from Leibig's 1840 Law. This states that whatever component is at a minimum or at a maximum tolerance for a species controls population size. Obviously, if a vital resource is missing the population is not viable in the long term. The classic case is the monarch butterfly *Danaus plexippus*, which shifts location *en masse* from hibernation to breeding sites (Ackery and Vane-Wright, 1984). Complementarity applies (see Box 3.4): if the full complement of conditions is not met, there is no habitat and a population cannot permanently be supported on the site. Individuals may migrate to and temporarily occupy the patch of resources but the resource base is deficient, unable to maintain a population through the year. A patch, maintained by immigration, is described as a **sink** (see Box 3.4). The second principle (**P3.28**) describes the outcome of observed conditions above this critical level; common sense tells us that if a population is small something must be limiting it, and therefore population size in many ways is a measure of resource abundance and suitability in content and spatial arrangement (Pywell *et al.*, 2004). It is often another matter to determine what is amiss when a population is small – the only solution is to do the fieldwork. So often assumed and rarely justified, particularly in metapopulation studies, the simplest and most direct relationship is between resource area and population size (**P3.29**). In this relationship, resources are typically measured as larval hostplant area. Any variance about this relationship evokes issues of other resources missing from the equation. Large populations provide founders for other, smaller, habitat patches and are called **sources**. At the other extreme, resource deficiency in all or any critical resource, as opposed to absence, will generally result in small population size. Such resource-deficient patches may furnish small populations; receiving rather than issuing individuals, they may appear to be sinks. They may, in fact, be **pseudosinks** (see Box 3.4) and immigration is not responsible for their maintenance in the short term.

The principle of supplementarity (**P3.30**) has an intuitive outcome (see Box 3.4); it is formalized in Brown's (1984, 1995) niche-based explanation for positive interspecific abundance–range size relationships. This proposes that species with broader niches are able to attain higher local abundances and wider distributions than are species with narrower niches. This relationship has been formally tested and found to hold for British Macrolepidoptera (Quinn *et al.*, 1997).

Box 3.4 **Some basic resource attributes that affect population size, density and status**

Figure B3.4 is a series of schematic diagrams illustrating contrasting situations for resource attributes and their impact on local populations.

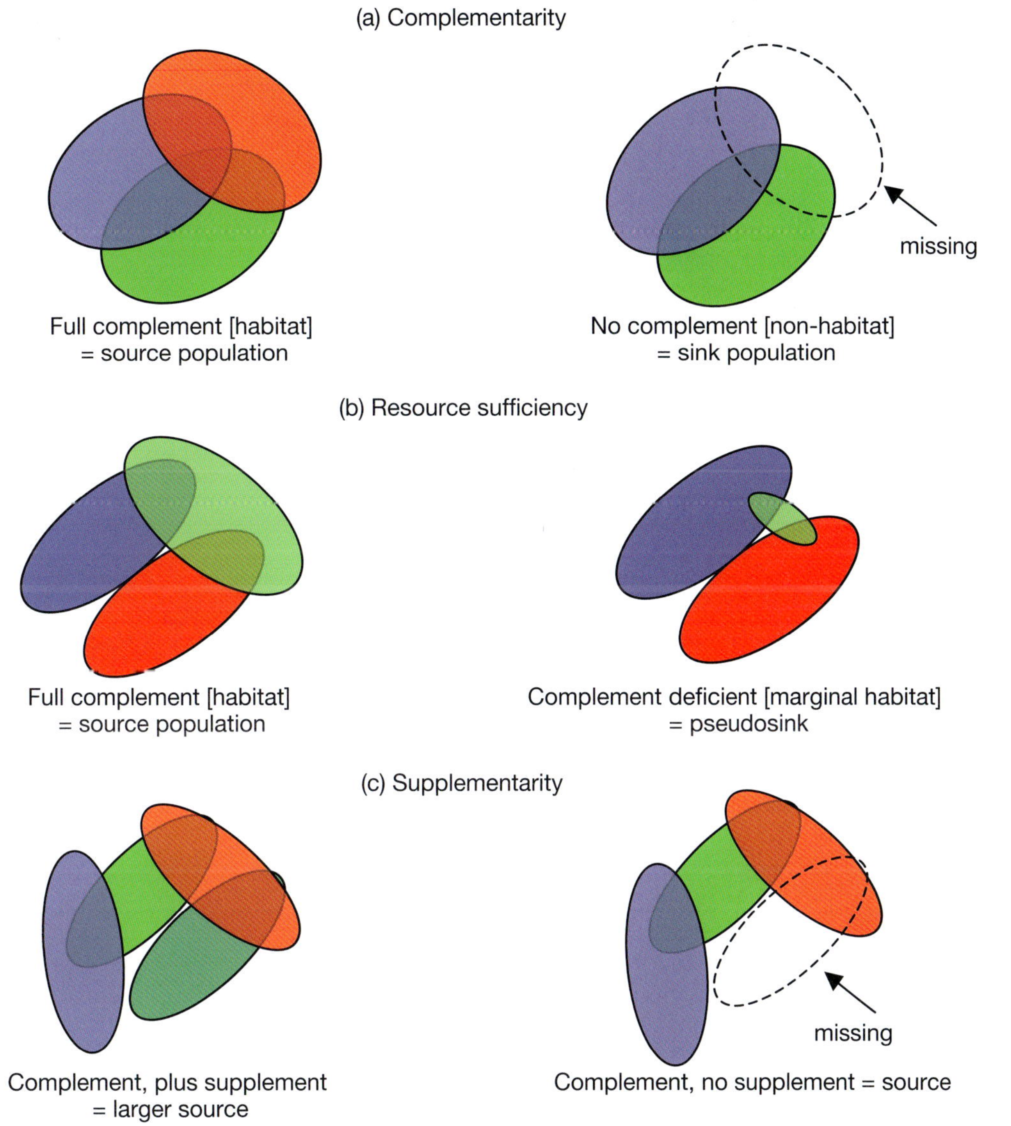

Fig. B3.4 (a) **Resource complementarity**: all resources are required for population persistence; the absence of a resource type may allow temporary existence of a 'population' which will be a pseudosink or sink. (b) **Resource sufficiency**: a resource type is too small or sparse to support a persistent population. (c) **Resource supplementarity**: two or more distinct outlets (e.g. hostplant species) over a site provide insurance against environmental changes and boost population size.

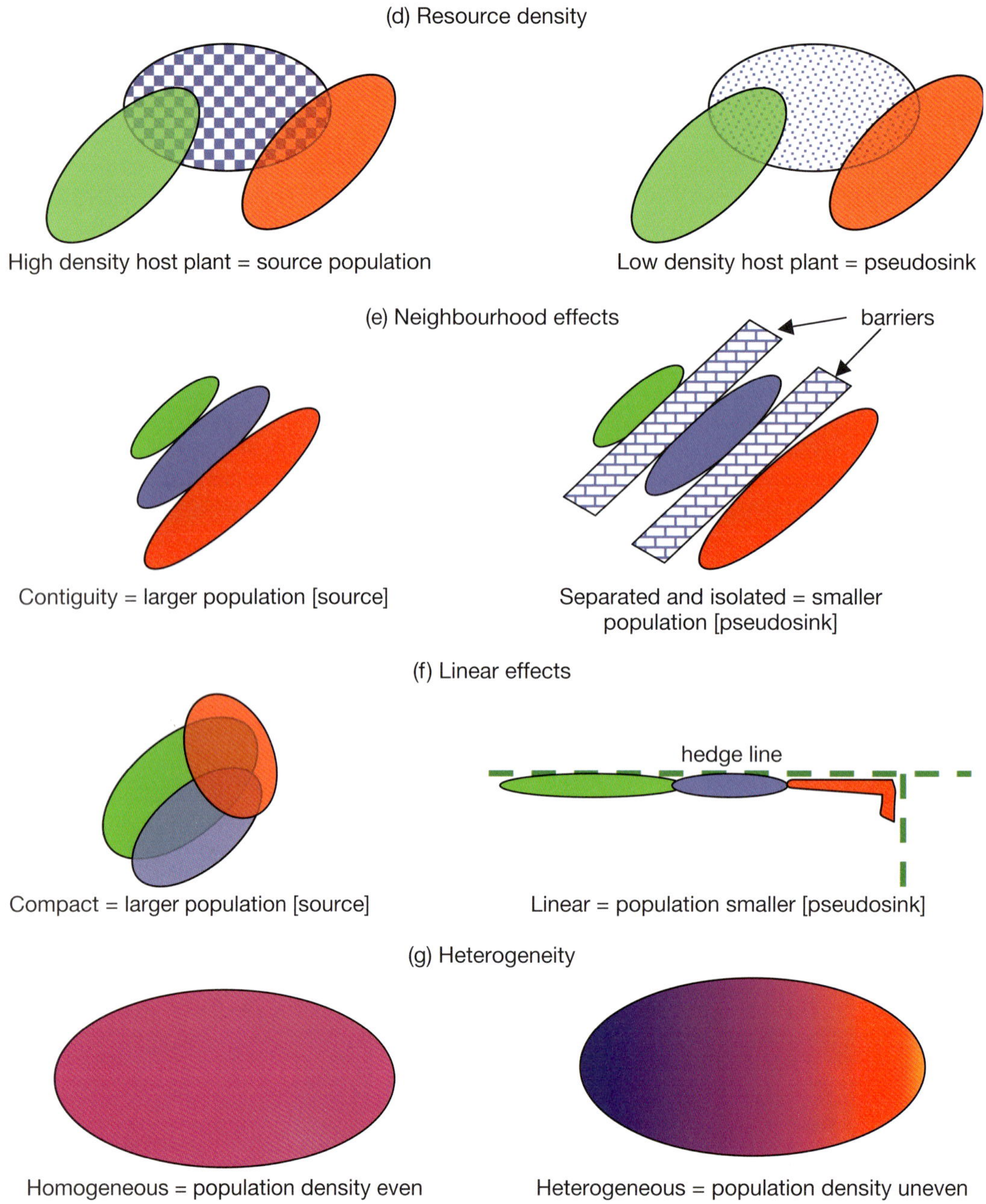

Fig. B3.4 (*continued*) (d) **Resource density**: a key resource (hostplants at higher density or in better condition within a site) supporting a larger population. (e) **Neighbourhood effects**: contiguity and absence of barriers ensure access and exploitation of all resources at a site and larger population growth. (f) **Resource package shape**: linear resources (e.g. hostplants) may well be of lower quality but also create more conspecific interactions and predation, thus depressing populations. (g) **Resource heterogeneity**: this results in an uneven density or quality of resources over habitats and movement out from high-density resource zones to lower density/quality areas, in effect sources and sinks within a habitat patch.

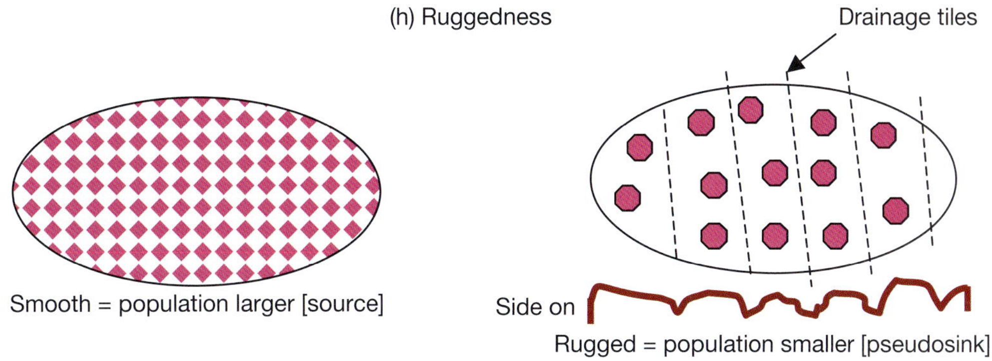

Fig. B3.4 (*continued*) (h) **Ruggedness**: an increase in resource spatial variance adversely affects population size and population persistence.

Ellipses are resources: green, larval hostplant; blue, adult nectar source; red, mate location site; purple, multiple resource distributions.

The outcome can be more testing. In the American silvery blue *Glaucopsyche lygdamus* (Doubleday), egg densities correlate significantly only with measures of host species' diversity. Patches consisting of a single host species, no matter how large, did not support high butterfly densities, but patches of multiple, equitably distributed host species did. The most likely explanation, in light of oviposition preference and larval performance data accumulated for this butterfly species, is that host species diversity is necessary for the persistence of *G. lygdamus* populations, because alternative host species buffer population losses during poor or unusual years. The dependence of both ovipositing butterflies and developing larvae on the ephemeral, young, hostplant flowers make the butterfly especially vulnerable to year to year variation in hostplant availability and quality (Carey, 1994; Hellman, 2002). Among British butterflies, an example is the occasional use by Isle of Wight late instar *Melitaea cinxia* of *Plantago coronopus* when the main hostplant *Plantago lanceolata* has been exhausted (Thomas and Simcox, 1982). Another example is the increase in the population of North Wales *Gonepteryx rhamni* since 1986 with the planting of alder buckthorn *Frangula alnus* along the roadsides and the introduced evergreen Mediterranean buckthorn *Rhamnus alaternus* (Gutiérrez and Thomas, 2000). However, there are also contrary indications; for instance, adaptations specific to particular hostplants can counter opportunities through supplementation (e.g., Aland islands *M. cinxia*; Kuussaari *et al.*, 2000; Hanski and Singer, 2001).

Principle **P3.31** describes one obvious way in which any key resource attribute may counter expectations under principles **P3.28** and **P3.29**. Denser hostplant areas have potentially greater productivity (larger populations for area) than sparser ones and reflect differences in composition via context. However, the relationship may be anything but simple and higher productivity on denser patches may be offset by density-dependent increases in predation and parasitization. Lower density areas of hostplant may imply the encroachment of denser contexts (matrix) and, depending on what the context is and the developmental stage it affects, this can influence productivity and survival in contrasting ways. Context, as explained earlier, can have multiple effects; it can influence access to resources (neighbourhood effects for larvae moving from exhausted to unused plants) as well as resource abundance and condition (Fig. 4.1). The effect is to generally depress butterfly populations, especially when context expands at the expense of a resource (Summerville and Crist, 2001). The decline in population can be linear or non-linear, in the latter case perhaps depicting response to thresholds of resource availability and matrix.

The last few principles (**P3.32** to **P3.34**) evoke a combination of the four basic effects described by Dunning *et al.* (1992) as well as earlier principles. Just how resources are arranged within a habitat space has implications for potential exploitation of supplementary and complementary resources evoking connectivity and thus neighbourhood effects. Linear habitats have been found in one Danish butterfly study to house significantly lower population abundances than non-linear habitats (**P3.32**) (Clausen *et al.*, 2001). Linear habitats, as mentioned earlier, tend to have different combinations of less well-connected resources than compact habitats (Dover, 1994, 1996); they also encourage emigration (T. G. Shreeve, personal communication). On the other hand, the spatial arrangement of linear resources in relation to environmental conditions (e.g., exposure to sunlight and wind) can have a major impact on how successful a collection of resources is in supporting a local population, and linear resources can be of critical importance for activities such as mate location and roosting (Sparks *et al.*, 1999).

Principle **P3.33** broadly explains heterogeneity in population density within habitats, that is, despite influences (e.g., mobility) and pressures (e.g., competition), tending to generate uniformity in density (Singer, 1972). Simply put, the greater the habitat heterogeneity, the less temporal population variability (viz., variance, coefficient of variation); Hanski (1991) demonstrates this nicely for bush cricket *Metrioptera bicolor* (Philippi). But, at a trivial level, a brief reflection discloses that variations in density within habitats is a given, *a priori*, for two reasons:

1 Each successive stage experiences losses, so density decreases from egg to adulthood.

2 Each stage or phase occupies different resources, so density changes amongst resource outlets with timescales of <1 day to the life cycle (Morris, 1992). This occurs because resources vary in dimensions in relation to conditions (seasonal climate, weather) as well as the space they occupy within habitats.

These issues underlie density and abundance differences among population units. But principle **P3.33** applies to a finer scale: variations in density for single stages can be brought about by either or both of differential access to and exploitation of, or differential mortality on or abandonment of, the various resources used and resource units exploited. At the simplest level there are shifts in individual density within the habitat space on a diurnal basis if only because there is a tendency for circadian rhythms in activity and behaviour to occur as well as shifts in resource suitability in relation to diurnal weather changes and sunlight. In Chapter 4, attention is given to reasons for variation in density on single resource patches.

The impact of spatial heterogeneity in resource distributions on population size and density within sites is no better illustrated than in the work on the mountain Alcon blue *Maculinea rebeli* by Jeremy Thomas and his colleagues (1998a). Specifically, they show that an increase in resource spatial variance (termed **ruggedness** by the authors) in the case of *M. rebeli* and host ant *Myrmica schencki* Emery and other ant species, adversely affects population size and population persistence. Just what the effects are likely to be for different species is not obvious without explicit spatial modelling. In the case of *M. rebeli*, the interactions between butterfly and ants on sites with variable substrates affects even the apparent relationship (coincidence) between ant species and hostplant (J. A. Thomas *et al.*, 1998a). Although less complex, the situation is likely to be similar for more straightforwardly phytophagous butterflies. Because of differences in mobility between larvae and adults, the ruggedness of the landscape drives larval resource exploitation more keenly than adult resource exploitation (**P3.34**).

It soon becomes clear that the four basic effects (complementation, supplementation, neighbourhoods, source–sinks) apply to a range of geographical scales – within a single habitat or site, as well as between habitats (see Box 3.4), including the next stage of multipatch situations of metapopulations and population mosaics. At the coarsest scale, complementarity becomes complementation and supplementarity becomes supplementation (Dunning *et al.*, 1992). It is inconceivable that there could be homogeneity for habitat quality among sites, though in metapopulation modelling this is usually assumed to be the case. Intrasite heterogeneity in population density can often be related to all four controlling effects operating at different space–time scales (Dennis and Bramley, 1985) (see Box 3.4). It needs to be emphasized that the outcome of these various effects, especially in combination, may be anything but obvious and can be completely contradictory to expectations. For instance, populations of habitats incorporating sinks or with adjacent sinks (part habitats) may be larger than those without, simply because individuals lost to the sink may be far fewer than those lost to competition without it. This has been nicely illustrated in work on European *M. rebeli* (J. A. Thomas *et al.*, 1998a).

RESOURCE DYNAMICS, POPULATION STATUS AND LIFE CYCLE STRATEGIES

Changes in the resource base have two major impacts on a butterfly species:

1 The size of the population.
2 The timing (occurrence, duration) of developmental stages and annual broods.

These can translate into larger, long-term trends for whole butterfly communities caused by, for instance, vegetation succession (see Box 5.3) manifested through different adaptations in different seres (Steffan-Dewenter and Tscharntke, 1997), biotope fragmentation, broad-scale transformations in landscape management and climate change (see Chapters 5 and 8).

Principles relating to population size and density

- **P3.35: As the size of the resource base continually changes, so does population size of dependent organisms.**
- **P3.36: The closer the synchronization between the suitable condition of a resource and butterfly stage, the larger the population will be.**
- **P3.37: The shorter the period of suitability of a resource, the smaller the population will be.**
- **P3.38: As spatial aspects of resources change continuously with time, there will be spatial variation in the incidence and abundance of the target organism with time.**

Although organisms are part of natural systems and therefore some degree of population stability is expected (see Box 3.3), populations can never be absolutely stable and population status will continually change in response to changes in the size of the resource base (**P3.35**). Resource dynamics occur in all time frames (<1 h, <1 day, within broods, between broods, over years). Such changes affect all resource types, but differently, and are generated by environmental conditions. In the short term this may cause fluctuations and in the long term trends, as with vegetation succession. Thus, populations will also change in status, for instance from sources, through pseudosinks to sinks and *vice versa* (Thomas and Kunin, 1999). Other aspects of resource dynamics will also influence population size. Population growth and population size should benefit from synchronization between the availability of a resource and a stage dependent on it. For example, asynchroneity between hostplant and early stages will tend to increase mortality (**P3.36**); e.g., *Favonius quercus* depends on the coincidence of hatching with oak bud burst. It follows that the shorter the period of time a resource is available, the smaller the population is likely to be (though the peak may appear to be larger), as synchronization becomes more critical to development (**P3.37**). This is also expected as the reciprocal of the supplementation principle (see above). Other factors may also be involved in this relationship; the more condensed in time is the flight period of one of two populations of equal size, the greater the population density (because of increased synchronicity) and therefore the greater the likelihood of density-dependent competition and predation.

On a finer scale, because of a continual flux in the status (quality) of resources in different parts of a habitat, population abundance and density of a butterfly population will also be spatially variable with time (**P3.38**). In a habitat, individual plants on which a butterfly can survive are never going to be equally suitable at any one given moment, therefore population density will vary over the space of a biotope even if plant geography remains static – which without management does not happen anyway. Such differences in resource status, some predictable, others unpredictable, occur on a range of timescales from diurnal changes in weather conditions, through broods, to gradual changes with vegetation succession. Switches in use between open conditions and wood cover by *Pararge aegeria* (Shreeve, 1984) nicely illustrate within-habitat changes in population density in relation to changes in the seasonal resource base, that of light and warmth. Short-term shifts (within a day) are characterized by the use of individual mole hills in a small field by territorial *Inachis io* (Dennis and Sparks, 2005). Other examples are the occasional utilization of annual hostplants, in addition to the usual perennials, in *Polyommatus icarus* (Dennis, 1985a) and *Aricia agestis* (Kemp, 1998), the first early in the season and the second late in the season.

Principles relating to stage appearance

Development stages are not expected to occur at times of the year when their critical resources are unavailable. Thus, adults are generally only on the wing when there is sufficient light and warmth and larvae are only actively developing when hostplants are growing. Also, butterfly species are likely to be more closely tied into the life cycles of their hostplants than of their nectar sources, as for these resources they are relatively specialists and generalists (Hardy *et al.*, 2007). Although generalist and specialist hostplant use tends to be reflected in a range of nectar exploitations (Tudor *et al.*, 2004), often butterfly species will use whatever nectar and other adult sources are accessible (Corbet, 2000) and most abundant during their flight season (Munguira *et al.*, 1997). Even so, there is considerable leeway as to when different developmental stages are found across butterfly species (e.g., overwintering occurs in all stages), as well as the number of broods occurring each year and their timing (see Appendix 5e) (Dennis, 1993a). Across species, relationships relating to stage appearance are expected to be strong. The focus, here, is on between-population (within-species) effects on stage appearance.

- **P3.39: The longer a suitable resource is available, the greater duration in the appearance of a dependent life history stage.**
- **P3.40: Species with supplementary consumable resources (polyphagy) will tend to have longer adult phases (flight periods).**
- **P3.41: The shorter the period a resource is suitable, the more synchronized the occurrence and development of a dependent life history stage or phase is expected to be.**

There is an expectation, under principle **P3.39**, that the longer the period of time a resource is available, *in a suitable condition*, the greater the duration of the life history stage using it. Across species, this relationship is evident for egg placement. Thus, eggs placed on woody tissue will outlast those placed on non-woody tissue and the larval stage will be longer on renewable foliage such as grasses than on those exploiting flowering organs. Between populations (within species), the principle is illustrated by resources that can vary between habitats. For instance, in higher latitudes there is likely to be biased use of sunspots used as male territories with southern aspects compared with sunspots having northern aspects; consequently, habitats with sunspots having southern aspects are likely to be used for longer than those with sunspots having northern aspects. Some resources may also be expected to influence later developmental stages. For instance, butterflies using hostplants suitable for herbivory over long periods (e.g., *Aglais urticae* on nettles), or using hostplants appearing at different times of the year (e.g., *Celastrina argiolus* on holly and ivy), would be expected to have increased broods and longer brood duration. This is not quite the same thing as expecting broods of butterflies on perennials to exceed in number those dependent on biennials and annuals, as the crucial phrase 'suitable condition' may not apply. In fact, voltinism increases significantly with the use of annual and biennial plants, less but still significantly with short-lived perennials but not at all with long-lived perennials (Fig. 3.3).

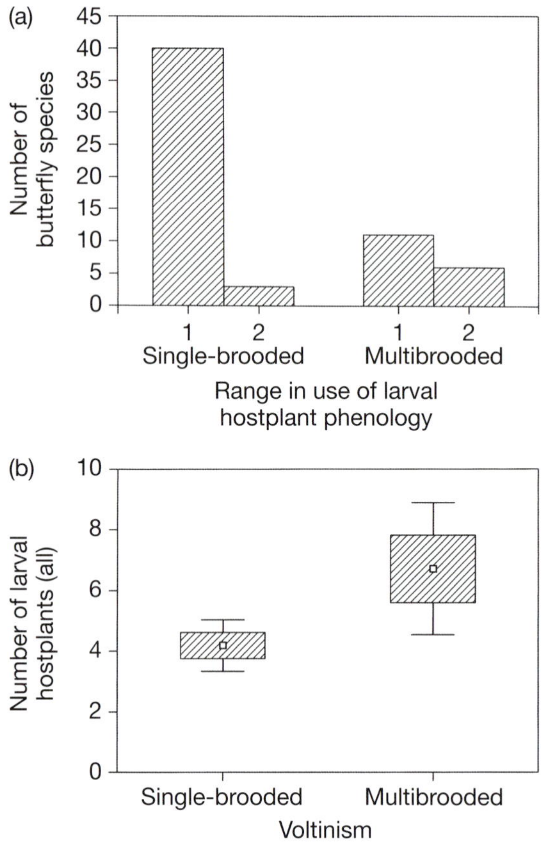

Fig. 3.3 Relationship between voltinism and larval hostplant use in British butterflies. (a) Relationship between the numbers of different larval hostplant life forms used by butterfly species and their annual numbers of broods (1, ≤2 life forms used; 2, ≥3 life forms used – combinations of annuals, biennials, short-lived perennials and long-lived perennials; Fisher exact test, $P = 0.018$). (b) Number of hostplants (all) used by British butterflies differing for number of annual broods (mean (small square) with 2 standard errors (box) and 1 standard error (whiskers)). Voltinism increases with the number of hostplants used (γ tests: all hostplants, $Z = 2.86$, $P = 0.004$; core hostplants $Z = 2.27$, $P = 0.02$), with the use of annuals ($Z = 4.91$, $P < 0.0001$), biennials ($Z = 4.36$, $P < 0.0001$) and short-lived perennials ($Z = 2.7$, $P = 0.005$), but not with long-lived perennials ($Z = -1.91$, $P = 0.056$).

In the last group the early larval stages are timed to use young fresh growth; mature foliage, with lower nitrogen and high tannin levels, is avoided (e.g., hairstreaks). This observation ties in closely with distinctions for qualitative versus quantitative plant defences (Cizek *et al.*, 2006) (see Chapter 5). It is not surprising that voltinism increases as the range in hostplant phenology increases, as it is probably related to access to increasing numbers of hostplants (Fig. 3.3). An extension of this, under principle **P3.40**, is the prediction that butterflies with access to more hostplants will have more broods each year (Garcia-Barros, 2000). This is supported by a significant correlation between numbers of foodplants and voltinism at the between-species level (Fig. 3.3). It would be interesting to determine whether this has relevance at the scale considered here, the between-populations (habitat), within-species level. The inverse of these relationships may also hold. Few resources, as well as resources condensed in time, are expected to produce short-lived, more highly synchronized phases or stages (**P3.41**). However, this relationship is unlikely to be simple, but is expected to show a geographical gradient in synchronization for any species from southeastern Britain to the northwest.

RESOURCES, MOVEMENTS AND DISPERSION PATTERNS INSIDE THE HABITAT

Above, attention is focused on resources as physical or plant structures in the habitat. Unsurprisingly, we established that changes in the density of a butterfly within habitats occur as a response to changes in resource distributions. The situation is, however, infinitely more fluid than this, in perpetual flux we find, when examining changes in butterfly density in response to the location of individuals – not just individuals of other species (enemies, competitors) but of conspecifics. Two fundamental principles here are:

- **P3.42: Since conspecifics represent both resources as well as competitors for landscape resources, the location of individuals affects movement and dispersion patterns.**
- **P3.43: Increased motility will lead to increased contact with resources but losses from the habitat (emigration).**

Individual butterflies move in response to the distribution of other individuals, both conspecifics and other organisms, and in doing so affect the dispersion patterns of the butterfly species within the habitat (**P3.42**). These movements and the resulting changes in distribution patterns occur because conspecific individuals represent a resource to each other (i.e., potential mates) or flag up the existence of potential resources, or are competitors for resources (e.g., egg-laying sites, nectar), whereas individuals of other species can be enemies (e.g., predators, parasites), competitors or symbionts. The consequences of such interactions, the resulting dispersion patterns, can be compared with resource distributions within the habitat; and the spacing between individuals can be different though described in the same way as for resources (i.e., regular, random and contagious (clumped)) (see Box 3.1). These patterns are explored in the next chapter in relation to specific resources.

The factors underlying the distribution of individuals in habitats are undeniably complex. It is well known that the spatial patterns of individuals of any one species within a habitat depend on the community of competing species and predators; for instance, a second competing species can reverse the effects of intraspecific competition within a target species, inducing habitat selection which would otherwise not exist (see Rosenzweig, 1991). At the core of this are resource distributions; their complexity within the habitat governs insect distributions (their spatial variation) and this, in turn, is considered to determine population stability. Arising from this is the concept of **minimum risk distribution** (MRD) (Floater, 2001). MRD is defined as the spatial distribution of individuals that results in the minimum number of premature deaths in a population given the distribution of mortality risk in the habitat (therefore leading to maximized population growth). This concept has been applied to 29 populations of the processionary caterpillar *Ochrogaster lunifer* (Herrich-Schäffer), where divergence from the MRD appears to be caused by high levels of background vegetation interfering with female host searching, resulting in the placement of egg batches in locations with high mortality risk (Floater, 2001).

Movement and dispersion are linked in a perpetual feedback loop; dispersion patterns generate movements that produce new dispersion patterns and so on affecting duration of speed and direction of movements (**P3.43**) (Boughton, 2000). The scale of movements inside habitats relates largely to the scale of resource distributions and their arrangement, which change on hourly to seasonal time fames, as many autecology studies have demonstrated (Brakefield, 1982a; Dennis, 2004b; Dennis and Sparks, 2005). Movements inside habitats, defining the 'home range', are often referred to as **trivial movements** (Johnson, 1969; Baker, 1969); however, only against the backcloth of mass broad-scale migration can they be thus described. Van Dyck and Baguette (2005) suggest the term **routine movements** – distinct from directed movements – but dispersal between resource patches and habitat units is, for some species (e.g., *Pieris napi*, *Anthocharis cardamines*), as routine as are within-resource patch movements for other species (Dennis and Hardy, 2007). Such 'trivial' movements for larvae and adults incorporate searching for resources and the avoidance of enemies and antagonists. Movements within habitat space are inextricably linked to migration generally, as would be expected with a resource-based definition for habitats, and can lead to movements beyond habitat and biotope boundaries (Dennis and Hardy, 2007); this subject is explored in Chapter 6. Here, it is sufficient to note that individuals of species are capable of recognizing habitat components and non-habitat components and adjust their movements accordingly. They do this to a different degree, thus individuals of different species (e.g., regal fritillary *Speyeria idalia* Drury, *Danaus plexippus*) recognize the same non-resources, such as roads and tree lines, as being obstacles of different magnitudes (Ries and Debinski, 2001). Some boundaries may be implicit rather than physical; individuals modifying their movements and returning to the habitat core metres short of a physical boundary, referred to as **edge-mediated behaviour** (e.g., lupine blue *Icaricia icarioides* Fenderi; Schultz and Crone, 2001). Often the cue is other individuals, as in the case of Rocky Mountains small apollo *Parnassius smintheus* Doubleday (Roland *et al.*, 2000). One generalization emerging from studies of movement within habitats is that physical obstacles decline in significance with increased motility. This is partly because faster movement brings individuals into contact with boundary features more frequently (Boughton, 2000) but also because motility is often generated by more suitable conditions for movement.

Chapter 4

EXPLOITING INDIVIDUAL RESOURCES

Sit down before fact as a little child, be prepared to give up every preconceived notion, follow humbly wherever nature leads, or you will learn nothing. (T. H. Huxley)

PATTERNS AND AGENTS IN RESOURCE USE

Now that multiple resources within habitats have been considered, we can move on to examine the finer influences of individual resources, their geography and composition, on the distribution of individual butterflies. In doing so, it is convenient to drop down in scale from coarser to finer detail. At least three divisions are suggested, from the scale of habitat to the part of a substrate used (see Fig. 2.2). A further division is useful for investigating structures created by butterflies, to include:

1 The placement of a resource type (e.g., nectar sources, hostplants) and its exploitation by members of a butterfly population within a habitat.

2 The placement of butterfly individuals on single resource patches (i.e., selection of part of patch and plant).

3 The placement of individual butterflies on single resource items (e.g., on plant, leaf).

4 The creation of microstructures (e.g., webs, leaf folding).

In examining these aspects of single resource use at different scales, it is convenient to do so by considering each developmental stage and the type of resource required by the sexes linked to behaviour (e.g., oviposition, nectaring, mate location, roosting, resting, predator escape, hibernation, etc.). Two observations are perhaps useful at the outset. First, individuals can only select from what is available in a habitat space. Because something is used most frequently from a choice of items or structures does not mean that it is optimal for use at any time. Second, current emphasis in explaining alleged adaptations in relation to 'hard' aspects of butterfly biology (e.g., mature egg number on eclosion, wing loading, thermal biology) is valuable in explaining broad patterns of resource use between species, but it is well to have in mind that variability in individual behaviour and preferences are themselves geared by inheritance. These points focus attention on selection from what is accessible in a variable environment in combination with differences in choice among individual butterflies due to experience (i.e., learning) and age (Resetarits, 1996). A final brief section for the chapter is given over to foraging decisions by individual butterflies, relating to different activities such as nectaring, oviposition and mate location. Foraging strategies attempt to explain the distribution of individuals in relation to maximizing their reproductive and survival benefits in the face of competition for resources, resource limitations and competing resource needs. A resource patch may seem to be uniform to a human observer but foraging behaviour can rapidly cause heterogeneity in resource quality (e.g., by resource depletion), and subsequent competition for what remains will result in different spatial patterns of individuals.

SOME PRINCIPLES RELATING TO SINGLE RESOURCE USE

As before, it is useful to try to focus on some generalizations common to species. These largely apply to the three prominently different levels in single resource use identified above and they are broadly placed into three groups to do with spatial variation, temporal variation and individual preferences or behaviour related to biology.

A Resource-Based Habitat View for Conservation, 1st edition. By Roger L. H. Dennis. Published 2010 by Blackwell Publishing.

Principles relating to spatial variation in a resource type

- **P4.1: The status of a single resource type (rewards/punishments) varies not only over the habitat but also over any single resource patch.**
- **P4.2: Location and situation of a resource type within a patch as within a site are key to its composition and constitution.**
- **P4.3: Spatial variation at fine scales presents opportunities for individuals to compensate for deficiencies in resources at coarser scales.**
- **P4.4: Aggregations in non-gregarious butterflies on resource elements is typically indicative of the limited availability or spatial and temporal concentration of resources of appropriate quality.**
- **P4.5: The boundaries between two vegetation zones (edges) often form sharp gradients in the suitability of a resource type.**

Variability in a resource is normal. A resource may simply be in a suboptimal position or in a position where it cannot be used (**P4.1**). A survey of Brereton Heath, Cheshire, UK in 1986 revealed that 48% of specimens of alder buckthorn *Frangula alnus* were under dense shade of trees (33% >2 m in height were dying) and were therefore not suitable for oviposition by the brimstone *Gonepteryx rhamni* (Gill, 1987). A survey on Alderley Edge National Trust property (Cheshire, UK) demonstrated that only 8% of nettle patches were suitable for egg laying and development of peacock *Inachis io* and small tortoiseshell *Agais urticae* (Fig. 4.1) (Dennis, 2008a). Classic cases exist where hostplants require contrasting conditions to adults egg laying on them. For example, cliff cabbage *Brassica oleracea* on the Great Orme in North Wales is largely restricted to shaded scars and cliff faces with north and east aspects, locations that lack sunlight and warmth for adult female large white *Pieris brassicae* and small white *P. rapae*. This is very different from its distribution on the deeper soils of the chalk cliff tops along the Dorset Purbeck coast in the UK (R. L. H. Dennis, personal observation). The status of resources is never even, if only because of the spacing of elements (e.g., shoots, flowers) and clusters of elements and the costs and benefits for individual insects exploiting each item in relation to foraging for other resource types. Thus, hostplants may receive more or less eggs in the vicinity

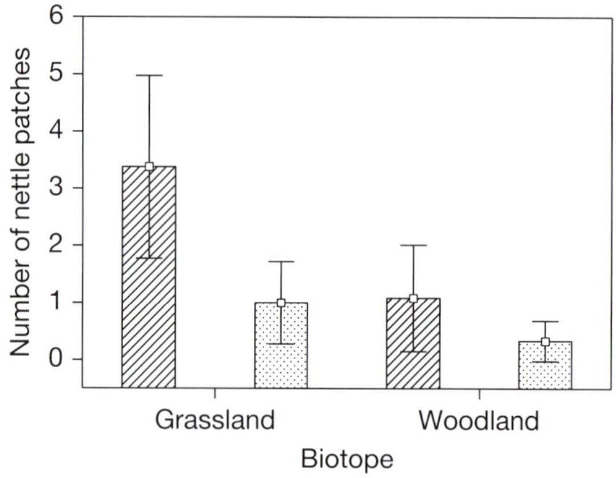

Fig. 4.1 Redundancy of nettle patches for the nymphalids peacock *Inachis io* and small tortoiseshell *Aglais urticae* in woodland and grassland on Alderley Edge National Trust (NT), Cheshire, UK. Bars, numbers of nettle patches (standardized for 100 m squares) occurring of adequate size (>200 × 200 cm) for egg laying in the two butterfly species (cross-shaded, in shade; stippled, exposed to sunlight); whiskers, 2 standard errors; based on a stratified, random sample of 24 ha over the NT property (18% of area), each square occupying >50% area within NT or Site of Special Scientific Interest (SSSI) boundaries. Comparison of nettle patches in grassland under shade or exposed to sunlight: $t_{(23)} = 2.98$, $P = 0.007$; comparison of patches in woodland: not significantly different. Both butterflies concentrate egg laying on larger nettle patches exposed to sunlight in the open or on edges with southerly aspects. (From Dennis, 2008a, courtesy of Entomologist's Gazette).

of nectar sources than those further away depending on other factors such as mobility and interference (i.e., male harassment). This can lead to variation in the distribution of individuals as a consequence of foraging decisions; alternatively, differential mortality (e.g., losses from predation and parasitoids) subsequent to placement on resource items can create spatial variation in individual placement. Thus, foraging may lead to a virtually random distribution but mortality of the same or later stages can result in spatial bias.

However, spatial variation in individuals on a single resource type can be more deterministically controlled (**P4.2**). Key to this variation is location, the situation of a resource type in relation to environmental agents. Some of these agents are physical and evident to the observant, such as slope, aspect, cover and substrate exposure. Clues to other agents may be contained in changes in vegetation composition. A closer inspection of plants reveals biased associations to different soil water retention characteristics, soil water pH, nitrogen and other nutrients, soil textures and temperatures. Invisible as these agents are to casual inspection, they nevertheless determine the suitability of hostplants for butterflies. A good example is the biased use of purple moor-grass *Molinia caerulea* on nitrogen-rich substrates by ovipositing chequered skippers *Carterocephalus palaemon* (Ravenscroft, 1994a, 1994b).

Variation at finer scales provides opportunities for individuals to compensate for deficiencies in resource attributes (viz., low nutrient levels, toughness of plant types, high toxin levels, lack of heat or light) at coarser scales (**P4.3**). The value of a single resource type to a butterfly will depend on its situation in the landscape (habitat), its actual site or ground conditions and composition (e.g., which of a number of nectar or hostplants, if available, are preferred at a specific site). Individuals selecting for them may do so on the basis of a hierarchy of criteria, as in the selection of molehills by peacocks *Inachis io* for territorial perches, the key driver being the thermal environment (Box 4.1) (Dennis, 2004c; Dennis and Sparks, 2005). Aggregations of individuals on larger resource units within a habitat are an inevitable condition for gregarious organisms, as they will obviously need more resources to develop (e.g., *Agais urticae* on nettle patches; Dennis, 1984b). However, apart from butterflies with gregarious larvae,

Box 4.1 Hierarchical exploitation of mate location resource use at different spatial scales by nettle-feeding nymphalids on the Carrs, Wilmslow, Cheshire, UK

Males of nettle-feeding Nymphalid butterfly species, the peacock *Inachis io* (L.), small tortoiseshell *Aglais urticae* (L.) and comma *Polygonia c-album* (L.) defend territories, limited areas, to acquire females (Baker, 1972; Dennis and Shreeve, 1988; Rutowski, 1991). Many of the features of these territories are shared by the three species; typically, they are in direct sunlight and have some vertical edge in the sun, typically a line of trees, a wall or hedge. In spring there is greater emphasis on selection of warm sites than in summer. Warmth is critical for butterfly activity. For sustained flight, butterfly species with moderate to fast wing beat frequencies typically require thorax temperatures in the range 28–40°C and for vigorous flight 33–38°C (Kingsolver, 1985), temperatures well above those experienced in central Britain during spring.

A study of territories using transects and mapped observations in the Carrs Public Park (March/April 2004/2005) (Dennis, 2004c; Dennis and Sparks, 2005) revealed a selection of territorial perch sites at three spatial scales (Fig. B4.1a): macroscale (landforms, top of slopes), mesoscale (field corners, 'fields' of earth mounds or molehills created by mole *Talpa europaea* L.) and microscale (features on the molehills including depressions, slopes, cols between peaks and grass surrounds). At macro- and mesoscales, preference is for the top of slopes and bare ground, including molehills, rather than the base of slopes and live vegetation. The three species used similar perch substrates; only *A. urticae* had any tendency to have the larval hostplant near its territories. There is a hierarchy in the selection of territorial perches by nymphalids, with macro-landforms (slopes) taking precedence over mesoscale features (mini-landforms, molehills), which in turn dominates choice of microfeatures. Thus, molehills at the base of slopes were not used as frequently (fields 5 and 7), if at all (fields 4 and 6) as those at the top of the slope (fields 1 and 2). The sites and features at all scales were warmer than their surroundings; thus the tops of slopes selected were warmer than the bottom of slopes, and earth mounds and bare ground were warmer than grass areas. The factors influencing the use of molehills were explored for two sites (molehill fields 1 (287 observations for 84 individual *I. io*) and 2 (30 observations for 21 individual *I. io*)); the key factors in molehill use in both molehill fields were area and shape (scored −1 depression to +3 complex peaked structure).

Butterflies changed their mean position over sites at all scales in response to weather changes, particularly wind direction, wind speed and sunshine/cloud. With switches from southerly to northerly winds, the mean location of individuals at site E shifted significantly downslope below a break in slope (Fig. B4.1b). Perch locations at this site, and on the concrete seat base at this site, were significantly influenced by wind speed and direction and changing sun angle during the day. On individual molehills, the features used for perching were warmer than those not used and the features used at any moment in time depended on sunlight, wind speed and ambient temperatures. On molehills, perching behaviour is strongly linked to microlandform selection to maintain warmth (Fig. B4.1c). Selection of perch sites for warmth dominates selection for visibility; thus, during cold weather, perch sites were adopted deep in the hollow within territory A (Fig. B4.1a), which greatly restricted the arc of detection, and conspecifics were sought by regular patrols of the area.

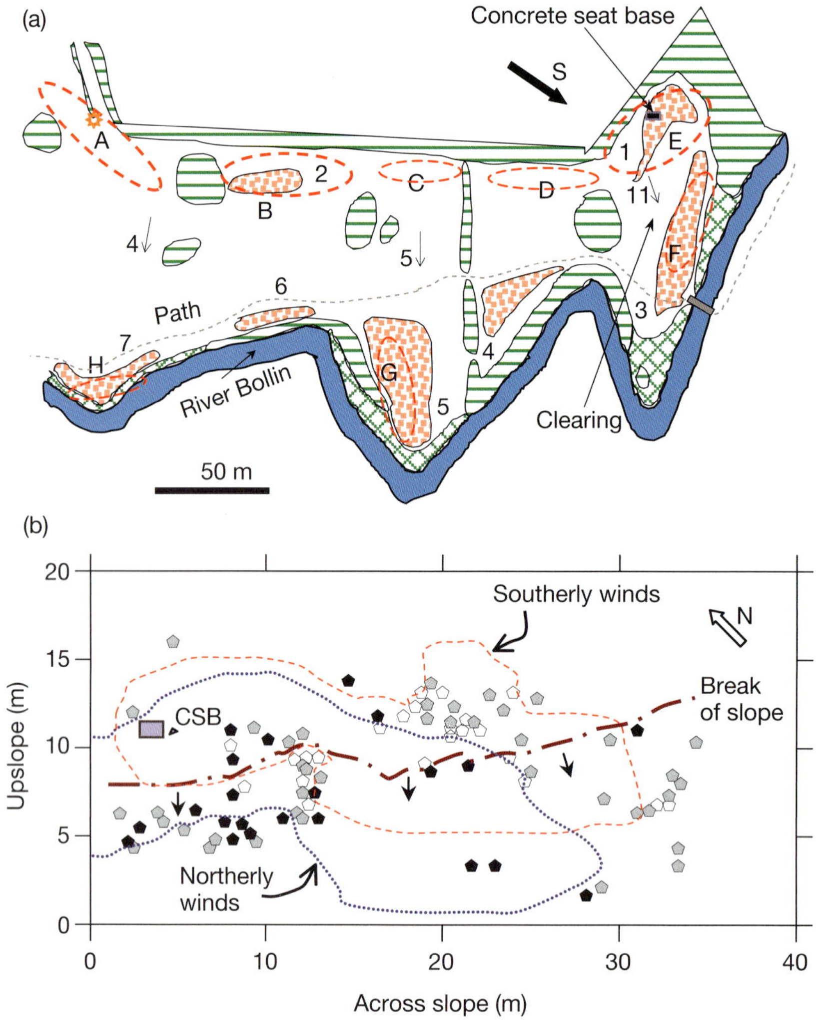

Fig. B4.1 (a) The Carrs, Wilmslow, Cheshire (340 × 110 m), which comprises a southwest-facing slope above the River Bollin, bounded by the river's north bank at the foot of the slope and by a narrow woodland separating the Carrs from housing at the top of the slope. In 2004/2005 molehills were concentrated in seven areas, two at the top of the slope (2004: 127 molehills) and five at the base of it (2004: 835 molehills). Site E includes a concrete rectangle (116 × 167 cm), a disused seat base, 5 m from the top of the slope and cut horizontally into it. The mean diameter of molehills in 2004 was 50.1 ± 1.3 cm with a total area of 193.7 m^2, occupying 1.54% of the study area. The main nymphalid territories (sites B and E) were concentrated at the two higher areas of molehills and on a hollow (>50 cm deep) located on the apex of a bank, the previous site of a tree trunk, at an abrupt reflex angle in the wood edge (red star). Molehill field attributes studied included size, shape, surrounding shelter, location across and upslope, and isolation from other molehills weighted by their size. Shading: brown stippling, molehills; green shading, wood; green cross-hatching, willow carr and tall herbs; blue, river; red pecked lines, butterfly territories (main ones heavily pecked). (b) A plot of the molehills in site E illustrating the shift in territories with winds from the north and south. The bold dashed line is the break of slope with the arrows indicating direction of slope. The finer pecked (red and blue) lines give the median number of territories (standardized) for different wind directions. The symbols illustrate molehills in three size categories (white, smaller than lower quartile; grey, between lower and upper quartile; black, larger than upper quartile). CSB and triangular object to right are the concrete seat base and remains.

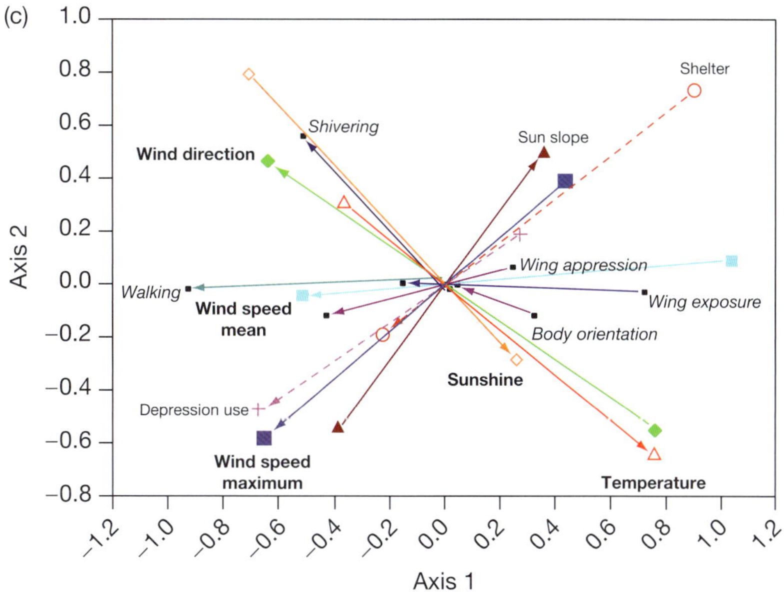

Fig. B4.1 (*continued*) (c) A multiple correspondence analysis plot describing simultaneous associations between behaviour, microfeature use and weather conditions for individual observations at site E, molehill field 1. Variables are simplified into binary states (insolation: 0 cloudy, 1 sunny; temperatures: 1 <16.0°C, 2 >16.0°C; mean wind speed: 1 <3 m s^{-1}, 2 >3 m s^{-1}; maximum wind speed: 1 <6 m s^{-1}, 2 >6 m s^{-1}; wind direction: 1 <270°, 2 >270°). Behavioural variables (open wing exposure, body orientation to sun, appression of wings to substrates (see Fig. 4.3a), shivering and walking were binary coded as present or absent) were entered as supplementary to the analysis. Variance of axes = 46%. Of 20 possible associations, 13 are significant ($P < 0.05$). With lower temperatures, cloud, stronger winds and more northerly winds, there is an increase in the use of sheltered fringes of molehills (see Fig. 4.3b) and depressions on the surface, as well as shivering, wing appression and walking to sheltered microsites. With increasing sunshine there is greater use of exposed sun-facing slopes on the molehills. (a, c, from Dennis and Sparks, 2005, with kind permission of Springer Science and Business Media.)

aggregations of butterflies (any stage) on a resource are either the result of a limited resource (suitable resource concentration to a small area) or selectivity for some attribute (i.e., quality, prominence) of the resource (**P4.4**). For example, orange-tips *Anthocharis cardamines* will load eggs onto large plants, despite cannibalism by larvae, when large crucifers are in short supply (Fig. 4.2) (Dennis and Hardy, 2006). Typically, the boundary between two vegetation units often forms a zone (edge) of butterfly aggregation (**P4.5**) (e.g., woodland or scrub abutting herb grassland) for any developmental stage. This often explained by the distinctive local climates of such boundary zones; mate location sites are typically formed along such boundaries where these are sheltered and sun-facing (e.g., *I. io* and *A. urticae*; Baker, 1972). But marginal sites can present changes other than to

Fig. 4.2 Excessive egg loading by orange-tip *Anthocharis cardamines* on a cuckoo flower *Cardamine pratensis* flower head. (Courtesy of Peter Hardy.)

local climate or microclimate, for instance facility of access, availability of resource elements and escape from enemies or from herbivory. Thus, crucifers on river levées can be more accessible to ovipositing *A. cardamines*, where winter over-bankful discharge has removed competing vegetation, than hedgerows at field margins which may nevertheless provide the same butterfly with better pupation sites. Much work remains to be done on butterfly perception and use of different surfaces (T. G. Shreeve, personal communication).

Principles relating to temporal variation in single resource types

- **P4.6: The spatial variation in every resource attribute changes over time.**
- **P4.7: The status of a resource type (rewards/punishments) for a butterfly varies over time for any single locality.**
- **P4.8: Distinctive temporal patterns occur in the use of some resources.**

The spatial variation in any resource will not remain constant but will change over a range of timescales from seconds (e.g., from changes in weather) to years (e.g., from vegetation succession or management) (**P4.6**). This is an extension (subdivision) of the same principle for habitat (multiple resource) composition and structure. Changes in weather conditions are sufficient to change the pattern of rewards for dependent organisms, as for territorial nymphalids. An example of this is provided by the shift in suitable molehills for territorial perches by *Inachis io* over several days. With a change in wind direction from south to north was observed a shift in molehill occupancy in a cluster of molehills relating to shelter from the wind (Dennis and Sparks, 2005). At a finer scale still (e.g., from second to second with passage of clouds and changes in wind speed), the butterflies were observed to change their position on molehills from exposed to sheltered locations in hollows or the shelter of the grass fringe (Fig. 4.3; see Box 4.1).

Following this is the principle that resource quality at a site never remains constant (**P4.7**). A simple illustration relating to *I. io* using molehills is that the numbers and positions of the earth mounds change from week to week, year to year. Experimentation with a plastic bag in a territory of the large skipper *Ochlodes sylvanus* indicated that even small human influences

(a)

(b)

Fig. 4.3 Bare ground resources for peacock butterflies. (a) Peacock butterflies *Inachis io* using a flattened molehill for mate location. (b) Territorial peacock *I. io* perching and basking on a molehill by engaging wing appression in a mini-depression and using the shelter of a grass surround.

can redirect resource use (Fig. 4.4) (Dennis and Williams, 1987). Changes can, however, be more subtle; a resource patch can remain much the same in dimensions over years but suddenly provide a very different 'resource' in quality (e.g., from changing water table levels, nutrient and toxin (herbicide) inputs). Thus, a nettle patch on Alderley Edge National Trust property used by nymphalids annually for 28 years was discarded in 2007 following illicit herbicide treatment of

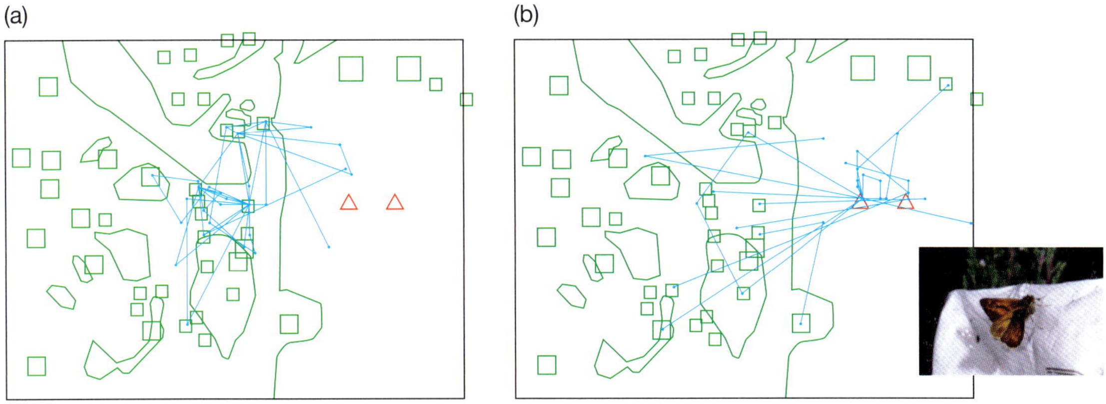

Fig. 4.4 Influence of the placement of a plastic bag on perch sites adopted by large skippers *Ochlodes sylvanus* (inset) within the centre of a territory on Lindow Common, Wilmslow, Cheshire, UK. Adopted perch sites were alternated four times over 10 min periods ($\chi^2_{(1)} = 51.6$, $P < 0.001$): (a) without the presence of plastic bags, and (b) with the presence of a plastic bag in one of two positions shown. The area shown is 8 × 6 m. Triangles, plastic bag; squares, birch seedlings; polygons, other vegetation boundaries; blue lines, lines of movement. (Dennis and Williams, 1987, courtesy of the Lepidopterists' Society, USA).

the unimproved pasture by the tenant farmer (Dennis, 2008a); a female *Aglais urticae* was observed to inspect the site closely for 30 minutes and then abandon it. Silver-studded blues *Plebejus argus* occupy varying portions of the upper slopes on the Great Orme headland from day to day, from one season to the next, despite continuity and relative stability in the status of the larval hostplant rock rose *Helianthemum nummularium* (see Box 3.2). This varying exploitation is associated closely with weather and seasonal climate, the upper slopes being used increasingly in hotter, sunnier and calmer conditions (Dennis and Sparks, 2006).

A distinctive feature in resource use is the occurrence of temporal patterns (**P4.8**). Many are obvious and associated with the seasons and stages in development (e.g., larval feeding, hibernation). But some temporal patterns in resource use are less obvious; there is a shift in what is used in the habitat space over time even when the physical resources appear to remain constant (e.g., grayling *Hipparchia semele* and speckled wood *Pararge aegeria*; Shreeve, 1986, 1990). Recently, it has been shown that the height and type of substrates used for resting (settling) varies from morning to dusk and from spring to autumn in a rather predictable manner (Hardy and Dennis, 2007). Interestingly, the height of settling differs from beginning to end for the daily and seasonal cycle (Fig. 4.5). Settling height is at a minimum during the mid part of the day, whereas throughout the season there is an increase in settling height from

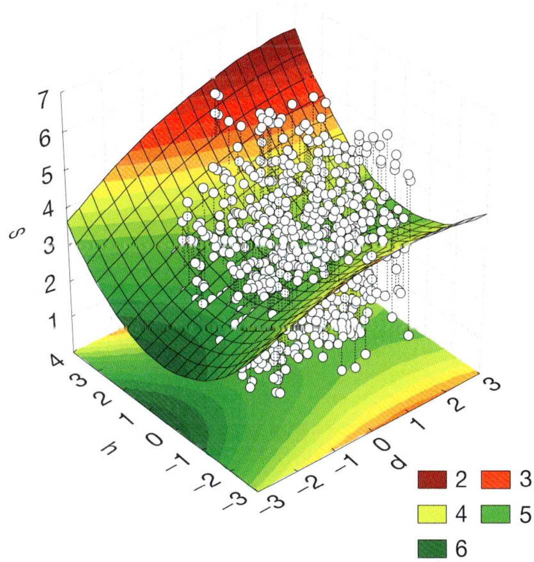

Fig. 4.5 Height above ground (S, scores) of settled butterflies in Greater Manchester in relation to time of day (hours of day, h) and season (days in year, d). Time of day and season have been standardized to the same scale with zero mean and unit variance. Height scores 1 = <2 cm; 2 = 2–10 cm; 3 = 10–50 cm; 4 = 0.5–1 m; 5 = 1–2 m; 6 = >2 m. Individual observations are shown as open circles. Significant seasonal and/or diurnal patterns occur for eight of the 13 species studied for which $N \geq 30$. (From Hardy and Dennis, 2007, with kind permission of Springer Science and Business Media.)

spring to summer followed by a levelling off in autumn. Distinctions in settling height are affected by thermal environment and are related to two aspects of thermoregulation behaviour, basking and wing appression, both having seasonal components. These findings demonstrate the wide variety of surfaces required by butterfly species as habitat components over time.

Principles relating to individual preferences and behaviour

- **P4.9: Different (supplementary) outlets of a resource type differ in their suitability for individuals in a population, in terms of reproductive success, development and mortality, and can be arranged in a suitability hierarchy.**
- **P4.10: Individual preferences differ in threshold and variability, thus range of use, quantitatively and ordinally (in rank).**
- **P4.11: That only a fraction of any resource is useful and used by a butterfly is indicative of selectivity for particular resource attributes.**
- **P4.12: Bias in the placement of any stage of a butterfly on single resource items is an indication of fine-tuning in exploiting the resource.**
- **P4.13: A hierarchy of spatial scales exists for selectivity of any resource attribute.**
- **P4.14: Resource attributes used are not constant but vary with the growth of individuals and changing conditions.**
- **P4.15: First instar larvae nearly always depend on tender young growth to feed on regardless of the requirements of later instars.**
- **P4.16: Butterflies engaged in gregarious activities require larger resource items (individuals, packages, patches) than those carrying out activities singly.**

A great many generalizations have developed around aspects of individual behaviour and preference. They are based on the observation that outlets for a resource type (e.g., larval hostplant species, nectar flowering plants) at a site can be arranged in a hierarchy of suitability for butterfly individuals potentially using them (**P4.9**). In this way Wiklund (1981) distinguished primary, secondary, novel and unsuitable hostplants for a butterfly species at any location. Investigating preference for resources and resource elements (e.g., oviposition sites, nectar flowers) has proved extremely elusive, primarily because it is difficult to eliminate the influence of the human observer and human influence in experimental design (Singer, 2004). A crucial matter is that both status of resources and preferences of individuals are not constant. There are also conflicting aspects of preference; these are better considered when discussing foraging activity (below). Even so, some pointers are available for consideration. Generally, individual preference displays a variability in threshold and order and quantitative differences (**P4.10**) (Singer and Lee, 2000). Different individuals often show variation for their order of preferences and the degree to which they prefer a resource type, and, more importantly, for the threshold in acceptance of it. This distinguishes individuals as **generalists** – those that have low thresholds in accepting a number of distinct resource outlets – and **specialists** – those that have high thresholds for all but perhaps one resource outlet. A typical example is oviposition preference for hostplants in species that use multiple hosts. The classic study is that of *Papilio machaon* by Christer Wiklund (1981); but then refer to the experiment carried out on the Glanville fritillary *Melitaea cinxia* with regard to egg laying on ribwort plantain *Plantago lanceolata* L. and spiked speedwell *Veronica spicata* L. to get a full appreciation of just how complex performance requirements can be to assess (Singer and Lee, 2000).

There is no reason why the basic concepts on generalism and specialism generated by Wiklund should not apply to all resource types and not just larval hostplant choice (see Box 2.3) (e.g., nectar flower foraging in the green-veined white *Pieris napi*; Goulson and Cory, 1993). Undertaking experiments on the same species, Wiklund (1975) demonstrated two distinct shifts in preference. First, he found changes in preference between a female and her offspring, a substantial difference in hostplant range in the adults and larvae of *P. machaon*, indicative of links to separate gene complexes. Second, he isolated differences in preferences between newly hatched larvae and those in their third instar (Wiklund, 1973), with the latter preferring species on which they had been reared. In his research, Wiklund (1974a) was able to refute one principle, **Hopkin's host selection principle**. This postulates that a memory of the larval food predisposes the adult females of phytophagous insects to oviposit on the same plant as that upon which they were reared. Bearing in mind the experimental difficulties alluded to, interpreting preferences (in terms of individual success and population fitness) is a difficult matter (Janz *et al.*,

1994). A good hostplant for the larvae is not necessarily a good one for eggs or pupae or for emerging adults (Reavey and Lawton, 1991). Janz *et al.* (1994) impress on us that preferences, as for oviposition but also perhaps for other resource use (e.g., nectar feeding), should be assessed against total offspring fitness and not any single performance measures such as growth, development rate of larvae, or emergence times of adults, which in any case may not be highly correlated (Thompson, 1988; Dennis, 1993a).

From the vantage of resource use in habitats, only a fraction of a basic resource (e.g., hostplant, nectar flowers) is used by a population of individuals. This is observed, however small the population, and this above all else forces the conclusion that quality of resources is crucial in butterfly biology (**P4.11**). Apart from capacity to locate resources, bias in the placement of any butterfly stage on single resource units may be indicative of selectivity and fine-tuning in selection (**P4.12**). Classic cases are the choice of dorsal or ventral leaf surfaces for oviposition (e.g., *Aricia* species; Asher *et al.*, 2001) or leaves in partial shade rather than sunlight (e.g., white admiral *Limenitis camilla*; Fox, 2005), and destinations chosen by pre-pupal larvae on smooth stems or rough twigs (e.g., *Pieris* and *Papilio* species; Smith, 1978, 1980). The reasons are usually evident, as in the case of selection of small or large plants by *Pieris napi* and *P. brassicae*, respectively (needs of a gregarious brood in the latter), or flowering plants for egg laying by *Anthocharis cardamines* (larvae feed on the developing pods). Often a hierarchy in selectivity is evident (**P4.13**) (e.g., selection of site dominates the selection of a mini landform for perching by territorial *Inachis io*; Dennis and Sparks, 2005; see Box 4.1). Selectivity can also change with growth, as for instance as larvae pass through instars (**P4.14**) (e.g., different conditions required by early and late instars for feeding and survival in the swallowtail *Papilio machaon*; Wiklund, 1973).

Typically, if for mechanical reasons alone, first instar larvae require young growth, the reason why hatching is synchronized with bud burst in so many Lepidoptera; they can usually move on to coarser plant material as they become older (**P4.15**) (e.g., *Aglais urticae*; R. L. H. Dennis, personal observation). Inevitably, butterflies that engage in aggregations, or are gregarious, prefer larger resource items (**P4.16**); mortality would be high for gregarious batches running out of food on small patches or shoots. In many cases, increased numbers on larger resource items may simply reflect the amount of resource available, i.e., a linear or log linear increase in butterflies occurring with resource size (e.g., butterflies nectaring on bramble bushes; *Plebejus argus* roosting on cotoneaster bushes; Dennis and Sparks, 2006) (Fig. 4.6). What would be of interest are unexpected relationships between exploitation and resource dimensions, i.e., numbers of individuals on resource items being greater or fewer than expected. Possible examples of excess numbers compared with those expected are aggregations (leks) of territorial butterflies on one of several tall trees (purple hairstreak *Favonius quercus*; Crawford, 1996) and egg batches on plant patches in particularly suitable locations or having unusually advantageous structural characteristics (e.g., *A. urticae* batches on nettles; Dennis, 1984b (Fig. 4.6b).

DISTRIBUTION OF INDIVIDUALS IN RELATION TO THE DISTRIBUTION OF RESOURCES

Regarding the distribution of individuals of a population over a single resource type, two things can vary within a habitat space or site:

1 The resource itself.
2 Individual butterflies on the resource.

They may do so exclusively or in combination.

Resource distributions fall into two broad groups:

1 Those that can occur virtually anywhere in a habitat space, in the biotope or vegetation unit(s) forming the template for the habitat.
2 Those that have primarily edge distributions and are usually linear; the edge may be marginal to the habitat space or cut through part of it.

The distribution of the former group, including most of the resources listed in Table 2.2, are dictated by environmental (e.g., light/shade, moisture, etc.) or other conditions (e.g., management) within the habitat space. Such resources usually occupy amorphous areas (e.g., hostplants, nectar source zones), but can assume shapes and can also occur in edge positions because of association with a vegetation unit or some structure such as a fence or ditch (e.g., hostplants such as hedge species *Alliaria petiolata*, referred to as Jack-by-the-hedge or hedge garlic, for *Anthocharis cardamines* (Dennis, 1983a). The latter group which have primarily, or invariably, edge distributions can be further divided (without suggesting mutual exclusion), mainly on the basis of utility, into exploitation of the following features:

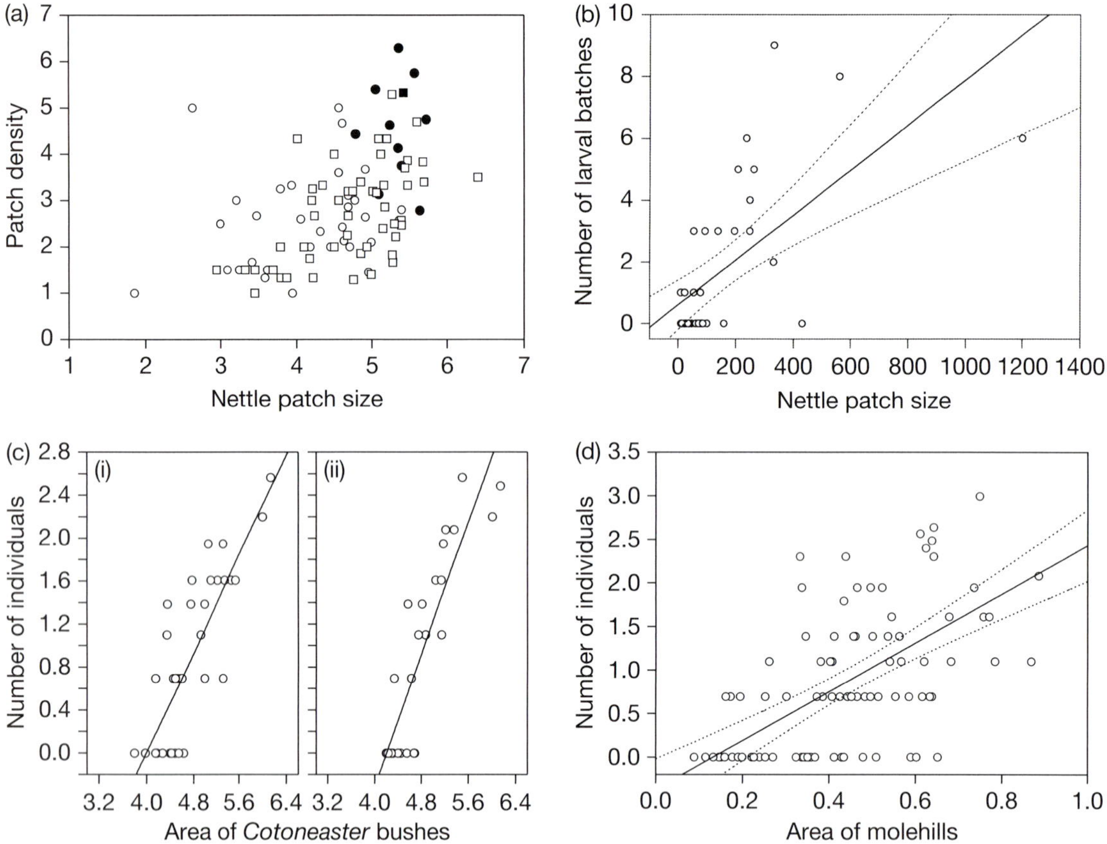

Fig. 4.6 Selected butterfly number–size relationships in resource exploitation. (a) Small tortoiseshell *Aglais urticae* egg batches on nettle patches on Alderley Edge National Trust, Cheshire showing the distribution of nymphalid larval batches on nettle patches for size ($S = \log_{10}(1*b*\pi/4)$, density ($\log D$, where D is the number of nettles per 20 cm quadrat along length of patch) and shade (0, none; 1, partial/full) caused by trees and shrubs (circles, no shade; squares, shade including partial shade; empty symbol, no nymphalids; black symbols, nymphalid larval batches). (b) *A. urticae* larval batches on nettle patches (number of nettles) in a field in the Bollin Valley, Cheshire in 1983 ($F_{(1,35)} = 24.0$, $P < 0.0001$, $R^2 = 41\%$). (c) Numbers of resting (log numbers day roosting) silver-studded blues *Plebejus argus* on cotoneaster bushes (log area, cm^2) on the Great Ormes Head, North Wales: (i) 24 June 2004, after 11 a.m., ambient temperature 12.5°C, wind speed 8.5 m s^{-1}, $R^2 = 67\%$; (ii) 2 July 2004, after 11.30 a.m., ambient temperature 15.9°C, wind speed 7.6 m s^{-1}, $R^2 = 79\%$). (d) Log number of visits by territorial and thermoregulating peacocks *Inachis io* on molehills (log area, cm^2) at the Carrs, Wilmslow, Cheshire (Pearson $r = 0.57$, $R^2 = 33\%$, $P < 0.001$). Regression line with 95% confidence limits are shown in (b) and (d).

- Physical structures, often boundary features such as ditches, hedges, fences, walls and rock faces, and boundaries between vegetation units.
- Local climate edges, typically with sun-facing aspects.

Numerous examples fall into the first category of resources, edge or boundary physical structures (e.g., pupal hibernation sites of *Pieris brassicae* on walls; egg laying locations under hedges in *Lasiommata megera* (Dennis, 1983c); adult feeding for mineral salts along river edges or beach tide lines, often witnessed in tropical settings), and into the second category of local climate-driven resources (e.g., the level of light needed for egg laying as in the case of the partial shade required by *Limenitis camilla* using honeysuckle at woodland edges (Pollard, 1979); basking sites and mate location territories on south-facing woodland edges in nymphalid

Table 4.1 Factors influencing oviposition bias on hostplant patches within a habitat space.

- Plant taxa/type (forms)
- Plant geography (density, abundance, spacing)
- Plant size/growth form (synchronization and timing) – note not just size of a plant but size of individual patches
- Plant architecture including colour (Kelber *et al.*, 2001)
- Plant composition/chemistry (nitrogen, tannin, etc., soil water condition, nutrient supply, colour) (Bink, 1992)
- Plant nutrients and defensive chemicals (N, Ca, Mg – nitrogen is the major limiting factor for growth) (Ravenscroft, 1994a, 1994b)
- Plant accessibility/context (overtopping), growth stage synchronization (bud burst) (Emmet and Heath, 1990)
- Plant environment (light, temperature, moisture, exposure, drowning in water)
- Plant juxtaposition (neighbouring resources: nectar, predators)
- Inhibitors/degree of exploitation (competitors, predators, parasitoids)
- Symbionts (ants for Lycaenidae); proximity of egg laying related to feeding strategy: cuckoo versus predatory (Thomas and Elmes, 1998)
- Plant edges or bare earth edges, plant bare ground (Bourn and Thomas, 1993)
- Factors that affect larval growth rate (e.g., plant toughness)
- Egg avoidance: visual and chemical (e.g., oviposition deterring pheromones) (Schoonhoven, 1990)
- Enemy avoidance (e.g., ants in non-myrmecophilous butterflies)

Note that individual preferences are not included.

butterflies (Baker, 1972)). Other examples, very likely the majority, may involve both a physical component and the advantages of a local climate, such as roosting in *L. megera* (Dennis, 1986a).

The following are some basic principles of individual distribution:

- **P4.17: The distribution of individuals for a specific activity in any developmental stage within a site is largely governed by the spatial pattern of the 'host' resource.**
- **P4.18: The spatial pattern (density) of individuals typically contrasts with that of a 'host' resource at the same or at a finer spatial scale.**

The distribution of individuals at any moment in time is generally developmental stage activity – 'host' resource dependent (**P4.17**), that is, linked to the relevant resource on which the activity is based for that stage. But, in detail, the density of individuals over resource patches can be complex. Spatial patterns of butterflies on single resources within the habitat space are governed by innumerable agents that depend much on the developmental stage involved (**P4.18**). To give some idea of just how varied factors can be, a listing is provided of a range of factors that can influence the distribution of eggs on more or less ubiquitous hostplant cover within a habitat space (Table 4.1). Much the same variation in factors can be isolated for other resource types. Clearly, edge resources determine edge distributions in the butterflies dependent on them, though usually the rule of fractional resource still applies; that is, only part of the edge structure available is used, and the reasons can be one of access (e.g., distance of walls for pupation from hostplant patches; first plants to be contacted on entry to a habitat, as in egg laying in *Anthocharis cardamines* on marginal patches for a hostplant resource; Dennis, 1983a, 1983b) or local climate (sunny, sheltered positions versus shaded, exposed positions; Dover *et al.*, 1997).

The observation that both resource and butterfly geography have some independence leads to two questions. First, how far does the spatial pattern or position of a single type of resource at a site (habitat) influence the spatial pattern of individuals and their population status? Second, when a resource is apparently homogeneously distributed, is spatial patterning in butterfly individuals equal and, if not, then what factors result in spatial patterning of butterfly individuals? These questions, left open at this stage, may appear somewhat teasing but some clues have been sown above in mentioning foraging behaviour in relation to competition for resources. Suffice it to say that aggregations of individuals may occur because a resource is limited or aggregated in its own right, but aggregations can also occur when a resource is apparently ubiquitous if only because the fraction used differs in some way that is important for exploitation. The solution may not be obvious to an observer. Answers to these types of question invariably have to consider scale effects. For instance, resource patterns may be apparently random at a broad scale but the condition of the resource may be greatly biased to factors at a finer scale (e.g., in plant size, nutrient status, etc.). Different kinds of patterns are referred to in distributions and these are illustrated in Box 3.1. Often, a number of distinctive pattern gradients are evident for a resource type, or butterfly individuals using them, within habitats: habitat edge to centre, biased aspects, and substrate floor to canopy ceiling. Underlying these may be a number of obvious explanations, such as differences in resource density, exposure and orientation, and access as suggested for instance by covered versus exposed resources or resource recesses versus protuberances (see Zalucki, 1993). But other factors may be more elusive and dependent, for instance, on substrate (soil) differences or the pattern of the age of a resource (e.g., nectar, with plants in patches flowering at different times) in relation to say substrate differences such as soil water and previous foraging. Perhaps, depending on the developmental stage and behaviour, spatial variation of individuals on resource outlets may occur from choice (i.e., selectivity by adults foraging, larvae foraging) or from differential selection against individuals (any stage); it should not be forgotten that enemy-free zones (predators and parasites) comprise crucial resources (Sanz, 1998; Shaw, 2006; Shaw *et al.*, 2009).

DISTRIBUTION OF INDIVIDUALS ON SINGLE RESOURCE PATCHES

Much as butterflies can be distributed in various ways over multiple (supplementary) resource patches within habitats, so too can they be over single resource patches. However, single resource patches, particularly those of a hostplant, are atypical inasmuch as characteristic edge distributions have been described for butterflies loading on their margins, to date invariably of eggs on hostplant margins. At least three distinctive edge distributions have been identified, with generic connotations, for the distribution of early stages (usually eggs or larvae) on hostplant patch margins, but there is no reason why the concepts should not be extended to include adults. These have been labelled **density effects**, **edge effects** and **recess effects** (Courtney and Courtney, 1982; Dennis, 1983a). Density effect is not strictly an edge effect phenomenon; it describes the excess loading of eggs on low-density plants in a habitat space (Courtney and Courtney, 1982; Dennis, 1983b, 1985b). It properly belongs to the previous section dealing with distribution of butterflies on patches within a habitat space but is dealt with here for convenience. Edge effects *per se* describe butterfly loading on the margin of a single resource patch. Confusion emerges from the fact that edge distributions can occur as a consequence of individuals experiencing changes in realized density in a resource on contact with the margin of a resource patch. Density effects tend to get associated with edge effects as small patches often lie outside large patches and these are the first and last patches with which females come into contact on entering and leaving a resource zone (Fig. 4.7). But density effects imply cause through changes in resource density and edge effects can include many more possible reasons for marginal situations. Recess effects describe distinctive structural associations in a variety of situations, including edge and marginal locations and indicate the occurrence of multiple factors in causing bias in butterfly distributions on a resource patch.

The following principles summarize marginal distributions on a resource patch or patches:

- **P4.19: Distributions of individuals associated with low or high density of a resource suggest advantages of resource opportunity or quality linked to resource density.**
- **P4.20: Edges of resource patches provide rapid transitions in conditions and attract biased distributions of individuals.**
- **P4.21: Use of physical recesses in resources by individual butterflies implicates the need for appropriate access, protection, cover or concealment.**
- **P4.22: Edge distributions increase the density of butterflies and thereby increase the potential for infestation by parasitoids and pathogens.**

Changes in resource density implicate changes in quality of a resource that individual butterflies may well exploit (**P4.19**). Patches, whether singular or multiple, are defined by margins and margins involve sudden changes in conditions and attract peripheral

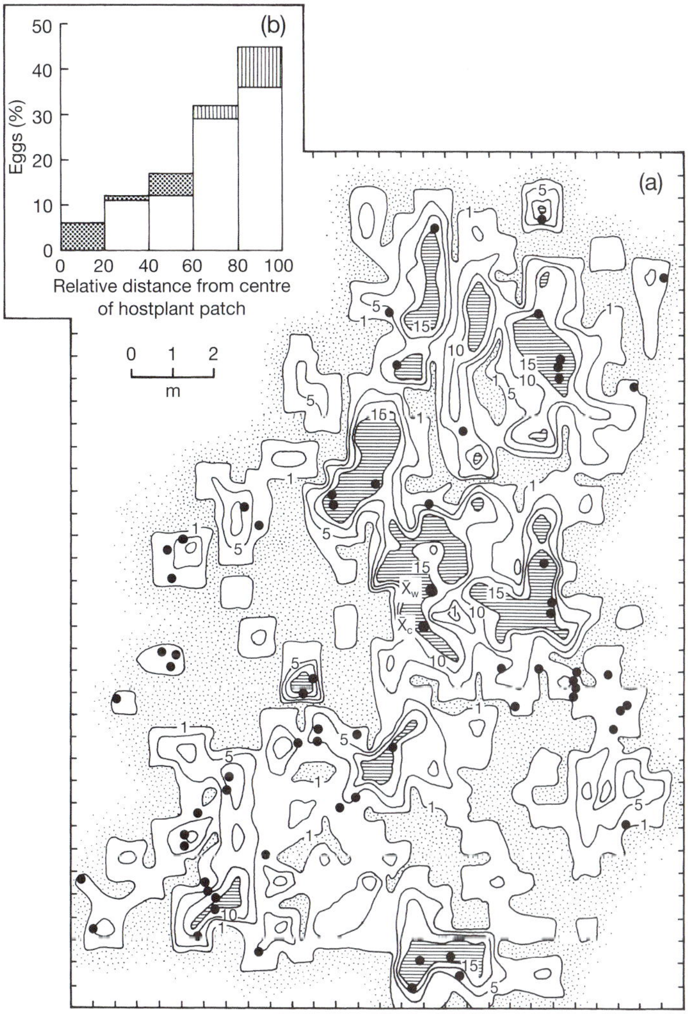

Fig. 4.7 Density effect in egg distribution of orange-tip *Anthocharis cardamines* on cuckoo flower *Cardamine pratensis* in the Bollin Valley, Cheshire, UK. (a) Eggs tend to have a peripheral distribution on satellite patches around a large hostplant patch. Symbols X_c and X_w give the centre (unweighted and weighted according to plant density) for the whole patch. Eggs and larvae are shown as black circles. Isolines give flower density at intervals of 1, 5, 10 and 15. (b) A significant bias in distribution of eggs to the periphery of the hostplant patch is illustrated in a plot of the percentage of eggs located in five annuli extending from the centre of the hostplant patch. The proportion of eggs expected in each annulus is given by the height of the bars and is based on the number of flower heads in each annulus. Vertical hatched areas, proportion of eggs exceeds that expected; stippled areas, deficit of eggs occurs from that expected by the model. (Dennis, 1983a, 1992, courtesy of Oxford University Press.)

distributions; one of these changes is that of the density of the resource (**P4.20**). Small patches amidst matrix (trivial space) outside larger, thus denser, patches are experienced as sudden increases in density by butterflies, say ovipositing females. Thus, what may appear to be an inverse density effect is in reality a direct density effect for the butterfly approaching the resource! Physical recesses less often occur around the entirety of a resource patch and their use suggests more definite reasons for their use (**P4.21**). However, although protection, cover, concealment and suitable microclimate come readily to mind as reasons, this is unlikely to complete the list of influences.

There are at least two clear, structurally distinct forms of edge distribution of plants that may help in distinguishing recess, density and edge effects: (i) a key distinction is that with density and edge effects, butterflies can be found around the entire margin; and (ii) with recess effects butterflies tends to be concentrated within a small fraction of the margin and, even if lacking bias in aspect, typically include further structural association. Distinctions may not always be clear, though, for two reasons. First, individual resource patches can be of any size or shape and it is more difficult to determine patterning over small patches and, indeed, over immensely large ones. In the first case

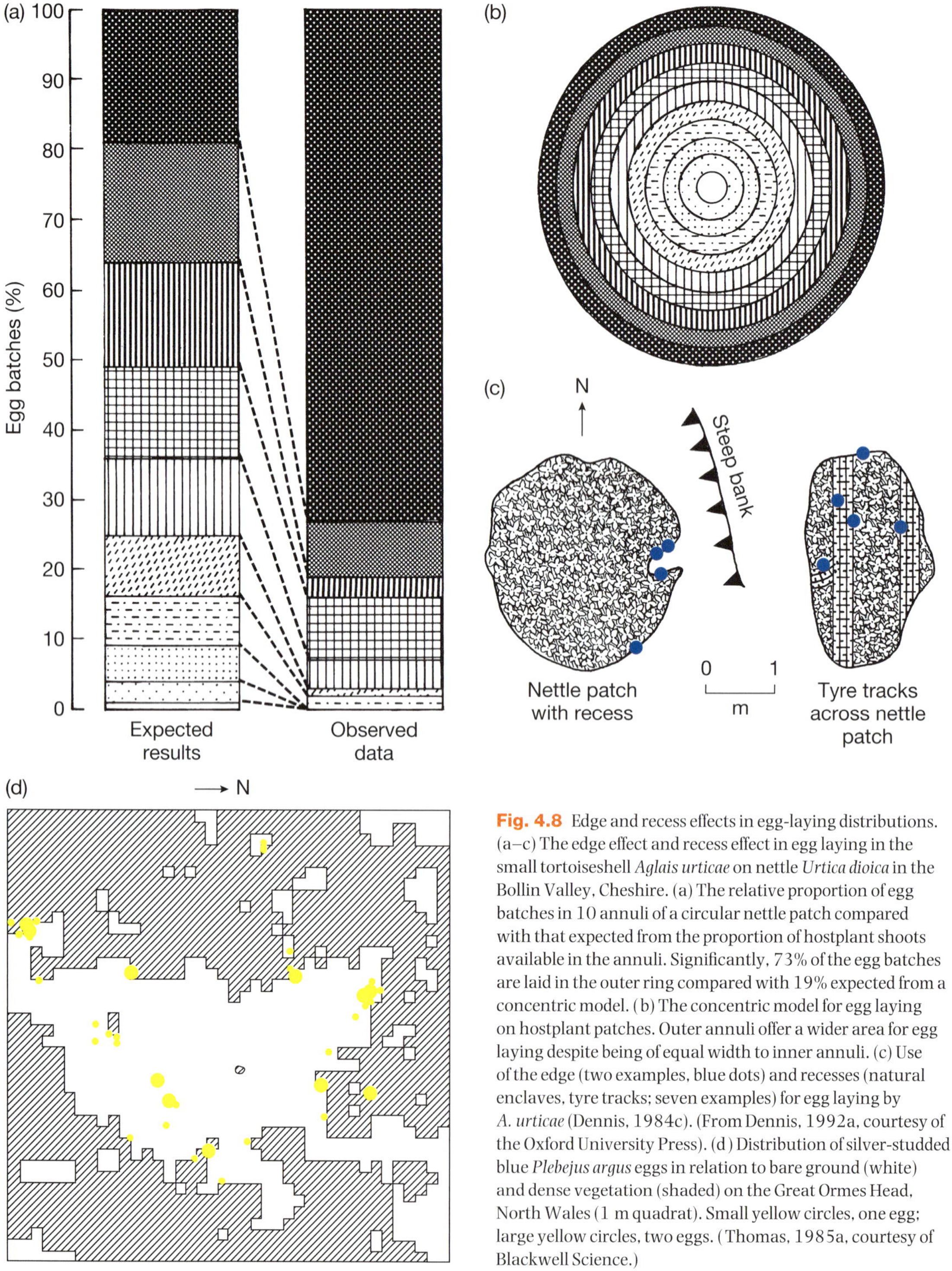

Fig. 4.8 Edge and recess effects in egg-laying distributions. (a–c) The edge effect and recess effect in egg laying in the small tortoiseshell *Aglais urticae* on nettle *Urtica dioica* in the Bollin Valley, Cheshire. (a) The relative proportion of egg batches in 10 annuli of a circular nettle patch compared with that expected from the proportion of hostplant shoots available in the annuli. Significantly, 73% of the egg batches are laid in the outer ring compared with 19% expected from a concentric model. (b) The concentric model for egg laying on hostplant patches. Outer annuli offer a wider area for egg laying despite being of equal width to inner annuli. (c) Use of the edge (two examples, blue dots) and recesses (natural enclaves, tyre tracks; seven examples) for egg laying by *A. urticae* (Dennis, 1984c). (From Dennis, 1992a, courtesy of the Oxford University Press). (d) Distribution of silver-studded blue *Plebejus argus* eggs in relation to bare ground (white) and dense vegetation (shaded) on the Great Ormes Head, North Wales (1 m quadrat). Small yellow circles, one egg; large yellow circles, two eggs. (Thomas, 1985a, courtesy of Blackwell Science.)

reaction may be to the whole patch and the margin not be relevant, and in the second case the patch may be too large to respond to as a single entity. Second, edges and recesses can also exist *within* a resource patch as well as in conjunction with the patch boundary (e.g., *Plebejus argus* egg laying where bare ground abuts the hostplant; Thomas, 1985b).

To be most useful for ecologists, marginal distributions on plant patches should not convey generic connotations. It is apposite that terms describe the patterning, and then an explanation is sought for it, as explanations can be multiple and not mutually exclusive. There are perhaps three distinctive patterns:

1 Edge bias: individuals having a biased distribution towards the edge or margin of the patch but not limited to the margin (Figs 4.7 and 4.8a, b).

2 Edge limited: individuals are concentrated exactly at the edge or margin (Fig. 4.8c, d).

3 Recess: individuals are concentrated at the outer edge (or an interior edge of a resource patch) but biased to some smaller structural feature limited to part of the resource patch (Fig. 4.8c).

In referring to edge distinctions, it should be understood that they may relate to exterior or interior patch margins. Patches are frequently not entire and may therefore have internal margins that can be exploited. Two basic reasons can be envisaged for edge-type (marginal) distributions on resource patches:

1 A sudden change from non-resource to a resource zone or change from one resource type to a complementary resource (e.g., from nectar to hostplant patch).

2 A specific resource element associated with the resource type is found, or is of higher quality, at the margin.

It is important to realize that the three patterns given above are not necessarily mutually exclusive; two or all three of these egg patterns may be observed for some butterfly species on separate hostplant patches (e.g., egg batch distributions on nettle patches in *Aglais urticae*; Dennis, 1984b). In detail, a wide range of suggestions has been made for such distributions. For the first group, the prime reason is resource stimulus, i.e., in the case of egg laying, the female carries an egg load and the host patch stimulates egg laying (e.g., *Pieris rapae*, *Anthocharis cardamines*; Jones, 1977; Courtney and Courtney, 1982). Regarding the second group, reasons suggested include: ease of access (e.g., *A. urticae*; Dennis, 1984b), concealment or protection (from enemies and conspecifics; e.g., *Polyommatus icarus*; Dennis, 1984a; wall brown *Lasiommata megera*; Dennis, 1983c), higher quality of a resource (e.g., *A. urticae*; Dennis, 1985b) or limitation of a specific resource element (e.g., ants for *Plebejus argus*; Thomas, 1985a), and suitable microclimates for development or activity (e.g., Adonis blue *Polyommatus bellargus*; Thomas, 1983a). The selection of edge sites for this reason evokes the earlier principle (**P4.3**) that adjustments to spatial variation at fine scales compensate for deficiencies (e.g., low nutrient levels, toughness of plant types, high toxin levels, lack of heat or light) of resource at coarse scales. Much needs to be done on this issue; edge distributions have frequently been described (Table 4.2) but unequivocal reasons for their occurrence in different species, just for egg patterns alone, are deficient.

Edge and recess types of distribution are open to experiment. Occasionally experiments can be the unintentional result of management. A neat example is the use of a 150 m row of stones for egg laying by

Table 4.2 Species observed producing edge-type distributions on hostplant patches in British butterflies.

Edge type	Species	Reference
Edge bias	*Pieris rapae*, *P. napi*, *Anthocharis cardamines*	Jones, 1977; Courtney and Courtney, 1982; Courtney, 1986
Edge limited	*A. cardamines*, *Aglais urticae*	Courtney and Courtney, 1982; Dennis, 1983a, 1984b, 1985b
Recess	*Hesperia comma*, *Plebejus argus*, *Polyommatus icarus*, *P. coridon*, *P. bellargus*, *A. urticae*, *Lasiommata megera*, *Coenonympha tullia*?	Dennis, 1983c, 1984b; J. A. Thomas, 1983a; C. D. Thomas, 1985a; Thomas *et al.*, 1986; Melling, 1987

Pyrgus malvae at Ryton Wood Meadows, Warwickshire, UK. This was constructed by volunteers to reduce damage to mowing machines; subsequently, eggs were found to be biased to this row of stones away from the ditch at the site where egg laying was previously concentrated (Slater, 2007). Regarding egg distributions there is the issue of just how important edge distributions are for this stage rather than for subsequent stages. Unexpected exceptions have been found to resource margin placement and this gives clues as to priorities among influential factors (Dennis, 1985b); also, later stages are often found to move away from the marginal placement on resources. Thus, for example, *A. urticae* egg batches are frequently biased to edges and recesses on nettle patches (Dennis, 1984b); they are also found in the centre of nettle patches and larvae often move in from the edge to the centre as they grow in size (Dennis, 1985b). In this case, it is considered that the availability of new growth in old nettle patches is more important to egg laying *A. urticae* than recess and edge locations. Crucially, it is expected that the pattern results in higher survival, more rapid development and ultimately in increased reproductive success, but this remains to be tested.

Edge-type distributions have yet to be considered on other types of resource patches, but may be found in larval overwintering (e.g., *Coenonympha tullia* on *Eriophorum vaginatum* tussocks; Melling, 1987) and would be expected in adult nectaring and roosting. From the vantage of understanding habitat components, edge distributions are an argument for the importance of habitat quality. This is supported by the notion that any increase in the density of organisms encourages discovery and infestation by pathogens and parasitoids (**P4.22**) (Dwyer, 1991). Selection of sites that increase density without the advantages gained from gregarious defences, as in *A. urticae*, *Inachis io*, the marsh fritillary *Euphydryas aurinia* and *Pieris brassicae* (Porter, 1992), imply that advantages exist in the selection of edge sites.

PLACEMENT OF INDIVIDUAL BUTTERFLIES ON SINGLE RESOURCE ITEMS

Bias in the placement of any stage of a butterfly on single resource items is an indication of fine-tuning in resource use. This is the level of the single resource unit (e.g., an individual plant in a hostplant patch, a particular flower on a flowering bush); in this situation a whole bush or tree may be envisaged to be more akin to that of a host patch, as butterflies are probably distributed over single shrubs and trees much as they are over patches of perennial or annual herbs, and the 'item' would then be a branch or twig. It also includes subdivisions of these resource elements. Thus, any part of a plant may be used, i.e., stem, leaves, flowers, and finer divisions still (e.g., eggs may be placed on leaf uppersides, undersides, the leaf edge or midrib, the leaf petiole or stem). Variability in selection of the actual site or surface for an activity at any stage is most evident at interspecies (or even higher) level. The classic case is that for egg placement described earlier where eggs may or may not be placed on hostplants (see Chapter 2). Although there are distinct interspecies preferences for resource elements, classically for size (large plants in flower, *Anthocharis cardamines*; small plants, *Pieris napi*), variability also prevails at the intraspecies level. Use of leaf upper- or undersurface can differ for closely related species (e.g., *Aricia artaxerxes* and *A. agestis*; Kemp, 1998; Asher *et al.*, 2001) and among individuals of a species (e.g., purple emperor *Apatura iris*; Dell *et al.*, 2005).

Just how important a resource(s) element and resource element microsurface is for a butterfly depends on two things:

1 How crucial they are to development and survival.
2 On their abundance.

In the case of *A. cardamines*, eggs are almost always placed on plants in bud or flower, usually taller plants in bud; the butterfly searches using hostplant flowers as a cue (Wiklund, 1978). Just how critical this can be depends on how many hostplant individuals are in bud or flower. If few, there can be egg loading on single plants and many of the eggs will be eaten by the first larva to emerge (Dennis and Hardy, 2006). The ability to fine-tune resource use depends, therefore, on the availability of suitable resource elements; this is illustrated in perch microsite use by *Inachis io* – impressing that this level of resource use is common to areas other than consumables such as hostplants and nectar sources (see Box 4.1). In the absence of the most suitable microsites, other sites, apparently subordinate in quality, are used. Just what effect this has on productivity, survival or development is largely unknown, particularly as there is a dearth of information on the use of resource elements and microsites. We need data on egg sites, larval eating sites (just what is eaten at different stages), larval resting positions, and surfaces used for roosts, basking, mate location, pupation, hibernation and other activities (see Appendices 5 and 6). After centuries of observations on British butterflies, it would have been thought this information would be readily available; it is not. For instance, Gillam (1998) found white-letter hairstreak *Satyrium w-album* eggs to be laid consistently at the base of buds of wych elm *Ulmus glabra*, as close as possible to the bud the larva will use, and found only old egg cases from the previous year on girdle scars; this conflicts with data given in several texts. There is still much to learn!

MANIPULATION OF THE MICRO-LANDSCAPE: MICRO-ARCHITECTURE

As increasing amounts of fine resolution work are carried out on the mechanisms of environmental manipulation by larvae, no doubt a number of important findings will be made, as in the case of the long period of research into *Maculinea* species (Thomas *et al.*, 1989, 1998a) and white admiral *Limenitis camilla* (Fox, 1996). Here, one principle will suffice:

- **P4.23: Microstructure building by butterflies can be critical for development and survival and depends on the existence of suitable conditions for the resource elements exploited.**

An important addition to the level of microsite selection is the modification of the immediate site for activity, that is, the creation of microstructures by butterflies (**P4.23**). Examples are larval webs, leaf folding, larval feeding platforms, larval shelters, latrines, cocoons, etc., and examples are given in Appendix 6b. Silk feeding platforms may allow larvae to evade the chemical defences of plant glandular trichomes (Rathcke and Poole, 1975). Some of these features have currently been classified to facilitate identification (e.g., larval shelters; Greeney and Jones, 2003). These structures have multiple applications, including predator (parasite) escape and basking to speed up development (e.g., larval webs in *Euphydryas aurinia*; Porter, 1982; Fig. 4.9), and devices to manipulate and exploit the hostplant and overcome physical or chemical defences. Egg chains (e.g., European map *Araschnia levana* (L.); Bink, 1992) and frass chains (e.g., larvae of the nymphalid butterfly *Eunica bechina* (Hewitson), which create frass chains to combat ants on the leaves of pequi *Caryocar brasiliense* Cambess; Freitas and Oliveira, 1996) are also suggested to protect against predators, usually arthropods such as ants. Feeding architecture may also help to retain nutrient concentrations as in the creation of notches in grass blades by *Carterocephalus palaemon* (Ravenscroft, 1994b) and feeding platforms in other Hesperiidae (see Appendix 6b).

Fig. 4.9 Marsh fritillary *Euphydryas aurinia* third instar larval web. Inset is an *E. aurinia* adult. (Courtesy of Keith Porter.)

Fig. 4.10 Micro-resources and features of the white admiral *Limenitis camilla* butterfly habitat. (a) The complex structure of the larva feeding platform and abandoned latrine. (b) Larva aerial latrine; line shows actual size. (c) Honeysuckle leaf surface showing sticky secretion from glandular trichomes which seriously compromise larval feeding. (d) Early stages in the larval hibernaculum. (e) Later stages (post winter) of the larval hibernaculum with a close up of the silk threads. (a–e, courtesy of Barry Fox.)

The essential thing about such structures is that they modify or control environmental conditions for the organisms producing them. Their importance is obvious; what should not be underestimated is that the conditions for successful construction may be very precise. This is no better demonstrated among British butterflies than the construction of sculptured leaf tips, piers, larval platforms, latrines and hibernacula on the leaves of *Lonicera periclymenum* by *Limenitis camilla* (Fig. 4.10). The butterfly is able to build the pier extension, on which it rests between feeding bouts, on plants in shade but is overwhelmed by sticky exudates from the leaves when the plant is in direct sunlight; the research done on this by Barry Fox (1996, 2005) is simply fascinating to read.

FORAGING: THEORY AND PRACTICE

Just what resources or resource elements individual butterflies use in a habitat space depends on foraging behaviour. This is an enormous subject area which lends itself to mathematical modelling and experiment – much of it beyond the compass of this introduction to habitats. Suffice it to say, that just how a butterfly forages over supplementary resource spaces, and between complementary resource spaces, depends on its daily requirements under changing environmental conditions, as well as on competitors, both conspecific and interspecific, and interactions with enemies. A large number of generalizations have emerged on the study of foraging behaviour; many of these depend on the quality or quantity of the resources or their spatial arrangement. A fundamental observation of models on foraging behaviour is that they generally result in patterning in resource use despite the absence of any obvious heterogeneity in resource geography and composition. A few foraging models are mentioned briefly below. This is followed by an outline of some principles that illustrate confounding features in resource use to impress on the reader not to adopt an overly deterministic view of foraging and resource use patterns.

Above, we discussed how bias in resource use can be the outcome of differences in resource elements or larger units. However, bias in resource use may not have anything to do with any noticeable quality of the resource elements. A number of dispersal models, collectively known as foraging models, have been developed to account for the spacing of individuals. An assumption is usually made that habitat selection is an optimal foraging process. Two models briefly introduced here – now widely referred to – are the **ideal free distribution** (**IFD**) and the **ideal pre-emptive distribution** (**IPD**) models (Fretwell and Lucas, 1970; Poysa *et al.*, 1998). These models recognize that the distribution of individuals is dependent not just on resource condition and status but also on competition among conspecifics for limited resources. The IPD model assumes that individuals are capable of 'defence' and that the first animal to occupy a site can pre-empt it; later individuals are forced to occupy different sites and their spatial pattern will be one of over-dispersion. This pattern is observed in the egg laying avoidance behaviour of *Anthocharis cardamines* and was once considered (erroneously) to be entirely the case in the mate location behaviour of speckled wood butterflies *Pararge aegeria* (Davies, 1978) but territory occupancy has now been shown to depend on the size and flight capacity of territory holders (see references in Wickman, 2009). The reverse of this is conspecific attraction where individuals are attracted by conspecifics to occupied sites, rather than repulsed by them, or where defence of the resource collapses – for example, lekking mate location behaviour provides an example as in a number of Theclini around tree tops (e.g., *Favonius quercus* (Crawford, 1996), *Satyrium w-album* (Thomas and Lewington, 1991)) and in *Ochlodes sylvanus* on flowering bramble bushes (Dennis and Williams, 1987). Care is needed to distinguish the causes of aggregations that may reflect underlying resource quality, as with nectar or roost sites, rather than the attraction of conspecifics. Lack of physical competition may result in IFD, which assumes that individuals do not defend resources and will distribute themselves, and continually redistribute themselves, until every individual has the same expected fitness, that is, access to the same amount of resources. It is based on notions of optimal foraging behaviour and implicates levels of knowledge and learning. With IFD large patches may simply have more individuals because they have more resources than do small patches (thus the interest in non-linear and non-allometric relationships where this breaks down), and be the outcome of passive sampling (Golden and Crist, 1999; Summerville *et al.*, 2002). Care needs to be taken in assuming that spacing is caused by competition. Animals will often move between resources for reasons other than simply to maximize resource payoffs even given perfect information about resource quality (Hugie and Grand, 2003).

The following principles consider some of the complicating issues:

- **P4.24: Learning (resource quality) is prominent in foraging but is not equally applied to all resource types.**
- **P4.25: Discrimination within a resource outlet can mask or confound discrimination among resource outlets.**
- **P4.26: Optimal choices in resource use may not have an optimal outcome.**
- **P4.27: Resource selectivity may not always be apparent because of competing pressures but the condition of individuals provides a clue as to relative success.**
- **P4.28: For species that occupy complex vegetation structures, spatial bias in vegetation structure with height from floor to canopy may influence the pattern of resource use near the ground.**

A variety of observations underlie these few basic principles and many more principles exist, and will be found as work on foraging proceeds, that account for exploitation patterns on individual resources. For instance, foraging is fundamentally affected by physical limitations of butterfly species and stages (i.e., larvae, adults). This is well illustrated by nectar use patterns (e.g., small, short-tongued butterflies do not visit deep flowers; Corbet, 2000; Hall and Willmott, 2000; Krenn *et al.*, 2001; but see Tiple *et al.*, 2009) and fragment patch size use in butterflies (e.g., high fliers avoid small units; Summerville *et al.*, 2002) but there are exceptions as in the case of *Pieris rapae* which has been found to be unaffected by scale of fragmentation of its resources (Banks, 1998). Foraging behaviour for the same kind of resource is also affected by sex (e.g., male and female *Polyommatus bellargus* exploit different nectar zones; Rusterholz and Erhardt, 2000) and age (e.g., age classes in puddling; Sculley and Boggs, 1996). If these are not enough reasons to apply caution to interpreting spatial patterns in resource use, then consider, too, reasons for foraging, which may vary and not always be that of maximizing net energy gain rate (Hainsworth and Hamill, 1993). For instance, costs of processing information may be prominent (Janz, 2003), thus the ability to assess costs of processing, as well as the capacity for learning and the part played by memory in resource use (**P4.24**). Learning is a distinctive feature of nectar use (Goulson *et al.*, 1997a, 1997b; Goulson, 2000) but not of host use (e.g., *Melitaea cinxia* (Schops and Hanski, 2001), *Euphydryas editha* (McNeely and Singer, 2001)); underlying this observation is the impact on fitness. Assessment, learning, training and memory have been shown to relate to vision (i.e., shape, colour, polarized light and distinctive spots; Kandori and Ohsaki, 1998; Kinoshita *et al.*, 1999; Kinoshita and Arikawa, 2000; Kelber *et al.*, 2001), as well as to odour (e.g., blue admiral *Kaniska canace* L. and Asian admiral *Vanessa indica* Herbst using sap; Omura *et al.*, 2000).

All these observations should perhaps make us aware that we cannot expect patterns observed in one species in one location to be matched in another location. Confounding observations occur (**P4.25**) such that discrimination within a resource outlet can mask or confound discrimination among resource outlets, as has been shown in oviposition preference experiments in *M. cinxia* on *Plantago lanceolata* and *Veronica spicata* L. (Singer and Lee, 2000). Thus, Goulson (2000) found that pollinators, in this case a bee *Bombus lapidarius* L., visit proportionally fewer flowers in larger patches. Following near optimal behaviour, it uses simple departure rules, two successive encounters with empty inflorescences encouraging departure from a patch. This behaviour is likely to be affected by differential apparency of patches contrasting in size and by the 'history' of nectar exploitation in a patchwork of nectar flowers. For instance, it can be affected by constancy in nectar flower use (Goulson *et al.*, 1997b). Where IPD 'rules' apply it is inevitable that what may be optimal for a butterfly in terms of accessing the resource type may not be optimal, even sufficient, in terms of provision (e.g., food quality, nutrients, mates) (**P4.26**). This is usually because tradeoffs occur among life history traits, the existence of counter effects at some other stage. A good example is provided by work on sooty copper *Lycaena tityrus* (Poda) by Fischer and Fiedler (2000). This study examined the effects of increased leaf nitrogen in natural foodplants on oviposition, pre-imaginal survival, growth and adult size of this butterfly. Females did not discriminate between leaves of high and low nitrogen content and, as in many previous studies, higher growth rates and concomitantly decreased development times were associated with high nitrogen intake. However, because of high pupal (and larval) mortality (total 73.0%) as well as a reduction in adult size (*c.* 8%) the impact of higher nitrogen levels was, overall, not beneficial to the butterflies. Thus, their results were not consistent with the broad interspecific trend that insect herbivore performance is positively correlated with leaf nitrogen and the findings undermine the general applicability of the **nitrogen**

limitation hypothesis. It is not difficult to find analogies in human society where survival determines a course of action that is nevertheless far from optimal for many individuals.

A feature of foraging processes as envisaged by IFD or IPD is that competing pressures can go some way to concealing selectivity (**P4.27**). In a study of alcon blue *Maculinea alcon* Schiffermüller, Van Dyck *et al.* (2000) demonstrated that ant-related oviposition on hostplants, which had previously not been observed in highly specialized *Maculinea* butterflies, could be counterbalanced by intraspecific competition and oviposition deterrency when hostplants already carried several eggs. From this result, it is clear that careful interpretation is required of the absence of a correlation between egg load and the presence of host ant nests at the end of the flight period. To understand this conundrum, the condition and status of individuals can be most revealing. It was interesting that in hill-topping *Lasiommata megera* on Corfu Castle, Greece, that those forced to occupy sites on the lower walls were slow, tattered and in poor condition whilst those occupying the summit parapets were vigorous and in good condition (Dennis, 1987).

A further complicating issue has just emerged. Recently, from casual observation it has become evident that speckled wood *Pararge aegeria* can establish territories higher in the canopy than previously realized (5–6 m or more; Dennis *et al.*, 2009; H. Van Dyck, personal communication). Davies's (1978) original work on *P. aegeria* mate location behaviour, which involved canopy observation, described territories at low levels in woods (1–2 m high above the ground) and patrolling in the canopy by individuals unable to obtain a territory. Following the first observation of this high territory behaviour on 1 August 2008, many more observations have followed. At the moment, it is not known just how common this behaviour is and whether it is related to population density or conditions (e.g., wet ground substrates may encourage use of higher territories). The fact remains that a map which records territories below 2 m in height may well differ from one which takes in all territories and will be influenced by vegetation structures within an area (**P4.28**). To understand the foraging strategies of this butterfly the third dimension to the canopy cannot be ignored.

All these observations suggest that identifying a habitat space is one thing, but determining an optimal arrangement and combination of resources for managing a butterfly in the space is quite another, more testing, endeavour. The reader may now have some sympathy for site managers when events do not go to plan and populations decline. There should at least be appreciation of the need for obtaining a thorough understanding of a target species' ecology for the purpose of managing butterfly populations.

Chapter 5

BUTTERFLY HABITATS: SEARCHING FOR ORDER

Science consists in grouping facts so that general laws or conclusions may be drawn from them. (C. R. Darwin, from Darwin, 1888)

As each butterfly species has a unique habitat, much as it fills a unique niche, it will tend to occupy different vegetation units (biotopes), an observation that underlies the basic failing in a vegetation-type definition of habitat. This is a familiar experience when moving from one location to another where changes in vegetation types can be subtle (e.g., ringlet *Aphantopus hyperantus* on tall-herb grassland in the Cotswolds, England and on *Juncus*-rich grassland on raised beaches in western Scotland; R. L. H. Dennis personal observation) or dramatic (e.g., silver-studded blue *Plebejus argus* on calcareous heath, and wet and dry ericaceous heath in North Wales; Thomas, 1985b). Earlier, we explained that this is also a basic observation for single populations of one species occupying a single habitat (e.g., *P. argus*; Dennis, 2004b; Dennis and Sparks, 2006). One butterfly species will tend to share biotopes with other butterfly species. Even so, it may not share all the same biotopes occupied by another species, again something that is typically observed locally as well as regionally (see Appendix 7). This chapter investigates the biotopes occupied by butterflies, shared biotopes, and the common factors that link butterflies using the same or different biotopes. An inevitable objective is the search for pattern in ecology and this chapter examines what we know currently of such patterns in butterflies.

First, there is the need to consider the language of multiple species groups in ecology: communities, guilds, ensembles and assemblages (Fauth *et al.*, 1996). These terms regularly arise in discussing multiple species groups, and are most easily envisaged in the space of a Venn diagram comprising three sets describing geography (places), phylogeny (taxonomic units clustered according to common descent) and resources (Fig. 5.1). Communities are simply groups of organisms (that inevitably interact between and within trophic links) living in the same place at the same time; as long as a boundary can be placed around the study site a community can be circumscribed. A community is not

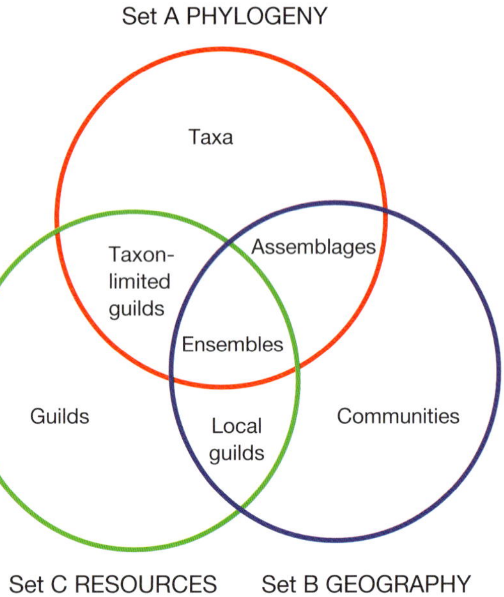

Fig. 5.1 Collective terms in community ecology arise from the intersection of three sets (A, phylogeny (taxonomic distinctions); B, geography (places); C, resources) in which populations of species are elements. No term is available for the intersection of resources and taxa; such taxonomically restricted groups are usually denoted by a compound description (e.g., grassland butterflies). (Modified from Fauth *et al.*, 1996, reproduced by permission of the University of Chicago Press.)

A Resource-Based Habitat View for Conservation, 1st edition. By Roger L. H. Dennis. Published 2010 by Blackwell Publishing.

restricted by phylogeny or resource use. Restriction by resource or phylogeny, or both, is described by alternative terms. **Guilds** are groups of species without regard to taxonomic position that exploit the same class of environmental resources in a similar way. The use of a resource by one species potentially makes it unavailable for use by others, thus guild members engage within a trophic level (Root, 1967). Phylogenetically related groups within a community are termed **assemblages** (e.g., butterflies within a vegetation unit). Species that share a common resource and occur in the same community comprise a **local guild** (e.g., grass feeders such as Hesperiidae and Satyrinae within one meadow). The intersection of all sets is occupied by an **ensemble**, which is a phylogenetically bounded group of species that use a similar set of resources within a community (e.g., pierid butterflies using the same crucifers in a meadow or fritillaries using violets within a wood). Organisms within communities function within what are called **ecosystems** – units of space–time containing living organisms interacting with each other and with their abiotic environment by the interchange of energy and materials (Odum, 1963). For species living within the same site there is potential for contact and competition for resources; and therefore an important principle:

- **P5.1: Competition among individuals can distort the apparent bounds of a species' habitat.**

Competition can be a significant factor in determining the arthropod composition of biotopes and therefore the bounds of habitats (**P5.1**). Kunte (2008) has recently demonstrated that by removing two *Anartia* butterfly species in a Costa Rican rainforest, not only was competition for nectar flowers reduced and nectar quality increased, the experiment facilitated an increase in butterfly community diversity, resulting in significantly less skewed species' abundance distributions and expanded frequency distributions for both body size and proboscis length of butterflies. Theoretically, it is clear that species using butterflies as prey (e.g., birds, lizards, parasitoids) can have a similar impact, though this should never simply be assumed but demonstrated to be the case. This leads to the concepts of a **realized habitat** (the resources a species is shown to use at a particular time) and a **potential habitat** (resources that a species could use in specific, if not ideal, conditions). These concept terms mirror and underpin realized and potential geographical ranges (see Chapter 8).

BIOTOPE DISTINCTIONS AMONG BRITISH BUTTERFLIES

'Habitats' of British butterflies have been classed in a number of ways, in terms of vegetation categories (biotopes) occupied and environmental conditions associated with biotopes, those underlying vegetation types (viz., geology, soils, human influences and thus management regimes) as well as measures relating to the vegetation itself (e.g., light environment, nutrient status, pH status, soil moisture availability). This has been done rather loosely without reference to butterfly resource use or precise designation of vegetation types. The most notable of the vegetation-type 'habitat' or biotope classifications is the field manual for the Joint Nature Conservancy Council, *Handbook for Phase 1 Habitat Survey* (Anon, 1990; www.jncc.gov.uk) – an invaluable basis for vegetation comparisons and one broadly applied to the 1981–1982 butterfly WATCH census (see below).

Biotope associations

Data on biotope associations among British butterflies are woefully inadequate. Even so, their importance is undoubted; basic regional biotope data have been used to predict butterfly distributions and relative densities in the Creuddyn Peninsula in North Wales (Cowley *et al.*, 2000). Even here, however, not all biotopes were covered that provide resources (habitat components) for butterflies (e.g., cliffs as mate location sites for the grayling *Hipparchia semele* and hostplants for the large white *Pieris brassicae*) and problems exist in this approach that were not fully considered in this study (Box 5.1).

Four direct sources of biotope data exist for butterflies at a national scale:

1 The 1981–1982 WATCH census (J. A. Thomas, 1983b). The WATCH census accumulated data, countrywide, using Pollard transects on 672 sites, covering 10 commonplace biotopes (Table 5.1). The largest number of species was associated with unimproved grassland, hedges adjoining unimproved grassland and abandoned railways, contrasting with low numbers for improved grassland and their associated hedges and manicured urban gardens.

2 Incidences of butterfly species for 24 basic, broad-scale biotopes established from the literature (Dennis, 1992a). This is updated in Appendix 7.

Box 5.1 Use of biotope data to assess the distribution and density of butterfly species on the Creuddyn Peninsula, North Wales

In an important study, biotope data have been used to predict butterfly distributions and relative densities in the Creuddyn Peninsula in North Wales (Cowley *et al.*, 2000). The peninsula comprises three Carboniferous limestone ridges including the two prominent headlands of the Great Orme and Little Orme with precipitous scars, cliffs and scree slopes, the sandy tombolo on which Llandudno is built, still persisting as dunes on the west side at Deganwy, and the intervening lower ground including boulder clay, deglacial outwash and marine deposits. Thus there is a mixture of semi-natural grassland, scrub and woodland biotopes, farmland, urban cover and linear features such as ditches, verges and lanes. Here, transect data for 16 habitats (= biotopes) were applied to predict the incidence of 30 butterflies and day-flying moths using logistic regression for 140 grid cells of 500 m covering an area of 35 km^2. The basic approach for testing the use of biotope data to predict butterfly incidence over an area or region involves:

1 Obtaining transect records throughout the flight season, from randomly placed transects, of the density of butterfly species over the full range of biotopes.

2 Mapping butterflies over the target area or region for the selected grid (500 m grid used in the Creuddyn study).

3 Mapping biotopes over the same region; this can be done remotely using geographical information systems (GIS).

4 Relating the incidence (presence/absence) of butterfly species over the grid to density estimates from the transect study using logistic regression (in which the dependent variables are scored as absent or present). A jack-knife procedure is used, in which predictions are made for each square in turn using data for the other squares.

For species on the headland, the percentage correct classification was found to vary between 58 and 98% (mean 83%, $n = 26$ species). This approach was found to be significantly more successful than just using larval hostplants (6–100%, mean 65%, $n = 30$ species).

The problems with this approach include the following:

- It is essential that biotopes assessed by transect are representative of the region for which predictions are to be made, including the all-important local climate contrasts associated with topography. Thus, it is not possible to translate the findings of transects collected over the Creuddyn Peninsula to wider regions in North Wales, for instance, as the geology and soils are unrepresentative of the North Wales region.
- An essential part of linking transect data to mapped areas is resource transference. There is an assumption that transects with butterflies include all the vital resources for the butterfly species found on that transect; in a resource-based habitat approach, it would be necessary to demonstrate that this is the case.
- A comparison with the presence/absence of larval hostplants is not really a rigorous enough test; it is both necessary to assess the quality of hostplants as well as their quantity in biotopes.
- Transect data and mapped records on butterflies may well be recording species from neighbouring biotopes (Dennis, 2001). There are problems with spatial autocorrelation, effectively accounting for neighbourhood effects and vagrant individuals (see Chapter 7). Part of this can be overcome by determining the status of the butterflies (sex, numbers, behaviour indicative of breeding) observed.

This approach, used with care, particularly if combined with resource data such as larval hostplants, has the potential for 'mapping' (assessing distribution and abundance of) butterflies over the British Isles and larger regions. Essentially, it can provide information on losses from habitat degradation and fragmentation (see examples of this for the small copper *Lycaena phlaeas* and the common blue *Polyommatus icarus* for the Creuddyn; León-Cortes *et al.*, 1999, 2000), including the reasons for extinctions (e.g., Duke of Burgundy fritillary *Hamearis lucina*, small blue *Cupido minimus*; León-Cortes *et al.*, 2003a, 2003b). The main proviso of this approach is that transects are fully assessed for butterfly resources and that the variability in resources is determined for biotopes on transects over the range of distinct substrates, topography and thus local climate.

3 The Millennium Atlas databank (Butterfly Conservation UK, Butterflies for the New Millennium (BNM)). This comprises coding for 22 broad-based biotopes of squares (100 m^2) in which butterflies have been recorded (unreported and not yet analysed).

4 The Butterfly Monitoring Scheme (BMS) databank at Monks Wood, UK. Vegetation types at two scales for sections and transects of >100 transect routes throughout the UK are being determined (J. N. Greatorex-Davies, personal communication). The number of sites, inclusive of biotope and resource data, covered by the BMS survey techniques is in the process of being greatly increased (T. M. Brereton, personal communication).

Table 5.1 Biotope associations of butterflies for transects on 672 sites in the 1981–1982 WATCH census. (Data from J. A. Thomas, 1983b, courtesy of the Institute of Biology.)

Biotope	Number of sites	Number of counts	Mean number of butterflies*
Wood (deciduous)	72	454	12
Grassland (unimproved)	125	979	28
Grassland (improved)	90	383	8
Hedge (unimproved grassland)	57	344	25
Hedge (arable; improved pasture)	35	224	7
Hedge (road verge)	58	449	16
Abandoned railway	26	179	26
Wasteland	36	256	14
Garden (neat urban)	92	660	5
Garden (mature)	88	676	11

* Mean number of butterflies of all species recorded per 30-minute transect in the 10 different biotopes. Butterflies were divided into: whites, peacock, red admiral, small tortoiseshell, brimstone, browns, small copper, blues and skippers.

Further indirect biotope data sources exist. For example indirect data sources are being compiled for plants that include most hostplants of British butterflies – e.g., data for seven biotopes of butterfly hostplants (*Comparative Plant Ecology* (CPE), Grime *et al.*, 1988; see Dennis *et al.*, 2004) and the National Vegetation Classification (NVC) database which records butterfly hostplants as elements in vegetation categories (Rodwell, 1991–2000).

Problems exist with all of these biotope sources. Each is limited as to the number and variety of biotope (vegetation) classes used. The WATCH census provided data for only five species and, otherwise, only for broad classes (subfamilies) of butterflies. The Millennium Atlas data are seriously flawed as records have been coded up for biotopes in squares, of which there may be a number, and not for the biotope(s) in which the butterfly was actually observed.

From this work, however, a number of generalizations (principles) linking butterflies to biotopes emerge.

Principles of biotope properties

- **P5.2: Each biotope (vegetation unit) is compositionally and structurally unique. Actual biotopes are never identical, but similar, to other biotopes.**

- **P5.3: Biotopes are recognizable, hierarchical vegetation/substrate classes that can be isolated at different levels in the hierarchy.**
- **P5.4: Biotopes vary in composition, physiognomy and connectivity, much as do resource patches for butterflies.**
- **P5.5: Butterfly resources are contained within vegetation units or biotopes and the finer the division of biotopes the more likely it is that resources are scattered among different biotopes.**
- **P5.6: The broader the biotope class, the greater its heterogeneity in resources.**

Two things make vegetation difficult to classify and, therefore, the interpretation of biotope use by butterflies. Vegetation units are unique and variation is gradual, even imperceptible to inexperienced human observers (**P5.2**), except where 'boundaries' have been imposed by land use and physical barriers. Thus, classification is necessarily based on resemblances (similarity, differences) and relationships tend to fall into a hierarchy as with the NVC database (**P5.3**). In this approach, based on similarities for abundances of different plant species, lower order vegetation categories tend to cluster into higher order groups. Sight should not be lost of the fact that whereas a number of biotopes may be regarded as semi-natural, if not natural, and thereby

plesiotypic (ancestral), many are unnatural, and thus **apotypic** (derived) biotopes (Wiklund, 1977). Vegetation has three dimensions and is extremely intricate in vertical stratification and spatial structure; therefore, vegetation units vary in physiognomy as well as in composition and similar units vary in degree of isolation (**P5.4**). However, not all vegetation classification schemes encompass an obvious vertical dimension of stratification, especially within vegetation types that do not encompass shrub and tree components. The location of butterfly resources within vegetation units depends on the degree of stratification of vegetation; the finer the stratification – the more homogenous the unit of vegetation is in composition and structure – the more dispersed among biotopes will be a butterfly's resources (**P5.5**). This occurs if only because a finite space is being subdivided into smaller parcels. Obversely, the broader the biotope class or vegetation unit, the greater the heterogeneity in vegetation and substrates and the more likely it is that it includes a greater variety of a butterfly's resources and more species of butterfly as a result (**P5.6**).

A wide range of different techniques has developed to describe this complexity. Techniques range from the use of relatively simple coding schemes for appraising plant physiognomy and assessing floristic content to painstaking quantitative assessments of the frequency, density and cover of individual species (Box 5.2). Several important sources of data exist for vegetation and substrate cover over Britain. The most important are the NVC database (Rodwell, 1991–2000), land use survey maps (Coleman *et al.*, 1974), remote land use data (Fuller, 1993; Cherrill, 1994; Cherrill *et al.*, 1994, 1995; Fuller *et al.*, 1994a, 1994b; Bunce *et al.*, 1996), the Scottish land use survey (MLURI, 1993; refer to www.mluri.sari.ac.uk) and CORINE land class data (Anon, 1991, 1992; Fuller and Brown, 1994, 1996; Brown *et al.*, 1996; see also the Centre for Ecology and Hydrology (CEH) website, www.ceh.ac.uk). These are briefly described in Box 5.2.

Box 5.2 British vegetation and land use: terms, sources and techniques in plant geography

Two different approaches are used to describe and classify plants and vegetation:

1 Physiognomic: the emphasis is on distinguishing life forms (e.g., trees, shrubs, herbs, bryoids, epiphytes, lianas), seasonality (deciduous, evergreen), size, leaf shape and texture, and vegetation cover (barren/sparse, discontinuous, tufts/groups, continuous) (after Raunkiaer, 1934; Dansereau, 1957; Peterken, 1967).

2 Floristic: detailed taxonomic (species) accounts of plants comprising vegetation.

Both can be determined by rapid (subjective) assessment techniques describing frequency, abundance,

Table B5.2 The Braun-Blanquet and Domin cover scales typically used for rapidly describing the presence of individual plant species or elements.

Scale value	Braun-Blanquet scale	Domin scale
+	<1% cover	A single individual, no measurable cover
1	1–5% cover	1–2 individuals; no measurable cover; individuals with normal vigour
2	6–25% cover	Several individuals but <1% cover
3	26–50% cover	1–4% cover
4	51–75% cover	5–10% cover
5	76–100% cover	11–25% cover
6		26–33% cover
7		34–50% cover
8		51–75% cover
9		76–90% cover
10		91–100% cover

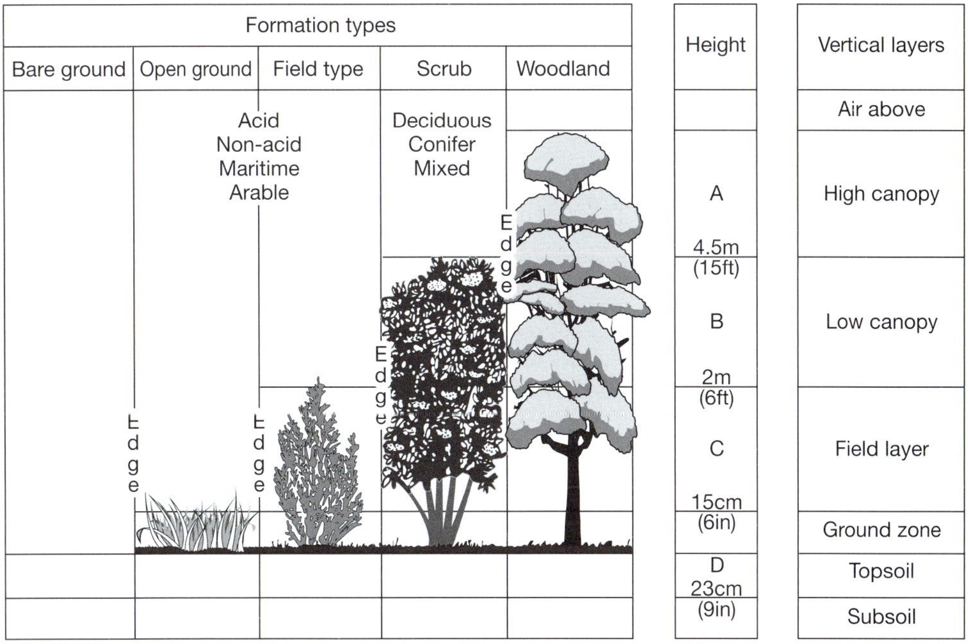

Fig. B5.2 Terrestrial biotope classes illustrating basic strata within woodlands and edge situations beyond woodland fringe. (Modified from Elton and Miller, 1954; redrawn and further modified from Vincent, 1990, Fig. 4.1, p. 77, courtesy of Routledge and Taylor and Francis.)

density and/or cover (Braun-Blanquet and Domin scales; Table B5.2) or by more precise (quantitative) surveys involving either complete mapping or sampling (see Kershaw, 1973; Kellman, 1975; Pears, 1977; Gilbertson *et al.*, 1985). Sampling using quadrats is fraught with problems related to quadrat size.

Early descriptions of vegetation recognized that plants frequently recur in association with other plants (e.g., Tansley, 1916). From this, a language of plant 'sociology' or **phytosociology** has developed to recognize vegetation associations (Tansley, 1935). Vegetation can be described by its vertical stratification, with three or four strata typical of woodlands (canopy trees, understorey of saplings and shrubs, field layer and ground layer; Fig. B5.2). Some ecologists treat each stratum in a complex vegetation as a minor community or **synusia** (a social aggregate of a few closely related life forms that have similar ecology). **Dominants** are plants that strongly influence the environment of other species present and are typically, but not always, the tallest plants in a community; a number do so by releasing chemicals, a process referred as **allelopathy**. An **association** is a plant community in which two or more species form the **codominants** and a **consociation** is a community in which one species is dominant. Where vegetation is advancing, plants of lower strata advance ahead of those of higher strata and edge biotopes (called **ecotones**) occur (Fig. B5.2).

The British vegetation has been described twice over, first by Tansley (1939) and more recently as the **National Vegetation Classification** (NVC) (Rodwell, 1991–2000). The NVC began its work in August 1975 and completed the task in 1989. The objective was to provide a comprehensive coverage of all natural, semi-natural and major artificial habitats. The method was based on quadrat sampling units taken from natural tracts of vegetation, whose floristics are relatively homogeneous in species and structure, called **stands** (vegetation homogeneous to the eye; groups of several members of a single species occur together). The communities were based on multivariate data processing of some 35 000 relevés (samples) from more than 80% of the 10 × 10 km grid squares on the British mainland and islands. A computer program (MATCH) to aid the assignment of vegetation data to the communities and subcommunities of the NVC is run by the University of Lancaster, Unit of Vegetation Science.

Vegetation categories in the NVC cover but a fraction of the land surface. A wider documentation of land cover is provided by successive land use surveys. This not only embraces natural and semi-natural vegetation but other land uses such as farmland, industry, settlement and transport, water bodies and bare ground. The first was completed before World War II under Professor L. Dudley Stamp; this achieved coverage of the entire British Isles on some 20 000 six-inch field maps. Unfortunately, the plates were destroyed during the war. The second national survey, based on volunteer recorders, dates to the 1960s and 1970s (Coleman *et al.*, 1974). This comprises 843 maps at a scale of 1:25 000 (200 km^2). As with soils, ground survey has proved to be too expensive and slow. More recent land surveys at the scale of the British Isles depend on remote survey using so-called geographical information systems (GIS, such as Landsat and aerial photography) (Fuller, 1993; Cherrill, 1994; Cherrill *et al.*, 1994, 1995). Extensive information on a variety of land uses is held by the Countryside Information System (CIS; website, cis-web.org.uk; see also www.ceh.ac.uk/data). The fundamental building block of these data services is the resource for 32 land classes recorded for Great Britain at a resolution of 25 × 25 m squares (Bunce *et al.*, 1996). Two land cover maps of Great Britain (LCMGB) have been produced for 1990 and 2000 by the Centre for Ecology and Hydrology (CEH); the map for 2000 also covers Northern Ireland (Land Cover Map 2000 (LCM2000); http://www.ceh.ac.uk/sections/seo/lcm2000_home.html). These provide data at two resolutions: 25 land classes at 25 m square resolution and 17 'key' cover types at 1 km resolution. The 25 classes consist of sea and inland water, beaches and bare ground, developed and arable land, and 18 types of semi-natural vegetation (www.ceh.ac.uk/data); data are available as **vector** (i.e., management units) and **raster** (grid) datasets. The land classes were determined by the Landsat satellite Thematic Mapping Sensor (Fuller *et al.*, 1994a, 1994b). Resolution is accurate to 2 pixels (0.125 ha); smaller features with strong spectral signatures can be discriminated, such as roads, farms and tracks (Fuller *et al.*, 1994a). Verification by ground survey indicates at least >70% accuracy and perhaps as much as 80–85% (Barr *et al.*, 1993; Fuller *et al.*, 1994a; Wyatt *et al.*, 1994). A more detailed land cover survey has been made of Scotland by the Macaulay Land Use Research Institute (MLURI) for the Scottish office. This identifies 126 separate land cover classes from 1:24 000 panchromatic aerial photography with a maximum resolution of 10 m (MLURI, 1993). Separate environmental data, obtained by a range of survey techniques, are available for administrative areas, designated areas, agriculture, climate, geology and topography, soils, ancient woodlands, hedges, verges and individual species at the same data facility.

The British Isles has tended to adopt its own land cover classification rather than apply CORINE land cover classes as on the continent under the direction of the European Environment Agency (EEA) (Anon, 1991, 1992). The objective of CORINE (Coordinating Information on the European Environment) is to map the whole of Europe at a scale of 1:100 000 for 44 land cover classes; the database now covers some 3.5 million km^2 for 19 countries in the European Union and neighbouring regions. In Britain the CEH is currently developing techniques for converting the existing land cover map of Great Britain to the CORINE land cover vector format using visual interpretation and manual digitizing of satellite images (Fuller and Brown, 1994, 1996; Brown *et al.*, 1996; see www.ceh.ac.uk/data). The fundamental problem with CORINE is that the minimal mapping unit is 25 ha. There is also a Mediterranean bias in land cover nomenclature unsuitable for Britain and only 35 of the 44 CORINE land classes occur in Britain. It is no substitute for the NVC as there are 'gaps' in the natural vegetation classes and a lack of clear rules to define vegetation classes (Rodwell *et al.*, 1991–2000; Cruickshank and Tomlinson, 1996). Finally, there is the European Nature Information System (EUNIS) habitat classification, a pan-European system, which was developed between 1996 and 2001 by the EEA in collaboration with experts from throughout Europe. It covers all types of natural and artificial habitats, both aquatic and terrestrial (see www.jncc.gov.uk).

Principles linking butterflies to biotopes

Despite biotopes not being habitats, they have immense relevance for butterflies; some important principles link the two:

- **P5.7: Biotopes may only comprise a part habitat for a butterfly species.**
- **P5.8: Butterflies resource seek, and have resources, in specific biotopes that lack their larval hostplants.**

- **P5.9: Biotopes function much as metapopulation patches in the sense that the incidence of a butterfly in one biotope patch containing its resources will depend on patch size, isolation and quality of the resources.**
- **P5.10: Evenness in the occurrence of a butterfly species across biotopes depends on its mobility in addition to the degree of resource ubiquity.**
- **P5.11: Just how large an assemblage (ensemble) of butterflies there is in a biotope,**

and just how related butterflies appear to be to one another ecologically, depends on the position in the hierarchy that a vegetation unit is defined.
- **P5.12: The species' richness of linear biotopes reflects on their physical structures as well as on their substrate and vegetation components.**
- **P5.13: The importance of linear biotopes for butterflies largely reflects the status of contiguous biotopes.**
- **P5.14: In providing habitats, biotopes contrast in the length of time butterfly resources and thus the habitats are available.**

A fundamental point to make from the above list is that specific biotopes may only comprise a fraction of the resources required by a butterfly species (**P5.7**), an observation that depends much on the level in a vegetation hierarchy to which a specific biotope is allocated; the finer the designation of vegetation units, the more likely is this to be the case. Failure of single biotopes to provide all resources for a butterfly species has been clearly demonstrated for *Plebejus argus* at one location, where it uses both grassland and adjoining scrub (Dennis, 2004b); the generalization extends to distinct substrates (e.g., grassland and crags for *Hipparchia semele*). It is necessary to distinguish breeding biotopes from incidences of adult butterflies in biotopes. Butterflies will be observed to be resource seeking (engaging search flight) in biotopes apparently lacking in larval hostplants (**P5.8**). A studied example is that of pierids (e.g., *Pieris brassicae*, small white *P. rapae*) which use scrub and woodland for mate location and roosting (Dennis and Hardy, 2007). A great deal remains to be discovered by observing butterfly behaviour in different biotopes (Dennis, 2004a).

The absence of a butterfly species may reflect on the lack of a key resource (e.g., hostplant) or of several resources. Even so, it is well to bear in mind that vegetation units (biotopes) function much as metapopulation patches in the sense that the incidence of a butterfly in one biotope patch containing its resources will depend on patch size, isolation and the quality of the resources (**P5.9**) (see Chapter 6). This explains the ease and success with which metapopulation studies have been able to treat vegetation units as habitats. It deserves emphasis that a biotope may have the correct consumer resources but still lack a target species because of insufficient warmth (range margin effects) (Shreeve *et al.*, 1996a). Evenness in cover by a species over different biotopes is partly explained by resource availability but also substantially by migratory ability (**P5.10**). Naturally, a number of mobile species will be found regularly in a variety of vegetation units, even small vegetation patches, if only as vagrants in search for resources. By extension, it also follows that some species, owing to poor mobility, may simply be absent from vegetation units because they have been unable to migrate to them. This is illustrated by range extension in speckled wood *Pararge aegeria* which has been some 42% slower in extending its range in areas where woodland cover is <25% of the landscape (Hill *et al.*, 2001).

With this in mind, some generalizations can be suggested for the distribution of species within biotopes. The absence of certain breeding species from basic biotopes usually reflects on the absence of key resources, usually hostplants or nectar, for those species. For example, linear biotopes, such as grass verges and hedges, tend to be suitable for butterflies dependent on hostplants with ruderal (e.g., *Pieris rapae*) and competitive (e.g., white-letter hairstreak *Satyrium w-album*) strategies but are not so suitable for butterflies dependent on stress-tolerant hostplants (e.g., chalkhill blue *Polyommatus coridon*) (see Box 5.4 and the section below on classification). The occupation of the same biotopes by a number of butterflies occasionally follows from their sharing hostplants as well as other resources such as nectar and features for mate location or roosts. Even some widely different butterfly taxa share hostplants (e.g., Heteropterinae, Hesperiinae and Satyrinae use grasses). Biotope bias, in the absence of size effects, for species' richness and incidence is most likely an indication of resource variation and range in different biotopes; in the WATCH survey unimproved pasture was the dominant biotope (Thomas, 1983b). In this, scale and context matter. Butterfly species' incidence and richness in biotopes depends on the level in the hierarchy of plant communities that the biotope encompasses, that is, its composition in terms of plant communities (**P5.11**). Broad divisions of biotopes will produce higher measures of butterfly diversity and vice versa. Unique associations are exposed; for instance several species are found almost exclusively on **calcicolous grasslands** (growing on soil rich in calcium salts) owing to restrictions in their larval hostplants (e.g., Lulworth skipper *Thymelicus acteon*, silver-spotted skipper *Hesperia comma*, small blue *Cupido minimus*, *P. coridon*, Adonis blue *P. bellargus*; Dennis, 1977; Warren, 1993). Much finer classifications of vegetation, for example using the NVC, should improve uniformity in the data with increased homogeneity of the vegetation and concomitantly lower diversity of butterfly species.

Species' richness of linear structures (e.g., hedges, banks) can depend on structural features as much as on consumer resources (**P5.12**). Local climate is a key here; cuttings and embankments, hedges and banks, benefit from aspect in relation to winds and sun angle providing shelter and higher insolation levels (Dover, 1994; Sparks and Parish, 1995; Dover *et al.*, 1997). The context of a target biotope affects biotope admixture, especially for linear biotopes (**P5.13**). For example, owing to neighbourhood effects, hedges separating unimproved permanent pasture are likely to have greater species' richness than those dividing improved pasture or arable (Thomas, 1983b; Saarinen *et al.*, 1998). Such edge-type biotopes will tend to receive greater interference from neighbouring biotopes than zonal biotopes. A comparison of linear and zonal biotopes, in particular, introduces a further principle of biotopes: the lifespan of resources for any species differs among biotopes that a butterfly occupies (**P5.14**); that is, biotopes have unique dynamics and offer appropriate conditions for different lengths of time for butterflies through their resources, especially their hostplants. Linear and zonal habitats may well contrast for biotope dynamics owing to their neighbourhood associations.

Principles relating to observations made in biotopes

Another influence that needs to be considered is the impact of recording and observation on distribution data and diversity assessments.

- **P5.15: Butterfly species in distinct biotopes may differ in their apparency to human observers.**
- **P5.16: Fidelity (stay times) of butterflies within biotopes depends on the resources for them and the timing in exploitation.**
- **P5.17: Recorded species' richness of a biotope continues to increase towards the regional total with time and does not attain an asymptote.**

It is well to understand that records of butterfly species for vegetation units or biotopes can be affected by bias related to their apparency (size, colour and behaviour) to observers (**P5.15**) (Dennis *et al.*, 2006a). The same species, different sexes even, are not equally apparent to the same observer in different biotopes (e.g., *Plebejus argus* in calcareous heath and scrub; Dennis and Sparks, 2006). Butterfly species may also differ in their fidelity to biotopes in relation to resource use, timing in exploitation and survey timing (**P5.16**) (Harker and Shreeve, 2008). Stay times (potential observation period) can depend on time of day (e.g., peacock *Inachis io* territorial sites are typically set up after midday), daytime conditions (e.g., effect of temperature and cloud on *Ochlodes sylvanus* (Dennis and Williams, 1987) and *Ypthima asterope* (Klug) (John *et al.*, 2010)) and the type of resources (e.g., small skipper *Thymelicus sylvestris* does not feed for long on garden buddleia; R. L. H. Dennis, personal observation). Eventually, if recording continues in a specific biotope over long periods of time, the number of species recorded will gradually rise to approach the regional total for species found in all surrounding biotopes, owing to the constant flow of vagrants over the countryside (**P5.17**) (Dennis, 2001). In that time, however, even species breeding in the biotope will change owing to environmental changes (e.g., vegetation succession, climate change); during the 1980s and 1990s on Alderley Edge woodlands, in Cheshire, UK, wall *Lasiommata megera* was replaced by speckled wood *Pararge aegeria*, gatekeeper *Pyronia tithonus* and *T. sylvestris* (Dennis, 2000).

Biotopes, environmental conditions and niche parameters

A more penetrating way of assessing biotopes as suitable templates for butterfly habitats is in terms of environmental variables, particularly in connection with their larval hostplants. Principles that apply are:

- **P5.18: Underlying suitability of vegetation units for butterflies are key niche variables such as light, heat, moisture, soil pH and fertility.**
- **P5.19: Vegetation units (plants) are strongly associated with distinct environmental conditions and this is reflected in plant adaptations, which in turn influence butterfly biology and life history.**
- **P5.20: Different butterfly stages and activities (e.g., mate location, roosting) require different environment conditions in the same (or different) vegetation units or biotopes.**

- **P5.21: The climate of a region (viz., oceanic position of Britain) influences the tolerance of butterflies to the range of environmental parameters.**
- **P5.22: Dry, open biotope-dwelling species are adapted to higher temperatures than shade biotope-dwelling species with greater life history opportunities in terms of higher fecundity and longevity.**

Key environmental variables describing the condition of biotopes for butterflies are those used for assessing butterfly niches: such as light and heat, soil moisture and pH, and fertility (nitrogen) levels (**P5.18**). The nature of the environment is reflected in the plants growing in biotopes, each of which exists within a range of light, heat, moisture, pH and other conditions. Among these variables, only the light environment of a restricted number of biotopes, woodland glades and clearings, has been specifically measured for assemblages of British butterflies (Warren, 1985c; Greatorex-Davies *et al.*, 1993; Sparks *et al.*, 1996). For substrate conditions we are largely forced to rely on surrogate measures for the hostplants (**Ellenberg values**; Table 5.2; Fig. 5.2; see Appendix 2c) (Ellenberg, 1988; Hill *et al.*, 1999), though direct measurements on the plants are accumulating (Dennis *et al.*, 2004; J. G. Hodgson, personal communication). Four of the biotope conditions (light, soil moisture, pH, nitrogen) are described by Ellenberg values (Dennis *et al.*, 2004). The importance of environmental attributes is underlined in the clear distinctions that have been recorded for them; for instance, most butterflies require high levels of sunshine under the British climate, yet a number of species depend on shade (e.g., *Pararge aegeria*, wood white *Leptidea sinapis*, white admiral *Limenitis camilla*, orange-tip *Anthocharis cardamines*, holly blue *Celastrina argiolus*; Greatorex-Davies *et al.*, 1993; Fox, 2005).

The link between vegetation type and environmental conditions provides a key to understanding contrasts in butterfly habitats. Vegetation units are strongly associated with distinct environmental conditions and are reflected in plant adaptations, which in turn influence butterfly biology and life history (**P5.19**). Loram (2004) has explored the connection between plants (other than hostplants) and butterfly incidence at sites and found that some provide markers for suitable habitats. It should not be forgotten, though, that different stages and activities require different conditions

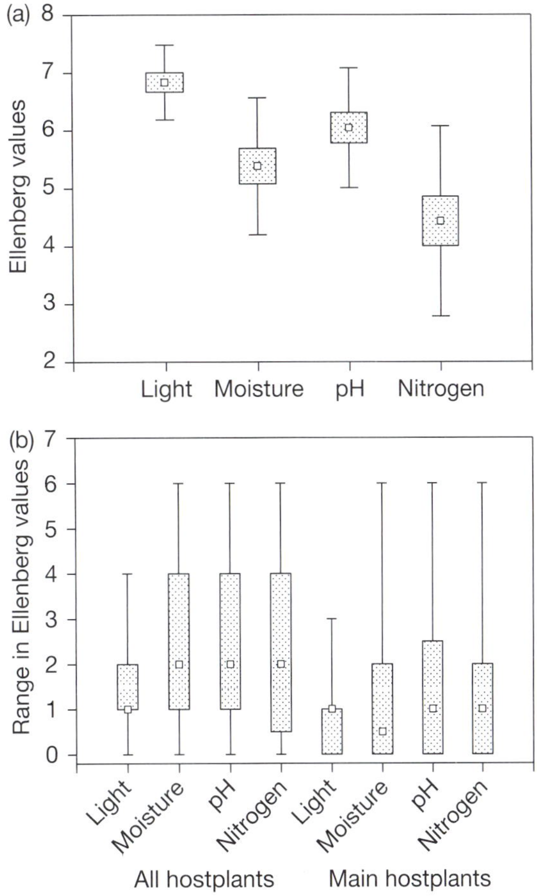

Fig. 5.2 Niche comparisons for British butterflies. (a) Comparison of average Ellenberg values for all larval hostplants (points, boxes and whiskers illustrate means, 2 standard errors and 1 standard deviation, respectively); there is no significant distinction between mean values for all and main larval host plants. (b) Comparison of the ranges in Ellenberg values for all versus main (core) larval hostplants (points, boxes and whiskers illustrate medians, quartiles and minimum/maximum values, respectively). The ranges for all hostplants significantly differ from those for main (core) hostplants: Wilcoxon matched pairs tests: light $Z = 4.46$, soil moisture $Z = 4.46$, soil reaction pH $Z = 4.11$, soil nitrogen $Z = 4.29$, all $P < 0.0001$. (See Table 5.2 for an explanation of Ellenberg values.)

(**P5.20**); this, in itself, accounts for their use of different vegetation types and occupation of contrasting biotopes. To date we lack quantitative measures of butterfly niches taken over many localities. Even so, Ellenberg values of hostplants provide cues linking butterfly

Table 5.2 Extreme values and range in Ellenberg values* for the main larval hostplants of British butterflies.†

Variable	Upper extreme (bright, wet, basic and/or fertile)	Lower extreme (shady, dry, acid and/or infertile)	Wide range of values
Light	[7] *P. brassicae* *P. rapae* *P. napi* *C. rubi* *C. minimus* *P. argus* *A. agestis* *P. coridon* *P. bellargus* *M. arion* *V. cardui* *A. urticae* *B. euphrosyne* *E. aurinia* *P. aegeria* *M. galathea* *H. semele* *P. tithonus* *M. jurtina* *C. pamphilus* *C. tullia*	[4] *S. w-album* *C. argiolus* *N. polychloros* *P. c-album* *P. aegeria* [5] *P. napi* *A. cardamines* *H. lucina* *L. camilla* *M. athalia*	[4] *P. aegeria* [3] *P. napi* *P. c-album*
Moisture	[10] *P. napi* L. dispar [9] *P. machaon* *P. c-album* *B. selene* *B. euphrosyne* *A. aglaja* [8] *C. palaemon* *O. sylvanus* *L. sinapis* *G. rhamni* *A. cardamines* *P. argus* *A. iris* *E. aethiops* *C. tullia*	[2] *H. semele* [3] *T. acteon* *P. argus* *P. coridon* *P. bellargus* *E. aurinia* *L. megera*	[6] *P. napi* [5] *P. argus* *B. euphrosyne* *A. aglaja* [4] *L. sinapis* *P. c-album* *B. selene* *E. aurinia* *H. semele* [3] *C. palaemon* *O. sylvanus* *G. rhamni* *L. megera*
Soil reaction (pH)	[8] *T. acteon* *S. w-album* *P. argus* *P. coridon* *P. bellargus*	[2] *C. rubi* *P. argus* *M. athalia* *L. megera* *H. semele* *C. tullia*	[6] *P. argus* *L. megera*

Table 5.2 (*continued*)

Variable	Upper extreme (bright, wet, basic and/or fertile)	Lower extreme (shady, dry, acid and/or infertile)	Wide range of values
	V. cardui *N. polychloros* *P. c-album* *A. adippe* *A. aglaja* *E. aurinia* *L. megera* *E. aethiops*	[3] *C. palaemon* *O. sylvanus* *F. quercus* *B. selene* *B. euphrosyne* *A. aglaja* *E. aethiops* *E. epiphron*	[5] *C. rubi* *A. aglaja* *E. aethiops* *H. semele* [4] *O. sylvanus* *M. athalia* [3] *C. palaemon* *L. sinapis* *A. adippe* *E. aurinia* *P. tithonus* *M. jurtina* *C. pamphilus*
Soil fertility (nitrogen)	[8] *P. brassicae* *P. rapae* *P. napi* *A. cardamines* *V. atalanta* *A. urticae* *P. c-album* *I. io* [7] *S. w-album* *A. iris* *N. polychloros* *P. aegeria* *H. semele* *P. tithonus* *A. hyperantus*	[1] *P. argus* *H. semele* *C. tullia* [2] *C. palaemon* *H. comma* *O. sylvanus* *E. tages* *L. sinapis* *C. rubi* *C. minimus* *A. agestis* *A. artaxerxes* *P. icarus* *P. coridon* *P. bellargus* *M. arion* *B. selene* *B. euphrosyne* *A. adippe* *A. aglaja* *E. aurinia* *E. aethiops* *E. epiphron* *M. jurtina* *C. pamphilus*	[6] *H. semele* [4] *O. sylvanus* *P. napi* *A. cardamines* *P. tithonus* *M. jurtina* *C. pamphilus* [3] *C. palaemon* *L. sinapis* *A. agestis* *P. c-album* *L. megera* *A. hyperantus*

* Ellenberg (1988; Ellenberg *et al.*, 1992) defined a set of seven indicator values (**Zeigerwerte**; four outlined here: light, moisture, soil pH and soil nitrogen or fertility) for the vascular plants of Central Europe, the basis of which is the realized ecological niche. These are on an arbitrary scale (plants are given values from 1 to 12). M. O. Hill *et al.* (1999) have calibrated values for 1791 plant taxa (1479 native species) in Britain. There are plans to develop ones reflecting climatic attributes (temperature, rainfall). Upper and lower extremes record the maximum and minimum values for the main hostplants of different species. Numbers in square brackets indicate the Ellenberg values and ranges (see Appendix 2c).

† For full species names, see Appendix 1a.

species and environmental conditions. Ellenberg values for hostplant species reflect known biotope conditions where butterfly adults are typically observed (e.g., shade for *P. aegeria*; basic mires for large copper *Lycaena dispar*; dry sunny basic conditions for *Polyommatus bellargus* and *P. coridon*). They give interesting perspectives on substrates used by early stadia (see Appendix 2c). As the modal values indicate, most of the hostplants of British butterflies occupy bright, moist, weakly acid, infertile or moderately fertile environments (see Fig. 5.2). Bright conditions are generally important for British butterflies as insolation and temperatures tend to be depressed under oceanic climatic conditions (Dennis, 1977, 1993a); the regional climate determines the tolerance of species to conditions (**P5.21**). A feature of butterflies in Britain is that since their arrival in the Late Glacial (16k years BP), they have become increasingly specialist (extreme) in environmental conditions compared with conspecific populations on the European continent (Dennis, 1977). This process is closely linked to the cooler, wetter British climate. As populations in different regions have become extinct, so too has there been a decline in biotope variation associations for British butterflies.

The light environment associations for butterfly adults corresponds broadly with that for butterfly hostplants; very few British species are not adversely affected by shade in the adult stage (e.g., *P. aegeria*, *L. sinapis*, *L. camilla*, *A. cardamines*, *C. argiolus*), all these exceptions bar *L. sinapis* are associated with low hostplant Ellenberg light values (Greatorex-Davies *et al.*, 1993). A recent study on the luminance range over which the eye pupil mechanism operates in 11 non-British butterfly species strongly suggests that butterflies are adapted to light environments. Three groups of butterflies were found differing in pupil sensitivities that match specific types of activity pattern in bright sunshine, shaded places and marginal times of the day (Jonson *et al.*, 1998).

Moist soil conditions are requisite for hostplant growth and larval development whereas pH levels are linked to nutrient availability, and high fertility increases competition from non-hostplants. The distribution of soil nitrogen for hostplants is bimodal at scores of 2 (infertile) and 5 and 6 (weakly fertile). Some species are clearly shown to occupy contradictory conditions (e.g., Scotch argus *Erebia aethiops* on wet acid and drier basic substrates; *Hipparchia semele* on dry acid and dry basic substrates); that is, they exploit different main hostplants with contrasting substrate affiliations. Also, as would be expected, species with more hostplants occupy a wider range of environmental conditions. A valuable corollary of this is that access to subsidiary hostplants permits a widening of exposure to the range of substrates that can be exploited when environmental conditions become suitable (see Chapter 8) (Kemp *et al.*, 2008).

Data, as yet, are deficient for the thermal environment; ambient (screen) temperatures at species' boundaries and measured in relation to butterfly behaviour and resource use fail to determine actual thermal conditions required by species or to classify species on the basis of microclimates suitable for persistence (Shreeve *et al.*, 2001). Karlsson and Wiklund (2005) have found that fecundity and longevity in four butterfly species that occur in Britain are influenced by ambient temperatures. Both exhibit bell-shaped curves in response to temperature regimes. Open-dwelling species (e.g., *H. semele*, *Coenonympha pamphilus*) have higher fecundity and longevity at higher temperatures than shade-dwelling species (e.g., *Aphantopus hyperantus*, *P. aegeria*) (**P5.22**). The difference in life history traits suggests either that dry and relatively warm, open biotopes provide life history opportunities in terms of higher fecundity and longevity, which are closed to butterflies adapted to cooler temperatures, or that life in dry, open biotopes actively selects for higher fecundity and survival as a result of increased offspring mortality. Of course, differences in longevity can also relate to predation rates which may be expected to be higher in shadier biotopes (Dennis *et al.*, 1986).

Principles relating to biotopes over time

Biotopes rarely remain static for long and, if they do, it is then usually because of human management or repeated action of geomorphic agents (e.g., cliff falls, scree movement, river flooding). Any changes in vegetation, of life forms and plant species' composition, is of importance to butterfly biology, as there will be an impact on butterfly resources, thus on butterfly population dynamics. Vegetation changes are of two basic types: **fluctuations** and **trends** (Burrows, 1990; Glenn-Lewin and Van der Maarel, 1992; Huisman *et al.*, 1993); vegetation fluctuations are regarded as reversible with no stable end-product whereas trends are directional (Miles, 1979). The distinction is often much a matter of scale, in time and space, as a fractional part of a fluctuation may appear as a trend and a trend, if extended, may be part of a fluctuation. A case in point is that of Holocene vegetation changes following deglaciation (ice melt) at the end of Devensian glaciation (18k years BP) to the climatic climax of 8000 years BP (see Chapter 8). A prominent successional trend on this timescale is but part of an ongoing cycle when viewed against the backcloth of the entire Pleistocene (Dennis, 1993a).

The following section deals with changes in biotopes over the medium term: vegetation succession from bare ground to climax forest and the regeneration complex (Box 5.3). Several principles are well established:

- **P5.23: Biotopes develop with vegetation succession to increasingly complex life form structures unless hindered by environmental or human agents.**
- **P5.24: Species' composition of autotrophs (plants) changes rapidly at first and then more slowly with vegetation succession.**
- **P5.25: Species' diversity of autotrophs increases initially and then becomes stabilized or even declines in older stages.**
- **P5.26: The organic structure changes with vegetation succession to produce increases in total biomass and non-living organic matter.**
- **P5.27: With vegetation succession, the number of heterotrophs (animals using organic material as food) increases; food chain relationships become more complex.**
- **P5.28: At a fine scale there is continuing cycling of plant species and life forms with the death of plants, particularly with the fall of forest-dominant trees.**
- **P5.29: Not all biotopes advance through stages to more complex life form structures but remain in flux owing to geomorphic processes.**
- **P5.30: Human intervention is key to halting succession and creating distinctive biotopes.**

Unless hindered by human intervention, geomorphic agents or high toxin status, bare ground is rapidly colonized and the vegetation advances steadily over decades through increasingly complex life form structures (**P5.23**), in the British Isles to a temperate forest or upland sphagnum mosses. This takes the form of distinctive climax woodland species in different parts of Britain in relation to regional climates and geology. Bennett (1989) has reconstructed this regional pattern of climax vegetation for the Holocene climatic climax at 8000 years BP (see Box 8.5). Composition of plant species rises rapidly at first but then more gradually (**P5.24**). At the same time, diversity of plant species increases rapidly, but then slowly and often declines as tall trees dominate the vegetation (**P5.25**), determining what persists under the canopy and open spaces with timber fall. For example, beech forests are so dense that very little light reaches the forest floor in summer months; consequently, the ground flora and shrub storey are restricted in development and much bare ground is maintained. Other changes accompany the transformations in life form with succession; organic structure increases, both living and dead biomass (**P5.26**) and, with this, there is increasing diversity of organisms and complexity in ecosystems (**P5.27**), and thus food web relationships. This process, the build up of dead as well as live organic matter, may not seem to be that relevant for British butterflies but it is important as a latent niche component and as a resource (see Chapter 7). It is vital for many other arthropods as well (e.g., longhorn beetles, hoverflies; Fayt *et al.*, 2006) and if a resource-based definition is to have any value at all, it must make sense and be to the advantage of all organisms. In energy terms, the early stages of succession are marked by an excess of production over respiration, but this balances out in late successional stages.

Vegetation change does not come to a halt with the development of forest cover. With the death and fall of forest tree dominants, new spaces are created below the canopy, light reaches the forest floor and there is rapid colonization by pioneering plants. With competition there is a 'succession' of plant species in the new space, by herbs, shrubs and saplings, and the emergence of new dominants that may differ from the previous incumbent. This cycle of vegetation change on the forest floor, the filling of a space made available by the fall of a forest-dominant tree, is referred to as the **regeneration complex** (**P5.28**). Of course, biotopes do not always advance through the succession of stages to high forest. They may be halted naturally by geomorphic agents (e.g., active scree slopes, flood plains) or extreme substrate (e.g., highly permeable gravel soils) and climatic (e.g., tundra) conditions (**P5.29**). A phase of the cycle over large areas can also be re-started by extreme events, as in the case of the October 1987 storm which decapitated many a woodland canopy in southern Britain. In fact, most biotopes are 'frozen' as **plagioseres** by human management; fields of pasture and arable, hedges and gardens, are all biotopes prevented from developing into more complex successional stages (**P5.30**). Plagioseres include highly valuable

Box 5.3 Vegetation changes: cycles and succession

This box provides an introduction to vegetation cycles and successional trends, both of which greatly affect butterfly populations.

With the exception of toxic substrates (e.g., mining and industrial spoil), an area of bare soil or regolith – exposed by fire, flood or clearance – does not remain bare for long but is rapidly colonized by a sequence of increasingly taller plants (Fig. B5.3a, c). This commonest of trends is termed **succession** (Clements, 1904; Miles, 1987; Andel *et al.*, 1993). Succession is described as occurring in a number of developmental stages (**seres**) until it attains equilibrium with climate and substrate conditions. Distinct sequences or stages have been described for different starting conditions (Fig. B5.3b), termed a **hydrosere** for freshwater, a **halosere** for salt marshes, **psammosere** for sand dunes and **lithosere** for bare rock. A reversal of this process is termed **retrograde-succession**, as occurred with developing glaciation over Britain (cf., Dennis, 1993a). Vegetation succession involves three components (Andel *et al.*, 1993):

1 Factors or causes, the agents responsible for change, usually abiotic (e.g., climate, fire) but also biota other than plants (e.g., herbivores, disease).
2 Mechanisms, effects and interactions among plants.
3 Pathways or stages of temporal change.

The most important agents driving succession determine the ultimate type of vegetation in an area; such are climate, geology and topography. Succession may start on bare substrate or from disruption of previous vegetation (e.g., by farming, deforestation), respectively referred to as **primary succession** and **secondary succession**. The mechanisms of succession describe the impact of plants themselves on the substrate and on each other, generally alluded to as distinct phases of activity: **nudation** (production of bare ground), **migration** (movement in of organisms, seeds, etc.), **ecesis** (successful establishment of colonizing plants), **competition** (interaction among colonists), **reaction** (modification of conditions allowing other plants to enter) and **stabilization** (homeostasis with regeneration cycles) (Clements, 1916). The pathways depend much on the environment where succession takes place, classic cases being those of glacier retreat (Crocker and Major, 1955), dune development (Olsson, 1958) and fall in lake base-level (Rydin and Borgegård, 1991). The literature is extensive on concepts and terms for vegetation succession, on the factors involved (**autogenic succession**, caused by intrinsic factors, versus **allogenic succession**, caused by extrinsic factors) and the nature of the climax vegetation, whether unitary (**monoclimax**, single climax vegetation type) or multiple (**polyclimax**, variety of climax

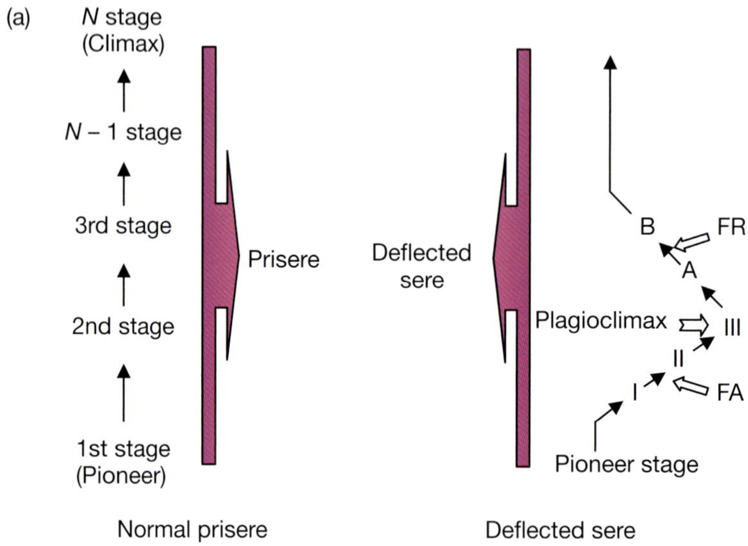

Fig. B5.3 (a) Development of a plagioclimax compared with that of a normal primary sere (prisere). FA, factor applied (factor creating plagiosere, e.g. grazing or mowing); FR, factor removed; I and II, plagiosere stages; III, plagioclimax; A and B, subsere stages. (Modified from Pears, 1977, Fig. 4.6, p. 56, reproduced by permission of Pearson Educational Ltd.)

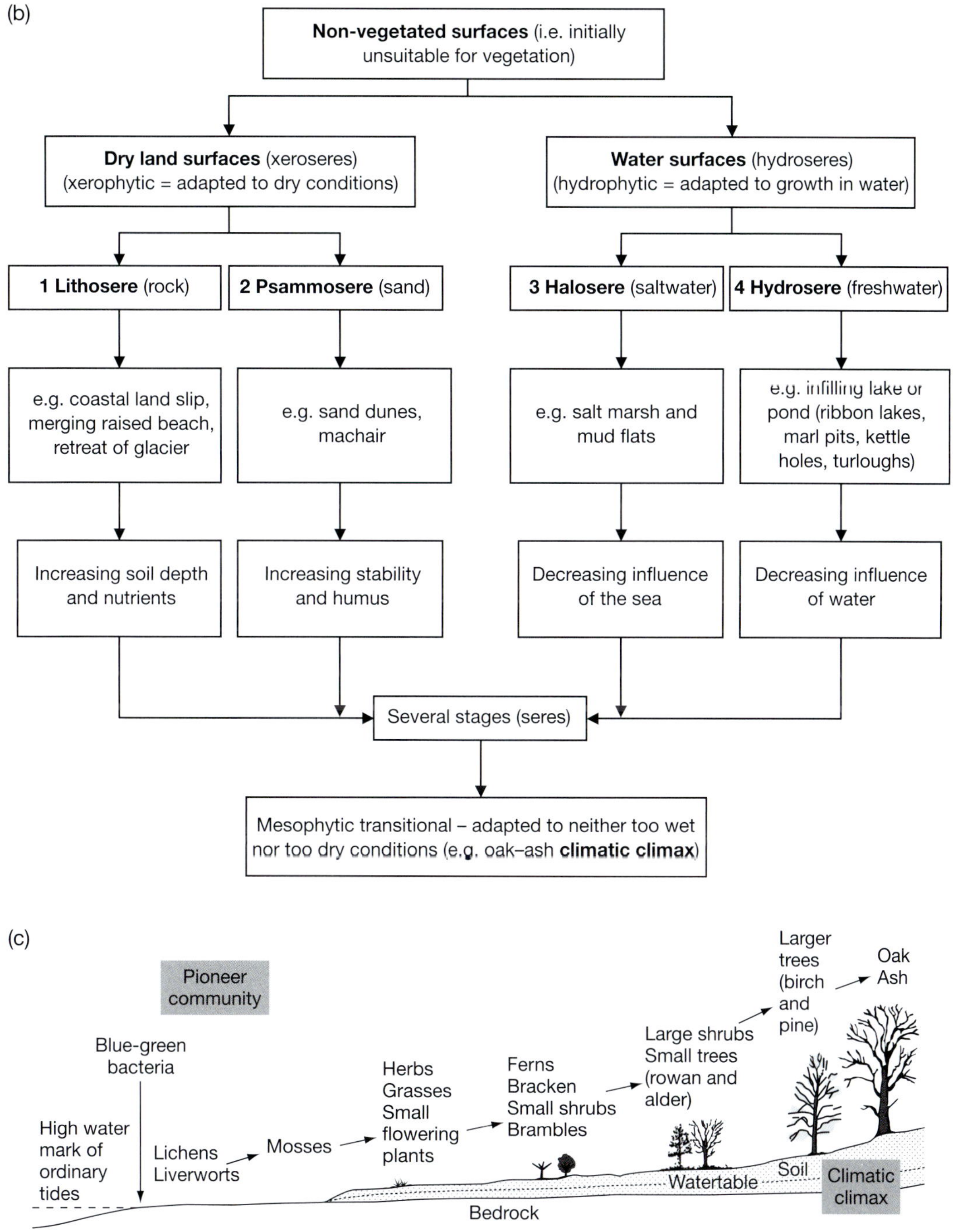

Fig. B5.3 (*continued*) (b) Development of a climatic climax vegetation. (c) Stages in a lithosere developing on a raised beach along the west Scottish coastline. (b and c, reproduced and adapted with the permission of Nelson Thornes from *Geography. An Integrated Approach*, 3rd edn, by David Waugh (ISBN 978-0-17-44706-1, first printed in 2000).)

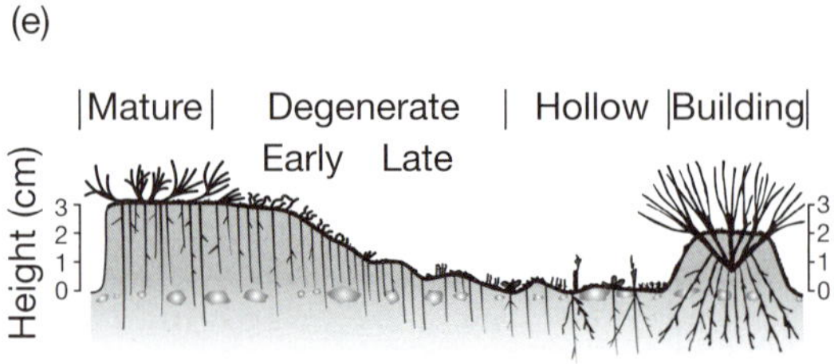

Fig. B5.3 (*continued*) (d) Woodland plagiosere (coppice with standards) management and illustrating vegetation before and after cutting. (Redrawn from Bennett and Humphries, 1974.) (e) Regeneration cycle at a microscale in grassland (*Festuca ovina* erosion cycle illustrating 'fossil' shoots in soil and mini-landform development and erosion. (From Watt, 1947, courtesy of Blackwell Science.)

vegetation types), and arrested development (**plagioclimaxes**, **disclimaxes**) from human activity (Fig. B5.3d) (Pears, 1977). There is still controversy over the direction of succession on different substrates and the nature of climax vegetation. Distinct changes occur with succession. From the vantage of butterfly biology, plant community (number of plant species) and ecosystem characteristics (e.g., biomass, non-living organic matter, complexity of food web, production/respiration ratio) change dramatically at first with vegetation succession, and slower later on (Odum, 1963); plant species' richness tends to decline towards climax vegetation development. Such changes have wide-ranging implications for butterfly biology, directly through plant dependency and indirectly via the impact on other life history resources. Succession is thus a key concept in butterfly ecology of increasing prominence as biotopes for butterflies decline, especially as farming is abandoned in mountain areas in Europe.

Both cyclic fluctuations and successional trends involve an orderly process of plant community changes, directional and to some extent predictable, and both are able to modify the physical environment. Even though vegetation, such as woodland, may appear to all extents and purposes unchanged for long periods, vegetation is never static. Cyclic vegetation changes represent fluctuations about a mean state of conditions and have been recognized in a number of contexts. Examples range from very fine resolution hummock and hollow cycles (e.g., *Calluna–Erica–Eriophorum* cycle on raised bogs (Godwin and Conway, 1939), sheep's fescue *Festuca ovina* cycle on grassland (Watt, 1947; Fig. B5.3e) to patchwork dynamics in forests. All may be described as a **regeneration complex** (a mosaic or patchwork of plants), each patch a part of the regeneration cycle – a cycle of pioneer plant invasion, building phases and maturity to degeneration and reinvasion. In the case of the examples of vegetation cycles on raised bogs and grassland, hummocks develop during the plant building phase that eventually become eroded to form hollows in late maturity and degeneration. The vegetation cycle produces a cycle of microtopography or micro-relief important for species such as large heath *Coenonympha tullia* (Dennis and Eales, 1997, 1999) and possibly silver-spotted skipper *Hesperia comma* (Thomas *et al.*, 1986) in cooler phases when bare ground is more important. It can also produce variation in plant quality in perennial plants, where new marginal growth outperforms older central growth (e.g., nettle *Urtica dioica*, cotton grass *Eriophorum* sp.; Kershaw, 1973), which may explain the use of marginal plants on nettle patches by nymphalids for egg laying. The basic feature of such cycles is that over a vegetation community these events are replicated many times; as such, all phases of the regeneration cycle may be expected to occur at any one time and the frequency of stages will remain much the same too. For instance, as pristine patches emerge with new tree fall in woodlands, other older patches at a later stage are closed under the canopy of a new dominant tree (Watt, 1947), a process that mirrors metapopulation dynamics of habitat colonization and extinction. In effect, this is succession occurring on a miniature scale, or micro-succession, in which a succession of species leads to a relatively stable end-product, a canopy tree. Much remains to be understood about the processes and product of regeneration from the processes involved such as seed production, seed predation and germination, tree root competition, browsing and over-canopy shading (Harmer, 1994a, 1994b, 1995, 1999, 2001).

biotopes for the maintenance of many species (but see below); a classic example is the woodland coppice cycle for the heath fritillary *Melitaea athalia* (Fig. 5.3).

Principles relating to vegetation succession and regeneration cycles

- **P5.31: The composition of butterfly species, as with all heterotrophs, changes with vegetation succession.**
- **P5.32: The diversity of butterfly species increases with vegetation succession but may decline with shrub and forest cover.**
- **P5.33: Without intervention, the regeneration cycle in climax vegetation is crucial for the maintenance of butterfly species requiring resources associated with early successional stages in their life cycles.**
- **P5.34: Species' composition and diversity in plagioseres relates more to autotroph diversity than life form structure.**
- **P5.35: The pattern of life histories will change with vegetation succession reflecting on changes in the mix of species and patterns of larval and other resource use in distinct seres.**
- **P5.36: There will tend to be taxonomic bias among butterflies in different seres reflecting patterns of resource use.**
- **P5.37: Butterflies associated with pioneer stages will more commonly have ruderal tendencies whereas those of later stages will tend to be adapted more to competition and stress tolerance.**

Although various resources for individual butterfly species occur amongst different vegetation or substrate types, these tend to be associated with specific successional stages. Consequently, the composition of butterfly species changes with vegetation succession, and with vegetation succession some species are lost from locations and new ones enter (**P5.31**), a pattern that can be experienced by observing changes by moving from areas of bare ground through contiguous zones of short herbs, tall-herb grassland, shrubs and woodland. Pollard (1982) developed a technique to identify trends in butterfly populations on sites subject to vegetation changes, either from succession or management. This involves comparing population data for a target site with similar data for sites within the same region over the period of interest. In this way he was able to demonstrate that management in the Monks Wood National Nature Reserve, Huntingdon, UK, had indeed influenced trends in several butterfly species (e.g., increasing trend in the meadow brown *Maniola jurtina*, decreasing trend for *Aphantopus hyperantus*). A note of caution is appropriate. Not all changes of butterflies within sites, even over long periods of say 20 years, are associated entirely with vegetation succession; they can also arise from changes of climate (e.g., *Pararge aegeria*) (see Chapter 8). Yorkshire fog *Holcus lanatus* grassland and bramble scrubland were abundant enough in north Cheshire during the 1970s but only in the mid 1980s were these sites, respectively, adopted by colonizing *Thymelicus sylvestris* and *Pyronia tithonus* (Hardy *et al.*, 1993). The change in species with vegetation succession is implied in the tabulation of breeding biotopes (see Appendix 7). However, the broad view of breeding biotopes over Britain does not list biotope exploitation of all resources and can conceal important resource–successional stage associations. Thus, *Hipparchia semele*, which is characteristic of rock faces and dunes, can also use tree trunks in wooded areas for territories in hot, calm, sunny weather (R. L. H. Dennis, personal observation) and *M. jurtina* takes to laying eggs on mossy tree trunks (Dennis, 2003). Similarly, purple emperor *Apatura iris* – a high forest butterfly with oak tree-top territories and *Salix caprea* shrub oviposition sites – also exploits mud and carrion as food on the floor of sunlit clearings and glades. This impresses on a further feature of successional vegetation categories: late stages in autotrophic succession may have elements, such as bare ground, characteristic of pioneer stages, but not vice versa.

Expectations are that with vegetation succession, faunal composition changes will be accompanied by increases in species' richness (diversity) (**P5.32**). Pre-climax limits have been found for trajectories in the richness of species in successional stages for both plants and animals (Odum, 1963) and findings for butterflies are not expected to differ. Just which seral stage has most species in any faunal group depends on at least two things. First, what basic adaptations are presented by the taxon as a whole in the region; and second, the opportunities for species in terms of key resources for early seral stages, and their capacity to access these resources in regeneration complexes and surrogates presented by human management within later seral stages of forest cover. The majority of British butterflies

Fig. 5.3 Coppiced woodland biotope for the heath fritillary. (a) An early stage in the coppice cycle of chestnut *Castanea sativa* Mill., suitable for the larval hostplant common cow-wheat *Melampyrum pratense* of the heath fritillary *Melitaea athalia* (inset) in Blean Woods, Kent, UK. (Courtesy of Martin S. Warren; inset courtesy of Jim Asher.) (b) The larval hostplant common cow-wheat *M. pratense* of *M. athalia*. (Courtesy of Caroline Bulman.)

use herbaceous larval hostplants and therefore it would be of little surprise if species' diversity declined with the vegetation characteristics of later seral stages (woodland). Of course, in the modern day, highly unnatural, context of forest, there are 'built in spaces' – glades, rides, clearings – which can be occupied by species with herbaceous hostplants. These greatly increase the number of species found in modern forests.

In the natural context of the British Holocene climatic climax, forest development had the potential for isolating species dependent on herbaceous hostplants and affecting their distributions (Dennis, 1977, 1993a). Three outlets prevailed in the forest for species requiring resources in more open biotopes: the regeneration complex, herbivore activities, and human clearances for hunting and agriculture (**P5.33**). The regeneration complex functions much like the original concept of the metapopulation model (Levins, 1969; see Chapter 6) inasmuch as new clearings were continually being created and overgrown but the number of them over the extensively forested area of Britain at 6000 years BP would have remained much the same. As Neolithic landnám clearances advanced with slash-and-burn techniques for cultivation plots following the Mesolithic firing of vegetation (Troels-Smith, 1960), the open spaces increased to a point by the Bronze Age from which they would never subsequently decline. The danger now is that owing to a decline in woodland management in the 20th century, forest and woodland blocks lack open spaces at metapopulation scales for open seral stages (see Chapter 6). Thus, many woodlands, even under public ownership, lack resources for many species of early seral stages. In the National Trust Alderley Edge woodlands (Cheshire, UK), there are suitable larval hostplant resources for only very few butterflies (e.g., *Favonius quercus*, *Pararge aegeria*, *Pieris napi*), though a number of butterfly species use other resources (e.g., *Vanessa atalanta* uses sun spots as territories, *Maniola jurtina* uses the woods for daytime roosts in hot summers) and many species pass through as vagrants (Dennis, 2000). These woods are now being opened up (Dennis, 2009c).

A naive expectation is that with so much 'open' space provided by farming there would be extensive opportunities for butterflies of earlier seral stages. Indeed, this must have been so in Medieval Britain but is obviously no longer the case. Butterfly species of early seral stages are largely confined to unimproved pasture and heath, and to land unit margins. Much of the landscape under improved pasture and arable is – from the butterfly viewpoint – largely a green desert **plagioclimax** (a stable stage in succession maintained by humans). 'Improvement' has resulted in a severe reduction of plant diversity and the accumulation of toxins and excess nutrients (NPK fertilizers). Nitrogen deposition is now such a problem that it is no longer one restricted to agricultural landforms, but is recognized as having an impact on almost all ecosystems. In fact, increasingly, most vegetation communities in Britain are part of **brotions**, ecological successions resulting from human activity. Phytophagous butterflies dependent on herbaceous larval hostplants simply reflect the diversity in plants (**P5.34**), their distributions in these early seral stages forced to refuges on residual patches and networks bounding land units.

With the changes in vegetation life forms, transformations in the mix of life histories are to be expected. Butterflies are closely tied into their larval hostplants and different hostplants tend to be characteristic of distinctive successional seres (**P5.35**); this observation is closely linked to life history strategies dealt with in the classification section below. Because of the phylogenetic association in larval host use, taxonomic bias is part of the change in the mix of species (**P5.36**); thus, for example, pierinae and satyrinae dominate in early herb seres and theclini dominate later shrub and tree phases. Phylogenetic associations are only part of taxonomic transformations with vegetation succession. Another part of it is a change in life history strategies in butterfly species matching those of plants (**P5.37**). Pioneer stages in vegetation succession tend to be dominated by plants, and thus butterflies, with ruderal tendencies, are geared to vagrancy and rapid development. Later stages in vegetation succession cycles are associated with plants and butterflies having competitive and stress-tolerant strategies, depending on environmental conditions, with the butterflies having fewer broods. The links occurring between butterfly and plant strategies are explained more fully below.

However, just what trends occur with vegetation succession depend largely on the sequence (range) of vegetation communities (the vegetation life form spectrum) analysed, from bare ground pioneer conditions to climax forest, and the regional context. Spurious relationships are likely to emerge if only part of a successional sequence is studied; studies in different parts of the European continent will certainly produce different taxonomic composition of butterflies with succession and very likely different patterns in species' richness too. In Britain, as with large parts of continental Europe, there is little or no original climax forest and forest depleted of plant communities may be equally depleted of butterflies, as in the case of the Alderley Edge woodlands mentioned above. Contradictory associations of life history attributes with vegetation succession, as in the case of hostplant specialization, have also been found in different studies (Futuyama, 1976; Steffan-Dewenter and Tscharntke, 1997; Kitahara and Watanabe, 2003). The resulting changes in life history patterns will also depend on just what life history strategies the plants, and thus dependent arthropods, in successional seres portray. Below, it will become evident what life history changes may be expected with vegetation succession: for example, increasing species' diversity, larval duration time and adult hibernation, but decreasing broods, are expected to occur with succession but not necessarily increasing egg size and egg clustering or decreasing body size (Dennis *et al.*, 2004). But trends in life history with successional sequence have also been found to differ from expectations (Steffan-Dewenter and Tscharntke, 1997). To reiterate a point of caution: what is found can depend on what is studied. Moreover, life history profiles for phytophagous insects in seres will not be exclusive, even if consideration is restricted to breeding species within a sere; adaptations are not that inflexible and there is variety in plant strategy and resource exploitation within a sere that deserves further study.

Communities, niches and invasibility

A key issue associated with living communities is the capacity of new organisms to join them, that is, to colonize sites where they previously did not exist. This is of particular interest when considering alien (non-native) species. In Britain there are many plants, animals and insects that fall into this group (Clement *et al.*, 2005; Hill *et al.*, 2005 lists 2723 organisms) but there have been few butterflies within historical times. Ford (1957) described the attempted introduction of the European map *Araschnia levana* into Britain but also its subsequent elimination. Of course, most British plants and animals have in effect been aliens at some stage when considered against a backcloth of the last deglaciation (18k years BP), thus an arbitrary date is typically selected to categorize alien status, usually in historical times before global sea navigation. The capacity for colonizing new sites or regions is referred to as **invasibility** and this notion draws heavily on the concept of the realized niche (see Box 2.2). This is a topic in which the fine distinction between niche and habitat becomes most useful, as will become apparent below. First, to draw on some important issues on invasibility:

- **P5.38: The richer the environment (site), the more alien species can invade it.**
- **P5.39: Species-rich communities are more resistant to invasion by alien species than species-poor communities.**
- **P5.40: Vacant niche spaces exist within communities (biotopes, sites, ecosystems).**
- **P5.41: Increasing disturbance promotes invasibility and increasing stress decreases invasibility.**
- **P5.42: Successful invasibility increases with propagule (migrant) pressure.**
- **P5.43: The enemy release hypothesis (EHA) states that an invading species will generally experience a decrease in regulation by enemies.**
- **P5.44: The evolution of increased competitive ability hypothesis (EICA) states that as alien species do not have to invest energy and resources in defence against enemies, they can invest these in increased competitive ability.**
- **P5.45: Different biotopes comprise contrasting numbers and proportions of alien species.**

The ability of organisms to invade communities depends on natural drivers, such as resource use, life history attributes (e.g., fecundity, migration capacity) and community invasibility (dependent on metapopulation dynamics), and on human influences such as biotope modifications, new biotope creation, transport systems and intentional introductions (see Kühn and Klotz, 2007, for a recent review). It is clear that the number (proportion) of alien species is spatially (and temporally) scale dependent – at a regional scale invasibility relates more to environmental heterogeneity, whereas at a local scale it depends increasingly on neighbourhood processes such as competition (Shea and Chesson, 2002). The ability of organisms to invade (colonize) a site depends on there being appropriate resources and a capacity to withstand competition from closely related organisms and enemies. It is expected then that the richer the site, in terms of potential resources (consumables and utilities), the greater the likelihood of invasion by alien species, since potential empty niche space is greater than in a poorly endowed site (**P5.38**). On the other hand, for any type of biotope (e.g., unimproved pasture, wood, garden), species-rich communities may be more resistant to invasion by alien species than species-poor communities (Elton, 1958); in niche terms, there is less niche space available for entry (**P5.39**). Where two species have very similar resource demands, there is a greater likelihood of competition and one species being excluded. This is formally referred to as Gause's hypothesis (**Gause's exclusion principle** or Grinell's axiom): two species with the same ecological requirements cannot coexist in the same place indefinitely; this may well be the case for some closely related butterflies (e.g., Essex skipper *Thymelicus lineola* and small skipper *T. sylvestris*). An unfortunate example among mammals is the replacement of red squirrels by grey squirrels. Clearly, where alien species converge on the niches of incumbent species, there will be disruption of interactions in the community.

From metapopulation studies (see Chapter 6), it is appreciated that potential sites (habitat) for a butterfly can at times be vacant or colonized; all the resources may be present but no butterfly is evident. Yet, the site may have been colonized in previous years and may be occupied in subsequent years. Absence often has something to do with the quality or size of resources, but it can also occur from stochastic processes, extinction caused by random events such as a wet, cloudy season coinciding with egg laying. In effect we have a vacant physical (habitat) space on the ground and an empty niche (the hyper-dimensional space) described by attributes (resources, environmental measures, competition and enemy-free space) in the community of organisms at that site (**P5.40**). In this way we may envisage a site that fails as a habitat for any current butterfly species in Britain, but nevertheless has adequate resources (a potential habitat but a vacant niche) for one not currently in the islands. Good examples exist of this among British mammals (e.g., mink, coypu). Several studies indicate that invasibility increases with disturbance of sites and decreases with stress (low nutrients, temperature extremes, salinity, etc.) (**P5.41**) (Alpert *et al.*, 2000; Davis *et al.*, 2000). The ecology of butterflies in relation to disturbance and stress is examined closely below. Invasibility is also likely to increase with an increasing number of migrant individuals (propagules) (**P5.42**), if only because a greater 'rain' of individuals increases the chance of a successful mating and colonization, and introduces greater genetic variability to withstand environmental uncertainties. Two other principles introduce two hypotheses based on plant communities, but which may well also apply to butterflies. The first (**P5.43**), the **enemy release hypothesis** (ERH), states that alien species should increase in abundance and distribution as they are unlikely to have been accompanied by their suite of enemies; they enter, for a time at least, enemy-free space (Keane and Crawley, 2002). They may nevertheless find themselves exposed to a new suite of enemies and threats (E. John, personal observation). The second (**P5.44**), the **evolution of increased competitive ability hypothesis** (EICA) is also based on the notion of aliens entering enemy-free space – the energy that would have been expended in defence can now been expended in growth, development and reproduction, i.e., competitive ability (Blossey and Nötzold, 1995).

From a butterfly viewpoint, there are two aspects to invasibility:

1 Invasion by butterfly species themselves.
2 Invasion by their resources, particularly consumer resources (larval hostplants, nectar flowers).

Resident butterfly species are continually invading new sites; much as there is perpetual local extinction, so too is there local colonization; niche issues of interaction among organisms are continually in flux (see Chapter 6). The issue is more interesting when alien species enter the country, either by natural processes of migration and colonization or by human introductions (see Chapter 8). Immigration and introductions of alien butterfly species have occurred regularly (Salmon, 2000), and only the long distance migrants have been able to establish footholds for any length of time (e.g., clouded yellow *Colias croceus*, *Vanessa atalanta*). Far more prominent is the invasion of alien plants, more of which have become nectar sources (50% are alien) for British butterflies than larval hostplants (22% are alien). Inevitably, different biotopes have contrasting proportion of aliens (**P5.45**). Gardens are prominent in receiving the bulk of introduced plants used by butterflies; as might be expected, an increasing use of alien nectar sources occurs among species that have larval hostplants in gardens (Hardy and Dennis, 2008).

ECOLOGICAL CLASSIFICATION OF BRITISH BUTTERFLIES

Links between butterflies, their hostplants and associated vegetation and substrates have been exploited to obtain classifications of butterfly habitats. Two approaches have been attempted:

1 Indirectly through species hostplants and their life history strategies (Hodgson, 1993; Dennis *et al.*, 2004).
2 Directly through ecological attributes of butterfly species (Shreeve *et al.*, 2001).

Hostplant strategies and butterfly habitats

A number of principles apply:

- **P5.46: Plants acquire sets of life history traits that reflect three contrasting groups of environmental pressures: competitive exclusion, chronic stress and severe disturbance (C-S-R).**
- **P5.47: Plants also acquire life history traits (physical and chemical defences) in response to herbivory, contrasted as quantitative and qualitative in effect on herbivores.**
- **P5.48: As butterflies are herbivores, ecological (habitat) distinctions among butterflies are expected to relate to contrasting hostplant life history strategies linked to C-S-R environmental pressures and plant defences.**
- **P5.49: Ecological distinctions among insect herbivores will be reflected in contrasting life history traits.**
- **P5.50: An increasing tendency of a butterfly's hostplants to a particular C-S-R strategy biases that butterfly species to distinctive life history attributes and resource breadth and type.**
- **P5.51: Butterflies feeding on quantitatively protected woody plants and grasses (S and C strategists) have fewer annual generations than those feeding on qualitatively protected forbs (R strategists).**

In this explicit approach to classifying butterfly habitats, a link is established between life histories and ecological parameters of butterflies on the one hand and ecological strategies already determined for plants, specifically their hostplants (Grime *et al.*, 1988; Hodgson *et al.*, 1999), on the other. There is an expectation that the ecological and life history adaptations of butterflies, because they are herbivores, are keyed into their hostplants and hostplant environments (Price, 2002; Hunter, 2003). The **C-S-R strategy** model for plants (Grime, 1974) is argued to form the foundation of herbivore (butterfly) adaptations to habitats; in effect, plants acquire life history traits that reflect adaptation to three distinct groups of environmental pressures: competitive exclusion, chronic stress, and severe disturbance (**P5.46**) (Grime, 1974, 1979). Grime *et al.* (1988) list an array of 20 morphological, life history and physiological attributes for plants corresponding to the different strategies. A key feature in these adaptations is that plants also differ in mode of defences to herbivores; defence is achieved either by synthesizing qualitative compounds or by protection through quantitative macromolecules, leaf toughness and low water and/or nutrient content (**P5.47**) (Cizek *et al.*, 2006). The palatability of the former remains relatively constant during a growth season but the palatability of the latter decreases with leaf age.

The C-S-R strategies were described initially for plants (Grime *et al.*, 1988) but clearly also affect herbivores such as butterflies dependent on them (**P5.48**); the butterflies occupy the same environmental conditions and are subject to adaptations acquired by plants in response to their environments. Thus, herbivores are expected to acquire sets of traits that reflect the same pressures, those experienced by their hostplants; the 20 traits acquired by plants are mirrored by 28 significant life history and resource attributes in butterflies (Box 5.4; **P5.49**) (Hodgson, 1993; Dennis *et al.*, 2004). The background details to this topic go well beyond the remit of this book; readers who wish to follow it up should refer to the original papers and the text book on plant strategies (Grime *et al.*, 1988). Many organisms (e.g., plants, butterflies) are expected to occupy intermediate conditions corresponding to particular equilibria between stress, disturbance and competition. The C-S-R equilibrium will vary spatially and temporally at different scales, and communities will often contain species of widely differing strategies, if only because of delays in adjustment to changing environmental conditions (Grime *et al.*, 1988) and opportunities in regeneration complexes. Butterfly species, as with plants generally, are found to occupy distinct zones in the ternary co-ordinates describing plant strategies (Box 5.4). The message is clear: hostplant strategies significantly affect butterfly biology and, as we shall see later, all aspects of butterfly geography and status.

In essence hostplant strategies influence three aspects of butterfly biology: development rates, mobility and resource range (**P5.50**). **Development rate**, in turn brood number and length of flight period, relates to fertility and length of feeding opportunity. Butterflies with C and R hostplant strategists develop relatively rapidly in fertile and disturbed environments and those with S hostplant strategists relatively slowly in the comparatively infertile, skeletal soils of scarplands (e.g., *Polyommatus coridon*, *Cupido minimus*, *Hesperia comma*), the cold conditions of mountains (e.g., mountain ringlet *Erebia epiphron*), acid conditions of blanket bogs and mires (e.g., large heath *Coenonympha tullia*, small pearl-bordered fritillary *Boloria selene*) and the shaded woodland floor (e.g., pearl-bordered fritillary *B. euphrosyne*, high brown

Box 5.4 C-S-R life history strategies for British butterflies

As butterflies are phytophagous, there is an expectation that contrasting hostplant attributes hold the key to changes in butterfly status because insect life history traits link to characteristics of their hostplants (Price, 2002; Hunter, 2003). The **C-S-R strategy** model for plants (Grime, 1974) proposes that organisms acquire sets of life history attributes that reflect three distinct groups of pressures: **competitive exclusion**, **chronic stress** and **severe disturbance** (Grime, 1974, 1979). Each threat occurs under particular types of environmental conditions and life history attribute sets confer a selective advantage in these extreme conditions. Under this model, the habitat that organisms occupy forms the template for life history adaptations. Southwood (1977, 1988) recognized five sets of adaptations acquired by organisms in their habitats: to physical conditions (e.g., heat, water supply), predation, food availability, mating and lethal conditions (e.g., escape by migration and diapause). The strategies correspond to the gradient of what is referred to as *r*- and *K*-selection (MacArthur and Wilson, 1967):

- ***r*-selected species**: characteristic of variable or unpredictable environments, typically with rapid development, high innate capacity for increase (*r*), early reproduction and small body size – they are thus opportunistic species.
- ***K*-selected species**: characteristic of relatively constant or predictable environment, typically with slow development, late reproduction and large body size – they thus have relatively high competitive ability.

The three-strategy system developed with the recognition of a type of ecological specialization not accounted for by *K*, and lying beyond *K*-selection, termed *S*-selection (Grime, 1974) or **adversity selection** (Whittaker, 1975).

The strategies were described initially for plants (Grime *et al.*, 1988). They arise as the outcome of four permutations of high and low stress and high and low disturbance. Stress consists of phenomena that restrict production and rates of growth, such as extremes of light and water, temperature and perhaps most importantly, mineral nutrient deficiency. Disturbance includes agents that destroy biomass, such as fire, floods, cliff falls, drought, frost, wind damage and human interference (e.g., ploughing, military ranges, bike scrambling). One of the permutations is discounted as organisms fail to persist where both disturbance and stress are high. The three remaining extreme strategies are **ruderals** (R: high disturbance, low stress), **stress-tolerators** (S: high stress, low disturbance) and **competitors** (C: low stress, low disturbance). Fundamentally, as applied here, the strategy of a butterfly's hostplant distinguishes two components: the nutritional quality and duration of larval food (i.e., C: food quality moderate to good, length of feeding opportunity intermediate (spring/autumn); S: quality poor, length of feeding opportunity long (all year round); R: quality good, length of feeding opportunity short (many short-lived hostplants)). C plant strategists are also monopolists and hence where they occur they do so in abundance; whereas R plant strategists are smaller (and transient).

Many organisms are expected to occupy intermediate conditions corresponding to particular equilibria between stress, disturbance and competition. The C-S-R equilibrium will vary spatially and temporally at different scales, and communities will often contain species of widely differing strategy, if only because of delays in adjustment to changing environmental conditions (Grime *et al.*, 1988). Grime and colleagues list an array of 20 morphological, life history and physiological attributes for plants corresponding to the different strategies. They have also classified plants in the Sheffield region to one, or combinations, of seven basic strategies describing their mean strategy. This model is particularly relevant for butterfly conservation since the three axes correspond to the three key directions of land use change (C = abandonment; S = inverse of eutrophication; R = mechanized disturbance; see Chapter 8). Elements of C-S-R strategies have been investigated among British butterflies (Hodgson, 1993; Dennis *et al.*, 2004). The basic findings for butterfly hostplants are tabulated in Table B5.4; as explained in the text, butterflies seductively follow the C-S-R strategy model (Fig. B5.4a, b). An obvious distinction is that ruderal species tend to be generalist hostplant users whereas competitors and stress-tolerators tend to be more specialist in hostplant exploitation. However, one big difference is the scarcity of ruderal species; even *Pieris* species do not exclusively use ruderal (annual) hostplants (see Kemp *et al.*, 2008).

An increased understanding of plant and herbivore strategies is advancing with the development of quantitative measures of both plant and animal attributes

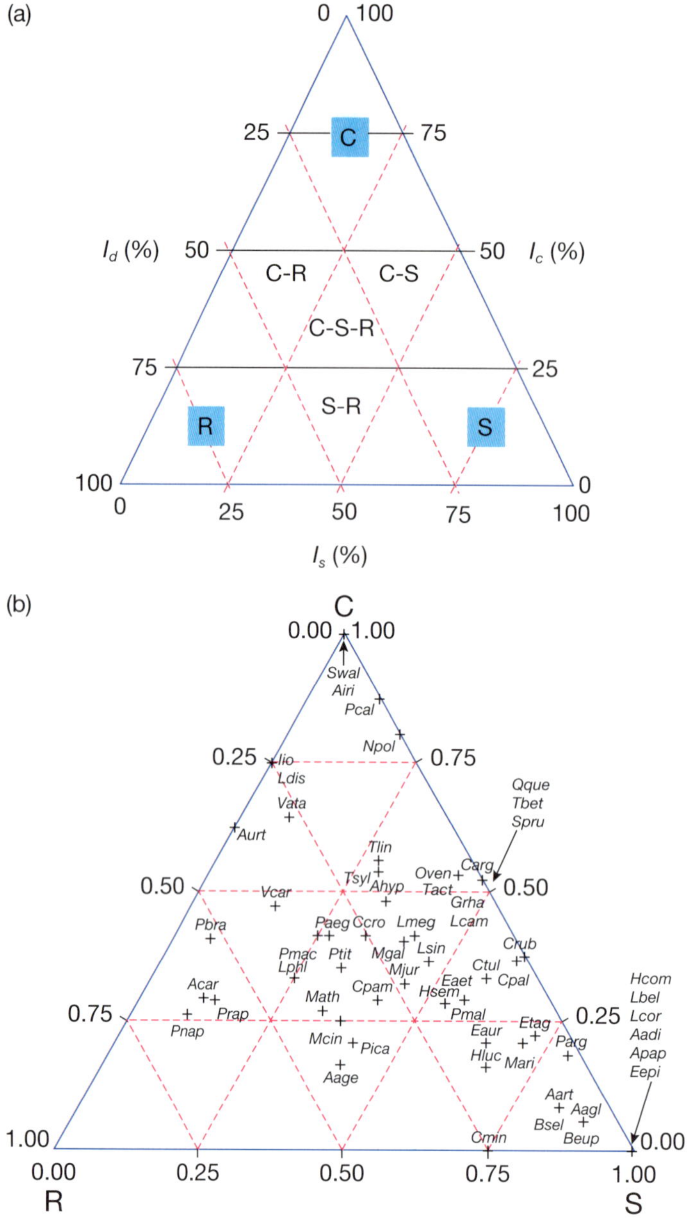

Fig. B5.4 (a) A model describing the various equilibria between competition, stress and disturbance in vegetation and the location of primary and secondary strategies. C, competitor; R, ruderal; S, stress-tolerator; C-R, competitive–ruderal; C-S, stress-tolerant competitor; C-S-R, 'C-S-R strategist'; S-R, stress-tolerator–ruderal. The three axes show the relative importance of competition (I_c), stress (I_s) and disturbance (I_d). (b) Distribution of mean butterfly species scores in ternary axes for C-S-R strategies based on all hostplants. Abbreviations for species comprise the first letter of the generic (subgeneric) name and the first three letters of the species name. (From Dennis *et al.*, 2004, courtesy of Blackwell Science.)

(J. G. Hodgson, personal communication). Work to date on butterflies has largely been reliant on Ellenberg values for plants. These are not entirely suitable. First, because they have not been measured where each butterfly species necessarily occurs. Second, because plants are given single values whereas we know that each is associated with a range of values (Grime *et al.*, 1988). Third, butterfly species are often restricted to a narrower band of hostplant conditions as in the case of the Scottish chequered skipper *Carterocephalus palaemon* which is biased to more nutrient-rich areas of purple moor – grass *Molinia caerulea* (Ravenscroft, 1994a, 1994b).

Table B5.4 Tendencies and comparisons in British butterfly species' biological and geographical traits associated with increasing C-S-R strategy scores.* (From Dennis *et al.*, 2004, courtesy of Blackwell Science.)

Biological trait	C	S	R
Wing expanse	Larger	**Smaller**	*Larger*
Proboscis length	Longer	**Shorter**	*Longer*
Oviposition behaviour	*Relatively precise*	**Relatively careless**	*Relatively precise*
Pre-oviposition period	*Relatively longer*	*Various*	**Shorter**
Egg load (total production)	*Various, generally higher*	**Lower**	**Higher**
Egg size	*Relatively larger*	*Relatively smaller*	*Relatively smaller*
Egg batch size	*Relatively larger*	*Relatively smaller*	*Relatively larger*
Larval growth rate	*Relatively faster*	**Slower**	*Relatively faster*
Larval duration	**Shorter**	**Longer**	Relatively shorter
Duration of early stages	*Relatively shorter*	**Longer**	Relatively shorter
Adult lifespan	**Longer**	Relatively shorter	Relatively shorter
Hibernation stage (egg)	*Several*	*Relatively more*	**Fewer**
Hibernation stage (larva)	Generally less	**Prominent**	*Various*
Hibernation stage (pupa)	*Various*	*Various*	**Increasing**
Hibernation stage (adult)	**Increasing**	**Decreasing**	*Generally decreasing*
Adult hardiness, tolerance of ambient conditions	Decreasing (except overwintering adults)	Increasing	Increasing
Symbiotic (relationship with ants)	Decrease	**Prominent (Lycaenidae)**	**Decrease**
Voltinism	*Various*	**Increasingly univoltine**	**Increasingly multivoltine**
Number of hostplants	*Various*	**Fewer**	**More**
Mean phagy score	*Various*	Relatively monophagous	**Relatively polyphagous**
Number of biotopes	*Various*	**Fewer**	**More**
Annual hostplants	*Lacking*	**Fewer**	**Increasing**
Biennial hostplants	*Lacking*	**Fewer**	**Increasing**
Short-lived perennial hostplants	**Generally fewer (exceptions Nymphalidae)**	Generally increasing	**Increasing**
Long-lived perennial hostplants	*Usual but various*	*Various*	**Fewer**
Hostplant phenology: range	*Generally narrowing*	Narrower	**Increasing**
Nectar sources	*Increasing*	**Fewer**	**Increasing**
Utility resources for pupae and adults	Generally increasing	Fewer	*Various*
Hostplant phenology	*Longer lived plants*	*Longer lived plants*	**Short lived plants**
Hostplant growth form	Relatively taller and more prominent plants	*Short herbs and shrubs*	**Short plants**
Plant life form association (immature stages)	**Typically tall herbs, shrubs and trees**	**Typically shorter turf to ground substrates**	*Typically tall herbs or below*
Plant life form association (adult stage)	**Taller vegetation, typically trees**	**Shorter vegetation to ground substrates**	**Relatively shorter vegetation**
Minimum habitat space occupied	Relatively larger minimum habitat size	Tendency for minimum habitat size to be smaller	*Various sizes*

Table B5.4 (*continued*)

Biological trait	**C**	**S**	**R**
Minimum population density	Tendency for typical minimum density to be lower	*Tendency for typical minimum density to be higher*	*Typical minimum densities vary*
Population structure	**Increasingly open structure**	**Increasingly closed structure**	*Open structure expected*
Mobility	**Higher**	**Lower**	**Higher**
Geographical range	*Various, including large*	**Increasingly restricted**	**Increasingly larger**
Metapopulation type	**Decreasing (none to patchy populations)**	**Increasing (distinctive and typically Levins)**	*Decreasing (none to patchy populations)*
Distribution cover	Relatively dense cover	**Sparse cover**	**Dense cover**
Incidence on offshore islands	Variable incidence	**Lower incidence**	**Higher incidence**
Change in population status over past 200 years	**Slower losses, persisting and expanding**	**Declining towards extinction**	Slower losses, persisting and expanding
Recent losses to distribution cover	**Relatively fewer**	**Increasing losses**	Relatively few
Conservation status	**Low priority**	**High priority**	**Low priority**

* **Bold**, consistently significant correlations for all and main hostplants; normal print, significant for one or other of all and main hostplants; *italics*, no significant correlation.

fritillary *Argynnis adippe*). **Mobility** relates to disturbance levels; appropriately, R hostplant strategists are significantly more mobile than S hostplant strategists. Increased mobility is a necessity for ruderalists (Dennis *et al.*, 2004) whose resources are ephemeral. S strategists are more likely to stay put and adapt to adverse, but predictable and persistent, conditions; compatible with this suggestion is their smaller wings. However, mobility is also strongly associated with C strategy scores and this very likely relates to the abundance and density of C strategy hostplants (Dennis *et al.*, 2004). Movement in hostplant C strategists reflects opportunities for colonization as their hostplants have been historically abundant; they have significantly larger wings for flight to exploit these resources (Dennis *et al.*, 2004). **Resource range** relates to disturbance and opportunity. Butterflies increasingly associated with R strategist hostplants, typical of disturbance and unpredictable conditions, are polyphagous; R plants are smaller and transient, making monophagy less tenable. As such, they also utilize hosts with varied phenology and exploit a wide range of biotopes; this also exposes them to a wider array of nectar sources. The opposite extreme is expected and found in butterflies increasingly dependent on S strategist hostplants that are generally adapting to adverse, even severe, conditions; such adaptations typically lead to specialization, monophagy or limited oligophagy (site monophagy) on longer lived hostplants. Such circumstances also provide opportunities for the evolution of specialized defence (e.g., symbiosis with ants). In the case of more limited nectar use this may result partly from the mechanical limitations of the smaller proboscises of smaller butterflies (Corbet, 2000) as well as exposure to fewer nectaring opportunities. C strategist hostplant feeders use prominent, ubiquitous, abundant, relative permanent resources, mainly trees, shrubs and persistent perennials, the larvae invariably making use of new growth high in nitrogen (e.g., *Inachis io* and *Aglais urticae* (Pullin, 1986, 1987), *Satyrium w-album* (Asher *et al.*, 2001)), again fitting expectations.

Other features are also disclosed that may be functionally linked to hostplant strategies, i.e., productivity, egg laying behaviour and hibernation. The biology of hostplant R strategists shows further pressures of continual disturbance, including high productivity, rapid egg release and hibernation in later stages; few opportunities are available for egg and larval hibernation among ruderalists on ephemeral plants with short lifespans and less predictable geography. S hostplant strategists in adverse, but long lived, environments are the converse. Along with slow development, the long larval period and restricted broods are low productivity, larval hibernation and, among Lycaenidae, specialized defence in symbiotic associations with ants. C hostplant strategists also hibernate in later developmental stages; they are typically associated with woodland in which adults can find niches in which to overwinter.

A recent twist to understanding life history strategies in relation to C-S-R is the disclosure that European butterflies feeding on quantitatively protected woody plants and grasses (S and C strategists) have fewer annual generations than those feeding on qualitatively protected forbs (R strategists) (**P5.51**) (Cizek *et al.*, 2006). These patterns relate only to foliage-feeding species. This supports the long held notion that plants defended by synthesizing qualitative compounds differ from those protected by the accumulation of quantitative macromolecules, leaf toughness and low water and/or low nutrient content. While the palatability of the former plants remains relatively constant during the growth season, the palatability of the latter group decreases with leaf age. As the authors point out, this relationship may explain why large butterflies tend to have fewer generations and feed on structurally complex hosts, and why some species remain univoltine although they are not restricted by a short summer growth season. There is clearly much yet to discover for the determined researcher.

Searching for ecological order in butterfly life history and resource use

Identifying groups of butterflies for hostplant strategies provides valuable insights into their biotope associations, immediately appreciated by geographers who understand the links between rock, soil, vegetation and climate. But, this line of reasoning, concerning life history strategies and multiple adaptations in organisms, can be carried further. As organisms are constrained by their resource base – entreated to adapt to it – it would be expected that their biotope affiliations (ecological differences) would be reflected in their life history strategies and ecological parameters without reference to their specific hostplants. A classification of species, then, based on their life history and ecological parameters carries expectations:

1 Species will tend to cluster or order for ecological attributes along lines of vegetation categories and biotopes.

2 Species belonging to ecological groups will also tend to occur together more frequently in vegetation units than species belonging to different groups.

3 Closely affined butterfly species in the classification will tend to share specific ecological characteristics and life history strategies.

4 Inevitably, species having a common ancestry, especially those sharing hostplant taxa, will tend to cluster in such a classification, but so will unrelated species that share hostplant taxa or which exploit different hostplants that tend to co-occur within vegetation associations.

5 Similar ecological groups or subdivisions of groups should also emerge from clustering and ordination of butterfly species on niche parameters, such as light, heat and substrate attributes.

These expectations are largely met in a classification of British butterflies based on 128 binary state ecological attributes, describing all stages of butterfly life cycles (Shreeve *et al.*, 2001). This classification avoids the use of variables that explicitly identify or describe biotopes, such as vegetation seral stage, soil type and specific hostplants. Both multivariate clustering and ordination identify ecological clusters of species that occupy distinctive vegetation types (Fig. 5.4) though the correlation with the simple list of biotopes (Dennis, 1992a; see Appendix 7) is poor. This is probably due to biotope divisions being overly broad and including a wide variety of vegetation categories, for instance, grass communities in woodland glades. Some agreement occurs relating the ecological groups identified with adult co-occurrences on BMS transect sections, despite the fact that the latter does not strictly identify breeding species associated with specific transect vegetation types. As would be expected, the classification reveals a phylogenetic component to the grouping of species. However, this is stronger in some families (e.g., Lycaenidae, Hesperiidae, Pieridae, Heliconiinae) than others (Satyrinae), tending to link congeneric taxa. Even so, members of all families are distributed throughout the different ecological groups. An example is the association of *Coenonympha pamphilus* with three members of the Pyrginae.

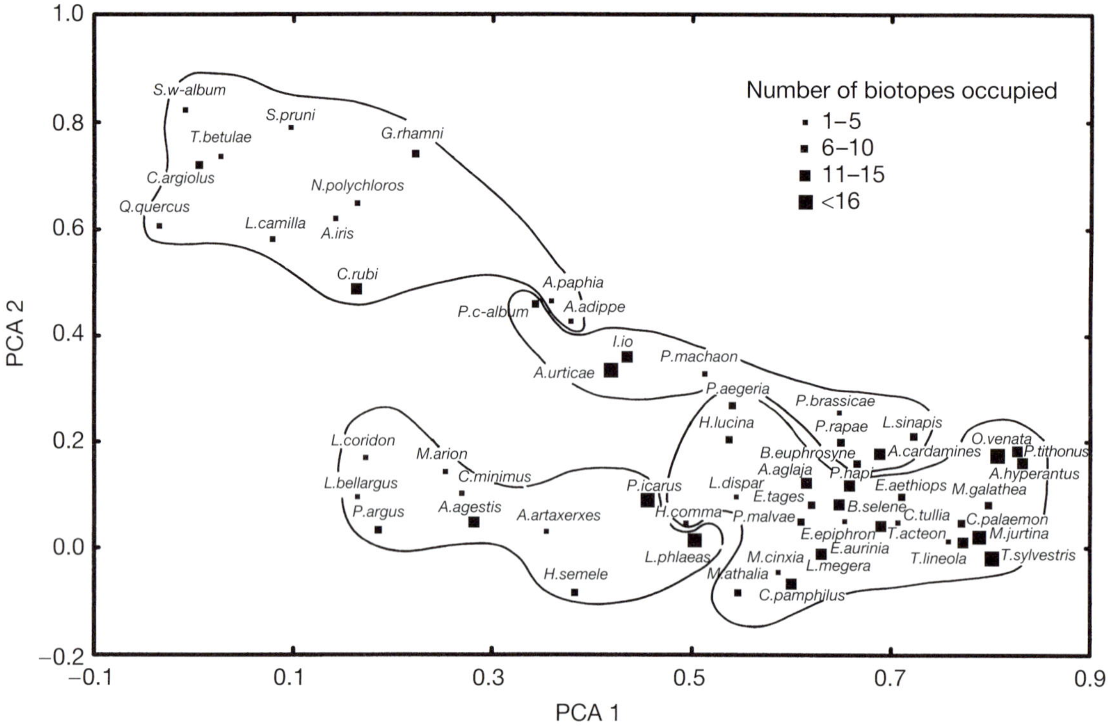

Fig. 5.4 Ecological clusters of British butterflies. An ordination plot (principal components) of 57 butterfly species based on 128 ecological (resource) attributes. Four groups of butterflies are highlighted by clusters (see text). (After Shreeve *et al.*, 2001, with kind permission of Springer Science and Business Media.)

Usefully, these analyses relate the ecological groups identified to key ecological attributes. What becomes clear is that no single factor or suites of related factors adequately describe the requirements and thus ecological affiliations of species. Even so, there is a tendency for some variables to be more prominent in determining group membership, for instance voltinism and attributes describing the physical structure of the egg, larval and pupal milieu. What is particularly interesting about this ecological classification is that it reproduces a number of the basic life history strategies associated with the habitat templet (Southwood, 1977, 1988) and the C-S-R strategies of Grime (1979), and it does so from a very extensive array of ecological variables. For instance, it distinguishes ruderal species (R strategists) associated with temporary habitats from those tuned to more competitive strategies (C strategists) found in longer lived, richer habitats such as tall grass and high forest. It also distinguishes species under stress found in relatively harsh or extreme environments, for example acid heathland and lime-rich heath (S or adversity strategists). The latter are generally members of the Lycaenidae with strong ant associations. As would be expected, that species grade between these extremes of conditions is in line with variation in conditions.

This work is just a start and is being extended by research at Oxford Brookes University (T. G. Shreeve, personal communication). Although, we perhaps know more about British butterflies than butterflies of any other part of the globe, these data remain a fraction of what is still unknown but required for a complete understanding of butterfly ecology, ecological relationships and associations.

Chapter 6

THE HABITAT CONTEXT FOR BUTTERFLY POPULATIONS

The metapopulation approach is based on the notion that space is not only discrete but that there is a binary distinction between suitable and unsuitable habitat types. If this does not fit one's idea of a particular environment, one is probably better off in using some approach other than the metapopulation approach. (Hanski and Gilpin, 1997, courtesy of Elsevier)

FROM POPULATIONS TO METAPOPULATIONS

Up to this point we have considered butterfly habitats largely in the context of single populations occupying single habitats on one site. Broadening this outlook we move to multiple interconnected but discrete populations occupying separate habitats on different sites (a **metapopulation**), each isolated by intervening ground supposedly absent of habitat called the **matrix**. It is useful for understanding the nature of matrix, and the status of populations, to see how one site can grade into many others. Populations, as identifiable systems, are defined largely on interactions (connectivity) among individuals (Dempster, 1989; den Boer and Reddingius, 1996:46). Connectedness depends on the capacity of individuals (e.g., propagules, gametes, seeds) to move and a butterfly's lifetime movement (**lifetime track**) is strictly limited by distance and time, so much so that for a population of most British butterfly species to be truly **panmictic** (viz., random breeding among individuals), it would necessarily occupy a small area (<10–50 ha) (Warren, 1992a). Populations of a small number of species with capacity to migrate across continents occupy vast areas (e.g., red admiral *Vanessa atalanta*; painted lady *V. cardui*). As the habitat (patch) on which a population is based becomes larger, it is less likely to be uniform in the type, quality and quantity of resources. Consequently, the population is increasingly likely to break down into areas of higher and lower population density, and thus local panmictic units or local populations, centred on areas where resources overlap or where there is a peak in resource quality, separated by larger areas lacking *obvious* resources (Fig. 6.1b). The population then becomes a **multipartite population** (Den Boer and Reddingius, 1996) comprising overlapping, interacting groups with indistinct boundaries. As habitat patches become increasingly distinct from surrounding ground, or their mutual isolation increases, interactions among them concomitantly decline and the populations fragment into distinct units (Dempster, 1991). Though disconnected into patches of suitable habitat, the patches may still be connected by regular transfers of individuals. In this situation, some distinction is achieved between the multiple population units as a system and its component parts (each local population); we have arrived at the typical metapopulation – a population of populations (Fig. 6.1c) (Levins, 1969, 1970).

There is not space here to recount the developments in metapopulation biology or its wide terminology (see Hanski and Simberloff, 1997). The basic concept is derived from **island biogeography theory** (MacArthur and Wilson, 1963, 1967) – at a single species level interacting populations on islands within archipelagos are a special case of metapopulations – and in effect it encompasses terrestrial island biogeography in which habitat patches are islands amid matrix, a sea of non-habitat (Box 6.1) (Driscoll, 2005). Underlying it, I am reminded, is den Boer's (1968, 1981) 'spread-of-risk'

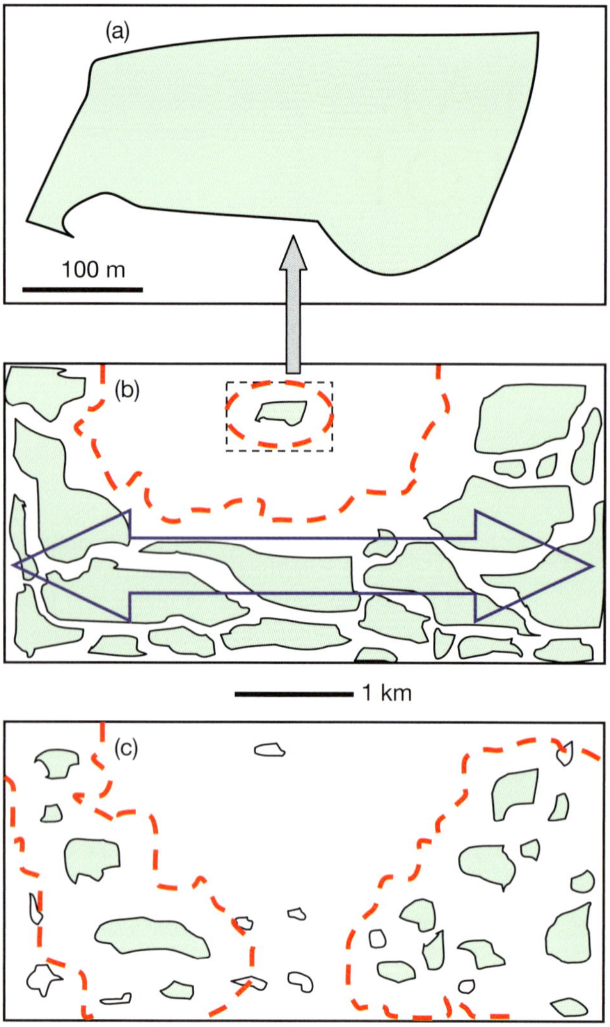

Fig. 6.1 Relationship of population structures to habitat components in multipartite populations and metapopulations. Local population (a) on the periphery of a multipartite population complex (b) in which there is movement throughout the system, and its degeneration to two metapopulations with habitat fragmentation (c). Green areas are populated habitat patches and white areas are unpopulated habitat (population persistence <5 years) amid matrix. Pecked lines indicate the bounds of most (e.g., 95%) of the individuals recaptured. A classic case of an outlying population being lost is that of the Scotch argus *Erebia aethiops* at Grassington in the Yorkshire Pennines, UK.

concept (A. C. G. Morton, personal communication). Key to these models is that the system, the metapopulation, is in equilibrium; that is, extinctions on some patches are matched by colonizations on others, so that the whole system is considered to be in homeostasis. Underlying this essential aspect of metapopulations are some simple principles (listed below). The focus of attention in this chapter is on how the view of a resource-based definition of habitat affects interpretation of butterfly metapopulations. In doing so, our attention is on spatially realistic metapopulations that may be mapped over the ground, discrete patches for butterfly species in real landscapes, often arising from biotope fragmentation, as opposed to mathematical models.

BASIC PRINCIPLES OF METAPOPULATIONS

The basic principles of metapopulations relate to habitat patch area and isolation of (distance between) patches; from these two variables can be modelled colonization, extinction, immigration and emigration. The principles below are made on the assumption that no distinction in habitat quality or resources exists.

Box 6.1 **Metapopulation models: structure and dynamics**

The simplest form of a **metapopulation** (Fig. B6.1a) envisages local populations occupying discrete patches in a uniform plane linked by migration. It is a landscape stripped down to its bare essentials of identically sized and shaped patches of consistently homogeneous habitat in an empty matrix, the matrix an isotropic plane with circular **isodapanes** around each habitat patch (effectively equal 'transportation cost' lines; the term from Alfred Weber who applied it in industrial and economic geography) in this case describing lines of equal movement (Wright, 1940; Levins, 1969; Hanski and Gilpin, 1997). It describes a persistent equilibrial population comprising a patchwork of infinitely numerous linked local populations. Levins modelled the fraction of local patches occupied as a density-dependent outcome (equilibrium) between the rate of successful colonization into empty patches and the rate of extinction in occupied patches. In his model, each patch has an equal probability of population extinction and receiving immigrants. As the number of occupied patches increases to unity then extinction in occupied patches exceeds the rate of colonization or recolonization into empty patches. Conversely, as the number of occupied patches approaches zero, then colonization into empty patches exceeds extinction in occupied patches. Levins (1970) was responsible for generating a set of metapopulation models of the general form:

$$\delta p/\delta t = i - e,$$

where p is the fraction of homogeneous patches occupied by a single species over time t, and i and e are common immigration (colonization) and extinction rates, respectively, measured across the entire set of patches. This model has been extended by Hanski (1982) and Gotelli (1991); their models, in turn, are shown to be extreme variants of a single model, which predicts the equilibrium fraction of sites occupied as a function of four parameters, two each for colonization and extinction (Gotelli and Kelly, 1993). The equilibrium differs for variants of the model and is not always singular (Hanski, 1982). A useful metaphor for Levins' single equilibrium model is of a city in which different lights are being switched on and off, but for which overall illumination remains the same (May, 1994). In this analogy, local populations turnover with a characteristic half-life (Gilpin, 1991).

Recently, Harrison (1994) argued that classical metapopulations are unlikely to be common and explains that an equilibrium condition between extinctions and recolonizations is improbable. To attain it, habitats must be uniform and occur in discrete patches, so that they can be occupied by local populations – small enough so that all local populations have a substantial risk of extinction, isolated enough to constitute separate populations, yet interconnected enough to permit recolonization. Local populations must also have asynchronous dynamics so that simultaneous extinctions are unlikely (Harrison, 1994; Hanski, 1997). Species fitting these conditions commonly occupy early successional habitats and have unexceptional dispersal ability (Harrison, 1994). Failing these conditions, a population is shown to fall into one of several other distinct categories in each of which persistence is more dependent on within-population than on metapopulation processes (Fig. B6.1b) (Harrison, 1991, 1994; Harrison and Taylor, 1997):

- Mainland–island model (Boorman and Levitt, 1973).
- Non-equilibrium model.
- Multipartite population model (den Boer and Reddingius, 1996).
- Patchy population (foraging) model.
- Transient population (migration) model.

These alternative models have been shown to be part of a continuum relating to variation in patch size and isolation. In the first case, the **mainland–island model**, at least one patch (the mainland) is large enough to resist stochastic (demographic and environmental) extinction. The balance between extinction and recolonization in the smaller patches is irrelevant for metapopulation persistence since persistence is assured by the existence of the mainland population. The multiple species case of the mainland–island model is the familiar equilibrium island biogeography model (MacArthur and Wilson, 1963, 1967), though a Levins-type model in which colonization depends entirely on a propagule rain (migrants) external to the metapopulation in question also equates with the island biogeography model (Gotelli, 1991).

Non-equilibrium [meta]populations and transient populations lie at extremes of isolation. In the former, local populations are virtually or completely isolated from one another. Population dynamics are generated largely by natality and mortality in each isolated unit. Consequently, ongoing local extinctions (extinctions > recolonizations) may result in a declining or retreating metapopulation (there is said to be **extinction debt**) and regional extinction. Another type of non-equilibrium, **transient populations**, occurs with progressive recolonizations (recolonizations > extinctions) and may produce a rejuvenated or advancing metapopulation associated for instance with range expansion. **Patchy populations** are akin to foraging models; the individual patches do not support separate local populations, transfers between resource patches exceed internal movements and the average individual's home range

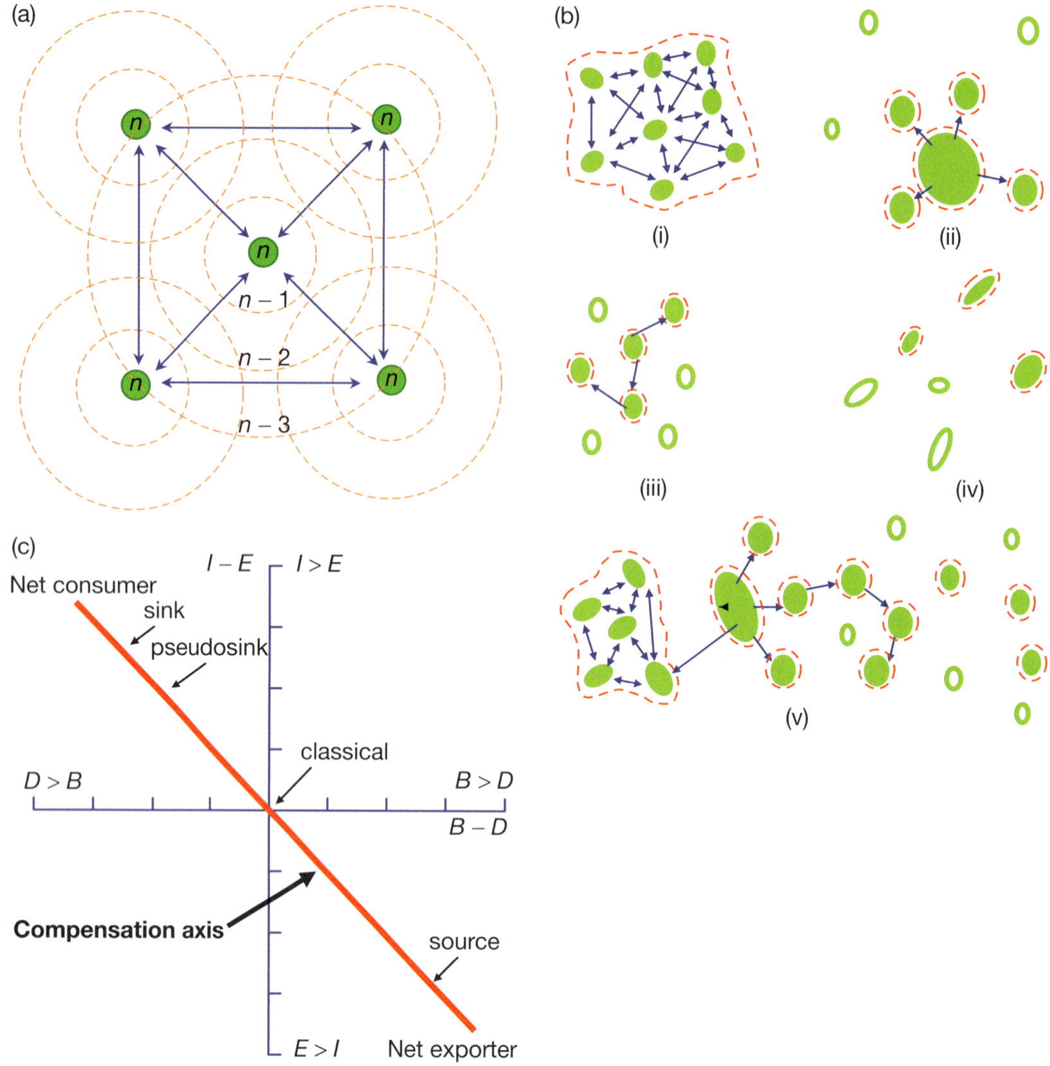

Fig. B6.1 (a) Basic features of metapopulations. Shown are equally sized and spaced habitat patches in an isotropic landscape in which movement is equal in all directions and over the entire surface. Shaded circles are patches, arrows show movements occurring in all directions and dotted lines illustrate equal number of movements from each patch. The dotted lines are also 'cost' lines inasmuch as individuals travelling further from their natal patches incur a cost (i.e., energy lost, fat reserves, etc.). The fall off in movement is exponential (see Fig. B6.4a(ii)). Thus where patches have a population of n individuals, the number migrating out is some fraction of n, $n-1$ and with the costs of migrating from a patch increasing, then $n-3<n-2<n-1$. (b) Different types of metapopulations: (i) patchy populations, (ii) mainland–island or Boorman–Levitt model, (iii) classical Levins model, (iv) non-equilibrium model showing extinction debt, and (v) combination (from left to right) of patchy populations, mainland–island, Levins and non-equilibrium situations as size, isolation and condition of patches changes. Solid patches, colonized; white patches, vacant; dashed lines, boundaries of local populations; arrows, exchanges. Note that no linear scale is provided and metapopulations are scaled by individual movements of organisms. (Modified from Harrison and Taylor, 1997, courtesy of Elsevier.) (c) The compensation axis for metapopulation units (local populations). B, birth; D, death; E, emigration; I, immigration; per capita rates; classical (Levins) situations have $I \approx E$ and $B \approx D$. (From Thomas and Kunin, 1999, courtesy of the British Ecological Society and John Wiley and Sons Ltd.)

Table B6.1 Population and habitat units. (From Thomas and Kunin, 1999, courtesy of British Ecological society and John Wiley and Sons Ltd.)*

	Birth (B), death (D)	(Actual or potential)	Immigration (I), emigration (E)	$I \Rightarrow$ zero†
Sites with populations				
Sources	$B > D$‡		$I < E$	Usually limited impact
Classical populations	$B = D$‡		$I = E$	Usually limited impact
Pseudosinks	$B < D$ ($B > D$‡ at lower density)		$I > E$	Decline
Sinks	$B < D$		$I > E$	Go extinct
Sites without populations				
Potential sinks	B < D			
Vacant habitat‡§	B > D		$I \approx 0$	
Sieves§	B > D	and $(E - I) > (B - D)$	$I < E$	

* See Pulliam (1988) and Watkinson and Sutherland (1995).
† Effect on local population size of reducing immigration rate to zero.
‡ Allee effects may result in $B < D$ at very low densities for populated areas (converting these areas into true sinks), and $B > D$ at higher densities, leading to alternative stable states.
§ When colonized, these habitats may become either net consumers or exporters.

inhabits more than one patch in a lifetime. Dispersal takes place on a scale greater than that of local events causing population fluctuations. As peaks in resource suitability will undoubtedly break up areas of apparent habitat into a mosaic of realized habitats, a **multipartite population** (see Fig. 6.1) may represent an intermediate situation between a non-equilibrium population and a patchy population (den Boer and Reddingius, 1996). Den Boer and Reddingius envisage a narrower definition for a multipartite population than that of Andrewartha and Birch (1954:670). In effect, the whole population uninterruptedly inhabits an area much larger than the average area occupied by a local population.

In transient populations, all the patches may be vacant at some stage during the year as a consequence of individuals migrating between seasonal resource patches. The patchwork may also be used unpredictably, colonized during one year but not the next. Each patch in the patchwork typically includes less than a full complement of resources required by the species during development. Classic examples of species belonging to this category engage in long-distance reversed migrations such as the nymphalids, the painted lady *Vanessa cardui* and red admiral *V. atalanta*. However, other species that use widely separated seasonal resources may also belong to this group (e.g., brimstone *Gonepteryx rhamni*).

The various spatial models have led to the concepts of source and sink populations, pseudosinks and the rescue effect (Table B6.1). **Sinks** are local populations that would go extinct in the absence of immigration from **sources** where growth rates at low density are positive (cf., Hanski, 1997, with Pulliam, 1988). **Pseudosinks** are patches that can support a local population without immigration, but where density would decrease relative to carrying capacity without immigration from source populations (Holt, 1985). Extending these concepts, the **rescue effect** refers to reduced extinction probability in a local population owing to immigration from a source population raising population size (Brown and Kodric-Brown, 1977).

It soon becomes clear that the existence of different types of metapopulations may simply be a matter of spatiotemporal scale or changes in migration capacity (highly vagile butterflies (e.g., small tortoiseshell *Aglais urticae*) integrate heterogeneity over greater extents than do sessile ones; Wiens, 1997). With changing size of patches and their isolation, metapopulations may present a mixture of model types, for example a continuum from patchy populations at one extreme to non-equilibrium metapopulations at the other, including an interdigitation of variably sized mainlands (Fig. B6.1b(v)). In the same way, as conditions change (e.g., climatic warming, changing management) different models may apply for the same organism over the same patchwork at different times, as in the case of silver-spotted skipper *Hesperia comma* on the North Downs in England (Hill *et al.*, 1996). This warns against casting a species into a single metapopulation mould. Thomas and Kunin (1999) have shown how each (meta)population system and part system (local population unit occupying a 'habitat') can be described by two axes based on four population parameters: per capita birth, death, emigration and immigration rates (Fig. B6.1c). This graph illustrates that the differences between sources, pseudosinks, sinks and sieves is a gradual one with no hard and fast divisions. In Fig. B6.1c, a population at a single site may be represented by a 'cloud' of points each taken at a specific point in time; as such, a population can move from being a source to a pseudosink to a sink or vice versa. The balance of these four processes describes the demographic space into which population units fall. At equilibrium, populations lie along a line defined by $(B + I) - (D + E) = 0$, since with equilibrium, factors increasing population size (birth and immigration) necessarily counter the forces decreasing it (death and emigration). This line is termed the **compensation axis**. Thomas and Kunin explain that the compensation axis serves as an attractor in demographic space; that is, any population that moves away from this line will be drawn back towards it (e.g., if immigration increases, there is tendency for a density-dependent response in which birth rate tends to decline and/or emigration to increase). Clearly population units may not be at equilibrium, and therefore stray away from the compensation axis, as is the case with transient populations and populations subject to deterministic changes to habitats (vegetation succession) that force them towards extinction. Empty 'habitat' units may occur where:

1 The unit is unsuitable (lacks key resources).
2 Suitable units are yet unoccupied.
3 Units are subject to Allee effects ($B - D < 0$ at very low numbers or densities; e.g., in species that mate after dispersal) or are **sieves** (leaky, isolated units where the drain from emigration greatly exceeds any immigration) (Thomas and Kunin, 1999).

Mathematical modelling of metapopulations is advancing rapidly in sophistication (Hanski and Gilpin, 1997; Moilanen and Hanski, 1998; Settele *et al.*, 1999), increasingly becoming spatially and temporally explicit, advancing to trophically dependent species and communities (Hanski and Gilpin, 1997) and gradually adopting habitat quality and matrix structure (Moilanen and Hanski, 1998). Expectations from theory on equilibria are that threshold numbers of local populations (**MVM**, **minimum viable metapopulation**) and threshold numbers of suitable habitat patches (**MASH**, **minimum amount of suitable habitat**) exist below which metapopulations are driven to extinction and above which they can persist for long periods (Hanski, 1997). Theory also shows that multiple equilibria can be generated for metapopulations (Gyllenberg *et al.*, 1997) and that spatial structuring of populations facilitates the coexistence of species, especially those in trophic relationships (Nee *et al.*, 1997). The models support a commonsense view in indicating that species at higher trophic levels are more constrained than those at lower trophic levels (Holt, 1997). For a deeper view of metapopulation theory and findings, see Ehrlich and Hanski (2004).

- **P6.1: The larger the patch size (habitat), the larger the population and the longer it is expected to persist.**
- **P6.2: The larger the patch size, the larger the number of emigrants expected to emanate from it as potential colonists for other patches. Smaller patches may nevertheless emit more emigrants *per capita* of their populations.**
- **P6.3: Larger patches, compared with small patches, are expected to receive more immigrants from the same source if equally isolated from it.**
- **P6.4: The larger the distances separating patches, the fewer individuals are expected to migrate successfully between them.**

- **P6.5: The persistence of populations within a metapopulation, and the metapopulation itself, depends on the number of patches and their spatial arrangement as well as on patch size and isolation.**

Persistence of populations in a habitat-homogeneous metapopulation is a function of patch sizes, isolation of patches, number of patches and their spatial arrangement. It is also dependent on evolutionary changes occurring within populations occupying the patches; for instance fragmentation may select for increased migration rate and body: wing size differences (Dempster *et al.*, 1976; Hill *et al.*, 1999b; Van Dyck, 2003; Van Dyck and Baguette, 2005; Merckx *et al.*, 2006). Larger patches have larger populations that persist longer (**P6.1**), emit most emigrants (**P6.2**) and, all things being equal, receive most immigrants (**P6.3**). However, small patches may emit more emigrants *per capita* for several reasons; prime among them is that individuals contact the edge of habitats more frequently in small patches, increasing their probability of leaving, and that lack of individuals – cues for the occurrence of mates – may more readily induce departure (cf., Allee effect). Unexpected relationships emerge in studying population units in metapopulations and emphasize the importance of a holistic analysis of explicit spatial attributes of the metapopulation. Thus, smaller patches may receive more *per capita* immigrants if only because they receive contributions from larger sources. Time to extinction is generally modelled as a simple function of patch area. Even if patches are treated as having equivalent habitat quality, extinction can occur in a patch from a number of causes related to small patch size (e.g., demographic stochasticity, emigration exceeding immigration, asynchrony in emergence of sexes, inbreeding (Saccheri *et al.*, 1998; Haikola *et al.*, 2001; Nieminen *et al.*, 2004)). Migration is distance and time dependent; it is typically modelled as a negative exponential function or inverse power function, with very few individuals covering large distances (**P6.4**; see Box 6.4).

In the negative exponential function, the probability (M) of an individual moving a certain distance D for locations $i \ldots j$ is:

$$M_{ij} = e_{ij}^{-kD} \text{ or } M_{ij} = \exp(-kD_{ij}).$$

Whereas in the inverse power function,

$$M_{ij} = D_{ij}^{-k},$$

in which k is a migration constant (describing the slope of the regression; i.e., it affects the rate at which movement declines with distance, with large k a more rapid decrease) (Hill *et al.*, 1996; Baguette *et al.*, 2000). These differ from a linear expectation in decay of movements:

$$M_{ij} = -kD_{ij}.$$

An important difference between these two models is that the negative power function has a longer tail, and a few individuals travel further. The metapopulation model differs from the island biogeography model, except perhaps extreme mainland–island versions of metapopulations, inasmuch as each patch provides a significant source as well as a destination for individuals (see Box 6.1). Thus, the persistence of the metapopulation depends on the number of patches and their spatial arrangement (i.e., shape, neighbourhood associations, etc.) (**P6.5**). To some extent, size of patches can be offset by number of patches for the same degree of isolation and similarly isolation of patches can be offset by the number of patches and their sizes. But, both modelling and empirical findings indicate that critical limits in size, isolation and number are reached below which the metapopulation starts to fail (Hanski, 1997; Hanski and Simberloff, 1997). *Expected* time to metapopulation extinction has been found to be closely related to the product of $p\sqrt{H}$, where p is the fraction of occupied patches (assuming homeostasis) and H is the number of suitable patches (Hanski, 1997). On the assumption that most patches in a patchwork are occupied, a minimum of some 15–20 well-connected patches are required for long-term persistence (equilibrium; Thomas and Hanski, 1997). This is termed the **minimum viable metapopulation** (**MVM**) but great caution should be used in relying on particular numbers of patches in specific cases (see below and Chapter 9). Supporting it there is the additional concept of the minimum number of habitat patches for metapopulation persistence (**minimum amount of suitable habitat**, **MASH**) (Hanski, 1997).

THE LINK BETWEEN STRUCTURE AND DYNAMICS IN METAPOPULATIONS

It soon becomes evident that there are numerous structural variants described for metapopulations (see Box. 6.1) and that multiple populations form a graded series, in which not all units are connected nor is the system in homeostasis, as in the case of the meadow brown *Maniola jurtina* (Shreeve *et al.*, 1996b) and ringlet *Aphantopus hyperantus* (Sutcliffe *et al.*, 1997a). Thomas and Kunin (1999) have shown how each (meta)population system and part system can be described by two axes based on four population parameters: per capita birth, death, emigration and immigration rates (see Box 6.1). This also defines the status of individual population units as **sources**, in equilibrium, **pseudosinks** or **sinks**. There is controversy in the literature as to how frequently classical (Levins type) metapopulations are expected to be found (see Box 6.1). Hanski and Kuussaari (1995) consider that as many as 57 of 94 resident Finnish species may belong to the Levins-type metapopulation, and Hanski and Thomas (1994) regard 50% of the 60 British species as conforming to this genre.

A word of caution is needed on the link between structure and dynamics: modelling and empirical surveys have been successfully combined to illustrate that the nature of colonization and extinction in butterfly metapopulations can do anything but follow the Levins model. For instance, both synchronized extinctions and colonizations can occur across population units (Sutcliffe *et al.*, 1997b) and warn against too great a reliance on numbers of populations units. Rohde (2006) would impress on us that systems in non-equilibrium are more widespread than those in equilibrium. As noted before hand (see Chapter 3), one crucial control in attaining at least a homeostatic, if not equilibrium, state is habitat heterogeneity; increased heterogeneity results in decreased population variability (e.g., bush cricket *Metrioptera bicolor* (Philippi); Hanski, 1991).

EMPIRICAL STUDIES OF BUTTERFLY METAPOPULATIONS IN BRITAIN: HABITAT QUALITY MATTERS

Empirical studies have been generated by the overriding significance given to patch size and patch isolation in metapopulation modelling. If these key variables control local population incidence on patches then they should explain a large part of the variance in patch incidence and provide accurate predictions of incidence. This approach involves application of logistic regression in which a binary response (dependent) variable (patch incidence: 0, absent; 1 present) is regressed against patch area and isolation. It has been carried out on a number of species (e.g., silver-studded blue *Plebejus argus*, silver-spotted skipper *Hesperia comma*, Lulworth skipper *Thymelicus action*, heath fritillary *Melitaea athalia* (Thomas and Harrison, 1992; Thomas *et al.*, 1992; Thomas and Jones, 1993)). In these studies, patch area is a zone of hostplant suitable for egg laying and development. Isolation of patches has been measured in two ways, to the nearest occupied patch and to the nearest patch occupied or not. The model, with isolation measured to nearest occupied patches, proved to be highly significant for all four species (variance accounted by Rho^2, 46.3–85.0%), though patch area was significant only for *P. argus* and *H. comma*.

It is clear that, for some species, isolation and area can explain a great deal of the number of occupied patches in a patchwork of potentially suitable habitat. It is possible that they may explain more variation with increasingly refined measures of isolation (see Moilanen and Nieminen, 2002, for a review of connectivity measures). Hanski (1994a, 1994b) recommends the use of a measure for isolation (I) that takes into account distances to all patches weighted by their areas (their potential for emigrants and colonists):

$$I = -\Sigma \exp(-\alpha D_{ij}) A_j$$

where $A_{j \neq i}$ is patch area, D_{ij} is distance between patches i and j and α is a constant describing the survival rate of migrants over distance D_{ij}. From this relationship, Hanski derives a measure of connectivity for populations defined in terms of the expected rate of immigration (number of individuals arriving per unit time) to a focal patch:

$$S_i = A_i \zeta im \Sigma_{j \neq i} \exp(-\alpha D_{ij}) p_j A_j \zeta em$$

where S_i is the connectivity of patch i, A_i is the area of patch i, p_j is the probability of occurrence of a population in patch j, $1/\alpha$ is the average migration distance of a species, and ζim and ζem are two parameters describing the scaling of immigration and emigration rates with the patch area. The isolation measure, I, is more typically used now in patch incidence studies (Hanski and Pöyry, 2007). In reality, measuring to patches beyond a certain distance adds very little to target patch status owing to the, at least, exponential decline in migration distances and the impact of intervening opportunities (Stouffer effect; see below), but the notion of incorporating potential 'influence' of individual sources is highly important. Even so, however apparently successful are such distance–area assessments, there is a question of how much of the variation accounted, and not accounted, is owing to habitat quality. Can we afford to ignore quality of habitats bearing in mind that resource quality varies spatially and is subject to constant changes with time?

Habitat quality has been regarded by the proponents of metapopulation models as being subsumed in patch area (Nieminen *et al.*, 2004). However, several, recent studies of metapopulations have demonstrated the importance of habitat quality (Dennis and Eales, 1997, 1999; J. A. Thomas *et al.*, 2001; Matter *et al.*, 2003; Valimake and Itamies, 2003; Auckland *et al.*, 2004). The first of these was carried out on the large heath *Coenonympha tullia* in Northumberland, a butterfly of oligotrophic mires (i.e., blanket bogs, raised valley bogs) associated with *Sphagnum*, hare's-tail cotton grass *Eriophorum vaginatum* L. and cross-leaved heath *Erica tetralix* L. (Box 6.2; Fig. 6.2). A habitat quality index for the butterfly was developed from 21 variables scored for individual sites (Box 6.2). In this study, habitat quality was found to be the best predictor of *C. tullia*

Box 6.2 Habitat quality and the Northumberland large heath *Coenonympha tullia* metapopulation

A study of the Northumberland metapopulation for *Coenonympha tullia* (Dennis and Eales, 1997, 1999; Eales and Dennis, 1998) (Fig. B6.2a) demonstrated that habitat quality ranked above habitat area and isolation in accounting for occupied sites, despite two measures being used to describe habitat area (both mire area and hostplant/nectar source area) and three measures for isolation:

$$I_1 = \log(D_{ij}) \quad \text{equation 1}$$

$$I_2 = -\sum \exp(-D_{ij}) \quad \text{equation 2}$$

$$I_3 = -\sum \exp(-D_{ij}) A_j \quad \text{equation 3}$$

where D_{ij} is the edge to edge distance (km) to the nearest neighbouring occupied site in equation 1 and to the five nearest neighbours in equations 2 and 3; $A_{j \neq i}$ is habitat area (100 m^2 units). I_2 was the most successful isolation predictor of the three isolation measures. The habitat quality index was the most successful predictor for the 1995–1996 data, for the 1997 data, and for the combined dataset. The index for sites is a sum of negative and positive scores for 21 site quality variables (Table B6.2a) based on the strength of their correlation with *C. tullia* incidence; site quality for *C. tullia* will vary if

Table B6.2a Ecological variables for *Coenonympha tullia* sites in Northumberland. (Adapted from Dennis and Eales, 1999, courtesy of Elsevier.)

Number	Description of attribute
Positive attributes	
1	Vigorous, dense growth form of *Eriophorum vaginatum*; flowering, no bare ground visible and without invasion from other grasses (e.g., purple moor-grass *Molinia caerulea*) or sedges
2	Dense, vigorous growth form of cross-leaved heath *Erica tetralix*; flowering, no bare ground visible below plants and lacking invasion from plants other than hare's-tail cottongrass *E. vaginatum*
3	Physical overlap of hostplant and nectar plants, i.e., compact habitat including an admixture of both adult and larval hostplants
4	Tussock development of *E. vaginatum*; plants raised on peat mound above watertable level
5	Wet site, existence of surface water
6	Sheltered site, with valley location, lee side of hill or tree shelter
7	Light grazing (1–2 animals ha^{-1})
8	Conservation action in progress beneficial for *C. tullia* (e.g., blocking drains on dry sites)
9	Low-intensity patch burning of site
10	Natural occlusion of existing drains
Negative attributes	
11	High-intensity burning of peat or low-intensity burning of whole site at one time
12	Overgrazing ($\geq$2 animals ha^{-1})
13	Severe trampling by stock or damage by vehicles causing crushing of vegetation
14	Drainage by narrow shallow drains
15	Drainage by extensive deep, wide land drains
16	Ploughing of site
17	Afforestation of site, usually with conifers
18	Site encompassed completely or almost entirely by forest
19	Advancing succession of grasses, bracken or trees; usually in association with either fencing or drying out of site
20	Active peat extraction
21	Chemical applications (e.g., liming, NPK fertilizers or herbicides); evident or indicated by landowner
Index	
HI	The value for each site is the balance of the binary positive and negative scores listed above added to an arbitrary value of 10, so as to avoid negative numbers

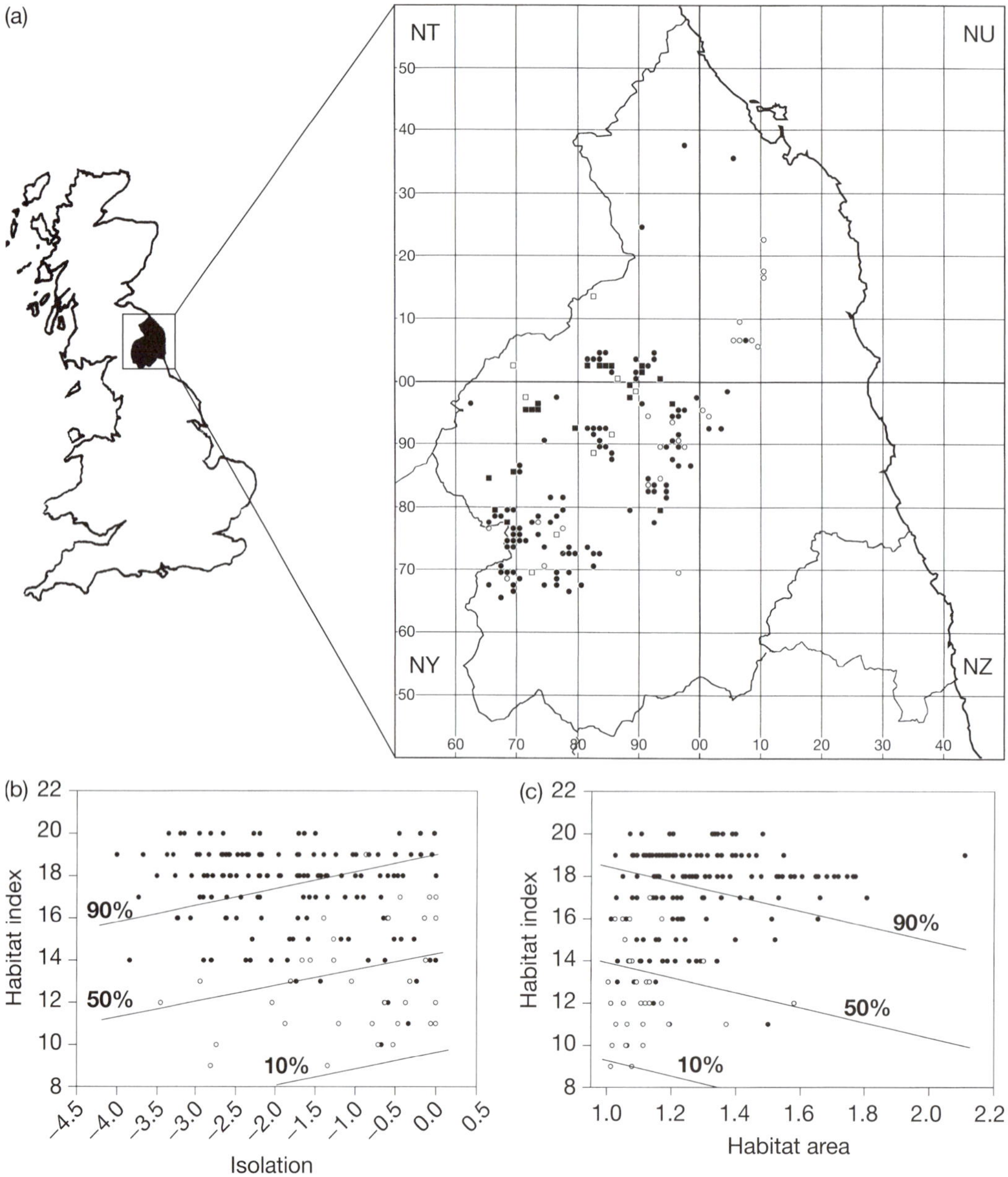

Fig. B6.2 (a) Distribution of *Coenonympha tullia* in County Northumberland during 1995–1997 (1 km squares): 1995–1996 (circles); 1997 (squares); vacant sites and unknown status (open symbols); and occupied sites (closed symbols). Inset map illustrates the position of Northumberland in the UK. (b) Distribution of sites in axes for habitat quality and isolation: colonized (black circles) and vacant (open circles) sites; and probabilities from logit regression of status on habitat quality index ($P < 0.0001$), isolation ($P < 0.01$) and habitat area (not significant). (c) Distribution of sites in axes for habitat quality and habitat area: colonized (black circles) and vacant (open circles) sites; and probabilities from logit regression of status on habitat quality index ($P < 0.0001$), isolation ($P < 0.01$) and habitat area ($P < 0.003$). Vacant sites > 400 m, surveyed in 1997, are excluded. (Adapted from Dennis and Eales, 1999, courtesy of Elsevier.)

Table B6.2b Frequency of habitat types for *Coenonympha tullia* sites in Northumberland.

Code	Biotope	Present	Absent
B1.1	Acid grassland/marsh unimproved	4	5
B1.2	Acid grassland/marsh semi-improved	1	0
B5	Marsh/marshy grassland	14	3
D1	Dry dwarf shrub heath	5	3
D2	Wet dwarf shrub heath	32	11
D6	Wet heath/acid grassland mosaic	17	11
E1.7	Wet modified bog	13	8
E1.8	Dry modified bog	5	1
E3.1	Valley mire fen	12	1
E3.2	Basin mire fen	5	2
E1.6.1	Blanket bog	61	13
E1.6.2	Raised bog	4	0
Total	All types	173	58

Multiple classification of sites are used. Habitat types are defined as in the *Handbook for Phase 1 Habitat Survey* (Anon, 1990).

only because habitats are found on different substrates with distinct vegetation associations (acid grassland, shrub heath, bog, fen) (Table B6.2b). The study was based on 181 sites; the 1995–1996 data comprised 148 sites, 122 colonized by the butterfly. A further 33 were added in 1997, the butterfly occurring on 23 of them. Results from the 1995–1996 study (Dennis and Eales, 1997) were used to predict site occupancy on the sites surveyed in 1997 (Eales and Dennis, 1998). Predictions were based on logistic regression and discriminant function analysis. Logistic regression equations take the form of:

$$Y = e^r/(1 + e^r)$$

where r describes the regression parameters ($b_0 + b_1X_1 + b_2X_2 \ldots b_iX_i$). Although the 1997 survey was carried out in a new area, significantly further north ($F_{1,198} = 5.4$, $P = 0.02$) and west ($F_{1,198} = 9.9$, $P = 0.002$) than previously, site data for the two year groups do not differ significantly for mean isolation, habitat area or habitat quality. However, they do differ in altitude and vegetation associations. Sites surveyed in 1997 are at a significantly higher altitude (mean 321 m) than those surveyed earlier (mean 275 m) ($F_{1,198} = 26.1$, $P < 0.001$) and occupied sites in 1997 are found significantly more frequently on wet and dry heath and somewhat less on marshy grassland and blanket bog than in 1995–1996 ($\chi^2_{(2)} = 9.1$, $P < 0.02$). Discriminant analysis based on three variables describing habitat quality (positive and negative habitat indexes (PHI, NHI); see Table B6.2a for positive and negative attributes) and habitat area (A) (Wilks' $\lambda = 0.5$, $F_{(3,144)} = 48.0$, $P < 0.001$) successfully classified 91% of the 148 sites surveyed in 1995–1996 and correctly predicted 79% of sites surveyed in 1997; the difference in the number of correctly classified sites in the two year groups is statistically significant ($\chi^2_{(1)} = 4.39$, $P < 0.05$). The difference related to three large (>70 ha) unoccupied sites at particularly high altitude (>400 m); two of the sites have the highest altitudes in the database (>450 m). The additional 1997 sites have the effect of switching the relative and absolute significance of habitat area and isolation. Isolation was the least successful predictor for the 1995–1996 data but still significant. However, habitat area was the least significant variable for the combined years datasets and lost significance. Significance was restored for habitat area by removing the three vacant sites at >400 m surveyed in 1997 from the equation (Table B6.2c). The relative importance of habitat quality, compared with habitat isolation and area, is evident in the distribution of probabilities for all sites over all years (Fig. B6.2b, c); Figure B6.2c illustrates the relationship between habitat quality and habitat area once the high level sites have been removed; area is still not as important as habitat quality. The three large unoccupied sites, apparently with suitable consumer resources for the butterfly, at elevations exceeding the group of sites on which the initial model was based, indicates that other habitat quality factors (e.g., thermal environment) may be involved that were not measured or that the butterfly has a biennial cycle at high altitude and this was missed during the 1997 survey.

Table B6.2c Logistic (logit) regression equations relating butterfly incidence at sites to habitat area (A, ha), habitat quality (HI) and site isolation (I_2, km). (Adapted from Dennis and Eales, 1999, courtesy of Elsevier.)

	Constant b_0	Habitat area ($\log_{10}A$)	Habitat index (HI)	Isolation (I_2)	Deviance	Significance of model (*P*)	R^2
Equations 1995–1996							
1	−12.92***	4.36**	0.66***	−1.10*	83.49	<0.0001	61.2
Equations 1995–1997							
2	−9.58***	0.67	0.60***	−0.81*	81.12	<0.0001	51.4
3	−20.40***	8.21*	0.73***	−0.83*	97.58	<0.0001	57.2

Parameters are calculated using a maximum likelihood function (quasi-Newton minimization algorithm). R^2 is obtained by a least squares estimation procedure.
* $P < 0.01$, ** $P < 0.001$, *** $P < 0.0001$.
Equations *N*: 1995–1996: 148; 1995–1997: 181; 1995–1997 (three highest sites excluded): 178.

Fig. 6.2 Meathop Moss, Cumbria, UK: a valley-raised bog biotope for the large heath *Coenonympha tullia* (inset). (Courtesy of Jim Asher.)

incidence in a habitat patchwork, accounting for as much variance (48%) as patch area and isolation jointly (46%). The findings suggested that the best conditions for the butterfly are provided by dense, vigorous, contiguous or overlapping growth of *E. vaginatum* and *E. tetralix* on sheltered mires, typically on peat (>0.5 m deep) with the watertable at or just below the surface. An important component of the mires is the tussock growth form of *E. vaginatum*; the tussocks may enable the butterfly to escape winter submergence that, experimentally, has been shown to cause high larval mortality (Joy and Pullin, 1997). It was subsequently found that habitat quality was the most prominent variable in correctly classifying 33 additional sites, surveyed in 1997, to the northwest of the 1995–1996 study area. Subsequent studies by J. A. Thomas *et al.* (2001) also demonstrate the importance of habitat quality at metapopulation level for three species, the Glanville fritillary *Melitaea cinxia*, Adonis blue *Polyommatus bellargus* and *Thymelicus acteon*. For those long-engaged in studying butterfly populations none of this should come as a surprise. A colleague, Ashley Morton, describes the situation well for small blue *Cupido minimus*. A large patch of kidney vetch may be of little use for ovipositing females requiring unopened flowers; the plant is **monocarpic** (living 1 year, flowering and fruiting and then dying) and it is not uncommon for large areas to be devoid of flowering plants in some years leading to local extinction of the butterfly (Morton, 1985).

These results were important for demonstrating that it is a mistake to ignore habitat quality. This does not mean that incidence cannot successfully be explained, on occasion, by isolation and patch area alone – in fact researchers will undoubtedly find that area, isolation and quality will assume varying importance as fragmentation increases, particularly isolation – but that there is now no excuse for ignoring measures of habitat quality in any study. A study by Gutiérrez *et al.* (1999) on the distribution of dingy skipper *Erynnis tages* on North Wales's Creuddyn Peninsula, illustrates how an orthodox approach of specific 'habitat' parameters – the quality of hostplants used in relation to management and shelter – linked to a metapopulation study can account for the incidence of the butterfly. Below, it is considered how patch quality may add to spatially realistic models based on patch area and isolation.

A number of principles of habitat quality for metapopulations can be made:

- **P6.6: Patches are not equal in resource composition, structure and connectivity (between resources outlets) and thus not equal in habitat quality.**
- **P6.7: Some resource other than consumables (larval hostplant and nectar) can be limiting for any life history stage. Utilities (e.g., roost sites, mate location sites) should be included in assessments of habitat quality.**
- **P6.8: Differences between local populations in basic inputs (topography, substrates, climate, land management) will inevitably cause differences in habitat quality.**
- **P6.9: Even patches with identical starting points will be unlikely to retain the same habitat quality trajectory owing to stochastic influences affecting inputs on sites.**
- **P6.10: Connectivity (between patches) can affect habitat quality as enemies and competitors are influenced (possibly differently to the target prey organism) by connectivity between habitat patches.**
- **P6.11: Although large patches more probably will have high-quality patches, this is not an invariable rule.**

Before examining aspects of patch quality, it is important first to consider the difficulties in patch identification. Patches can only be unequivocally designated when the habitat is identifiable as discrete homogeneous entities and as every fieldworker will appreciate the scale of variation depends much on the scale of measurement. As soon as there is some spatial variability in a resource in relation to vegetation units, or temporal variability in a resource used to define a habitat (e.g., a hostplant, weather conditions; Dennis and Sparks, 2006), there is difficulty in defining patch bounds. In most metapopulation studies it has been necessary to establish a set of arbitrary rules to define the patchwork. Typically, distinct patches are defined by a distance related to dispersal capacity (i.e., 25 m, *Hesperia comma* (Hill *et al.*, 1996); 20 m, *Plebejus argus* (Lewis *et al.*, 1997); 20 or 50 m, *Melitaea cinxia* (Nieminen *et al.*, 2004); 25 or 50 m, brown argus *Aricia* spp. (Wilson *et al.*, 2002), the alternatives based on the occurrence of scrub). Where patches are separated by smaller distances they are deemed to be part of the same patch. When habitats are described in terms of a range of consumable and utility resources, difficulties in patch designation increase substantially and habitat bounds inevitably must be based on an autecological survey of resource use. A procedure for doing so has been illustrated for the green hairstreak *Callophrys rubi* (see Box 9.2) (Vanreusel and Van Dyck, 2007; see also Dennis *et al.*, 2006b). It is likely that physical habitat bounds will generally be enclosed within 2 to 3 standard deviations of linear daily movement distances from the 'centre' or core of the habitat, data usually derived by applying mark–release–recapture of individuals (Shreeve, 1992b). When a vegetation-type surrogate is used for a habitat patch, then measures of habitat quality relating directly to distinct consumer and utility resource attributes should be included in the study.

Metapopulation studies typically designate patches on the basis of a single consumable resource: larval hostplants. Defined in this way, patch area makes no reference to the composition and structure of resources comprising habitat patches (**P6.6**) or to the connectivity amongst resources within patches (Dennis *et al.*, 2003, 2006b). Regarding composition of resources alone, variability in the presence or absence of supplementary resources will influence the viability of populations on patches and thus act as differences in overall habitat quality. If resource composition is identical in two habitats, but internal structure and connectivity differ, then so can the quality of resources for an array of reasons associated with how butterflies interact with one another and the resources' distributions. Compact and highly connected resources may well lead to very different density-dependent effects than diffuse and separated resources, influencing levels of mortality, natality, immigration and emigration. A large number of components and environmental agents influence potential differences in patch quality (**P6.7** to **P6.10**) and should be borne in mind in a study of multiple integrated populations units.

Patch quality has often been considered to be subsumed in patch area and it is important to see why this is and the consequences of this assumption (**P6.11**). A wide array of resource attributes may affect patch quality and population size (Table 6.1). Many of these attributes may be thought to correlate with patch size (i.e., resource density, complementarity, supplementarity, contexts, resource area, resource overlap, resource shapes, variation in resource comminution, connectivity, temporal variability). But, there is no firm reason why they should and the relationship between resource types and patch area has been little studied. The relationship between population density and patch size can therefore break down for very obvious reasons:

- A disproportionate increase in internal patch matrix with patch size (Matter and Roland, 2002; Matter *et al.*, 2003).
- Inequality in economic potential for land units of different size leading to increased intensity in the exploitation of larger units.
- Increased exposure – lack of shelter – of larger land units relative to boundary vegetation (Dover *et al.*, 2000).

Patches for target species in published metapopulation studies often range in size (area) over three or more orders of magnitude, usually from 0.1 to 100 ha. There can thus be a relationship between population size (or incidence) of a species and patch area if only because increments in population size will tend to be associated with increases in patch size despite any lowering in the quality of larger patches. The reason for this is that with big increases in patch size, resource quantity can partly offset reductions in resource quality. But, if differences in patch areas are in orders of magnitude but differences in their populations are not, and vice versa, then quality and composition of the resource base very likely lies at the root of the difference. If resource quality has no influence over the effects of patch area and isolation, then it would not add to the explanation in stepwise, logit regressions of species' incidence against area, isolation and quality measures as in the case of Northumberland *Coenonympha tullia* (Müller) (Dennis and Eales, 1997, 1999). Techniques are fast advancing in the study of factors influencing population incidence over patchworks (Jiménez-Valverde *et al.*, 2008; Schröder *et al.*, 2009); variation partitioning and hierarchical partitioning approaches promise a deeper understanding of causation, the prominence of habitat quality attributes, in explaining the occurrence and size of butterfly populations (Heikkinen *et al.*, 2005).

METAPOPULATIONS AND A RESOURCE VIEW OF THE MATRIX

Strict metapopulation models ignore the intervening matrix between patches; it has been treated as if it was sea, without content or structure, the effects of which are invariable on movements. This is one reason why traditional metapopulations may reasonably be referred to as patchworks of habitat but not to more sophisticated topologies such as networks of habitat. However, a number of features within the matrix render this view of it as a 'blank' entirely unjustified, and some principles of matrix quality for metapopulations can be established:

- **P6.12: The matrix is only distinct from habitat by degree and is rarely absolutely lacking in any resources.**
- **P6.13: The matrix is not an isotropic plane (a smooth level surface) with movements equivalent in all directions. Matrix structure affects the direction, distance and speed of movements.**
- **P6.14: Matrix resources affect survival, maintenance and reproduction within metapopulations.**

Table 6.1 Resource attributes affecting patch quality of metapopulations.

Composition of resources
- Resource complementarity (Leibig's law): all resources present or not
- Resource limitation: a resource in short supply or of suboptimal status
- Resource fraction: resource elements used by a butterfly (e.g., taller or shorter plants, flowering plants) are always less than resource type
- Resource density
- Resource context: suitability of surroundings (e.g., overtopping, shade) and substrates (e.g., dry or wet substrates) for all resources
- Resource substitutability and supplementarity: single outlets for more than one resource requirement and multiple outlets for one resource requirement
- Resource independence: matching of conditions (e.g., moisture, humidity, sunshine, temperatures) for different resource requirements

Physiognomy of resources
- Resource size (area) and variability in composition and physiognomy: do resource outlets match in target organism requirements for size and spatial attributes?
- Resource topography (slope, orientation, altitude) influencing context (e.g., hill tops or edges for mate location)
- Resource substitutability: multiple resource use (e.g., same plant providing hostplants and nectar) increases habitat compaction
- Resource shapes: effects of linear as opposed to compact resources
- Resource contagion: clumping, random, overdispersion (e.g., comminution and fragmentation of resource elements)

Resource connectivity
- Resource proximity and spatial variance (overlap, neighbouring, isolation, ruggedness) and effects on connectivity
- Substitutability and compaction of resources increases connectivity
- Resource abundance (density) or size and compaction increase connectivity
- Resource context and topography influence mobility and connectivity

Temporal aspects
- Resource base changes with time: affects composition, physiognomy and connectivity
- Resource time limitation: resources have limited availability and timing
- Resource independence: resources respond to different influences over time
- Resource supplementarity and resource lifespan
- Resource dimensions (area) and relationship to temporal changes
- Contextual influences on temporal changes in resources

Composition refers to the occurrence, density, multiplicity, variation in and context of a resource type. Physiognomy describes the geography of a resource type, its location, altitude, orientation, slope, area, height, shape, contagion and fragmentation. Connectivity describes contact and isolation of resource types from one another and involves adult and larval mobility.

Boundary issues between patch and matrix

If a patch (= habitat) is not a discrete, homogeneous entity, where exists a virtual 1 : 1 correspondence between resources and a vegetation unit (e.g., a host-plant with a national vegetation classification (NVC) unit; Rodwell, 1991–2000), then there will be difficulty in distinguishing patch from matrix (**P6.12**). Such particulate patch identity is more typical of industrial farming systems (e.g., East Anglian cereal farmland in the UK) which generate landscape fragmentation and simplification; the binary distinction increases with intensity of human activity. Where environmental conditions are described as 'semi-natural' or 'natural', the sheer difficulty in establishing rules for determining habitat patchworks is all too clear (Dennis *et al.*, 2006b). Matrix is an extension of non-resource space *within* habitats, as shown in a study of *Maniola jurtina* and gatekeeper *Pyronia tithonus* on Alderley Edge, Cheshire (see Box 2.1) (Dennis, 2004a), and demonstrated by

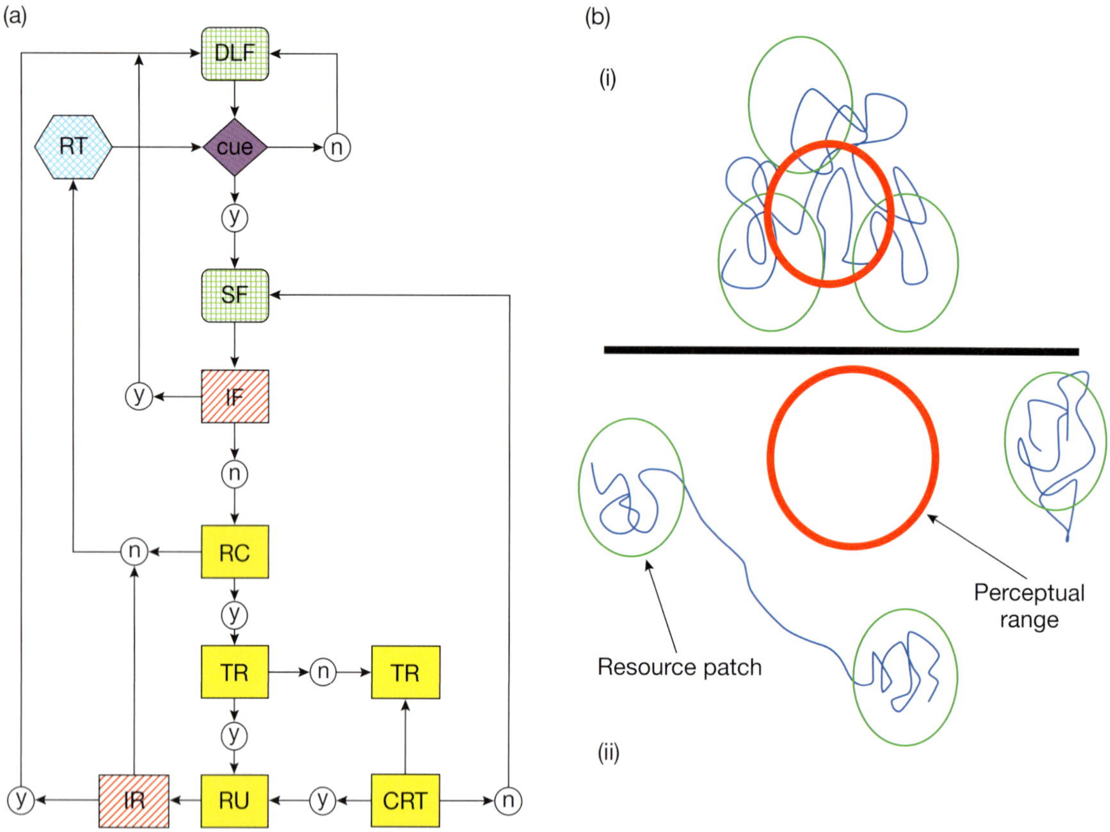

Fig. 6.3 Response of organisms to the landscape: resource seeking and migrating. (a) Flowchart of suggested switches in flight behaviour in response to resource cues in *Pieris* butterflies within both habitats and matrix. Blue boxes: RT, resource targeted; green boxes: DLF, direct linear flight; SF, search flight; yellow boxes: resource variables – CR, complementary resource; CRT, complementary resource targeted; RC, any resource contacted; RU, resource use; TR, targeted resource; red boxes: IF, interaction in transit (with butterfly or predator); IR, interaction on resource. Open circles: connectors, yes (y) or no (n); diamonds: proximate 'cues' including visual and scent stimuli triggering switch in flight types. (From Dennis and Hardy, 2007.) (b) Schematic representation of movements as shaped by the relative scales in the grain of resources within the landscape and the perceptual range of an organism: (i) when the grain of the landscape is of the same order as the perceptual range, dispersal should occur as a by-product of routine movements; (ii) when the grain of the landscape exceeds the perceptual range of the organism, dispersal bears a cost and response is to decrease the cost in transfer. (From Baguette and Van Dyck, 2007.) (a and b, with kind permission of Springer Science and Business Media.)

the frequent switches between search flight and direct flight of butterflies through so-called matrix (Fig. 6.3a) (Dennis *et al.*, 2006b; Dennis and Hardy, 2007).

Metapopulation modellers and empiricists have to face up to two uncomfortable axioms. The greater the fraction of the complement of resources that make up habitats and are used to define them, then inevitably the more resource types and elements will be found in the matrix. But, the fewer the resource types that are used to define habitat bounds, the more resources that should be included within the habitat space will be allocated artificially to the matrix (Box 6.3). Adherents of metapopulation modelling, whether they limit their definition of a habitat to a single resource or encompass the entire complement in the process, will find that they have resources defining a habitat for the organism dispersed throughout what they categorize as matrix. Inevitably, whatever our view of what is or is not a habitat, the matrix becomes a zone of resources. Another truth is that the fewer the resources used to

Box 6.3 Numbers of resources in habitat designation and landscape complexity

A butterfly's landscape is described by the number of resources used by the butterfly to designate a habitat. Three observations emerge in association with **P6.12**:

- **PI: With an increase in the number of resources used to define habitats there is an increase in habitat size and a decrease in the expanse of the matrix.**
- **PII: With an increase in the number of resources used to define habitats, more resources are found in the matrix.**
- **PIII: With an increase in the number of resources used to define habitats, there is the expectation of increased dynamics in the bounds observed for the habitat patches.**

When a butterfly habitat is described by a single resource (larval hostplant), a binary landscape is described for a butterfly; for any site there are just two possibilities, the presence or absence of a habitat. As the number of resources used to describe a habitat is increased then the combination of resource types occurring in the landscape increases. This is described by:

$$2^n$$

where n is the number of resources, or

$$2^n - 1$$

if situations where resources are completely absent (i.e., $n = 0$; a lake) are ignored.

Thus, five resource types result in 32 combinations and 10 resource types result in potentially 1024 combinations. With each additional resource item there is a doubling of combinations possible. With an increase in the number of resource items used to designate a habitat, there is an increasingly complex landscape in which some combinations match circumstances that would be considered suitable for a habitat. Such are a complete intersection or union of resources types (see Fig. 2.1a, d) or a disjoint distribution of resources (see Fig. 2.1e) where there is diurnal access to sequential resource requirements.

This situation can be constrained by placing a premium on one or more resource items. For instance, a habitat could be presumed to be non-existent in the absence of a larval hostplant. A distinction can then be made between overlapping resources where the hostplant is an ingredient and those where it is absent. Combinations with the key resource are described by:

$$\tfrac{1}{2}(2^n).$$

And combinations represented by any resource but the key resource are:

$$(2^n) - (2^{n-1}) - 1 \text{ or } \tfrac{1}{2}(2^n) - 1.$$

A landscape in which five resource types are considered as being suitable for habitat designation has 16 combinations of resources including the key resource, the larval hostplant. The landscape is a mosaic of resource distributions. For every additional resource there is a doubling of combinations. The landscape becomes more complex outside the habitat as it is occupied by other resource combinations lacking the key resource. To illustrate this point, these combinations for five variable types are illustrated in the framework of Edwards' Venn diagram (Fig. B6.3); the elements are units of landscape space and subsets are described by five resource items in which habitat is defined by any element with a hostplant. Thus,

$$\text{H (habitat)} \Leftrightarrow \text{l (larval hostplant)}.$$

The universal set (ξ) is divided into two complementary zones: H (habitat) and H′ (non-habitat), the latter of which contains the empty set ϕ.

The key hostplant is placed centrally. Although, to accept this limitation, intersection or union of the hostplant with any other resource type gives 16 combinations, if diurnal movement of the butterfly species were to be accounted the resource combinations 'occupied' would increase but of course not be acceptable as habitat in this strict set notation. There are large areas where the hostplant does not coincide with other resource types and in the strict definition of sets these zones are not the habitat set; they are, in effect, matrix, but matrix with resources.

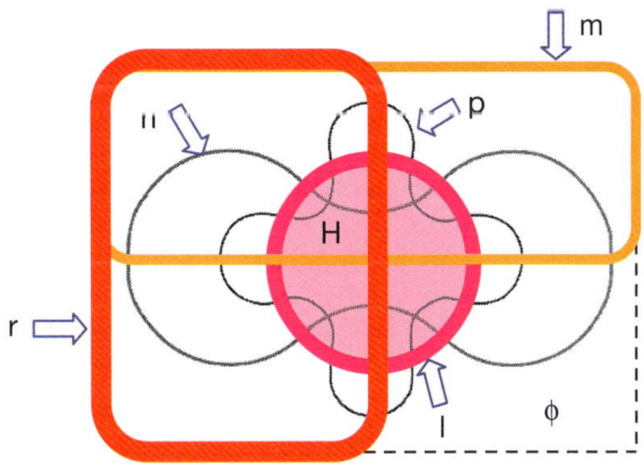

Fig. B6.3 Combinations (subsets) for five resource types (l, larval hostplant; n, nectar source; p, pupation sites; m, mate location sites; r, roosts). Habitat combinations (H) are shown pink in the figure: lnpmr, lnpm, lnpr, lpmr, lnmr, lnp, lnm, lnr, lpm, lpr, lmr, ln, lp, lm, lr, l; matrix combinations (H′): npmr, npm, npr, nmr, pmr, np, nm, nr, pm, pr, mr, n, p, m, r, ϕ.

define a habitat patch, the smaller it will be; simultaneously, the bigger becomes the matrix around it, and the more likely it contains resources. Thus, patch dimensions become truncated.

The matrix has been defined largely through human perception rather than perception of the target organism. Resources (or lack of them) occurring in the matrix not only depend on the definition of what constitutes a habitat patch, but on humans' inevitable tendency to filter out small objects; screening for size of resource objects can also arise from their inaccessibility to study or lack of survey resources. Units or packages below a certain size (<0.01 or <0.001 ha) tend to be ignored or merged into larger patches if sufficiently close to larger ones to fall prey to arbitrary rules of patch designation. But, this is what is distinctive about consumable resources, if not utilities, within the matrix: the resources are often in small, even tiny, packets and they are disparate (Fig. 6.4) (Dennis *et al.*, 2006b). Small resource elements and items in the matrix are frequently regarded either as below the scale to which insects respond or, if used, then as inflicting a cost (see below), slowing movement and acting as sinks in reproduction (Pulliam, 1988).

Patch and matrix bounds are further confounded by temporal changes (Wiens, 1996; Thomas and Kunin, 1999) and spatial (regional) variation. Just what appears to be a habitat patch changes on scales of seconds to decades. Those engaged in conservation practice are constantly faced with successional changes on sites as well as changes in conditions induced by human activities (Sheppard, 2002; Offer *et al.*, 2003; Underhill-Day, 2005). Changes on fine timescales have important implications for recognizing just what resources are important for organisms; there are also inevitable implications for habitat recognition. The habitat space used by *Plebejus argus* (Lycaenidae) on the Carboniferous limestone headland of the Great Orme (North Wales) oscillates upslope and across slope from the vicinity and shelter of scrub with changes in sunshine, temperatures and wind speeds; the warmer the conditions the larger the area used and the response is instantaneous on weather changes (see Box 3.2) (Dennis and Sparks, 2006). Changes with weather and seasonal conditions have also been recorded in mate location surfaces and elements in the peacock *Inachis io* (Nymphalidae) at three different spatial scales: the landscape, surface substrates and in relation to microfeature topography (see Box 4.1) (Dennis, 2004c; Dennis and Sparks, 2005). Seasonal shifts are well known in vegetation and biotope occupancy in both sexes of the speckled wood *Pararge aegeria* (Shreeve, 1984, 1986, 1987) and for vegetation and hostplant use by different generations of the butterfly *Polyommatus bellargus* (Roy and Thomas, 2003). All this frustrates the distinction of habitat and matrix in the metapopulation view.

Matrix resources and movements

A steady stream of work is emerging which demonstrates that the matrix is not an isotropic plane with movements equivalent in all directions (**P6.13**) (Dover *et al.*, 2000; Roland *et al.*, 2000; Kindlmann *et al.*, 2004; Matter *et al.*, 2004; Ouin *et al.*, 2004; Ovaskainen, 2004). The matrix is made up of objects or structures (entities having dimensions as points, lines or areas), some of which can be ascribed to resources and, as with habitats, distinguishable in terms of composition, physiognomy and connectivity, and non-resource structures that nevertheless affect butterfly ecology. Structural components affect movements in three basic ways, through:

1 Composition in terms of life history resources (e.g., consumables, utilities).

2 Physiognomy and thus environmental conditions (e.g., weather, local climate).

3 Landscape connectivity (networks, flyways, visual lines and surfaces).

Matrix resources and structures influence various aspects of adult migration: numbers, frequency, direction, height, distances and speed all influence fields of movement (*supra cit.*) as well as having profound capacity for affecting survival, maintenance and reproduction in adults and early stages that may develop in the matrix (**P6.14**). Anything that enhances an organism's fitness in passing through the matrix is effectively a suitable resource and anything that detracts from it is not. This has less to do with the speed of transfer and more to do with individual condition (e.g., fat reserves, competitive ability) on attaining a new habitat patch. The smallest objects can create barriers and this is no better illustrated than in the experiment carried out by Dover and Fry (2001) on the effects of a tape drawn across part of a cereal field on adult butterflies in transit (see Fig. B6.4b, c). Physical structures greatly influence movements of adults in conjunction with local climate and weather. Sunshine, high temperatures and calm conditions or tail winds encourage or assist movements and shade/cloud, low temperatures and head winds

(a)

(b)

Fig. 6.4 Small-scale resources in the matrix: a fence line on Alderley Edge, Cheshire, showing summer (a) and early spring (b) conditions along fine-mesh (left side) and coarse-wire (right side) fence sections. Note, the denser grass in winter, suitable for hibernating satyrines, where there is a fine mesh in the fence.

oppose it. Structures (e.g., hedgerows, wood edges, fences, walls, rides, lanes, etc.) that enhance sunlight, warmth and tail winds, depending on the prevailing sun azimuth and wind speed, become important resources in the matrix, if only temporarily or for short periods during daylight hours (see Box 7.6). Thus, in transmitting between habitat patches, butterflies are regularly noticed to use flyways along sheltered, sunlit wood edges and hedges, lines of movement that change during the day (e.g., orange-tip *Anthocharis cardamines*; Dennis, 1982a, 1982b) and adhere to rides and corridors rather than over-fly woodlands (e.g., *Aphantopus hyperantus*; Sutcliffe and Thomas, 1996).

There is increasing evidence of resource use within the matrix (e.g., *Maniola jurtina* and *Pyronia tithonus*; Dennis, 2004a). Butterflies do not invariably engage direct linear flight to pass through an entire zone of matrix; it is more likely that they switch between direct linear flight and search flight in relation to an array of resources as do Pieridae (Dennis and Hardy, 2007) if not to the same degree (see Fig. 6.3a). The crucial factor seems to be the relative scale in the grain of resources within the landscape compared with the perceptual range of an organism (see Fig. 6.3b) (Baguette and Van Dyck, 2007). What we do not know yet is how the occurrence of resources within the matrix affects body maintenance, survival, reproduction, development and successful transfers within the metapopulation at large. Thus, even if small 'habitats' present as sinks (Pulliam, 1988) rather than as sources, they may nevertheless function as effective stepping stones. Many of the species studied thus far in metapopulations have few apparent and easily identified larval hostplant species, and these have been used to define habitat patches in which observations of behaviour have been concentrated. It is interesting that those butterflies that use a wider array of less apparent plants (e.g., Satyrinae feeding on grasses such as *M. jurtina* and *A. hyperantus*) display more complex metapopulation structures and dynamics (Shreeve *et al.*, 1996b; Sutcliffe *et al.*, 1997a). It is not widely enough appreciated that hostplants occur as matrix resources, even for those butterflies that have apparent hostplants used to define patchworks (e.g., *Plebejus argus*; Dennis, 2004b; Dennis and Sparks, 2006). Typically they exist in tiny units, their existence largely recognized by some arbitrary, unconsciously imposed, limit for inclusion during survey (see Fig. 6.4). Nectar sources usually abound in the matrix as do other resources such as resting sites, roosts, thermoregulatory sites and mate location sites. The study of matrix effects on metapopulation persistence has yet to be undertaken; it has been awaiting new techniques for observation and tracking transfers (Boiteau, 2001; Riley, 2002; Boiteau and Colpitts, 2004).

Although it seems reasonable to treat matrix effects as a constant for any metapopulation, there can be substantial spatial variation in influences on transfers and local population persistence owing to matrix structures and resources. Population interactions are typically treated as a function of population sizes and their mutual isolation, as modelled along the lines of Newtonian gravitation theory by Reilly (1929):

$$T_{ij} = P_i P_j (D_{ij})^{-k},$$

where T is the interactions (transfers) between the two populations, P is population size for patches i and j, D is the distance between them and k is a constant. For humans, Reilly suggested an exponent of −2, but this has been found to be variable for different phenomena (Haggett, 1965). It is not difficult to see how the exponent k may be affected by resources, conditions or structures within the matrix. Stouffer (1940) labelled this effect as **intervening opportunities**. For butterflies, it may simply be suitable looking biotope (vegetation) patches, but lacking a key resource such as a hostplant, for the migrants emanating from the two patches. Transfer time is increased by the apparent resource, which for conservation headlands Dover (1997) labelled as **distracting resources**. The influence of intervening opportunities, here called the **Stouffer effect**, may be formulated as:

$$I_{ij} = P_i P_j (R_{ij})^{-k},$$

where R is the number (or some measure of size) of resource outlets and k is a constant. In this case, the interactions, over a given geographical distance, are directly proportional to the product of population sizes, but inversely proportional to the number of intervening opportunities. Just how complex this equation can become depends on the number of resource types considered and the spatial and temporal variation within the matrix. There is no reason, if hostplant units within the matrix permit breeding, why the exponent should not be positive, i.e., the number of individuals increases with resource outlets owing to successful breeding. It will, at any rate, vary through the metapopulation and can be envisaged much as **isotims** and **isodapanes** in human geography (Longley *et al.*, 2001), cost 'contours' of equivalent movement. Such intervening opportunities can function as stepping stones; stepping stones potentially vary in status from single resource items, through sinks (intrinsic rate of population increase <1) to full habitats.

There is no reason why the Stouffer effect should not be incorporated into standard measures of isolation between population units. From the gravity model (Reilly, 1929) it is a simple step to describe movements by the Pareto-type equation (after Wilfredo Pareto, economist and sociologist) for the analysis of statistical distributions of specific populations in relation to some measure of size, e.g., impact of distance on the individual movements, as discussed above:

$$M = aD^{-k},$$

where M is the number of immigrants, D is the distance from potential sources, and a and k are constants (Kant, 1946). The two-dimensional (map) view of this is a **field of movement** (the area from habitat core through periphery and matrix within which movements of individuals and interactions among them occur) (Box 6.4).

Box 6.4 Fields of movement and barriers for butterfly populations

Butterfly species have such very different mobility profiles (the frequency distribution of adult individuals moving over increasing distances) to suggest clear distinctions for this life history trait. The author has developed a scale of mobility for British butterfly species (Dennis, 1993a) and these values (see Appendix 5e) correlate closely with the frequency of vagrants occurring outside resource (larval hostplant) zones in two regions contrasting in biotopes (Cook *et al.*, 2001).

Even so, different butterfly species have a similar *pattern* of movements. Accumulated movements of organisms within and beyond a habitat can be described as a **field of movement** (Fig. B6.4a(i)). In cross-section this approximates the form of a bivariate normal distribution but with a sharper peak and longer tails (Fig. B6.4a(ii)). Fields of butterfly movements for a population are rarely **isotropic** (movement same in all directions), but are easily influenced by structures in and outside a habitat, including random effects, and are thus **anisotropic** (movement varying with direction) (Fig. B6.4a(vii)). Figure B6.4a illustrates some of the patterns typically found in nature, associated with structures that act as barriers or corridors; the bounds for fields of movement are abstracted as a section cut through the density distribution parallel to the ground (plane of movement). Dover and Fry (2001) carried out a simple but ingenious experiment on butterfly behaviour on three groups of butterfly 'species' (scarce copper *Lycaena virgaureae*, heath fritillary *Melitaea athalia* and, together, the high brown fritillary *Argynnis adippe* and niobe fritillary *A. niobe*) in relation to two artificial structures in Norway (Fig. 6.4b, c): a sight-line (a narrow red-and-white builders warning tape) and a windbreak (green horticultural windbreak). These structures were drawn across a cereal field from one rocky outcrop (Åkerholme) to its neighbour, flush with and just overtopping the crop respectively. All three taxa were strongly affected by the windbreak, both as a barrier and corridor; in the latter case an increased proportion of butterflies moved along the windbreak, and further, compared with the control (no windbreak). Interestingly, the sight-line did not elicit a uniform response by the different taxa. Significantly more *A. adippe*/*A. niobe*, but not *M. athalia* or *L. virgaureae*, moved along the visual link, and further, than the control (no link); the sight-line had a slight barrier effect on *L. virgaureae*. The importance of this work is that it demonstrates that the same structure can be a corridor or a barrier, not just for different species, but different individuals of the same species. The responses seem to have been to the change in colours and structural contrast with the background. Dover and Fry explain the increased movement along the artificial structures (corridor effect) in terms of response to apparent resources:

- Foraging (red or white patches of tape as nectar flowers).
- Mate location (structures interpreted as locations for females).

Even though movement from one Åkerholme to another was not witnessed, it is clear that even the simplest structures can initiate oriented movement along it, analogous to linear features in the landscape (e.g., hedge). The barrier effect in all species could relate to:

- A rejection of the 'resource'.
- Interpretation of the structure as a limit to resource zone and the availability of conspecifics.

Previously, observations in the Bollin Valley, Cheshire between 1982 and 1984, revealed truncation in the fields of movement of two local populations north and south of a motorway bridge across the valley (Fig. B6.4e). Using mark–recapture, movements were found to be reduced by 91.8% compared with that between sections either side of the bridge (Dennis, 1986b). Direct observation of 132 'individuals' approaching the bridge revealed that the decision to fly along the embankment and to return was closely related to implicit barriers (local climate: sun, shade, wind speed) (Fig. B6.4d); only five individuals spiralled up to attain the level of the road and a further 13 did so by turning and flying along the embankment, most to return. The majority turned on their approach to the bridge, in sunshine (Fig. B6.4d(i)).

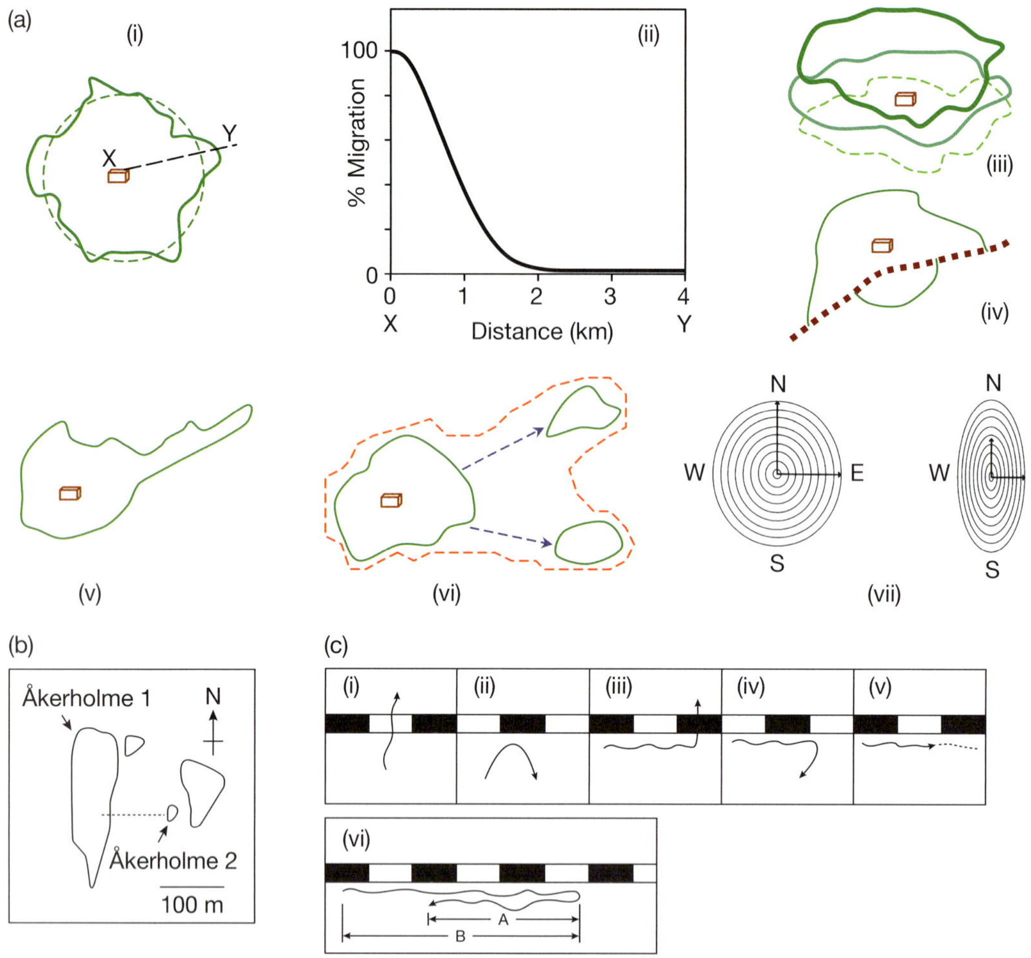

Fig. B6.4 (a) Alternative types of fields of movement: (i) circular field of movement influenced by random variation and minor barriers to gives lobes, waves and indentations (the line X to Y indicates the direction over which movement is measured over the *x* axis in (ii); (ii) typical pattern in the decay of movements away from the habitat (movement within habitat included); (iii) **displacement**, shifts in field of movement over time (e.g., with seasonal climate or weather conditions); (iv) **truncation**, field of movement influenced by a barrier (thick pecked line); (v) **distortion**, movement influenced by a corridor of habitat (e.g., field edge, narrow valley); (vi) **fragmentation**, (e.g., loss of habitat zone or alternatively budding of sinks and pseudosinks from the main habitat); and (vii) movement directionality with isolines describing the same value of movement decreasing from the centre: circular, **isotropic**; elliptical, **anisotropic**. The square illustrates the centre of the habitat; the solid lines illustrate bounds for 95% of movements; and the dashed lines indicate an earlier stage in the field of movement. (vii, from Fortin and Dale, 2005, p. 11, with courtesy of Professor Marie-Josée Fortin and Cambridge University Press.) (b) Field study site at Vestby, Norway. The pecked line indicates the position of the 'sight-line' and 'windbreak' structures for experiments between two Åkerholme (rock islands dominated by mixed woodland with clearings fringed with narrow, but species-rich, dry meadows) in a cereal field. (From Dover and Fry, 2001, with kind permission of Springer Science and Business Media.) (c) Behavioural responses noted at the Vestby experimental site: (i) butterfly crossed tape with no apparent response; (ii) approached tape but did not cross and turned; (iii) flew along line of tape and ultimately crossed; (iv) flew along line of tape but ultimately turned back; (v) flew along line of tape but was lost from view; and (vi) distances that butterflies flew along the tape included changes in direction (A + B) with maximum displacement from its point of initial observation (B). (From Dover and Fry, 2001, with kind permission of Springer Science and Business Media.)

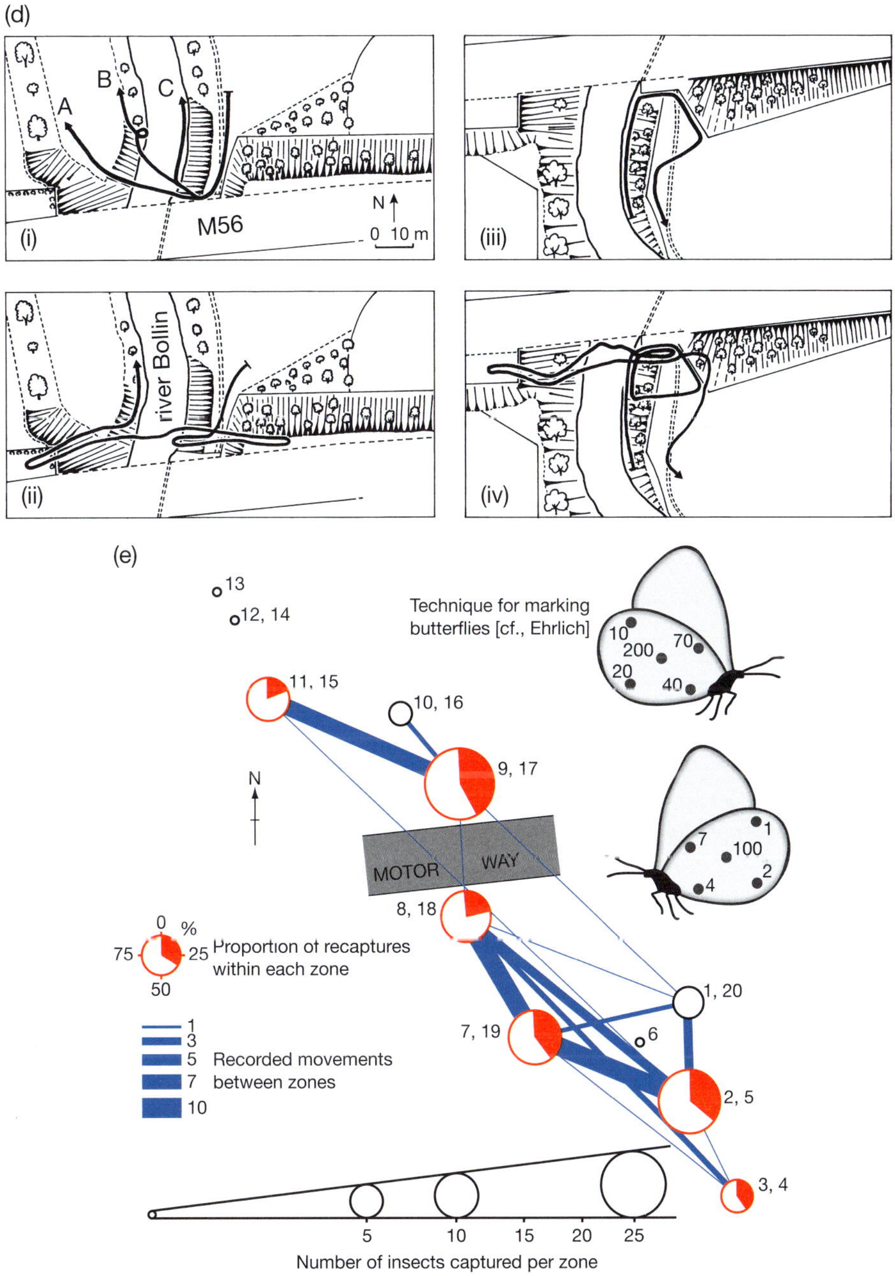

Fig. B6.4 (*continued*) (d) Examples of behaviour of orange tip *Anthocharis cardamines* males at a motorway (M56) bridge over the River Bollin, Cheshire: (i) and (ii) north side with typical tracks for different times of day shown in (i) (A, <12.00 h; B, 12.00–15.00 h; C, >15.00 h); (iii) and (iv) south side. (From Dennis, 1986b, courtesy of Amateur Entomologist's Society.) (e) Movements along the Bollin Valley in 1983 divided by the motorway bridge. (From Dennis, 1982b, courtesy of Vasculum and the Northern Naturalists' Union.)

A characteristic of fields of movements is that they are theoretically continuous distributions with a very rapid fall-off near their origin and a very slow, almost asymptotic fall-off at their outer limits. Because of the continuous nature of individuals' movements, the size and shape of the movement field cannot always be described, in practical terms, by the absolute limits of individual movements. However, useful generalizations can be made if some arbitrary limit (e.g., mean field or 2 standard deviations from centroid) is substituted for that of maximum field. In a patchwork embedded in a uniform matrix, the arbitrary line determining the field of movement would be expected to be approximately circular. Often, they can be amoeba-like owing to stochastic variation (see Fig. B6.4a(i)).

Of course, fields of movement and distribution of resource use are rarely continuous, smooth gradients in the natural world. Instead, there can be, and usually are, sharp transitions and distinctive shapes (i.e., funnels or corridors) in both resource use and movement. These typically relate to 'structures' in the landscape (see Chapter 7). Thus, it is not difficult to envisage modifications to the size and shape of fields of movement caused by variations in patchwork and matrix structures. For example, some barrier across the habitat or matrix, say a wide river, may result in a truncated field of movement (see Fig. B6.4a(iv)). Similarly, a patchwork of some, supplementary, resource such as nectar-producing flowers along a hedgerow, occurring in one direction from a habitat may cause a distortion in the field of movement (see Fig. B6.4a(v)). Just whether this results in more or less *net* movements in the direction of the resource patchwork depends on the type of resource and the overall effect on migrating butterflies. As in the case of examples used to illustrate the impact of intervening opportunities above, movement may be enhanced or impeded; this is evident from the revealing experiments carried out in Norway by Dover and Fry (2001) (see Fig. B6.4b, c). Real as opposed to model landscapes, even artificial ones under intensive human management, suggest more profound adjustments to gravity models and fields of movement (see Fig. B6.4d). When many factors are involved, the interaction (movement) between mass (population size) and distance may not be simple, but perhaps be better expressed in some more complex (e.g., quadratic) form, as has been shown to be the case empirically in human economic studies.

Topology for resource use and movement

In the last chapter we considered how resource use related to biotope composition. Biotopes also have structure and connectivity. These can be described using topological nomenclature: vertex (point), edge (line) and face (surface) (Box 6.5). Below, the impact of biotope topology in the matrix is considered on resource use and movement. Among butterflies (Baker, 1978; Dennis, 1982b; Shreeve, 1992b, 1995) research reveals two basic kinds of feature into which landscape is divided and over or along which movements occur or are impeded:

1 Linear features or **edges**, such as hedges, road verges, river banks, wood margins, walls and roads.

2 Areas or **surfaces**, such as fields, woods, lakes, moors and urban sprawl. A distinction between a point and an area in landscape is a matter of spatial scale.

Where movements predominantly follow linear features, these are often linked up as **networks** (e.g., field boundaries) (Fig. B6.5a, c). Both edges and surfaces provide 'channels' or **corridors**, along which movements occur and 'fences' or **barriers**, where they are prevented (Fig. B6.4b–e). To understand corridors, it is essential that their meaning is made clear. In a resource-based habitat view, a corridor is any structure that permits (increases numbers and frequency of) movement between two habitat units. As such, this can be a gap in a hedge linking two neighbouring habitats as much as a green lane that joins, or partially joins, two isolated habitats. The key feature (process) is that transfer exceeds that of equivalent units of neighbouring structures. Various resources for butterflies occur at distinct points, along edges as well as over surfaces. At their most inclusive they comprise, as indicated above, consumables, utilities, conditions and physical links (e.g., visual and other stimuli), all of which can affect movements. Surfaces with distinct resources for a butterfly species may be called **resource zones (resource patch)**; whereas, edges with resources are better referred to as **resource lines (resource edges)** (Fig. 6.5). Along edges, resources are typically sequenced and complementary. Often, resource types are concentrated at particular points along edges, usually at intersections, providing a **resource node** (Fig. B6.5a, d) (Forman and Godron, 1986; Noss and Harris, 1986). Nodes, where a number of resources coincide in a small area, virtually a point on a map, may also exist away from edges, for example in mid-field depressions (e.g.,

Box 6.5 Basic terms in landscape topology: nodes, edges and faces

Terms and usage

In a move to describe the landscape of resources for organisms, some terms have been borrowed from topology. These are **node** (**vertex**), **edge** (**arc**) and **region** (**face**). They roughly equate to point, line and surface in Euclidean geometry and are zero, one- and two-dimensional figures in two-dimensional (map) space (see Box 3.1). In organism geography these terms are not really applied as they would be in topology. Topology differs from Euclidean geometry in that it is primarily concerned with order and contiguity and less concerned, if at all, with distance, linearity and orientation. Thus, nodes, edges and faces (hereon **surfaces**; region is used differently in this book to describe large-scale areas of the countryside) cannot be applied in butterfly biogeography as they would be in a strict topological system where these features occur in strict ratios, as for instance in human dermatoglyphics (finger prints) where loops (defined by ridges) occur in strict ratio to the number of triradii (ridge meeting points). Even so, the terminology is extremely useful for butterfly landscapes, as resources can be conceived as occurring at nodes, along edges and over surfaces. Such structures have a distinct impact on butterfly behaviour, resource exploitation, mobility and consequently on population integrity and persistence. Moreover, rigid Euclidean geometry is not always appropriate in understanding resource use; barriers (e.g., dense conifer wood) and corridors (e.g., green lane) distort movements; the perception that a butterfly has of landscape is not Euclidean, even in flat map space, but a more elastic arena (Fig. B6.5d(ii)).

Herein, nodes, arcs and surfaces are treated as separate sets (Fig. B6.5a). Each of these components can be distinguished or defined in terms of:

- Landscape features, and/or
- Biota (target organism(s)),

but may be relatively persistent features of the landscape (e.g., rock outcrop) or dynamic entities (e.g., populations). Thus, nodes can be physical features (e.g., pond, pit, mound, hedge junction, molehills) or biological entities (e.g., aggregations of a species, community of organisms). The components are also distinguished in terms of:

- Form (composition, structure, features, resources).
- Function (population units, e.g., sources, pseudosinks and sinks; movements, e.g., conveyors, barriers, magnets, gateways).

As any element of each component (node, edge, surface) can have or lack resources for an organism, they may also display the full spectrum of population attributes, i.e., they can be sources, sinks or pseudosinks (see Box 6.1). In the same way they may also function as conveyors, barriers, magnets or gateways for organisms. Elements acting as conveyors and gateways permit the passage of organisms in transit, whereas barriers cause reflection, deflection and repulsion (Fig. B6.5b) and magnets cause attraction and increased length of stay. Each of these movement types is related to a **disruptor** (physical feature), such as resource availability or its absence (Table B6.5a).

A landscape is made up of elements of nodes, edges and surfaces in combinations that can be treated as potentially independent of type or number (Fig. B6.5a gives several examples), although obvious statistical

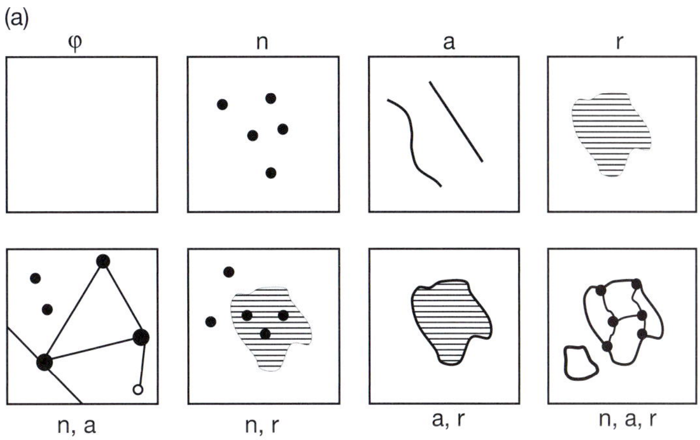

Fig. B6.5 (a) Sets of nodes, arcs and faces: φ, empty set; n, nodes; a, arcs; r, surfaces. Nodes are distinguished by size (example in the lower left box) with disconnected nodes (small black), terminator nodes (small white), two nodes (medium black) and three nodes (large black). Note the same or different features can be integrated or independent. (Modified from Cole and King, 1968, courtesy of Wiley-Blackwell.)

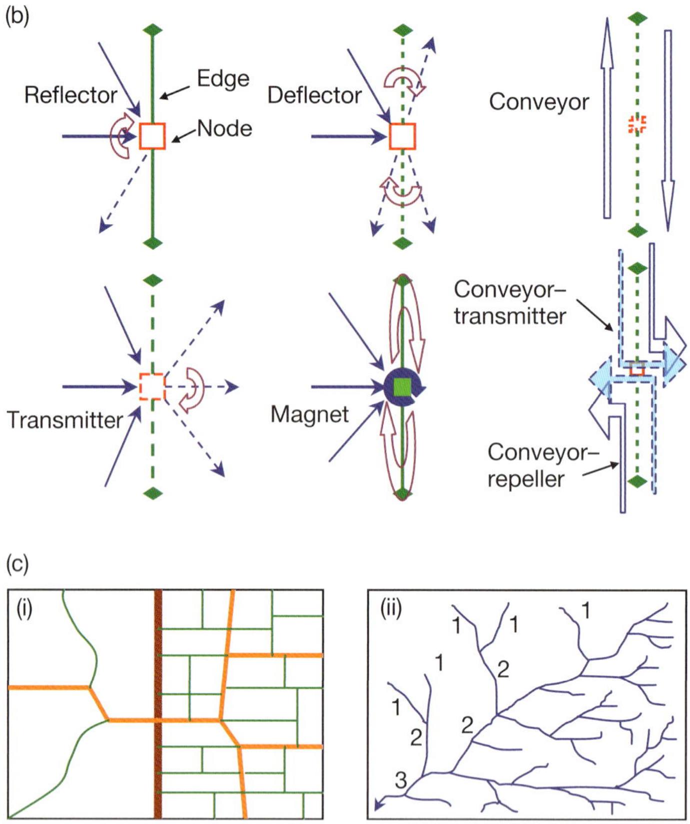

Fig. B6.5 (*continued*) (b) Response of butterflies in transit to node and edge margins. **Reflector**, entry and return at edge or node greater than critical angle; **deflector**, entry greater than critical angle but exit less than critical angle (switch to conveyor) at either node or edge; **transmitter**, entry and transmission at node or edge greater than critical angle; **magnet**, transit discontinued and remains locally within node or edge; **conveyor**, transit along edge (no returns) with movement away from edge less than critical angle; **conveyor–repeller**, disruptor (physical features causing response) causes movement away from edge at greater than critical angle; **conveyor–transmitter**, disruptor causes movement away from edge at greater than critical angle after transmission through edge. Angles are taken from the margin of the feature at the point of interception; the critical angle (>22.5°, >2 m of movement away from margin for every 5 m of edge length) is taken to be that which indicates an obvious movement away from the feature. **Barriers** are reflectors and conveyor–transmitters, and deflectors, conveyor–repellers and magnets when movement remains on the same side as that of contact; **corridors** are conveyors and deflectors; and **gateways** are transmitters and conveyor–transmitters where the rate of movement post contact is greater or equal to that prior to contact. In the right pair of diagrams, the node is a disruptor (solid, bottom) and non-disruptor (pecked, top). (c) Contrasting topologies for: (i) land boundary (rectilinear) and (ii) drainage (dendritic) systems; low density edges are on the left and high density ones on the right in each diagram. For streams, the Strahler stream ordering system is illustrated for the low-density half. As edges increase, so do the number of independent surfaces and nodes (junctions, confluences).

relationships between surfaces and edges (e.g., fields, hedgerows) and edges and nodes (streams and confluences, hedge intersections and hedges) will be prominent in strict (physically linked) networks (Fig. B6.5c). Thus, with an increase in fields there will be an increase in field boundaries (edges) and junctions (field intersections and corners); in stream networks, as stream density (edges) increases there will be more confluences (nodes). Each type of network tends to have different properties (see corridor effects in Table B6.5b).

(d)

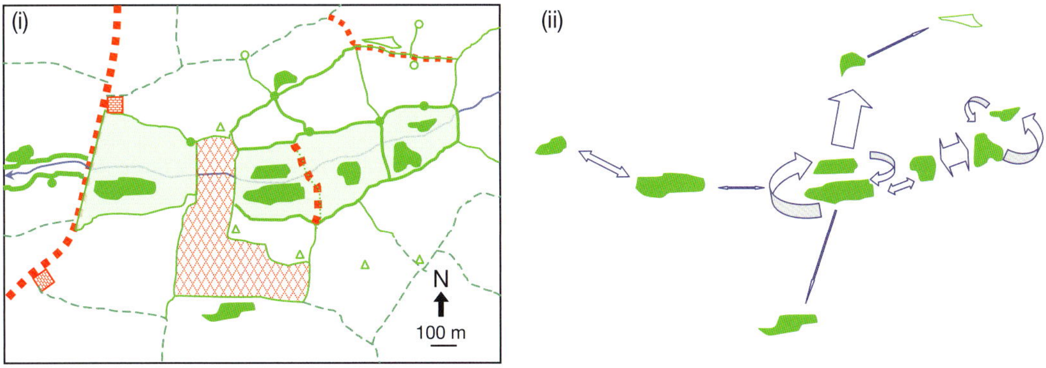

Fig. B6.5 (*continued*) (d) Schematic maps illustrating resource nodes, edges and surfaces associated with part of a metapopulation for a butterfly: (i) habitat patches, nodes, edges and surfaces; and (ii) habitat patches and ranked movements. Nodes: resource (open circle), habitat complement (closed circle), resource vacant (triangle), barrier (red square); edges: resource (thin green line), habitat (thick green line), resource vacant (dashed line), barrier (red interrupted line); faces: resource (pale green), habitat (dark green), resource vacant (white), barrier (cross-hatched red). Movements in (ii) are only shown between the main habitat patches; they do not necessarily coincide with patch proximity, but are influenced by corridors and barriers. The thickness of the arrows indicates the relative movement of butterflies.

Table B6.5a Association of resource status and movement types for contacts with nodes, edges and surfaces.

	Movement type on contact with landscape component				
Resource status	**Barrier**	**Transmitter**	**Deflector**	**Conveyor**	**Magnet**
Habitat	✗✗	✓✓*	✓	✓✓	✓✓
Habitat > resources > 0	✗	✓✓	✓✓	✓✓	✓
Resources = 0	✓✓	✓	✓	✓	✗

* Sources likely to act as a magnet whereas sinks will act as transmitters.
Double sign emphasizes link.

It should be noted that individual mathematicians, and biologists, often give distinct and different meaning to node, arc, edge and surface. For example, Noss and Harris (1986) apply the term node to any scale in the biological and spatial hierarchy; thus, whole nature reserves can be envisaged as nodes, and quite rightly – at a scale of 1:500 000 or 1:1 million they are but points on a map. What it does mean is that the terms require clear definition before they can be used effectively in any study. Their application is more flexible at finer resolutions as there are more opportunities for distinguishing subsets of node, edge and surface (vacant of resource, resources less than complement for habitat, habitat complement, barrier, etc.) at scales <1:50 000; an imaginary example is illustrated to show how this might be applied (Fig. B6.5d(i)).

Scale and survey

There are inevitable questions. One obvious one involves spatial scale: how is a node or edge to be distinguished from a surface? A resource unit appearing as a point or a line on a map of scale 1:50 000 will take on more the appearance of a surface at a scale of 1:500. It is well to appreciate that 50 and 20 m on land at a relatively coarse scale (e.g., 1:50 000 map: 1 and 0.4 mm, respectively) are still, to all extents and purposes, points at a finer resolutions (e.g., 1:10 000 map: 5 and 2 mm, respectively).

Table B6.5b Attributes of corridors associated with rectilinear and dendritic networks. (Modified from New, 2006, and Forman, 1991, courtesy of Surrey Beatty & Sons Pty Ltd, Chipping Norton, Australia.)

Human landscape features (rectilinear)	Drainage (dendritic or trellised)*	Both types of networks
Circuitry	Stream order (Strahler)	Corridor width and length
Variable links per node	Tributary density	Connectivity
Mesh size	Matrix types (quality)	Node distribution
Surface (patchwork) types (quality)	Singular directionality	Angles of links at nodes
Variable directionality		Curvilinearity of links
		Variable patch/matrix size

* Anastomosing stream systems (e.g., deltas) also have a degree of circuitry.

At the outset of surveys, the simplest solution is to select nodes on the basis of landscape features (e.g., edge junctions and termini, hedge intersections, hedge terminations, ponds or pits, etc.) as then no *a priori* decision is made about their impact on butterfly biology. Their comparative value for resource use or movements may then be hypothesized and tested; for instance, ditches may differ from hedges in species' diversity, as may tall uncut hedges from short cut ones. Blind terminator nodes at the end of a hedge may be found to have fewer resources for butterflies than two-node features along a hedge (see below) but lead to greater aggregations of butterflies. Nodes will inevitably be small when compared with surfaces (e.g., fields, source habitat patches) and part of their selection will usually be that they form a fractional component of surfaces (they are not habitats or expanses of so-called matrix) or edges (they are not prominently linear). The status of biotic aspects (resources, populations, movements) of particular nodes, edges and faces becomes evident with the survey. Clearly the resources comprising a habitat can, within reason, be of any scale but the function of the unit can change with scale, as we have seen (see Box 6.1), from a source to pseudosink, and ultimately to a sink. The problem is that the change of scale may not be a simple one and will vary for different species. All this evokes the issue of what is or is not a habitat (resource) patch in the landscape – therefore what is ignored in butterfly biology – a vital point aired in the text.

Using this approach, one thing is abundantly clear: the metapopulation landscape with patch and matrix comes alive with resource geography, nodes, edges and surfaces of resource; factors underlying differential movements among habitat patches are then exposed (Wiens, 1989). Noss and Harris (1986) describe their importance for conservation in the development of **multiple-use modules** (**MUMs**); these consist of inviolable or well-protected core areas, which they call nodes of diversity, surrounded by buffer zones outwards of increasing utilization by humans.

marl pits) – a **disconnected node** (Fig. B6.5a). Where all the required resources for a particular species exist over an area (surface), these are **habitats** or **habitat patches**. Where they occupy a linear feature, they are called **linear habitats** (Fig. 6.6), compared with which if they occur concentrated at points (viz., field corners, hedge intersections) along edges they are **habitat nodes**. Remember, if the intrinsic rate of increase for the population in a habitat unit is <1 (mortality > births) it is a sink, and small size of the resource concentration alone may make a habitat node a sink or pseudosink. Thus, the landscape over which butterflies move comprises a hierarchy of resource nodes, resource lines and resource zones, some natural and some human in creation. Just how important these distinct features with their resources are for movements may be gauged from the contrasting expectations of metapopulation dynamics and landscape ecology. Metapopulation theory postulates that landscape features do not bias movements over space, whereas landscape ecology predicts biased movement patterns that relate directly to landscape features. Direct comparisons have yet to be made. Even so, considerable evidence is accumulating that landscape structures critically influence butterfly movements (Dover and Settele, 2009).

Fig. 6.5 Dry-stone walls with bilberry *Vaccinium myrtillus*, a larval hostplant for green hairstreak *Callophrys rubi* (inset) in the Pennines Macclesfield Forest, UK, overlooking Tegg's Nose, a site of a large colony of the butterfly. (Inset courtesy of Paul Kipling and Peter Eeles.)

Principles of movements and resource use in real matrix situations

Landscape structures in the matrix influence three basic processes which cause movements or prevent them outside main habitat patches:

1 Corridors: lines, zones and conduits of more movements than expected for a unit of distance.

2 Barriers: lines and zones of fewer movements than expected for a unit of distance.

3 Aggregations: collection points of no net movement producing (permanent or temporary) accumulations of otherwise moving individuals.

The phrase 'than expected' indicates comparison with (relativity) neighbouring structures and spaces. Some general points are outlined for these first and then, below, each is considered in more detail.

- **P6.15: An increase in matrix structure results in increased opportunities for corridors, barriers and aggregations.**

- **P6.16: Both edges and surfaces (faces) can act as corridors (flyways) and barriers as well as generating aggregations. Just how structures function depends on the direction of movement of individuals and their resource associations.**
- **P6.17: Edges and surfaces have profoundly different effects on resource use and movement in the matrix depending on the balance of corridors, barriers and collection points (nodes) causing aggregations that individuals encounter.**
- **P6.18: Edges provide more contrasting visual lines, local climates and resource availability than surfaces.**
- **P6.19: Although collections of individuals (aggregations) are most likely to be caused by a resource(s) concentration, they can also be caused by its absence, barriers and edge termini (blind edges).**

(a)

(b)

Fig. 6.6 Human-made landscape structures providing linear habitats and flyways for butterflies and other organisms. (a) The Middlewood Way (a landscaped and wild-flower-seeded, disused railway line) between Bollington and Poynton in Cheshire, UK, a biotope for small skipper *Thymelicus sylvestris* (top inset), meadow brown *Maniola jurtina* (middle inset) and gatekeeper *Pyronia tithonus* (bottom inset). (Insets courtesy of Peter Eeles.) (b) The Macclesfield Canal, running parallel to the Middlewood Way between Bollington and Poynton. The path and hedgerows provide valuable biotopes for speckled wood *Pararge aegeria* (inset). (Inset courtesy of Andrew Burns and Peter Eeles.)

If matrix is understood to be regions through which organisms must pass on emigrating from one habitat to colonizing another, then matrix structure, and its associated resources, operates to enhance or deter these movements. Matrix structure essentially provides corridors for movement, barriers against it, and places where individuals collect, if only temporarily. As structure increases in the landscape, so opportunities increase for corridors (flyways), barriers (obstacles to movement) and places where individuals may aggregate (**P6.15**). Clearly, as structure increases, there is a stronger likelihood of resource clumping, thus habitat creation; in this case, the aggregations may be viable populations in habitat patches. Just whether structures operate as corridors, barriers or places of aggregation depends on the direction that individuals are taking and the resources required by them (**P6.16**). This is highly species-specific, with strong taxonomic bias. In many species, although individuals show preferred directions in migration, different individuals usually have different preferred compass directions and the preferred direction for populations may be variable (Baker, 1978). The latter pattern contrasts with the high directionality of other species, typically long-distance migrants with seasonal reversals of migration (e.g., *Pieris* sp., *Vanessa atalanta*, *V. cardui*). The impact of matrix structure will reflect on these individual and population preferences. Structure can also act differently for individuals with the same directional tendency; a barrier for one individual is not a barrier for another, as in the case of individual *Anthocharis cardamines* attempting to cross motorways (Dennis, 1986b). Among those for which the motorway embankment operated as a barrier, for a fraction it provided a flyway, effectively a corridor, and they adopted a path along the sheltered sunny side of the embankment, using resources (e.g., nectar flowers) in transit.

In this way, edges and surfaces adopt very different roles in resource use and movement among species providing distinctive patterns of corridors, barriers and collection points (**P6.17**). The role of structure is species-specific. Even so, edges provide more contrasting structures in terms of visual lines, resources and thermal climate than open surfaces (**P6.18**), and it is at edges that the most profound distinctions will be found in behaviour among conspecific and interspecific individuals. The tendency to form aggregations is a particularly important collective behaviour involving numbers of individuals, usually of one, but occasionally more than one, species. Movement has ceased or is being curtailed, if only temporarily. It is important as – unless the collection point or edge has all the resources a butterfly needs (= habitat proper) – it can lead to higher mortality or reduced fecundity despite an increase in the number of matings, particularly if the aggregation attracts predators. Aggregations can evidently be created by a concentrated resource but may also be caused by a barrier (physical obstacle) including an unconnected flyway (blind corridor or terminal vertex) (**P6.19**). The latter situation should increase with fragmentation where hedgerows are being uprooted and other linear features truncated and erased.

Corridors, barriers and aggregations

It is vitally important that when considering corridors, barriers and aggregations, a distinction is made between different aspects of movement, namely, numbers, frequency, direction, distance travelled and speed of movement. A corridor, for example, may indeed enhance numbers and frequency of movement (by definition), but may adversely affect the speed of movement and distance travelled depending on the type of resources within it. Fewer butterflies may take a direct track across a uniform surface such as a lake, but the movement may well be unimpaired by intervening resources or obstacles (Shreeve and Dennis, 2009). Circumstances may also be envisaged where movements are frequent but the matrix conditions passed over are apparently uniform and resource-less to the human eye. These situations are in need of urgent study as are the influence of different corridor conditions on different aspects of transfers. There is the question: what is perceived to be a resource-less matrix to a butterfly in transit?

Some general observations can be made on corridors and barriers:

- **P6.20: A corridor effect is enhanced by zones of resources, whereas a barrier is imposed by an absence of resources.**
- **P6.21: A corridor rarely involves a single resource, so it is unlikely to have a single effect in the population dynamics and movement of a species.**
- **P6.22: The greater the contiguity among resources, the more likely it is that a zone will function as a corridor.**

• P6.23: Edges and marginal structures that lack resources other than enhanced local climate for activity encourage rapid transfers between habitats.
• P6.24: Even though surface corridors may have more numerous and varied resources than edge corridors, they may well be less effective in transmitting movements.
• P6.25: Edges provide stronger visual lines for movement than broader zones and provide a wider range of local climate conditions to modify weather and seasonal climate.
• P6.26: Larger, denser structures are key edge attributes facilitating movement along them but also for preventing movement over them, particularly when wide.
• P6.27: Biotopes that simulate the structure (life forms and texture) of patches comprising habitats stimulate increased numbers of transfers, whereas those that differ counter it.

The concept of corridors has risen to prominence with advancing habitat fragmentation (Hobbs, 1992; Simberloff *et al.*, 1992). Even so, for butterflies, the effectiveness of corridors in enhancing movements has yet to be demonstrated, as has what constitutes an effective corridor (Davies and Pullin, 2007). It would seem obvious that a corridor for a particular species, whether strictly linear or zonal, should comprise suitable resources (e.g., strips of wet heath and nectar sources encourage movements in Dutch Alcon blue *Maculinea alcon* (Schiffermüller); Wynhoff *et al.*, 1996) (**P6.20**). But just which resources, and how their spatial arrangement (i.e., whether juxtaposed or sequenced), may affect movement remains largely unknown. As corridors may very often include more than one resource for a species and usually these are not audited, the outcome of different combinations of resources can be complex (**P6.21**) involving different inducements to 'stay' or 'leave', or even 'return'. Indications are from studies in farmland that the presence of resources along hedges encourages movements, especially if continuous, as individuals can forage using trivial flights among the resources much as in habitats (Dover, 1996). Observation and research suggest that linear corridors enhance population abundance and mobility of the least mobile species (Merckx *et al.* 2008), but even individuals of species that are not well known for being mobile occasionally adopt cross-matrix transfers rather then use linear features (Shreeve and Dennis, 2009). Evidently, the crucial factor is the distance separating vital, sequentially required, resource components for a particular species (**P6.22**). However, movements may be faster, if fewer, when surfaces are resource vacant depending, of course, on conditions and structures enhancing local climate for flight (**P6.23**). The virtual absence of resources on lakes, cereal fields and herbicide-treated pastures has been shown to induce butterflies to fly rapidly, in straight lines, usually without stopping (Baker, 1978, 1984; Dennis, 1982b; Shreeve and Dennis, 2009; but see Dennis and Hardy, 2007); the effect may be much the same on resource vacant edges. The broader the corridor (surfaces), the greater the potential quantity, if not also variety, of resource types for different species (**P6.24**). However, narrow structures (edges) are more likely to form corridors for species than surfaces or fields, if only because there is less likelihood of resource continuity and maintenance over broad swathes of land with increasing land management and surface homogeneity than along boundary features. Nevertheless, it is important to distinguish between transfers per unit area; those along a hedge may exceed those over a field per unit area, but the bulk of transfers (numbers) may still be across the field.

Edges may have other advantages as corridors associated with local climate and because they also provide physical continuity and visual lines (flyways) for individual butterflies to follow (**P6.25**) (Dover and Fry, 2001). Local climate is expected to have profound effects on movements. As butterflies function most efficiently within narrow thermal limits, movement is facilitated by warm and calm conditions (Dennis, 1993a). Edges provide contrasting conditions (e.g., sides of a hedge), for sunshine, warmth, moisture and wind speeds, depending on aspect to the sun and exposure to prevailing winds. Thus, movement will tend to concentrate along distinct linear features rather than in the open in weather conditions that are marginal for activity. Examples of the use of edges as 'flyways' for local climate reasons are known in *Anthocharis cardamines* (Dennis, 1982b). This species changes its flight path from one side of the hedge to another in response to diurnal shifts in sun angle; a number of species (e.g., wall brown *Lasiommata megera*, *Maniola jurtina*) have been shown to do the same in response to wind direction (Dennis and Bramley, 1985; Dover *et al.*, 1997). Dover (1997) has made a study of the selection of edge attributes in relation to wind speed and direction. He found that structures of edges matter; as would be expected, dense, high hedges facilitate activity (movements) more than sparse, low hedges (**P6.26**). Open areas tend to be more exposed and cooler in high latitudes. This very probably restricts movements or at least limits times when extensive movements are possible, as on hilltops (Wickman, 1988). However, individuals caught up in high winds and travelling downwind may be able to cover greater distances in a shorter time (Shreeve, 1992b), but cold, strong winds may simply ground butterflies altogether (Dennis and Bardell, 1996).

Conditions contrary to producing corridors establish barriers (**P6.20**). Thus, for *Aphantopus hyperantus* woodland constitutes a barrier for the butterfly in Monks Wood, Cambridgeshire, UK (Sutcliffe and Thomas, 1996). The effectiveness of an obstruction to movement may be measured in the same way as corridors, by its permeability – the relative numbers of individuals approaching a barrier that successfully cross it. The capacity of an organism to cross a barrier depends on:

1 The dimensions of the feature, height, breadth and density (**P6.26**).

2 Its context, its location amid landscape components, climatic conditions and topography (Fry and Robson, 1994).

Surfaces lacking resources and visual cues typically tend to be barriers, as are surfaces and edges having marginal conditions for flight (**P6.20**, **P6.25**). Not all linear features present suitable local climates for flight. For instance, farm tracks can be exposed or act as wind tunnels limiting flight activity (Dover, 1996). The context is clearly important in determining the effectiveness of barriers. Ford (1964) considered that areas of cropped grassland and even hedges constitute barriers for *Maniola jurtina* in the Isles of Scilly and Cornwall. Yet, elsewhere in the UK, as in Cheshire, these are by no means absolute barriers; nor are woodland, lakes or large buildings (Shreeve *et al.*, 1996b). Evidently, the costs and benefits of moving and crossing barriers depend on the different elements that make up a landscape mosaic. Moreover, a corridor for one species may well represent a barrier for another. In landscapes of semi-natural vegetation, the seral stage of species' hostplant habitats is expected to have a bearing on the coincidence among species for barriers and corridors (**P6.27**). Seral stages of vegetation differing from those in which hostplants occur might be expected to cause a reduction in number of movements, though they may encourage greater speed of movements.

Although autecological research using mark–release–recapture of individual organisms has led to the identification of barriers to movement in many species (Thomas, 1983a, 1984), very few studies have been conducted of particular obstacles and how they affect movements. Yet, the impact of different surfaces and edges on movements is particularly important in view of increasing habitat fragmentation and urbanization. For instance, road surfaces not only provide minimal resources (e.g., thermoregulation, rest), but also are associated with extreme weather conditions and an additional potential mortality agent in traffic flow. Numbers of butterflies are often found squashed on roads as, for example, along the western side of Marine Drive, around Great Ormes Head, North Wales in July and August, the victims specimens of the two local races of the grayling *Hipparchia semele* and *Plebejus argus* (Dennis, 2008b). A study of the M56 motorway in Cheshire demonstrated that it greatly restricts movement of *Anthocharis cardamines* along the Bollin Valley (see Fig. B6.4d, e). A close study of the movement patterns in the vicinity of the motorway showed that most butterflies were turning back at the side of the obstacle and only very few ventured onto the road itself (Dennis, 1986b). Observations suggested that cross-motorway flights may be limited by deep shade cast by the motorway bridge across the valley (none flew under it), and higher wind speeds and vortices caused by traffic above it, all causing low temperatures. However, not all roads present the same local climate barriers to butterfly movements (Munguira and Thomas, 1992).

Some general observations regarding aggregations are as follow:

- **P6.28: Aggregations at an edge or node are facilitated by an increase in resource types.**
- **P6.29: Prominent edge features lend themselves to aggregations of individuals usually in relation to concentrations of a resource(s).**
- **P6.30: Aggregations will tend to occur where there exists a refuge from exposed, cold climate or weather conditions.**
- **P6.31: The context of aggregations is essentially one of neighbouring supply (source) zones for individuals.**
- **P6.32: A favoured resource is often a key contributor to the formation of aggregations.**
- **P6.33: Aggregations can be caused by barriers or blind links (edges), called terminal vertices.**

A noticeable aspect of mobile individuals is their tendency to gather in aggregations at collecting points (nodes) or edges, especially where numerous resources (i.e., larval hostplants, nectar, roosts, mate location sites, pupation sites, hibernation sites, shelter, etc.) occur (**P6.28**). This is the case even despite the fact that some of the important components that make up a habitat may be missing at these collecting points (e.g., aggregations on abundant nectar sources such as bramble *Rubus* sp.). Aggregations are more noticeable at edge features (e.g., scallops in vegetation) and intersections (nodes) than surfaces (e.g., fields), as these structures lend themselves to resource concentration, refuges from more intensive land uses in adjacent fields (**P6.29**) (Sparks and Parish, 1995). Aggregations of *Maniola jurtina*, *Pyronia tithonus* and *Aphantopus hyperantus* tend to be larger where hedges are higher, where verges are wider and where boundary features comprise key nectar sources (e.g., *Rubus* sp., Compositae; majoram *Matricaria* sp.) (Dover, 1996). They also occur where there is direct insolation and shelter from wind (e.g., scalloped edges) (**P6.30**). Context is important as it often accounts for the source of individuals; for instance, aggregations are associated more with pasture and uncultivated biotopes, less with arable (**P6.31**) (Dover, 1994, 1996). Sparks and Parish (1995) made similar findings for a larger group of species. Of course, part of the reason for aggregations occurring at edges may be the presence of tiny pockets of breeding habitat or specific resources (e.g., *Inachis io* mate location sites; Baker, 1972). In Britain, as many as 24 species may breed in arable field margins (Dover, 1994). Some edge features create aggregations of particular species but not others, as in the case of *P. tithonus* which tends to cluster on old man's beard (**P6.32**) (Dover, 1996). The most notable collecting points for aggregations are nodes, such as sheltered field corners, hedge intersections, depressions on field boundaries and pockets of higher than usual floral diversity; these are points least likely to be disturbed and likely to have suitable local climates. Edge aggregations (nodes) are less ephemeral phenomena and involve more species than aggregations over wider surfaces; this reflects on the greater permanency of the edge features, the bias some species have in using linear features for movements, and advantages of local climate. An example of aggregations occurring outside habitats associated with restricted resources on open ground is provided by cereal fields undersown with turnip *Brassica rapa*. In Cheshire, UK, this can produce aggregations of *Pieris* species, but these conditions are temporary and do not provide nectar sources. The butterflies (green-veined white *Pieris napi*, small white *P. rapae*) typically stay for at most several days, breeding and ovipositing, before moving on; very few of the progeny may survive subsequent grazing, trampling and ploughing (Dennis, 1982b).

Aggregations of individuals, outside habitats, properly belong to movement phenomena as they are temporary (e.g., *Vanessa cardui*; John, 2001). The collecting point forms only part of a resource set required for completing a life cycle and therefore individuals must eventually move on to locate other resources. Individuals are initially attracted to a particular resource or have their activity restricted to specific locations by matrix structures (e.g., edge terminus), physical barriers or environmental conditions (**P6.33**). Aggregations outside habitats have connotations for fields of movements. Collecting points could act as temporary refuges for moving individuals, but then they could also be hazards for them. As aggregations characteristically form in atypical biotopes (e.g., edges and intersections), individuals may be more vulnerable to greater levels of predation than in their usual habitats. Hedge intersections, woodland edges and corners provide particularly favourable hunting ground for insectivorous birds. Thus, nodes, which may benefit moving individuals through the provision of resources, could act as **ecological traps** and be responsible for high mortality and, consequently, reduced movements through landscapes.

Just how edge resources affect movement is not known exactly, but the indications are that although movement may be delayed by resource concentrations along edges compared with flight over fields, movement along edges may enhance survival and generate wider movements. Further research is needed. As part of this, the costs and benefits of nodes along networks remain to be investigated. They will clearly vary with time as do the use of corridors and the effectiveness of barriers. The overall message is that attention should be given to the matrix (surrounding landscape) and the resources it can usefully provide, as well as the obstacles it presents, for a target species, and not to expect this measure to be a metapopulation constant in space–time.

FROM METAPOPULATIONS TO AN ENTIRE LANDSCAPE APPROACH

With increasing fragmentation of the British landscape, a metapopulation approach obviously has an important part to play in the study and conservation of butterfly populations. Metapopulation models are clearly more appropriate to what are termed LC (low coverage) as opposed to HC (high coverage) landscapes. With the former, suitable habitat fragments cover a relatively small fraction of the total landscape; the latter more closely approaches semi-natural and natural landscapes. In dealing with LC landscapes, the metapopulation view of nature is complex enough; regarding HC landscapes it appears to be a poor metaphor for how landscape ecologists view reality, but then this was clearly stated by Hanski and Gilpin (1997:3; see chapter head). As such, Wiens (1997) has called for common ground to be struck between metapopulation and landscape ecology. Even so, for proponents of metapopulation analyses not to draw spurious conclusions, there is an urgent need to incorporate both habitat quality and matrix quality along with patch area and isolation. In doing so, isolation has Stouffer elements to absorb, both in terms of habitat patches and matrix elements. In effect, patch and matrix quality modulate variation in patch area and isolation, and equations can be generated to incorporate aspects of habitat and matrix quality. Before this, there is exploratory work to be done as to how patch area, isolation, patch quality and matrix measures relate to one another and influence population incidence in different situations. A valuable approach is to apply data reduction techniques such as principal components analysis to identify dimensionality in agents influencing populations within the metapopulation; this can easily be done by coding populations up as supplementary variables in analyses on their geographical coordinates, much as done in more traditional biogeographical studies (Maes *et al.*, 2003).

A crucial act in furthering the understanding of resource use is to advance studies on butterfly behaviour in relation to biotope, vegetation and other substrates. Individual dispersal behaviour affects metapopulation connectivity and patch viability (Heinz *et al.*, 2006). This is a necessary step if we are to divorce human observer perception from butterfly perception of landscape and its resource distributions. In the business of defining patch and matrix, very little attention is actually paid to what arthropods perceive and respond. Pertinent questions are: do arthropods actually experience the environment polarized as patch and matrix or as landscape with variable resource distributions? How much does size matter when it comes to resource recognition and use? Empirical studies of just what arthropods do in landscapes can reveal how behaviour is related to biotopes, vegetation units and substrates; it is information that can easily be acquired during either transect surveys (e.g., Butterfly Monitoring Scheme; Pollard and Yates, 1993a) or when recording for mapping programmes (e.g., Butterflies for the New Millennium Programme; Asher *et al.*, 2001; Fox *et al.*, 2006) (Dennis 2004a; Hardy and Dennis, 2007). It is expected that, with a typical vegetation patch (= habitat)–matrix model, movements will be of two basic types:

1 Routine searching, sinuous flight.
2 Direct linear flights (Van Dyck and Baguette, 2005).

Direct linear flights will dominate what is supposed to be the matrix. However, two studies on butterflies reveal that they can treat the matrix as comprising resources. In the matrix – defined either on the basis of traditionally accepted unsuitable biotopes or very low density of the target organism – *Maniola jurtina* and *Pyronia tithonus* engage more in resource searching and resource using than they do in direct linear flights, typical of dispersal and escape (Dennis, 2004a); non-random movements of *M. jurtina* in the matrix have since been confirmed (Kindlmann *et al.*, 2005). Subsequently, pierid butterflies were found to switch between direct linear flight and search flight in response to resource cues across a variety of substrate or vegetation surfaces (see Fig. 6.3) (Dennis and Hardy, 2007). Close attention to resource attributes indicates that movements are affected by more than just interpatch distances, by at least five aspects of resource geography and timing (Fig. 6.7) (Dennis *et al.*, 2003). This lies at the root of the occurrence of both types of movements observed in *Pieris* species across landscapes; frequent switches between the two types are expected. The key thing is that hostplants occur in the matrix but are often so small that they are missed by human observers. But they are found and used by arthropods even when they are not visible to the observer (e.g., hostplants in *Pieris napi* (Dennis, 1985c; Courtney, 1988)). There is a serious lack of data on interactions between behaviour and matrix components, yet such data are critical to understanding the potential of the matrix and the functioning of species in Britain's industrial farming landscapes. There is an urgent need to know more about the scales of resources used and distances over which resource elements can be sensed and the part played by vision and olfaction in resource tracking (Vane-Wright and Boppré, 1993; Cant *et al.*, 2005).

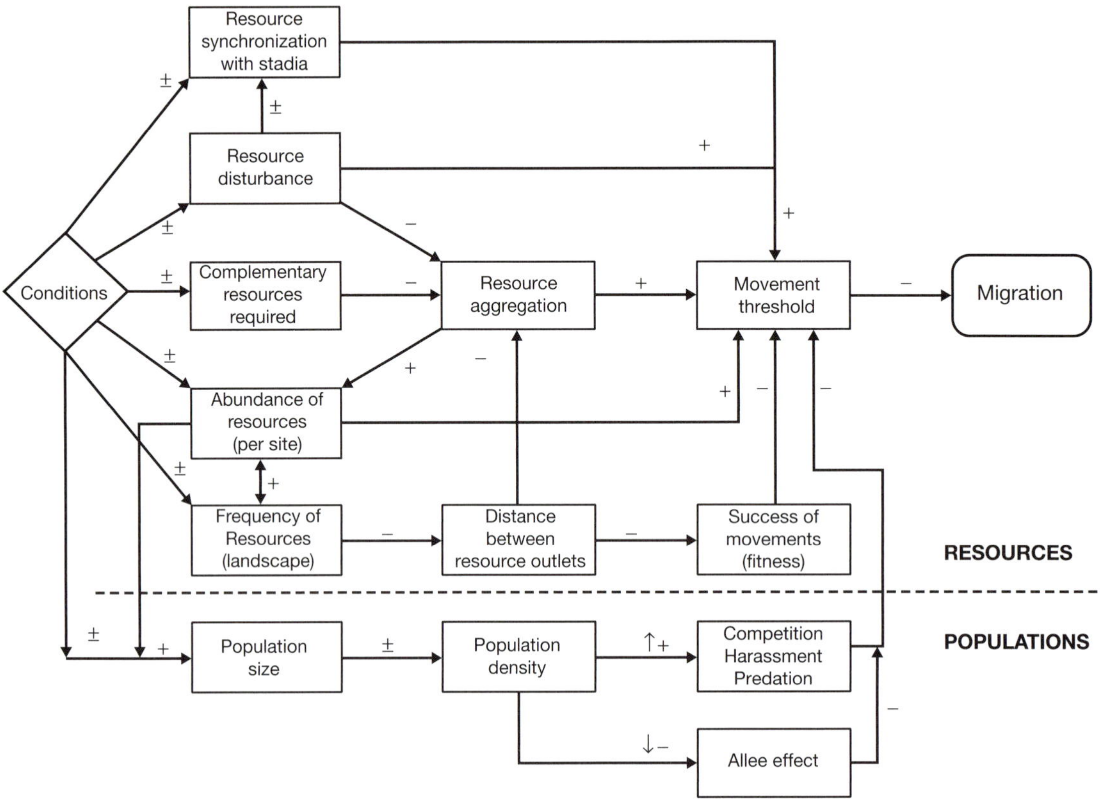

Fig. 6.7 Resource variables affecting movements in the landscape. Initial conditions include a variety of agents (e.g., vegetation succession, human management, weather and climate). These impact on five basic attributes of resource distributions that, in turn, over different space–time frames, influence the tendency to move over the landscape. Resource disturbance refers to vegetation changes (Grime, 1974) and hostplant dynamics (e.g., generation time). Complementary resources refer to non-substitutable resource outlets (Dunning *et al.*, 1992). Some degree of dependence occurs among the resource variables (illustrated). No attempt is made to expand on individual (e.g., lifespan) and population influences on movement and migration distances in this simple process–response model other than to indicate a link to resources, nor on the direct influence of conditions (e.g., weather) (Dennis and Bardell, 1996; Dennis and Sparks, 2006). In Baker's (1978) initiation factor model, the probability of an individual initiating migration depends on its migration threshold being exceeded. According to this model, both population density and environmental conditions interact to affect individual mobility, including changes to resources generated by individual resource use (e.g., the ideal free distribution; Calow, 1999). (From Dennis *et al.*, 2003, courtesy of Blackwell Publishing.)

Chapter 7

LANDSCAPE INFLUENCES ON BUTTERFLY HABITATS

I have steadily endeavoured to keep my mind free so as to give up any hypothesis, however much beloved (and I cannot resist forming one on every subject) as soon as facts are shown to be opposite to it . . . I cannot remember a single first formed hypothesis which had not after a time to be given up or be greatly modified. (C. R. Darwin, from Darwin, 1888)

LANDSCAPE-SCALE STUDIES

Studying butterflies at the landscape level involves a shift up in scale from that of foraging models within habitats (Krebs and Davies, 1993) and metapopulations (Hanski and Gilpin, 1997). Although it encompasses these, it extends to the scale of multiple metapopulations and to that of species' geographical ranges. Landscape ecology, which began with Troll (1939; see also Wiens, 1997), emphasizes the links between landscape structure, function and change, particularly the impact of structure on process (Forman and Godron, 1986; Scott *et al.*, 1991; Yip *et al.*, 2004). The roots to this approach date back to the geomorphologist W. M. Davis (1909) and his focal concern with structure, process and stage (i.e., the form structures have attained with time) in understanding landform evolution. The difference is that contemporary landscape ecology has a much deeper appreciation of interdependency, particularly the scale interdependency of landscape structure and function, evolving from an ecosystem approach (Odum, 1963; Turner, 1989). The basis for this distinct level of investigation is the existence of scale-specific processes. In metapopulation models, no constraints are visualized on the distribution of patches (habitats), which can theoretically occur anywhere. However, habitat patches, let alone distinct metapopulation units, do not, in reality, have an equal probability of occurring everywhere as resources comprising their habitats are dependent on specific conditions imposed by landscape components. Shifts in landscape type counter the more typical, intuitive, observation, that of similarity among landforms in adjacent spatial units (e.g., fields, grid squares) due to the effects of contiguity. Relationships among neighbouring space and time units are referred to as autocorrelation, on which Tobler's 1970 First Law of Geography is based (Box 7.1) (Longley *et al.*, 2001).

This chapter continues with a brief introduction to landscape agents and proceeds with an account of how key variables, separately and in combination, as well as landscape elements can potentially influence the distributions of butterfly habitats. Attention is then given to more formal treatments of landscape influences on butterfly habitats and distributions, inherent problems with surveys, and a brief account of some recent studies relating butterfly distributions to landscapes.

Box 7.1 Autocorrelation, contiguity and Tobler's First Law of Geography

Contiguity and autocorrelation are basic concepts for any observations or measurements made on adjacent units in space or time. Time series are one-dimensional and cause has a single direction, forward in time. Spatial series (viz., adjacent grid cells, natural regions, habitat patches) are more complex; they may be one-, two- or three-dimensional, the latter taking into consideration topography, thus changes in altitude. Causation for changes can take place in any direction. A **contiguity effect** occurs when adjacent observations differ from

A Resource-Based Habitat View for Conservation, 1st edition. By Roger L. H. Dennis. Published 2010 by Blackwell Publishing.

randomly paired observations (e.g., the probability of a field having a marsh fritillary *Euphydryas aurinia* colony is increased by an adjacent field having one; or resources for butterflies in one grid square correlate with resources for butterflies in adjacent squares). Contiguity effects in space or time are referred to as **autocorrelation**, **spatial autocorrelation** for geographical observations and **temporal autocorrelation** for time. Measures of spatial autocorrelation (e.g., correlation coefficients) quantify the degree to which near and distant objects are interrelated; measures of temporal autocorrelation do the same for objects over time units. It is important to appreciate that there are two distinct reasons underlying autocorrelation (Legendre *et al*., 2002):

1 Pseudoreplication: this is autocorrelation *sensu stricto* and relates to sampling issues, such as the recording of an individual butterfly in more than one cell of a grid, owing to vagrancy.

2 Spatial dependence: this is generated by the joint presence of a resource for a butterfly in adjacent cells linked to distribution of an environmental agent such as geology.

It is the second type, spatial dependence, that underlies heterogeneity (spatial independence) in the distribution of habitats and metapopulations. There can be quantum jumps in the distribution of agents (e.g., landscape processes) that go counter to expectations arising from autocorrelation; in fact such sudden changes are expected with say sharp transitions in lithology, with rock outcrops and soil chemistry (see Box 7.2 illustrating Surrey geology and its impact on butterfly distributions). In this box, the focus is on landscape agents controlling the distribution of suitable habitats, which in turn influence butterfly distributions. Two useful principles emerge:

- **PI: Tobler's law. Although some phenomena violate the law, varying with extreme irregularity from place to place, nevertheless, as a general rule, spatial data exhibit an increasing range of values (increasing heterogeneity) with increased distance.**
- **PII: Owing to contrasts in landscape agents and landscapes, butterfly habitats and metapopulations do not have an equal probability of occurring everywhere.**

My geography tutor in Durham, Ian S. Evans, made the memorable point that geographical observations are always, to some extent, autocorrelated: a land surface that is not autocorrelated would be very uncomfortable to travel across and may well be inconceivable! Thus where there are abrupt changes in geology, soils and vegetation and butterfly resources, there will nevertheless be leakages of butterflies (vagrants) into adjacent zones as well as complementary/supplementary resources required by a butterfly species in those contiguous, but distinct, units. The issue then arises: is correlation of a species' occurrence across spatial units owing to pseudoreplication, spatial dependence or both?

The presence of a species in surrounding units is a powerful predictor of its likely occurrence in a target unit; logistic regression models that include neighbourhood variables are called **autologistic models** (e.g., prediction of individual butterfly species in French departéments; Dennis *et al*., 2002; Dennis and Shreeve, 2003). An array of spatial statistics now exists to measure this autocorrelation (Haining, 1990; Legendre and Legendre, 2000; Fortin and Dale, 2005). These measure changes in autocorrelation over space or time. The correlation in the incidence of a butterfly species or its resources across grid squares is unlikely to be homogeneous over a landscape; if it is, then the measure of autocorrelation is said to be **stationary** (independent of absolute position in space or time). Often spatial dependence varies in different directions over a landscape; it is said to be **anisotropic**. As autocorrelation can vary in relation to the interval taken (100 m, 1 km, 2 km, 10 km, etc. grid units; 1 year, 5 years, 10 years), it is necessary to test for autocorrelation for different units. Thus, in correlating for numbers of a butterfly occurring through time, correlations are typically made between x_i and x_{i+t}, where t is the 'spacing' and i is the position in the series; the correlations are made for different values of t (or s for space) to assess the rate at which autocorrelation varies over time or space (it usually declines) with increasing 'distance' (separation). Some spatial statistics summarize the strength of autocorrelation over a landscape. One of the most widely used spatial autocorrelation measures for areal units (e.g., grid units, fields) and interval scale attributes (e.g., butterfly numbers) is the Moran index, another the Geary index (Longley *et al*., 2001). At one time, the abundance of a species, and the causes underlying it, were determined without reference to autocorrelation – no longer. Now, demonstrating cause for the incidence of organisms over landscapes involves a careful assessment of autocorrelation among predictor and dependent variables (e.g., Belgian butterflies; Maes *et al*., 2004). A range of techniques is now available for interpolating values for items (i.e., organisms or resources) over space, based on values for data points in that space. A lucid summary of these methods (e.g., inverse distance weighting, kriging) is given in Burrough (1986), Burrough and McDonnell (1998) and Longley *et al*. (2001) as well as the terms (e.g., semivariogram, sill, nugget, kernel) associated with them.

LANDSCAPE COMPONENTS AND THEIR INFLUENCE ON BUTTERFLY HABITAT DISTRIBUTIONS

There is insufficient space to provide detail on landscape processes in physical geography, but some key concepts in landforms, geology and soils, together with UK source material, are boxed for easy reference (see Boxes 7.2–7.5). Attention is focused on key variables that affect butterfly habitats and how these are distributed unevenly over the country. Three basic principles apply linking butterflies to landscape heterogeneity:

- **P7.1: Butterfly habitats and metapopulations are influenced largely through five variables: minerals, soil moisture, substrate exposure, vegetation and climate.**
- **P7.2: A hierarchy of distal to proximal factors is involved in the incidence of butterfly habitats (viz., geology and climate tend to dominate and drive natural systems of landform processes, soil formation and vegetation succession, though human influences are rapidly taking over in prominence).**
- **P7.3: Similar associations of butterfly resources with environmental variables (e.g., soil moisture or minerals, substrate exposure, warmth) can be repeated over scales varying from a few metres to tens of kilometres within the same broad region.**

Landscape is the spatially varied product, and undoubtedly a complex one, of several basic interlocked, interacting 'spheres' of activity; from ground base upwards they are stacked rock types and structure (geology), landforms and water bodies, soils, and biota (plants and animals), atmosphere and climate. Human constructions and landscape modifications present a further layer of structural complexity. These agents impact on one another and on the factors that influence resource distributions comprising butterfly habitats (Fig. 7.1).

Autecological research on butterflies (Thomas, 1984), supported by research on C-S-R life history strategies described earlier (see Chapter 5) (Grime *et al.*, 1988; Hodgson, 1993; Dennis *et al.*, 2004), has disclosed that five variables, operating separately or in combination, are more directly responsible for the distribution of butterfly resources (**P7.1**):

1 Substrate minerals or nutrients (soil acidity or soil pH).
2 Substrate moisture (or drainage).
3 Substrate exposure (or vegetation cover).
4 Vegetation association (or degradation).
5 Local climate (light, heat, shelter).

As we have seen in Chapter 5, these conditions are usefully summarized by Ellenberg values on plants (Oostermeijer and van Swaay, 1998; Hill *et al.*, 1999; Dennis *et al.*, 2004). They vary substantially across different landscape types, for different spatial scales from broad regions to single slopes, and reflect on interactions among different **endogenetic agents** (i.e., geological lithology, structure and tectonics) and **exogenetic agents** (i.e., geomorphological processes, biota, climate, humans) (Fig. 7.1). At a regional scale, the variation disclosed by these agents is usefully summarized by the natural areas map of England dating to 1996 (Fig. 7.2). This map appropriately identifies 97 terrestrial biogeographical zones that reflect the geological foundation, natural systems and processes over England. It does so appropriately because geology and climate determine the basic inputs into natural systems affecting landforms, soils, vegetation, biota and human activity, thus resource distributions. At any scale, both the spatial dimensions and shape of butterfly distributions – for instance whether linear (e.g., fault-line scarps) or curvilinear (e.g., outcrops of folded strata; volcanic ring dykes) – are indicative of constraints associated with these variables being placed on resource geography.

Substrate chemistry and butterfly habitats

Nutrient status and soil acidity (soil water pH reaction) are determined by mineralogical content, climate (precipitation : evapotranspiration ratio), soil processes and, increasingly, anthropogenically derived aerial deposition. Basic mineralogical status depends on solid and drift geology (Box 7.2), the latter including deposits from human activity. Soil processes (e.g., leaching, eluviation, illuviation, gleying), drainage (e.g., free-draining, impaired drainage) and biotic activity (e.g., ionic exchange, nutrient recycling) can enrich or deplete soils of nutrients (Box 7.3). In Britain, a number of butterfly species are particularly sensitive to nutrient or mineral status but there is no example to match the exotic association of the ridge checkerspot *Euphydryas editha bayensis* to serpentine substrates in California (Harrison *et al.*, 1988). However, some species are confined largely to larval hostplants dependent on lime-rich substrates (e.g., Adonis blue *Polyommatus bellargus*,

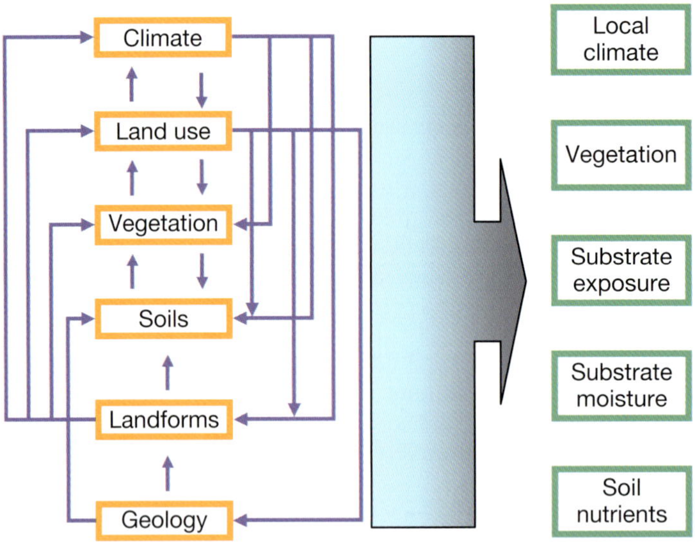

Fig. 7.1 Broad interactions between landscape-forming agents showing basic cascades and feedbacks in natural systems affecting butterfly habitats. Landforms, soils and biota are ultimately the product of two very different forces: (i) **endogenetic agents**, driven by radiogenic energy from the Earth's interior (e.g., mountain building with plate tectonics); and (ii) **exogenetic agents**, powered by solar radiation and operating on the Earth's surface, etching tools of land surface denudation – these are atmosphere driven, generated from heat, wind and precipitation, operating under gravity (e.g., coastal waves, movement of materials on slopes under gravity, running water, flowing ice, wind blasting). Thus, climate and geology dominate natural systems, though human activity (a prominent exogenetic agent) has grown exponentially in the last two centuries to encroach on all natural systems. A simple way of conceiving landscape is as a function of these influences:

$$L = f(r, w, s, c, p, a, m)\Delta t$$

where L is land or landscape, r is rock, w is water, s is the shape of the terrain or landform (topography), c is climate, p is plants, a is animals, m is man, t is time, f is 'function of', and Δ denotes 'variation of' or 'change in' (Zonneveld, 2005).

chalkhill blue *P. coridon*) or nitrogen-rich soils (e.g., small tortoiseshell *Aglais urticae*). The substrates may be allochthonous (derived) not autochthonous (*in situ*), and human in origin rather than natural. In the Manchester conurbation, the common blue *Polyommatus icarus* is largely associated with imported or derived lime-rich resource patches (e.g., bird's-foot trefoil *Lotus corniculatus*) on wasteland (e.g., tips, derelict building sites, rail yards) or reclaimed land (e.g., topsoil from the Peak District Carboniferous limestones) (Hardy, 1998). Soil nutrient status can have an impact on hostplant preferences affecting other aspects of life history and biology such as fecundity, larval development and chemical defences. For instance, in the case of the American buckeye *Junonia coenia* (Hübner), preference for plants on higher nitrogen substrates results in choice for plants with lower chemical defences (Prudic *et al.*, 2005). Distinctions at the landscape level in soil nutrient status can be substantially affected by fertilizer applications with bleak, abhorrent, implications (green desert) for butterfly persistence amid British farmland (see Chapter 8).

Substrate moisture and butterfly habitats

Soil moisture is affected mainly by precipitation levels and evapotranspiration rates, substrate permeability and topography. Areas with high precipitation and low evapotranspiration, impermeable substrates, concave (water collecting) and foot slopes and level ground, tend to have high soil moisture, and areas with low precipitation and high evapotranspiration, steep slopes, especially convex (water shedding) slopes and permeable substrates, tend to have lower soil moisture. Permeability does not simply reflect the perviousness or porosity of rock; it is also affected by the height of

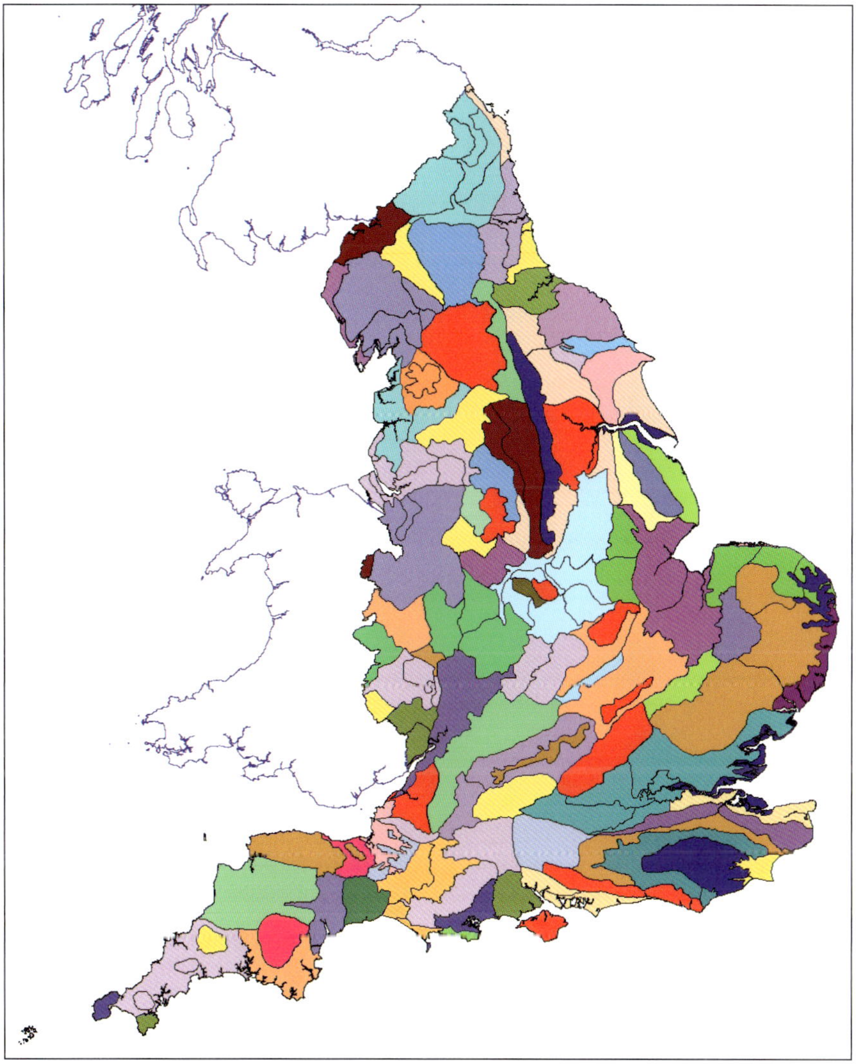

Fig. 7.2 Natural areas map of England for 2008 (see www.naturalengland.org.uk for the key to the areas). With increasing recognition of the importance of the whole countryside in nature conservation, replacing the traditional preoccupation with individual sites, has come the division of the landscape into 120 **natural areas** by Natural England (Idle, 1995; Gray, 2001). These areas, in effect a biogeographical zonation of England, are based on the regional coherence, the coincidence, of distinct geology, landform, soils and plants all of which have implications for animal distributions. Note that the colours do not infer any relationship between areas (see http://www.english-nature.org.uk/science/natural/na_search.asp for names). The biogeographical zonation of Scotland into 10 discrete zones, although based on different variables including 12 climatic and four topographic variables for each 10 km square, has much the same design, to reflect species' distributions and form a basis for conservation (Carey *et al.*, 1995). (Courtesy of Keith Porter; © Crown copyright. All rights reserved Natural England 100046223 2008.)

Box 7.2 **Geological source material for the UK**

Geology forms the foundation for butterfly habitats, the basis for landscape and a major input into soils and vegetation types. It includes solid rock and unconsolidated material, referred to as drift deposits. Geological source material can be obtained from the British Geological Survey (BGS; website http://www.bgs.ac.uk) (Bailey, 1952). The whole country is mapped at four scales (1:10 000, 1:50 000, 1:250 000, 1:625 000) and some areas of special interest (e.g., Snowdonia; see Condry, 1973) are mapped at a useful scale of 1:25 000. The BGS has produced a set of 20 regional handbooks on the geology of Britain; the data are also available digitally. Mapping in Ireland has had a long history (Davies, 1983). Themes include solid and drift geology (unconsolidated sediments, mass movements and artificial ground).

Rocks fall naturally into three groups on their formation: igneous (from molten magma), sedimentary (from pre-existing surface materials, eroded, weathered or life forms, and deposited in layers) and metamorphic (previous rocks altered by intense heat and pressure) (Woodcock, 1994; Maltman, 1998). A basic classification is given in Table B7.2a. These groups differ substantially in structure, mineralogical composition and texture, therefore in their susceptibility to breakdown from weathering and erosion and the material they provide for soil formation. Geology, through rock type and structure, is a major determinant in the evolution of landforms, and is reflected in landforms. Holding geomorphic processes constant, the impact of rock type on landscape depends on:

1 The relative resistance of different rocks to denudation.
2 The disposition (exposure) of different rocks to denudation.
3 The time rocks have been exposed to denudational processes (Sparks, 1972; Woodcock, 1994).

In turn, the resistance of rocks to denudation depends on rock hardness, including chemical composition, and permeability. Rock hardness is a complex issue and is very difficult to measure entirely. It involves properties such as particle compaction and crystal bonding, shear strength of minerals and planes of weakness such as bedding, fractures and jointing. Chemical composition determines the rates of chemical weathering and corrosion (erosion by solution activity). Permeability measures the rate of water transfer through rocks, which may pass through pores within the body of the rock (porosity) or down joints and fractures and along bedding planes (perviousness). As a general rule, harder rocks form more upstanding relief than do softer rocks. Thus, the hard pre-Cambrian gneiss of the Malvern Hills stands out abruptly from the surrounding Devonian and Triassic sediments. Permeable strata are an important exception. As erosion of land surfaces is principally caused by running water, rocks that reduce surface runoff are usually more resistant to denudation than less permeable strata, even when the latter are harder and less soluble. Limestones (e.g., Carboniferous limestone and chalk) are particularly permeable and typically form upstanding relief. Just how rock type affects landscape depends largely on their disposition, their juxtaposition with other rocks of differing properties and their degree of exposure to denudation. The same rock type sandwiched between more resistant rocks in one place and between less resistant rocks in another may form a trough (valley) in the first instance and a ridge in the second case. Thus, the Carboniferous limestones of the north Pennine block stands above the softer Triassic sandstones of the Vale of Eden in northern England, whereas in the Brecon Beacons, South Wales, Carboniferous limestone forms subdued relief against the quartz conglomerate-rich millstone grit.

Much of the contrast in British landscape and the potential for landscape evolution is controlled by geological structure. Folding, faulting and volcanic forms determine both contrasts in relief and varied surfaces for denudation to work on. The geology of Britain has the impression of being built up in layers from northwest to southeast. At this scale, the regions of dramatic relief and landforms in northern and western Britain are not only areas of old hard rocks, but zones of past intense folding, faulting and volcanic activity; some very ancient volcanoes, much reduced, still project above the English countryside as prominent features (e.g., Wrekin to Ragleth Hills, Shropshire). The gentler, subdued relief and landforms of south and east Britain, in contrast, are on softer, younger rocks; folding has been relatively gentle, faulting sparse and concealed beneath superficial deposits, and surface volcanic forms generally absent. Close attention to specific case areas reveals that all rock types have unique influences on landscape. Occasionally, rock or structural properties produce strikingly distinctive landforms that have direct relevance for organisms found on them, their biology and geography. The impact is typically through geomorphic agencies described below and the combination of substrates, slopes, aspects and drainage. Such, for example, are the scars, pavements and solution hollows found on Carboniferous limestone, the coombes and dry valleys on scarp and dip slopes of chalk (see Fig. 7.10), and the tors of granites.

Table B7.2a Basic rock types and examples.

	Texture or chemistry		
Formation	**Acid**	**Intermediate**	**Basic or ultra basic**
Igneous rocks			
Extrusive	Rhyolite	Andesite	Basalt
Minor intrusive or hyperbyssal	Quartz porphyry	Porphyrite	Dolerite
Major intrusive or plutonic	Granite	Diorite	Gabbro Peridotite Picrite
Free quartz (amount %)	Oversaturated >10% SiO_2	Saturated 5–10% SiO_2	Undersaturated <5% SiO_2
Prominent minerals	Orthoclase feldspar, quartz, plagioclase feldspar, micas	Plagioclase feldspar, quartz, orthoclase feldspar, hornblende	Plagioclase feldspar, augite, olivine
Sedimentary rocks			
Mechanically formed	Rudaceous (coarse texture, e.g., conglomerates, breccias, scree) Arenaceous (moderate texture, e.g., gritstone, sandstones, sltstone) Argillaceous (fine texture, e.g., mudstone, marl, clays, shale)		
Organically formed	Calcareous (chalk, limestone (shelly, coral)*) Siliceous (diatomaceous earths) Ferruginous (limonite) Carbonaceous (peat, lignite, coal) Phosphatic (guano and bone beds, coprolites)		
Chemically formed	Evaporites (rock salt, gypsum, anhydrite)		
Metamorphic rocks			
Thermal (heat dominant; e.g., contact with magma, causing recrystallization)	Shales → slates → hornfels (minerals grow in haphazard texture) Sandstones → quartzites Limestones → marbles		
Regional (enormous compression and heat; e.g., typical of subduction zones in plate tectonics)	Schists (finely foliated layers) Gneisses (thicker foliated rocks of distinct mineral composition) Minerals develop parallel orientation; occurrence of porphyroblasts such as garnets For example dolerite/basalt → chlorite-schist → hornblende schist → eclogite		

* Jurassic limestone is really chemically and mechanically formed.

Drift (superficial) geology covers much of the solid geology of the countryside; often this cover bears little relationship to the underlying solid geology, the material being allochthonous (alien) as is the glacial till overlying Triassic sandstones in Cheshire (see Clark *et al.*, 2004). To understand the landscape, soils and vegetation overlying it, it is necessary to become familiar with the variety of drift deposits, material laid down by various geomorphic agents and human activity. Drift deposits are laid down by rivers (e.g., alluvium), slope wash (e.g., colluvium), glaciation (e.g., till, outwash gravels), periglaciation (e.g., solifluction deposits), the sea (e.g., marine clays, beach deposits) and the wind (e.g., loess). Each of these deposits has a characteristic texture and structure, though mineralogy may be inconstant, that give them unique qualities and permits identification. Extensive areas are also covered by organic remains from the accumulation of decaying vegetation (e.g., peat). The geology of England is nicely reflected in the natural areas map (see Fig. 7.2).

The importance of rock types to British butterflies is demonstrated by the number of species that are either entirely found on a single rock type or which have the majority of their populations linked to specific lithology. Limestones, whether soft as in chalk, or harder limestones, are key rock types for British butterflies (Dennis, 1977, 1992a). Species dependent on them for their habitats in the British Isles are noted in Table B7.2b. An example is provided of substrate-limited habitats for butterflies in the county of Surrey; a cross-section and map illustrate the geology (Fig. B7.2a, b); Table B7.2c lists butterfly species associated with the main rock types and their associated soils and vegetation types.

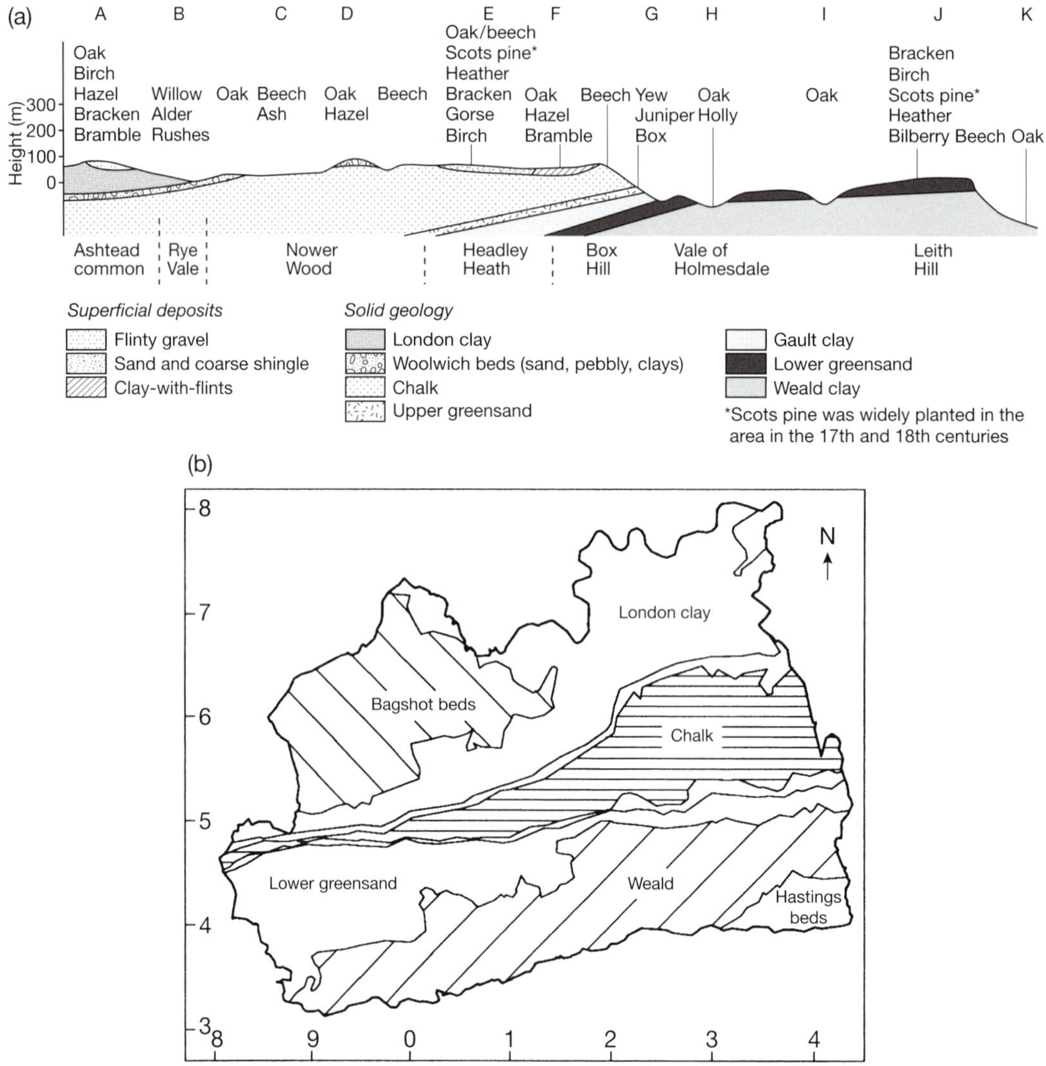

Fig. B7.2 (a) North to south cross-section of the geology from Ashtead Common to Leith Hill across the Weald and North Downs. (Adapted from Briggs *et al.*, 1979, courtesy of Oxford University Press.) (b) Basic geological divisions of Surrey. The distance between grid numbers is 10 km. (From Collins, 1995, courtesy of Surrey Wildlife Trust, School Lane, Pirbright, Woking, Surrey GU24 0JN, UK.)

Table B7.2b Butterfly species dependent on limestones in the British Isles.

Butterfly species	Limestone
Limited (or almost so) to limestones	
Lulworth skipper *Thymelicus action*	Chalk, Portland limestone
Silver-spotted skipper *Hesperia comma*	Chalk
Chalkhill blue *Polyommatus coridon*	Chalk, Portland limestone, oolitic limestone
Adonis blue *Polyommatus bellargus*	Chalk, Portland limestone, oolitic limestone
Glanville fritillary *Melitaea cinxia*	Chalk
Main populations on limestones	
Small blue *Cupido minimus*	Chalk, Portland limestone, oolitic limestone, Carboniferous limestone
Brown argus *Aricia agestis*	Chalk, Portland limestone, oolitic limestone, Carboniferous limestone
Duke of Burgundy *Hamearis lucina*	Chalk, oolitic limestone, Carboniferous limestone
Pearl-bordered fritillary *Boloria euphrosyne*	Ireland Burren, Carboniferous limestone
Wood white *Leptidea sinapis*	Ireland Burren, Carboniferous limestone

Table B7.2c Butterfly species with significant substrate bias in the county of Surrey. (Modified from Collins, 1995.)

	Limestones	Sandstones, pebble and flint beds	Clays
Geology	Chalk	Hastings beds Lower greensand Upper greensand (Woolwich beds) Bagshot beds	Weald clay Gault clay London clay
Soil type	Calcareous brown earth (dip slope) and rendzina (scarp slope)	Podsols	Gleys and gleyed soils
Vegetation	Calcareous grassland	Heathland	Woodland
Butterfly species*	*Hesperia comma* (*Erynnis tages*) (*Pyrgus malvae*) (*Callophrys rubi*) *Cupido minimus* *Aricia agestis* *Polyommatus coridon* *Polyommatus bellargus* *Hamearis lucina*† *Argynnis aglaja* *Melanargia galathea*	*Plebejus argus* *Hipparchia semele*	*Leptidea sinapis* (*Thecla betulae*) *Boloria selene* *Boloria euphrosyne*

* Unbracketed species are almost entirely restricted to the substrate; bracketed species have colonies/records on other rock types in the county. An additional 25 resident butterfly species occur widely over the different substrates. Elsewhere, some of these species may be found off limestones (e.g., *Aricia agestis*; Kemp *et al*., 2008).

† Extinct on limestones in 1997 (G. and S. Jeffcoate, personal communication).

Box 7.3 **Sources for UK soil types**

Classification and mapping of soils in mainland Britain is carried out by the Soil Survey and Land Research Centre at Cranfield University (SSLRC, England and Wales; www.cranfield.ac.uk, see also www.cis-web.org.uk) and the Macaulay Land Use Research Institute (MLURI, Scotland; www.mluri.sari.ac.uk). In Ireland it is the Teagasc (Johnstown Castle Research Centre, Wexford; www.teagasc.ie/research). Soils are described from three-dimensional vertical sections (**pedons**) usually accessed by digging pits or by taking cores. These sections or cores typically reveal a quasi-horizontal 'layering' similar in appearance to that seen in sediments but are quite different in origin; that is, not the result of deposition as strata. Instead, these layers arise from movement of materials – water, humus, nutrients, minerals – from the activity of organisms and changes in soil pore composition of gases and water. Collectively these soil-forming processes are referred to as **pedogenesis**. They include:

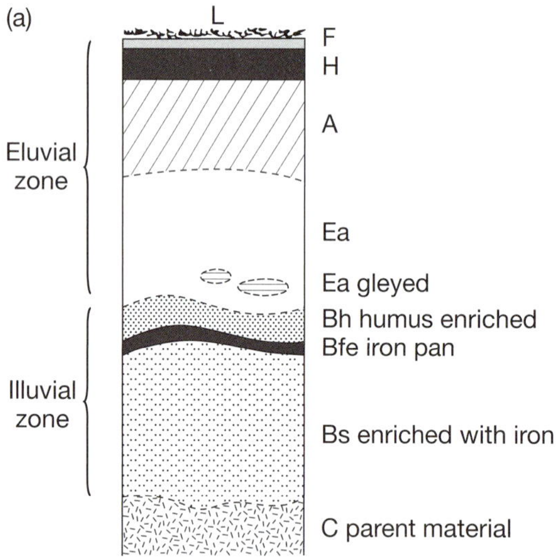

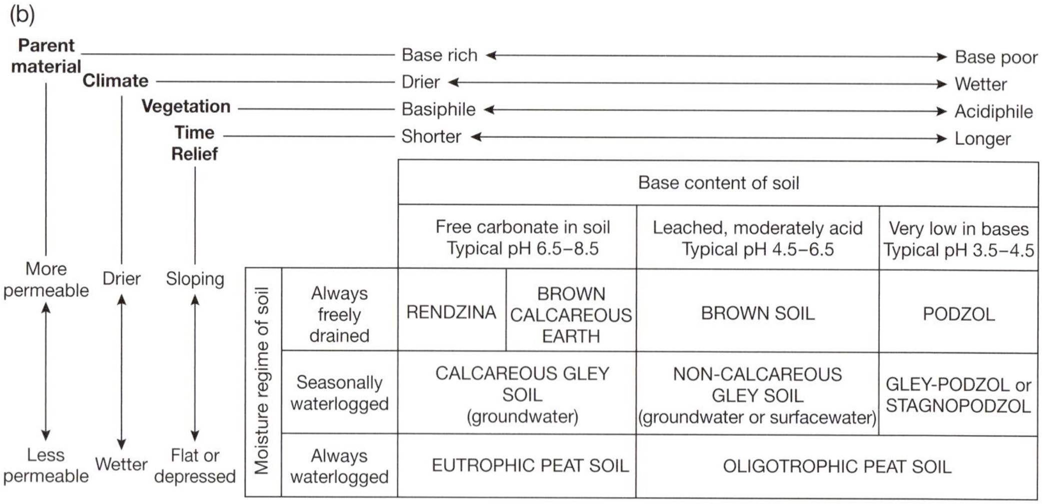

Moisture regime of soil	Free carbonate in soil Typical pH 6.5–8.5		Leached, moderately acid Typical pH 4.5–6.5	Very low in bases Typical pH 3.5–4.5
	Base content of soil			
Always freely drained	RENDZINA	BROWN CALCAREOUS EARTH	BROWN SOIL	PODZOL
Seasonally waterlogged	CALCAREOUS GLEY SOIL (groundwater)		NON-CALCAREOUS GLEY SOIL (groundwater or surfacewater)	GLEY-PODZOL or STAGNOPODZOL
Always waterlogged	EUTROPHIC PEAT SOIL		OLIGOTROPHIC PEAT SOIL	

pH figures refer to natural soils (unlimed)
Arrows indicate tendencies not all of which apply to a given situation

Fig. B7.3 (a) Schematic diagram of a soil profile (a podsol), illustrating the notation for indexing soil horizons using capital and lower case letters (not all the following are illustrated): L, little decomposed litter; F, fermenting layer, with comminuted litter showing some plant structure; H, humus layer (decomposed organic layer often with some mineral particles); O, organic layer typical of peat forming under poor drainage conditions; A, mineral surface horizon with well-incorporated organic matter; E, eluviated horizon depleted of clay or iron and/or aluminium sesquioxides; B, horizon altered by downward translocation of materials; C, parent material, solid or unconsolidated rock; G, gleyed horizon; h, added humus; s, added sesquioxides; t, added clay; g, gleying; fe, iron pan; a, ash-like or ashy grey. (b) Some effects of formative agents in British soil types; arrows indicate general tendencies which may not apply in a given situation owing to the influence of other factors.

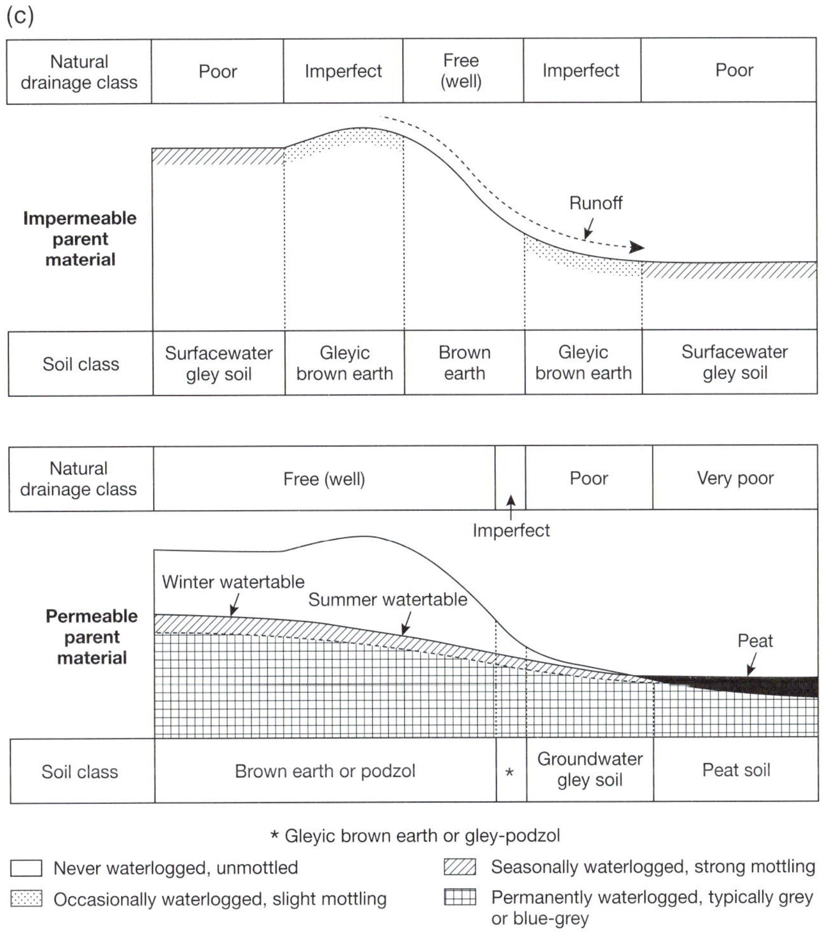

Fig. B7.3 (*continued*) (c) Effects of relief and hydrology on soil and drainage for similar slopes on impermeable rocks (top) and permeable rocks (bottom). (b, c, from Burnham, 1980, courtesy of the Field Studies Council, Preston Montford, Montford Bridge, Shrewsbury SY4 1HW, UK.)

(i) the breakdown of organic matter (**humification**); (ii) the mobilization of anions and cations; (iii) the downward and downslope translocation of materials including **leaching** (losses of N, Ca, Mg, Na and K in drainage water) and **eluviation** (losses of clay particles and iron and aluminium sesquioxides in drainage water), and their precipitation and redeposition (**illuviation**) which ultimately causes **podsolization** (formation of podsol soils, Fig. B7.3a) and in extreme conditions **pan formation** (mineral crust development); and (iv) aeration or the reverse in waterlogging (**gleying**). Soil organisms (e.g., worms, moles) form a vital component in the movement of mineral and humus in soils. In effect, soil is a living layer; British soils may comprise more than 2000 distinct organisms at a single site and as much as 2 tonnes of bacteria per hectare. Movement can be upwards by capillary action as well as downwards or laterally downslope. As more generally with slope processes, these various activities describe soils as active systems with internal modification from incessant inputs and outputs (Courtney and Trudgill, 1976; Trudgill, 1977).

In the UK, soil classification has four levels and is based on the **soil profile** (Butler, 1980). At the highest level are 10 major groups (two types of immature soils, shallow soils, clay soils, well-drained brown soils, podsols, two types of seasonally waterlogged soils, artificial soils and peats) (see Avery, 1973). These groups are further subdivided so that there are 43 groups and at the lowest level there are some 700 named soils (Hodgson and Thompson, 1985). Soil maps, including explanatory booklets, are available at scales of 1:250 000, 1:50 000 and 1:25 000 but only approximately 25% of England and Wales are mapped at the higher resolutions. In Ireland, 44% of the country has been mapped based on the 10 main Great Soil Groups of the United States Department of Agriculture (USDA) 1938 system (Gardiner and Radford, 1980), specifically: browns earths, podsols, brown podsolics, grey brown podsolics, gleys, skeletal stony soils (lithosols), rendzinas, immature soils (regosols) peats and raised bog. A generalized soil map is available at a scale of 1:575 000 and the aim is to produce a map at a scale of 1:126 720. Each organization has now moved away from intensive soil survey to using environmental data and predictive techniques to fill in gaps and to produce digitized data and soil information databases.

Soil types are influenced by parent material, topography (slopes), climate and vegetation (Fig. B7.3b); regional variation in soil types over the British Isles is striking and is illustrated in colour in Burnham (1980) at a scale of 1:2 000 000. The major soil groups divide along three lines involving distinct processes: the degree of development, acidification and drainage (waterlogging). Immature and shallow soils can arise for one or more of four reasons; either the bedrock releases very little insoluble residue (e.g., limestones, especially chalk), the bedrock is extremely resistant to weathering (e.g., quartzites), there has been insufficient time for soil formation, or soil erosion exceeds new soil development. Bare rock surfaces are increasingly common in the north of Britain where glacial erosion has removed much of the regolith as recently as 10 000 years ago. Acidification increases with downward translocation of water. This may arise because of high levels of precipitation, permeable substrates and low soil water pH. Low soil water pH levels generally reflect on the high carbon to nitrogen content (C : N ratio) of plants (e.g., conifers and ericaceous species such as heather *Calluna*) and have consequences for the mobilization and availability of soil nutrients, conditions for life in soils and soil formation. In acid conditions, downward translocation of materials is exacerbated since organic activity is reduced and so, consequently, is redistribution of minerals and nutrients throughout the soil profile. Conversely, horizon development in soil profiles tends to be reduced in soils with increased organic activity. Waterlogging of soil profiles occurs for three, often combined, reasons: high water availability (from precipitation or surface runoff), water collection on near level surfaces or concave profiles, and low infiltration capacity affected by clay textures and collapsed soil structures (Fig. B7.3c). The result ranges from gleying, caused by deoxygenation of horizons, to peat accumulation in extreme circumstances. Gleying and seasonal waterlogging can also occur where duricrusts (iron pans) have developed in previously excessively drained, nutrient-depleted regolith; the pan, occurring from the downward translocation of iron and aluminium sequioxides, sustains an artificial watertable.

Regional contrasts in soil types form one end of the **pedocal–pedalfer** sequence across Europe; in the former, upward translocation of soil moisture dominates formation and in the latter downward movement is dominant. At this scale climate is the main controlling factor of soil types. Britain belongs to the region of pedalfer and waterlogged soils (e.g., gleys, organic soils) and this alone suggests how marginal environments are for many butterflies. Even so, there is much variation in temperature and precipitation throughout Britain, producing contrasts in soil organism activity and the translocation of minerals and nutrients in soils. Consequently, there are distinctive geographical trends in soil types in Britain; organic soils, gleyed soils and podsols dominate in the west and north and brown earths in the south and east.

What is particularly intriguing is that contrasts similar to those occurring between north and south Britain can be found within much smaller areas, affected by altitude, slope, aspect and geology. This occurs primarily because contrasts in water balance and temperatures affected by local conditions can match regional climates. Landform influences are mediated through parent material. Especially important properties of geology are permeability, base content and susceptibility to weathering, including the texture and mineralogy of the regolith. These various factors determine water infiltration rates, aeration, organism activity and nutrient status. In these relationships the impact of solid geology tends to be indirect; most soils in Britain develop on young superficial (drift) deposits, not all of which are evident on geological maps. Nevertheless, the sequence of soil types differs for permeable and impermeable underlying rock types and soil sequences tend to be dominated by geology to the extent that **catenas** (a chain) of soil types mirror repeated exposures of strata (Avery, 1973). Relief and topography control soil temperature and water supply. Soils at higher altitudes are wetter and colder than those at lower altitudes, as are those with northern aspects compared with others having southern aspects. Slope shape and position determine whether soils are in water-shedding or water-receiving environments (Fig. B7.3c). Where the slope geology is varied, with contrasting beds of clays, sands and calcareous strata, very complex influences on soil type and character can result.

local and regional base levels (watertables, spring lines) and soil conditions. For instance, vegetation cover, especially deep rooted trees in woodland and bracken on hillsides, can facilitate drainage, whereas underlying impermeable soil horizons (iron pans in podsols) impede it. The classic case of a species dependent on specific wetland patches in Britain is that of the large heath *Coenonympha tullia* on *Eriophorum vaginatum–Erica tetralix* mires, typically as raised bogs over boulder clay, or marine clay-lined hollows in water-collecting lowlands or blanket peat expanses over relatively impermeable rocks such as gritstone and shale on moorlands subject to high precipitation (Fig. 6.2). Other species can be affected by moisture levels, if more subtly: for example, the host red ant *Myrmica sabuleti* Meinert of the large blue *Maculinea arion* is susceptible to drought and in Britain seems only able to maintain high densities on sites with deeper, moister soils (Asher *et al.*, 2001). Other species require dry soil conditions (e.g., grayling *Hipparchia semele* for pupation; Dennis and Bardell, 1996).

Substrate exposure and butterfly habitats

The degree of substrate exposure depends on rock type and endogenetic (tectonic) and exogenetic (geomorphic) agents (see Fig. 7.1). Bare substrates are typically associated with indurated (compacted) rock that, with denudation, releases little in the way of insoluble regolith (e.g., Carboniferous limestone) or inert (e.g., Shropshire Stiperstones and Sharpstones orthoquartzites) weathering products, and unconsolidated drift deposits that are continually in flux as a result of slope movements referred to as mass wasting (e.g., talus slopes, fresh moraines). Slopes exceeding 45°, termed 'free faces', because they undergo continual active mass wasting of falling debris, are produced by a variety of tectonic activities (e.g., earthquakes, volcanicity) and geomorphic processes (e.g., cliffs of marine, glacial and fluvial erosion). The end product can be very different, depending on the combination of lithology and process; thus, the contrast between so-called coastal soft cliffs and hard cliffs and their distinctive, associated insect faunas (Howe *et al.*, 2008). Disturbance agents also expose bare ground and may have natural (e.g., fire) or human (e.g., mining, quarrying) origins. Many butterfly species in Britain are attached to biotopes with bare ground (e.g., silver-spotted skipper *Hesperia comma*, *Polyommatus bellargus*) that need warm microclimates for early stages (Thomas 1993), but a number of butterflies also use bare ground directly as a resource (e.g., peacock *Inachis io*, wall brown *Lasiommata megera* and *Hipparchia semele* for mate location and thermoregulation; see Box 4.1). Factors that maintain exposed substrates, such as steep slopes, also generally limit vegetation succession and retain early seres (see Chapter 5). Along the Welsh border, away from the Carboniferous limestone screes at Llangollen, *H. semele* is restricted largely to sites of human origin, prehistoric earthworks, such as hillforts, and spoil heaps from quarries and abandoned lead mines (Joy, 1996a; Loram *et al.*, 2003).

Vegetation succession and butterfly habitats

Very little, if any, of the natural Holocene climax vegetation persists today (see Box 8.5) (Bennett, 1989). Vegetation cover is largely controlled by humans. Human influences, mainly through intensive farming an urbanization, are particularly prominent in arresting vegetation succession, through specific land uses, and even more so in degrading and fragmenting biotopes. Examples abound of butterfly species where patchwork distributions are restricted to plagioclimax vegetation associated with grazing (e.g., *P. bellargus*; Thomas, 1983a) or mowing, and past coppicing practices (e.g., heath fritillary *Melitaea athalia*; Warren, 1987c). Subtle changes in practices over time, such as mowing regimes, can have serious consequences for population persistence, as in *Maculinea* species (Johst *et al.*, 2006). In all the cases, the controlling factor influences the availability of a key resource or resources in suitable condition for the butterfly, usually the hostplant but also nectar, but not necessarily simply the incidence or the abundance of the resource. The classic case is the dependence of *Maculinea arion* on grazing levels for the condition of the hostplant (wild thyme *Thymus polytrichus*, wild marjoram *Origanum vulgare*) and presence of the symbiont (*Myrmica subuleti*); the former plant may thrive even when grazing levels are inadequate for the ant (Thomas, 1984).

Light, warmth and butterfly habitats

Butterflies are highly dependent on heat, light (sunshine) and moisture levels and are therefore particularly vulnerable to agents of weather and climate (see Chapter 2) (Dennis, 1993a). Experiments have shown that butterflies tend to have optimal temperature requirements related to open (e.g., *Hipparchia semele*, small heath *Coenonympha pamphilus*) and shaded (e.g., ringlet *Aphantopus hyperantus*, speckled wood *Pararge aegeria*) conditions (Karlsson and Wiklund, 2005). Weather and climate involves a complex of factors that may affect butterfly biology, directly (Shreeve, 1992a) or through their resource distributions, and associated with regional location in Britain, topography (e.g., altitude, hills, valleys, slopes) and vegetation features (e.g., open areas, edge features), at scales varying from that of the landscape (Dennis, 1993a) to that of a single plant or molehill (Dennis and Sparks, 2005). Coastal distributions of butterflies towards the north of Britain, as for *H. semele*, would suggest some climatic constraints that can be tested locally for different aspects on headlands and hillsides such as the Great Orme in North Wales (Dennis, 1992a; see below).

Hierarchy and scale recurrence in factors influencing butterfly habitats

Although some species may show considerable tolerance for environmental variables (e.g., meadow brown *Maniola jurtina*), others are inordinately influenced by one of the key variables, or may show bias for more than one of the following: soil nutrients, soil moisture, substrate exposure, vegetation status and local or regional climates (Oostermeijer and van Swaay, 1998) and therefore with agents influencing these variables. It is sometimes difficult to extricate which factor among many is uppermost in accounting for habitat distributions. Thus, *Coenonympha tullia* is limited to substrates that are not only waterlogged but mineral deficient (acid peat) (Dennis and Eales, 1997, 1999). Butterflies restricted to limestones in Britain (e.g., *Hesperia comma*) not only occupy unique rock chemistry, but tend also to occur in drier and thinner soils, on distinctive topography (scarp slopes, coombes, dry valley sides) with characteristic local climates, relating to aspect and insolation levels, and land uses. Influences are often evident in the local distribution patterns of species; both the spatial dimensions and shape of distributions are indicative of constraints being placed on resource geography. Clearly, potential patchwork geography (metapopulation geography) may vary for different species dependent on the same basic resource, such as a hostplant, as demonstrated by Gutiérrez *et al.*, (2001) for the dingy skipper *Erynnis tages*, *Polyommatus icarus* and *Zygaena filipendulae* on *Lotus corniculatus*. The degree to which their patches may overlap depends on the extent to which resource demands coincide for the five controlling niche variables.

Examination of butterfly distributions reveals that two further important principles apply in habitat distributions. First, in terms of causes for habitat incidence, there is a distinction between distal and proximal causation – factors operating earlier or later in the chain of cause to effect, respectively; thus a hierarchy in causation exists (**P7.2**). Geology and climate can dominate the outcome of other variables and this is an important issue for comparative studies that attempt to determine causation for butterfly distributions (see below). Thus, in the case of the limestone example above, permeable limestones and associated geomorphic processes produce characteristic landscape forms of cuestas – scarps and dip slopes – in turn influencing soil type, soil exposure, soil moisture and vegetation type, and distinctive local climate, but not vice versa. A study of an organism such as *Polyommatus bellargus* over the whole of Britain will likely identify geology as a key, if remote, variable but another study restricted to its distribution will identify variables lower in the hierarchy.

Over the past half century humans, increasingly, have become the dominant influence in the distribution of butterfly habitats (Heath *et al.*, 1983). This is evident from the losses in distributions at the 10 km scale (Asher *et al.*, 2001; Fox *et al.*, 2006); the scenario is even clearer with finer scale mapping at 1:25 000 scales or higher resolutions (Thomas and Abery, 1995) and indirect assessments based on transect counts over different biotopes (Cowley *et al.*, 2000) (see below). The basis for it is the intensification of agriculture and growing urbanization (Fox *et al.*, 2006). Agricultural land under pre-industrial management, which evidently provided habitats for butterflies requiring grassland and herb biotopes, is now increasingly a biodiversity desert. It is a sobering reflection that this open farmland was initially created by humans, first by hunting practices in the Mesolithic and then by encroachment on the Holocene forest cover by farming from the Neolithic to Medieval times, when during the latter period such biotopes would have been optimal for most British butterflies (Dennis, 1977, 1992a, 1993a). At an alarming rate farming, in its industrial mode, is taking away what it has given in traditional and subsistence practice.

As a further principle, environmental effects on resources can recur over varying spatial scales (**P7.3**). Geological distinctions (across geological periods) and climatic contrasts affecting butterfly resources over a broad scale from southern to northern England can be repeated over a single slope occupying a few hundred metres in soil **catenas** related to rock outcrops and local climate gradients (see Boxes 7.2 and 7.3). The typical example is the mirroring of latitudinal climatic gradients on butterflies and their resources, including hostplants, by those associated with increasing altitude, repeatedly for different localities, within limited areas. These limits suggest common agents in resource limitations, but the impact can be a multiple one. Thus, much as range limits rarely extend to the same latitude on either side of the country, they will undoubtedly be found to vary with altitude for different mountains and hill regions (see Chapter 8).

INFLUENCE OF LANDSCAPE AND LANDFORM ELEMENTS ON BUTTERFLY HABITATS AND RESOURCES

Landform and landscape features

Landscape is made up largely of two basic components describing form and structure: **landforms** and **landscape features**. Without imposing a rigid distinction between the two, landforms determine the basic shape of the land, formed by underlying rock and regolith, whereas surface structures or objects on landforms can be referred to, if only for convenience, as landscape features. Both landforms and landscape features can be natural (glacial, periglacial, fluvial, marine, aeolian, etc.) or human in origin and are often provided with terms describing their form or structure (Box 7.4) (Sparks, 1972; McCullagh, 1978; Hart, 1986). In the case of landforms these terms may be specific, conveying genesis (e.g., oxbows of rivers, cirques of glaciated uplands, the spits and bars of coastal deposition), or have more general application and include words (e.g., hill, slope, knoll, plateau, cliff – which may be coastal, a river bluff or quarry) used in everyday communication, indicative of uncertain or complex origins. Even so, all landforms regardless of presence or absence of suitable terms can be described using the notation of slope elements (Box 7.5). Landforms often have mixed origins, as in the case of those created by changes in base (sea) level (e.g., raised beaches) and by changes or conflict in agents (e.g., river terraces cut in glacial deposits). Whatever the agents of action, landscape features are the result of distinct processes: erosion (e.g., roche moutonnée) or deposition (e.g., drumlins) or both (e.g., crag and tail) – to take examples from glaciation – and vary considerably as to size (e.g., rabbit scrape and molehills to scarplands and mountains; single brick to a conurbation), degree of permanence and their impact on landscape and, potentially, on butterfly habitats and resources.

Specific landforms are recognizable, and so named, by having distinctive physical characteristics. Particularly important in distinguishing landforms is their shape, orientation, size and location, and prominence relative to other features, but details of slopes and the texture of materials are also important (see Boxes 7.4 and 7.5). A number of components determine their value as sites for butterfly resources and habitats; among these are:

- Location in Britain (longitude and latitude).
- Altitude.
- Degree of permanence (how long they persist in historical and geological time).
- Size of the landform (in three dimensions).

Box 7.4 Natural landforms

Landscape is not a fixture; it is continually changing. Land raised above sea level by endogenetic forces (volcanicity, folding, faulting) is in the process of continual modification and transformation by exogenetic agents of denudation. New landforms and land surfaces are produced by weathering (modification *in situ*) and erosion of the original surface and by deposition of debris and regolith (weathered material). Denudation achieves two ends. The first is the reduction of the land surface towards sea level, as material is removed and the height of the land is reduced. The second involves the smoothing of land surfaces, a process to which the term **grade** applies; erosion occurs in zones of surfeit energy and deposition where there is a deficit in energy. The consequence is that rock surfaces (slopes, river and glacier profiles) become smoother or graded. The agents of erosion are water (e.g., rivers, slope wash, wave action at the coast), ice (e.g., glaciers, frost heaving) and wind, alone or in conjunction with detritus. These agents, the individual processes of erosion and deposition, and the resulting landforms have distinctive geography over a region the size of the British Isles. During the last two million years, and as recently as 10 000 years ago, northern Britain has been dominated by glaciation and southern Britain by periglaciation (see Chapter 8), processes associated with permafrost (permanently frozen ground); with the Holocene world sea level rise, valleys have been drowned by sea – glacial troughs in Scotland and river valleys in southwest England. Isostatic rebound (rise) of highland Britain relieved of overlying ice, has left raised beaches as far south as Wales, reminiscent of higher marine platforms during higher sea levels in

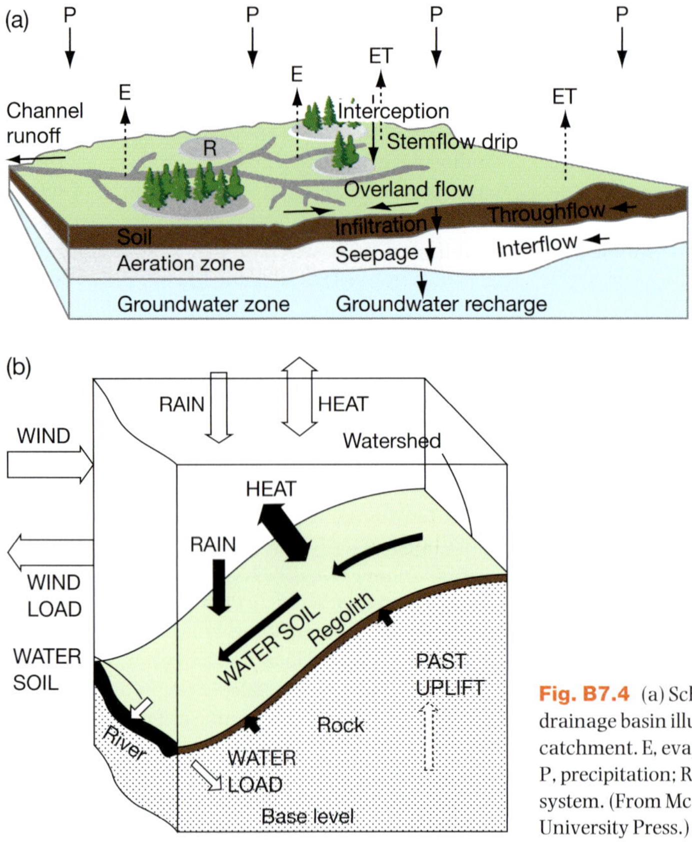

Fig. B7.4 (a) Schematic diagram of a third order drainage basin illustrating inputs and outputs in a river catchment. E, evaporation; ET, evapotranspiration; P, precipitation; R, surface runoff. (b) A valley slope system. (From McCullagh, 1978, courtesy of Oxford University Press.)

the Tertiary (Wooldridge and Linton, 1955; Hart, 1986). Superimposed on glacial and periglacial surfaces are the landforms of rivers and coastal waves. These agents bias to erosion or deposition, accompanied by distinctive landforms, in different regions depending on the energy environment, for example, fluvial erosion in highlands and deposition on lowlands. Geomorphological processes, the same as well as different denudational agents, have operated on different parts of the landscape with different intensities; over time, the same agent varies in intensity at different times in the same place and the evolution of the landscape is characterized by an orderly sequence of landforms. This suggests that landscape has a simple basis. It is perhaps less obvious that each land surface has been subject to very different agents over geological time, producing a complex of surfaces and features (Sparks, 1972; Hart, 1986). Thus, some gently folded escarpments have been planed by wave action (marine planation) in the past (e.g., Portsdown and Littlehampton chalk anticlines in the Tertiary; Wooldridge and Linton, 1955) whereas others have been overwhelmed by later glaciation (e.g., Magnesian limestone of Durham).

Despite the great age of much of the bedrock, little of the country's topography is older than the Tertiary and most is no older than the Pleistocene; these are the surfaces relevant for current living organisms. Landscapes older than the Quaternary are generally ones that have been exhumed, but even then new landforms, for example lines of drainage, have usually been superimposed on the older surface (e.g., the radial drainage pattern in the Lake District that developed on a Tertiary dome of folded sediments now cuts across underlying older Ordovician sediments and volcanics). Even one set of glacial deposits is superimposed on past ones and landforms created by glacial erosion are regenerated. There is now a detailed map illustrating the glacial landforms of the Devensian glaciation; this is accessible via a website in Clark *et al.* (2004).

A close study of modern landscapes discloses their dynamism; drainage catchments facilitate the division of landscape into a nested hierarchy of systems, each

Table B7.4a Some metrics of drainage basins. (From Hart, 1986, Table 7.1, p. 82, courtesy of Allen and Unwin and Taylor and Francis.)

Perimeter for drainage basin:	p
Area of drainage basin:	a
Maximum elevation within drainage basin:	H_{max}
Minimum elevation within drainage basin:	z
Relief of drainage basin:	$r = H_{max} - z$
Circularity of the basin:	C
Total length of stream of order u:	lu
Total length of all streams:	$\sum lu$
Drainage density:	$D = \sum lu/a$
Bifurcation ratio:	$R_b = n_u/(n_u + 1)$
Hysometric integral:	$I = \int_0^{100} da/dr$

Table B7.4b Prominent nominal landforms associated with geomorphic agents and British examples of their use as butterfly resources and habitats.

Landforms	Examples*	Status†	Cover‡
Coastal landforms			
Cliff (hard)	Great Ormes Head, North Wales: *Pieris brassicae*, *P. rapae*	3	1
Cliff (soft)	Chalk on the Isle of Wight: *Melitaea cinxia*	1	1
Undercliff (rotational slip segment)	Chalk and lias clay in Devon/Dorset: *Leptidea sinapis*	1	1
Spit	Spurn Head, Humber Estuary: *Lasiommata megera*, *Coenonympha pamphilus*	1	3
Raised beach	Easdale, West Scotland: *Aphantopus hyperantus*, *Maniola jurtina*	3	1
Sand or shingle bar and offshore bar	Loe Bar, Cornwall: *Polyommatus Icarus*	3	3
Tombolo	Llandudno, North Wales: *Hipparchia semele*, *Polyommatus icarus*, shingle on Chesil Beach. territorial perches for *Vanessa cardui* in March 1985	2	3
Beach and cuspate foreland above high water mark	Dungeness, Kent: *Lasiommata megera*, *Coenonympha pamphilus*	1	2
Dune complex including slacks	Ainsdale, Lancashire: *Hipparchia semele*; Penhale Dunes and Gwithian Dunes, Cornwall: *Plebejus argus*	1	1
Fluvial (or fossil fluvial) landform			
Interlocking spurs; valley slopes	Longmynd, Shropshire: *Coenonympha pamphilus*	1	1
River bluff/incised meander	Bollin Valley, Cheshire and Wear Valley, Durham City: *Pararge aegeria* (former), *Anthocharis cardamines* (both)	3	2
Erosion terrace	Bollin Valley, Cheshire: *Pieris napi*	4	2
Alluvial fan	Starbotton fan on Cam Gill Beck (early Holocene fan), off River Wharfe, Yorkshire: *Aricia artaxerxes*?, *Coenonympha pamphilus*?[a]	4	3
Point bar	(no known examples, but minimal resources for resting or water intake in hot weather)	4	3

Table B7.4b (*continued*)

Landforms	Examples*	Status†	Cover‡
Oxbow lake and cutoff meander	Short- to tall-herb grassland/meadow butterflies, e.g., Bollin Valley, Cheshire: *Anthocharis cardamines*	3	2
Levées	River Ouse near Cambridge: *Thymelicus lineola*, *Coenonympha pamphilus*	2	2
Floodplain	Meadow grassland butterflies; usually lost to intensive agriculture; e.g., Dolgarrog, Conway Valley, North Wales: e.g., *Aphantopus hyperantus*, *Maniola jurtina*	2	1
Lacustrine delta; drowned lower course valleys/fen	Norfolk Broads: *Papilio machaon*	1	2
Glacial and periglacial landforms			
Trough end and glacial trough	Glen Lyon, Grampians: *Coenonympha tullia*	1	1
Cirque (without tarn)	Ben Lawers, Grampians: *Erebia epiphron*	1	2
Arête (slopes)	Ben Lawers, Grampians: *Erebia epiphron*	1	2
Hanging valley	Cwm y Llan, Snowdon: *Coenonympha pamphilus*	1	2
Roches moutonnées and rock bar	Camas Rathaid, Ballaculish, Glen Coe near Oban and Upper Deeside: resources – thermoregulation and mate location in nymphalids[b]	4	3
Drumlin	Kendal to Lancaster, Lancashire: *Maniola jurtina*	3	2
Terminal (recessional, lateral) moraines	Coast edge between Cley-next-the-Sea and Mundesley, North Norfolk: *Hipparchia semele*	1	2
Esker	Castlesampson, Co. Roscommon, Ireland: *Erynnis tages*, *Hipparchia semele*, *Cupido minimus*, *Lasiommata megera*, *Leptidea reali*	2	3
Kames and outwash plains	Breckland, Suffolk[c]: *Erynnis tages*, *Aricia agestis*	1	2
Kettle hole (infilled as hydrosere)	Wynbunbury, Cheshire: *Callophrys rubi*	3	3
Pingo	Breckland, Suffolk: *Erynnis tages*, *Aricia agestis*[c]	1 or 2	3
Solifluction terrace/bench	South Downs, Sussex: *Lysandra bellargus*	1	2
General landforms of slopes			
Rock wall/free face	Allt Wen, Conwy, North Wales: *Hipparchia semele*	4	1
Scree slope	Allt Wen, Conwy, North Wales: *Hipparchia semele*; Tegg's Nose, Macclesfield, Cheshire: *Lasiommata megera*	1	1
Col	Honister Pass, Cumbria: *Erebia epiphron*; *Coenonympha tullia*	1	2
Ridge	Kerridge Ridge, Bollington, Cheshire: *Lasiommata megera*	1	1
Monadnock (residual hill)	Longmynd range: *Coenonympha pamphilus*[d]	1	2
Peak (summit of mountain or high hill)	Shutlingsloe, Cheshire: *Vanessa cardui*, *V. atalanta* (hill-topping)	4	1
Karst landforms (in chalk and limestones)			
Dry valley and gorge	Cressbrook Dale, Derbyshire: *Aricia artaxerxes*	1	1
Scarp slope of cuesta	South Downs, Brighton: *Cupido minimus*, *Polyommatus bellargus*	1	1

Table B7.4b *(continued)*

Landforms	Examples*	Status†	Cover‡
Coombe	Devil's Dyke, South Downs: *Cupido minimus*, *Polyommatus bellargus*	1	2
Pavement (clints/grikes)	Silverdale, Lancashire: *Boloria euphrosyne*	1	2
Doline/swallow hole	Malham Tarn area: *Aricia artaxerxes*?, *Coenonympha pamphilus*, *Polyommatus icarus*	1–3?	3
Volcanic landforms			
Volcanic cone (extinct, ancient)	Caer Caradoc and Ragleth Hill, Shropshire: *Lycaena phlaeas*	3	3
Sill (laccolith/ phacolith)	Great Whin Sill: *Aricia artaxerxes*?[e]	?	3
Dyke	Ardnamurchan peninsula ring dyke complex: *Erebia aethiops*?	?	3

* Where references are cited, they are indicated by superscript letters: a, Coulthard *et al*., 2002; b, Sugden *et al*., 1992; Glasser, 2002; c, Clark *et al*., 2004; Evans *et al*., 2005; d, N. Stone, personal communication; e, Dunn and Parrack, 1986.
† Typical status of examples indicated: 1, highly important as habitat for Biodiversity Action Plan (BAP) species; 2, sites of habitats for numerous species; 3, habitat for a few generalist species; 4, site of butterfly resources though not full resource complement making up habitats; ?, no record.
‡ Areal cover and frequency in landscape: 1, total area over countryside extensive or feature sufficiently frequent to cater for independent metapopulations; 2, frequent features; 3, sparse or rare features. Examples could not be found for all features (e.g., a pediment). Classification is based on the highest status observed. Features of no significance for butterflies (e.g., waterfalls, rapids, tidal wavecut platform) were omitted from the table.

of which is characterized by distinctive inputs, outputs and changes to the system. Thus in drainage catchments without human abstraction of water supplies:

River discharge = precipitation − (evaporation + transpiration) ± Δ storage (rocks, soils, vegetation, etc.)

The route for water to pass through to discharge at gauging stations is a complex one involving movement of water through soil and rock (Fig. B7.4a). Similarly, each slope segment within a drainage catchment can be treated in the same way (Fig. B7.4b). Underlying habitats and biotopes on which butterfly species depend, ecosystems, are subject to the interaction of such geomorphic agents. The end product (landforms) can be mapped either as distinct land units (areas similar in construction such as valley floors, valleys sides, plateau tops, rock outcrops) or as individual landforms. There are two provisos: first, not all named landforms may be obvious features to casual observation (e.g., glacial lateral moraines, kames) of the land surface as many become modified with time; second, deposition landforms may differ in composition (e.g., eskers are usually on sands rich in quartz as in Co. Tyrone, Northern Ireland and poor in plants and butterflies; in contrast, those of central Ireland are mainly of gravelly limestone, rich in flowers and butterflies such as the dingy skipper *Erynnis tages*, small blue *Cupido minimus* and wall brown *Lasiommata megera*). Cooke and Doornkamp (1974) provide a detailed legend of geological and geomorphological symbols to be used in mapping landscapes (see also Hart, 1986). An array of measures has also been devised for comparing different landscapes. Some for drainage basins are given below and very likely have relevance for broad resolution comparisons of butterfly diversity (Table B7.4a); others for slopes, together with morphological mapping symbols for slope features, are noted in Box 7.5. An array of measures has been applied to grid units; direct comparisons using them, as for northern England, indicate that landscapes have many contrasting features beyond the incidence of nominal landforms (Cole and King, 1968; King, 1976) and it would be surprising if landscape metrics did not underlie butterfly diversity and species' associations. This is suggested to be the case by the contrasting potential of specific landforms for butterfly habitats and resources (Table B7.4b).

Box 7.5 **Slopes, morphometric mapping and digital elevation models**

An important observation, one we shall consider in greater depth below, is that all landscape comprises surfaces, that can be divided into **slope segments**, on which weathering, surface wash and mass wasting operate. These slope segments are named and associated with different processes (Fig. B7.5a). **Weathering** describes processes leading to the *in situ* breakdown of rock and minerals and their movement downslope is referred to as **mass wasting**. All slopes, however gentle in gradient, undergo continual change, some very slow involving almost imperceptible creep of soil and regolith, others extremely rapid caused by the slumping of slope segments. The speed of movement depends much on the nature of the material (bedrock, regolith), the amount of material under movement (mass), water content (lubrication) and slope angle. Water moves downhill orthogonal to the contours with the result that concavities cause convergence of water (marshy ground) and convexities divergence of water (drier ground) (Fig. B7.5b). Some highly important landforms for butterfly habitats are formed by landslips, for instance rotational landslips along coastlines (Fig. B7.5c), features that in miniature are reproduced as terracettes on steep slopes. Distinctive slope segments are produced in different environmental settings (e.g., arctic, temperate and tropical; coastal and inland) by different agencies and as such they can be considered as equifinal products of different processes. For example, cliffs (**free faces**) – bare slopes with gradients >45° – are the product of numerous agents, endogenetic (e.g., faulting, volcanicity) and exogenetic (e.g., rivers, wave action), the latter including human activity (e.g., quarries). The universality of slopes and the contrasting impact of different segments on biota, elevates their study and the understanding of mass-wasting agents to one of prominence in butterfly ecology.

Landscapes are distinctive enough in features for many of them to be named or termed. Even so, such features, varying as they do in scale and form, render comparisons difficult. Instead, variables are needed that can be applied over all land surfaces regardless of form and scale. Work by geomorphologists over the past 40 years reveals that variation in land surfaces can be expressed as statistical measures (mean, variance, skewness, kurtosis) based on five basic attributes of landscape: **altitude**, **slope angle**, **slope aspect**, **profile convexity** and **plan convexity** (Fig. B7.5d, e) (Speight, 1968). Four are derivatives of altitude, height above sea level. Altitudes are typically point measurements taken in the field or from **digital elevation models** (**DEM**). DEMs are raster representations in which each grid cell records the elevation of the surface, providing a field of elevation values. These are typically the elevation of the central point or the mean elevation of a cell. The measures are derivatives since they can be calculated directly from heights of neighbouring (nine) data points (Fig. B7.5d). Thus, from this figure:

$$b = (z_3 + 2z_6 + z_9 - z_1 - 2z_4 - z_7)/8D$$

$$c = (z_1 + 2z_2 + z_3 - z_7 - 2z_8 - z_9)/8D,$$

where b and c are tan (slope) in the x and y directions, respectively; D is the grid point spacing (i.e., between z_1 and z_2), and z_i denotes the elevation at the ith point (Fig. B7.5d). The equations give the diagonal neighbours of z_5 only half the weight of the other four neighbours in determining the slope at z_5.

$$\tan(\text{slope}) = \sqrt{(b^2 + c^2)},$$

where slope is the angle of slope over the steepest section, and

$$\tan(\text{aspect}) = b/c,$$

where aspect is the angle between the vertical and the direction of the steepest slope, measured clockwise. As aspect varies from 0 to 360, an additional test is necessary that adds 180 to aspect if c is positive, and 360 to aspect if c is negative and b is positive (see Wilson and Gallant, 2000; Longley *et al*., 2001).

Further quantitative definitions are given in Krcho (1973), and Evans (1972, 1979) computerized the calculations. These measures allow the influence of different landscape components on biota to be determined despite the uniqueness of the regional land surface (Evans and McClean, 1995; Evans, 1998, 2001; Evans and Cox, 1999). Their strength lies in providing a three-dimensional description of surfaces, as shown in Fig. B7.5e, which is essential if processes affecting biota are to be fully accounted. It has long been known that these variables affect mass wasting, soil development and their dependent biota (Aandahl, 1948; Curtis *et al*., 1965; Lanyon and Hall, 1983; Pennock *et al*., 1987). Altitude, aspect and slope have major impacts on climate, particularly on insolation, temperature and water availability (see Box 2.4). Profile convexity (concavity) affects the speed of movement of water downslope, whereas plan convexity (concavity) directs the divergence or convergence of flow.

Owing to the importance of slopes in understanding physical process in landscape and changes to ecosystems and thus butterfly habitats, it is often useful to be able to map slope features. For this purpose, a number of morphometric mapping symbols have been devised; these are distinct from the geomorphological mapping symbols discussed in Box 7.4 and are illustrated in Fig. B7.5f. A range of techniques, varying in sophistication are available for surveying slopes (Gilbertson *et al*., 1985). In addition a number of measurements characterize slopes that influence plant and animal communities including butterflies (Table B7.5).

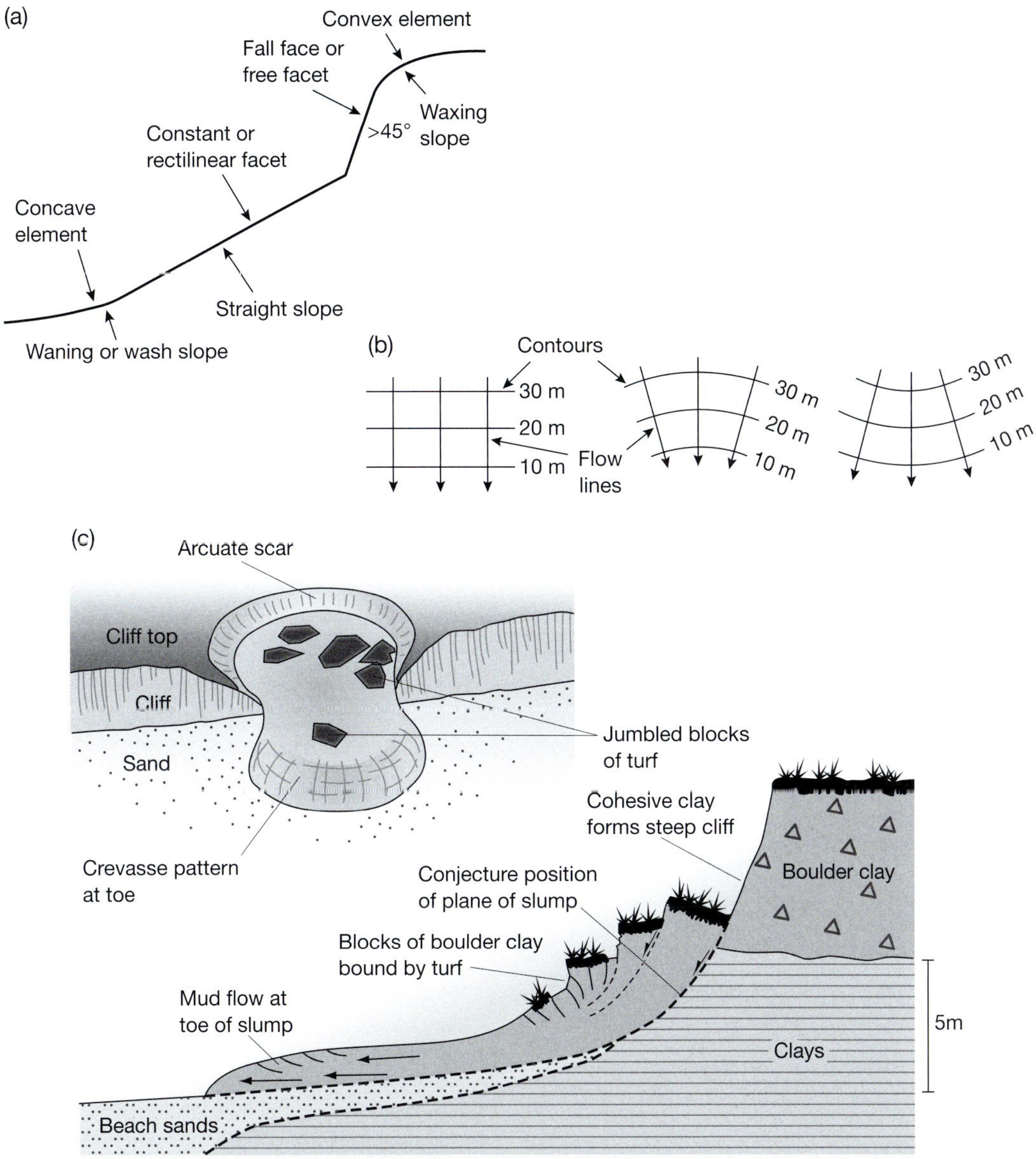

Fig. B7.5 (a) Standard terms for elements of a slope. (From McCullagh, 1978, courtesy of Oxford University Press.) (b) Direction of water movement in relationship to contour lines for rectilinear, concave and convex slopes (left to right, respectively). (From Clowes and Comfort, 1983, courtesy of Pearson Education Ltd.) (c) Undercliff landforms formed from coastal cliff landslips on the Norfolk coastline. Top, plan view; bottom, profile. (From Clowes and Comfort, 1983, courtesy of Pearson Education Ltd.)

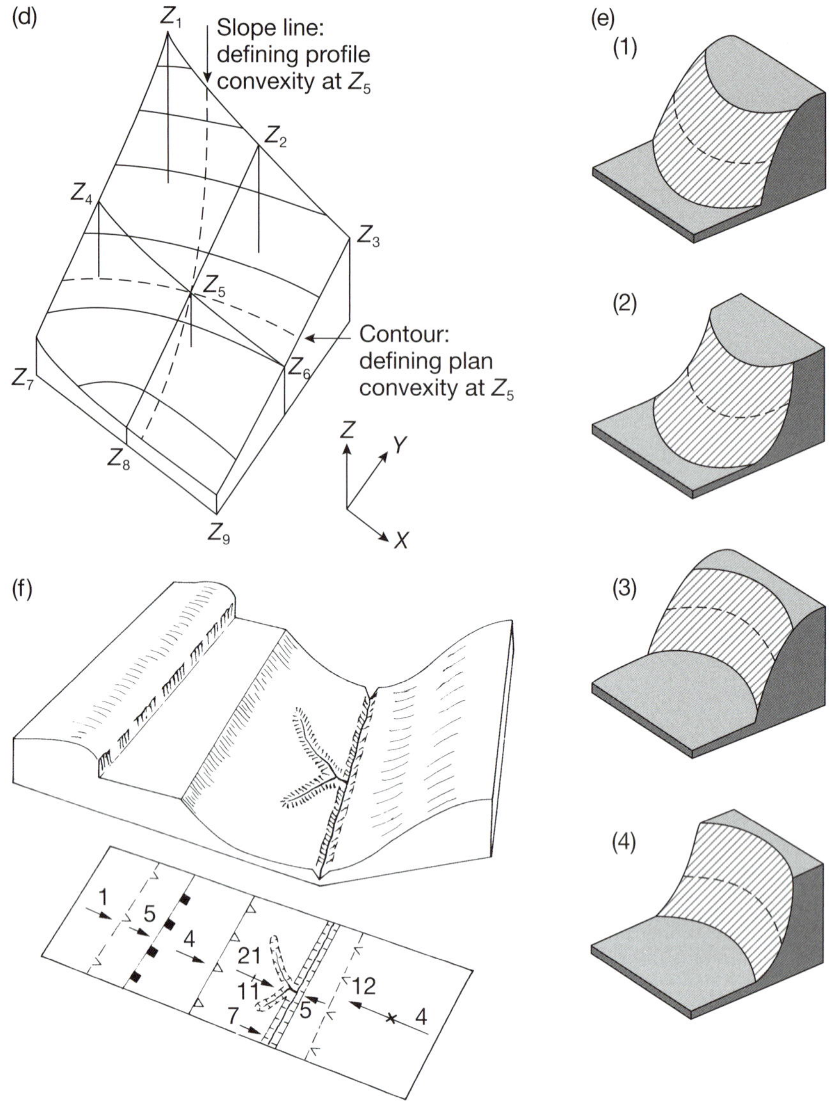

Fig. B7.5 (*continued*) (d) Key altitudinal points in the determination of profile and plan convexity (concavity). (After Evans and Cox, 1999, with kind permission of Springer Science and Business Media.) (e) Shapes of slopes (second derivatives of spot altitude measurements) showing four combinations of profile and plan convexity: 1, convex radial convex contour; 2, concave radial convex contour; 3, convex radial concave contour; 4, concave radial concave contour. (f) Morphometric mapping symbols illustrated in plan and map view; distinctions are made between breaks and changes of slope. (From McCullagh, 1978, courtesy of Oxford University Press.)

Table B7.5 Some properties of slopes relevant to butterfly communities. (From Hart, 1986, Table 7.2, p. 82, courtesy of Allen and Unwin and Taylor and Francis.)

Maximum angle of slope:	β
Average angle of slope:	$\sin \alpha = Ht/L_g$
Aspect of slope:	*Az*
Height of slope profile:	*Ht*
Length of slope profile:	L_g
Depth of soil or weathered regolith:	*W*
pH value of soil or regolith:	pH
Moisture content of soil or regolith:	% dry weight, $S\mu$
Porosity or void ratio of the soil or mantle:	V_R
Organic content of soil or regolith weight:	W_o
Surface cover of plants:	%, S_c
Height of vegetation:	H_v

- Composition (hard rock or unconsolidated material).
- Frequency or incidence in the landscape.
- Rate of regeneration.
- Potential for human exploitation.

Larger, more abundant, more permanent or readily regenerated, low-lying landforms at lower altitude, comprising softer rocks, are likely to make more important butterfly habitats than smaller, scarce, temporary, infrequently generated, montane landforms in harder rocks. Butterfly diversity in Britain is greatest in the south and at lower elevations and thus landforms with southern distributions provide more opportunities as a base for resources and habitats. Landforms may obtain permanency in two ways, by being durable in the face of geomorphic processes, or by being constantly regenerated. Permanency of landform ensures that suitable resources become established together with the butterfly populations that use them. Increased size and frequency raises the probability of metapopulations (e.g., on scarps and coastal cliffs) and not just local populations or sites with few resources (e.g., infilled oxbow lakes along lower river courses; tors on Dartmoor granites). Landforms in unconsolidated substrates provide opportunity for continual regeneration of slopes and dependent vegetation and soils (e.g., sea cliffs, scarp slope soils) (see Box 7.4), but tend also to present varied opportunities for human exploitation (deep boulder clay soils on drumlins versus skeletal soils on dunes). Thus, coombes and spurs on escarpment slopes provide more important habitats for butterflies than cirques and arêtes in mountain Britain, though the latter may be important for single species such as the mountain ringlet *Erebia epiphron* (Fig. 7.3). Soft cliffs in Cretaceous chalk similarly present greater opportunities for butterfly habitats than hard cliffs in Carbonferous limestones. Deposits such as coastal dunes and spits equally house more butterfly habitats than glacial eskers, kames and drumlins distributed largely north of the Severn to the Wash. Thus, it is well to appreciate that landforms have biased locations and this bias is occasionally associated with unique environments for butterflies. Such are the raised beaches of northern Britain linked to land rebound after the last (Devensian) glaciation (see Box 7.4). These provide valuable sites for *Maniola jurtina* and *Aphantopus hyperantus* habitats near Oban, Scotland (Dennis, 1977). The importance of landforms as sites for habitats is closely linked to potential, or lack of it, for exploitation; for example, as the midlands of Ireland are generally heavily farmed except where there are bogs, fens, lakes or wooded estates, eskers (and the often associated quarries) provide one of the few areas of unimproved grassland, and so form an important reservoir of biodiversity for diurnal Lepidoptera (I. Rippey, personal communication).

Individual landforms readily lend themselves to being treated as systems, much as ecosystems mentioned in the preface, with energy and mass inputs and outputs and systematic transformations of morphology. This is a valuable concept, since it underlines the integrity of landscape and the dynamism of landscape which directly affect butterfly habitats or resources on a range of timescales from seconds to millennia. Examples of natural systems, a slope profile within a river catchment, are illustrated in Box 7.4. The current British landscape, as any other, is characterized by a nested hierarchy of such natural systems, ecosystems

Fig. 7.3 Glaciated mountain scenery featuring the cirque Nethermost Cove, between Nethermost Pike and Helvellyn, in the Lake District biotope for the mountain ringlet *Erebia epiphron* (inset). (Courtesy of Ian S. Evans; inset courtesy of Paul Kipling and Peter Eeles.)

locked into drainage basins, described by stream order links (see Box 6.5). Inputs into ecosystems (biotopes) can affect whole catchments and the organisms inhabiting them. Thus, management is essentially required at a catchment scale as changes to upland ecology (burning, pollution deposits, drainage) have consequences for landforms, biotopes and organisms at lower elevations (Crowle, 2007) and consequently butterfly habitats at all levels (Dennis, 1992a:234). Even so, landforms in drainage basins depend on their location in the hierarchy. Landforms are more distinct in low order catchments (interlocking spurs, V-shaped valleys, valley side slumps, steep slopes) than in higher order catchments (open valleys and floodplains, valley side bluffs, terraces, oxbows, levées, gentle slopes); so too is human land use (e.g., rough pasture amid walled fields and moor tops versus arable farming in massive fields with cropped hedges or ditches). Thus, **interfluves** (the higher ground between two streams or rivers that is part of the same drainage basin) make for poor butterfly habitats in the lower courses of a river compared with those in the upper courses. But, as butterfly diversity is greater at lower, warmer altitudes, this is the reverse of what is needed for conserving the butterfly fauna. Often, butterfly species in the plain courses of rivers hang precariously to levées along river banks (e.g., Essex skipper *Thymelicus lineola* in the Cambridge Fens), the floodplain occupied by crops intensively treated with herbicide and fertilizer (Figs 7.4 and 7.5).

In every respect the human impact on natural systems (vegetation, soils, landforms) has been exponential, matching population growth (Goudie, 1993). Humans act as any other exogenetic geomorphic agent of landform transformation, where quarries and road cuttings are produced by erosion and pit heaps and embankments are deposits. Although the impact of vegetation changes on butterfly habitats is well appreciated, that of landforms and landscape features is less so. There is very little of the British countryside that has not been exploited by humans in one form or other and few sites that reveal no archaeological evidence of their habitation or use (Mitchell, 1965). Examination of Ordnance Survey maps from around the country reveals numerous features, many dating to prehistory that are likely to affect the incidence of butterfly habitats. Pre-Norman conquest, there are hill forts, their ramparts and ditches, henge monuments, burial mounds, mines and quarries, ancient settlements, walls, banks and ditches bounding fields and estates, tracks, sunken roads and ponds to mention but a few features (Hoskins, 1969; Rackham, 1986). At a finer scale too, evident on aerial photographs, are the ridges and furrows of the Anglo-Saxon open field system and the outlines of abandoned villages (Mitchell, 1965; Roberts, 1977). Even such small-scale

Fig. 7.4 Embankments along the lower course of the River Great Ouse in Cambridgeshire, biotope for Essex skipper *Thymelicus lineola* (top inset). Note, that the river is higher than the surrounding floodplain. In this part of the river the natural banks (levees) have been enhanced and artificially strengthened. The bottom inset of the river shows the importance of this sinuous biotope for butterflies amidst the intensively farmed landscape. (Courtesy of J. N. Greatorex-Davies; inset of *T. lineola* courtesy of Peter Eeles.)

Fig. 7.5 Interfluves and valley sides of an upper course drainage network in the Macclesfield Forest of the southern Pennines, biotope for the small heath *Coenonympha pamphilus* (inset). The fields on steeper slopes are increasingly left for coarse grazing, abandoned to scrub, bracken and rushes or given over to forestry. *C. pamphilus* occupies the abandoned pasture, green-veined white *Pieris napi* the hillside flushes of cloughs, and speckled wood *Pararge aegeria* the forest margins and rides. (Inset courtesy of Peter Eeles.)

differences in topography as the almost imperceptible line of ancient house walls in the turf can make for blatant distinctions in soil quality, microclimate and conditions for butterfly hostplants. That is, they can make the difference between what is traditionally described as the basis for a suitable butterfly resource, or not. Often, with these human-created landforms, the features of erosion and deposition, thus wide extremes of conditions for organisms, are closely juxtaposed; the material for the rampart or bank (deposition) comes from the ditch or ditches (erosion). New slopes are created varying in substrate exposure, aspect and hydrology. Ridges and furrows of open fields involve the same processes at a finer scale. Some of the prehistoric landforms are sizeable, for instance the hill fort ramparts of Maiden Castle near Dorchester, or extensive as in the case of the flint pits of Grimes Graves in Norfolk. As an extreme example, the henge or burial monument of Silbury Hill in Wiltshire, built more than 4500 years ago, is a conical hill 40 m high and a likely spot for hill-topping butterflies (see below).

Post-Norman conquest landform transformation has been little short of spectacular. This may not be immediately apparent until it is realized that housing, buildings and whole settlements of villages, towns, cities and conurbations are but deposits of rock, ergo landforms; they are simply more sophisticated in design than the spoil and slag heaps of mining and industry and the ramparts of fortifications. To butterflies, a building simply presents a vertically sided hill, somewhat sterile perhaps though occasionally useful for thermoregulation, mate location and roosting (e.g., red admiral *Vanessa atalanta*, *Hipparchia semele*; Dennis and Asher, 2009), with sharply contrasting local climates depending on aspect. Roads are exposed rock surfaces – and tarmac heats up quickly – but are inadvisedly used for thermoregulation (e.g., *H. semele*; Dennis, 2008b). Although a large variety of novel types of human landforms were introduced post-conquest, typically associated with new communications, some landforms were a continuation of earlier activity. For example, the rash of field boundaries associated with the 18th and 19th century Enclosure Acts, current mining of new raw materials and the impact of high explosives on Ministry of Defence land. Other new land uses simulated old patterns from different activities. For example, Anglo-Saxon ridges and furrows have structural parallels with patterns of drainage lines following the development of clay tile drains in 1800, furrows developing over the drains with continuous soil loss. Modern features also produce a matching of erosional and deposition features, for example quarries and their spoil heaps as in the excavation of copper, tin and china clay in Cornwall (Bradshaw and Chadwick, 1980; Johnson *et al.*, 1996) and the cuttings and embankments of canal, rail and motorway. All these features have an impact on butterfly resources and habitats – though contrasting in scale – from the single small feature such as a household brick or a temporary pile of gravel used by *Lasiommata megera* as a territorial perch site (Dennis, 1982–1983) and artillery craters (e.g., Salisbury Plain, Wiltshire where craters develop rings of devil's-bit scabious *Succisa pratensis* for the marsh fritillary *Euphydryas aurinia* and contain horseshoe vetch *Hippocrepis comosa* supporting *Polyommatus bellargus*; M. S. Warren, personal communication) to whole cities (e.g., habitats for the holly blue *Celastrina argiolus* in the suburbs).

Principles of landform and landscape influences on butterfly biology

- **P7.4: Specific landforms and landscape features contrast in the provision of resources and/or habitats for butterfly species.**
- **P7.5: Landforms and landscape features have an impact on the incidence of butterfly habitats and resources largely through five basic measures of form: altitude, slope gradient, slope aspect, profile convexity (concavity) and plan convexity (concavity). The impact is through soil minerals, soil moisture, substrate exposure, local climate and vegetation type/stability.**
- **P7.6: Landform components affect the behaviour of butterflies.**
- **P7.7: Landforms and landscape features generally differ from the surrounding land in their vulnerability to human exploitation and in the resources provided for butterflies.**
- **P7.8: Landscape components produced by humans have both direct and indirect positive, as well as often quoted negative, influences on the incidence of butterfly habitats. Sites of antiquity can prove to be valuable foundations for butterfly habitats.**
- **P7.9: The larger the landscape element, the more likely it is to provide (or exclude) entire habitats for butterflies rather than just specific resources.**

Landscape components influence butterfly habitats differently depending on their composition, form and location (**P7.4**). Landforms may not offer butterflies any obvious resources such as hostplants and nectar flowers but may still have a prominent part to play in resourcing butterflies, as in the case of flyways and hill tops (see below). Different landforms similar in overall structure may have a similar or contrasting impact on butterfly resource availability and use; for instance, topographic features such as hills and valleys may both provide 'virtual corridors' for movements in areas of apparent biotope homogeneity (Brunzel *et al.*, 2004; Pe'er *et al.*, 2005). One example will suffice to illustrate contrasting use. Drumlins and eskers/kames are both upstanding landforms providing a variety of slope elements produced by glaciation, but the first is a till deposit from a glacial advance and the second is the product of fluvio-glacial outwash (permeable sands and gravels) during glacial melt. These features contrast substantially in substrate permeability, subsequent erodibility and vegetation associations. Not in Britain, but in Finland, *Maculinea arion* is virtually confined to ridges of past glacial outwash deposits where substrate exposure has been maintained by fire and the instability of the deposits, both apparently essential for the ant symbiont, allegedly *Myrmica lonae* Finzi (Kolev, 1998). On a finer scale, it has been shown that resources differ for vegetation patch types on human-created landforms (road verge, lane, hedge bank, grazed, mown and unproductive meadows) in a study of *Maniola jurtina* in western France with concomitant effects on behaviour and resource use (Ouin *et al.*, 2004).

Landforms can be described by five slope measures that can be derived from altitude (see Box 7.5) and these affect butterfly habitat and resource status through the five variables discussed above – soil minerals, moisture, exposure, vegetation and climate (**P7.5**). The importance of slope elements is examined in a separate section below. Landform and landscape feature components, particularly structure, can affect butterfly behaviour (**P7.6**). A clear example is the distinction in mate location behaviour in *Pararge aegeria* between woodland and fragmented farmland; in the latter landscape males showed higher levels of aggressive, fast take-offs (an indicator of territorial perching), but also higher levels of displacement (an indicator of patrolling) (Merckx and Van Dyck, 2005). Matrix situations can lead to distinct differences in resource use and movement behaviour in butterflies, some of it unpredictable (e.g., Rocky Mountain parnassian *Parnassius smintheus* Doubleday flies less frequently, for shorter distances, and at lower rates in forest edge matrix biotopes than in habitats within meadow biotopes; Ross *et al.*, 2005). Thus, adjacent landscapes contrasting in habitat and matrix can expect to generate contrasts in butterfly behaviour and life history including resource use. Observations on contrasts in landscapes and landforms and their impact on the balance of available resources are germane to understanding the impact of landscape components as corridors for movement (see Chapter 6). Corridors have been demonstrated to enhance movement in butterflies through pollination experiments (Townsend and Levey, 2005). Two ideas persist concerning how corridors function. The **traditional corridor hypothesis** posits that corridors increase immigration and emigration by functioning as flyways between patches. The **drift fence hypothesis** posits that corridors function by 'capturing' organisms dispersing through the matrix, redirecting them into associated habitat patches. Just how they function depends on landscape and landform components and the availability of species-specific resources (see Chapter 6).

A further important way in which specific landforms and landscape features affect the incidence of butterfly habitats is through their relative insusceptibility to further human exploitation compared with the surrounding countryside (**P7.7**). For example, a wide collection of features of human origin (e.g., hill forts, long barrows, henge monuments) in time become valued and protected by ancient monument legislation (Ancient Monuments and Archaeological Areas Act 1979; see English Heritage: http://www.english-heritage.org.uk/); sadly by no means all valuable sites seem to have this protection. Many human-created landforms effectively provide refuges for butterfly habitats (e.g., Maiden Castle hill fort from agriculture and Woolsbarrow Fort from forestry, both in Dorset). A good example is on Martin Down, Dorset, where *Hippocrepis comosa* for *Polyommatus bellargus* is confined to ancient earthworks with thin soils (Bockerley Dyke) surrounded by flat chalk grassland (M. S. Warren, personal communication). A fascinating example of the value of ancient sites as a resource for Lepidoptera is the use of overhangs on the standing stones of the Ring of Brogar, Stenness (Orkney Isles) as pupation sites by the Magpie moth *Abraxas grossulariata* (L.) (Waring, 2006). Very likely, Pennine field walls form hibernation and roost sites for nymphalid butterflies such as *Aglais urticae* and *Inachis io* (R. L. H. Dennis, personal observation). However, the degree to which a feature provides a refuge can depend on composition. Thus, hard and soft limestones provide cliff refuges for cliff cabbage *Brassica oleracea* and the large white *Pieris brassicae*, but on the soft cliffs (chalk) of Dorset this refuge is on the cliff tops, a margin avoided by heavy machinery, whereas on the Great Orme in North Wales – where land use, goat and sheep grazing, abuts the cliff edges – this is on the inaccessible cliff ledges which are more resistant to erosion and cliff collapse than chalk. There are thus distinctions in soil depth and moisture availability for these cliffs and the plants differ for these locations, having large thin leaves on the chalk and small tough leaves on the Carboniferous limestone.

Much of human exploitation has a negative impact on butterfly habitats, but by no means all (**P7.8**). In an extensive study of 29 species on 308 prime butterfly sites in south central England, Warren (1993) found that 42% of the prime sites on calcareous grasslands had key resources on man-made features, including ramparts of ancient forts, ancient trackways, lynchets and banks, disused quarries and chalk pits, recent earth banks, and railway cuttings. In the case of brown argus *Aricia* species, disused quarries, road and rail cuttings and embankments, and tumuli, all provide suitable habitats (Wilson *et al.*, 2002). Even so, it is well to recognize that a specific feature may be negative or positive in impact on habitats or resources depending on the attributes described above as well as management; for example, a hedgerow may provide suitable resources for butterflies if left untrimmed and untreated, but not if severely cut, with the under-flora affected by herbicide spray. Similarly, a fence between two intensively used fields is more likely to provide resources for butterflies (e.g., suitable hostplants, larval retreats, pupation sites and nectar for *Maniola jurtina*; Dennis, 2004a) if the lower section is closely cross-fenced by wire than if left open and accessible to grazing (see Fig. 6.4). The importance of what is occasionally referred to as rural furniture and architecture for butterfly resources and diversity is considered in a separate section below. Inevitably, the larger the feature, the more likely is it to ensure larger and more varied habitats (**P7.9**) but the same comments on size apply as for metapopulation patches.

Butterfly landscape divides at the British Isles scale

Landscapes fall into larger divisions at scales greater than perhaps those of natural regions. Thus, Britain divides into domains that concentrate on arable (east and south) and pastoral (west and north) farming, the latter with a further tier in hill sheep farming country; this division is linked to topography and altitude, and evokes a division between highland and lowland, with strong geological contrasts of older, indurated rocks compared with younger, softer and unconsolidated strata. There are also fast growing divisions between urban and countryside with expanding conurbations (see Chapter 8). These divisions implicate type and intensity of land use and provide different opportunities for wildlife, including butterflies, in effect refuges where butterflies cannot be considered in isolation from other organisms (see Chapter 9).

Landscape refuges for butterfly habitats

The concept of landscape components providing **refuges** for butterfly habitats (specifically, from contemporary human impacts) is an important one and can be extended into a wider principle:

- **P7.10: A variety of landscape features, human and natural, provide opportunities for complementary butterfly resources, thus habitats. The essence of such refuges is landscape heterogeneity.**

A wide range of features provide habitats and resource outlets for butterflies in Britain; they occur at different spatial scales from individual landforms or landscape features to distinct landscapes. There is not room here to discuss individual features, natural and human-made, in any detail – an obvious exercise would be to document their value for butterflies. An example of such a landscape is the Arnside Area of Outstanding Natural Beauty (AONB), north Lancashire. The key to landscape as a refuge is landform heterogeneity and such a landscape presents a wide variety of form, feature, composition and substrates, which can be complementary in linking up habitats for butterflies between regions distinguished by geology and major changes in land use (**P7.10**). It is important to appreciate that human-created features play a vital part in this process, including a wide range of structures, from abandoned quarries and golf courses (New, 2005) to way-leaves under power lines simulating forest breaks (e.g., chequered skipper *Carterocephalus palaemon*; Douglas, 2003; Fig. 7.6); even fences can provide valuable habitat refuges and resources for butterfly species (see Fig. 6.4) (Dennis, 2004a).

Fig. 7.6 Way-leaves through coniferous forest in Scotland, providing habitat for the chequered skipper *Carterocephalus palaemon* (inset). (Courtesy of Paul Kirkland; inset courtesy of Peter Eeles.)

It becomes evident that a number of distinct features, both natural and human in origin (see Table B7.4b), influence each butterfly species' resource distributions. An example can be taken from the orange-tip *Anthocharis cardamines* in Cheshire where the butterfly uses a few crucifer hostplants (i.e., cuckoo flower *Cardamine pratensis*, hedge garlic *Alliaria petiolata*, large bitter-cress *C. amara*). *C. pratensis* and *C. amara* are typically found in damp sites; among natural features providing suitable conditions are valley floors, especially river terraces, cut-offs, abandoned channels and spring lines, whereas human-constructed landforms include roadside ditches and clogged-up drainage channels. An important landscape feature used over Cheshire interfluves are the margins of rain-infilled, excavated marl pits (Fig. 7.7). In this county, some areas abound with marl pits which are biased to areas of boulder clay; the pits arose with the exploitation of the clay and lime subsoil used as a fertilizer on adjoining fields. They now form invaluable ponds for freshwater organisms such as great crested newts (*Triturus cristatus* Laur.), dragonflies and water beetles (Boothby and Hull, 1994). *Alliaria petiolata* is found in drier sites along river levèes (natural feature) and road verges and boundary banks (human features) (Dennis, 1982a, 1982b). Together the distribution of these host plants provides an example of landscape complementarity (see below).

CASE EXAMPLES OF THE IMPACT OF LANDSCAPE FEATURES ON BUTTERFLY RESOURCES

Although it is not possible to explore in detail all the varied ways that landscape can impact on butterfly resources and habitats, three examples are selected below to illustrate how important natural landforms and human landscape features can be for butterfly resources.

Hill tops and hill-topping: a special case?

- **P7.11: Hill-topping, exploiting peaks as a resource, occurs among a number of butterfly species during their life cycles, though hill tops tend to lack permanent conditions (complementary habitat resources) for population maintenance.**
- **P7.12: Not all hill-topping butterflies are engaged in mate location on hill tops, but may be using peaks during migration.**
- **P7.13: The physical form of hill tops (peakedness, elevation, ancillary landforms) affects the frequency of hill-topping on them.**

Fig. 7.7 A rainwater-infilled marl pit above the Bollin Valley, north Cheshire, the site of crucifer hostplants for the orange-tip *Anthocharis cardamines* (top inset) and green-veined white *Pieris napi* (bottom inset). (Insets courtesy of Paul Kipling and Peter Eeles.)

Hill-topping presents an unusual form of resource use in butterflies inasmuch as tops of hills tend to be bare, lack resources and are cold, and thus seem unlikely places to find butterflies, the more so in the British islands where conditions are substantially cooler, less sunny and windier than at low elevations (**P7.11**). Until 2005, the records of butterflies on British hill tops were limited to a few observations (see Dennis and Dennis, 2006). Hill-topping was first described by Shields (1967) in North America and has been thoroughly studied since then, mainly in the United States (e.g., Baughman *et al.*, 1988, 1990; Brown and Alcock, 1990) but it has also been observed elsewhere (e.g., McFarland, 1976; Pinheiro, 1990; Tennent, 1995). The behaviour is far from being fully understood (Ehrlich and Wheye, 1986; Brown and Alcock, 1990; Singer and Thomas, 1992). Fundamentally, hill-topping behaviour is considered to facilitate mating in populations occurring at low density (Scott, 1968; Pe'er *et al.*, 2004). Modelling the behaviour indicates that topography essentially channels movement; in landscape terms it enhances connectivity (Pe'er *et al.*, 2006). Males typically occupy the summit and await the arrival of females that, once mated, descend to lower elevations to lay eggs. Males either patrol the summit or establish territories that they defend. Although the key to the activity may well be low population density, clearly suitable conditions are required to occupy hilltops, especially if at any great altitude. Sunny, hot, calm conditions are particularly propitious for hill-topping, and the opposite – cloudy, cool and windy – conditions are likely to deter the behaviour, if only because males will be too cool to locate and acquire mates at the summit. As such, it is very possible that the limited observations on hill-topping among Lepidoptera in Britain reflect the comparative rarity of the behaviour in this region because of poor summer weather conditions. Even so, butterflies are observed on mountain peaks (e.g., *Vanessa atalanta* on Moel Llyfnant summit, North Wales, at 751 m on 4 February at 14.00 h; Bailey, 2007) and increasing numbers of hill-topping records are being made (Dennis and Dennis, 2008).

From observations in 2005 it is now evident that hill-topping relates to at least two, if not three – not necessarily mutually exclusive – activities in Britain (**P7.12**):

1 Mate location behaviour.
2 Migration.
3 Hibernation.

In 2005, hill-topping in *Lasiommata megera* for mate location was observed on a number of millstone grit peaks and ridges in the West Pennine Peak District between Macclesfield and Buxton (Shutlingsloe at 506 m, Tegg's Nose at 330 m and the Kerridge Ridge hog's

Fig. 7.8 Shutlingsloe hill top, Cheshire, above the moorland plateau used by the red admiral *Vanessa atalanta* (top inset), painted lady *V. cardui* (bottom inset) and wall brown *Lasiommata megera* for territorial hill-topping. (Insets courtesy of Andrew Burns and Peter Eeles.)

back at 290 m; Fig. 7.8) at heights where the climate lies outside the conditions experienced within the geographical range of resident populations. *L. megera* qualifies as a hill-topping candidate for mate location purposes. It occurs at low density and the butterfly was engaged in territorial perching, patrolling and conspecific skirmishes during both broods (Dennis and Dennis, 2006) as recorded elsewhere in Europe (Dennis, 1987; Wickman, 1988). Since the early 1980s, when it was virtually ubiquitous throughout Cheshire in abandoned sand pits and sandstone quarries and along the edges of arable and pasture fields, adhering to the hedgerows (Dennis, 1982–1983; Dennis and Bramley, 1985), it has become extremely scarce in the county (Shaw, 2005). To test for prevalence of this behaviour, further observations were carried out in a different region, Devon, in 2006; there it was found to be hill-topping on seven Dartmoor granite tors and one coastal peak (Fig. 7.9), but absent or scarce at lower elevations (Dennis and Dennis, 2007). Both on the Pennine peaks and Dartmoor tors, other, nymphalid, species (i.e., *Vanessa atalanta*, painted lady *V. cardui*, *Aglais urticae* and *Inachis io*) were observed to be engaged in territorial perching on hill tops (Dennis and Dennis, 2007, 2008). The records for hill-topping in *L. megera* indicate that less strong-flying butterflies (Satyrinae) are capable of the behaviour in a region where conditions are usually marginal for activity in uplands.

Hill-topping also probably occurs in species undertaking long-distance migrations (**P7.12**). Pierids (i.e., green-veined white *Pieris napi*, small white *P. rapae* and *P. brassicae*) and nymphalids (i.e., *V. atalanta*, *I. io* and *A. urticae*) have been seen making their way in a direct line over Pennine peaks (Shutlingsloe, Shining Tor). The presence of pierids and the non-breeding generation of nymphalids (*I. io*, *A. urticae*) simply flying over the summits of these Pennine peaks, even against a strong breeze, would suggest that other cues than mate location are also involved, likely landmarks for migration. In the case of pierids seen on Dartmoor in May the movement was northwards whereas the nymphalids in September were flying southwards (Dennis and Dennis, 2006, 2007) as expected from previous observations on migration for different broods (Baker, 1969, 1984).

As the frequency of territorial species and individuals differs among hill tops, it is clear that peaks of hills do

Fig. 7.9 Granitic tors (Haytor) in Devon, mate location sites for the wall brown *Lasiommata megera* (inset) and nymphalids. (Inset courtesy of Paul Kipling and Peter Eeles.)

not present equivalent resources and therefore not equal opportunities (**P7.13**) (Dennis and Dennis, 2008). Peaks differ in elevation, ruggedness and in variety and positioning of landforms (crags, scree slopes, slope form) and observations suggest that where there are matching elevations, a peaked hill top with crags on the sheltered side will receive more territorial males than one where the peak is ill-defined and crags face into the prevailing wind (Dennis and Dennis, 2008). Higher peak elevation is accompanied by higher wind speeds and cooler conditions. As such, *L. megera* will tend to congregate around alternative prominent landforms and use lower slope tops, rock outcrops and other physical features, when conditions are cooler and windier (Hardy and Kinder, 2007; Dennis and Dennis, 2008). As in the case over the Cheshire plain in the 1980s, where sheltered, sunny edge sites were used for territories, *L. megera* is generally found on sunny aspects and on Devon tors was found more frequently around the base and lower sides in shelter rather than on the summits themselves (Dennis and Dennis, 2007).

Recently, a third reason for hill-topping has been raised by Keith Bland (2009). He describes large numbers of *A. urticae* (≥200 individuals) regularly hibernating in an unheated byre of a hunting lodge at Feahar (525 m) in Perthshire, distant from adequate larval resources. He also observed movement of *I. io* in a northwesterly direction at 600 m up a burn on Meall Greigh, Perthshire, feeding intermittently on honeydew on 4 September 2008. There is clearly much yet to be learnt about the way butterflies use upland resources. In this case, hibernating nymphalids may select upland sites to ensure that diapause is not broken too early as Bland (2009) suggests.

The significance of slopes and their aspects for butterfly habitats and resources

It is a simple but important observation that all landforms comprise surfaces called slope segments. The universality of slopes, the contrasting impact of different segments on soils and biota, elevates their study and the understanding of slope processes (fluvial erosion and mass-wasting agents, pedogenesis) to one of prominence in butterfly ecology. In a number of different ways, through these processes and the interrelationships amongst different agents, slope form and position affect nutrient status, soil moisture, substrate exposure and soil depth, vegetation cover and type, local climate and human land use – the main factors controlling butterfly resources. This is illustrated with respect to soil types in Box 7.3 and local climate in Box 2.4. In Box 7.5 an outline is given of the geometry of slope forms, slope nomenclature, basics of slope systems and factors in slope form generation.

The following principles relate to the importance of slopes for butterfly ecology:

- **P7.14: Changes and breaks in slopes, and changes in aspect, provide opportunities for complementary and supplementary resources, dependent on different conditions (e.g., geology, climate), required by butterfly species.**
- **P7.15: Different combinations of slope parameters, altitude and substrates can produce suitable conditions for a butterfly resource(s), which is then an equivalent product of different agents.**
- **P7.16: Conditions provided by slope parameters may conflict for a key resource required by a butterfly species and the butterfly itself, excluding the butterfly from a site.**
- **P7.17: Slope and aspect combine to have compound effects on butterfly resources reflecting agents operating over varying scales.**
- **P7.18: Slope parameters provide a variety of conditions over short distances enhancing the potential for butterfly species' diversity at a site.**
- **P7.19: Different slope segments provide varied ground conditions – thus refuges – for butterfly resources subject to climate fluctuations and trends.**

With variation in slope form (breaks and changes in slope angle, slope shape, aspect) very different conditions – for soil depth, nutrient status, moisture, substrate exposure and local climate – occur over short distances. This provides opportunities for the development of different soil types, vegetation cover and land uses and consequently for very different resources – supplementary and complementary resources – comprising habitats for butterflies (**P7.14**). A simple example is a free (rock) face above scree slopes. *Hipparchia semele* uses the former for mate location, establishing defended territories, and the latter for egg laying where tufts of sheep's fescue *Festuca ovina* grow on screes that have started to stabilize. In a similar way, *Polyommatus icarus* will lay eggs on *Lotus corniculatus* and lesser trefoil *Trifolium dubium* on drier convex slopes but will use *Trifolium* spp. and greater bird's-foot trefoil *Lotus pedunculatus* on moist, convex slopes. It would be a mistake to imagine, however, that resources are bound to precise slope forms. Because of the interactions among slope processes and agents, particularly underlying soil and drift geology, similar conditions can be produced by different combinations of parameters (**P7.15**). For instance, moist soils can be produced by steeper slopes in a cooler aspect, or where spring lines break the surface, and where mass wasting or erosion has removed much of the topsoil in the past on gentler slopes as with limestone pavements above Malham Cove, Yorkshire, UK. Thus, the same end conditions for butterfly resources can be produced in different circumstances as for example in the occurrence of *Cardamines pratensis* on both narrow valley (clough) sides and on valley floor mires; both are used by *Pieris napi* in the Peak District of Cheshire and Derbyshire. These are sites of excess soil water, the presence of which is revealed by rush *Juncus* species, the first relating to spring lines, impermeable bedding and surface runoff and the second to a permanent high watertable and peat accumulation. Another example is the occurrence of *Hipparchia semele* on precipitous rock scree and undulating dune, which in North Wales are close to one another at Llandudno and Conway (Dennis, 1972a).

However, conditions can also conflict for a key butterfly resource and its other resource requirements (**P7.16**) as is evident from the observation that butterflies rarely use up the total area of an apparent resource type. This is illustrated on the Great Orme, North Wales, for *Pieris brassicae*. Conditions coincide for nasturtium *Tropaeolum* spp., cultivated *Brassica* and the butterfly on the south-facing, sheltered gardens of Cwlach on the headland, but cliff cabbage *Brassica oleracea* is largely restricted to eastern and northern cliff ledges, where there is just sufficient moisture for the plant but where shade and exposure to winds make for cool conditions for egg-laying females. Here, larval development occurs on plants with thicker cuticles than found in gardens and allotments. A more critical situation is apparent for the Duke of Burgundy *Hamearis lucina*. This once occurred on the Creuddyn Peninsula (pre 1944) of which the Great Orme forms a part. However, on the Orme, where the hostplants primrose *Primula vulgaris* and cowslip *P. veris* abound on northern and eastern slopes, shaded for much of the day and subject to cold winds, conditions are entirely unsuitable

Fig. 7.10 The scarp slopes with coombes (short dry valleys) and solifluction terraces of the South Downs, Sussex, an essential biotope for butterflies such as the Adonis blue *Polyommatus bellargus* (top inset) and silver-spotted skipper *Hesperia comma* (lower inset) in southern Britain. This photograph was taken above Charleston Farmhouse and Alciston near Beddingham Hill (grid ref TQ46 06). (Courtesy of Crispin Holloway, who also monitors the species and breeding trends of butterflies at Malling Down Nature Reserve, Lewes; insets courtesy of Peter Eeles and Peter Hardy.)

for the adult butterfly let alone developing early stages (León-Cortés *et al.*, 2003a). Even so, northern slopes exist that have sufficient complexity of slope element aspects as to make invaluable butterfly sites (e.g. South Downs; Fig. 7.10).

A distinctive feature of landforms with different slope segments, facets and elements, is that a variety of different conditions are provided over a fine scale; the importance of this observation is evident in the case of two butterflies, the Lulworth skipper *Thymelicus acteon* (Fig. 7.11) and the Glanville fritillary *Melitaea cinxia* (Fig. 7.12) at the southern edge of Britain. In combination, different slopes and aspects of a landform can have a compound effect on butterfly resources reflecting the operation of agents over varying time-scales (**P7.17**). This situation is usefully illustrated for regular landforms (e.g., conical hills, basins) where all aspects (see Box 2.4) and a wide range of slope angles are available. A suitable example is again presented by the Great Ormes Head in North Wales; this near symmetrical headland of approximately horizontally bedded Carboniferous limestone, with marginal scars and screes, has slopes of much the same profile varying in aspect. Here, two local butterflies, known for their dwarfed adults and early emergence (*Plebejus argus* and *Hipparchia semele*; see Fig. 3.2) do not occupy the entire headland (Dennis, 1972a, 1972b, 1977), but tend to concentrate around the southern and western sides. In effect geology and landforms (topography) are held constant; but local climate, soils and vegetation are not. The different aspects in relation to sun angle and exposure to winds produce strikingly contrasting local climates that have a profound effect on immediate conditions as well as on long-term ones via vegetation and soil development. The headland is subject to daily massive variations in sunlight and wind speeds that can affect the two butterflies directly (e.g., adult thermo-regulation, mating, egg laying). Over seasons these conditions profoundly impact on soil temperatures and moisture, influencing plant types, vegetation growth and accumulation and pedogenesis. This inevitably has consequences for the development of early stages as

Fig. 7.11 Lulworth Cove with chalk cliff biotopes used by the Lulworth skipper *Thymelicus acteon* (inset). (Courtesy of Jim Asher; inset courtesy of Peter Eeles.)

Fig. 7.12 Chines, a soft cliff biotope for the Glanville fritillary *Melitaea cinxia* (inset) on the Isle of Wight. (Courtesy of Robin Curtis; inset courtesy of Peter Eeles.)

well as for the prevalence and suitability of hostplants. The contrasts are evident to anyone walking around the Orme from the north to south toll house around the Marine Drive; one passes through palpably different local climates and observes plants associated with damp, cool, shaded conditions (e.g., dog's mercury *Mercurialis perennis*, *Primula* spp., *Salix* spp., cliff *Brassica oleracea*) on the east and north faces and then those associated with warm, dry, bright conditions (e.g., *Helianthemum* sp., fennel *Foeniculum vulgare*, *Cotoneaster* sp.) on the west and south. Higher soil moisture levels on the east and north screes lead to the swamping of hostplants for species by bracken and denser calcareous grassland. On the summit, as on other gentle gradients, geology and land use once again dominate, where areas of till replace the limestone, and the butterflies are effectively excluded (Dennis, 1992a, 1993a). The significance of slope aspects for British butterflies is such that even in south central England there is substantial bias for southerly aspects of prime butterfly sites for species on calcicolous grasslands (BUTT, 1986; Warren, 1993). Not every species has this bias (e.g., *Hamearis lucina*, *Euphydryas aurinia*), reflecting perhaps on the tradeoff between the needs of the butterflies for warmth and the hostplants for moisture.

Variety of slopes and aspects not only produces complementary conditions for a single butterfly species but a variety of resources for different species (**P7.18**). Slopes varying in form provide refuges for butterfly species in different conditions and enhance butterfly diversity. The first situation is illustrated by the use of taller, denser grassland on deeper soils during warmer conditions and droughts (e.g., *Hesperia comma*; Davies *et al.*, 2005). The second situation is well illustrated by Pennine slopes in northern England (Fig. 7.13; see also Box 7.3) where a succession of species is found occupying different conditions: *Coenonympha tullia* on plateaux peats above 300 m, green hairstreak *Callophrys rubi* on bilberry *Vaccinium myrtillus*-covered immature soils over screes and thin podsols on steep slopes, small heath *Coenonympha pamphilus* over *Festuca*-dominated deeper podsols and acid brown earths, and *Pieris napi* on crucifers among *Juncus* growing on waterlogged gleys and peaty gleys forming valley floor mires. Where such slopes terminate in more level lowlands, raised bogs once again provide peats for *C. tullia*.

Different slope segments also provide a variety of different conditions for the same key resources (i.e., hostplants, nectar), potential refuges for different seasonal and weather conditions, and more importantly for trends in climate change (**P7.19**; see Chapter 8). Change of aspect (less southerly and southwesterly in Britain) and slope angle (less orthogonal to the sun's azimuth) become important with drought and hot weather, sheltered hollows and lee slopes in high winds; butterflies adopt these situations as they do locations with more cover (e.g., hedge lee and open scrub) when so exposed. Both the silver-studded blue *Plebejus argus* and *Hipparchia semele* employ these tactics on Great Ormes Head in different conditions (Dennis and Sparks, 2006).

Rural 'architecture' and 'furniture' and their impact on butterfly resources

- **P7.20: Field boundaries over agricultural land provide both resources and complete habitats for butterflies.**
- **P7.21: Field margin landscape features enhance conditions for butterflies on the neighbouring land.**
- **P7.22: Local climate (e.g., shelter, temperature, shade) is often the most important attribute for butterfly species in field margins, not just for exploiting resources but in enhancing flyways.**

Very little of Britain is covered in semi-natural biotopes typically regarded in metapopulation studies as suitable for butterflies. Indeed, some 80% of Britain is under agriculture (21% arable) divided into fields, compared with 0.8% of the area taken up by nature reserves (Barr *et al.*, 1993). Even so, Dover (1994) identified some 32 butterfly species in arable land using resources found in field margins and later listed 39 of 62 species (63%), with 26 breeding in hedgerows (Dover and Sparks, 2000); Warren (1992b) previously noted 24 species breeding in field margins. Dover indicated that field margins may provide surrogates for biotopes typically occupied by butterflies, such as unimproved pasture and woodland coppice (**P7.20**). Some outstanding issues exist for future research. One is the extent to which numbers of butterflies in field margins reflect resources in margins as opposed to adjoining land. Field margins differ in composition and structure, therefore in quality, amount and density of resources, local climate, etc., from biotopes to which they may be most closely related. An important question remains as to how suitable they are in providing habitats, as opposed to resources, for butterflies. A second issue is the interaction effect. Instead of just reflecting adjoining land use, or the resources of the field margin

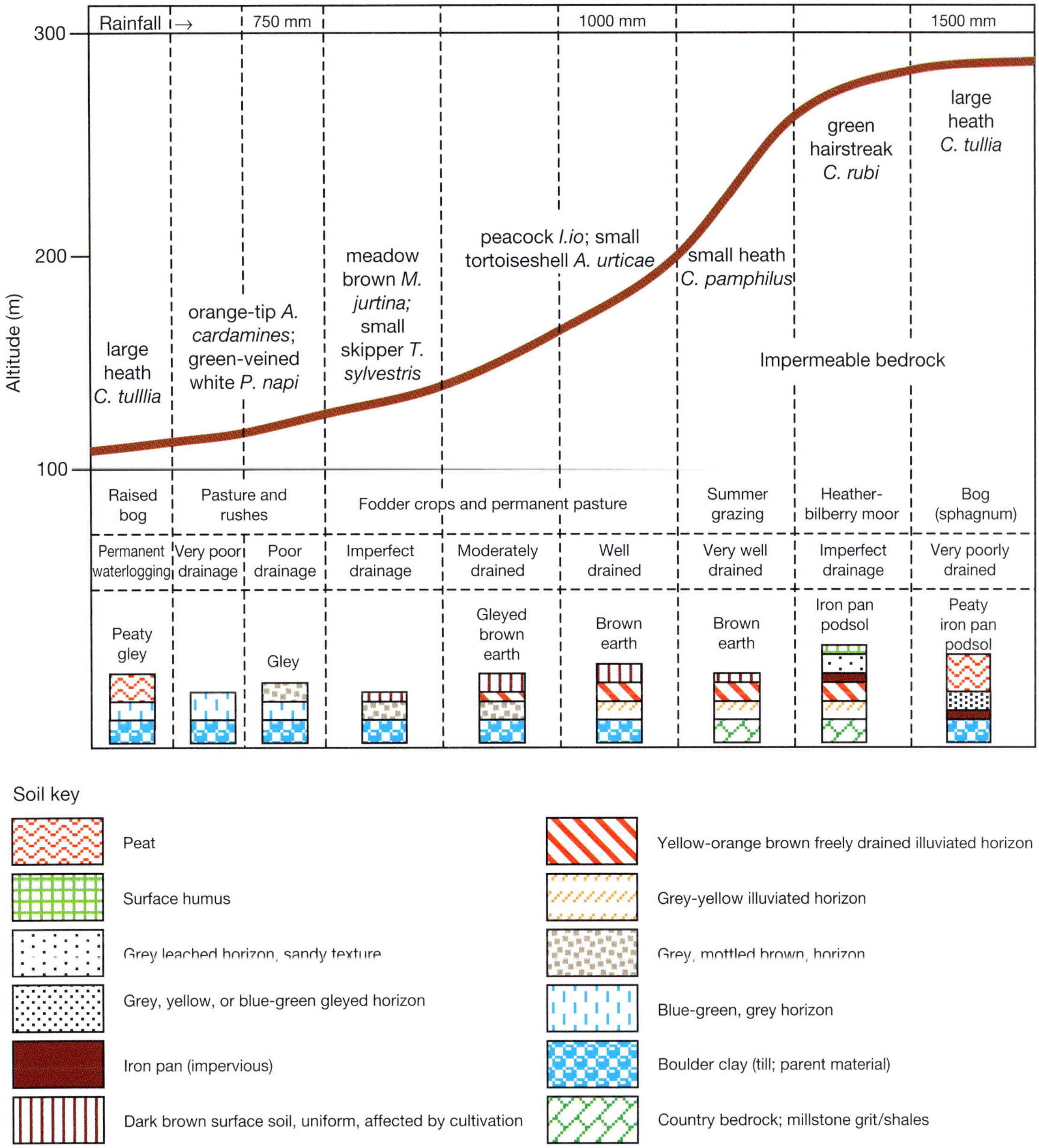

Fig. 7.13 Butterflies associated with changing altitude, slopes, bedrock exposure, soils and land uses of a highland zone in northern Britain (e.g., Pennines, Forest of Bowland). A distinctive zonation of butterfly species typically accompanies changes in conditions over short ranges in altitude. Large heath *Coenonympha tullia* are found on sphagnum bogs at high and low altitudes (blanket bogs and raised bogs, respectively); green hairstreak *Callophrys rubi* on bilberry moor sides where the country bedrock, typically gritstones and shales become exposed; small heath *Coenonympha pamphilus* on sheep's fescue *Festuca ovina* overlying better drained, deeper soils on steep hillsides; nymphalids (e.g., *Inachis io*, *Aglais urticae*) occupy nettle patches on lower pastures over improved soils enriched with nitrogen; and more warmth-tolerant species (e.g., *Maniola jurtina*, *Thymelicus sylvestris*) are found at lower elevations on moist ground where land is free of fertilizer and herbicide applications. Where drainage is impeded, increasingly crucifer-associated pierids such as orange tip *Anthocharis cardamines* and green-veined white *Pieris napi* occur. How the various factors integrate to influence butterflies above the improved zone is illustrated in Dennis (1992a:234). (Based on material in JMB Advanced Geography Syllabus Paper 1, 1972, question 10.)

biotope, butterfly numbers in field margins may actually arise because the marginal features enhance conditions in the adjoining biotope (**P7.21**); resources in the field may not be fully exploitable without the field margins (Dover, 1988; Dover *et al.*, 1992, 2000). From the latter point arises the question of what actually constitutes a boundary and the concept of a '**footprint**', the zone of influence of the boundary feature (Box 7.6).

These are not simple questions to answer. Field margins entail a wide variety of features, mostly linear structures, including hedges (a hedgerow includes the whole field boundary structure, including single trees), banks, walls, fences, tree lines, ditches, ditch banks, verges, wood edges, green lanes and tracks – the latter two may also be raised or sunken below adjoining fields (Helps, 1994). Nodes occur in field corners and field intersections as well as other larger features or zones (copses, relict areas, soil dumps, pits, ponds). These features vary immensely in form, reflecting regional and local practices, management and topography, making them difficult to compare for butterfly resources. Thus, individual features vary extensively for width, height, density, life forms (trees, shrubs, herbs, grasses), nectar flowers, hostplants, adjacent land uses, management (e.g., herbicide spraying, fertilizer applications, uncultivated margin width, physical disturbance such as close ploughing and uprooting, cutting) and local climate (shelter, direct sunlight, temperatures, humidity). A number of comparisons have been made for these features with respect to butterfly numbers. Distinctions are not always found, for example between short and tall hedges or between hedges and grass banks (Dover, 1996), but certainly occur between hedges and green lanes (Sparks *et al.*, 1999; Dover *et al.*, 2000; Croxton *et al.*, 2005). Green lanes are superior to other linear farmland biotopes (hedgerows, grass banks, wall lines), except wood glades, for butterfly abundance and species' richness; they have more resources (nectar) and are more sheltered. There is a question of whether like is being compared with like in such studies. Nature rarely presents neat experiments where the impact of the many potential confounding factors is nullified. Ancillary research on other factors in different settings should be consulted (e.g., agrochemicals (Rands and Sotherton, 1986; Davis *et al.*, 1991, 1993), shade (Sparks *et al.*, 1996)); these studies indicate what is to be expected for field boundary features.

Only two extensive comparative studies have been carried out on the complex effects of field margin features on butterfly numbers, one in Swavesey Fens, Cambridgeshire (Sparks and Parish, 1995) and a second in North Hampshire (Dover, 1996), both in Britain. In Swavesey Fens, a study of field margins covered some 26 km of field boundaries in 131 transects each of 200 m length; these were divided into three groups: bounding pasture, arable and both land uses. Butterfly abundance was found to be enhanced in those boundaries with large hedgerows, grassy areas as verges and banks, and by flower abundance (nectar sources) – attributes found to be associated with pasture. The negative aspect of management using agrochemical applications also emerged as it did for the Hampshire study (Dover *et al.*, 1990). In Hampshire, Dover (1996) found shelter to be the dominant variable in accounting for numbers of *Maniola jurtina* (except females), *Aphantopus hyperantus* and gatekeeper *Pyronia tithonus*, though results differed for species and sexes. Key structural pluses were shelter, insolation, width of hedge bank or grass verge and adjacent uncultivated land. Key consumable pluses were specific nectar sources, bramble (*Rubus fruticosus*), thistle-like composites, wild marjoram (*Origanum vulgare*) and mayweeds (*Matricaria*). Distinct negatives were cow parsley (*Anthriscus sylvestris*) and old man's beard (*Clematis vitalba*) which were not used and had a **swamping effect** on butterfly resources, much as nettle seemed to have for Swavesey. The latter is an interesting finding and indicates that one butterfly's hostplant may exclude other species' hostplants. The distinctions for field margin attributes in butterfly diversity and abundance of individual species suggests distinctions in habitat and resource capacity but it is extremely difficult to extricate the influence of individual variables owing to high correlations among them, as is clear from both these studies. For example, in the Swavesey study, larger hedgerows and *Rubus* are more typically associated with pasture than adjacent arable land uses.

These two studies demonstrate that there are great opportunities for additional research on the effects of field margin features on butterflies in different landscapes and different conditions. Some key studies have already been carried out that test for single features or attributes, on the influence of shelter and the impact of linear features on movements. Indications are that enhancement of local climate is, when taken over the entire country, the most important attribute of field margins (**P7.22**). Dover *et al.* (1997) found that more sedentary species declined in unsheltered areas and increased in sheltered areas with increasing wind

Box 7.6 **Shelter belts, footprints and scoring shelter for butterfly studies**

Shelter forms a crucial part of the environmental conditions for butterfly populations and for butterfly individuals exploiting resources. John Dover (1996) has been instrumental in developing a method for scoring shelter along hedgerows, along lanes and field corners. An index is constructed based on wind flow affecting an imaginary butterfly from eight potential compass directions (Fig. B7.6).

This system can be modified to account for the pattern of prevailing winds on such structures. In the scheme described for Fig. B7.6a, shelter from easterly winds would obtain the same score (3) as that illustrated, coming from the west. A possible solution is to multiply the proportion of winds from a particular direction by 8. Thus, in Fig. B7.6a, where 60% of the wind comes from the west (25% from northwest, 20% from west, 15% from southwest), the score would be $8 \times 0.6 = 4.8$, whereas in the reverse of this situation, with 25% of the wind coming from the east (10% from northeast, 10% from east, 5% from southeast), the score would be $8 \times 0.25 = 2$. The period that prevailing winds are measured over would depend on the study; scoring such features for a biogeographical study would consider prevailing winds over a long period (e.g., period of atlas records), whereas a behavioural study on population movements would determine the pattern of wind directions for the study period. Note, only wind directions affecting a site would be considered as illustrated in the diagrams.

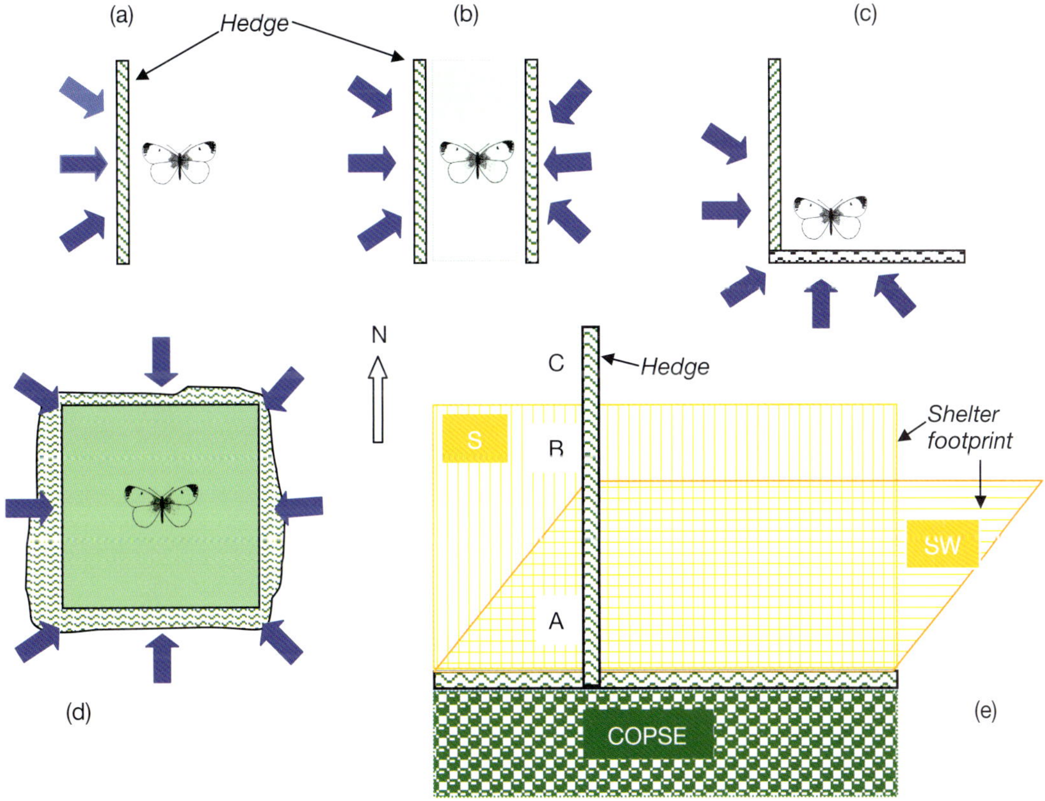

Fig. B7.6 (a) A butterfly protected by a single unbroken hedge, windbreak, wall or copse edge receives a score of 3. (b) A butterfly in a double hedged or walled green lane or similar double-sided structure receives a score of 6, provided that the two sides are not separated from one another to the extent that lee effects do not overlap. (c) An L-shaped junction (e.g., field corner) obtains a score of 5, as would scallops in wood edges. (d) A small clearing in a woodland, where the lee effects overlap, receives a score of 8. (e) Schematic representation of shelter footprints from a copse along a field margin with a hedge abutting it at 90°. 'A' receives a score of 5 with shelter from the hedge (score 3), the southwest and the south, 'B' a score of 4 with shelter from the hedge and the south, and 'C' a score of 3 with shelter only from the hedge. Footprints represent lee effects ×10 the height of the copse (Lewis, 1969). (From Dover, 1996, courtesy of the author and Blackwell Science.)

speeds and that this depends on hedge structure (see Box 7.6). Studies on movements have shown that the impact of linear rural features on movements can be complex (Fry and Main, 1993). Hedges can clearly be barriers, as was suggested to be the case in the appearance of the reversed cline in *Maniola jurtina* hindwing marginal lunulation at a hedge boundary (Creed *et al.*, 1970; but see Shreeve *et al.*, 1996b) and from autecological work on sedentary species such as *Polyommatus bellargus* (Thomas, 1983a). This is supported by experimentation using tapes (Dover and Fry, 2001) and tarpaulins (Fry and Robson, 1994). Even so, work by Dover (1989, 1990, 1991, 1994, 1996; Dover *et al.*, 1992) indicates that butterfly movement is largely restricted to uncropped areas of open farmland – 98% of butterflies' movements are along field boundaries as opposed to across the crop. Within some of these movements may well be examples of **trap-lining** (Erhardt and Mevi-Schütz, 2009), a behaviour described for *Heliconius* butterflies (Gilbert, 1980), where particular resources, if not identical elements of those resources, are used repeatedly. These findings corroborate suggestions in the chapter on metapopulations (see Chapter 6) that networks are critical structures for transfers, but there is much room for study of 'gaps' in linear features on movements in butterfly species (Dennis, 1982b, 1986b; Munguira and Thomas, 1992) as on birds (Bosschieter and Goedhart, 2005).

Studies of the impact of rural features on butterfly resource use are in their infancy and it is as well to briefly outline the pitfalls. One has already been mentioned, the often high **collinearity** (linear correlation) among variables in such studies. Ways of removing the impact of correlations in regression studies are not entirely successful, but there are data reduction techniques that assist in determining clusters of variables affecting butterfly numbers. A second problem is autocorrelation (see Box 7.1). Numbers in sampling units (transect sections, squares) will tend to be more similar to those of neighbours than distant units. One way around this is not to use adjacent units and another is to calculate for (remove the effect of) autocorrelation (see below). A third issue is that findings will depend much on the regional context, dominant influences over land use and management such as geology, topography, altitude, markets, etc., and that studies should account for spatial variation in landscape (Kleijn and van Langevelde, 2006). A fourth problem is that not all important variables may be measured (e.g., humidity, temperature, shade) or perhaps not measured satisfactorily. Shade is a predonant factor in woodland rides and this is likely to be matched for hedgerows (Sparks *et al.*, 1996). Shelter measures based on structure are not really a substitute for actual measures of wind speed. Local climate can now be directly modelled for sites using artificial neural networks (Bryant and Shreeve, 2002). Part of the measurement process is determining the bounds of influence of the features (the footprint) (see Box 7.6). Fifth, there is a real problem of finding suitable controls for features (e.g., for headlands (Feber *et al.*, 1994) and green lanes (Dover *et al.*, 2000)) owing either to availability or access to land. Finally, for comparisons to be meaningful they need to relate butterfly resource use with features. Merely recording the presence of a butterfly does not inform about what it is doing and using (Dennis, 2004a).

TRANSLATING CONCEPTS FROM THE HABITAT TO A LANDSCAPE SCALE

From the above it is evident that a number of themes distinguish landscape ecology from a metapopulation approach to organism biogeography (Wiens, 1997); for instance:

- Resource quality is continuous in space–time rather than discrete; landscape patches are mutable.
- Edges and boundaries have potentially important consequences for within-patch and between-patch population dynamics.
- The degree of connectivity among components in a landscape mosaic has major consequences on patch dynamics and interactions as well as affecting population dynamics of the target organism.
- Patch context matters. Events within a patch depend on its location relative to the structure of the surrounding mosaic of land uses.

To the two spatial dimensions of the metapopulation landscape is added a third, vertical, dimension, giving height and depth to landscape features. The sheer complexity in structure is well illustrated by the variety of measures needed to describe specific landforms, namely: position, orientation and direction, area and size, height and depth, gradient, shape and context (cf., Evans, 1987, for landforms). In effect, there is a translation of complexity, found within habitats, to the landscape scale. Much as resources within habitats could be described in terms of composition, physiognomy and connectivity, so can landscape components (Taylor *et al.*, 1993). The distinction with habitats is that instead of these measures being made directly on resources, they can also be made of 'deeper' structures within landscapes underlying resource distributions. At a fundamental level, processes in landscape ecology do not differ from those within habitats (patches) or in metapopulation-scapes. Distribution patterns ultimately depend on transfers, colonizations and extinctions. At the landscape scale, these processes are brought about by four relationships (Dunning *et al.*, 1992):

1 Landscape complementation.
2 Landscape supplementation.
3 Source/sink relationships.
4 Neighbourhood effects.

These occur, and have already been described, for the habitat and metapopulation (see Chapter 6) (Ouin *et al.*, 2004). At a coarser landscape scale, the relationships may be conceived to have regional analogues describing species' distributions and ranges. The ingredient that defines the nature of the relationships is the movement of organisms. Within habitats, it is daily (routine) movements between resource outlets. At the landscape scale, it is migration of vagrants between habitat units or resource outlets. Thus, nearby distinct landscapes, say on different rock types, may hold complementary resources (e.g., hostplant, roost site), supplementary resources (e.g., different hostplants or different nectar sources) and have neighbourhood structures that enhance or deter connectivity. Neighbourhood resources have been demonstrated to have a prominent influence on diversity in adjacent biotopes and landforms, such as abandoned quarries (Novak and Konvicka, 2006) and urban areas (Snep *et al.*, 2006). An adjacent landscape may function as a sink for one species whilst acting as a source for another; in which case retreat from sink landscapes results in relict distributions and refuges in the source landscape.

LANDSCAPE-SCALE STUDIES ON BUTTERFLIES

The basic idea of landscape studies is to test for effects of landscape composition, structure and connectivity on organisms while treating the landscape components as potentially continuous. To understand distribution patterns of organisms such as butterflies in landscapes – the ecological processes responsible for these distribution patterns – we need not only information on obvious aspects of composition such as habitat patches, where these are recognizable, or more credibly resources such as hostplants and nectar sources, but also data on the pattern or structure of the land cover that influences resource distribution and connectivity (Herzog and Lausch, 2001).

Cautionary principles for landscape-scale studies

Before reviewing the few studies that have attempted landscape-scale assessments of butterfly distributions in Britain, it is well to consider some cautionary principles relating to such studies:

- **P7.23: Factors influencing the distribution of species' habitats or distributions depend on the scale of studies (extent/mesh) and location of study.**
- **P7.24: Measures of landscape influences are often inadequate surrogates for true landscape features. The latter are difficult to measure, index and quantify; the difficulties increase inversely with scale (mesh) of study.**
- **P7.25: Butterfly studies are affected by biased survey. Recorders tend to focus on 'home' squares and hotspots.**
- **P7.26: The status of individual butterflies recorded (vagrants, residents) is crucial for landscape studies especially for fine scales (mesh). Progressive bias in species' status is symptomatic of fine-grained mapping units subject to repeated sampling.**

Inevitably, the factors found to be responsible for distributions will change with location and scale (**P7.23**) (Bergman *et al.*, 2004; Kivinen *et al.*, 2006), particularly the grain or mesh of study (MacNally *et al.*, 2004). This is indicative of the heterogeneity in landscape – the quantum shifts in resource geography with geology, climate and land use. Coarse-scale (countrywide) studies are influenced by dominant gradients (e.g., climate) and fine-scale studies (extent of a few kilometre squares or less) are limited by specificity in landscape components. Thus, at the scale of the British Isles, land uses expected to enhance butterfly incidence, such as steeper slope angles and forest cover, were negatively – not positively – associated with species' richness simply because squares that score highly on these attributes are biased to northern regions with lower butterfly diversity (Dennis and Williams, 1986).

Quantification of landscape influences is in its infancy (**P7.24**) if only because of the expense of surveying landscape or the often inadequate nature of surrogates (e.g., geographic information systems (GIS) measures from remote sensing such as the Landsat Thematic Mapper; Seto *et al.*, 2004). It is nevertheless proving to be highly successful in determining patterns in butterfly species' diversity, as in northeast Spain (Stefanescu *et al.*, 2004), and for producing conservation strategy maps based on relevant habitat data as in the case of *Melitaea athalia* in Kent (Holloway *et al.*, 2003). Furthermore, the techniques are improving in sophistication (e.g., airborne light detection and ranging or LiDAR providing a resolution of 10–20 cm on a vertical axis). Landscape pattern is a mixture of natural and human-managed patches, varying in size, shape and arrangement, the result of complex interactions of physical, biological and socioeconomic forces (Forman and Godron, 1986; Krummel *et al.*, 1987; Turner, 1987). Different cover classes for these phenomena can have very different geographical patterns, as in the case of forest components in Snowdonia, Wales (Gkaraveli *et al.*, 2001). Quantification is an essential part of relating distributions of organisms, the persistence of populations and species' richness, to land use cover and patterns (Fahrig and Merriam, 1985). In this process, links between landscape components and butterfly resources, the quality side of patchworks and habitats, has yet to be bridged. Increasingly, the approach is the construction of landscape indices based on measurable attributes of cover classes, such as the number, size, shape and edge length of individual areas and the distances between areas. Large numbers of landscape indices have been developed for patchworks and new ones are continually emerging, all of immense importance with the continuing fragmentation of biotopes suitable for butterfly species. A few examples and sources are outlined in Box 7.7.

Box 7.7 Metrics for describing the landscape

Whether habitats or resources are associated primarily with rock types, soils, landforms or vegetation units (biotopes), they can all be considered as occupying 'patchworks', despite the fact that many may be very small or linear. There are three basic aspects to these patchworks: composition, structure and connectivity. At a landscape scale, connectivity can be conveniently subsumed in structure. The distribution of organisms such as butterflies depends on both, such that:

$$D = fL_c + fL_s,$$

where D is the distribution of organisms, L_c is land cover types, L_s is land structure and f is 'function of'.

Metrics of landscape pattern are primarily concerned with land structure and these are discussed below. Quality is often subsumed in a binary categorization of landscape (i.e., habitat and matrix) and in time measures will be developed that cater for land composition, that is, for habitat quality.

A key aspect of understanding population persistence is the measurement of patchwork; measurements allow comparisons over time and between areas (regions). One of the outstanding problems has been what and how to measure the patches in a patchwork. They may vary in size (area), number, shape (perimeter, degree of fragmentation), isolation, location, orientation and in many other ways. Many indices that have been developed measure much the same phenomena. In one study, 55 metrics of landscape pattern and structure for 85 maps of land use and land cover were boiled down by factor analysis to six factors explaining 87% of the variation. Only 26 variables were needed of the original 55 and the authors suggested the use of simple univariate substitutes (Ritters *et al.*, 1995). Giles and Trani (1999) also suggest that there may be as few as six important factors for describing a mapped area: area, classes, proportion of dominant class, number of polygons, polygon size variance and elevation range. Increasingly, computer programs are available for calculating these indices (e.g., FRAGSTATS (Tischendorf, 2001), SPAN (Miller *et al.*, 1997)) and spatial statistics are becoming part of statistical packages (SYSTAT, 1999, see www.spss.com).

Much research effort is being expended at the moment on the evaluation of landscape pattern metrics (e.g., Hulshoff, 1995; Schumaker, 1996; Chuvieco, 1999; O'Neill *et al.*, 1999), often using simulated landscape patterns (Haines-Young and Chopping, 1996; Traub and Kleinn, 1999; Jaeger 2000; Bowersox and Brown, 2001). They are often accompanied by highly original labels (e.g., index of lacunarity to measure habitat contagion; McIntyre and Wiens, 2000). Indices vary in robustness and sensitivity (Saura and Martinez-Millan, 2001). They are affected by map scales, both of extent and mesh size (Cullinan and Thomas, 1992; O'Neill *et al.*, 1996; Grossi *et al.*, 2001; Saura and Martinez-Millan, 2001), GIS data transformations and data misclassification (Bettinger *et al.*, 1996; Wickham *et al.*, 1997) as well as sampling issues (Hunsaker *et al.*, 1994) and stages of fragmentation (Jaeger, 2000). One of the basic problems is poor consistency in relationships between landscape and ecological variables over different landscapes (Tischendorf, 2001). From a detailed comparison of perception and metrics, D'Eon and Glenn (2000) point to the lack of meaningful quantitative measures which encourages reliance on intuitive human perception for appraisal and management. The need for standardized procedures is increasing, particularly with the explosion in use of remote images as databases (Herzog and Lausch, 2001). For instance, numerous techniques are now available for measuring biotope fragmentation (Bowen and Burgess, 1981; Krack-Roberg *et al*., 1995) which together with patch size is a crucial attribute of landscape pattern. These techniques form the basis of the metapopulation approach to butterfly populations with intuitive consequences for movement and colonization (Vos *et al*., 2001).

Recently, there have been advances in finding single key measures of landscape patterning (e.g., Φ (phi) is a standardized summary for habitat area, perimeter, number and isolation founded on Euclidean distances) (Bogaert *et al*., 2000). Perhaps the most useful of these indices are those describing fragmentation, as this is key to butterfly habitat loss in Britain (Jaeger, 2000). The measures described by Jaeger (degree of coherence *C*, degree of landscape division *D*, splitting index *S*, splitting density *s*, net product *N* and particularly effective mesh size *m*; Table B7.7 and Fig. B7.7a) are not only simple to use and mathematically elegant, but improve on previous measures in having low sensitivity to very small patches, greater monotonicity in reaction to fragmentation phase (Fig. B7.7b), and distinguish better among spatial patterns. They have been compared against *n* (number of patches),

Table B7.7 Measures of fragmentation. (From Jaeger, 2000, courtesy of Jochen Jaeger, with kind permission of Springer Science and Business Media.)

Description	Equations
Degree of coherence (*C*): the probability of two animals placed in different areas within the study region finding one another	$C - \sum_{i=1}^{n}\left(\frac{A_i}{A_t}\right)^2$
Degree of landscape division (*D*): the probability that two randomly chosen places in the study region are *not* situated in the same undissected area	$D = 1 - \sum_{i=1}^{n}\left(\frac{A_i}{A_t}\right)^2$
Splitting index (*S*): the number of patches obtained from dividing the total region into parts of equal size but producing the same degree of landscape division as obtained for the '*n*' subset of patches	$S = \frac{A_t^2}{\sum_{n=1}^{n} A_i^2}$
Effective mesh size (*m*): the size of areas obtained from dividing the region into S areas (each of the same size A_t/S) with the same degree of landscape division as for the '*n*' subset of patches	$m = \frac{A_t}{S} = \frac{1}{A_t}\sum_{i=1}^{n} A_i^2$
Splitting density (*s*): when a study region is characterized by the splitting index (*S*) then the number of 'meshes' per unit area is given by the splitting density	$s = \frac{S}{A_t} = \frac{A_t}{\sum_{i=1}^{n} A_i^2} = \frac{1}{m}$
Net product (*N*): product of the effective mesh size, (*m*) and the total area of the study region	$N = m \cdot A_t = \sum_{i=1}^{n} A_i^2$

Jaeger's measures of fragmentation first require the 'construction' of a binary categorical map divided into two sets of 'habitat' patches and fragmenting features (lines, areas). The set of *n* patches of a landscape is denoted by:

$$\Phi = \{A_i \mid i = 1, \ldots, n\}.$$

The total area of the study region is given by:

$$A_t \geq \Sigma i = 1\, A_i,$$

where A_i is the size of the patches ($i = 1, \ldots, n$) and A_t is the total area of the region.

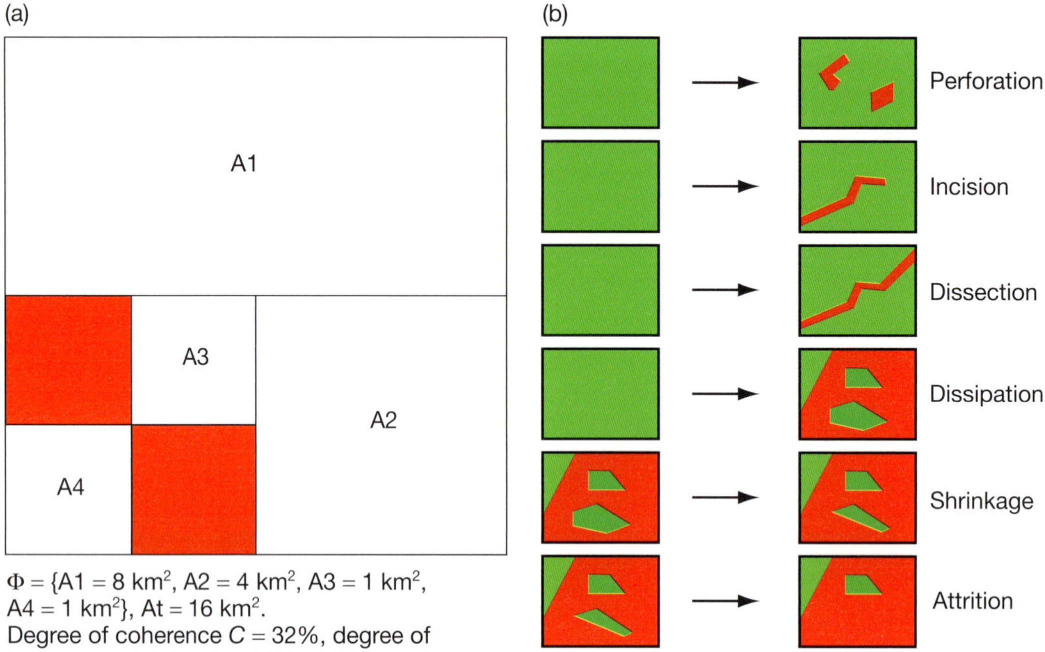

Φ = {A1 = 8 km^2, A2 = 4 km^2, A3 = 1 km^2, A4 = 1 km^2}, At = 16 km^2.
Degree of coherence C = 32%, degree of landscape division D = 68%, splitting index S = 3.1, effective mesh size m = 5.1 km^2, splitting density s = 0.2 1/km^2, net product N = 82 km^4

Fig. B7.7 (a) Area configuration to illustrate the calculation of fragmentation measures. (b) Phases of the fragmentation process, distinguished according to geometric characteristics. (From Jaeger, 2000, courtesy of the author and with kind permission of Springer Science and Business Media.)

$\bar{A}$ (average area), l (line density = L_j/A, where L_j is the length of the network and A is the regional area), the landscape dissection index LDI (Bowen and Burgess, 1981) and the relative partitioning index PI_{rel} (Deggau *et al.*, 1992, in Jaeger, 2000). Some measures (e.g., D, S and n (number of patches)) are not **intensive** (i.e., do not have constant value when the study region is multiplied (enlarged)) and are suited only for comparison of regions with the same total area or comparison of the same area over time. However, the measure m is both intensive and **area-proportionally additive** (the value for regions can be combined). Jaeger (2000) suggests that the number of patches (n), average patch size ($\bar{A}$) and relative partitioning index (PI_{rel}) should be replaced by S, m and D, respectively, and that in the case of landscape incision (Fig. B7.7b), road density (l) or the landscape dissection index (LDI) should also be applied. Recently, effective mesh size m has been modified to solve boundary problems (i.e., where patches are truncated at study boundaries) (Moser *et al.*, 2007).

Butterfly distributions are affected by biased survey. Inevitably, owing to costs and time available, recorders focus attention to their home areas and to hotspots further afield (**P7.25**) (Dennis *et al.*, 1999; Dennis and Thomas, 2000). Other aspects of biased recording involve the apparency of different butterfly species to recorders; these largely affect comparative study, but can also affect understanding of resources used by sexes that differ in apparency to recorders (e.g., *Plebejus argus*) (Dennis and Sparks, 2006; Dennis *et al.*, 2006a). One part of this bias is the repeated sampling of squares (**P7.26**). This will have the effect of inflating the number of species in fine-scale (mesh) studies, as with progressive sampling an increase in the number of vagrant species will be recorded (Dennis, 2001). It is important to record the status of species in squares, which can be done by noting behaviour of individuals in relation to substrates (Hardy and Dennis, 2007).

Data on recording intensity can be used for selecting geographical units (Maes *et al.*, 2003) or adjusting records for diversity and species' incidence (Dennis and Hardy, 1999; Hardy and Dennis, 1999).

Empirical findings of remote landscape-scale butterfly studies

Landscape studies on butterflies in Britain are scarce and still tend to be limited in their remit. There are two basic approaches; a distinction is typically made between direct and indirect approaches. The first relates species' occurrences to landscape components directly; the second samples the link between landscape components and butterfly incidence or abundance and projects the findings to a wider area. The latter has the great advantage of projecting the relationships assessed at finer resolutions over coarser scales as long as they both relate to the same landscape (Cowley *et al.*, 2000).

A direct approach study of 21 butterfly species for 17 variables (11 land variables, six climate variables) at tetrad scale (1-in-100 sample) over nine counties in southern England, disclosed a number of valuable insights into land use associations of butterfly species' incidence (Sparks *et al.*, 1997). Species' richness is shown to increase in tetrads with a number of attributes (i.e., low wind speed, smaller cultivated edge scores, lower frequency of improved grassland and larger amounts of tilled edge) indicating the importance of field margins in the arable dominated biotopes of southern England. Regarding individual species' incidence, many of the significant agents were found to be climatic. Particularly influential is the negative impact of summer wind speed and relative humidity indicating preference for shelter, wooded, lower ground away from coasts. In this study no adjustment was made for sampling biases and problems are evident in measuring land attributes. Data on land uses, some 10 attributes, were available at high resolution (25 m), but remain limited in scope, being available only in a binary form (present or absent) and lacking measures of quantity and quality.

Substantial improvements are expected in the near future in both butterfly and environmental data collection, particularly in the resolution and ground truthing of data (Fuller, 1993). Methodological improvements are also required. There is a need to compare associations tested at different scales. A start has been made using data from the Manchester Butterfly Atlas, part of the Butterflies for the New Millennium (BNM) scheme (Hardy, 1998; Asher *et al.*, 2001). This database demonstrates a number of important developments. The region has been mapped at three different scales. Recording effort has been measured for squares, and data have been obtained on both environment and specific resources for species. This survey usefully focuses on the impact of fast developing urban cover on species' incidence, and is important inasmuch as it demonstrates that urban growth may not necessarily be detrimental to butterfly populations (Hardy and Dennis, 1999), a finding confirmed by a more intensive subsequent study (see Box 8.4) (Dennis and Hardy, 2001).

Broad-scale landscape studies on butterflies in Britain have yet to measure or account for the effects of connectivity (see Box 7.1). An essential aspect of this is spatial contiguity; squares are more likely to record a species if adjacent to squares also having the species than to those lacking it. One study, based on French departéments has revealed just how important is this effect. Predictions of incidence for species in departéments were found to be related closely to a neighbourhood measure based on incidences in contiguous departéments (Dennis *et al.*, 2002) and confirmed by subsequent records (Dennis and Shreeve, 2003).

The indirect approach of sampling landscape components for butterfly incidence and abundance, and testing this over wider areas holds great promise. This has been done on the Creuddyn Peninsula in North Wales, where sampling over transects for 16 biotopes has been compared with records obtained for 140, 500 m grid cells with known biotope cover (see Box 5.1) (Cowley *et al.*, 2000). Agreement was very high (83%). This has a vital application for future monitoring of changes in butterfly distributions over regional scales to back up full atlas mapping, which logistically has limitations.

These types of survey, linking butterflies and landscape, are producing some valuable insights into factors influencing butterfly distributions at local and regional scales and will clearly burgeon in the future. On the continent of Europe they have revealed, by way of example, the importance of small farms over large farms, organic over conventional farms (Belfrage *et al.*, 2006) and the relative merits of particular grazing and mowing regimes (Saarinen and Jantunen, 2005). Other studies identify key resource needs, such as nectar, and key vegetation associations and combinations for burnet *Zygaena carniolica* (Scopoli) and pearly heath *Coenonympha arcania* (L.) (Binzenhofer *et al.*, 2005). However, they have their limitations and caution is required in interpreting the results. The smaller the scale (extent) of surveys, the less transferable the results are likely to be, as survey context, for instance of landscape and conditions, is clearly an important influence on associations. Repetition of a familiar cautionary note is that much stronger links need to be forged between butterfly records and landscape, by recording behaviour in relation to context and substrates (Dennis, 2004a). Nevertheless, such survey designs are essential if we are to develop an understanding of the influence of landscape components on butterflies, an obvious example being that of the effect of landscape features on butterfly movements.

LANDSCAPE MODELLING APPROACHES

The need for realistic landscape models is borne out of the observation that metapopulation models, despite their acknowledged sophistication, are but poor metaphors for actual patterns and processes. Modelling, for all its complexities, is essential, as it is impossible to survey both the entire landscape and all species. The models grow in realism by incorporating data from sampling landscape influences on butterflies and the behavioural responses of butterflies to the natural environment. Several requirements are recognized for the new generation of models:

- To relate population dynamics to actual landscape components (i.e., distinct resource distributions) not just area.
- To relate movements to landscape permeability (i.e., percolation) not just isolation and frequency distributions of movements.
- To incorporate behavioural responses of butterflies to agents and landscape components (e.g., modelling of emigration in terms of migration thresholds; Baker, 1969, 1984).
- To incorporate the effects of extrinsic agents on growth, mortality and movement (e.g., aggregations, predation, parasitization, competition).
- To build in a dynamic base where all data layers can change with time and in response to one another.
- To incorporate contiguity effects (neighbourhoods), where values of data layers integrate horizontally to develop meaningful spatial contexts.

Simulations of growth, decline, extinction and movements are typically conducted in spatially explicit mosaics consisting of lattices, a grid, where the cell size is equivalent to the smallest scale at which an organism recognizes spatial heterogeneity. Thus, landscape structures of greater extent appear within clusters of cells and can be attributed connectivity, other than contiguity, as well as homogeneity. In this grid, the *X* and *Y* coordinates describe two-space and the array of variables, including elevation, are effectively stacked as a series of *Z* coordinates. Each variable is potentially mutable, undergoing changes intrinsically, as in vegetation succession, or extrinsically, as driven by the environment. The data requirements for the study frame (extent and mesh) are theoretically enormous. Provision, increasingly, is being facilitated by remotely acquired data analysed in a GIS system. Techniques are advancing rapidly that allow the determination of incidence and abundance for species over whole areas from sample data (e.g., desktop GARP; Ball and Henshall, 2006; www.nhm.ku.edu/desktopgarp/). These can produce estimates either directly, for example using neighbourhood techniques (e.g., kriging; Burrough, 1986; Liebhold *et al.*, 1993) or indirectly by applying regression modelling techniques which allow predictions to be made from a set of independent, environmental variables (see below).

Dynamic models necessarily incorporate movements of individuals as well as population changes within patches. In landscape mosaics, distances between patches do not scale directly with distance (Euclidean), but more complex functions of permeability at boundaries and of differential viscosity within patches are needed (Wiens, 1997). Movements have been simulated in a number of different ways; some algorithms treat movements as if they obey simple laws of physics (e.g., random diffusion; Okubo, 1980; Johnson *et al.*, 1992) and others fail to properly represent bias in direction (e.g., correlated random walk; Kareiva and Shigesada, 1983; Turchin, 1989). Reaction diffusion algorithms carry more promise, inasmuch as they incorporate spatially varying diffusion rates, in which movements may be made responsive to any kind of resource (Corbett and Plant, 1993); movements occur between cells where some critical threshold is exceeded. Yet, advances still need to be made to model behaviour of organisms in terms of migration thresholds to conditions within cells as well as to conditions in adjacent cells. Fry (1995) illustrates the systems applications for a reaction diffusion model to changes in dispersion of the scarce copper *Heodes virgaureae* (L.) from a meadow in a Norwegian field system. From such beginnings, we shall undoubtedly experience an exponential growth of increasingly complex models that, without care, are in danger of conveying less useful information than current models to those that manage landscapes for organisms, particularly if based on single species.

Chapter 8

HABITAT ISSUES IN BUTTERFLY GEOGRAPHICAL RANGES

The most notorious source of logical fallacy is probably post hoc ergo propter hoc, *which attributes that: since that event followed this one, that event must have been caused by this one.*

COMPONENTS OF GEOGRAPHICAL RANGES

Geographical range is perhaps the most important attribute of a species determining its status. But what the term entails is not readily appreciated. Because it can be conceived and measured in different ways it is not a unitary concept. Following a brief overview of the concept, attention is focused on how ecological factors (resources, habitat conditions) first influence contemporary butterfly ranges and then secondly range changes, in the past on a geological scale and in recent times with projections for the immediate future.

Definitions of range and distribution

Range describes the geographical extent of a species, in two dimensions longitudinally and latitudinally, and in three dimensions with the addition of altitude. A species rarely occupies all the habitat patches available for it; thus, there is a distinction between **potential range** and **realized range** (Box 8.1). To the extent that a line (an epidermis) can be drawn linking perimeter records for a species, the **range margin**, it is possible to distinguish between unoccupied exterior and interior space within the potential range, and occupied space from core to periphery within the realized range. Putting aside human error in recording, the key to understanding occupied and unoccupied space is to appreciate that ranges are not static but reflect population parameters which in turn mirror conditions in habitats. The range margin is not invariably a physical barrier (e.g., coastal edge), so there is no reason why an empty space inside the range margin should necessarily differ from another outside it, but it is expected from the distribution of environmental agents that there may well be distinctions in number, type and condition of resources available between range core, periphery and exterior (Shreeve *et al.*, 1996a). In habitat terms, a species may be recorded in variable circumstances – where there is a habitat, or just a single resource, and even where no resources exist (e.g., at sea). Similarly, it can also be absent where there is a habitat, fewer resources than the complement for a habitat, and obviously where there exist no resources at all (Box 8.1). The potential for species to appear at a site with less than the full complement of resources comprising a habitat depends on its life history and ecology as well as on contiguous conditions and records. Species with extensive migration capacity can more readily occupy sites without resources (no habitat), if only temporarily, as well as those sites with single resources. Examples of the latter are species such as the brimstone *Gonepteryx rhamni* with isolated breeding and overwintering sites (i.e., disjointed habitat resources; see Fig. B2.1f).

The magnitude of problems of species' status for sites is also inversely related to spatial scale. The finer the scale of mapping, the greater the likelihood of butterfly records being made where habitats are absent or incomplete (Dennis, 2001). Thus more vagrants will

A Resource-Based Habitat View for Conservation, 1st edition. By Roger L. H. Dennis. Published 2010 by Blackwell Publishing.

Box 8.1 Components of geographical ranges

The concept of a geographical range is not a simple one and this box considers some basic points. Measures of geographical ranges fall broadly into two groups:

1 Extent of occurrence.

2 Area of occupancy.

Extent of occurrence describes the spatial limits of an organism's distribution (e.g., distances separating latitudinal, longitudinal and altitudinal limits) whereas **area of occupancy** describes the distributional cover of the organism (e.g., number of grid squares occupied on a map). **Geographical range** describes the bounds of spatial cover whereas **distribution** describes geographical location; range as extent of occurrence is not affected by size of **mesh** (the units used for mapping), but range as area of occupancy may well be and so are measures of distribution.

In the measure of geographical range, there are problems with both extent of occurrence and area of occupancy. Inevitably, measures of extent of occurrence include areas that are 'empty' or vacant; they may have been occupied by the species but are no longer occupied during the period of survey. On the other hand, measures of areas of occupancy are affected by grid mesh size, i.e., the size of the sampling unit chosen to represent species over the geographical region studied.

Figure B8.1 illustrates some components of geographical ranges. In Fig. B8.1a, sites with suitable habitats for an imaginary butterfly species are shown as colonized

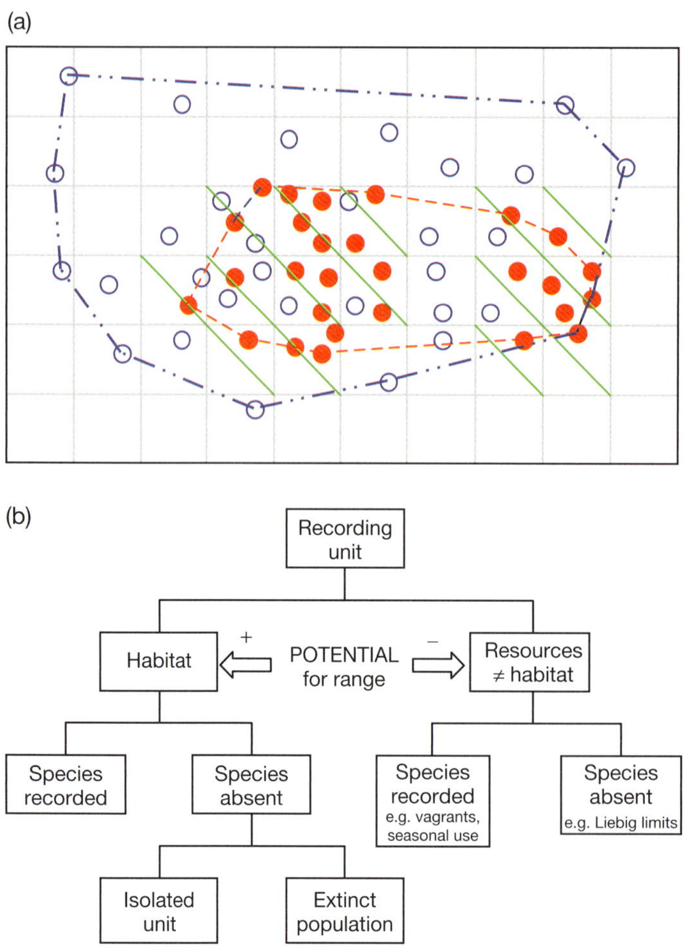

Fig. B8.1 (a) Habitat sites in a grid, which are either vacant (open dots) or colonized (black dots). The extent of occurrence (range) is illustrated as two regions drawn as convex polygons around the outlying sites: the realized (actual) range (pecked lines) and the potential range (bold pecked lines). The realized distribution is illustrated by black dots within the realized range, and potential distribution by open dots within the realized range. The area of occupancy (distribution) is illustrated by diagonal lines across squares that include the centre of black dots, indicating colonized sites. In this example, the range is actually broken into two areas of distribution enveloped by the line determining the extent of occurrence. (Modified from Gaston, 1994.) (b) Dendrogram of relationships illustrating the alternatives for sites having habitats and lacking them. Unoccupied habitats are effectively part of the potential range whereas those sites having resources that fail to make a complement (habitat) are not. Understanding ranges involves important issues of habitat. How can a site be determined to have an adequate habitat for a butterfly? If the site is no longer deemed suitable as a habitat, has this always been the case? How do we interpret consistent records of species that occupy grid squares when habitats do not appear to be available, only some resources?

Table B8.1 Distinctions of potential and realized ranges and distributions in relation to habitat parameters (resources and suitable environmental conditions).

		Environmental conditions	
		Absent	Present
Habitat resources	Absent	00 OR	01 PD
	Present	10 PR	11 RR, RD

OR, outside range; PD, potential distribution; PR, potential range; RD, realized distribution; RR, realized range. Although the realized range and realised distribution are distinguished by having sufficient conditions and resources for a species, sites exist within the realized range that fail to meet the conditions for the realized distribution.

or vacant. This helps to distinguish the realized (actual) range of a species from the potential range and the potential distribution from the realized distribution. Potential range and potential distribution identify zones that could possibly be occupied. Table B8.1 distinguishes these four terms (and a fifth one, outside range):

1 **Realized distribution**: sites where an organism is confirmed to have a population, clearly comprising both habitat (consumer and utility resources) and sufficient conditions of existence.

2 **Realized range**: the geographical bounds linking marginal sites occupied by a species within which occur sites with populations of a species.

3 **Potential range**: the bounds lying outside the realized range (for altitude as well as latitude and longitude) but which have suitable habitats (utility and consumer resources).

4 **Potential distribution**: sites lying within the realized range that may or may not have adequate habitats for a species.

The significance of distinctions between realized and potential distributions and ranges for explaining species' incidences is covered in an important paper by Jiménez-Valverde *et al*. (2008) as well as guidance given to the techniques that should be used.

Figure B8.1a also illustrates the distinction between extent of occurrence and area of occupancy; it is noteworthy that if the grid mesh was halved, the number of dots counted, depicted by diagonal lines, would increase, but the extent of occurrence would remain unchanged. Figure B8.1b illustrates some of issues underlying both measures. An assumption is made that recording accurately reflects the consistent presence or absence of species over a period of atlas construction (see Box 8.6 to see how this may often not be the case). Over a potential range, from prior autecological survey, suitable habitats (or suitable biotopes, e.g., raised bogs for large heath *Coenonympha tullia*) may be occupied or vacant. When vacant, this is often because of factors disclosed by metapopulation studies; that is, the sites have been colonized but through stochastic processes and lack of migrants from source populations have become extinct. This process is expected to be typical of the region within the actual range, caused by fragmentation of suitable biotopes and thus loss of habitat. The sites may also have never been colonized; this is more likely to be the case in areas outside the realized range but within the potential range. These are areas that could become colonized with climate change (warming). A brief consideration of potential ranges in the light of Holocene climate and vegetation changes (see Box 8.5) reveals that this is by no means certain, that large parts of the potential range may indeed have been colonized (Dennis, 1977, 1993a). The other side of Fig. B8.1b illustrates alternatives for the absence of entire habitats at sites; resources may be present but insufficient in complement to form a habitat. A species is expected to be absent. But it can be present, either as frequent vagrants or indeed because a seasonal resource is being used (e.g., a wood for hibernation by brimstone *Gonepteryx rhamni* which lacks the larval host shrubs *Frangula alnus* and *Rhamnus cathartica*). Records such as these fill the area of occupancy and are consistent with the extent of occurrence.

be recorded for a smaller region (e.g., Wiltshire) mapped for a grid of tetrads than for a larger one (e.g., Britain) mapped for 10 km squares.

Some basic principles of butterfly ranges can be enumerated:

- **P8.1: Breeding ranges of species (excluding vagrants) delineate the zone where resources and conditions comprising habitats co-occur within migration capacity.**
- **P8.2: Resources for species exist well beyond range margins and where populations are not found within range margins.**
- **P8.3: Habitats (complements of resources) for any one species can occur beyond the range margin but are not occupied owing to one or more of several reasons.**
- **P8.4: Not all the interior space within range margins will ever have had suitable habitats.**
- **P8.5: From range core to margin it is expected that the gap between suitable habitats will increase and also that migration capacity will decrease.**
- **P8.6: Range boundaries are not static; they can only be regarded as relatively static in relation to the scale of mapping: the coarser the scale of mapping the more static they appear to be.**
- **P8.7: The highly contrasting climatic (environmental) conditions at northern and southern (similarly, eastern versus western) range margins suggest that different combinations of conditions or different types of resource loss impinge on the species at different points along the range margin.**
- **P8.8: All resources for any butterfly species vary in composition and relative abundance over its geographical range. Habitat is a multivariable species-specific trait.**

The range of species, excluding vagrants, encompasses the minimal zone within which habitats occur for species (**P8.1**). Not only do resources such as hostplants exist beyond the range margin (**P8.2**), but so can entire habitats (**P8.3**). Several reasons, separately or taken together, explain why they may not be occupied; these include:

- The inability of individuals to access the site.
- The site provides too limited a habitat (e.g., area limited) or climate conditions (e.g., heat) for long-term persistence.
- An extreme event (e.g., fire, drought, rainstorm, frost) has removed the population.

Only a variable fraction of interior space, however this is scaled, within attained range margins may ever have had suitable habitats at any one time; this is expected from metapopulation dynamics (**P8.4**). This fraction will also vary among species, but on range expansion may necessarily be sufficiently close for colonization. Just what fraction of sites has resources less than the complement making up a habitat will depend on the scale of mapping (mesh size) compared with movements by individuals utilizing the resources – recall that habitat bounds are defined by an organism's mobility (see Box 2.1). It is expected that habitats will become more isolated towards the periphery of ranges than towards the 'core' (**P8.5**) inasmuch as the margin heralds conditions adverse to population persistence. But the core may not be the geographical centre (mean centre) of the range particularly if the area considered is only part of the entire realized range. Range margins are not static (**P8.6**). This should now be a familiar concept following the section on metapopulations; habitat distributions change as do, of course, occupied and vacant habitats. However, changes are not expected to match for rate or degree of change at different points along the margins as these points experience very different combinations of conditions and changes in conditions (**P8.7**). Thus, the same apparent resource (e.g., nectar flowers) may be influenced by different environmental and land use factors at northern and southern range margins. All this leads up to a crucial point, a rather frustrating one for conservation. Butterfly resources (the habitat) vary in composition (type) and abundance over a species' geographical range (**P8.8**) (J. A. Thomas *et al.*, 1999); knowledge of habitat cannot simply be acquired from studies of a single site. A wide sample of sites is required, representative of environments over which a species is distributed; only when this kind of work has been done do we start to properly understand what a species requires in different conditions. Readers can demonstrate this for themselves by making a study of larval hostplants and nectar flower sources used in different locations on different substrates.

Measuring geographical ranges

Measurement of range and exploring potential causative agents depends foremost on whether the whole global

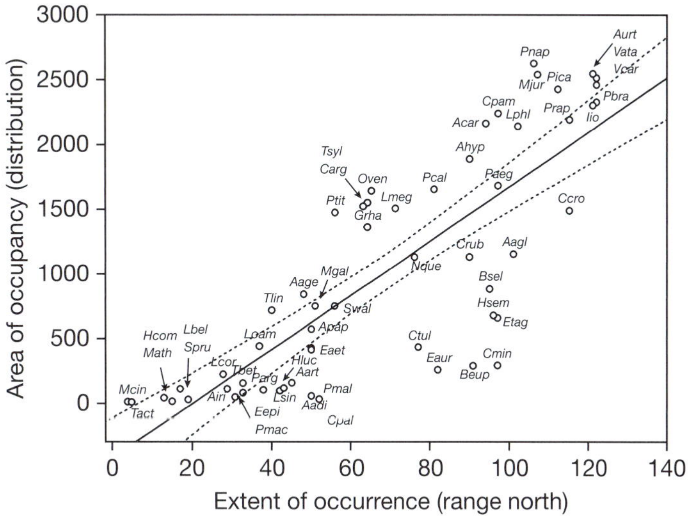

Fig. 8.1 Relationship between area of occupancy (distribution, 10 km squares) and extent of occurrence (geographical range, distance north in 10 km squares) in British butterflies (regression line with 95% confidence limits). $F_{(1,55)} = 107.1$; $P < 0.0001$; $R^2 = 66.1\%$; $N = 57$.

range of a species is being considered or just some fractional part. There is a significant correlation between butterfly ranges in Britain and Europe but this is not perfect (area of occupancy (Butterflies for the New Millennium, BNM, atlas square cover and Mapping European Butterflies, MEB, atlas 'squares'): $r = 0.74$; number of hostplants of butterflies in Britain and Europe: $r = 0.73$; both $P < 0.0001$). By focusing on a region, Britain in this case, realized ranges are being truncated and the prominence of causative indicators for range will depend on the 'selection' of species. Mapping for the BNM atlas (Asher *et al.*, 2001; Fox *et al.*, 2006) makes no distinction as to the status of the records (i.e., breeding, overwintering, vagrant, introduction) other than that between species that regularly breed in Britain and those that do not during the recording period. Ranges considered here are of 60 species (see Appendix 1) excluding those that rarely migrate to Britain (e.g., long-tailed blue *Lampides boeticus*, Camberwell beauty *Nymphalis antiopa*, pale clouded yellow *Colias hyale*).

Ranges can be measured in two distinct ways, by **extent of occurrence** and **area of occupancy** (Gaston, 1994), which figure prominently in the **International Union for the Conservation of Nature and Natural Resources** (IUCN) assessment criteria (www.iucnredlist.org/static/categories_criteria_3). The former typically measures the spatial bounds, using longitude and latitude; the latter the distributional cover. Gaston advises use of the area of occupancy to avoid artefactual correlations with environmental agents, but it is expected that the two measures will correlate across species (Fig. 8.1). In fact, these two approaches to species' geography measure different things, as Wilson *et al.* (2004) have ably demonstrated in relating the fractal dimension D_{ij} (which describes the slope of the range–distribution relationship over different spatial scales) in British butterflies to losses and gains. But, a word of caution; as the variation in common (explained) is 66% ($r = 0.82 - 0.96$, $P < 0.0001$; Dennis *et al.*, 2004), one should avoid the temptation to consider that range and distribution can be attributed entirely different causes. Extent of occurrence reflects more the limits to which a species has spread, and area of occupancy more subsequent fragmentation. But current extent of occurrence may well be an underestimate, possibly a substantial one, of historical maximal range extension, whereas the area of occupancy can obviously be affected by subsequent recolonization of sites within the range margin. It is important to understand that saturation of the range is unlikely but comes closest to fruition as a transitory phenomenon during range expansion for the reasons discussed below (Dennis, 1977).

ECOLOGICAL FACTORS UNDERLYING RANGES AND DISTRIBUTIONS

These measures of extent of occurrence and area of occupancy have been used to explore the factors underlying butterfly ranges and this section briefly outlines how the basic findings relate to butterfly habitats. Analyses have investigated ranges from the vantage of species' richness and for individual species for climate and for other factors (i.e. altitude, islands, urban cover) that have distinct geographical trends (Dennis, 1993a).

Species' richness: trends and climate

British butterflies display a distinct latitudinal trend; two independent studies based, respectively, on climate stations (Turner, 1986; Turner *et al.*, 1987) and evenly spaced grid intersections (Dennis and Williams, 1986; Dennis, 1992a, 1993a) as sampling points revealed this to correlate strongly with summer temperatures and sunshine, confirming an earlier correlation based on more limited data (Dennis, 1977). This is a tentative finding since it was not possible to fully account for spatial bias in recording which also has a latitudinal trend (Asher *et al.*, 2001), although bias was probably less in the study based on mean values for neighbouring 10 km squares at 50 km grid intersections (Dennis and Williams, 1986). A more serious problem is that many other environmental variables (e.g., altitude, islands, urban) also tend to have spatial trends – south–north and east–west – and the effect of these is extremely difficult to extricate statistically from climate influences as a number of these are linked to climate (e.g., altitude; Dennis, 1993a). Turner *et al.* (1987) attributed the latitudinal trends in species' richness to the **energy hypothesis**. This is an alluring hypothesis; ecosystems are ultimately controlled by climate (Scriber and Lederhouse, 1992), particularly solar radiation and ambient temperatures, but the reasons may be more complicated and include other aspects of climate including increased wind speeds northwards associated with the greater impact of depressions and fronts. A full account is given in Dennis (1993a:77) which suggests an increased **hazard effect** northwards as an additional burden for butterfly species at northern and western boundaries, an idea supported by more extreme population fluctuations at northern margins (Thomas *et al.*, 1994). An indication, without any attempt to be complete (Dennis, 1993a), of the ways that the heat environment may influence species' richness throughout mainland Britain is given below:

- Less adult activity (shorter flight period, mate location, oviposition), thus lower reproduction, with lower temperatures and less sunshine (Courtney, 1986).
- Adults are unable to migrate as efficiently (slower, shorter distances) in cooler, less sunny conditions, thus colonization is impeded (Dennis and Bardell, 1996).
- Slower development, thus fewer broods and susceptibility to disease, occurs with lower temperatures and sunshine (Dennis, 1993a).
- Increased torpidity with lower temperatures and less sunshine in all stages renders butterflies more vulnerable to enemies (predators and parasitoids) which may not be any less active (Pollard, 1979).

Some principles can thus be summarized:

- **P8.9: Common factors (e.g., cold, heat, moisture) are likely to impinge on different species at range boundaries, especially at extremes of latitude, longitude and altitude.**
- **P8.10: Climate is likely to be the primary factor in geographical ranges (extent of occurrence); thermal conditions are key to all butterfly activities and development. However, human factors (fragmentation) are taking over control of ranges (area of occupancy).**
- **P8.11: The key consumer resources, larval hostplants and nectar flowers, extend beyond the geographical ranges of most butterflies and rarely act as limits to range size *per se*.**

The general trend in declining species' richness northwards suggests that a common factor affecting all ecosystems and biotopes is responsible (**P8.9**). The link between decreasing species' richness northwards and summer ambient temperatures and sunshine suggests that a general factor such as climate limits populations towards the northern margin (**P8.10**). Cooler conditions (i.e., lower temperatures and sunshine and higher wind speeds and surface moisture) have multiple effects on individuals that limit reproduction, movement and survival, which in turn limit butterfly populations and makes them more prone to extinction. In Eurasia other, similar, climatic limits for butterfly species' richness are experienced to the east and south where conditions become drier and hotter (Dennis, 1993a; Hawkins and Porter, 2003). In Britain richness has been steadily declining in the eastern counties of England and this is suggested to be the consequence of intensive agriculture and chemical inputs (Warren, 1992b), a broad-scale factor that simply erases semi-natural biotopes and essential resources for butterflies (Thomas, 1984; Dennis, 1992a). One thing is clear, many consumer resources extend beyond species boundaries; for nearly all British

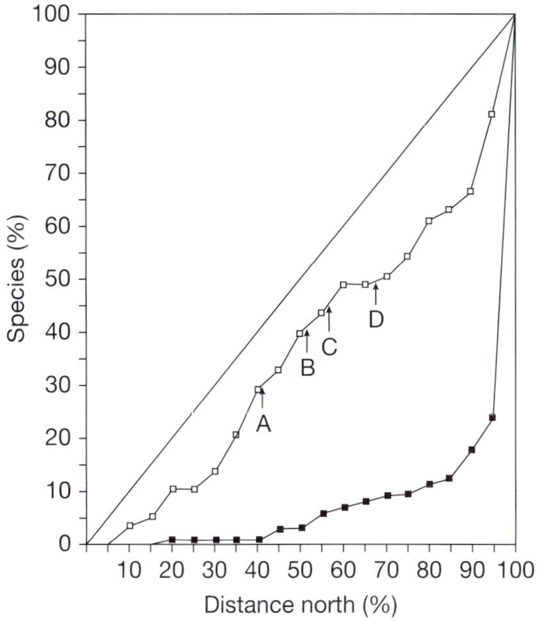

Fig. 8.2 Relationship between the geographical extent of butterflies and their larval hostplants in Britain showing Lorenz (cumulative frequency) curves comparing the two features. Open squares, butterflies ($N=57$); black squares, larval hostplants ($N=95$). The diagonal line describes an even decline in the range of species. Butterfly species decline more severely than their hostplants, most of which extend to the Shetland Isles. Only *Peucedanum palustre* (for *Papilio machaon*) (A), *Hippocrepis comosa* (for *Polyommatus bellargus* and *P. coridon*) (B), and *Frangula alnus* (for *Gonepteryx rhamni*) (C) and *Brachypodium pinnatum* (for *Thymelicus acteon*) (D) would limit the range extensions (occupation of habitats) of butterflies dependent on them into northern Scotland. Other plant species that have southern limits to their ranges are subsidiary hostplants. Note that hostplant quality is not considered. (From Dennis and Shreeve, 1991, courtesy of Elsevier.)

butterflies the key resource, larval hostplants, have much larger geographical ranges than the butterflies dependent on them (Fig. 8.2; **P8.11**) (Dennis and Shreeve, 1991; Dennis, 1993a), as do nectar sources.

In seeking solutions to factors affecting ranges of species, when considering species' richness, it is as well not to lose sight of the fact that butterflies, even in small regions such as the UK, also have southern boundaries. The latter evokes factors other than climate, or at least not the same climatic factors, in explaining ranges and suggests a more profitable approach, that of comparing data for different species. However, condition of a resource does not necessarily equate with its availability (see below).

Local population abundance–range size rule

There are few useful 'laws' in the biological sciences, but one relationship that has been elevated to prominence is that linking local population abundance and range size of organisms:

- **P8.12: The population abundance–range size rule states that locally abundant species tend to be widespread and locally sparse species tend to be narrowly distributed.**
- **P8.13: The population abundance–range size rule weakens as the fraction of species' ranges considered is increased, intimating increasing historical influences.**
- **P8.14: Range size is an inherent species' attribute relating in some way to past or present population size (abundance, density) and mobility, and reflects on species' ecology, life history and historical opportunities.**

The first issue in understanding species' range sizes is the population abundance–range size rule (**P8.12**). For a given taxonomic assemblage, there is a robust positive interspecific abundance–range size relationship (a significant, though modest, correlation) (Hanski, 1982; Brown, 1984; Gaston, 1994). Therefore, it has seemed logical to look for causes that link local population abundance for species to their geographical ranges. Gaston *et al.* (1997) examined eight explanations, two artefactual and six causal (range position–abundance relationships, resource breadth, resource availability, 'habitat' (=biotope) selection, metapopulation dynamics and contrasts in vital growth rates), the latter dynamically linking abundance to increased site occupation. In view of the modest correlations involved, there is, however, no absolute reason why current species' ranges should relate closely (dynamically) to their local population abundances although clearly the spread of individuals is enhanced by population numbers. An increase in local population size makes available, stochastically, a proportional increase in potential long-distance colonists. It would also be difficult to envisage the post-glacial spread of species unaccompanied by a burgeoning of local populations (Dennis, 1977).

Three reasons have been advanced why the range size and local population abundance may not be so closely linked now. First, there is the likelihood of non-equilibrium between species' ranges and population abundances, reflecting on the vastly different temporal scales each potentially represents. Contemporary ranges in British butterflies are the outcome of historical events, for some species possibly dating to the early Holocene (Dennis, 1977) and reflecting on ecological opportunity, but local abundances may be expected to relate more to current resource availability and the immediate carrying capacity of sites. Second, species' ranges may be affected by ecological variables that influence migration, colonization and population persistence directly (e.g., long-distance migrant species with extensive ranges are able to persist at lower population density; Bink, 1992) and independently (e.g., a species can spread if its resources are ubiquitous without necessarily the resource base being able to generate large local populations). Finally, the relationship is likely to have been affected substantially by the exponentially increasing human impact on landscape and local populations during the last 200 years, causing declining status of core populations (see Chapter 9). What is climatically marginal for butterflies can also be marginal for humans – range margins may remain largely intact where human impact has been less, whereas local population sizes may have declined at the core. The relationship between local abundance and range could be transformed, of course, with projected climate changes, as a warming climate should result in virtually simultaneous population increases and range expansions, assuming the presence of suitable habitat beyond the margins of the existing range (see below). If there is a historical component to the relationship then correlation between range size and local population abundance should decrease as larger parts of the realized range are considered (**P8.13**). This is exactly what occurs over five shifts in range size from local to global ranges in British butterflies (i.e., coefficients of determination, R^2: local 0.52, regional 0.59, national 0.23, European 0.05, while the global relationship is negative; Cowley *et al.*, 2001a). A major stumbling block is the lack of adequate measures of comparative population size; Butterfly Monitoring Scheme (BMS) data on abundances cannot be used 'raw' for interspecies comparisons as they are biased for species' apparency (Dennis *et al.*, 2006a). All that can be said with any certainty at present is that range size is an inherent species attribute relating in some way to past or present influences, but that species contrasts for range size are unlikely to be transmitted entirely through present differences in population size (**P8.14**).

Contrasts for species' geographical ranges: ecological and life history influences

From this brief view of the population abundance–range size relationship, it is to be expected that there is no simple reason for the contrasts in range size. Several studies have investigated the potential factors (e.g., Hanski *et al.*, 1993a; Quinn *et al.*, 1997; Dennis *et al.*, 2000a, 2004, 2005; Cowley *et al.*, 2001a, 2001b). These include both life history and resource variables underlying habitat requirements.

Some principles linking life history and resource requirements to geographical range size are:

- **P8.15: Species having more larval hostplants are expected to have larger ranges but butterflies with few larval hostplants do not necessarily have limited geographical ranges.**
- **P8.16: Resource variables are prominent in accounting for taxonomic differences in geographical ranges.**
- **P8.17: The capacity to migrate is a key variable in interspecies contrasts for geographical range.**
- **P8.18: Climate is a significant controlling factor of individual species' ranges.**
- **P8.19: Ecological factors and range size; beware chicken and egg situations in reasoning – the cautionary biogeographical principle.**
- **P8.20: Hostplant C-S-R strategies have relevance for the size of butterfly ranges and distributions.**

It has generally been found that hostplant and biotope generalist butterfly species have significantly larger ranges in Britain (Hanski *et al.*, 1993a; Quinn *et al.*, 1997; Dennis *et al.*, 2000a, 2004, 2005). But the most recent study demonstrates that the number of hostplants actually accounts for a small amount of the variation in range size (main hostplants 14% and all hostplants 33%) (**P8.15**) (Dennis *et al.*, 2005). Some species with limited numbers of hostplants (e.g., nettle-feeding nymphalids) requiring fertile substrates are clearly encouraged by modern agriculture. In fact, a variety of resource types (consumer and utility resources), as well as different resource attributes (e.g., hostplant distributions, density, abundance and temporal availability), figure prominently in range size (Box 8.2; **P8.16**) (Dennis *et al.*, 2004; 2005). A number of life history variables also correlate with differences in range size. With one exception, these mostly contribute small amounts of explained variation (e.g., wing span (Quinn *et al.*, 1997), voltinism and development times (Dennis *et al.*, 2004), flight period and overwintering stage (Dennis *et al.*, 2005)). The exception is migration capacity, which is perhaps a key variable (**P8.17**), the first to be entered in stepwise regressions against species' range sizes (Box 8.2). It is important to understand that mobility is not dependent on wing expanse (which alone explains little of range size, 2% in Quinn *et al.* (1997); wing size with mobility, $r^2 = 28\%$), but is

Box 8.2 Ecological and life history influences on the geographical ranges of British butterflies

Although butterfly species and their resources (habitats) cannot persist where conditions (e.g., temperatures, sunlight, moisture) are inadequate – and the factors underlying butterfly diversity suggest the ultimate influence of climate (Turner *et al.*, 1987; Dennis, 1993a) – simply ascribing cause to climate alone, or some other broadcast agent (e.g., C-S-R strategies; see Box 5.4) does not help in building a complete picture of ecological and life history factors that are proximally linked to geographical ranges. Emerging from studies of ecological associations of geographical range it is becoming evident that a complex group of factors are involved. As is evident in Fig. 8.2 (Dennis and Shreeve, 1991; Quinn *et al.*, 1997, 1998), butterfly distributions are not restricted by the geographical ranges of their key consumer resource, larval hostplants, as these extend much further north than do butterfly species. The likelihood is that hostplant quality is of critical importance. The objective of a recent study was to distinguish between the potential influences of different resources linked to hostplants and to compare them with key life history attributes (Dennis *et al.*, 2005). An important question focused on the extent to which ranges could be explained by the number of larval hostplants used by each butterfly species. Although butterfly distributions are known to be positively correlated with the

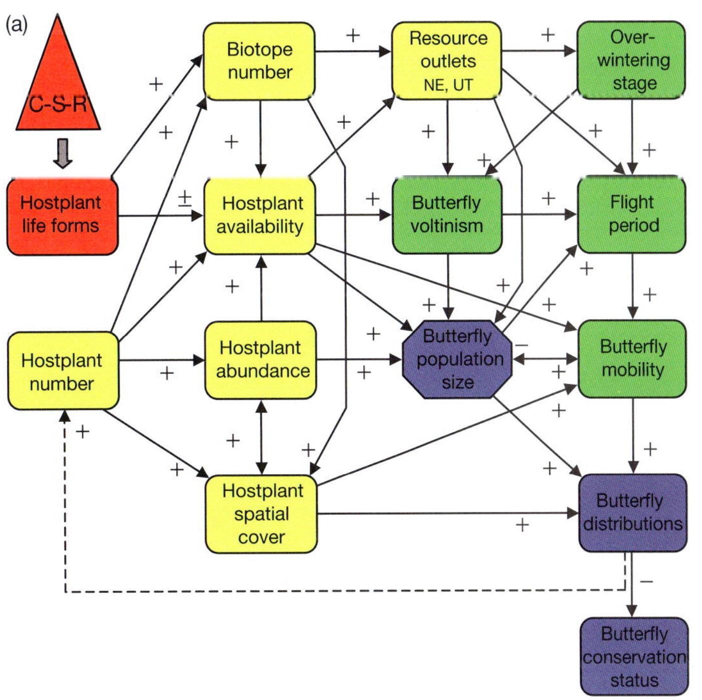

Fig. B8.2 (a) Potential causative links between hostplant generalism, resources, life history, population abundance and butterfly distributions. Signs indicate directions of expected correlations and influences; see Table B8.2a for variables. The C-S-R strategy proposes that plants acquire sets of life history attributes that reflect three distinct groups of pressure: competitive exclusion, chronic stress and severe disturbance. Red boxes, life forms (plant phenology, plant lifespan, leaf toughness, leaf nitrogen, etc.) associated with C-S-R influences (triangle); yellow boxes, resource variables associated with number of host plants; green boxes, life history variables; blue boxes, population, distribution and conservation variables. Resource outlets are complementary resources to host plants including nectar (NE) and utilities (UT). Pecked line indicates feedback between distributions and resource use. (From Dennis *et al.*, 2005.)

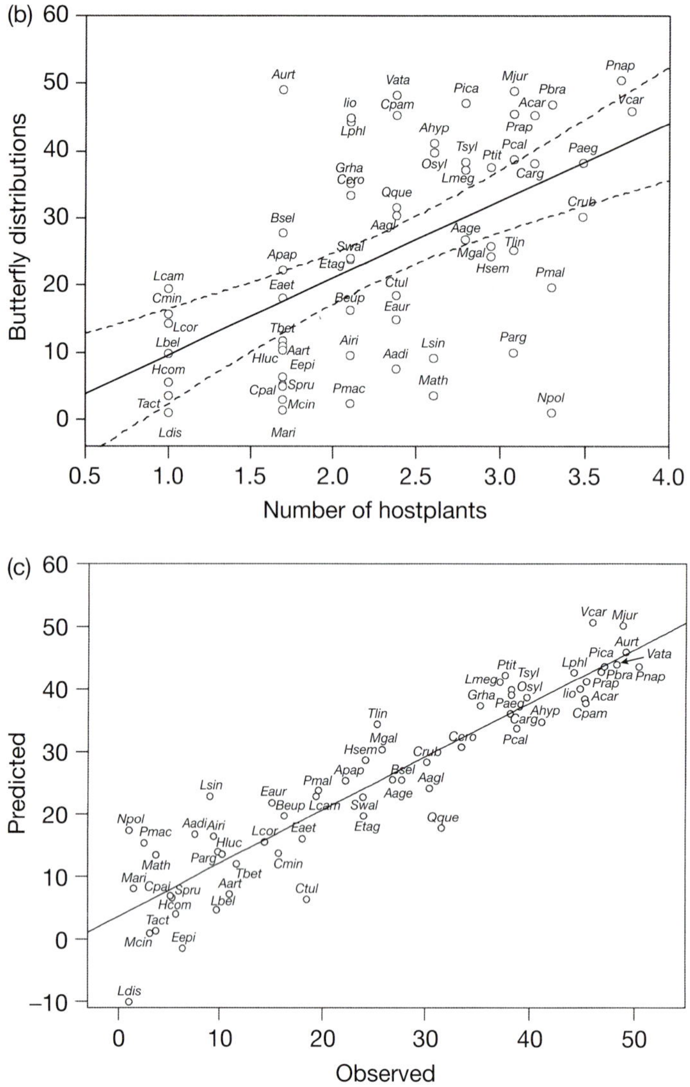

Fig. B8.2 (*continued*)
(b) Relationship between butterfly distributions on hostplant number (all hostplants). Abbreviations for species comprise the first letter of the generic (subgeneric) name and the first three letters of the species name (see Appendix 1a). Butterfly distributions are the number of occupied 10 km squares (square root transformed) against the number of hostplants (log transformed); regression line with 95% confidence limits are illustrated. (From Dennis *et al.*, 2005.) (c) Relationship between observed and expected butterfly distributions (measured as in (b)) based on resource and life history variables (all hostplants). See Table B8.2b for the parameters for the regression line. (From Dennis *et al.*, 2005, with kind permission of Springer Science and Business Media.)

number of larval hostplants used (Dennis and Shreeve 1996; Dennis *et al.*, 2000a, 2004), it was not known to what extent larval hostplant number uniquely influenced butterfly distributions and the extent to which any effects are indirect through other variables. This issue is central to understanding the part generalism and specialism in host use play in organism persistence, particularly butterfly conservation.

To answer this question, a model was constructed illustrating the connections between number of hostplants used by butterflies and their distributions, and the links between these variables (Fig. B8.2a). It identified the key variables that connect the number of hostplants used by butterflies and the size of butterfly distributions (Table B8.2a). Among a number of life history and resource variables, different attributes of hostplants were investigated: their density, abundance, spatial cover and temporal availability. Access to more hostplants was shown to affect a number of resource and life history variables impinging on butterfly population abundances and butterfly distributions. Use of partial regression analysis demonstrated that the unique contribution of

Table B8.2a Descriptions of ecological and life history variables.

Variable type	Variable
Hostplant number	Number of hostplants (main and all hostplants)
Hostplant distribution	Hostplant spatial cover (number of 10 km squares occupied; main and all hostplants)
Hostplant density and abundance	Hostplant species' density (records for 10 km squares: squares recorded/ unique squares occupied; main and all hostplants)
	Hostplant abundance in eight biotopes (weighted by percent cover in UK) (main and all hostplants)
Hostplant availability (time)	Active larval feeding period (months)
Biotopes	Number of breeding biotopes for butterfly species (main and all hostplants)
	Number of vegetation types (NVC communities) (main and all hostplants)
Nectar sources	Number of nectar sources used by butterfly species
Utility resources	Number of different utility resources used by butterfly species (e.g., mate location, roosting, pupation, hibernation structures etc.)
Hibernation stage	Hibernation stage (egg, larva, pupa or adult) (coded 1 to 4; 2.5 for *Pararge aegeria*)
Broods per year	Voltinism score (maximum number of annual broods; coded 1 to 4: single, single + part, double, multibrooded)
Flight period*	Length of total annual flight period (mean length of flight period in days)
Mobility†	Mobility (mobility score: sum of ranked scores for nine variables)
Local population abundance‡	Mean national density (BMS for three periods 1978–1982, 1980 and 1990, and mean for years) (Cowley *et al*., 2001a, 2001b)
Butterfly range size	Distribution (distribution cover: number of 10 km squares 1995–1999)
Distribution losses	Percentage of 10 km squares lost since 1982
Butterfly conservation status	Butterfly Conservation Priority (coded 3 to 1: high, medium, low) and UK Biodiversity Action Plan status (coded 3 to 1: priority, conservation concern, unclassified).

BMS, Butterfly Monitoring Scheme; NVC, National Vegetation Classification.
* Flight period data are restricted by BMS scheme starting date of 1 April and completion by 1 October in each year; data based on 59 species.
† Mobility variables included are: ex-habitat vagrants, garden records, urban central business district records, at-sea records, mass movements, range expansions, overseas migration from continent to Europe, regular reversed long-distance migration, and over-ocean (Atlantic) migration.
‡ Local population density data based on 47–49 species.

hostplant number to butterfly distributions is actually relatively small (R^2 = 14–33%) and indicated that the number of hostplants has a limited, direct impact on distributions (Fig. B8.2b). Butterfly distributions are largely accounted for (R^2 > 81%) by a set of resource and life history variables linked to numbers of hostplants: mobility is most important, but biotope occupancy, nectar sources used, utilities (the number of physical structures used by each life cycle stage) and hostplant abundance also have significant influences (Table B.2b and Fig. B8.2c). The inference drawn from the study is that, owing to independence of most of the resource and life history variables, ranges of butterfly species are not limited by the number of plant species that represent their primary consumer resource, larval hostplants. Thus, specialist butterfly species have a number of distinct outlets, via other resource and life history variables, to compensate for the lack of supplementary larval hosts enjoyed by generalist butterfly species.

The expected scenario runs as follows (see Dennis *et al*., 2005, for an account of each link in the chain): contact with new resources is enhanced by distribution size, but size of distributions will also be influenced by how many resources are used. Use of more hostplants

Table B8.2b Stepwise regression of butterfly distributions on resource and life history variables.

Regression model	Hostplant status	Variable	β	SE	R^2	CumR^2	*P*
Distribution on resources	Main hostplants	Intercept					<0.0001
		Nectar sources	0.40	0.08	49.6	49.6	<0.0001
		Utilities	0.37	0.07	14.8	64.4	<0.0001
		Hostplant abundance	0.34	0.09	7.7	72.1	<0.001
	All hostplants	Intercept					<0.0001
		Hostplant abundance	0.48	0.08	57.1	57.1	<0.0001
		Utilities	0.37	0.07	15.9	73.0	<0.0001
		Nectar sources	0.29	0.08	4.9	77.9	<0.001
Distribution on resources and life history	Main hostplants	Intercept					<0.0001
		Mobility	0.41	0.08	62.5	62.5	<0.0001
		Nectar sources	0.28	0.08	10.7	73.2	<0.001
		Utilities	0.25	0.07	5.0	78.2	<0.001
		Hostplant abundance	0.23	0.07	3.3	81.5	<0.01
	All hostplants	Intercept					<0.0001
		Mobility	0.35	0.07	62.5	62.5	<0.0001
		Number of biotopes	0.19	0.09	16.2	78.7	0.03
		Nectar sources	0.18	0.07	2.6	81.3	0.02
		Utilities	0.21	0.07	1.8	83.1	0.002
		Hostplant abundance	0.24	0.09	2.2	85.3	0.006

(generalist strategies) provides access to more biotopes and use of more biotopes provides access to more plants. Thus, increased resources permits wider distributions, and species with wider distributions will come into contact with more varied resources. This relationship evokes the **encounter–frequency hypothesis** (Southwood, 1961) in which the probability of an insect species encountering a potentially suitable new hostplant depends on the frequency of exposure to the host, with widespread plants receiving greater exposure. The relationship between host use, resource use and distribution will be influenced by life history variables: voltinism, overwintering stage, flight period and mobility. An increase in host use exposes a butterfly to a wider seasonal resource base, facilitating increases in the number of broods and therefore the range of resources and utilities that can be used, promoting wider resource use still and greater potential distributions. For example, flight period length increases the range of resources encountered (temporal component) promoting generalist strategies and widespread distributions. Mobility is influenced by use of hostplant numbers via a number of routes (see Fig. B8.2a) (Dennis *et al.*, 2003). Host use generalism should promote mobility as resources for generalists are expected to be more abundant and widespread and therefore less isolated from each other. Thus generalists, by their resource use, will tend to have open population structures. By contrast, specialists, with limited resource use, will have fewer opportunities within the landscape matrix and tend to have closed population structures (Pollard and Yates, 1993a; Dennis *et al.*, 2004).

The findings are a salutary caution against assuming that butterfly distributions, at any scale, depend purely on the number of larval hostplants or any single factor. Among species, larval hostplants are never equivalent in the resource they provide. Different butterfly species use different parts of plants, plants differing in life forms that vary in biomass, availability and larval preference. Butterflies are unable to use all hostplant areas in a landscape; some are clearly not in a suitable condition or in suitable environmental situations. Consequently, different butterfly species have access to differing fractions of their primary consumer resources.

affected by number of broods and flight period and is evolutionarily linked to resource availability in space and time. Altogether, resources alone explain 72–78% of variation in range sizes; with mobility this reaches 82–85% (Box 8.2).

It is as well to understand that investigation of taxonomic contrasts in range size presents real analytical difficulties. To mention a few: first, many of the variables are closely correlated and although procedures exist for distinguishing precedence, changes in the way variables are measured can affect results; range, as we have seen above can be measured in different ways. Second, a number of variables are difficult to measure or adequate data are as yet unavailable (e.g., temporal availability of resources types). We have no measurement of predation on butterfly species, but a high density of parasitoids has been linked to small ranges (Quinn *et al.*, 1997). A key variable(s) for individual species involves thermal ecology and climate limits, suggested to be prominent in species' richness analysis (**P8.18**; see Box 8.2). Climate (three variables) and habitat (biotope) have been shown to be limiting for individual species in studies of ranges, initially for the speckled wood *Pararge aegeria* (see below) (J. K. Hill *et al.*, 1999a; 2002). This remains one area difficult to model appropriately in comparative studies of species' ranges, as different variables will almost certainly be limiting for different species. Third, phylogenetic relationships should strictly be accounted as species do not constitute independent data points in analyses; relationships may be generated by evolutionary links rather than through individual adaptations. This may be less of a problem than initially expected as results are very similar in analyses applying phylogenetic techniques (e.g., CAIC; Purvis and Rambaut, 1995) or ignoring phylogeny. We still await an accurate phylogeny based on molecular techniques. Fourth, none of the studies is spatially explicit. Although data have been accumulated from British space, they do not represent specific sites and analysis has not been carried out on site criteria. This is crucial, as, for instance, the number of hostplants available throughout Britain may not match the number used within sites (habitats). Fifth, although it is possible to follow what appear to be plausible causative pathways linking life history, resource and population variables to range size, there are chicken and egg situations in which it is possible to reverse the logic (**P8.19**) and yet appear to make very good sense of the reasoning! For instance, an expectation is that for butterflies having small ranges and few (single) hostplants, the hostplants are limiting range size. However, butterflies may have few hostplants in Britain simply because their range has been fragmented; hostplants have been lost with loss of sites. A view on their wider continental ranges becomes essential. Finally, it is important to bear in mind that these studies look at taxonomic *differences* in range size; a butterfly's range is crucially dependent on its resources, all of them.

Digging deeper, we find that butterfly species' ranges are linked to hostplant C-S-R strategies – thus conditions affecting hostplants – via life history characteristics and resources either directly or through population status (**P8.20**; see Box 8.2 and Chapter 5) (Dennis *et al.*, 2004). Species with wider ranges (extent of occurrence) and denser distributions (area of occupancy) have C and R hostplants particularly with biased scores to rapid development, short early stages, multivoltinism, long flight periods, early seasonal emergence, higher mobility, polyphagy, wider resource availability and biotope occupancy. In population terms they have open, areally expansive, patchy population structures. On the other hand, species with narrow ranges and sparse distributions have S hostplants. These feature slow development, fewer broods, short flight periods, smaller wing expanse, lower mobility, monophagy, restricted resource exploitation and biotope occupancy. Their populations are closed, areally limited and have typical metapopulation structures. These associations are tabulated in Box 5.4.

Relationships between range size and C-S-R strategies of hostplants have a logical basis that hinges on development rate, resource exploitation and mobility and the relationship between these characteristics. Increasing S strategy is associated with limited, specialized resource use and low mobility; both these attributes encourage the development of closed populations, described as having typical metapopulation structures (Hanski and Gilpin, 1997). The low productivity, slow development rates and long development times also restrict broods and flight periods, thereby limiting opportunities for adults to exploit conditions for colonizing new sites. R and C hostplant strategists appear to have extensive distributions for different reasons. C hostplant strategy is associated with higher mobility linked to hostplant abundance. R hostplant strategists have increased resource opportunities and increased productivity, numbers of broods and increased length of flight period to exploit the more extensive, but shorter lived, resource opportunities. Mobility increases with longer flight periods (flight period and mobility, $r_s = 0.58$, $P < 0.0001$), if only because of the greater likelihood of suitable weather windows for moving greater distances occurring more frequently. This matches the need to exploit more numerous and less predictable resources, related to vagrant behaviour and patchy population structures (Cook *et al.*, 2001). These different outlets both ensure wider distributions and insure against extinction. In the case of C hostplant strategists this is reinforced by their having open population structures.

Altitudinal limits: upper and lower limits of tolerance

Change in altitude has the most dramatic effect on regional climates. With increasing elevation in Britain there is a drop in temperature and sunshine and increase in frost, cloud cover, rainfall, snow cover, humidity and wind speed. But these changes are not uniform with aspect and substrate (Box 8.3); contrasting environments (e.g., soils and vegetation) occur with changing substrates and slopes, all of which have an impact on butterfly diversity and species' incidence (see Fig. 7.13). A number of principles are evident:

- **P8.21: All species have upper, and a number have lower, altitudinal limits in their habitats associated with changing environmental conditions.**
- **P8.22: Mountains and uplands are associated with high species diversity associated with greater variation in environmental conditions and characteristically lower intensity of land use.**
- **P8.23: Different altitudinal thresholds (range limits) for species' habitats occur with aspect and latitude up mountain and upland ranges.**
- **P8.24: Rapid shifts in altitudinal range are facilitated by butterfly mobility but changes to the resource base (habitat development) are much slower.**
- **P8.25: Mountain areas provide potential refuges for butterfly species from more intense selection pressures and human activity at lower elevations.**

As already noted in Fig. 7.13, all butterfly species in Britain have upper range limits in mountains and uplands (**P8.21**), though records may be made of butterflies on mountain tops (i.e., hill-topping; see Chapter 7). These limits are typically associated with sharp changes in environmental conditions; the factors controlling upper range limits may, however, be complex and involve not just a change in climate but also in geology, slopes, soils, aspect and management. A few species also tend to have limits to their lowest elevations; the obvious example is the mountain ringlet *Erebia epiphron* which is the clearest example of a butterfly occupying unique upland conditions in Britain. Most species rise up upland slopes to reach a limit, but the habitats of some can have a bimodal distribution in elevations, as in the case of the large heath *Coenonympha tullia*, which occupies blanket bog on moors and raised bogs in valleys and lowlands. With the proximity of different environments up mountain sides, and thus biotope diversity, is expected a high diversity in butterflies (**P8.22**). On the continent, the highest butterfly diversity in Europe occurs in the Alpine mountain range (Dennis *et al.*, 1991). One observation is expected, and that is the upper range limit for species varies among upland ranges (**P8.23**), typically declining with latitude and contrasting with aspect in the same upland range. Clearly, slopes with different aspects receive different quanta of energy (see Box 2.4), but more than that, windward and lee sides of mountains experience different moisture and heat environments in relation to the Föhn effect (see Box 8.3). However, no work has been carried out to determine whether the effects of drier, warmer slopes on the lee (east) side of mountains are countered by the effects of increased levels of solar radiation on windward (west) sides of mountains; nor has a simple comparison been made of range limits on eastern and western slopes and margins associated with climate oceanicity and continentality.

One advantage of range extensions to a higher elevation compared to that of latitude is that equivalent climatic zones are closer together up mountains. Thus butterflies have ready access to new environments up mountain sides and are able to reach mountain tops (see Chapter 7). However, their capacity for colonization depends on changes to environments. In many cases, all but sufficient heat may be available to make higher sites suitable (e.g., *Coenonympha tullia*; Dennis and Eales, 1999). However, in the majority of cases, colonization may also require upward migration in larval hostplants and structural resources (e.g., shrub cover) – a much slower process (**P8.24**). A key feature of mountain and upland areas is that chemical inputs

Box 8.3 **Climatic conditions and changing butterfly incidences with higher altitudes in Britain**

Range extensions of butterflies are expected to be particularly noticeable in upland regions owing to:

- The ease with which adult butterflies can move up hill sides.
- The rapidity with which some biotopes can be converted to suitable habitat with climate change.

Uplands are associated with distinct climatic and environmental gradients (see Fig. 7.13). Below, some details are briefly provided on:

- Climatic changes over uplands and mountain sides.
- Changes in butterfly species' incidence in northern England from plain to high moorland in the Pennines.

Upland climate

Upland areas are associated particularly with lower temperatures and sunshine, and higher cloud cover, precipitation, snow cover and wind speeds. The drop (**lapse rate**) in temperature with increased elevation varies between region and with season, but in Britain is of the order of 7.1°C km^{-1} (White and Smith, 1982). With increasing altitude, the drop in air temperature relates to decreasing air density and reduction in heat retention and not to solar radiation, which can increase depending on whether or not the sky is clear. Greater cloud and precipitation is caused by increased condensation of water vapour as rising air is cooled over mountain areas. The relationship between altitude and precipitation is not a simple one; windward sides of mountains are noticeably wetter and cloudier than lee sides, which are often described as being in a '**rain shadow**'. Put simply, condensation on the windward side of mountains releases latent heat, energy added to the heating up of the air mass on descent, from compression, on the leeward side. This can lead not only to a drier lee side, but also to temporary warmer conditions at comparable elevations on the lee, termed the Chinook (snow-eater) effect in Canada and the **Föhn effect** in the European Alps (Fig. B8.3a). Aspect in relation to sun angle is still of importance for mountains as for hills in explaining thermal differences for locations at similar elevations as well as the number of snow days on mountain sides; this is strongly suggested by the northeast bias in the distribution of British cirques (i.e., cwms, corries), massive glaciated depressions, in the Lake District, North Wales and Scotland (Evans, 1977).

Regional wind speeds increase dramatically with higher elevation, some 7–9% per 100 m increase in elevation (Caton, 1976), and greatly reduce temperatures through **wind chill**. Mountains also have their own distinctive air movements relating to diurnal heating and cooling (**anabatic winds** – ascending air up hills sides during daylight hours; **katabatic winds** – descending air down mountain sides at night time). The descent of cold dense air down mountain sides at night can lead to inversion temperature layers and severe valley frosts, with potentially serious consequences for plant and animal life, including butterflies, at susceptible times in the life cycle. The classic case on the European continent is the Gstettneralm sinkhole near Lunz in Austria (Geiger, 1965). However, similar effects are known in Britain (e.g., Houghall Hollow, Durham), and even in the southernmost parts of England (e.g., in Kent) orchards are planted higher up south-facing slopes not just to catch the direct sun's rays but to avoid damaging frosts from cold air sinking down the slopes at night. In these situations, slope shape matters; the general rule is that at night, concave surfaces are cold and convex surfaces are warm. Fog also tends to develop in deep valleys (e.g., Church Stretton rift valley, Shropshire) as falling temperatures in the descending air reduces it to its dew point and the cold air cools the warm air displaced above it. These effects, together with marked variation in seasonal conditions, will persist even with climate change and should not be overlooked when considering range changes by butterflies in uplands (Dennis, 1993a). There will be shifts in the position and abundance of soil types and vegetation zones up hill sides with climate change, but zonation will still be evident. Even so, some biotopes (blanket peats, loss of sphagnum species) may well come under threat (Crowle, 2007).

Upland refuges and limits for butterflies

Conditions in upland areas impose altitudinal limits on most butterflies, in effect by marginalizing their habitats and limiting their resources. Thus, for example, Gardiner and Gardiner (2008) recently observed a limit for small heath *Coenonympha pamphilus* on the north side of Snowdon at 498 m. The limit on the southern side is likely to have been at a higher elevation, and the limit on other uplands will differ in relation to exposure and changing plant ecology. However, mountain areas can also be important refuges for butterflies and a number have minimum, as well as maximum, altitudinal limits. The classic case is the mountain ringlet *Erebia epiphron*, which is not found much below 500 m in the Lake District and 350 m in Scotland (Asher *et al.*, 2001). Variability in climate and ecology within relatively small areas provides the link between biodiversity and upland areas (Dennis, 1993a). Particularly valuable refuges in upland areas are provided by valley floors and valley sides, land that receives less chemical treatment than adjacent plains. Thus, the Goyt Valley, that runs parallel to the

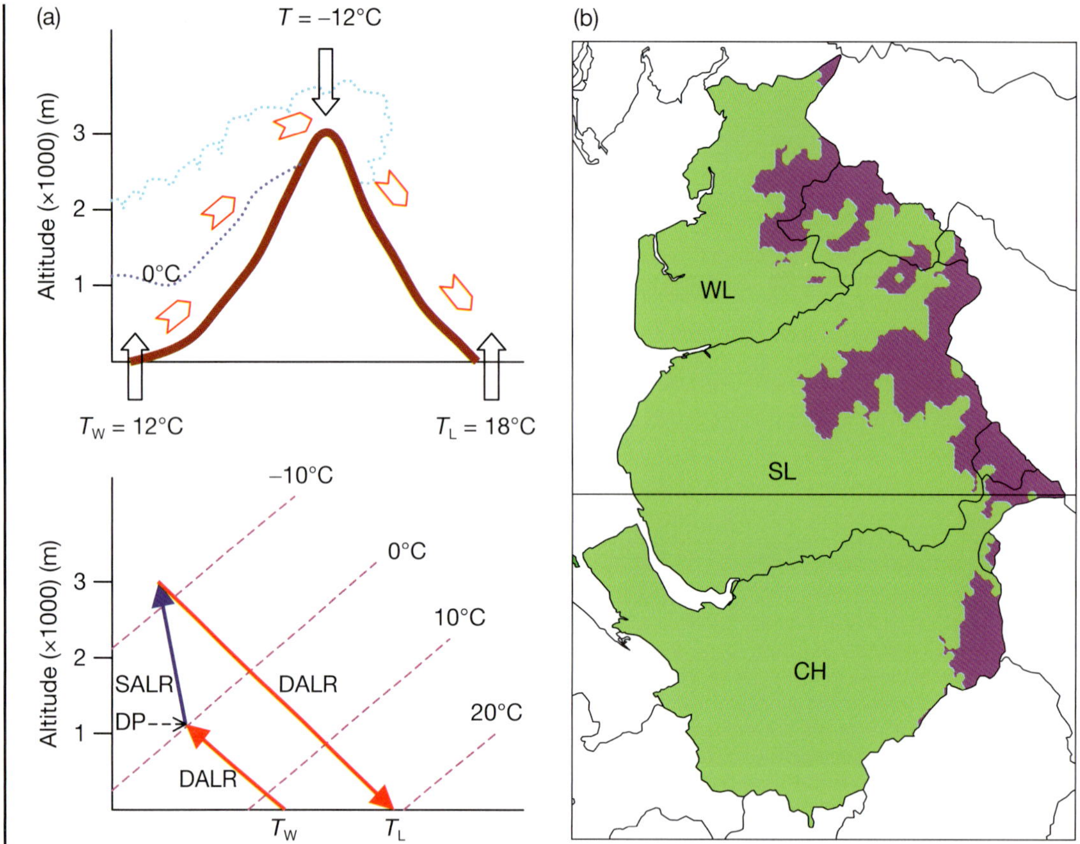

Fig. B8.3 (a) The Föhn effect over mountain ranges taken from an example over higher mountains than in Britain. DALR, cooling of mass at the dry adiabatic lapse rate (9.8°C km^{-1}); DP, dew point at condensation level; SALR, cooling at saturated adiabatic lapse rate; T_L temperature on leeward side; T_W, temperature on windward side. The rate at which temperature decreases in a rising, expanding air parcel is called the adiabatic lapse rate. Above the level at which water vapour condenses, the air cools at a lesser rate, when latent heat is liberated, at the SALR. Thus, lee mountain sides tend to experience warmer and drier conditions with cross mountain airflow than windward sides. (Redrawn from Barry and Chorley, 1982, Fig. 3.10, p. 144, courtesy of Methuen and Co. Ltd and Taylor and Francis.) (b) Northwest England showing areas under and over a minimum elevation of 150 m. Vice counties are marked: WL, West Lancashire (VC60); SL, South Lancashire (VC59); CH, Cheshire (VC58). (Courtesy of Peter B Hardy and Loren Hardy.)

west Pennine margin close to the Cheshire–Derbyshire border contains a number of species (25 species; Mallon, 2004; R. L. H. Dennis, personal observation) and has rich butterfly pastures. It also has some unusual species for upland areas; in 2008, the author recorded both ringlet *Aphantopus hyperantus* and white-letter hairstreak *Satyrium w-album* in the valley.

The distribution of butterfly resources with altitude over different uplands has yet to be studied in Britain. Yet, we will not be able to make sound predictions for future climate changes until this is done. A taste of the significance of upland areas for butterflies can be gleaned for the accompanying data on northwest England (Fig. B8.3b). Figure B8.3c illustrates changes in butterfly numbers (predictions, based on 20 visits) for two altitudinal bands, two regions (100 km grid squares SJ and SD divided by grid northing 400) and three periods. Figure B8.3d shows actual numbers for

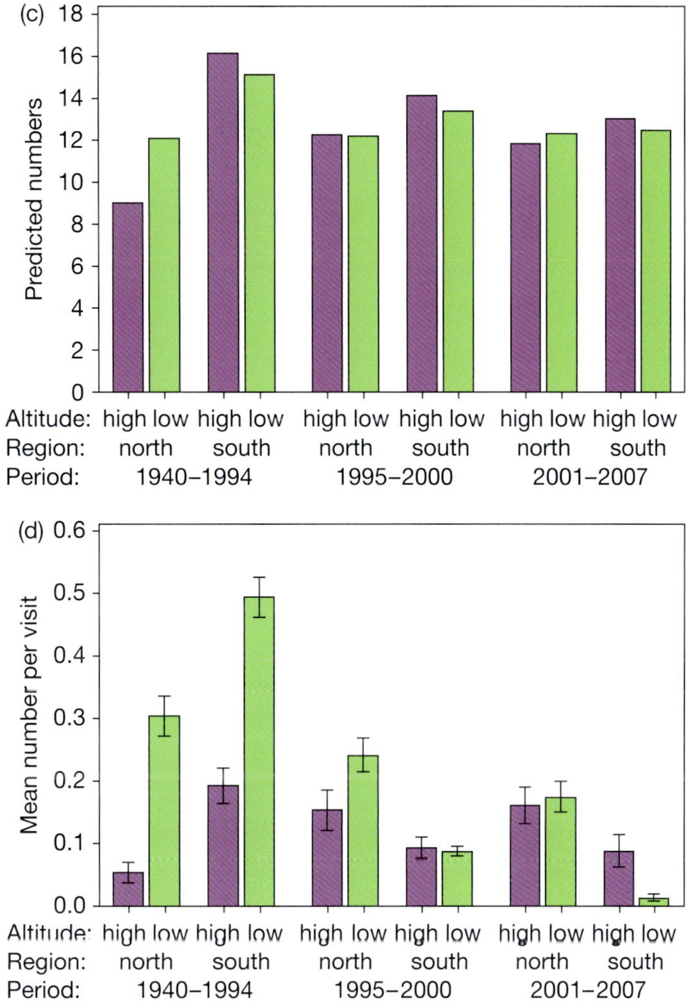

Fig. B8.3 (*continued*)
(c, d) Predicted numbers of all butterfly species (c) and actual numbers for wall brown *Lasiommata megera* (d) for upland and lowland areas (minimum altitude 150 m above sea level) north and south of the 400 northing (just north of the Mersey Valley) for three periods of recording. Predicted values are based on the number of visits (squares with less than three visits are excluded). (Data courtesy of regional recorders for Cheshire, Lancashire and Greater Manchester; analysis courtesy of Tim H. Sparks.)

20 visits for one species, wall *Lasiommata megera*, for the same classes. For the region to the north of the line it is predicted, typically, that more species will be found at lower than higher elevations in the Pennines. Interestingly, the reverse is the case for the region to the south of the line. For all three periods, more species are predicted for higher than for lower ground. The differences in the absolute numbers for the three periods may well reflect different protocols (time period surveyed and methods; Barbour, 2007) as well as recording success for visits during the three periods. The pattern is mirrored for the now rare wall butterfly but with several striking differences. First, the butterfly was far more abundant on low ground than high ground in both regions during the first period. Second, numbers for the butterfly drop, often dramatically, in later periods at all elevations except on high ground where they increase (north region) and stabilize (latterly north and south region). The bias for higher elevations in the southern region suggests a degree of stasis in upland areas, but certainly of smaller losses in upland than lowland areas of the Cheshire Plain. This is manifestly an **upland refuge effect**, a geographical bias in differential survival rates.

(e.g., nitrogen, herbicides) are generally much lower, even on lower slopes (J. G. Hodgson, personal communication). This is linked to substantial difficulties in exploitation (fields are steep and waterlogged). Thus, upland valleys and hill sides can become a refuge from human abuse (**P8.25**) (cf., Goyt Valley in the Peak District with the north Cheshire Plain; R. L. H. Dennis, personal observation). Having said that, some human constructions can be of great importance for butterfly survival at higher elevations, as in the case of outbuildings for hibernating nymphalids (Bland, 2009).

The importance of range for sourcing islands with butterfly species

The number of species found on islands has long been related to island area and isolation (MacArthur and Wilson, 1963, 1967; Simberloff, 1974; Williamson, 1983); **isolation** describes the capacity for immigration, which decreases with distance to a source for species, whereas **area** dictates potential for colonization and persistence – capacity increasing with island size. In many respects area is a surrogate for variety of biotopes and habitats and the number of habitat patches. These relationships have been explored for butterflies on British offshore islands (Dennis and Shreeve, 1996, 1997; Dennis *et al.*, 2000a, 2004) and present an interesting case in island biogeographical studies. The key variable accounting for species' richness is not area or isolation, but the size of the faunal source (62% of variation); the influence of both area and isolation are weak, jointly accounting for less than 43% of the residual variation once the contribution of faunal source has been extracted. Of course, just how effective area and isolation are likely to be depends on the range of variation for these geographical attributes. Then, unexpectedly, long-range colonizations are indicated from data on some species (e.g., grayling *Hipparchia semele*; Dennis *et al.*, 1998a, 1998b). The key variable, size of faunal source on the nearby 'mainland' large islands of Britain and Ireland, relates closely to all measures of geographical range which in turn correlate highly with ecological variables and conditions (e.g., heat, sunshine) as disclosed above. It is hardly surprising then, that the incidence of each species on the 73 islands investigated correlates closely with ecological variables describing their potential for migration (migration capacity, voltinism, flight period) and colonization (hostplant taxonomic breadth, hostplant abundance, biotopes occupied) (Dennis and Shreeve, 1997). An interesting ecological twist to butterflies on British islands, with implications for habitats, is that there is evidence for the long-term survival of species on British offshore islands. Part of the explanation for this rests on the size of patchworks and constant immigration 'rain' from nearby sources. But, islands provide unusually stable situations for some open biotopes that form the basis for butterfly habitats. Compared with average conditions on the adjoining mainlands of Britain and Ireland they are relatively immune from human impact (chemical inputs); they are also more exposed to high winds that maintain better the open seral-stage biotopes required by most species. One particular vegetation type, raised and blanket bogs – the essential biotope for *Coenonympha tullia* – is an ancient climatic climax component of many Hebridean islands (Dennis and Shreeve, 1996).

Butterfly species in cities and conurbations: changes in diversity and incidence

Breeding populations of butterfly species have long been known to penetrate to the heart of large cities (e.g., World War 2 bombsites in London; Owen, 1949, 1951), an observation confirmed by early atlas mapping (e.g., Sheffield (Garland, 1981), London (Plant, 1987)). Studies of plants in urban areas, as in Sheffield and Birmingham, have developed concepts and a language to describe the association and tolerance of plants to the human impact of urban settings (M. O. Hill *et al.*, 2002). Urban situations (presenting fertile biotopes, high disturbance, abundant nectar flowers and warmer local climates) contrast dramatically with what we regard as naturalness, often expressed as 'nativism' (Peretti, 1998), and consequently butterflies in towns have been given less attention compared with those butterflies of the so-called 'natural' countryside. Even so, there can be **urbanophiles**, organisms common in cities (e.g., plants such as rosebay willowherb *Chamerion angustifolium* L. and nettle *Urtica dioica* L) but that also occur in the open countryside. Thus, organisms can be related to **hemeroby** (measures of human impact), or more specifically with **urbanity** (measure of association with urban structures and the opposite of **ruderality**). Among plants, species that do well in cities are typically **annuals** (thus **annuality** is important; Kemp *et al.*, 2008) and/or **neophytes** (recently introduced plants, as opposed to older (i.e., <1500 AD) introduced plants or **archaeophytes**), which derives from the concept of **xenicity** (e.g., particularly alien nectar and larval hostplants; Hardy and Dennis, 2008). All these terms are immediately accessible to butterfly studies (see M. O. Hill *et al.*, 2002).

As there are many fewer butterfly species than plants, there is unlikely to be an equal range in patterns of urban affiliation among butterflies, but urban butterfly geography is becoming increasingly important as rural areas continue to be degraded. Immediate questions are: what is the pattern of species' richness in urban areas? Which species tolerate cities or benefit from resources in cities? How are these relationships affected by the location of cities? Some of these answers are evident on turning the pages of the Sheffield and London butterfly atlases (Garland, 1981; Plant, 1987). One expects diversity to decline with increased urban cover and for some species to do better than others. A simple calculation for London shows that species' richness at a 10 km scale declines sharply with urban cover (Spearman $r_s = -0.60$, $P < 0.001$; Hardy and Dennis, 1999). However, the pattern is complicated by scale, not just by map extent (area covered), but also by grain (mesh, scale of mapping or grid size). This is evident from the mapping of butterflies across Manchester (Box 8.4); three scales were selected: the whole of Greater Manchester (tetrad scale), southwest Manchester centred on Sale (7 × 5 km zone, 1 km scale) and a slice across the Mersey Valley (3 × 2 km zone, 100 m scale). This mapping has been unusual not just for its variety of approaches but also for the 'resource' or 'habitat' data accumulated at the two finer scales (hostplants, nectar sources, biotopes) as well as urban measures (Hardy, 1998; Hardy and Dennis, 1999). Three other features make it unusual. First, visits to squares that produced nil returns (i.e., **blank visits**) were recorded so that sampling bias could be assessed. Second, rates of change in richness and incidence with urban cover were tested for thirty 1 ha squares based on complete sampling (a transect covering each square walked repeatedly throughout a season). Finally, mapping of resources enabled a distinction to be made between breeding populations and vagrant individuals.

Within the two smaller areas of Greater Manchester, species' richness was found to decline with urban cover, 0.65 species per 10% increase in urban cover for southwest Manchester and 0.19 species per 10% urban cover for the Mersey Valley. The latter was affected by highly variable recording and the more detailed study using transects for 30 squares revealed species' loss rates of the same order as that for southwest Manchester (>0.67 species for every 10% increase in urban cover). With a more stringent measure of urban cover that treated gardens as non-urban, the loss rate increased to 0.9 species) (see Box 8.4) (Dennis and Hardy, 2001). What was interesting is that for Greater Manchester as a whole, species actually increased with urban cover (0.38 species for every 10% urban cover). Once sampling intensity was accounted for, this figure was found to relate to altitude, much of the bleak moorland surroundings within Greater Manchester having lower species' richness. The point is made: rates of change and differences in diversity within conurbations depend on just what is compared (Hardy and Dennis, 1999). In the case of London, the map surrounds include downland with high species' richness. The pattern of incidence for individual species varied within Greater Manchester. Many declined with urban cover but some increased along with their hostplants (e.g., holly blue *Celastrina argiolus* with holly and ivy; large white *Pieris brassicae* and small white *P. rapae* with *Brassica* crops and nasturtium *Tropaeolum* sp.) and may be true urbanophiles. There is certainly potential for urban species; city gardens in the UK increase with

Box 8.4 Factors influencing butterfly distributions in urban Manchester

The effect of urban development on butterfly species' richness and incidence has been tested for the Greater Manchester conurbation (310 tetrads) and two sample areas (southwest Manchester, 35 × 1 km squares; Mersey Valley, 600 × 1 ha units) mapped at finer scales within the conurbation (Fig. B6.4a). The study included measures of bias (recording effort), and mapping at the two finer scales included measures on biotopes and resources for butterflies, specifically nectar and hostplants (Hardy and Dennis, 1999). The impact of urban development (see text) was found to differ for the three different scales, despite the smaller areas forming nested units within the conurbation. The rate for southwest Manchester was found to describe a loss of 0.65 species for every 10% increase in urban cover, significantly higher than that for the Mersey Valley (0.19 species per 10% increase in urban cover). Two factors were considered to be influencing this difference: recording bias and failure to

Table B8.4 Regression parameters for butterfly species on urban cover for 30 sample squares within the Mersey Valley.

	Suburban cover		
Species	**β ± SE**	***R*²%**	***F* and *P***
Thymelicus sylvestris	−0.79 ± 0.115	62.7	47.12***
Ochlodes sylvanus	−0.73 ± 0.129	53.3	31.93***
Gonepteryx rhamni	−0.33 ± 0.179	10.6	3.32
Pieris brassicae	0.39 ± 0.174	14.9	4.88†
Pieris rapae	0.54 ± 0.160	28.7	11.29*
Pieris napi	−0.70 ± 0.136	48.4	26.29**
Anthocharis cardamines	−0.60 ± 0.151	35.9	15.68**
Lycaena phlaeas	−0.22 ± 0.184	4.8	1.40
Polyommatus icarus	−0.43 ± 0.171	18.5	6.35†
Celastrina argiolus	0.49 ± 0.164	24.4	9.03*
Vanessa atalanta	0.36 ± 0.176	13.0	4.19†
Vanessa cardui	0.20 ± 0.185	4.0	1.15
Aglais urticae	0.13 ± 0.187	1.8	0.50
Inachis io	−0.32 ± 0.179	10.1	3.16
Polygonia c-album	−0.22 ± 0.184	5.0	1.47
Parage aegeria	−0.62 ± 0.148	38.6	17.59**
Lasiommata megera	−0.32 ± 0.179	10.2	3.19
Pyronia tithonus	−0.26 ± 0.182	6.9	2.09
Maniola jurtina	−0.80 ± 0.115	63.2	48.13***

† $P < 0.05$, * $P < 0.01$, ** $P < 0.001$, *** $P < 0.0001$.

distinguish breeding from vagrant species in squares. The effect of differences in recording effort was directly assessed using stepwise regression and was found to be important. The importance of vagrancy was suggested by the higher correlations between species' incidence and nectar sources than between species' incidence and their hostplants.

In 1999 a more detailed study was undertaken within the Mersey Valley to assess the impact of urban biotopes and surfaces on species' richness, vagrancy and species' incidence. Systematic transects were made uniformly covering the entire areas of 30 ha squares, ranging from open areas to dense urban cover, squares that previously had been visited a minimum of 20 times (mean 121 visits) (Dennis and Hardy, 2001). Vagrancy was defined conservatively on the presence or absence of larval hostplants; thus, a butterfly habitat was assumed to be present if its hostplant was present in a square regardless of plant condition or suitability. It was shown that loss rates were as high as those recorded previously at lower resolution over wider areas in southwest Manchester (0.67–0.68 species for every 10% increase in urban cover). Loss rates are also influenced by the definition of urban cover. More stringent definitions of urban cover, i.e., excluding gardens, were associated with higher losses in butterfly diversity (0.90 species for every 10% increase in urban cover). The results linking numbers for each butterfly species to urban cover were not reported. The results are illustrated in Fig. B8.4b and detailed results summarized in Table B8.4; data for atlas records and the 1999 survey were amalgamated as no significant differences were found in rates of decline between them (Dennis and Hardy, 2001). Total number of species, and breeding species, hostplants and nectar sources, decline significantly with urban cover whereas the proportion of vagrant species increases with urban cover. As previously shown to be the case, the abundance of most butterfly species also declines with urban cover, but there is confirmation of increases with urban cover for holly blue *Celastrina argiolus* and large and small whites *Pieris brassicae* and *P. rapae*. Results for 11 of the 19 species have significant association with urban cover. Six butterfly species have a positive relationship with urban cover, whether it includes or excludes gardens, four significantly so. It is particularly interesting that the abundance of hostplants for three of the four butterfly species that increase with urban cover also significantly increase in urban areas (nasturtium for *P. brassicae* and *P. rapae*: $\gamma = 0.75$, $P = 0.03$; holly and ivy for *C. argiolus*: $\gamma = 0.48$, $P = 0.02$). In the case of *Vanessa*

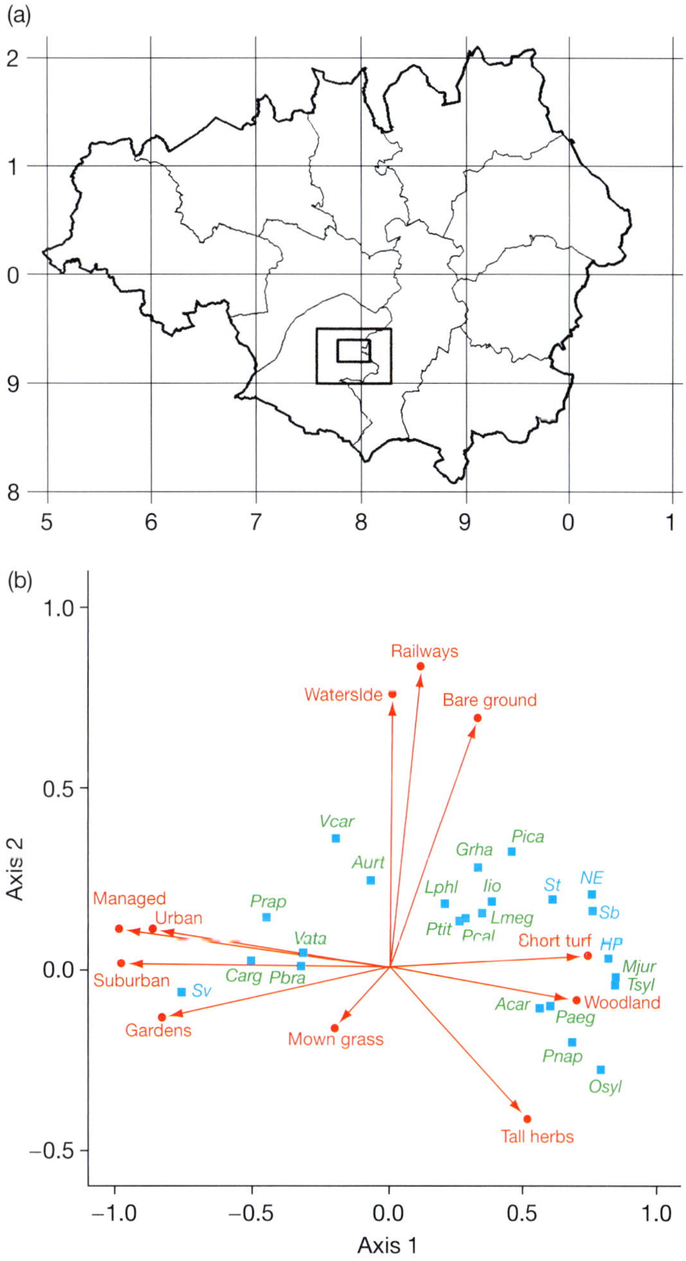

Fig. B8.4 (a) Greater Manchester, illustrating district boundaries and the two areas mapped at higher resolution: southwest Manchester (7 × 5 km area) and the Mersey Valley (3 × 2 km area) within a 10 km grid (Hardy and Dennis, 1999, courtesy of Elsevier). (b) Principal components analysis plot of the first two dimensions (axis 1: 44% variance; axis 2: 18% variance) of butterfly species within urban biotope types for 30 ha squares in the Mersey Valley. Suburban is distinguished from urban by including gardens. Butterfly species entered as supplementary variables are noted by the first generic letter and the first three letters of the species name. Data are a combination of atlas data pre-1999 and 27 transects covering the entire squares in 1999. *HP*, number of host plants; *NE*, number of nectar sources; *Sb*, breeding butterfly species; *St*, total number of butterfly species; *Sv*, percentage of species that are vagrant and lack hostplants in squares. Insets show two butterflies that have adapted well to the city environment: holly blue *Celastrina argiolus* (top) and large white *Pieris brassicae* (bottom) (courtesy of Peter Hardy).

atalanta and *V. cardui* there is a significant negative relationship with urban cover (nettles: $\gamma = -0.47$, $P = 0.005$; thistles: $\gamma = -0.64$, $P < 0.0001$), suggesting that increases into urban areas are vagrants associated with use of nectar sources on which they are more apparent. The results are particularly gratifying as most species are closely associated with their known biotope preferences (e.g., common blue *Polyommatus icarus* with bare ground, large skipper *Ochlodes sylvanus* and speckled wood *Pararge aegeria* with woodland and tall-herb grassland).

urban and human population density and occupy 21.8–26.8% of urban areas (Loram *et al.*, 2007). In resources they tend to be heterogeneous (Loram *et al.*, 2008a, 2008b) and can provide potential links between larger areas of semi-natural biotopes (Gaston *et al.*, 2007). Even so, the incidence of most butterfly species declines with urban cover and analysis of resources shows that generally urban development reduces the area of semi-natural biotopes and the abundance of both nectar sources and hostplants (Hardy and Dennis, 1999). Interesting finds are the degree of vagrancy among butterfly species within urban Manchester and the extent to which even tiny resource pockets (hostplants and nectar flowers) are used by species (e.g., small copper *Lycaena phlaeas* has been found egg laying in a tiny garden of a terraced house in Sale; P. B. Hardy, personal observation).

The findings for Manchester are paralleled by those of a study on the distribution of four species (green-veined white *Pieris napi*, meadow brown *Maniola jurtina*, gatekeeper *Pyronia tithonus*, small heath *Coenonympha pamphilus*) in relation to genetic structure within the West Midlands conurbation. An ability to move between patches for these butterflies is indicated but also limitations related to the availability of suitable habitat (Wood and Pullin, 2002). This has been tested further in the West Midlands conurbation on four additional butterflies of grassland and scrub (common blue *Polyommatus icarus*, *C. pamphilus*, dingy skipper *Erynnis tages*, green hairstreak *Callophrys rubi*; Loram, 2004); for all four species habitat quality accounted significantly for the greatest variance in distribution, connectivity very little and patch area not at all. In view of these results it is interesting that more extreme specialists are notable for their absence from city regions – specialist butterflies (S strategists) such as chalk downland blues (e.g., Adonis blue *Polyommatus bellargus*, chalkhill blue *P. coridon*) and woodland fritillaries (e.g., small pearl-bordered fritillary *Boloria selene*, pearl-bordered fritillary *B. euphrosyne*) (Plant, 1987; Fox and Williams, 2006). One final crucial point: urban environments are no more stable as butterfly habitats than rural areas; a dismal trend is the concreting over of front gardens for parking spaces and the infilling of gardens with new housing (Loram *et al.*, 2007, 2008b), the use of weed killer and excessive zeal for the 'outside' room. A recent study in Prague in the Czech Republic (Kadlec *et al.*, 2008), has demonstrated how, although species' diversity may not differ significantly over time in urban environments (30 years in their study), the butterfly community may be transformed. They found a shift towards species of taller grasslands and xeric scrub, particularly in larger reserves with a higher proportion of natural as opposed to urban perimeter in the city. This was interpreted as the result of gradual successional changes that affect the reserves despite conservation management, a common feature of rural settings.

RANGE CHANGES BEFORE RECORDS

Mapped distributions of butterfly populations and their habitats provide time slices of their geographical status, but the quality of information we have on butterfly distributions changes over time. Range changes in British butterflies fall into three broad time periods of interest:

1 Past changes before records were made.
2 Historical to current period of records.
3 Future projections from present knowledge.

Origins

Details of the reconstructed palaeohistory of British butterflies are available elsewhere (Box 8.5) (Dennis, 1977, 1992a, 1993a). Here, attention is focused on principles touching on long-term, past changes to butterfly habitats and their value for making **retrodictions** (predicting past events) underlying, as they do, species' recent geological origins and re-establishment following glacial advances. We need look only at the last glacial advance (*tabula rasa*) and retreat, the Late Devensian glaciation (18k years BP), during which the polar desert conditions were severe enough to exclude all current butterflies from present-day Britain. The basic idea in retrodictions for organisms without adequate fossil material is to match data on current habitats (resources and conditions) with those for past space/time slots; conditions for the latter are suitably assessed using data from other (fossil) organisms with contemporary populations and other environmental markers. This process is plausible because extensive data on organisms (e.g., plants, Coleoptera; West, 1969; Evans, 1975; Roberts, 1989; Chambers, 1993) with fossil and current populations reveal there to have been massive and dramatic transformations in conditions (i.e., climate, biotopes) over the past 20 000 years and equally vast changes in species' ranges.

The following principles for retrodictions can be summized:

Box 8.5 **Changing butterfly habitats during the Late Devensian glaciation and Holocene Britain**

The graphs in Fig. B8.5 cover the period from the height of the last glaciation to the modern day. During the Devensian glaciation, ice sheets extended south to South Wales and north Norfolk – the region to the south being under permafrost. With the melting of the glaciers, vegetation returned to Britain together with the butterfly species. It is likely that many butterfly species occupied the islands in the early Holocene (e.g., *Erebia* species) that were lost with subsequent forest development (Dennis, 1993a).

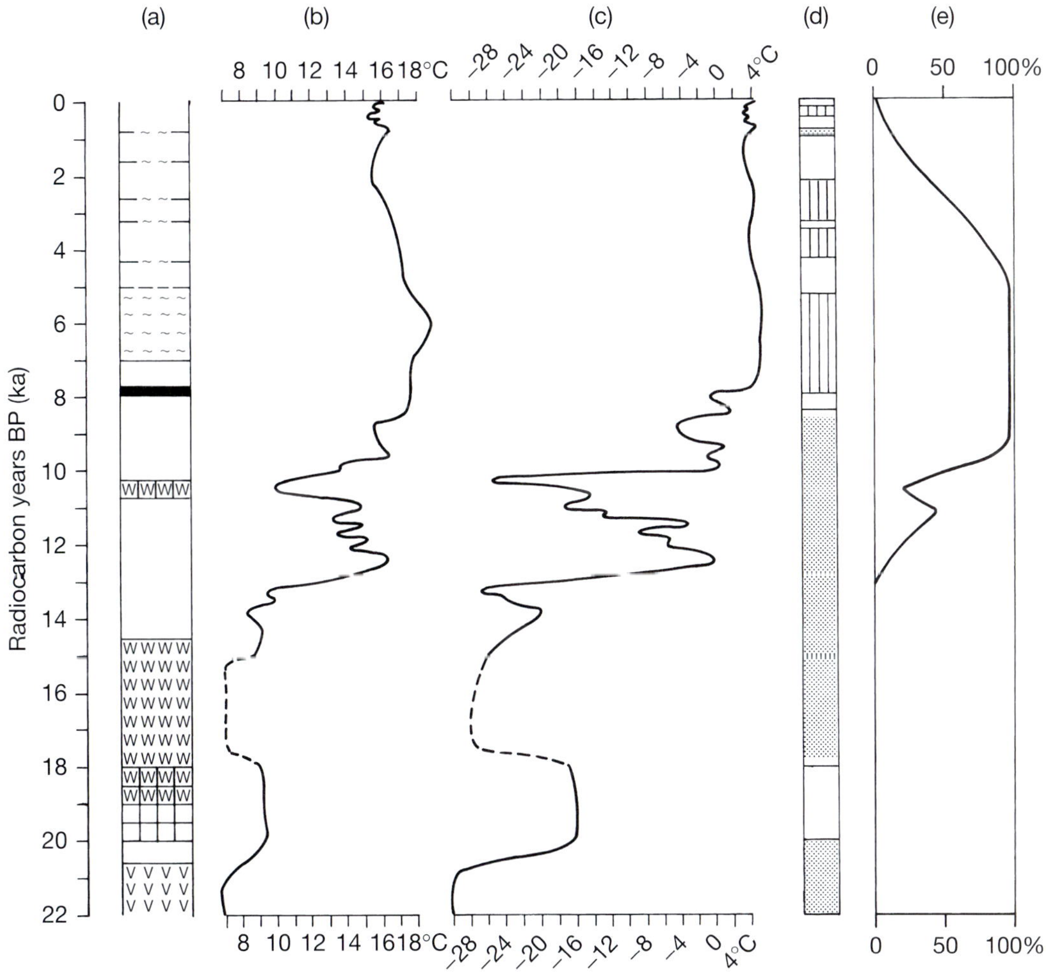

Fig. B8.5 The colonization, extinction and evolution of butterflies in Britain associated with habitat changes over the past 22k years. (a) Geomorphological events: small squares, glacial advances in Britain; V, large ice wedges; W, small ice wedges, involutions and pingos; thick bar, severance of land connection with continent; ~ ~ ~ ~ , development and expansion of ombrogenous peat and recurrence surfaces (peat regrowth) ideal for large heath *Coenonympha tullia*. (b) Mean temperature of the warmest month for central England. (c) Mean temperature of the coldest month for central England. (d) Annual precipitation. Grey shading, drier than period 1940–1970; vertical shading, wetter than period 1940–1970. (e) Forest cover.

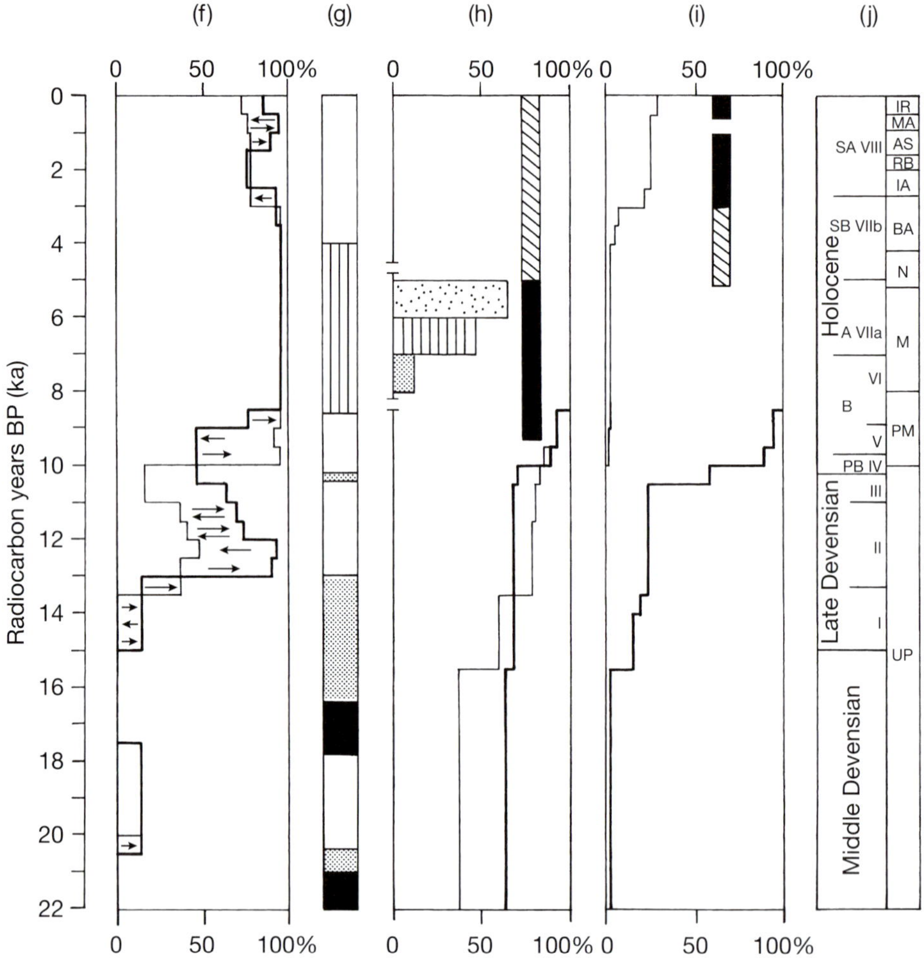

Fig. B8.5 (*continued*) (f) Tolerance curves for known resident British butterflies of mean screen temperatures for the warmest month in central England. Lower (thin) curve, for lowest temperatures per unit time; higher (thick) curve, for highest temperatures per unit time (500-year units based on 59 species). Arrows indicate direction of changes in increases or decreases in butterfly species. (g) Thermal limits for British butterflies. Vertical shading, tolerance of conditions in central England based on the mutual climatic range of species currently resident in lowland Britain (based on 55 species); grey shading, mean conditions for the coolest month probably critical for >50% of current British butterfly fauna; black, mean conditions for coolest month critical for >90% of the current British butterfly fauna. (h) Hostplant and biotope limits for current resident butterflies within Britain. Left (thin) curve, presence of habitat structural components (not symbionts), as known, for resident British butterfly species in central England; right (thick) curve, presence of hostplants for resident British butterfly species within Britain. Hostplants are assumed to persist for the entire pollen zone (Fig. B8.5j) in which they are found. The bar chart shows the species vulnerable to forest climax development: stippled, butterfly species with hostplants not a typical part of the field layer or high strata within forest and requiring open biotopes; vertical shading, species whose hostplants attain a suitable quality in the regeneration cycle of forests but require the continuous production of open spaces; grey shading, species requiring both tree or liana hostplants and biotopes. The divided bar shows the constraints for specific groups: black, period of isolation, range contraction and microevolution for species dependent entirely on open conditions; diagonal shading, population extinction and range contraction for species ($N = 29$ species) dependent on forest biotopes.

Fig. B8.5 (*continued*) (i) Arrival sequence and climatic constraints. Thick curve, proposed cumulative earliest date for the continued presence of British butterfly fauna north of the southern English coast based on climate, hostplants and biotopes (500-year periods); thin curve, cumulative curve for species experiencing upper and lower critical thresholds for mean screen temperatures of the warmest month in central England during the late Holocene ($N = 59$ species). The black bar shows the period of microevolutionary and ecological changes relating to the lower critical threshold summer thermal limits. This period may have started earlier (diagonal shaded zone) depending on the extent to which local and regional climates were influenced by human agents (i.e., deforestation, grazing, loss of cover, shelter and wind chill). (j) Chronology and human cultures. Left part, Blytt and Sernander post-glacial periods and Godwin pollen zones (PB, Pre Boreal; B, Boreal; A, Atlantic; SB, Sub Boreal; SA, Sub Atlantic); right part, cultures (UP, Upper Palaeolithic; PM, Proto Maglemosian; M, Mesolithic; N, Neolithic; BA, Bronze Age; IA, Iron Age; RB, Romano-British; AS, Anglo-Saxon; MA, Medieval age; IR, Industrial Revolution).

- **P8.26: As butterflies leave few fossil remains, historical projection of ranges requires spatial reconstruction of basic resources (i.e., hostplants, biotopes, climate) from independent fossil organisms or alternative markers for different time slots to determine probable occupancy.**
- **P8.27: A fundamental assumption of retrodictions for organisms lacking fossil evidence is that resource thresholds for butterfly species have not changed significantly during the period investigated.**
- **P8.28: Historical reconstructions should be based on resources available throughout the whole current range for the species as regional adaptations may have resulted in the loss of 'global' resource associations (e.g., British *Papilio machaon*).**
- **P8.29: Climate is likely to be the primary factor in geographical range and range change; thermal conditions are key to all butterfly activities and climate change can both be rapid and regionally extensive.**
- **P8.30: Most species are expected to shift their distributions to keep track of climate changes rather than adapt *in situ*.**
- **P8.31: Geological and archaeological principles apply (e.g., failure to find a resource in a deposit is not absolute proof of its absence).**
- **P8.32: Butterflies may have occupied biotopes (vegetation categories) and climates (combinations of climate variables) no longer currently available.**
- **P8.33: It is expected that during expansion of species' ranges (post-Devensian) the capacity to migrate will increase with range expansion, thereafter decreasing when a limit has been attained. These changes are driven by selection and population parameters.**
- **P8.34: Range expansion in a species will have broad biotope associations inasmuch as key resources tend to be linked to different seres (vegetation succession stages).**
- **P8.35: Range changes can be instigated by non-climatic factors such as management and land use, since these directly affect vegetation content and structure, thus the resource base, and with broad economic changes which can affect extensive regions simultaneously.**

It is inevitable that the reconstruction of past ranges and distributions prior to records being available is based on a number of assumptions. The first is that current habitats (resources, conditions) for butterflies can be reconstructed for past space–time frames from data on other, fossil organisms and environmental markers (e.g., deposits, fossil landforms such as ice wedges) and can be adequately dated. Ideally, retrodictions would be based on all resource requirements, but as these are poorly known in many cases and lack transference back in time, three basic resources are assessed, viz., thermal conditions (summer and winter temperatures), hostplants and biotopes, the latter a surrogate for other resource requirements (**P8.26**). The errors involved depend much on the resolution of time units chosen; the latest and most detailed reconstruction (Dennis, 1993a) applies 500-year units, which adequately embraces errors around C^{14} dating estimates back to 10k years BP, calibrated as they are using dendrochronology. A second assumption is that habitat criteria and condition thresholds for butterfly species, taken 'globally', have not changed significantly over the period investigated (**P8.27**) – the last glacial stage and Holocene. For instance, it would be a mistake to base the history of the swallowtail *Papilio machaon* on its current resource limitations within the British Fenland (Dennis, 1977). Thus, retrodictions are suitably founded on resource use and adaptations throughout the current range of each species (**P8.28**); more conservative estimates could now be made as for future predictions based on logit regression analysis (cf., J. K. Hill *et al.*, 1999a).

Although three types of resource have been used in reconstructions, climate is regarded as the most important variable for several reasons (**P8.29**):

• From work on beetles *most* species are expected to shift their distributions to keep track of climate changes rather than adapt *in situ* (**P8.30**) (Coope, 1987; see also the next section which is a contradiction of this principle).

• Sunshine and temperatures are key to all butterfly activity, survival, reproduction and development (see above).

• Data on specific hostplants are susceptible to error and to the basic geological principle that failure to find a hostplant in a deposit (macroscopic remains) is not evidence of its absence (**P8.31**), nor is the presence of pollen proof of its local occurrence.

• Butterfly species may have occupied different biotopes (vegetation assemblages) and climates in the past; habitats could have been incorporated in different vegetation associations (**P8.32**).

Undoubtedly, the magnitude of changes occurring in conditions lie outside our recorded experience. Dennis (1977) explained how migration capacity may well have been much greater with de-glacial range expansion – an accelerating front (**P8.33**), because of two factors:

1 Hostplants lay well to the north of species' climatic limits.

2 Shifting range margins would have been occupied by individuals having the greatest migration capacity.

This is now evident from recent empirical work comparing species in intact and fragmented landscapes (e.g., Merckx *et al.*, 2003; Schtickzelle *et al.*, 2007) and range edge and range centre situations (e.g., Komonen *et al.*, 2004). The latter study in Finland demonstrated a lower migration rate in species that have restricted niche breadth, low resource availability and those on the northern edge of their geographical range where resources are now sparse. In such species, selection against mobile individuals is likely to result from the decreased probability of finding another habitat patch suitable for egg laying. With saturation of habitat space, the advantages of large migration capacity would be lost as migrants would be arriving at already occupied habitats and competing for exploited resources. There are likely to have been exceptions to this range expansion model. Colonizations can only operate as fast as there are resources to support them. Although herbaceous hostplants occurred well to the fore of the post-Devensian butterfly migration front (Dennis, 1977), shrub and tree hostplants were well in arrears. In the case of species dependent on plants of later successional stages (seres) (e.g., purple hairstreak *Favonius quercus*), these species would have been more constrained by the occurrence of their hostplants, and where hostplants have important vegetation associations holding other resources (i.e., for roosting, mate location) butterfly species might be more limited by whole vegetation communities (e.g., purple emperor *Apatura iris*) (**P8.34**) (Dennis, 1993a).

Even so, not all range changes are instigated by climate change (**P8.35**). Massive range retractions with fragmentation of habitats in eastern England is a prime example (Dennis, 1992a; Wilson *et al.*, 1999), as is the lag in range expansion where suitable habitats are fragmented (e.g., *Pararge aegeria*; J. K. Hill *et al.*, 1999a). Massive biotope changes accompanied Late Glacial and Holocene environmental changes (see Box 8.5), undoubtedly generating large range shifts with forest development and subsequent forest removal starting with Neolithic landnám clearances and agriculture. During the last 50 years, particularly, the decimation of all wildlife with modern industrial farming and rapid urbanization is the newest and most pernicious phase.

Establishment

Owing to population growth, dispersal, panmixis and widespread recombination of genes, colonizing species were probably highly varied (Dennis and Schmitt, 2009). This would have allowed them initially to exploit the fast developing landscape of north Europe and subsequently to adapt to changes. Establishment after colonization involved adaptation to changing conditions. Some principles apply:

> • **P8.36: Habitat bottlenecks occur following arrival (a wave of colonizations) with the establishment of organisms resulting in site and regional specializations.**

• P8.37: Products of habitat bottlenecks are distributional changes, ecological adaptations, life history changes, and changes to genetics and phenotypes.
• P8.38: Extinctions at different scales (mass extinction, taxonomic extinction, regional extinction, population loss) have been consequent on habitat bottlenecking.
• P8.39: Local and regional adaptations (i.e., life history, ecology, distributions, phenotypes, genetics) provide clues as to the nature of habitat bottlenecks for species after arrival.
• P8.40: Changes to populations within habitats (genetics, phenotypes, life history, host use, distributions) can be extremely rapid, occurring in a matter of years or broods.
• P8.41: Human exploitation of landscapes throughout the Holocene is key to understanding a number of habitat bottlenecks experienced by butterflies. The impact is now exponential and threatens regional extinction of the butterfly fauna.

Changes in Late Glacial and Holocene Britain are discussed in detail elsewhere (Dennis, 1977, 1992a, 1993a). The issue here is that just what happened depended on bottlenecks in habitats. Bottlenecks to habitats occurred very rapidly following colonization of the British landmass in the Late Glacial and Holocene, caused by changes in climate, in ecosystems (i.e., biotopes and associated resources), sea levels (marine transgressions) and human activity (**P8.36**). Following multispecies' mass range expansions, populations adopted specializations in relation to changes within habitats as they adapted to changing conditions, which become more severe towards the end of the interglacial stages. In effect, what happened was a loss in the number and size of habitats, with changes to their composition, expanse, structure and connectivity ultimately affecting species' ranges, distributions and adaptations (**P8.37**). During the Devensian maximum glaciation, these changes to habitats were large enough to result in regional extinctions and even species extinctions (**P8.38**) (Dennis *et al.*, 1991, 1995, 1998c, 2000b). The most severe bottlenecks are illustrated in Box 8.5. Three climatic bottlenecks occurred for warmth-loving butterfly species:

1 The Devensian maximum glaciation around 18k years BP.
2 The Younger Dryas or Loch Lomond Readvance about 11–10k years BP.
3 Climatic deterioration for a period following 4000 years BP.

Three bottlenecks also occurred for butterflies of cold-associated biotopes (e.g., *Erebia epiphron*):

1 The early Holocene (post 10k years BP).
2 The Medieval period.
3 The current period of climatic warming.

A habitat bottleneck for species of open biotopes occurred with forest closure (vegetation succession) after 9000 years BP (see Bennett, 1989, for forest cover types in Britain at 5000 years ago) and a bottleneck for forest floor and canopy butterfly species (human activity) followed the Norman Conquest and the decimation of forests and abandonment of ancient woodland practices in recent history (see Dennis, 1977, 1993a, for details of refuges). A further bottleneck was created by sea level rise and the fragmentation of the present British landmass into islands; islands reduced in size lose habitat patches and resource variability and increasingly depend on repeated colonization from mainland sources (Dennis and Shreeve, 1996, 1997).

Each bottleneck tends to leave its fingerprint; various adaptations, be they distributional, ecological, life history, phenotypic or genetic, provide clues to the nature of the bottleneck (Dennis, 1977) when assessed in the context of a species' resource requirements and environmental tolerances (**P8.39**). Casually speaking, the appropriate *modus operandi* is first to determine the period over which a species could have persisted within the area and then to search for events that approached or broached thresholds for the attributes presented (Dennis, 1977). Habitat bottlenecking can occur with or without apparent significant reductions in ranges or distributions. The gradients in phenotypic modifications among species in northwest Britain (Dennis and Shreeve, 1989) indicates changes to 'resources' and adaptation to them despite distributions being more or less continuous (e.g., *Maniola jurtina*; Thomson, 1987). These changes are linked to contemporary climate gradients (Dennis, 1977:201) and were subsequently modelled in relationship to three pressures on butterflies: successful thermoregulation, mate location and predator escape – the outcome depending on resting posture and wing energy absorbance and reflectance (Dennis and Shreeve, 1989; Dennis, 1992a, 1993a). This model describes compromises necessary for enhancing these activities in habitats with variable heat and sunshine. It has been tested and substantially refined for *Polyommatus icarus* (Howe, 2004) and applies well to wing phenotypes for *Coenonympha tullia* (Dennis *et al.*, 1984, 1986). Gradients arising from habitat bottlenecks undoubtedly exist in other attributes, in ecology (e.g., chequered skipper *Carterocephalus palaemon*, hostplant and nectar use) and life history (e.g., *Coenonympha pamphilus* and voltinism; ringlet *Aphantopus hyperantus* and size; *Pieris napi* and migration; Dennis, 1977, 1993a), but ecological aspects have yet to be studied in any detail.

Habitat bottlenecking is often accompanied by an assortment of specializations, among them characteristic distributions and ranges. The most distinctive distributional and range retractions occur on offshore islands (e.g., *Maniola jurtina cassiteridum* on the Isles of Scilly) and within ecological islands, for instance on mountain tops (e.g., *Erebia epiphron*), and on distinctive substrates (e.g., *Erynnis tages bayensis* on the Burren limestone), the former from rising sea levels and the latter from forest encroachment and climate change. Distribution changes are accompanied by ecological associations generated, for instance, by geology, topography and vegetation succession with restrictions to extreme substrates (e.g., large copper *Lycaena dispar* to Fenland peats; Duffey, 1968), slopes with southerly aspects (e.g., Glanville fritillary *Melitaea cinxia* on the Isle of Wight; Thomas and Simcox, 1982) and persistently maintained early seres (disclimaxes, plagioclimaxes) (e.g., large blue *Maculinea arion* to rabbit-grazed downlands; Thomas 1977a, 1977b, 1978, 1980a, 1980b).

An extraordinary example where habitat bottlenecking has impinged on a number of attributes is that of the Great Ormes' *Hipparchia semele* and silver-studded blue *Plebejus argus*. These two taxonomically unrelated butterflies are dwarfed, emerge early in the year and occupy similar distributions on the Welsh headland despite having very different hostplants and occupying different biotopes. Originally, these local populations were thought to be distinctive subspecies and glacial relicts (Beirne, 1943, 1947; Ford, 1945) but from matching current habitats with past conditions, arguments for more recent origins have been advanced pointing to the contemporary significance of their adaptations relating to the topography, geology, local climate and land use of the Orme and its Holocene history (Dennis, 1971, 1972a, 1972b, 1977, 1992a). Habitat bottlenecks have had a profound effect on the British butterfly fauna; there is evidence that British species which belong to the widest faunal element (extent unit) in Europe (Dennis *et al.*, 1991) have become specialized compared with conspecifics on the continent (e.g., comma *Polygonia c-album*; Nylin, 1988) using fewer hostplants and occupying narrower biotopes (Dennis, 1977). Over the years that British butterflies have been investigated, perhaps the greatest shift in perception has been in relation to how rapidly changes in butterfly biology can occur with habitat changes (**P8.40**). Prior to 1971, it was thought that significantly distinct phenotypes indicated differentiation at subspecies level (Beirne, 1943, 1947; Ford, 1945, 1964; Heslop Harrison, 1950). But in 1971 it was demonstrated that substantial, highly significant, differences in butterfly populations can occur within 30 years (Dennis, 1972b). These differences were demonstrated in populations of *P. argus*, 90 individuals of which had been transferred from the Great Orme to the Dulas Valley over 14 km away near Abergele in 1956 (Merchant, 1956) (see Fig. 9.5).

Throughout the Holocene, butterfly habitats (resources and conditions) have been vastly altered by human activity (**P8.41**) (see Chapter 9). Humans are responsible for two major habitat bottlenecks in historical times, deforestation and what can be most simply coined as industrialization – recent industrial-scale farming and urbanization. However, early human activity may well have provided the conditions – short turf biotopes – that ensured the survival of many of our rare grassland species (e.g., *Polyommatus bellargus*, *Maculinea arion*) that required warmer habitats during the climatic downturn after 4000 years BP (Thomas, 1993). Through vegetation clearance, initially for hunting, later for farming, they provided the open biotopes for species dependent on herbaceous hostplants. The irony is that species whose survival depended on open habitats generated by human activity are now being exterminated by a combination of abandonment and more intensive techniques to exploit the same land.

PRESENT AND FUTURE DISTRIBUTIONS: CLIMATE AND LAND USE CHANGES

It is one thing to demonstrate range (or distribution) changes and another to explain them. It is, in fact, no simple matter to demonstrate a change in the geography of a species because the evidence is based on records, and both recording intensity and geography can change too. The nature of sampling protocols is particularly important in these studies. Even so, as over other parts of Europe (Van Swaay and Warren, 1999; Maes and Van Dyck, 2001), there has long been compelling circumstantial evidence for changing geography in British butterflies, some leading to extinctions (e.g., *Lycaena dispar*; Ford, 1945; Emmet and Heath, 1990), but some to range expansions (e.g., *Polygonia c-album*, *Pararge aegeria*, wall brown *Lasiommata megera*; Dennis, 1977; Emmet and Heath, 1990). More recently, evidence for substantial, rapid range expansions during the late 1980s and early 1990s has accumulated (e.g., small skipper *Thymelicus sylvestris*, *P. aegeria*, *Pyronia tithonus*; Hardy *et al.*, 1993) in relationship to what has been regarded as climatic warming (Dennis, 1993a). Outlying populations of *Aphantopus hyperantus* (e.g., Camore Wood, Dornock, Sutherland, UK; Patterson and Wells, 2006) have been found to support range expansion in other species. These observations have led to formal attempts to describe and predict losses (C. D. Thomas *et al.*, 2004; J. A. Thomas *et al.*, 2004; Franco *et al.*, 2006) and gains (Parmesan *et al.*, 1999; J. K. Hill *et al.*, 2002; Hickling *et al.*, 2006). In Britain, this has depended on processing (e.g., subsampling) the records owing to disproportionate 'sampling' of squares for different time periods. It is important to understand, then, that claims for numbers of species in grid squares, their extent of contraction and expansion, depends on how data are processed, complicated by the fact that protocols for recording have also changed with time, from one study to the next. With this proviso (Barbour, 2007), recent range and distribution changes in British butterflies are described in Box 8.6.

As a cautionary note, bearing in mind the importance of making correct predictions for the future, some of the key working principles are outlined as follows:

- **P8.42: Although range expansions are expected to be accompanied by distribution infilling and population increases within the same region, and the reverse, there may well be regional independence for these processes, and conflicting processes within regions owing to species' individual responses to environmental changes.**
- **P8.43: Rates of gain and loss in ranges and distributions may be affected by relative scale dependency of colonizations and extinctions.**
- **P8.44: Comparisons of atlas records are susceptible to misinterpretation owing to different levels of recording between time periods and between regions.**
- **P8.45: No protocol exists for dealing with 'zero' records (blank visits) or unsuccessful observations for squares. This adds to the uncertainly of negative records for species.**
- **P8.46: Means for climatic variables (e.g., monthly means) can only be crude surrogates for actual climatic agents affecting species' populations, distributions and ranges. Species may respond to short-term extreme events with long-term consequences.**
- **P8.47: Biotopes are surrogate variables for habitats and unlikely to be entirely satisfactory ones.**
- **P8.48: Modelling range changes is an inexact science because of data restrictions and difficulties in simulating reality.**
- **P8.49: Rates of range expansion for future climate changes cannot be predicted/projected accurately from past changes owing to poor recording/poor data on expansions and standstills, and conditions occurring outside those experienced.**

Box 8.6 The rise and decline in British butterflies under climatic change and habitat destruction

Currently British butterflies are subject to two potentially conflicting processes, shifts in geographical ranges and fragmentation of biotopes (habitats) (Warren *et al.*, 2001b). Summarizing the product of both processes over the British Isles, across different species, depends on comparing records obtained for two or more time periods. This can only be achieved directly if recording is consistent and unbiased over space and time or, at the least, correction can be made for any bias. Unfortunately, this has not been the case and attempts to correct for alleged differences in sampling intensity between earlier and later atlases have resulted in further bias, primarily because intensity of recording for the period prior to 1983 was not fully documented (Barbour, 2007). Records of species, and their habitats, for spatial units contain a number of potential errors. Even when a

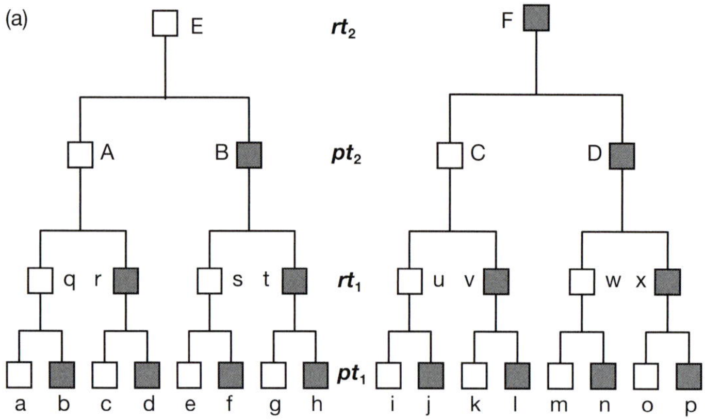

E. Species recorded as currently absent

A. Species actually currently absent

(a) NEVER PRESENT: species accurately recorded as absent in both time periods. Continual lack of butterfly habitat accurately recorded

(b) MISSED LOSS: species was there and missed. Inaccurate recording of past butterfly habitat.

(c) FALSE LOSS: species was not there but mistakenly recorded as present in earlier recording period. Habitat loss falsely recorded

(d) TRUE LOSS: species was present and accurately recorded as a loss. Real habitat loss recorded

B. Species actually currently present

(e) MISSED GAIN: species present in second recording period but missed as a gain during recording. Habitat development missed by recording

(f) MISSED CONTINUOUS RECORD: species there on both mapping occasions but missed. Continuous habitat for species unrecorded

(g) FALSE LOSS FALSELY RECORDED: species mistakenly recorded during first period is missed during a second period; a gain is registered as a loss. Habitat development missed

(h) APPARENT LOSS for a CONTINUOUS RECORD: continuous habitat registered as lost

F. Species recorded as currently present

C. Species actually currently absent

(i) FALSE GAIN: species never present but mistakenly recorded in second period. Habitat development falsely recorded.

(j) LOSS FALSELY RECORDED AS A GAIN: species missed in first period is mistakenly recorded during second period. A habitat gain falsely recorded for a habitat loss

(k) CONTINUOUS ABSENCE FALSELY RECORDED AS CONTINUOUS RECORD: habitat considered to be present where one has never existed (sounds amazing but happens!)

(l) MISSED LOSS: species loss by second recording period missed. Loss of habitat falsely recorded

D. Species actually currently present

(m) GAIN: species and habitat gain accurately recorded

(n) FALSE GAIN: species and habitat continually present but registered as a gain

(o) FALSE CONTINUOUS RECORD: species and habitat gain falsely recorded as continuously present

(p) CONTINUOUS RECORD: species and habitat accurately recorded as continuously present

Fig. B8.6 (a) The status of butterfly records and changing habitats. Butterfly records are affected by mistaken identifications as well as errors in data files. Comparison of records between two time periods, in any spatial unit, establishes 16 possible combinations of real to perceived records for species in any recording unit (grid square). In the figure t_1 and t_2 are atlases at times 1 (e.g., Heath *et al.*, 1983) and 2 (e.g., Fox *et al.*, 2006), *r* are records in the atlases, and *p* the true status of records (what is actually present or not present). A grey square denotes presence and a white square absence of a species. For simplicity it is assumed that the presence of a species indicates the presence of habitat for that species. Interpretations for the combinations of records labelled 'a' to 'p' for the status of species and habitats in squares are listed below. Errors are basically of two types: (i) failure to record a species when it occurs (b, f, j, n for t_1; e to h for t_2); (ii) recording a species when it does not occur (c, g, k, and o for t_1, i to l for t_2). The first is far more likely to happen than the second, owing to survey failures (inadequate sampling of squares in space/time, observer deficiencies, low population numbers during visits), but the second also occurs (identification errors, recording errors, database errors). Errors increase further if the records for butterflies are of vagrants – that is if observations on butterflies convey no information about the behaviour of individuals (e.g., females egg laying, mate location). Comparisons of records for any two time periods typically consider only a, m, d and p (Table B8.6).

one-to-one correspondence is assumed between a record for a butterfly species and the existence of its habitat, a number of mistakes can confound the interpretation of butterfly distributions and habitats. Some are illustrated in Fig. B8.6a; of 16 possible combinations of records for two time periods, registering assumed and actual presence, 12 errors are possible. The outcome is that butterfly habitats may occur and not be recorded and may not occur but be recorded. Some of these errors may appear to be fanciful; the depressing truth is that they occur more often than may be expected (P. B. Hardy, personal observation).

If records are not logical indicators of habitats, but of vagrants or migrants, then further error is possible and, as explained in the text, such errors are increasingly frequent the finer the grid used for comparison. In fact, with increased sampling, more vagrant butterfly species visiting a spatial unit will be observed and spatial units lacking habitats will gradually accumulate a list of species records found in units that have habitats. This is explained by the finite number of the regional species pool – there is a limit to what can be recorded over time. Although species are rapidly accumulated for units with habitats, after a certain number of surveys have been

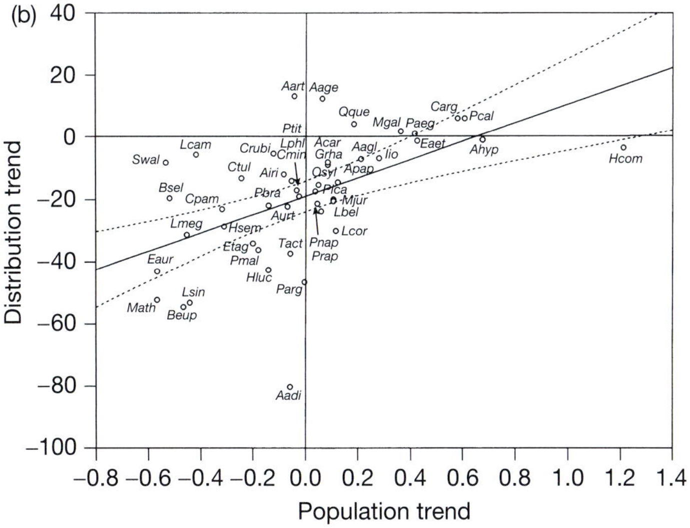

Fig. B8.6 (*continued*) (b) Relationship between long-term distribution changes and population changes in British resident butterflies. Distribution trends (regression line with 95% confidence limits) are given for 1970–1982 versus 1995–2004, and population trends back to 1976 (but differing for species; see Fox *et al.*, 2006). Distribution trends have been corrected for bias incurred during sub-sampling and populations trends are transformed for normality ($\log_{10}[(\text{population trend}/100)+1]$). Three migrants are excluded, as are species lacking BMS data; $F_{(1,44)} = 18.11$; $P < 0.0001$; $R^2 = 29.2\%$; $N = 46$. The cross lines indicate zones of decrease and increase for distributions and population trends. For only six species are both distributions and populations increasing (brown argus *Aricia agestis*, holly blue *Celastrina argiolus*, purple hairstreak *Favonius quercus*, comma *Polygonia c-album*, marbled white *Melanargia galathea*, speckled wood *Pararge aegeria*), three times this number are recorded as experiencing both population and distribution decline. Against the odds, northern brown argus *Aricia artaxerxes* has declining populations but increasing distributions, but many more (15 species) show population increases in the face of distribution losses.

conducted units without habitats begin to acquire more species than squares with habitats – the simple records for species between units with and without habitats start to look increasingly similar (Dennis, 2001). Among potential problems in mapping, distinguishing vagrants from residents (individuals with habitats) is particularly important. Ultimately, the objective of mapping is to illustrate records of species and their habitats. Thus, part of mapping should include the behaviour of individuals that signals the existence of habitat (egg laying, early stages, mate location behaviour) as well as numbers and sexes of species – this information provides means for distinguishing the status of records for spatial units.

As discussed in the text, biological theory draws on a strong relationship between population abundance and range size; one expects that butterflies that have growing populations will have expanding geographical ranges. However, a situation in which climatic warming is countered by habitat (biotope) fragmentation can dis-associate these two processes; the range margin of a butterfly may be expanding northwards while the core populations are declining with habitat loss. In this case, although a cross-species correlation is expected to occur between changes in range size and population size, the correlation will be weakened by antagonistic indicators of climate and human impact. In Fig. B8.6b is recorded the cross-species relationship between changes in ranges and populations. It has been corrected, as far as possible, for bias caused during the process of sub-sampling records (see Barbour, 2007) by regressing trends in the recent atlas (Fox *et al.*, 2006) on changes in raw numbers (Table B8.6). It is inevitable that randomly sub-sampling records across species, rather than for each species separately, will disadvantage those species whose presence in squares is based on fewer records (Dennis, 2009b), This evidently occurs as the residuals from the regression of atlas trends in Fox *et al.* (2006) are substantially accounted for by measures of species' apparency (Table B8.6). The scattergram illustrates that the few butterfly species that are actually extending their ranges also have growing populations. On the other hand, many species with contracting ranges also have net population declines. This is particularly worrying

Table B8.6 Changes in a species' distribution between two time periods, t_1 and t_2.

		t_2		
		0	1	Totals
t_1	0	00 a	01 m	a + m
	1	10 d	11 p	d + p
	Totals	a + d	m + p	n = a + d + m + p

In this contingency table, t_1 is time 1 (first atlas records) and t_2 is time 2 (second atlas records); 0 is absent, 1 is present, 'a' are joint occurrences, 'm' are gains, 'd' are losses and 'p' is continued occupancy (see Fig. B8.6a). The differences in distributions between the two time periods for any species based on complete records for the British Isles is:

$$D = [(m + p) - (d + p)]/n.$$

D varies between −1 (extinction) to +1 (colonization). Trends for sub-sampling are corrected by obtaining the predicted values from regressing the atlas trends reported in Fox *et al.* (2006) on *D* across species. The difference (residuals) between atlas trends and *D* is shown to be significantly affected by the number of records for species, and measures of butterfly apparency to observers, based on their colour, size, height of activity, resting posture and length of flight period: $F_{(1,52)} = 75.90$, $P < 0.0001$, $R^2 = 59.3\%$, $N = 54$ species.

as it demonstrates, again assuming a one-to-one link between a species' grid record and the occurrence of habitat, that habitat is being eroded (populations diminishing with resource loss) and lost altogether (populations becoming extinct). A habitat 'crunch' is exposed: although populations may be increasing for a number of species, and will do so periodically, this may not generate distribution gains if habitats are being lost and increasingly isolated (Fig. B8.6b).

As a brief introduction to the complexities of climatic warming and human impacts, it is worth considering the 'exchange' or compensation that might be perceived as occurring between the effects of climatic warming (new habitats in the north and west) and human impact (loss of habitat at the range core). Range margin habitats do not have the same status as range core habitats. Habitats gained at the geographical fringe with climatic warming can equally easily be lost with a single extreme season's weather event (e.g., from severe frosts or a cloudy, wet spell in the flight period) or from a periodic downturn in summer conditions; range core populations (metapopulations) provide vital source units for expansions. As it is, climatic warming is not always beneficial. It can obviously be accompanied by drought which can have serious consequences for species in wetland biotopes (e.g., large heath *Coenonympha tullia*). But there are more subtle issues involved. WallisDeVries and Van Swaay (2006) have shown how climatic warming can, in effect, result in critical seasonal *cooling* for many butterflies, particularly those hibernating as eggs or larvae and using herbaceous hostplants. With climatic warming, spring comes earlier to plants, driven by ambient temperatures, than to butterfly larvae that depend more on direct sunlight. Early plant growth is also exacerbated by atmospheric nitrogen deposition. The result is, they argue, that climatic warming causes microclimatic cooling in niches occupied by spring eggs and larvae of species using herbaceous hostplants. To date this is supported by statistical data and not experiments. It is also complicated by the fact that hibernation stage correlates with number of broods (the later the hibernating stage, the greater the number of potential broods) and length of flight period, both of which increase potential for population growth. Nevertheless, a salutary warning is delivered. Establishing distribution and range changes in species is no simple matter; explaining it is several orders of magnitude more complicated but a most enjoyable challenge!

A first difficulty in range change analysis is that processes may conflict, not just for different regions but for parts of the same region (**P8.42**), if only because species respond differently to environmental agents (WallisdeVries and Van Swaay, 2006). Gains and losses may also not occur at the same rate (**P8.43**). Extinction is area (population size) dependent, whereas colonization depends on migration capacity in relation to isolation between sites. To the extent that habitat area and isolation are disassociated, these processes will operate at different rates. Although range expansion should generally be accompanied by distribution infilling and increased population abundance, a process subject to reversal, distribution losses from one reason (habitat loss) can erode range expansion arising from another cause (climate warming). Thus, just how range change is measured matters and it is appropriate to use measures of both extent of occurrence and area of occupancy. Their joint use has been put to good use by Wilson *et al.* (2004) who demonstrated that colonization and extinction processes leave distinctive spatial signatures on species when ranges and distributions are compared over different spatial scales. A second difficulty is that all measures are influenced by recording intensity (**P8.44**). Records are 'samples' of species' occurrences, subject to errors (see Box 8.6), and intensity of recording (i.e., sampling) has massively increased from the early (Heath *et al.*, 1983) to later atlases (Asher *et al.*, 2001; Fox *et al.*, 2006). To overcome the problem of uneven sampling, a process of subsampling later records has been attempted (Fox *et al.*, 2006). There are indications that subsampling has been too aggressive (Barbour, 2007), but see Dennis (2009b). The difficulty here is that data have not been collected to an identical protocol during different periods, so it is not a simple matter of adjusting records for different recording intensity based on visits to squares. A further problem is that 'zero' records, failed searches (blank visits), thus negative observations in squares or biotopes, are less likely to be recorded than positive observations of species (cf., Box 8.4). This will increase the uncertainty of species' absences and is a problem if unreported visits have geographical bias, say to range centres or margins (**P8.45**). For instance, observation of species may be influenced by expectations of finding them and may lead to less effort and attention being expended in areas where few species or particular species are expected from past records. This counters a more general problem that colonizations are generally more easily detected than extinctions, especially over coarse spatial scales (Franco *et al.*, 2006).

A number of potential problems accompany representation of variables and their measurement, and to analytical models applied to predict range changes. Average conditions may miss the impact of significant extreme events (e.g., drought, storms, floods, fires) causing sudden declines (**P8.46**) (Dennis, 1993a; Dennis and Bardell, 1996). In this respect the decline of *Lasiommata megera* is most interesting; this broadcast decline, too rapid for land use changes, over much of lowland biotopes (e.g., Cheshire, UK) occurred after 1984 and it is possible that a chronic shift in weather conditions linked to a disease agent (e.g., a virus) during part of its development cycle may be responsible. The impact of insect pathogens is probably greatly underestimated, as are interactions with parasitoids (Hochberg, 1991; Dwyer and Elkinton, 1995; White *et al.*, 2000; Burden *et al.*, 2006; Creaser *et al.*, 2008). Models will always be affected by use of surrogate variables, such as the use of biotopes for finer vegetation categories, themselves surrogates for habitat resources (**P8.47**). In the study of range changes in *Pararge aegeria*, woodland inevitably used as a surrogate for habitat included conifer and deciduous woods and made no distinction for quality or amount of potential resources for the butterfly (J. K. Hill *et al.*, 1999a). Modelling for range changes is rife with problems (**P8.48**). In the same study on *P. aegeria*, data registered for mean elevation in squares is likely to have influenced predictions. The model was also a linear one though climatic limits for ranges almost always evoke quadratic or higher order functions having upper as well as lower bounds and probably involving interactions among variables. No measure of contiguity was applied to distinguish squares with and without neighbouring records and mainland from islands, and low thresholds for predictions (<0.5) were used. Matching of data with predictions provides support for a model and is suggestive of causal links but is never proof (see chapter head). In particular, it is unsatisfactory to rely on predictions that are modelled beyond the bounds of data (i.e., extending regression lines beyond data points) to establish those models (**P8.49**) or when models are not based on autecological findings.

Recent and future range and distribution changes in Britain: basic habitat issues

Once changes in range and distribution have been determined, the process of relating change to agents can begin. This again is no simple matter as different agents of change can, and almost certainly do, conflict – expansions from climatic changes countered by contractions from habitat losses or by effects of pollution – and the same agent can have diametrically opposite effects on different species (WallisDeVries and Van Swaay, 2006). Whilst viewing the current findings on range and distribution changes (see Box 8.6) it is first useful to consider some broad principles relating to habitats:

- **P8.50: Species will respond individually to agents of range and distribution change owing to distinct habitat requirements, particular conditions and life history associations.**
- **P8.51: Generalists (i.e., competitive and ruderal strategists) respond more readily to positive influences on populations and ranges, whereas specialists succumb more readily to negative agents.**
- **P8.52: Groups of species (e.g., northern, southern, coastal) having different geographical ranges within a region, thus different habitat conditions, will respond differently to agents of range change (e.g., climatic changes, warming or cooling).**
- **P8.53: Expansion of species' ranges is inversely related to the isolation of habitat patches and directly related to their migration capacity.**
- **P8.54: Distribution infilling and range expansion are consolidated with the development of metapopulation structures and enhanced metapopulation dynamics; distribution thinning and range contractions are subject to the reverse.**
- **P8.55: As conditions change (e.g., climate warming or cooling) the patchworks of habitat on which metapopulations are based may also change; that is, climate change *effects* changes in habitats.**
- **P8.56: Species with expanding range margins will tend to increase their breadth of resource use, and those with contracting range margins will tend to experience loss of resources.**
- **P8.57: Changes in a collection of attributes – demographic, morphological, physiological, biochemical and genetic as well as ecological (habitat) – are symptomatic of marginality, whether it is at the range edge or at the range core.**

It is expected that different species will respond individually to changes in any environmental agent owing to their contrasting resources, conditions for existence and contrasts in life history (**P8.50**). This will occur if only because different agents for change will impinge unevenly on different resources or life history characteristics. There is a geological precedent for this principle illustrated by idiosyncratic movements of plants over Europe during the late Quaternary into the Holocene (Huntley, 1991). This implies that variables successful for modelling range changes in one species are unlikely to be equally successful for modelling range changes in other species. Even so, some ecological patterns are found. Specialists (S strategists or stress-tolerators; see Chapter 5) are more likely to retract owing to limited resource associations and life history idiosyncrasies, whereas generalists (R and C strategists) are more likely to expand ranges, particularly as generalism is significantly linked to migration capacity (**P8.51**) (Dennis *et al.*, 2004). This has been confirmed with recent atlas data (Fox *et al.*, 2006) where it has been shown – though using more arbitrary criteria distinguishing species found in the farmed countryside and towns compared with those limited to discrete habitat patches – that specialist species' populations and distributions (populations: 78% of species declined; distributions: 35% decrease) have declined compared to generalist species (populations: 30% of species declined; distributions: 1% increase) (Fox *et al.*, 2006). It follows, from differences in life history and resource use, that species with unlike geography will differ in nature and degree of response to range and distribution-changing agents (**P8.52**) (e.g., northern and southern species to climatic warming; specialists and generalists to biotope fragmentation; see below).

But whether species expand or contract in response to an agent will depend on habitat area and isolation and migration capacity (**P8.53**). These variables determine their metapopulation dynamics, thus vulnerability to change and potential for persistence (see Chapter 6). For a constant migration capacity among species, a dense expanse of habitat, or potential habitat, permits rapid expansion and greater resistance to decline than sparse habitat distributions. For equivalent habitat arrangements, species with greater migration capacity can respond more rapidly to inducements or can better resist reversals in conditions. Thus, habitat distribution (fragmentation) is crucial in predicting species' responses to environmental changes (J. K. Hill *et al.*, 1999a) and only the most mobile species are expected to keep up with rapid climate changes (J. K. Hill *et al.*, 2002). The outcome of range expansions and contractions, or complementary distribution infilling or thinning, depends on the impact on metapopulation dynamics (**P8.54**), which respectively enhances the geographical stability of populations or undermines it (see Chapter 6). Range expansions can be most effective if they lead to recolonization of **semi-independent networks** (SINs; isolated metapopulations). The fact that these have previously been vacated indicates that without revitalization of the patchwork, as with enhancement of resources with climatic warming, they are unlikely to persist long term. An important concept here is the distinction between metapopulation dynamics (loss and gain of population units on habitat patches) and patch dynamics (expansion, contraction, changing shape, emergence or disappearance and shifting location of the habitat patches) (**P8.55**). Changing conditions may create new patchworks if species are able to exploit a new range of resources.

Consequently, an important feature of expanding range margins and distributions is that species undergoing them tend to acquire additional resources (**P8.56**). For example, the brown argus *Aricia agestis* has exploited additional hostplants (e.g., cut-leaved crane's-bill *Geranium dissectum*, dove's-foot crane's-bill *Geranium molle*) during range expansion (Kemp, 1998; Kemp *et al.*, 2008). Part of the explanation for this is stochastic (i.e., contact with a wider range of biotopes), but part is likely to be genetic and driven by selection (C. D. Thomas *et al.*, 2001). Genetic variability increases with population growth accompanying range expansion (Ford, 1964; Holt, 1990). The opposite is also suggested to occur – loss of resources with range contraction – for the same reasons, but in reverse, a phenomenon very likely accounting for the limited hostplant use by Fenland *Papilio machaon* (Dennis, 1977). This process may well explain the relationship between the breadth of nectar source use and geographical range in British butterflies (Hardy *et al.*, 2007). A feature of widening resource use during range expansion is that this is expected to reduce isolation, the 'new' resources, in effect, providing stepping stone patches between zones of previously isolated populations associated with less exotic resources (C. D. Thomas *et al.*, 2001). In fact, it is expected that a constellation of attributes is symptomatic of marginality, whether it is at the range edge or at the range core: demographic, morphological, physiological, biochemical and genetic attributes as well as ecological (habitat) (**P8.57**) (Shreeve *et al.*, 1996a). These were evident from range contractions during the Holocene (Dennis, 1977, for examples). It has also been shown – from changes to the introduced population of *Plebejus argus* in the Dulas Valley, Wales (Dennis, 1972b) – that such changes can occur in few (<25) generations; it is evident that such changes may be even more rapid (Merckx and Van Dyck, 2002; Merckx *et al.*, 2003).

Range and distribution changes: response to specific agents of change

Extensive range and distributional changes over the past two decades have increasingly been attributed to two basic causes: climate change and habitat loss (see Box 8.6). In fact, it is more accurate to apportion range changes broadly to atmospheric and terrestrial inputs; some of these are 'natural' in origin, but the main driver is exponential increase in human activity. There is much interaction between air and land-based inputs (see Dennis, 1993a, for details), so to simplify matters the sections below consider separately the impact on habitats leading to distribution and range changes of (i) climate change and (ii) habitat destruction, pollution and modification. Each of these processes can influence the broad-scale geography of species – increases as well as decreases in distributions and ranges – and do so in relation to:

1 The scale (spatial) of influence of agents of change.
2 The frequency of such processes.
3 The rate of input and the magnitude of change.

Isolating causes of range change is no simple matter and depends largely on the principle below:

> • **P8.58: In their impact on butterfly range changes, different agents tend to convey distinct habitat (resources and conditions) occupancy or abandonment signatures.**

Different agents are expected to carry distinct environmental signatures in effecting range changes (**P8.58**). In some cases, these agents may be obvious; for instance, species spreading northwards in Britain across what is accepted as a deteriorating (fragmenting) landscape are almost certain to be responding to climate warming, but northern species declining at their southern boundaries could equally well be responding to habitat or climatic change (Franco *et al.*, 2006). Climatic warming can cause habitat deterioration (e.g., impact of drought on habitat

Table 8.1 Factors in losses at southern boundaries for four British butterflies. (Data from Franco *et al.*, 2006, courtesy of Blackwell Publishing.)

Species	1 km sites surveyed	Butterfly incidence	Total extinction (%)	Extinctions due to hostplant loss (%)	Sites with hostplants	Extinctions on sites with hostplants (%)	Main extinction agent	
							Habitat loss	Climate
Aricia artaxerxes	111	56	49.5	25.2	83	24.3	✓	✓
Erebia aethiops	136	103	24.3	6.6	102	17.6	✓	✓
Erebia epiphron	89	56	37.1	1.1	88	35.9	✗	✓✓
Coenonympha tullia	85	41	51.8	28.2	61	23.5	✓✓	✗

Based on sites re-surveyed from records in Heath *et al.* (1983) and Asher *et al.* (2001). Key factors from stepwise logistic regression of survival in squares with hostplants: northern brown argus *A. artaxerxes* – vegetation height, latitude and an index of climatic suitability based on current European range in Kudrna (2002); scotch argus *E. aethiops* – latitude; mountain ringlet *E. epiphron* – elevation; large heath *C. tullia* – hostplant occurrence. The main extinction agents are here indicated simply by dominance of habitat loss (viz., loss of larval hostplant) and climate change. The only habitat components considered are hostplant occurrence, abundance and vegetation height. On this basis, habitat loss was implicated in most sites for *C. tullia*, some of those in *A. artaxerxes* and *E. aethiops*, and in just one site for *E. epiphron*.

abundance and quality) as well as enhancing habitats. In most cases, however, the agents causing change are by no means obvious – as in the case of southern species' margins; for example, complex interactions occur among atmospheric pollutants and terrestrial ecosystems with conflicting outcomes for butterfly habitats and their populations (Dennis, 1993a). The explanation (unexplained variation) must necessarily be 'partialled out' using statistical techniques, typically by applying regression models. In this way, Franco *et al.* (2006) were able to very neatly attribute percentage causation of survival, thus losses, at the southern boundaries to climate and habitat loss in four species and to identify the specific climatic and habitat variables responsible (Table 8.1). A note in Box 8.6 from research by WallisDeVries and Van Swaay (2006) illustrates how climatic warming can result in habitat decline for a group of species that overwinter as eggs and larvae and how caution is required in seeking explanation.

Principles associated with climate change

- **P8.59: Where there has been no change in the consumer and utility resources of habitat patches, rapid adoption of biotopes along a broad front with changing (cooler or warmer) conditions suggests the influence of a climatic agent(s).**

- **P8.60: Many species are expected to shift their distributions to occupy new biotopes, thus new types of habitat patches with different resource conditions, with climatic change.**
- **P8.61: With climatic warming scenarios, species are expected to become extinct/adapt to new habitat conditions at lower latitudes and altitudes and expand to occupy new habitat patches at high latitudes and altitudes.**
- **P8.62: Species will respond differently to climatic warming at northern and southern range margins owing to the impact of different influences on their habitat resources and conditions.**
- **P8.63: With climatic warming, generalists are more likely to colonize new locations than specialists whereas specialists are more likely to be lost from habitat patches than generalists.**
- **P8.64: Species' richness changes lag behind climate change owing to specialist butterfly species failing to respond as rapidly as generalist butterfly species.**
- **P8.65: Responses to climatic warming range changes can be rapid – in ecology (to habitat criteria such as vegetation structures, hostplant use, aspect and biotopes) and in morphology, physiology and genetics, etc. (e.g., wing size in crickets and butterflies).**

- **P8.66: As the adoption of new hostplants or habitat patches in different biotopes during range expansions is not generally the result of the sudden influx of consumer resources within sites, their adoption is likely to be due to ecological factors (e.g., microclimate conditions) that enhance population growth.**
- **P8.67: The adoption of novel habitats (alien hostplants or unusual utility structures) provides the conditions for adaptation to new conditions by selection.**
- **P8.68: Evolutionary changes to organisms during range expansions increase the capacity of those organisms to find new habitat patches.**
- **P8.69: The colonization of migrants of novel butterfly species, as well as their enemies, is expected with climatic warming.**

A large number of signals now support the notion of climate change; in Britain the phenology of communities is shifting forwards in time by some 30 days in the year (Sparks, 2006), including fungal mycelium (Mattock *et al.*, 2007), indicative of advancing decomposition rates. Climatic signals are perhaps most obvious when there is a rapid shift in species' distributions occurring during apparent stasis in habitat consumer and utility resources at sites. Rapid adoption of biotopes along a range margin, whatever the spatial scale, across distinctly cooler or warmer, drier or wetter gradients, suggests the influence of a climatic agent (**P8.59**). A classic case is the adoption of northern slopes and thicker turf by the silver-spotted skipper *Hesperia comma* in recent years (C. D. Thomas *et al.*, 2001; Davies *et al.*, 2005, 2006). Care is required to eliminate the influence of enemy-release from pathogens.

As noted above, data on fossil beetles and plants has led to the suggestion that most species are expected to shift their distributions in response to climate change – to colonize new sites and abandon old ones where conditions, respectively, make these suitable and unsuitable (Coope, 1987; Huntley, 1991) – rather than adapt to new conditions (**P8.60**). But this depends on the degree and the type of change in respect of habitat status and just what land use shifts are induced by climate changes (J. G. Hodgson, personal communication). For example, although site extinctions have occurred along the southern margin for the scotch argus *Erebia aethiops*, the most southerly site at Arnside, Cumbria (UK) remains intact. Above we noted how many British butterfly species have survived climatic and biotope changes during the Holocene and in doing so have acquired changes in morphology, physiology, genetics, life history and ecology (habitats) (Dennis, 1977, 1992a). Climatic warming to the end of the current century confronts British butterflies with changes to climate and environment at unprecedented rates, and the outcome is not easily predictable, affecting as it does the condition of resources as well as butterflies directly in *all* habitat patches (Dennis and Shreeve, 1991; Dennis, 1993a). Even so, butterfly species are expected to expand northwards and to higher altitudes and retract (become extinct) or adapt at southerly margins and at lower altitudes in Britain (**P8.61**), shifts which involve adopting new habitat patches and abandoning old ones. A contrast in response is expected of northern (with southern range limits in Britain) and southern species (with northern range limits but occupying all southern England); southern species have room to expand but northern species have much less leeway. Northern species are expected to contract northwards and to higher elevations and southern species to expand northwards (Dennis, 1993a; Franco *et al.*, 2006). A contrast in response to climatic change is also expected for any species between its northern and southern range margin in Britain (**P8.62**). Theoretically, climatic warming will have the effect of lifting species from lower climatic limits at the northern margin but of exceeding upper limits at the southern margin. The impact is expected to be greatest at northern margins as there is a question of how close to climatic limits are the southern boundaries, which may be limited more by other factors (e.g., community interactions) than by climatic variables (Davis *et al.*, 1998).

An important relationship is the likely contrasting response of generalists (C and R strategists) and specialists (S strategists) to climatic change (**P8.63**). Generalists are more likely to colonize new locations with climatic warming than specialists (J. K. Hill *et al.*, 2002); by definition, they have more resource outlets, but also have greater migration capacity (Dennis *et al.*, 2004). Specialists are more susceptible to habitat loss and are associated with distribution losses (Dennis *et al.*, 2004). For these reasons species' richness changes lag behind climate change (**P8.64**) (Menéndez *et al.*, 2006).

The remaining 'principles' single out important attributes linking the response of insects to changing habitat conditions during climatic change. A feature of such adaptations to climatic change, as to habitat loss, is the rapidity of response (**P8.65**). This response is not just to site ecology – occupation of sites with different vegetation structures and biotopes (e.g., *Hesperia comma*; Davies *et al.*, 2005) and to new hostplant use (e.g., *Aricia agestis*; Kemp *et al.*, 2008) – but can also involve morphology, physiology (e.g., contrasts in wing loading between woodland and agricultural landscape *Pararge aegeria*; Merckx *et al.*, 2006) and behaviour including dispersal (Van Dyck and Baguette, 2005). A number of these changes, especially response to novel habitats, involve ecological processes (**P8.66**). As the adoption of new hostplants, or usual hostplants in new settings (aspects) or in different growth forms, is not the result of their influx to sites, their adoption is probably due to climatic changes impacting on resource conditions, such as resource microclimates, that enhance conditions for population growth (viz., reproduction, fecundity, development or survival). This effectively switches patches from sinks to potential sources. The occupation of what in effect are novel habitats provides the conditions for improving adaptation, by selection, to the new climatic conditions (**P8.67**) (C. D. Thomas *et al.*, 2001). Climate change may affect resources in subtle ways (e.g., nectar and amino concentrations of nectar flowers; Rusterholz and Erhardt, 1998) and impinge on tolerances. Such evolutionary modifications developing during range expansions increase the capacity of organisms to find new habitats (e.g., increased fractions of longer winged, dispersive individuals among bush crickets *Conocephalus discolor* (Thunberg) and *Metrioptera roeselii* (Hagenbach)) (**P8.68**) (C. D. Thomas *et al.*, 2001) and survive in them (e.g., changes in marginal wing ocellation relating to predation by birds in *Pyronia tithonus*; Merckx and Van Dyck, 2002). Once habitat patches in new regions have been occupied (space-infilled), specialists and non-dispersive forms may be selected for, especially if there are trade-offs between dispersal and reproduction.

A distinctive feature expected of climatic change with enhanced greenhouse warming is the invasion of the British islands by novel (alien) species (**P8.69**), including the return of butterflies that used to occur in the British Isles (e.g., large tortoiseshell *Nymphalis polychloros*). Many Lepidoptera visit British shores as overseas migrants (Box 8.7) (Sparks *et al.*, 2005); some butterfly species are unlikely to persist in Britain without the annual input of migrants (e.g., red admiral *Vanessa atalanta*, painted lady *V. cardui*, clouded yellow *Colias crocea*). Another feature of current warming is that more individuals have been reported to overwinter in Britain and Ireland (Asher *et al.*, 2001; Dennis *et al.*, 2006c; Fox *et al.*, 2006). As yet, although many immigrant moths have colonized southern Britain, there have been no additions to the list of British butterfly species. The occurrence of numerous records for the swallowtail *Papilio machaon* in southeast England, found feeding on a wider range of larval hostplants than the Fenland race, indicates the potential for novel colonizations. The current lack of colonists into the British mainland has been questioned (Leverton, 2007), an issue that can be contrasted with Ireland where new colonists have been confirmed (e.g., comma *Polygonia c-album* (Fox *et al.*, 2006); Essex skipper *Thymelicus lineola* (www.wexfordnaturalists.com/gaggle.php; I. Rippey, personal communication)). Their absence probably reflects on the poor state of biotopes in southern Britain for colonization, lying at the root of butterfly habitat fragmentation and losses (Warren *et al.*, 2001b), rather than any inability to migrate – which is clearly increasing (Sparks *et al.*, 2007). Even so, phenological responses to climate change can be complex and may prevent colonization of Britain by novel species (Sparks *et al.*, 2006). Yet Leverton (2007) raises a valid point that a number of migrants may be overlooked, especially the hairstreaks. This is an important issue. The recent discovery of *Aphantopus hyperantus* in Cheshire and Manchester has led to closer inspection of passing dark satyrines such as *Maniola jurtina* (P. B. Hardy, personal observation); no doubt this will lead to further finds of the ringlet. Along with migrant Lepidoptera must be expected their migrating enemies, particularly parasites. An example is the arrival in 1999 of *Sturmia bella* (Meigen, Tachinidae; Diptera), a parasitoid of the nymphalids small tortoiseshell *Aglais urticae* and peacock *Inachis io*. Since then it has spread across England and Wales, and there have been reports of this parasitoid fly attacking these nymphalids and causing high mortality (Ford *et al.*, 2000; John, 2004; Lewis, 2008).

Principles associated with habitat destruction, pollution and accident

A wealth of autecological studies on British butterflies dating back to the 1960s demonstrate that it does not take much to render a site unsuitable for a butterfly (Thomas, 1984); the hostplant and other resources may appear to be intact but the patch is abandoned. The balance of conditions required for sustaining butterfly populations is precarious; this is witnessed time and again in unsuccessful attempts to manage sites. Unhappily,

Box 8.7 Changing status of migrant Lepidoptera in the British Isles

With climatic warming, as predicted from enhanced greenhouse gases emissions, more migrant Lepidoptera, including butterflies, are expected to arrive in the British Isles. More migrants are also expected to enter Britain from human agencies (i.e., in food stuffs, with nursery plants, timber products, accidental or intentional releases, etc.). Increased numbers of migrant Lepidoptera in Britain are certainly being recorded with higher temperatures (Sparks *et al*., 2005, 2007). A number of migrant butterfly species have long been able to breed in Britain during the summer months (e.g., long-tailed blue *Lampides boeticus*) and to overwinter in the islands (e.g., red admiral *Vanessa atalanta*) during a series of years. With climatic warming, it is expected that more species will be able to breed and to overwinter, ultimately becoming part of the resident fauna. Figure B8.7 illustrates the alternatives for species in the face of climatic change. With time, species long able to migrate to Britain, but unable to breed or overwinter, will be able to breed, overwinter and do both, becoming long-term residents (see Appendix 1). Prospective candidates are clouded yellow *Colias crocea* (already breeding on arrival and overwintering in Dorset), *V. atalanta* and painted lady *Vanessa cardui*, but it is thought that many more species might be involved. Although continental swallowtail *Papilio machaon* has been found breeding in Kent, the migrants have not yet colonized the southeast of Britain, breeding and overwintering over several successive years without a 'rescue effect' of further migrants.

Undoubtedly, there will continue to be immigrants for species that become residents and these species are distinct from a further group of residents that are never recorded as having immigrants. It should be noted that this latter group (D) may well include moth and butterfly species that are also immigrants (that is, they should be in C) but the immigrants are insufficiently distinct in morphology to be identified. Genetics (examining DNA) has a part to play here. It should also be noted that, for simplicity, no distinction has been made between species that migrate here naturally and other species that are assisted by a human agency. If such a distinction were to be made, then the zones marked as an empty set (φ) would be represented by species making their way to Britain through human influences.

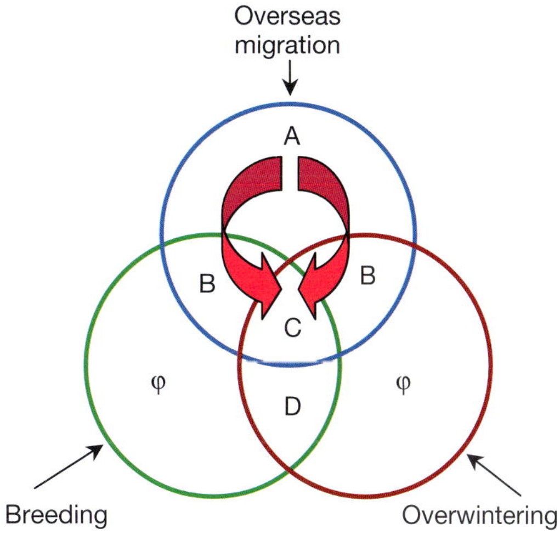

Fig. B8.7 (a) Classification of migrant and resident species. Three sets are identified including species that migrate into Britain (by whatever means), breed or overwinter in Britain. Butterfly and moth species are elements of the sets. Five options emerge: A, migrants that neither breed nor overwinter; B, migrants that breed or overwinter but do not do both; C, migrants that are also residents, breeding and overwintering; D, residents that do not include migrants; φ, empty set. The arrows indicate changes expected with climate warming with more species in categories A and B becoming components of C.

Group C has several potential subgroups:

- Congruence of migrants and residents: (i) current residents exist outside the migration zone (sampling point); or (ii) residents exist within the migration zone.
- Recency of residency: (i) residents are extinct or believed extinct (A or B apply); (ii) residents are of long standing (pre-study period); or (iii) colonization (residency) occurs during study period.
- Source and sink status for B and C: (i) residents are substantial populations (sources generating vagrants); or (ii) resident populations are sinks (small, with occasional failure of either or both of breeding or overwintering and cannot persist).

the bulk of human activity and intentions is not even this benevolent but weighs heavily against wildlife and the effect on butterfly habitats has been literally devastating. It has been estimated that the approximate flight areas occupied by 20 of 45 more sedentary British butterfly species now cover only some 1.44% of the land surface within occupied regions (Cowley *et al.*, 1999). Even so, it is a difficult matter attributing cause to loss of species in a scientific way when what has been lost leaves little imprint in the landscape. It is perhaps most effectively demonstrated by studies that apply current butterfly–biotope associations to comparisons of what is residually available of that biotope against what has been available historically, even in relatively recent times (see Box 5.1) (Cowley *et al.*, 2000). The findings should be alarming to us all.

The following principles link butterfly habitats to human influences and accidents:

- **P8.70: Direct destruction and deterioration of habitats by human activity is the primary cause of distribution losses and range retractions in Britain.**
- **P8.71: Pollution by human airborne and land-delivered chemicals can adversely affect butterfly species' geography by undermining habitat suitability for species; toxification and fertilization (eutrophication) are the main agents.**
- **P8.72: Atmospheric changes from enhanced greenhouse gases (carbon dioxide) and pollution can modify resource quality and thus suitable resource availability with concomitant effects for butterfly biology and geography.**
- **P8.73: Human activity is capable of directly or indirectly, inadvertently or intentionally, creating butterfly habitats and thus expanding species' geography (e.g., hemeroby).**
- **P8.74: Abandoned sites of even the most intensive human activity can provide valuable butterfly habitats, thus extending species' geography.**
- **P8.75: Well-intentioned management of butterfly habitats have furthered the decline of some species because habitat criteria have not been fully appreciated.**
- **P8.76: Human impact on other wildlife can influence butterfly habitats with consequences for butterfly species' geography.**
- **P8.77: Introduction of alien species (plants, animals) can greatly affect butterfly habitats, and resources, and thus their populations and distributions.**
- **P8.78: Naturally occurring catastrophic events can change butterfly species' geography directly through habitat resources.**

The dominant trend in butterfly geography is decline (see Box 8.6) attributed firmly to human impact on habitats; readers are directed to the latest atlases for details associated with the principles listed above (Asher *et al.*, 2001; Fox *et al.*, 2006). Here, just sufficient examples are given to illustrate each point. Part of the process is simply that of direct destruction of semi-natural biotopes with the development of intensive, mechanized agriculture requiring larger land units and increased production, but also with industrialization (chemical inputs), urbanization and the growth in infrastructure (e.g., housing, transport facilities) accompanying population growth and demand (**P8.70**). During the 20th century some 32 new towns were created in the UK, not all as well provided for as Milton Keynes with its 20 million trees (www.englishpartnerships.co.uk/miltonkeynes.htm). These activities involve direct erasure of habitat resources, both consumer and utility, but can create others (see p. 228). However, not all geographical changes are the result of biotope erosion. Abandonment of old land management practices, which fortuitously had been beneficial to many species with the arrival of agriculture in Britain (Dennis, 1992a), also have a large part to play in declines. The classic example has been the abandonment of coppicing practices, which in Blean Woods, Kent, has had serious consequences for the heath fritillary *Mellicta athalia* (Warren, 1984a, 1985a, 1985b; but see also Warren *et al.*, 2001a). This process is different from outright destruction in the impact on habitats as many, if not all, resources may still be present, but not in sufficient abundance or in suitable condition for butterfly species, despite the survival of the basic biotope though not the strict vegetation categories.

If abandonment of traditional practices is subtle in effect then the impact of airborne and land-delivered pollution from humans is more so, insidious as well as pernicious (**P8.71**). The effect is twofold, acidification and/or fertilization; the consequence is simply toxification of biotopes and unsuitability of resources in them for butterflies and most other wildlife. This

Fig. 8.3 Herbicide treatment of nettle and dock patches along a public footpath on Alderley Edge, Cheshire, inimical for breeding nymphalids such as the peacock *Inachis io* (top inset) and small tortoiseshell *Aglais urticae* (bottom inset). (Insets courtesy of Peter Eeles.)

process even affects our highest moorlands (Freeman *et al.* 2001; Caporn *et al.* 2005; Crowle, 2007). It is most blatant where 'weeds' are sprayed (e.g., nettles and thistles with serious consequences for nymphalid butterflies; Dennis, 2008a) (Fig. 8.3), but it also explains why expanses of grassland are vacated of satyrines. Soil water pH has a major impact on nutrient availability to which butterfly larvae are particularly sensitive, as with leaf nitrogen levels affecting growth rates and development, but also other aspects of life history such as diapause induction (Bink, 1985, 1986; Bink and Siepel, 1996). Enhanced greenhouse gases and pollutants can have complex effects on butterfly resources (**P8.72**) (see Dennis and Shreeve, 1991; Dennis, 1993a, for a wider discussion). Whereas increased carbon dioxide can increase growth and the C : N ratio in plants, plant quality is not thought to be offset by increased biomass; larvae grow more slowly and thus may be exposed for longer periods to enemies (Loader and Dalman, 1991). Loss of quality may, however, be offset by other pollutants (nitrogen compounds). Just how growth and nutrient status of plants and butterflies is affected depends on an intricate relationship between atmospheric pollutants and climatic change, both temperature and moisture. Larval nutrient status can influence other aspects of butterfly biology, such as nectar dependency (Mevi-Schütz *et al.*, 2004), thus population parameters, and have consequences for distributions. But it is too early to predict the consequences of current atmospheric changes for specific times ahead; the most likely outcome is that C and S strategists will be more vulnerable than R strategists, the latter having higher mobility and potential rate of population increase (Dennis, 1993a).

Of course, it does not have to be like this. There are exemplary cases where human interaction, directly or indirectly, has been (or could be) beneficial to butterfly populations and geography at a variety of scales by improving biotopes for butterfly habitats (**P8.73**) (Haughton *et al.*, 2009). At the species level, nothing has been more spectacular or more welcome than the reintroduction of *Maculinea arion* into Britain – the creation of viable habitats and metapopulation patchworks for a butterfly with a complex life history by J. A. Thomas and Butterfly Conservation. This accomplishment is made more extraordinary by the encouragement of a major transport company to engage in habitat creation in Somerset (see Fig. 9.4) (Simcox and Bourn, 2006). The result is that this species now again has a British range and distribution. Another example is the use of a 13 tonne, 360° slew bucket excavator to dig out scrapes (3 × 16 m to 30 × 10 m; depth *c.* 10 cm) over 0.2 ha on the Isle of Portland, to enhance egg-laying resources for *Plebejus argus* (de Whalley *et al.*, 2006). On a vaster scale, both in species and land area, are the developments through agri-environment schemes (Brereton, 2004) and landscape-scale programmes (e.g., Sandstone Ridge ECOnet Partnership or SREP in Cheshire). These projects are almost certainly arresting declines (e.g., high brown fritillary *Argynnis adippe* on Morecambe Bay limestones; www.ukbap.org.uk/UKPlans.aspx?ID=108). Habitat creation is no easy solution; the following three principles demonstrate just how easy it is to adversely affect species' distributions and ranges, as well as improve on them, depending on subtle habitat criteria. Not all human activity is deleterious; brownfield sites (abandoned industrial and urban sites) occupied by scarce species demonstrate what is possible (Wood and Pullin, 2002; Angold *et al.*, 2006). Such sites for *Erynnis tages* on the Creuddyn Peninsula form crucial patches in the metapopulation for this species in North Wales (Gutiérrez *et al.*, 2001; Gutiérrez, 2005); they provide vital patches and stepping stones for many butterfly species extending their distributions and ranges into conurbations (**P8.74**) (Hardy, 1998; Hardy and Dennis, 1999; Wood and Pullin, 2002; Loram, 2004; Angold *et al.*, 2006). In the East Midlands, the majority of sites for both *E. tages* and grizzled skipper *Pyrgus malvae* are brownfield sites, quarries, railways, pits and industry (Table 8.2) (Anon, 2007; Prater, 2007); Fig. 8.4 illustrates a notable site in the Pennines.

Unfortunately, management does not always work to plan, also management for one species affects others (see Chapter 9) and this can have deleterious consequences for species' distributions and ranges. Thus agri-environmental schemes and Sites of Special Scientific Interest (SSSIs) have not been as successful as hoped (Brereton *et al.*, 2005; Fox *et al.*, 2006); management for one group of species, those requiring short turf biotopes, can be deleterious for those requiring tall-herb grassland. These mistakes arise generally because of a lack of knowledge and failure to appreciate the details of butterfly habitats (**P8.75**) (Field *et al.*, 2005, 2007). Management for butterflies must necessarily understand entire ecosystems before undertaking changes; changes to one wildlife element of an ecosystem can have substantial consequences for butterfly habitats (**P8.76**), populations, distributions and ranges – as infecting rabbits with myxomatosis has had on the decline of *Maculinea arion* and other lycaenids (Thomas, 1984). Other examples include the introduction or removal of large herbivores (e.g., goats on calcareous grasslands, deer in woodlands) (Pollard and Cooke, 1994; Feber *et al.*, 2001). In all these situations a multitrophic approach is required to assess the full consequences for butterflies of different grazing stock types and densities (Stewart, 2001); imbalances in grazing density can clearly cause shifts in distributions.

An increasingly serious aspect of human intervention in butterfly habitats is the introduction of alien plants and animals (**P8.77**). Some 31.5% of vascular plants (1057 of 3354 species) are now regenerating, well-established (competing) aliens (Kent, 1992). **Baker's Law** predicts that self-compatible species, particularly those that do not need the services of pollinators, are more likely to establish and spread

Table 8.2 Brownfield sites for dingy skipper *Erynnis tages* and grizzled skipper *Pyrgus malvae* in the East Midlands. (Data from Prater, 2007, with kind permission of the author.)

Species	Quarry	Railway	Pit	Other industry	Semi-natural	% brownfield site
Erynnis tages	13	9	24	12	7	89.2
Pyrgus malvae	6	11	0	2	3	86.4

Based on survey of 132 sites in Nottinghamshire, Leicestershire, Rutland and lowland Derbyshire. At seven sites both species were present. The biotope was based on current or previous use.

Fig. 8.4 Abandoned mining sites bounding undisturbed farmland in Cressbrook Dale, Derbyshire, where northern brown argus *Aricia artaxerxes* (top inset), dingy skipper *Erynnis tages* (middle inset) and common blue *Polyommatus icarus* (bottom inset) are found. (Insets courtesy of Paul Kipling and Peter Eeles.)

after long-distance dispersal. This is supported by research in the USA based on 361 plants introduced from Europe (Van Kleunen and Johnson, 2007). The full list of alien species, animals and plants, is recorded in http://www.brc.ac.uk/resources.htm (English Nature's Audit Report No. 662 on non-native species; Hill *et al.*, 2005). Butterfly species use alien plants as larval hostplants and nectar sources; their use as nectar sources dominates over their use as hostplants (see Appendix 4). Alien plants are potentially important for both population dynamics and species' geography (see Chapter 2). Butterfly species can expand their ranges using alien hostplants introduced to novel biotopes (e.g., *Gonepteryx rhamni*; Gutiérrez and Thomas, 2000). However, alien hostplants can entice exploitation but be toxic (Graves and Shapiro, 2003). Key to alien nectar use by British butterflies are variables providing access to these sources (i.e., greater mobility, southern distributions, longer adult lifespan, hostplants in garden biotopes) which tend to resolve to fertile biotopes (Hardy and Dennis, 2008). The picture is very different for alien hostplants. Mobility still figures as the most significant variable for alien hostplants used (proportions and numbers). But other key variables differ for proportional use and numerical use of alien hostplants. Those butterfly species having a larger proportional use of alien hostplants are ones requiring woody hostplants with quantitative rather than qualitative chemical defences (Cizek *et al.*, 2006) – and butterfly species using a larger number of alien hostplants tend to be polyphagous and already have access to a greater abundance of native hostplants in gardens. Butterflies in decline and under threat (Biodiversity Action Plan (BAP) priority status) use fewer alien hostplants and nectar plants, though no study has been made on the extent to which butterfly species depend on alien resources in different localities (e.g., dependence on buddleia flowers in suburbia; *Pieris* species on nasturtium *Tropaeolum* species as a hostplant). Study should also be made of the extent to which some aliens (e.g., Indian balsam *Impatiens glandulifera* Royle; Japanese knotweed *Fallopia japonica* Houtt) swamp other resources, and nectar and hostplants of butterfly species (J. G. Hodgson, personal communication).

Finally, it is well to appreciate that naturally occurring extreme events – atmospheric or terrestrial – can

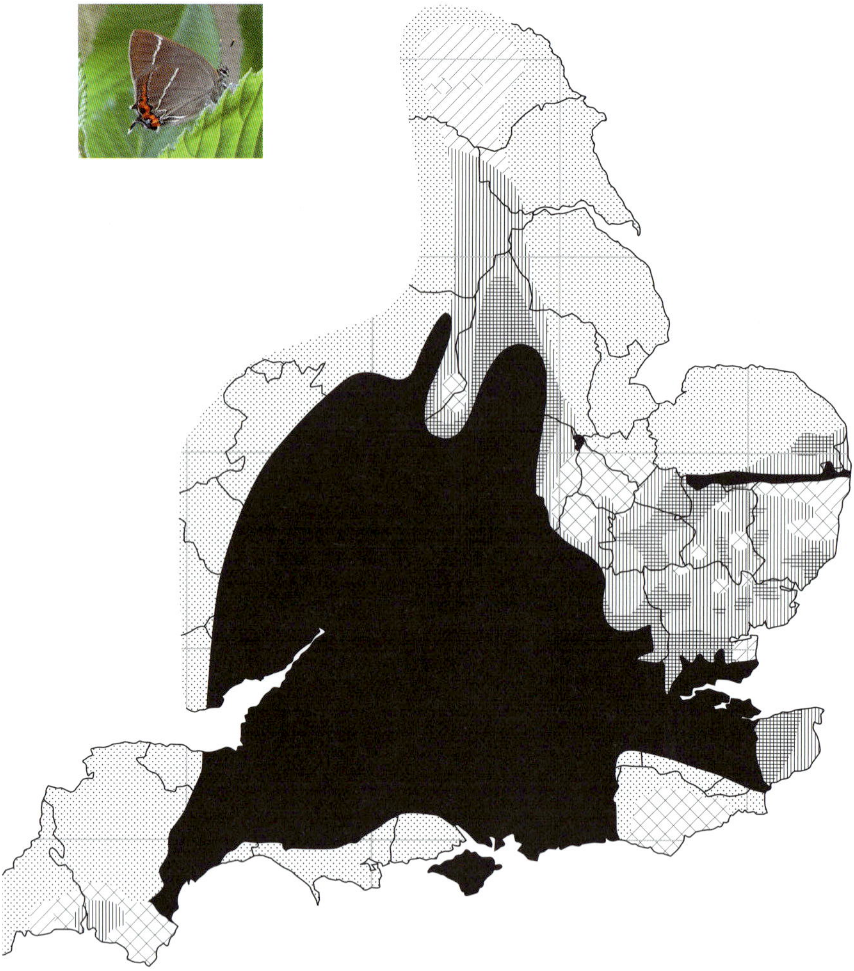

Fig. 8.5 Progress of the current epidemic of Dutch elm disease in 1983. The map illustrates the proportion of elms killed or severely diseased at the end of 1983 as a proportion of all elms >6 m high at the start of the epidemic. Shading indicates percentages from white to black (white <10%, diagonal shading 10–30%, wide cross shading 30–50%, vertical close shading 50–70%, close cross shading 70–90% and black >90%). Dense stipple indicates areas lacking information on elm decline. The inset shows a white-letter hairstreak *Satyrium w-album* which uses elm as a larval hostplant. (From Rackham, 1986; inset courtesy of Paul Kipling and Peter Eeles.)

affect butterfly distributions by direct impact on habitats (**P8.78**). Many extreme events tend to affect small areas (e.g., storms, severe frosts, floods, droughts, fires) (Dennis, 1993a), but some natural events can have widespread repercussions for butterfly geography. The classic example is the loss of 20 million mature elm trees in Britain since the 1960s caused by the Dutch elm disease fungus *Ophiostoma novo-ulmi* Brasier transmitted by Curculionidae beetles such as *Scolytus multistriatus* (Marsham) and *Hylurgopinus rufipes* (Eichoff) (Hubbes, 1999). The consequences for the white-letter hairstreak *Satyrium w-album* continue to be serious (Fig. 8.5) (Fox *et al.*, 2006) but steps are being taken, particularly in the management of woodland and hedgerows for elms, to counter the losses, as in the county of Cheshire (see www.cheshire-biodiversity.org.uk). Another British event operating virtually overnight was the storm, a deep depression, of 16 October 1987; this felled 15 million trees and opened the canopy of many woodlands and doubled the amount of dead wood to 23 m^3 ha^{-1} (Kirby *et al.*, 1998). Such devastating wind throws over wide regions have happened periodically in Holocene Britain (Allen, 1992) and the regeneration complexes provide open spaces for butterflies dependent on larval herbaceous hostplants. They also provide nectar sources for tree-dependent species.

All butterfly distribution and range shifts are initiated in habitats; that some resource within the habitat space has changed is evident in population crashes or explosions. Sometimes, the change is obvious, as at a local scale with the application of a skin of tarmac and concrete (Hardy, 1998), a flood or drought (Dennis and Bardell, 1996) or a severe storm as in October 1987 (Kirby *et al.*, 1998). The intensity of such events has a distinct periodicity; thus, storms are evident in the Holocene British fossil record (Allen, 1992). But changes are usually more gradual and a great deal more insidious, such that we are left with expanses of green deserts, grass or wood, in which very few resources are left for butterfly species (Dennis, 2000, 2008a).

Chapter 9

HABITATS IN BUTTERFLY CONSERVATION

The planet has long experienced episodes of white (polar) deserts and red (hot) deserts. Humans are unique among living organisms in generating green deserts.

APPROACHES TO CONSERVATION AND CONSERVING BUTTERFLIES

It is entirely fitting that examination of butterfly habitats should finish by considering their part in butterfly conservation. So substantial has the impact of humans been on global climate and ecosystems since the 18th century that Crutzen (in Crutzen and Streimer, 2000) has coined the term **Anthropocene** for this unique era – among other things this includes an accelerating deadly cocktail of climate change and habitat destruction and the threat of mass extinction (Samways, 2007). Conservation is the negative feedback response to damage to the environment initiated by the animal responsible for the damage. As long as there is a will for conservation and the logistic support for it, whatever these are, there will be controversy about the direction conservation should take and the approaches adopted. Even so, as Professor Tim New has impressed on me, a common ethical purpose provides a unifying theme and direction. This chapter focuses on how specific approaches to butterfly conservation are influenced by the interpretation of 'habitat', particularly the one adopted here, the functional or resource-based definition of habitat (Dennis and Shreeve, 1996), which emphasizes the spatial coincidence of resources in the right condition. Contrast is made with interpretation of habitat based on vegetation units or biotopes (Dennis *et al.*, 2003, 2006b, 2007). Conserving organisms requires understanding of the spatial and temporal dynamics of their resources within the landscape. When land areas are relatively undeveloped and large enough (>10^3 km^2), it is a simple and cost effective enough matter to rely on landscape, and thus resource, heterogeneity. At this scale, resource turnover via geomorphic processes and vegetation dynamics is sufficient to maintain heterogeneity and therefore species' persistence. However, in smaller regions such as the British Isles and parts thereof, long exposed to human impact, the persistence of resources becomes increasingly uncertain as semi-natural areas contract and become scarce, in which case there is a need to understand organisms' habitats and their disposition, and to intervene to prevent further loss. Intervention involves a kaleidoscope of potential approaches; these are considered in relation to a resource-based definition of habitat:

1 Species versus 'habitat' approach.
2 Habitat (= patch) versus landscape approach.
3 Single species versus multispecies approach.

The chapter then moves on to look at the habitat principles underlying conservation of butterfly species; this requires planning and action at three levels: (i) within sites and at levels above and below this; (ii) within resource patches at sites; and (iii) among sites over the wider countryside. Some of the basic principles in site and countryside conservation and restoration are then considered. The chapter finishes with a synopsis of the resource database for British butterflies and the position of butterflies as flagships and indicators for habitats of other arthropods. The overall message of the chapter is that:

Greater emphasis must be given to plant and animal communities; even when the objective is conservation of a sole species, attention needs to be given to whole landscapes and not just patches, and management for change not stasis.

A Resource-Based Habitat View for Conservation, 1st edition. By Roger L. H. Dennis. Published 2010 by Blackwell Publishing.

This book is not about 'political' issues but it is inevitable that they dominate conservation and a cautionary note accompanies this chapter: principles of human activity take precedence; humans inevitably and understandably give precedence to their own personal survival before other organisms – the more so in a resource-limited world. It is axiomatic but usually conveniently ignored that long-term human survival and that of plants and animals generally are interlocked in the maintenance of sustainable ecosystems. At a practical level, approaches are dictated by choices but often there is little leeway. Many theoretical alternatives (e.g., SLOSS – should we have single large sites rather than several small ones; or single sites versus patchworks) do not actually exist in reality for conservation organizations; they make do with what they can obtain – often indeed all that is left in suitable condition – and can achieve within spatial and financial constraints. Choice is crudely proportional to purchasing power and the labour available; the more cash available, the greater the likelihood of being able to address the theoretical ideas of ecologists and the practical needs of species. However, by no means does it all come down to cash. A sustainable economic system for humans has undoubted benefits for the natural world and other organisms' habitats; depleting the landscape of organism diversity has consequences for global ecosystem stability (Huby, 2001; Hodges, 2005; Barker, 2008). An intelligent organism will adopt sustainable solutions. Labour for conservation is separate; as many volunteers give their free time willingly to work for butterfly conservation, it would be insulting to them simply to list all labour under considerations of cash – ethics become paramount.

Species or 'habitat' approaches

• **P9.1: Without detailed knowledge, conserving a species by adopting vegetation types is only likely to be successful when the vegetation units selected are so broad as to include the array of necessary resources (as well as non-resources), or the landscape so severely fragmented that there is virtual 1:1 correspondence between resources and the vegetation unit surrounded by other units typically under intense exploitation (e.g., agricultural production) that lack any key resources.**

• **P9.2: A species approach and habitat approach are only identical in the ground they identify for conserving a species when a resource-based definition of habitat is applied.**
• **P9.3: For vegetation units to make suitable surrogates for resource-based habitats, there must necessarily be sound knowledge of a close association between the two entities.**

There is a continuing debate between the relative merits of species conservation and habitat conservation. Is it better to concentrate on individual species (e.g., Adonis blue *Polyommatus bellargus*, silver-studded blue *Plebejus argus*)? Or should we be looking after vegetation associations such as grassland or heathland with the implicit assumption that the latter will facilitate persistence of constituent and dependent species? There is, currently, no clear distinction between the two approaches. Management for a target species inevitably involves the management of whole sites or at least the appropriate parts of sites, whereas if the objective of management is the continued presence of a 'habitat' in its own right, the habitat is defined in species terms and the persistence of selected species is used as a measure of management success. Are these simply two ways of looking at the same situation interchangeable and, if not, does it matter? The outcome depends on how habitat is defined; the problems arise when 'habitat' is equated with vegetation type (Dennis *et al.*, 2007).

When habitat is regarded as being synonymous with a vegetation category or biotope (see Fig. B1.1b) it is mapped as if a vegetation unit or a biotope. This can, of course, be achieved by analysing a species' resource base and linking it precisely to vegetation components. More typically, it arises for two other reasons:

1 The vegetation units are so extensive, broad and vague in definition that they include resources of the target species and those of many more species.

2 The vegetation units are narrow and precise, usually because severe fragmentation leaves a small refuge for a species (**P9.1**) (e.g., an abandoned field corner cut off by road construction; see Fig. B1.1b) (Dennis *et al.*, 2003, 2006b).

As explained in Chapters 1 and 5, vegetation associations can be described at a hierarchy of levels, they are often 'idealized' statistical entities (e.g., National Vegetation Classification (NVC); Rodwell, 1991–2000), they can lack key resources for butterflies such as larval hostplants, and species often extend over a number of distinct vegetation types regardless of scale (Dennis *et al.*, 2003, 2006b). This situation has recently been described for *Plebejus argus* on the Great Ormes Head, a 3 × 2 km headland in North Wales (Dennis, 2004b; Dennis and Sparks, 2006). Here, the butterfly not only occupies shorter turf areas (<15 cm) of calcareous grassland (NVC CG1 and CG2 categories; Box 9.1) where its hostplants (rock rose *Helianthemum* spp., bird's-foot trefoil *Lotus corniculatus* L.) are found in association with ants of the genus *Lasius* (Thomas, 1985a; Thomas and Harrison, 1992), but also adjacent areas of scrub which it uses for adult feeding (e.g., *Cotoneaster* spp., bramble *Rubus* spp.), mate location, thermoregulation, daytime resting in cool, windy and cloudy conditions and roosting (see Fig. B1.2b and Box 3.2). These scrub areas also have small pockets of hostplant that are used for egg laying. Despite this, there is continued adherence to regarding habitats as occurring in distinct vegetation patches and the treatment of such patches as being uniform in composition, all of which may be to the detriment of extinction risk assessment and management.

So, are the species approach and 'habitat' approach as ploys in conservation interchangeable? In the case where a habitat is being considered for a single species and is defined on linked resources they are; in the situation where a habitat is regarded as synonymous with a vegetation unit they are not (**P9.2**). In a species approach to conservation, based on essential resources, this would require identifying and mapping the resources exploited; such resources, linked by individual movements, demarcate the habitat and this can be spread over a number of vegetation classes and biotopes (**P9.3**) (see Fig. B1.1c). A so-called habitat (= biotope or vegetation unit) approach would not necessarily include all the complementary resources for a target species and therefore not envelop the entire habitat for the species. It may well, of course, include both resources and entire habitats for other species, whereas the precise demarcation of habitat in a species approach for any target species is less likely to coincide with habitat bounds of another species. These may, at first, appear to be fine distinctions as the two approaches will inevitably involve some overlap. But, the difference is one of precision about resources, an understanding of habitat, and ultimately the part played by the wider landscape that becomes particularly relevant when multispecies conservation is considered. The distinction is brought home in the detailed comparison of resource use by a lycaenid, the violet copper *Lycaena helle* (Schiffermüller), and a nymphalid, the bog fritillary *Proclossiana eunomia* (Esper), which share the same biotope (peat bogs and wet hay meadows) and hostplant (*Polygonum bistorta* (L.)) in Fange de Pisserotte nature reserve, south Belgium (Turlure *et al.*, 2009). The functional habitats of these glacial relict species are quite different.

Two provisos for successful conservation practice hang over the above comparison. First, the single species objective versus 'habitat' (= vegetation unit) approaches are inextricably bound up with complexities in shifting scales from single sites to the wider landscape. Second, the single species objective versus 'habitat' (= vegetation or biotope) approaches confuse single as opposed to multispecies maintenance as the focus of management objectives.

'Habitat' (= patch) versus landscape approach

- **P9.4: Multiple habitat patchworks provide a basis for more stable long-term persistence of species than a focus on single habitat patches.**
- **P9.5: A total landscape approach based on patchworks is essential in the face of climate change and landscape fragmentation that induce both extinctions and colonizations of species.**
- **P9.6: Conservation of species is enhanced by a total landscape view, incorporating networks of resources in a joined-up landscape, rather than one just based on a habitat patchwork isolated by matrix.**
- **P9.7: The matrix is not beyond repair and is rarely without some resources; it can be resurrected into useful butterfly habitat.**

It is now well appreciated that, given the choice of actions, a species' persistence is generally enhanced by habitat patchworks compared with single sites. The SLOSS argument in favour of single large sites compared with several small ones is somewhat artificial inasmuch, given the restriction of matching area for area, the balance shifts with the following factors:

- The size of the units (e.g., SLOSS could be SHOMS: single huge or many small patches).
- The degree of isolation among units of a multiple patchwork.
- The increasing substrate and biotope heterogeneity and host use specialism/generalism (Steffan-Dewenter and Tscharntke, 2000; Tscharntke *et al.*, 2002).

Box 9.1 Vegetation and other substrates on the Great Ormes Head, North Wales, UK and breeding resources for butterflies occupying the Carboniferous limestone headland

An important issue is how much of the countryside is used as habitat by organisms. Although, clearly, there are biotope hot spots that have high biodiversity, it is argued that taken over all organisms, much of what is referred to as matrix is used for resources by some species (Dennis *et al.*, 2007). This is most obvious for birds but is also evident in butterflies. Here the example of the Great Ormes Head is taken. Butterflies have been mapped in detail for this headland by the Leeds University team (Cowley *et al.*, 2000, 2001a, 2001b) and, together with observations from the author, have been related to biotopes, substrates and vegetation units previously mapped by the Countryside Council for Wales (Fig. B9.1). Substantial differences emerge in the use of different vegetation classes and substrates, but there is not a single surface not used by a butterfly species as a resource at some stage, with the exception of open water, and even then the springs on the headland (e.g., Ffynnon Gogart on the west slopes) have associated crucifers accessible for the green-veined white *Pieris napi* (Table B9.1). These details indicate that a great deal more attention is needed as to what butterflies actually use; this can be achieved by observing their behaviour in relation to substrates and surfaces, rock, vegetation, human structures and all else. This can become simply part of casual observation on walks, or undertaken more formally during transect work (Dennis, 2004a) and during mapping for atlases (Hardy and Dennis, 2007). These observations form the basis for converting matrix to resources and habitat for butterflies and other organisms, countering the notions of a green desert (Dennis, 2009a).

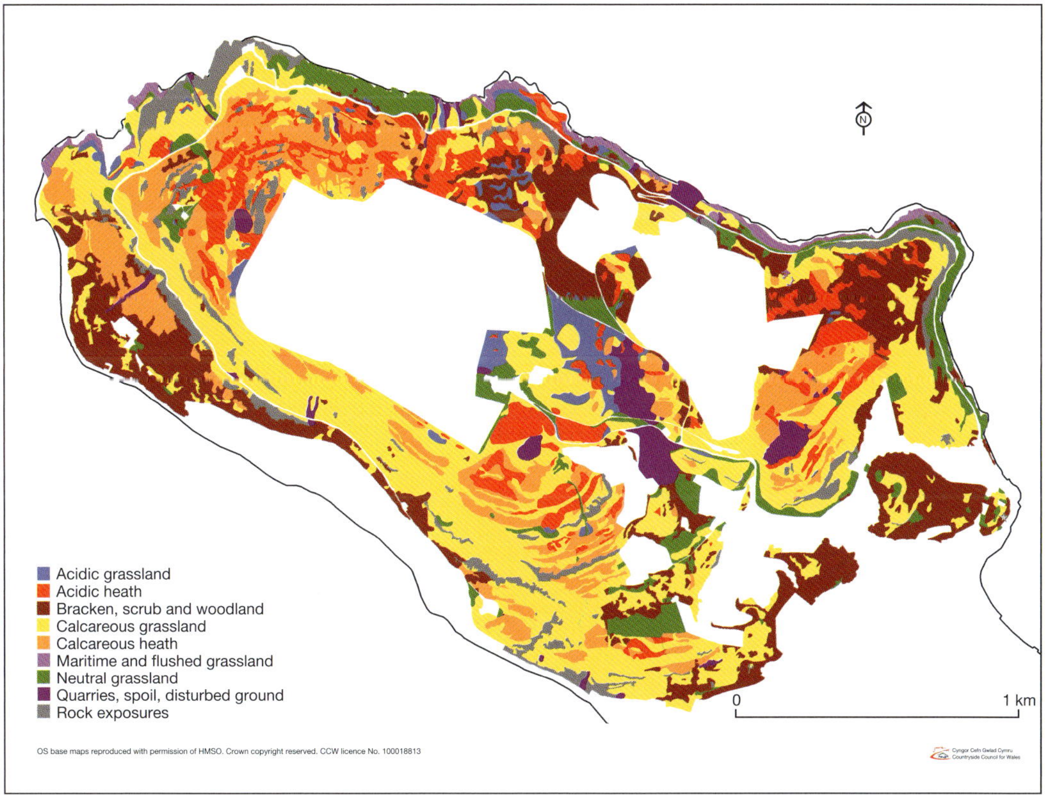

Fig. B9.1 Distribution of vegetation units on the Great Ormes Head mapped by D. G. Guest and S. L. N. Smith in 1994 (Countryside Council for Wales, Bangor; www.ccw.gov.uk/). (From Stevens *et al.*, 1995, courtesy of Adrian Fowles; © Crown copyright. All rights reserved. Countryside Council for Wales, 100018813 [1994].)

Table B9.1 Butterfly species associated with vegetation and substrates on the Great Ormes Head, North Wales. (From Dennis *et al.*, 2007, courtesy of the Royal Entomological Society.)

Vegetation and substrate class*	Butterfly species†
CG1 *Festuca ovina–Carlina vulgaris* grassland	*O. sylvanus*, ***C. crocea***, ***P. argus***, ***A. agestis***, ***P. icarus***, ***A. aglaja***, ***L. megera***, ***H. semele***, *P. tithonus*, *M. jurtina*, *A. hyperantus*, ***C. pamphilus***
CG2 *Festuca–Avenula pratensis* grassland	*O. sylvanus*, ***C. crocea***, ***P. argus***, ***A. agestis***, ***P. icarus***, ***A. aglaja***, ***L. megera***, ***H. semele***, ***P. tithonus***, ***M. jurtina***, *A. hyperantus*, ***C. pamphilus***
CG6 *Avenula pubescens* grassland	***T. sylvestris***, *O. sylvanus*, *C. crocea*, *A. agestis*, ***P. icarus***, *A. aglaja*, *L. megera*, *H. semele*, ***P. tithonus***, ***M. jurtina***, *A. hyperantus*, *C. pamphilus*
CG10 *Festuca ovina–Agrostis capillaris–Thymus praecox* grassland	*T. sylvestris*, *O. sylvanus*, *C. crocea*, *P. icarus*, *A. aglaja*, *L. megera*, *H. semele*, *P. tithonus*, ***M. jurtina***, *A. hyperantus*, ***C. pamphilus***
U4 *Festuca ovina–Agrostis capillaris–Galium saxatile* grassland	*T. sylvestris*, *C. crocea*, *P. icarus*, ***V. cardui***, *A. aglaja*, *L. megera*, *H. semele*, *P. tithonus*, ***M. jurtina***, ***C. pamphilus***
MG1 *Arrhenatherum elatius* grassland	***T. sylvestris***, ***O. sylvanus***, *C. crocea*, ***P. icarus***, *V. cardui*, *A. aglaja*, *L. megera*, *H. semele*, ***P. tithonus***, ***M. jurtina***, ***A. hyperantus***, *C. pamphilus*
MG6 *Lolium–Cynosurus* grassland (semi-improved grassland, cemetery)	*T. sylvestris*, *C. crocea*, ***L. phlaeas***, ***P. icarus***, *V. atalanta*, ***V. cardui***, *A. urticae*, *I. io*, *L. megera*, *H. semele*, *P. tithonus*, *M. jurtina*, *C. pamphilus*
MC4 *Brassica oleracea* maritime cliff-ledge community	***P. brassicae***, ***P. rapae***, ***P. napi***
MC8 *Festuca rubra–Armeria maritima* maritime grassland‡	*P. aegeria*, *H. semele*, *P. tithonus*, ***M. jurtina***, *C. pamphilus*
MC9 *Holcus lanatus* maritime grassland‡	*T. sylvestris*, *P. aegeria*, ***L. megera***, *H. semele*
M24 *Molinia–Cirsium dissectum* fen meadow²	***O. sylvanus***, ***M. jurtina***
H8 *Calluna vulgaris–Ulex gallii* heath	*O. sylvanus*, *A. agestis*, *P. icarus*, *A. aglaja*, *H. semele*, ***P. tithonus***, ***M. jurtina***, ***C. pamphilus***
Brachypodium sylvaticum grassland‡	***O. sylvanus***, *P. aegeria*, ***L. megera***, *A. hyperantus*
H10 *Calluna vulgaris–Erica cinerea* heath	*T. sylvestris*, *P. icarus*, *A. aglaja*, *L. megera*, *H. semele*, ***P. tithonus***, ***M. jurtina***, ***C. pamphilus***
'CGH' calcicolous grass heath	*T. sylvestris*, *C. crocea*, *P. argus*, *A. agestis*, *P. icarus*, *A. aglaja*, *L. megera*, *H. semele*, ***P. tithonus***, ***M. jurtina***, ***C. pamphilus***
U20 *Pteridium aquilinium–Galium saxatile* community (dense bracken)	***T. sylvestris***, ***O. sylvanus***, ***A. aglaja***, *H. semele*, ***P. tithonus***, ***M. jurtina***, *C. pamphilus*
Scrub (*Ulex europaeus*, *Rubus* spp.)	***O. sylvanus***, *L. phlaeas*, ***P. aegeria***, ***P. tithonus***, *M. jurtina*
Woodland	*P. c-album*, ***P. aegeria***
Exposed rock (cliffs, crags, pavement, erosion scars, scree, quarries, rock walls)	*P. napi*, *C. argiolus*, *V. atalanta*, ***L. megera***, ***H. semele***
Amenity (improved) grassland (playing fields, intensely used farmland)	***V. atalanta***, ***V. cardui***, ***A. urticae***, ***I. io***, ***P. c-album***
Urban and gardens	***G. rhamni***, ***P. brassicae***, ***P. rapae***, ***P. napi***, ***A. cardamines***, ***C. argiolus***, ***V. atalanta***, ***V. cardui***, ***A. urticae***, ***I. io***, ***P. c-album***, ***P. aegeria***, *H. semele*, ***P. tithonus***
Hedges, ditches, verges, tracks, paths, banks and springs	*T. sylvestris*, ***O. sylvanus***, ***P. napi***, ***L. phlaeas***, *C. argiolus*, ***A. urticae***, ***I. io***, ***P. aegeria***, ***L. megera***

* Mainly UK National Vegetation Classification (NVC) categories mapped for the headland by D. G. Guest and S. L. N. Smith in 1994 (Countryside Council for Wales, Bangor; www.ccw.gov.uk/; Stevens *et al.*, 1995). Note that some do not entirely match NVC classes (e.g., CG2, H8, *Brachypodium sylvaticum* grassland).

† Butterfly species recorded having hostplants in >50% quadrats. Bold type indicates species with suitable breeding biotopes, most supported by observations of egg laying and occurrence of both sexes (R. L. H. Dennis, personal observations). Nectar and utility resources are not disclosed but ensure wider use of vegetation and substrate on the headland than listed (see text). For full species names, see Appendix 1a.

‡ Vegetation units covering small areas.

Given the choice, a metapopulation structure is more stable than a single population (**P9.4**) (see Chapter 6). Conservation of organisms is enhanced by patch size, quality, proximity and number; even here there are provisos – proximity can have its downside and organisms do best when interpatch distances allow transfers of individuals but are sufficiently isolated to depress demographic synchrony and deter equal access by enemies (e.g., parasites) (Hanski *et al.*, 2004). Conservation is also enhanced by the occurrence of multiple metapopulations in distinct landscapes; vulnerability is reduced by independence of these systems and their disposition over varied environments, the approach becoming indispensable with pressures of climate change causing both losses at sites and colonization of new sites (**P9.5**). Conservation should not ignore resources and 'obstacles' occurring outside habitat patches in the wider countryside, in the so-called matrix, as these may enhance or deter interactions among patches. Their impact should be considered in developing and enhancing metapopulation structures from patchworks to networks of resources (e.g., Cheshire ECOnet project; www.cheshire.gov.uk/SREP/NHE_ECOnet_Ecointro.htm) (**P9.6**). The matrix is capable of repair and converting to habitat; again it is a matter of cost, but also of time and objectives (**P9.7**). Inevitably, a landscape approach will become a wider multispecies approach, if only because landscapes include more resources than single sites, and species require different resources: one species' matrix may well be another species' habitat. But attention to a wider landscape does not mean we can ignore what happens in single patches. As it is, theory provides the means for assessing the relative merits for improving conditions within single sites as opposed to investing greater effort into patchworks (Hanski *et al.*, 2004:279). All these issues are explored below in more detail.

Single species versus multispecies approach

- **P9.8: A single species approach is largely driven by special standing including rarity, decline or indicator status (flagship).**
- **P9.9: A multispecies approach is justified by substantial losses currently experienced by diverse components of the fauna, including butterflies.**
- **P9.10: Conservation effort on behalf of one (or a few) species tends to affect the resource availability and distribution for other species, including butterflies and other arthropods.**
- **P9.11: Homogenization of vegetation units with fragmentation of semi-natural biotopes tends to result in the coincidence of species' bounds but does not avert the issue of treating species in the residual landscape as requiring distinct resources.**
- **P9.12: As species require different resources, often contrasting in distribution, the multispecies concern equates with a wider landscape approach to conservation.**

With so many insects worldwide, let alone in Britain, there are insufficient resources to allocate the same attention to each species. But, as long as there is a will to conserve single species and resources are generated for this purpose there is no reason why conservation resources should not be directed to them (**P9.8**). British butterflies are sufficiently few in number to allow this approach. Priority is typically directed to species with special standing, particularly rare species (e.g., Lulworth skipper *Thymelicus acteon*), declining species (e.g., pearl-bordered fritillary *Boloria euphrosyne*), those under pending threats such as from climate change (e.g., Glanville fritillary *Melitaea cinxia*, northern brown argus *Aricia artaxerxes*), those lost from the fauna and in the process of being restored (e.g., large blue *Maculinea arion*) (Fox *et al.*, 2006; Simcox and Bourn, 2006), and also flagship species that convey delight and wonder to encourage further support from the wider public (e.g., swallowtail *Papilio machaon*, purple emperor *Apatura iris*). The list of special cases is growing alarmingly, as witnessed in the number of UK Biodiversity Action Plan (BAP) Priority Species (see Appendix 1a) (Fox *et al.*, 2006; Bourn, 2007). Attempts to resurrect two (large copper *Lycaena dispar* and chequered skipper *Carterocephalus palaemon* in England; Pullin, 1997; Asher *et al.*, 2001) have thus far failed.

The sad fact is that a large component of the British butterfly fauna is facing losses (41 species, 76% of 54 species; Fox *et al.*, 2006; Dennis, 2009a, 2009b). Some of this may not be evident, either because much of the loss has occurred prior to the period of detailed mapping in the 1970s (Heath *et al.*, 1983) (e.g., common blue *Polyommatus icarus*, small copper *Lycaena phlaeas*; León-Cortés *et al.*, 1999, 2000), because losses are at scales finer than used in countrywide mapping (Thomas and Abery, 1995) or because positive population indicators biased to special sites such as nature reserves conceal wider negative trends (Pollard and Yates, 1993a). With so many species under threat, a multispecies view assumes an increasing importance as it does anyway in definitions of rarity based on frequency distributions (**P9.9**) (Gaston, 1994): defined in this way (e.g., lower quartile), rare species are always with us. There is, in any case, little escape from a multispecies approach, even when the focus of conservation action is on a single species. Management for one species inevitably affects the resource base of other species (**P9.10**). The greater the difference in resource vegetation structural associations between a target species and other species sharing the same site, the greater the likelihood of adverse effects of targeted single species management. This can be the ecological equivalent of war-time collateral damage and is exactly what has happened with agri-environment schemes and management of Sites of Special Scientific Interest (SSSIs): species requiring varied vegetation structures (e.g., small blue *Cupido minimus* and Duke of Burgundy fritillary *Hamearis lucina* requiring shrubs with herb-rich grassland) have declined following management favouring short turf biotopes to enhance conditions for species such as *Polyommatus bellargus* and silver-spotted skipper *Hesperia comma* (Fox *et al.*, 2006). Unfortunately, the 'habitat' (= vegetation type) view, that species sharing small residual sites in a fragmented landscape belong to a single community, where uniformity of management is prompted by apparent homogeneity of vegetation, is entirely wrong (**P9.11**). This approach can only exacerbate the most serious declines currently on record of species inhabiting flower-rich grasslands, woodland clearings, heaths, dunes and bogs. Monitoring data show that agri-environment schemes have failed to halt the general decline of 40 butterflies assessed on farmland in England (Fox *et al.*, 2006). The resource review emphasizes the uniqueness of species' ecology and management geared to resource diversity. Single species management, in effect, becomes a multispecies vantage. Extending this view, where some diversity is left in the landscape, a multispecies approach is synonymous with a total landscape approach. To recount an earlier maxim: **one species' matrix is another species' habitat** (**P9.12**). For conservation to be doing its job properly, it has no choice but to adopt an ecosystem, entire landscape, approach in which different biotopes require different management regimes (see Ausden, 2007).

THE SINGLE SITE IN BUTTERFLY CONSERVATION

The first part of conserving butterfly species requires attention to sites and resources within sites (patches). In this section, some of the principles concerned with developing habitats for butterflies are considered, followed by guidelines on mapping butterfly habitats. It is axiomatic now that sites cannot exist long in isolation; issues for conserving butterflies at a landscape scale are dealt with in the following section.

Basic principles for within habitat conservation

- **P9.13: A key factor in site success is site area as it provides potential for greater resource abundance and variety.**
- **P9.14: Conservation of species within sites must necessarily consider all the resources required by the species, including utility resources as well as consumer resources for larvae and adults.**
- **P9.15: A resource at any one life history stage may be limiting but what this is may not be obvious to casual observation.**
- **P9.16: Supplementary resources on sites decrease the likelihood of population extinction.**
- **P9.17: Understanding the spatial structure of populations within sites, their resource use in different stages, is key to planning resource allocation within sites.**
- **P9.18: Conservation at the within habitat level must plan for environmental changes at scales ranging from the diurnal to those over decades; current management may not be geared for a species' current incidence or long-term persistence.**
- **P9.19: Topographic heterogeneity and other aspects of habitat diversity provide buffers and may critically influence the capacity of habitat patches to maintain viable populations.**
- **P9.20: Much as the distribution of other organisms on sites can enhance conditions**

for a butterfly species, others pose a threat to butterfly populations and need to be accounted in the management of sites.
- **P9.21: Site management must consider the impact on other species, including non-butterfly species.**

A number of clear principles emerge in conserving species at sites. These contain guidelines for the successful establishment of a species' habitat, the resources and conditions required for long-term persistence at that site without input from surrounding sites. The reader is directed to earlier chapters that deal with issues within the habitat (see Chapter 3) and resource patches (see Chapter 4); the present section sifts the main points. It goes without saying, perhaps, that political issues arise which are not dealt with here. For instance, ownership or tenure of sites, particularly a free hand to direct appropriate management – which is not always the same thing – is a key consideration, together with long-term site security for management to be effective (T. R. New, personal communication).

As a first point, it is inevitable from all we understand of island biogeography and population persistence that site area is of fundamental importance to population persistence (**P9.13**). The larger the area of the site, the larger and more varied the potential resource base and conditions, and the greater the room for management to manoeuvre fractionally over the site in developing suitable habitat. All this makes for a larger population and a bigger buffer against fluctuations that can lead to failure. **Buffers** are landscape structures (e.g., north- and south-facing slopes) that create variation in resources, soften edges and allow for unpredictable changing conditions such as seasonal weather. It is also important for the multispecies context, considered below. There is a counter to this principle as discussed earlier; land use intensity tends to increase with site area and conservation can thus become dependent on small residual sites, an issue that Tim New (personal communication) rightly felt deserved emphasis (Yip *et al.*, 2006). A site must also have all the resources and conditions required by a species (**P9.14**). Flexibility in the resource base requirements no doubt varies with resource attributes (e.g., roost sites may be more varied than suitable larval feeding sites), but a single attribute, deficient or in excess (for composition, physiognomy or connectedness), at any life history stage can be terminal (**P9.15**) and not necessarily obvious to casual survey (e.g., conditions for larval platforms in the white admiral *Limenitis camilla*; Fox, 2005) (Hanski *et al.*, 2004). Moreover, frustratingly, several years of data are often required to detect trends. Clearly, supplementary resources for any resource type such as multiple larval hostplants (e.g., hedge garlic *Alliaria petiolata* in addition to cuckoo flower *Cardamine pratensis* for the orange-tip *Anthocharis cardamines*) provides critical backup for species (**P9.16**), a belt-and-braces approach that can be vital for the persistence of species (e.g., use of buck's-horn plantain *Plantago coronopus* as a secondary larval hostplant by *Melitaea cinxia* in Britain; Thomas and Simcox, 1982).

This means that careful assessment is required, from autecological surveys conducted over a wide enough sample of sites, of consumer and utility resources needed by a species and the dependency of life history stages on environmental (e.g., thermal, moisture) conditions. Reliance on historical conditions at the site, always fragmentary and inaccurate, provides insufficient background. Crucial to success is an understanding of the spatial disposition of resources at two scales, attention to the distribution of complementary resources within sites and of life history stages on resource patches (**P9.17**); these two things encompass what is commonly referred to as resource quality. Just how complex attributes of resource quality can be was revealed earlier in Chapter 4. Variation in resources and conditions lies at the root of planning for environmental changes (**P9.18**). This includes diurnal changes associated with light and weather, seasonal changes associated with local climate, and yet others associated with vegetation succession, direct human impacts and indirect ones such as projected climate changes. It is tempting to consider that current management is responsible for a species' incidence on a site. It may well not be, nor may the pattern of current management in the surrounding region be a template for future management (Warren, 1993). Traditional management (e.g., grazing regimes, coppicing) became steadily uneconomic through the 20th century and became largely superseded by modern industrial practices. Not only is the effect of re-establishing past practices unlikely to produce instantaneous success but sound economic grounds are the only effective basis for their continuance. All sites go through changes on a range of timescales and these are part of normal site dynamics. This is why it is of particular importance to butterfly species, and arthropod species generally, to maintain botanical and structural variability in time and space.

Critical to site resource variability is topographic and structural variability (**P9.19**), long known from work on the ridge checkerspot *Euphydryas editha* (Ehrlich and Hanski, 2004); a hummock and hollow topography is a suitable substitute where natural landforms or earlier human earth works are absent. In the absence of detail on habitat components for many butterfly species, and with clear evidence that more species are declining than improving in status despite decades of conservation management, promotion of spatial heterogeneity is likely to have the greatest impact on maintaining habitats for species, simply because it is a strategy that is most likely to maintain a diversity of resources (Dennis *et al.*, 2007) and act as a buffer to population fluctuations from variable weather phenomena (Hanski *et al.*, 2004). Part of developing the resource base is an assessment of components that attract enemies or are in themselves mortality agents on the site (referred to as **traps**) and their disposition in relationship to other resources (**P9.20**). For instance, major nectar sources may become killing grounds for butterflies when close to tree cover (e.g., buddleia bushes in gardens; Roper, 1992, 1993). A classic case of a potential consumer resource acting as a killing agent is the demise of some six million large white *Pieris brassicae* trying to feed on, but becoming glued to, sundew *Drosera* on an island in Sutton Broad, a lake in Norfolk, in August 1911, during one of the butterfly's mass migrations (Williams, 1958). A survey of the plant base is vital – alien plants in particular can swamp butterfly resources (e.g., Japanese knotweed *Fallopia japonica* (Houtt.)) and clearly present a threat to site viability (Hanski *et al.*, 2004).

Finally, as a matter of course, management of sites should assess the impact of changes for a target species on others, not just butterflies but all plants and animals on the site (**P9.21**). Today's common species may become rare ones tomorrow; the number of examples among birds is alarming. Larger sites ensure greater space for all species, though conflict of interest is not inevitably dependent on site dimensions. With increasing size of site there is greater likelihood of variety in substrate and vegetation units and therefore greater potential in catering for diversity, among this rarity. On smaller sites there is greater likelihood of homogeneity in vegetation and substrate; thus, there may also be fewer species to cater for and greater coincidence of habitat *bounds*, but conflict can still arise from the variable dynamics of resources required by different species. These observations on site area are, of course, susceptible to human interest; sites inevitably attract attention from speculative developers in relation to their size and location.

Management and development of existing sites

Much of conservation effort is directed at the enhancement of sites that already contain a target species. These are often identified from the historic occurrence of the species at sites, their need for attention drawn by decreasing numbers at the site or distribution losses elsewhere. Where continuous monitoring of sites occurs (Butterfly Monitoring Scheme (BMS) sites), increasing variation in population numbers provides a good measure of the probability of site extinction (Hanski *et al.*, 2004). Once sites have been identified for enhancement or development, and before action is taken, a resource-based view of habitat suggests a number of issues be taken into consideration:

- Full consideration of resource data required by the species, consumer resources and utilities. A resource database is essential for this task (see below).
- In the absence of resource data, a study is required of resource use variability over a representative sample of sites. It is important to investigate past (extinct sites) and current sites in relation to the spatial array and use of the full resource base.
- Similar data are required of the species' capacity to move in different weather conditions and vegetation structures. Resources must be accessible.
- There is a need to consider allocation of appropriate resources for diurnal and seasonal changes in conditions and projected long-term environmental trends.
- To provide a suitable resource base it may well become necessary to factor in landscape modification or creation, to increase topographic (e.g., hummocks and hollows) and structural variation (e.g., hedges and wood edges as sun traps, wind breaks and movement conduits). Topographic variability enhances some conditions (i.e., light, heat, slope drainage) and moderates others (i.e., wind speed, vegetation succession).
- To evaluate threats to the current site (e.g., insufficient space, human impacts, butterfly enemies). There is little point in developing sites where failure is likely, renders other species vulnerable (T. R. New, personal communication) or continuous management is needed and expensive to maintain. For example, planning the bulk of readily used nectar resources in areas exposed to chilling winds may result in high predation rates.

• To establish threats to other organisms on the site through management objectives for the target butterfly. Will enhancement of the target species' resources impact on other species of conservation concern?
• Once site enhancement has been carried out, it is essential to monitor changes in the population and resource base (Spellerberg, 2005). Some resources will 'decay' faster than others. Testing the success of 'operations' is becoming common practice and methods are rapidly being developed for stricter protocols in comparative evidence-based conservation (Pullin and Knight, 2001; Sutherland *et al.*, 2004).

Currently, site management for most species lacks adequate data on habitats (resource use) to provide clear objectives for site design. There are notable exceptions. For instance, Willmott (1987) explicitly illustrates eight aspects of ideal woodland habitat creation for purple emperor *Apatura iris*, including mature wood edges with southerly aspects, canopy gaps for territories, widely distributed, condensed patches, north to south ride edge lines of mature sallows for breeding, and muddy tracks and stony roads for male nutrition. Management has to fall back to a broader base, that of variation in topography, substrates and structures, vegetation units and biotopes. Far more work is required to test what is most suitable for species in different circumstances and it is likely to become prominent with the development of evidence-based conservation (Pullin and Knight, 2001; Sutherland *et al.*, 2004). This may well require bolder methods than applied in the past, to act as substitutes for periodic, stochastic and 'catastrophic' events (e.g., floods, cliff falls, slope collapse, mass tree falls, forest fires), to achieve objectives, and to bring public opinion onside. Some basic principles for fixed site management apply; these are considered below in a section on landscape restoration.

Mapping and predicting butterfly habitats

It is by no means easy to recognize a habitat, but some situations make the task much easier than others. When habitat is regarded as being synonymous with a vegetation category or biotope, the outcome is that a habitat (or metapopulation patch) is mapped as if a vegetation unit, biotope or land use type. Although this can arise, usually because of severe landscape fragmentation (e.g., an abandoned field corner cut off by road construction; see Fig. B1.1b), it is more often than not highly inappropriate, as emphasized throughout this book (Dennis *et al.*, 2003, 2006b). To avoid mistakes, a working rule of thumb should be that **butterfly species extend over a number of distinct vegetation types, as well as being variably incident and abundant in the same vegetation type**.

When targeting single species, what we need are procedures for directly identifying habitats. Mapping of habitats is made easy where resources coincide and correspond to vegetation units. This is only likely to happen when biotopes have been so degraded, as in much of southern England, that what is left comprises small parcels (patches) of semi-natural vegetation amidst intensively used farmland, patches that can be easily identified by land use survey. In most cases, resources will be in differently sized and shaped parcels, isolated from one another by, what appear to be, non-resource zones – a situation illustrated in Fig. B2.1e. For small areas (*c.* 5 ha) it is possible to map the resources directly in a number of ways (Bennett and Humphries, 1974) and to assess their use either by direct measurement of movements using site mark–release–recapture (MRR) techniques (Henderson, 2003) or by following individuals (e.g., Cant *et al.*, 2005).

As *potentially* suitable areas of biotope increase in size for a target butterfly, and thus the areas of potential resources for it, this direct approach becomes impracticable. It does so also when multiple sites (multipartite populations) are involved, spread out over a large area. A recent, valuable piece of work in Belgium illustrates how the identification of habitats for a butterfly species may proceed for a large, single site (Vanreusel and Van Dyck, 2007). In this situation, identification of habitats requires a three-tier process (Box 9.2):

1 Smaller areas of study are used to identify a set of the butterfly's resources within the different vegetation zones and to determine the capacity for the butterfly to move between resource outlets using MRR.
2 A broader scale mapping programme is required of vegetation zones and resources within vegetation units.
3 Mapping of habitats is then based on the conjunction of resources buffered with daily movements.

Hence, it becomes possible to delineate functional habitat units or 'patches' that do not necessarily reflect physical patches or homogeneous zones in terms of vegetation. Moreover, functional units of butterfly habitat may cover different vegetation types relating to different requirements (e.g., roosting in trees and foraging in nectar-rich grassland). The recognition of the spatial scale at which different resources form functional units depends on our understanding of the behavioural ecology of movements and hence of resource tracking by individuals. For butterflies, this approach has so far only been carried out on one species, the green hairstreak *Callophrys rubi* (L.) Lycaenidae in the Belgian National Park, Hoge Kempen (Vanreusel and Van Dyck, 2007). Although this study did not apply all potential resources to habitat delineation, it indicates a suitable methodology for determining habitat based on resources and movements in line with a resource-based definition of habitats.

Increasingly, there is need to identify habitats for species over much larger areas (political states) and to predict their occurrences for different sites in distinct regions. Models for predictions are used widely in biogeography and ecology (Guisan and Zimmerman, 2000; Manel *et al.*, 2001; Guisan and Thuiller, 2005); accurate predictive models have enormous potential in conservation for identifying the following:

- Sites where there is a high probability of target (usually rare) species.
- Sites or areas that may usefully be restored for the target species in their absence.

Box 9.2 Mapping habitat for the green hairstreak *Callophrys rubi* in the National Park, Hoge Kempen, Belgium

Mapping a butterfly's habitat involves more than just identifying a field boundary that includes the larval hostplant, or just the hostplant itself. The habitat includes a variety of resources, referred to as consumables (i.e., larval hostplants, nectar sources) and utilities (e.g., mate location sites, roosts, hibernation sites. etc.), as well as suitable conditions (i.e., light, heat) where these occur (see Box 2.1). Wouter Vanreusel and Hans Van Dyck (2007) have demonstrated how the process of mapping a habitat(s) is carried out over a large area in the Hoge Kempen National Park of Belgium for *Callophrys rubi* based on a selection of its resources (for larvae (i.e., hostplant, microclimate) and adults (i.e., nectar plants, mate-locating sites, shelter, microclimate)) and the capacity of the butterfly for movement. The process involves building up layers of information on the butterfly, its density in different biotopes, its capacity for movement, as well as its use of different resources in different developmental stages. The habitat is the zone identified by the distribution of resources buffered by a butterfly's capacity for movement. A resource is not much use to a butterfly unless it can access it.

Identifying habitats properly requires painstaking work in collecting suitable data. For this study, the two authors mapped the entire study area (Fig. B9.2a) and delineated vegetation units as homogenous polygons, drawing the boundaries on aerial photographs, using additional GPS measurements. These polygons were digitized in Arcview GIS 3.2 and assigned a vegetation type using the standard vegetation classification for Belgium. Densities of the adult butterfly over different biotopes representing all vegetation types (heath, bogs, nutrient poor grassland, woodland and adjacent biotopes) were obtained by survey of a 784 ha portion of the study area in 2001 and 2002. Movements were studied in 2003 over a 44.6 ha zone (Neerharerheide) using mark–release–recapture (MRR). For each vegetation unit they estimated the status of the key resources selected for *C. rubi* larvae and adults using a simple and therefore easily reproducible scale from 0 to 2 (0, absent; 1, present but not abundant; 2, abundant), the final vector maps having a resolution of 5 m.

The figures illustrate some of the key layers of information built up on the study area for the butterfly, including vegetation types (Fig. B9.2a), heat environment (insolation; Fig. B9.2b), a habitat quality index (Fig. B9.2e) and the final habitat areas identified by buffering using the butterfly's capacity for movement (Fig. B9.2f). This clearly differs from habitat based on biotope (Fig. B9.2c) and the larval hostplant (Fig. B9.2d). Their functional, resource-based habitat approach to mapping a habitat selects those zones that comprise essential resources and conditions within an appropriate spatial window determined by capacity for daily transfers. Appropriate resource variables that were retained in a logistic regression model were used to calculate larval, adult and combined habitat indices in a geographical information system (GIS), taking into account thermal constraints on resource use. To group different (and sometimes scattered) ecological resources into functional habitat zones, the authors derived a measure of space use from MRR data. By least-cost modelling this spatial window was adapted to the nature of the vegetation between sets of resources. The habitat zones that were delineated using this approach matched the observed distribution of butterflies significantly better than did a classic approach based on vegetation types with hostplants only. Here, there is insufficient space to fully describe the details of the technique and readers are encouraged to read the original publication; it forms a vital step in fully understanding habitats of butterfly species.

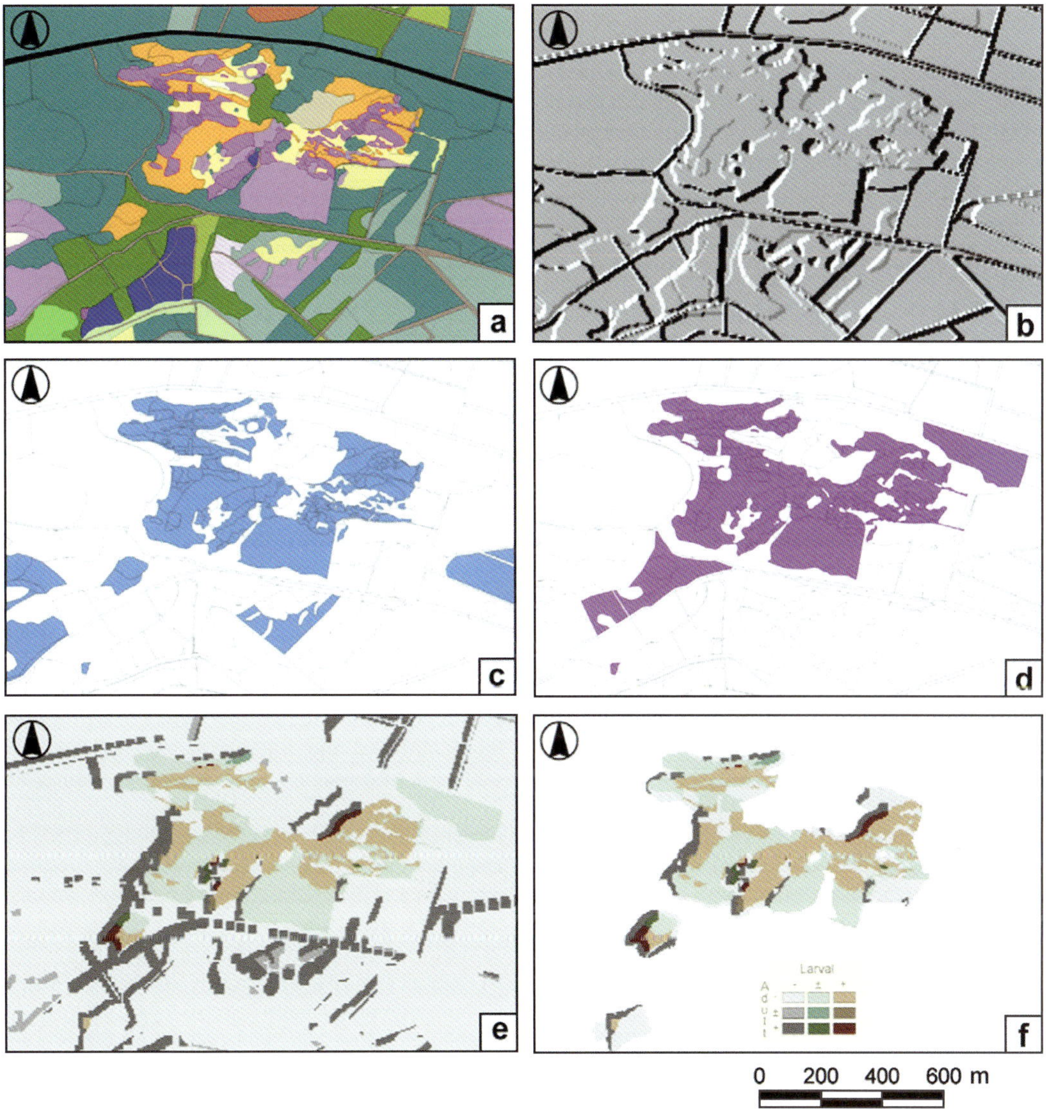

Fig. B9.2 Illustrations of the different information layers for part of the study area, Neerharerheide. (a) Vegetation map (dark and light turquoise: mixed woodland and conifer plantations; dark and light green: woodlands dominated by oak-birch and oak; pink: dry heath with small trees; mauve: dry heath; purple: wet heath; pale and dark yellow: degraded dry heath dominated by *Molinia* grass and small trees; pale brown: large woodland clearing with pioneer vegetation; dark blue, open water). (b) Insolation map (grey areas are flat surfaces with homogeneous insolation, dark areas are shaded and white areas have the highest insolation). (c) Distribution of habitat based on a biotope definition. (d) Distribution of habitat based on the distribution of the larval hostplant. (e) Calculated habitat quality. The larval index is represented by colours: (grey, unsuitable; green, suboptimal; brown, optimal) and the adult index is represented by lightness (brightest light, unsuitable; medium light, suboptimal; dark, optimal). (f) Habitat zones derived after applying buffering techniques. (Courtesy of Wouter Vanreusel and Hans Van Dyck and Elsevier.)

The main benefit is huge cost savings in labour-intensive surveying. Predictive modelling is fast developing for butterflies (Cowley *et al.*, 2000; Maes *et al.*, 2004; Vanreusel *et al.*, 2006). Currently, much of it is based on general environmental data (viz., topography, geology, soils, land use, climate) and applied to broad-scale mapping (>1 km resolution). Such low-resolution approaches have clearly been successful and are very useful for identifying broad areas for conservation (Maes *et al.*, 2003; Cabeza *et al.*, 2004). Key to these developments is testing of predictions; these are typically carried out:

- In the same areas.
- In different regions.

The latter is particularly important as otherwise predictive modelling has narrow finite spatial application.

Ideally, conservation managers require fine resolution (fine mesh) predictions for distinct regions (wide extent). In other words, the desire is to apply the resource-based habitat model to diverse sites in different landscapes from the original surveys. The process has recently been validated for two butterfly species (i.e., grayling *Hipparchia semele*, *Callophrys rubi*) for three areas in the same general region of north Belgium using data on hostplants, nectar sources, shelter and microclimates (Vanreusel *et al.*, 2006). This represents a distinct advance to using topography and vegetation units (Binzenhöfer *et al.*, 2005) and is a promising start. But, transferring resource data between regions raises a potential conundrum: translocating data from diverging (telescoping) scales of mesh (smaller) and geographical extent (larger). Can fine-scale resource data be transferred over wider geographical scales across quantum shifts in landscapes, lithology and climate? With such quantum shifts come major changes in biotopes, vegetation types, vegetation structure and hostplant use and other resource associations (Fielding and Haworth, 1995). For example, *C. rubi* uses different hostplants in different regions of Britain, which may be symbiont (ant) associated (see Appendix 5e) (Thomas in Dennis, 1992a), and also it is known to imbibe sap in the absence of nectar sources (H. T. Eales, personal communication). Two observations follow. First, in predicting habitats a wide database of resource use is a prerequisite (Dennis *et al.*, 2008). It is clearly not sound to predict the occurrence of a species from resource data obtained in a single region. In view of both spatial and temporal variability in the occurrence of resources used, it is necessary that a full understanding of a species' habitat is based on autecological studies in different settings, different conditions and different regions. Second, environmental variables provide a basis for analysing quantum shifts in resource use and can be placed into regression models as distinct factors. It should be pointed out that, in Britain, we are indeed fortunate to have such data. Other regions, even with a high standard of living (e.g., Australia) are not so fortunate and must of necessity use alternative approaches (New, 2006).

As a final point, neither of these fine examples of identifying and predicting butterfly habitats by Vanreusel and colleagues should negate attempts to seek simple vegetation surrogates for butterflies or other organisms over broad regions (Binzenhöfer *et al.*, 2005). The responsibility, of course, remains of demonstrating that such vegetation units hold the resources required by a species. That a real advance in conservation is possible using these approaches is demonstrated by Maes *et al.* (2004) for the Alcon blue *Maculinea alcon* (Schiffermüller). Using data on hostplant, biotope, mobility and colonization capacity, they identified three types of scale-dependent **functional conservation units** (FCUs): FCU-1 occupied patches buffered by 500 m, the maximum local movement distance; FCU-2 occupied patches buffered by 2 km, the maximum observed colonization capacity; and FCU-3 were potential introduction sites (recently extinct sites). They suggested different management and planning measures for each type of FCU. With increased knowledge on resources for species, such FCUs can become more precisely geared to conserving target organisms. Butterfly Conservation has 76 projects as of September 2008 which incorporate this approach (Fig. 9.1) and techniques are fast advancing for prioritizing landscapes for butterfly conservation (Moilanen *et al.*, 2005). These amalgamate functions for connectivity, thus population persistence, with species' complementarity and rarity (Arponen *et al.*, 2005; Moilanen *et al.*, 2005; Moilanen and Wintle, 2007).

MULTIPLE SITES IN SINGLE AND MULTISPECIES APPROACHES

It is now well established that a metapopulation approach, a patchwork of sites, is essential for butterfly conservation. The principles have been largely established for butterflies (e.g., *Melitaea cinxia*) by Ilkka Hanski (Helsinki University) from work in the Åland islands (Hanski and Gilpin, 1997; Ehrlich and Hanski,

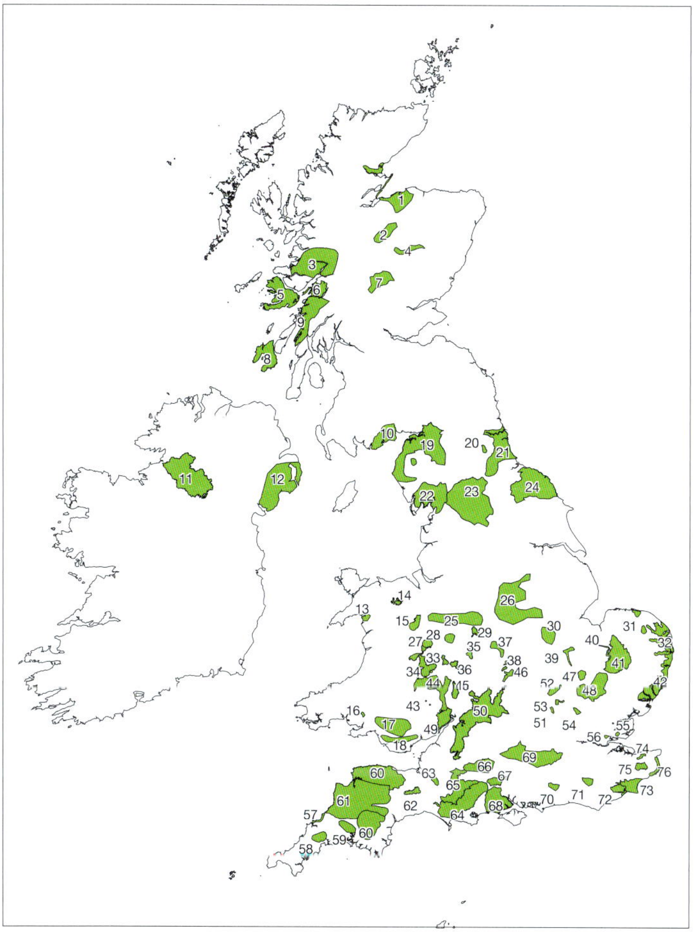

1 Moray Firth
2 Badenoch & Strathspey
3 Lochaber
4 Upper Deeside
5 Mull
6 Lorne
7 Highland Perthshire
8 Islay
9 Mid-Argyll & Knapdale
10 Solway Coast & Hinterland
11 County Fermanagh
12 County Down
13 Harlech
14 Clocaenog Forest
15 Oswestry Uplands
16 Mynydd Mawr
17 South Wales Coal Measures
18 High Brown Brackenlands
19 North Cumbria
20 Durham Coalfield Pennine Fringe
21 North East Brownfields
22 South Cumbria Low Fells & Morecambe Bay Limestones
23 Yorkshire Dales
24 North York Moors
25 North Shropshire & Staffordshire Mosses
26 Notts/Derby Coalfields Brownfields
27 South Shropshire Metal Mines
28 Telford & Wrekin Brownfields
29 Cannock Chase
30 Ketton Limestone Area Brownfields
31 Norfolk Heaths
32 Norfolk Broads
33 South Shropshire Wet Flushes & Rush Pastures
34 South Shropshire and North Herefordshire Woodlands
35 Black Country Brownfields
36 Wyre Forest
37 North Warwickshire Brownfields
38 Princethorpe Woodlands
39 Peterborough Brownfields
40 Cut off Channel
41 The Brecks
42 Suffolk Coast & Heaths (inc. Sandlings)
43 Herefordshire Commons
44 Herefordshire Woodlands
45 Malvern Hills
46 Southam Lias Grasslands
47 Cambridge Elm Belt
48 Cambs & Essex Chalk Grasslands
49 Forest of Dean
50 The Cotswolds
51 Bedfordshire Chilterns
52 Bedfordshire Brownfields
53 Hertfordshire Chalk Grassland
54 Hertfordshire Quarries
55 South Essex Woodlands
56 Langdon Hills, South Essex
57 North Cornwall Coast
58 Mid Cornwall Moors
59 Tamar Valley
60 Two Moors Project
61 Reconnecting the Culm
62 Blackdown Hills (Neroche Project)
63 Polden Hills
64 Cranborne Chase & Wessex Downs
65 Blackmoor Vale
66 Salisbury Plain
67 Tytherley Woods
68 New Forest
69 Thames Basin Heaths
70 Surrey/Sussex Woods West Weald
71 Ashdown Forest
72 Rother Woods
73 Dungerness/Romney Marsh
74 Blean Woods
75 Denge Woods
76 Sandwich Bay

Fig. 9.1 Landscape scale projects conducted by Butterfly Conservation in the UK. The map shows past, present and future landscapes that Butterfly Conservation is working on and includes some projects which were completed before 2008. (Courtesy of Butterfly Conservation and Caroline Bulman, December 2008. © Crown copyright. All rights reserved Natural England 100046223 (2008).)

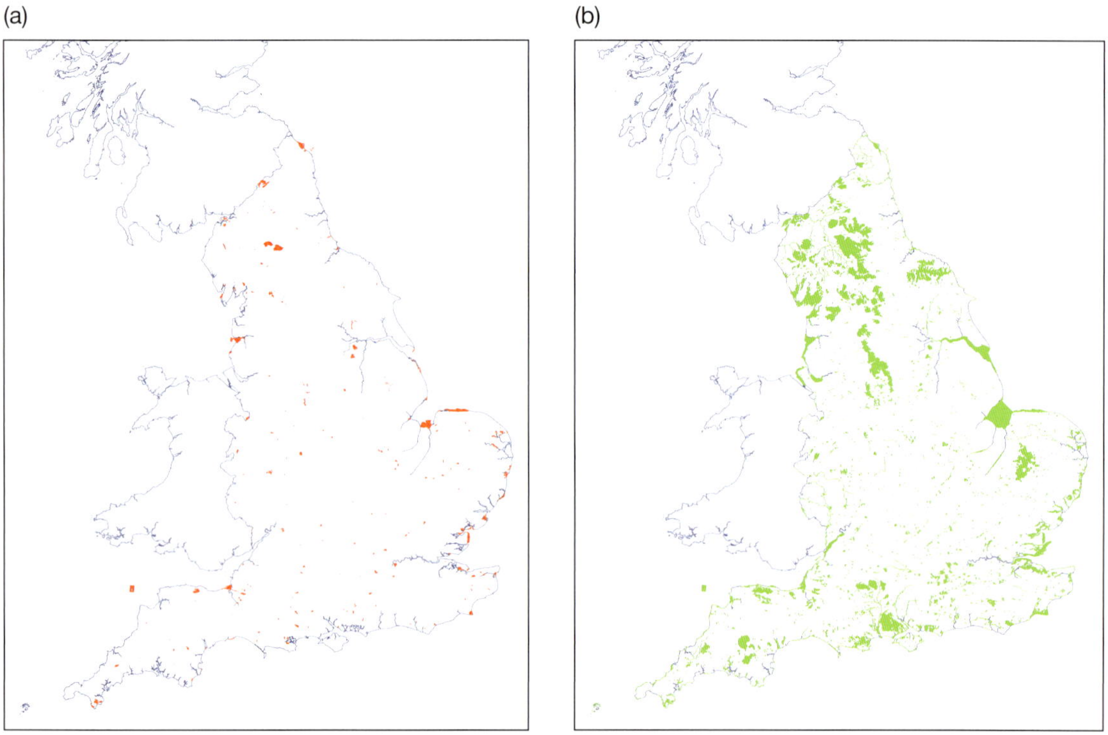

Fig. 9.2 Nature reserves form the bastion for a wildlife patchwork over Britain; these are areas where other organisms are given priority. A number of different types of nature reserves exist, some specifically sites owned by Butterfly Conservation where the focus of attention is butterflies. Key among nature reserves are the National Nature Reserves (NNRs), the most important sites for nature (a). They are sites owned and/or controlled by Natural England. By the end of September 2006, there were 222 NNRs in England. These sites, together with other reserves such as Sites of Special Scientific Interest (SSSIs) (b), form a dense patchwork of distinct biotopes (e.g., downs, heaths, bogs) over the country and are effectively population sources of biodiversity and rare species for sites not held as reserves. (© Crown copyright. All rights reserved Natural England 100046223 2008.)

2004). However, conservation cannot rely entirely on a patchwork of nature reserves, however densely distributed these appear to be on a map (Fig. 9.2). Increasingly, it is being recognized that this approach is not sufficient and that a view of the whole landscape needs to be taken – networking sites and attending to resources within the so-called matrix. When transferred from the case of a single butterfly species to the wider issue of butterfly diversity, and further into entire faunas and floras, this wider landscape view becomes inevitable. It is inevitable simply because of the limitations in available space for habitat patches, and the dependence of conservation on the encouragement of landowners to allow land management for the benefit of wildlife. At its simplest, to emphasize an earlier maxim, one species' matrix is another species' habitat. At the very least habitat bounds are likely to differ for species; as species join the list for concerned interest, so these bounds gradually consume the whole landscape (see Box 9.1). It may be recalled, too, that as more resources – and more small resource patches – for species are considered, so more of the matrix becomes habitat (see Box 6.3). Conservation of species then comes down to what is needed against what is available. If landscape resources are insufficient, there are two choices: more landscape resources must be found by whatever means (e.g., purchased land, land restored) or, inevitably, the long-term prognosis is that of local extinction. It is important to emphasize the long-term view; all sites, however apparently healthy, face extinction from either stochastic (random) or deterministic (e.g., vegetation succession, management changes) causes. The usually recognized objective of a landscape approach is attainment of dynamic homeostasis for species over large areas or regions through the integration of sites, habitats and resources. With such integration, the impact of adverse events is spread over numerous sites, each in turn acting as sources or sinks in relation to their site potential, typically linked to size and isolation. A key question, of course, is just how best to enhance landscapes for species. Where many different organisms (birds, mammals, plants, arthropods, etc.) are involved, indicator groups, comprising assemblages of key species, are valuable for identifying conditions, structures and resources (Webb and Lot, 2006). However, a focus on British butterflies justifies reference entirely to species' resources. For this is needed a butterfly resource database; this is now steadily being developed (Hardy *et al.*, 2007; Dennis *et al.*, 2008; see below). First, a view of what principles apply for butterfly landscape conservation.

Basic principles for conserving butterfly habitats at the landscape scale

- **P9.22: Understanding the spatial structure of populations is crucial to preserving patchily distributed species.**
- **P9.23: Metapopulation models combined with reserve site selection algorithms provide a robust framework for conservation of metapopulations.**
- **P9.24: Single, large, high-quality patches may play an important role in maintaining regional persistence.**
- **P9.25: A moderate degree of connectivity is generally most beneficial for long-term persistence.**
- **P9.26: The presence of an active metapopulation in the current landscape may be deceptive because metapopulation dynamics tracks environmental changes with a substantial delay.**
- **P9.27: Components and/or elements of the matrix stimulate resource use by individuals leaving habitat patches.**
- **P9.28: The composition, structure and connectivity of elements of the matrix influence movements and thus interactions among populations occupying habitat patches.**
- **P9.29: Species search for resources within the matrix, perhaps more so with increased structural complexity, providing a strong argument for mending the matrix.**
- **P9.30: Networks with broken or poor habitat connectivity, creating conduits of movement to 'dead ends' (terminator or end nodes), present priority areas for consideration.**
- **P9.31: The smaller the habitat patches supporting a metapopulation, the greater the attention needed to be given to the matrix resource base and structures.**
- **P9.32: An array of valuable resources for butterflies can be provided for butterfly species in the matrix based on available landscape structures.**
- **P9.33: Opportunities should be taken to expand habitat patches into the surrounding matrix wherever possible.**
- **P9.34: The impact of the matrix on a species will vary with time and conditions, all of which need to be accounted in a management programme.**

An underlying principle riveted into conservation by the Helsinki team (Hanski *et al.*, 2004) is that understanding the spatial structure of populations is crucial to preserving patchily distributed species (**P9.22**). As all species become patchily distributed with biotope fragmentation, in effect this includes all butterfly species. It just becomes more urgent for those with limited powers of dispersal. In concrete terms, Hanski *et al.* (2004) envisage that metapopulation models combined with reserve site selection algorithms provide a robust framework for conserving metapopulations (**P9.23**). As there is usually little room for choice of habitat sites, site selection in most cases is best reinterpreted as site design and management; this determines site quality and increased resource quality makes up for limitations of site dimensions. Hanski has shown how metapopulation theory can be used to predict how the structure of habitat patchwork influences metapopulation size and persistence. His group has advanced the notion that models developed for thoroughly studied, relatively common species provide a basis for understanding the dynamics, and thus conservation, of ecologically related rarer and threatened species. A proviso is worth bearing in mind from our resource-based habitat view, that 'ecologically related' does not imply being 'ecologically identical'; differences can have critical consequences, potentially, if this formula is treated as being axiomatic. Moreover, it is important to emphasize the dangers of assuming simplistic notions of some magical threshold patch number, size and isolation for attaining long-term persistence. Caution is required in following commonly quoted rulings of the sufficiency of arbitrary numbers of patches, typically 20 patches, based on limited metapopulation research into a few species in somewhat exotic (range edge and substrate-confined) circumstances as the Åland islands or *Euphydryas editha* sites on serpentine grasslands in California (Thomas and Hanski, 1997; Ehrlich and Hanski, 2004). Just how long metapopulations persist very likely depends on regional-scale deterministic processes such as vegetation succession, changes in farming practice and climate, or invasions of plants and animals. Notably, the initial study on the silver-spotted skipper *Hesperia comma* metapopulation (Hill *et al.*, 1996) failed to predict its later expansion associated with climate change (T. G. Shreeve, personal communication). It has long been known, however, that single, huge, high-quality patches play an important part in maintaining regional species' persistence (**P9.24**), as with *E. editha* at the Morgan Hill site (>1500 ha) for the butterfly in California (Hellmann *et al.*, 2004). What is less well understood is that an intermediate level of patchwork connectivity is generally more beneficial for long-term persistence (**P9.25**) (Hanski *et al.*, 2004). Greater isolation gradually reduces colonization to zero, whereas close interactions risk extinction owing to regional stochasticity and correlated environmental events (viz., high mortality from parasites). One should not forget that patchwork structure can select for biological attributes, such as migration rates and larval hostplant affiliation, which may further affect persistence (see Chapter 6). Nor should one lose sight of the fact of possible extinction debt; the presence of a metapopulation, as of a population at a single site, may conceal systematic 'decay' (**P9.26**). Worrying processes may be in train such as reduced immigration, excessive emigration or inbreeding depression (Saccheri *et al.*, 1998).

A number of crucial principles, which cannot be ignored in conservation, apply to what is going on beyond the targeted patchwork (habitat patches) and indeed that affect the dimensions and make-up of the targeted patchwork. The first of these is that components and elements of the matrix facilitate resource use by individuals leaving habitat patches (**P9.27**). Earlier (see Chapter 6) it was explained how a wide array of resources, mainly utilities, but also consumer resources (i.e., nectar, small clumps of larval hostplants) occur and are used in the matrix. The very structure of the matrix creates contrasts in local climate, crucially the thermal environment, to which migrants are subject and can exploit. It is not difficult to appreciate, then, how matrix components can influence movements between habitat patches, thus success of colonization and population exchanges and population numbers and turnover (**P9.28**). Other observations support attention to the matrix. Butterflies inevitably search the matrix for resources even when these do not exist (**P9.29**) and the intensity of search may well reflect the degree of matrix structural complexity. This searching behaviour can potentially have serious consequences for a metapopulation depending on how matrix structures affect connectivity. Low connectivity (e.g., truncated hedgerows) may create traplines and 'death pools' where individuals congregate and are subject to higher predation, clearly important structures for the attention of management (**P9.30**). There may well be a rule of thumb about matrix management that advocates increasing attention to the matrix the smaller the basic habitat patchwork (**P9.31**). Underlying this are expectations of the **Allee effect** – increased emigration expected to occur from sparser populations – which, if limitations of small patch size are not compensated for by increased patch quality, probably coincides with small sites having vulnerable, small populations.

With these points in mind, it is not difficult to envisage how landscapes can be greatly enhanced for butterflies. The key in what to do, always must be the study of what butterflies use and the effects landscape components have on movements and population (actually, metapopulation) parameters (see Chapter 6). Above all, it should be borne in mind that it is not always to a butterfly's advantage to maximize transfers between population units (see Hanski *et al.*, 2004). Even so, an array of valuable resources can be provided for butterfly species in the matrix founded on the *available* landscape structures such as hedges, verges, ditches, banks, fences, wood edges, walls, hollows, pond edges, gardens and so on (**P9.32**). In this process, clearly every opportunity should be taken for expanding the number of habitat patches (e.g., adding resources to promote sinks to sources) and enlarging and enriching smaller patches (e.g., converting biodiversity-poor biotope parts of patches, such as dense conifer plantations, to richer ones) (**P9.33**). Ultimately, sight should not be lost of the fact that the impact of matrix influences will vary with time and changing conditions, the consequences of which need to be accounted in any management programme (**P9.34**).

Broad principles for conserving landscapes for multispecies

From the above, a number of principles apply to conserving landscapes for butterfly communities:

- **P9.35: A resource-based view of butterfly habitats would suggest that a successful approach to the conservation of landscapes for butterfly communities should not just be based on broad vegetation units.**
- **P9.36: As species have distinct resources, their habitats are unique and their habitat bounds will differ, such that more of the matrix becomes butterfly habitat in need of management the more species that are considered.**
- **P9.37: As fragmentation of biotopes leads to equivalence of habitat bounds among species but not equivalence of resource distributions and resource lifespan within those bounds, attention needs to be given to resource allocation within biotopes.**
- **P9.38: A useful guide to managing landscapes for butterflies is provided by ecological groups reflecting species' associations for similar resources and conditions.**
- **P9.39: Landscape elements may conflict for different species in terms of both patch and matrix, and consequently management for landscape heterogeneity is advocated.**
- **P9.40: Common elements of landscapes are used by species, particularly thermal conditions, providing key management targets.**

When consideration is extended from single species to multiple species, as must be the case anyway when considering how management may affect other species, the resource-based view of habitat would suggest that the most profitable way forward would be to cater for the resources and conditions of all species likely to occur within the neighbourhood of sites (**P9.35**). Even when conditions appear to be homogeneous they rarely are, as Warren (1993) discovered for management over 308 prime butterfly sites in south central England: grazed, mown and actively managed calcicolous grasslands revealed substantial regional bias and local contrasts. The key consideration is that as habitats (and niches) for species are unique, the habitat bounds are unlikely to be identical and more of the landscape becomes relevant for consideration in any management programme as attention shifts from single species to biota (**P9.36**). Only in severely fragmented landscapes where the matrix has been homogenized (e.g., East Anglian patches in expanses of cereals) are species likely to share habitat bounds, but as their resources differ in kind and 'dynamics', conservation will need to consider the balance in resource allocation within the fragments and plan for networking the fragments for species transferring between patches (**P9.37**). A valuable guideline for determining commonality in resource requirements among species is available from research into ecological associations (**P9.38**) (see Chapter 5). Species broadly differ for biotopes, vegetation types, landforms and substrates and such basic distinctions in landscape provide a starting point to determine how specific resources may be allocated to species belonging to such divisions. The concern here is that landscape elements suitable for one species may be unsuitable for another species (**P9.39**); this does not imply insuperable conflict of interest and should not necessitate the removal of an entire landscape element, but management for contrasting landscape elements (e.g., light and shade; dry and wet biotopes; nutrient-rich and -poor substrates; grassland and shrub) in the same area. Not all will be conflict; many resources are used by several species (e.g., large nettle patches in open, sheltered spaces by the nymphalids peacock *Inachis io* and small tortoiseshell *Aglais urticae*) and some resources (warm local climates) are common to most if not all species in the British islands (**P9.40**). It has long been known that animal diversity is driven by biotope heterogeneity (e.g., McCoy and Bell, 1991; Tews *et al.*, 2004); conservation has the concern of all organisms, not just butterflies, and heterogeneity should be a key principle of site managers.

Management and development of existing landscapes

Most conservation is faced with one essential fixture – a given landscape, with fixed topography, landforms and human landscape features with butterflies largely dependent on semi-natural biotopes. Conservation is often 'politically' bound, limited to areas of local jurisdiction (e.g., National Trust) or administrative divisions (e.g., county boundaries). Just how the landscape should be managed depends crucially on whether the objective is for a single (rare) species or for a community of species, though it should be impressed on management that in conserving one species the impact should be minimized on other species. Some basic management issues for an area, which would typically involve one or multiple metapopulations for each butterfly species, can be listed; the issues already addressed for single sites above pertain:

1 A start should be made by accessing literature/data sources for the species and the status of species within the area under consideration – especial note being taken of the BAP and rare species (lowest quartile of distribution cover). If data are lacking then the first task is one of mapping distributions. The detail of this, as in everything, depends on the spatial scale (extent) and the number of surveyors. If resources for the task are limited then map organisms for distinct land units (vegetation types, biotopes, substrates, landscape features).

2 Obtain information on resources required by species and, if necessary, survey species to obtain information on specific resource and environmental associations of species over a representative sample of sites within the area. For single species, a resource-based approach can be made; for butterfly communities, different approaches are possible, including those that are taxon orientated (indicator species, key species assemblages; cf., Webb and Lott, 2006) and feature orientated (key landforms, biotopes). Multivariate techniques can be used to identify landforms, landscape features, conditions and resources that different groups of species share in common.

3 Assess how semi-natural biotopes can be enhanced for the species, not just rare ones.

4 Assess how the 'matrix' can be enhanced for species; what possibilities are there for resource nodes and corridors, flyways between habitat patches, etc? In this multispecies context, matrix is not so much considered as non-habitat land but as land outside 'control', often aggressively exploited, in which it will be necessary to apply considerable diplomacy and whatever benefits are at disposal to encourage suitable management.

5 Continuous monitoring is essential. Things do not stay the same and in particular vegetation changes rapidly. For example, it may be necessary to apply crude mechanisms to 'jump-start' or retard vegetation succession for species that require short turf and bare ground conditions.

The most realistic approach to managing landscapes for butterfly communities is a resource-based one, with the objective of creating ecological networks, ideally ecosystem networks founded on fluvial catchments (Box 9.3). In Britain, we are generally aware of what species occur within our neighbourhood and therefore it is a matter of planning for them using resource data. But the information required to carry this out is not readily available for all species, in which case resort to alternative approaches is often necessary. In case a reader has directly jumped to this section, it is necessary to reiterate that as species have distinct resources, their habitats will be unique and their habitat bounds are likely to differ as a consequence, the more so as resources differ. As semi-natural vegetation units become reduced in size and homogenized, there is an increased probability that species will share much the same habitat bounds. But there is still an issue of whether they share the same fine-scale substrates within single vegetation units, a level below any of the most detailed mapping programmes (e.g., 10×10 m). Each substrate or vegetation subunit has its own dynamics, and congruence in habitat boundaries is not synonymous with identity in resource use and lifespan. Patch and matrix bounds are further confounded by temporal changes (Wiens, 1996; Thomas and Kunin, 1999) and spatial (regional) variation. Just what appears to be a habitat patch changes on scales of seconds to decades (Dennis and Sparks, 2006). Those engaged in conservation practice are constantly faced with successional changes on sites as well as changes in conditions induced by human activities (Sheppard, 2002; Offer *et al.*, 2003; Underhill-Day, 2005). But heterogeneity is not equally accessible from every vantage; it is more easily achieved in later successional stages than in earlier ones, as woodland takes time to grow and bare ground can be created instantly within it but not vice versa. Change is integral for sites and a strong argument has been put forward for conservation to be geared to managing dynamics (Dennis *et al.*, 2007).

As part of promoting dynamics we also advocate heterogeneity and attention to the matrix (Dennis *et al.*, 2007). In this, it is unwise to ignore the potential of the general matrix for at least three reasons:

1 Resources are present in the landscape matrix for species (Dover and Sparks, 2000; Dennis 2004a) and can be promoted in the matrix.

Box 9.3 Components of ecological networks for conserving butterflies at the landscape scale

Conserving organisms at the landscape scale, even when given a free hand to manage at will, constitutes different considerations in the case of single species and whole butterfly (or more inclusively, all organism) communities. The reason is simply that management appropriate for one species can be entirely inappropriate for another. Nevertheless, at the landscape scale and working for diversity the aim is to build ecological networks (Fig. B9.3c). Three fundamental directives apply in the single species case:

- **Maximize the area presented by patches of semi-natural biotopes outside production**.
- **Enhance the quality of resources within semi-natural biotope (habitat) patches**.
- **Maximize the connectivity among sites**.

These three principles may be called the '**metapopulation trio**' (patch area, quality and isolation; Samways, 2007). By enhancing area of patches (e.g., removal of alien conifers such as spruce *Picea* sp. plantations in woodlands), populations of a target butterfly species can be increased on sites. The population is maximized by enhancing the quality of resources (Fig. B9.3b), ensuring that resource composition, utility structures and conditions for the species, and internal connectivity within sites for resource outlets, are appropriate as determined from autecology surveys (see Box 9.2). Finally, maximizing connectivity among sites (e.g., by enlarging vegetation types with semi-natural biotopes and acquiring neighbouring land; Fig. B9.3a), enhances viability of the patchwork as a metapopulation. In strict metapopulation terms, this can only be done by increasing patchwork number or area; both will reduce the distance between patches.

To these may be added other directives that move the landscape from 'patchworks' (Fig. B9.3d) to 'networks' (Fig. B9.3c, f). In doing this, we make the landscape not just fit for all organisms but enhance it also for the human animal too.

- **Increase biotope and substrate heterogeneity of the region (multiple patchworks of distinct biotopes, resources and substrates for different organisms)**.
- **Establish ecological networks through the matrix, linking up the patchworks (edges, nodes of resources)**.
- **Convert parts of the matrix to produce resources and 'mainland' for communities, particularly those within dispersal distance of already occupied sites, a process of providing buffers**.

These three objectives have the aim of increasing the variety of patchworks (e.g., different vegetation communities), a **mosaic**, and ensuring that they are linked up with transit routes. A variety of patchworks is required because different organisms have different habitats (Fig. B9.3e), that is, they require different resources. Ecological networks imply ecosystem linkages, to cater for the variety of organisms in transit between their distinctive patchworks (habitats), and not just a unitary type of link required for a single species (Fig. B9.3f). An example from butterflies will make the point: a trimmed, sparse, hedge-line fronted by bare ground, or a hard, fenced-off woodland edge verging fields, may be sufficient to enhance migration of a nymphalid such as *Aglais urticae* or *Inachis io*, but satyrines such as *Maniola jurtina* and *Pyronia tithonus* will benefit more from having grass verges, paths and banks, and set aside margins in arable fields abounding in nectar sources. All these organisms may well be able to cross arable deserts during hot, calm, fine weather, but in summers matching that of 2007 in Britain, such networking becomes critical for transits.

Networking requires making provision for a variety of resources – to match those resources in habitat patches – along edges in the landscape (river banks, motorway and road verges, rail embankments and cuttings, field margins, woodland surrounds, canal banks). This includes, where possible (e.g., at hedge junctions, crossroads, pits and hollows, hummocks, abandoned quarries, etc.), the development of nodes (mini-habitat patches). Although, studies of single species have called into question the validity of 'corridors' and stepping stones in a metapopulation context (Hanski and Gilpin, 1997), which can indeed become sinks or lines of increased mortality, there are much stronger arguments for ecosystem networking in the multiple organism case. 'Edges' and nodes through the matrix are habitats for many organisms, if not for particular butterflies, and increased physical structuring of the landscape simultaneously exaggerates and moderates local climates, to the benefit of overall diversity. **The objective should be arthropod and all-organism conservation, not just provision for butterflies**; for instance, ponds may not feature much in butterfly conservation, though they should for their margins (Dennis, 1982a), but are vitally important for other organisms (Biggs *et al.*, 2005). These principles are illustrated in Fig. B9.3a–f; Fig. B9.3g illustrates features of connectivity. In restoration ecology, connectivity is understood as physical links between reserves and should not be confused with the measure of connectivity between populations occupying patches (see Chapter 6). Clearly, reserves may be linked with various degrees of sophistication. Minimally, in a completely connected system, the number of vertices (arcs, a) is $(n - 1)$ where n is nodes (see Box 6.5). With increasing

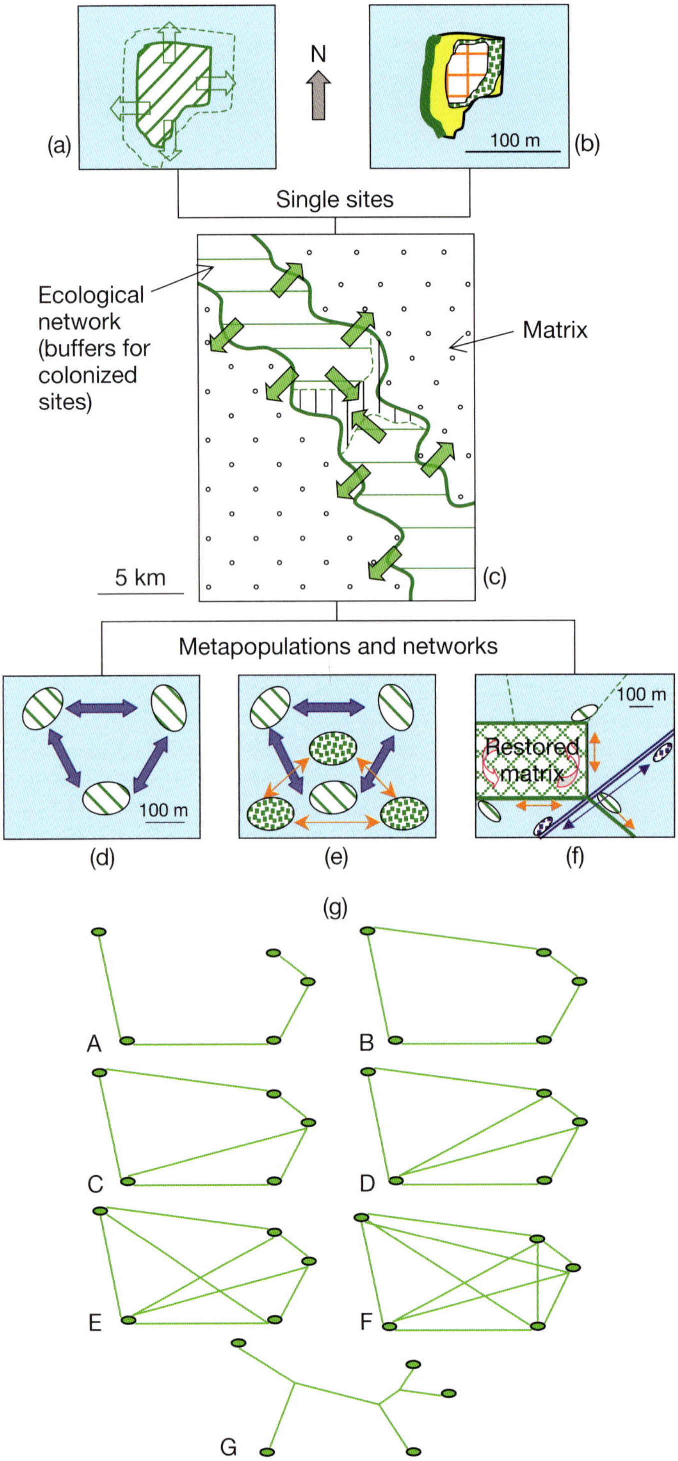

Fig. B9.3 (a) The size of single patches is key to conserving organisms. Acquiring adjacent land is one way of increasing patch size. (b) Habitat quality is a key issue in persistence of populations. This can be acquired by improving local climate (e.g., windbreaks: green band on west of patch) or improving resources (e.g., scrub clearance: green stipple, scrub; cross shading, cleared area). (c) Building ecological networks (horizontal shading) across the countryside involves taking in more of the matrix (stippled), often by donations of adjoining land (vertical shading) as in the case of the Cheshire ECOnet (e.g., National Trust's Bickerton Hill received land from ISH Dennis Estate). This is best achieved by taking in land within dispersal distances of occupied sites of target species. (d) Metapopulations (three patches, linked up by movement across the matrix) are a key feature in ecological networks. (e) Different organisms may require very different patchworks. Here two are shown linked up independently by vagrants across the matrix. (f) Networks for organisms illustrating two patchworks (stippled blue, marsh; cross shaded, tall-herb grassland) with field edges having low trimmed hedges (pecked lines) and tall dense hedges with banks (green lines differing in thickness for effectiveness of shelter), a canal and its bank (parallel blue lines). A restored matrix is shown cross-shaded. Arrows indicate movements of organisms with general movements through the restored matrix. The pale blue areas in (a), (b) and (d–f) are matrix. (g) Networks in landscape restoration. Parts A–F illustrate increasing links for a five-node network with a standardized β index rising from 0.4 to 1.0. Part G illustrates the shortest set of lines connecting all five nodes. Ellipses, reserves; lines, corridors or flyways. (a–f, substantially modified from Samways, 2007, courtesy of the Royal Entomological Society; g, modified from Bunge, 1962, in Haggett, 1965, courtesy of Hodder Education.)

links there is an increase in the standardized β index (Fig. B9.3g), where:

$$\beta = a/{}^{1}/_{2}n(n-1).$$

Where restoration is limited to an original field system when field integrity has to be maintained, increasing connectivity depends on re-establishing (renewing walls, hedges, banks and ditches) and improving structures of field bounds. However, in a landscape where restoration has a free hand but limited resources, reserves can be linked more directly (Fig. B9.3g(G)), a solution that can be determined analytically or mechanically (Haggett, 1965). Degree of connectivity depends on objectives and resources for the process.

Ideally, management for conservation should essentially be planned at a fluvial catchment scale, since most of the inputs into the land find their way over the surface and in groundwater to river channels, and many more organisms are thereby affected than butterflies. On a grand scale, including Britain, the European Union is seeking to ensure biodiversity by conserving natural habitats and wild fauna and flora in the territory of the member states. An ecological network of special protected areas, known as **Natura 2000**, is being set up for this purpose (europa.eu/scadplus/leg/en/lvb/l28076.htm). The network is given coherence by other activities involving monitoring and surveillance, reintroduction of native species, introduction of non-native species, research and education. The main act is Council Directive 92/43/EEC of 21 May 1992 on the conservation of natural habitats and of wild fauna and flora (amending acts followed, e.g., Directive 97/62/EC).

2 We know very little of the importance of matrix conditions for so many species.

3 Species clearly search for resources even in what is regarded as unprofitable biotopes (Dennis and Hardy, 2007).

On the Great Ormes Head in North Wales, the focus may be on a small selection of rare species and forms, but it is evident that there is not one pixel of the landscape unused by butterflies, let alone other organisms (see Box 9.1; Dennis *et al.*, 2007). Promotion of landscape heterogeneity is more likely to facilitate species' persistence in the face of climate change than focusing on specific locations, simply because it is most likely to maximize resource diversity (consumables and utilities) and heterogeneity.

An alternative to the resource-based approach for other arthropods, which is likely to become increasingly associated with using assemblage indicators, has been to focus on site features (i.e., vegetation units, specific substrates, microclimates) forming part of insect habitats considered to be rich in species or significant in some way with respect to their invertebrate fauna. Previously, such links have been influenced strongly by the knowledge and experience of the entomologists involved. In this approach, the site is dissected into component parts which hold different invertebrate interests and that require different management. A range of multivariate ordination techniques is, of course, now available (e.g., Ludwig and Reynolds, 1988; ter Braak and Smilauer, 2002) that takes the guesswork out of linking species with resources, substrates and structures (e.g., butterflies; Shreeve *et al.*, 2001; Dennis *et al.*, 2004; cf., Coleoptera, refs in Eyre, 2006). One cautionary note, a habitat view would suggest that there is a need to find all the features crucial for a target group of insects; discovery of clusters of a species at one time in one place does not ensure the presence of other vital resources not identified at the time of survey.

Sites (patches), in this section, are treated as 'given', as is the whole landscape, and consist of one or more vegetation units and/or substrates (e.g., biotopes: scree slopes, cliff sections, sand dune, garden). That is, although it is possible to modify the conditions within the patches and perhaps to their edges and the matrix by agreement with adjoining landowners, their fundamental location and the size and shape of landscape units, cannot be substantially altered. Such options, including specific landscape modifications, are left to the following section on landscape restoration.

GUIDING PRINCIPLES FOR LANDSCAPE RESTORATION

Much can be achieved when cost is no option, there is unlimited land to play with and there are no restrictions as to the action managers can take. Such is utopia for conservationists and largely imaginary. Even so, opportunities arise to restore landscapes for organisms and butterflies are no exception (Morris *et al.*, 1994); the ambitious vision for the Great Fen between Huntingdon and Peterborough, an area which once had habitats for *Papilio machaon* and *Lycaena dispar* is an

Fig. 9.3 The Great Fen Project (composite artist's impression) between Huntingdon and Peterborough is the restoration of an area which once had habitats for *Lycaena dispar* (top inset) and *Papilio machaon* (bottom inset). (Courtesy of Chris Gerrard, the Great Fen Project manager, the Wildlife Trusts for Bedfordshire, Cambridgeshire, Northamptonshire and Peterborough; project partners are the Environment Agency, Huntingdonshire District Council, Natural England and the Middle Level Commissioners; see www.greatfen.org.uk; insets courtesy of Peter Eeles.)

example (Fig. 9.3) (Bowley, 2007). Of course, when the objective of restoration is species specific, habitat creation is unique. In these cases the physical landscape template is usually intact and it is a matter of focusing on restoring resources through careful management; this may often require direct cooperation with business enterprises (Fig. 9.4). Cases in point are reintroductions of the large blue *Maculinea arion* (successful; Simcox and Bourn, 2006), the large copper *Lycaena dispar* and chequered skipper *Carterocephalus palaemon* in England (unsuccessful; Asher *et al.*, 2001). Here, attention is drawn to some basic 'rules' in landscape restoration for biodiversity emerging from landscape ecology (Forman and Godron, 1986; Turner, 1989; Forman, 1995a, 1995b; Fry, 1995; Dramstad *et al.*, 1996; Dover and Settele, 2009); these rules increase in sophistication and scale from that of a patch, to patchworks and ultimately to networks. As the cost of error can be high with 'real' management for particular species, it is necessary to reiterate the cautionary note applied to all principles in this text (Box 9.4). There is often an underlying assumption of the principles being derived from comparisons of large and representative samples or tests – which is rarely the case – and on the understanding of *ceteris paribus*; that is, in comparisons, all other factors are assumed to be equal, and like is being compared with like. In fact, clear exceptions are found and the general rules are regularly contested (Hanski and Simberloff, 1997:19). In viewing the points made below it is necessary to drive home to readers the need for careful thought in planning management objectives.

Size, shape and placement issues for single habitat patches

In habitat restoration the basic size, shape and condition of patches matters. Nevertheless, as noted in the previous chapter, even the most heavily abused land (brownfield sites; see Fig. 8.4) can become important sites for nature conservation. Some of the obvious principles that apply generally for the multispecies case are as follows:

- **P9.41: Large patches have more conservation potential than small patches.**
- **P9.42: One large patch is more effective for conservation than several small ones encompassing the same area – the SLOSS argument and the accountant's retort.**
- **P9.43: Circular-shaped patches are to be preferred for conserving species than elongated patches.**
- **P9.44: Neighbouring patches are to be preferred for conservation than isolated patches.**

Fig. 9.4 Business enterprises can get involved in the restoration of sites. Here, Network Rail helped in the restoration of a large blue *Maculinea arion* site in 2005 along the Taunton to London Paddington line near Green Down in Somerset. Near to Green Down the railway line is raised on an embankment that needed shoring up. Instead of simply replacing the existing topsoil after the work was completed, Network Rail worked with David Simcox and the Centre for Ecology and Hydrology to choose an ideal soil type and wild seed mix suited to the large blue (inset). The site is now five times larger than the previous one that had been threatened by the failure of a cutting. (Courtesy of David Simcox; inset courtesy of Peter Eeles.)

Box 9.4 Comparative patch topology for conservation: rules and exceptions

Over the years a series of 'rules' have emerged to direct landscape restoration. The assumption is that planners have a free hand to engage the process on a blank map as flat as the fens. A number of these rules emanate from biogeography theory and practice (Fig. B9.4a) and this gives them credence for enhancing biodiversity. However, in the case of a single species, and for particular higher taxa, there are clear exceptions to the rules. As indicated in the preface, there can be no excuse for assuming that these rules take on a law-like status. Students using this book should invest some time considering each of the principles on which these rules are based (**P9.41** to **P9.53**) and listing situations and organisms to which they might apply and where reversals to the rules are likely to be more important. The information in this box, then, is intended to build in the reader an inquiring approach to problems – restoration rarely is given an opportunity of a blank canvas; 'rules' provide guidelines but should not be adhered to rigidly. At the very least, the reader should be aware that these different designs are not independent of one another and conditions being compared may not be equivalent for a measure other than the one under consideration.

To provide examples for the first two sets of designs in Fig. B9.4a, one large reserve would generally be accepted as superior to several small ones occupying the same area of reserve (the SLOSS argument). But this rule may not apply to any one species in every case. For the speckled wood *Pararge aegeria*, one large wood may be superior to several small ones if the wood is of mature deciduous trees, but certainly not in the situation where the wood is densely packed with tall coniferous spruce. In the former case the butterfly will be able to occupy much of the deciduous woodland. However, in the latter case, the butterfly will be forced to occupy the margins, and the circumference of several small woods – thus usable habitat – will be greater than that of one large wood. Clearly, **reserve core species** will do better

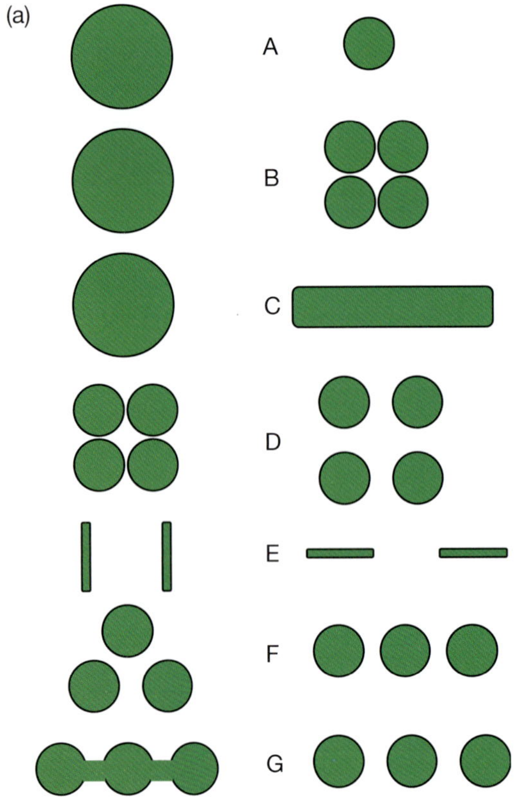

Fig. B9.4 (a) Rules for reserve design emerging from island biogeography theory. Each rule follows principles **P9.41** to **P9.47** (A to G, respectively). For each rule, the design on the left is regarded as superior to the alternative on the right; each symbol represents a single habitat patch or reserve. (Modified from Hanski and Gilpin, 1997; after Diamond, 1975; IUCN, 1980.)

on large patches and **reserve margin species** will be more successful in a situation of many small reserves. Moreover, several small reserves are not entirely comparable to one large reserve on a plain, for this butterfly or any other organism, as the space between them creates different conditions (wind speeds for adults, contrasting resource variability for early stages) for the margins of the smaller reserves than exists for the single large reserve, depending on their isolation from one another as well as reserve shape.

In Fig. B9.4b, a reserve with several biotopes is expected to be superior (for biodiversity) to another of the same size with fewer biotopes, or just a single biotope. However, fragmenting a reserve into distinct vegetation units effectively fractionates population size for any organism dependent on just one of those biotopes, making them more susceptible to stochastic extinction. Thus, a butterfly that depends on *Calluna* heath, such as the silver-studded blue *Plebejus argus*, may actually prosper better on the homogeneous reserve if this was heathland with all its consumer resources, than if the reserve also comprised equivalent areas of woodland, dense scrub and tall-herb grassland. Even so, if the homogeneous heath had no buffer from surrounding influences (high wind speeds), this situation may be reversed, as woodland and scrub could provide filters from external influences, both natural (e.g., weather conditions) and human (e.g., chemical inputs from agriculture) (see Box 3.2 and Fig. 7.6). In a multispecies situation, different species are favoured by different juxtapositions of biotopes; some are favoured by **convergence points** (junctions where three or more biotopes or vegetation units converge), others by **contrast adjacencies** (different combinations of adjoining biotopes) and others still by **biotope interspersion** (biotopes scattered rather than aggregated) (Dramstad *et al.*, 1996).

Nothing is simple in restoration ecology. For instance, there may appear to be no contest between having a larger reserve than a smaller one. But, relative size is affected by absolute dimensions of sites. For instance, in Chapter 6, it was shown that habitat quality was more important than patch size for Northumberland large heath *Coenonympha tullia*. It is feasible that absolute size alone could account for differences in population size, but *inversely*. Any specialist parasitoid will have a smaller population than its butterfly prey. Therefore there may be an optimal size of reserve for a butterfly species, large enough for medium-term population maintenance, but sufficiently small so that its parasitoid enemy experiences more regular stochastic extinction. We must be prepared for surprises in butterfly ecology; things do not always work out the way we expect.

Warren and Stephens (1989) rightly regard manipulation of vegetation structure in existing biotopes as a form of habitat creation. Networks and links, providing corridors (flyways and corridor links) are a vital component in woodland landscapes as well as open ones (see Fig. B9.3g). But, they also warn against destruction of butterfly habitat components in the process of restoration; it can be an easy matter to accidentally remove hostplants and nectar sources in ride widening, especially at the edges of coniferous woodlands, where these components may be restricted to the ride margins. In Fig. B9.4c a number of solutions (none to be taken as mutually exclusive) to providing open spaces through woodlands are indicated, expanding on this issue as illustrated in Fig. B9.4b(C). Again, solutions on the left may be presumed to be superior to those on the right for raising biodiversity (expanding the size

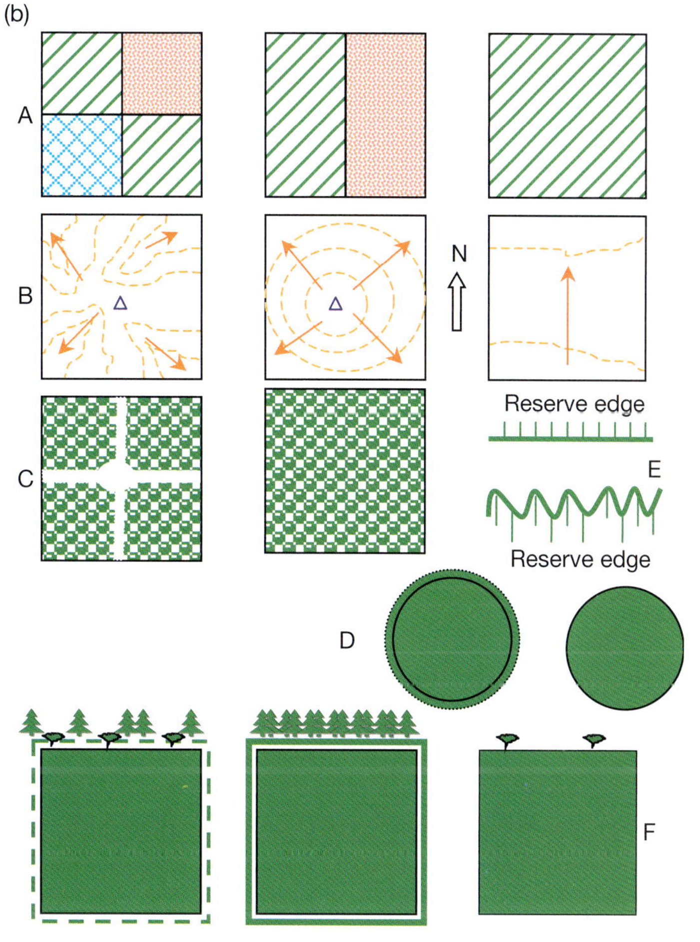

Fig. B9.4 (*continued*) (b) Rules for interior reserve design. Each rule follows principles **P9.48** to **P9.53** (A to F, respectively). For each rule, the design on the left is allegedly superior to those on the right of it except E in which the lower model is superior to the upper one. A, biotope variability in which each shade class indicates a different biotope, illustrating a convergence point (left) and contrast adjacency (centre); B, topography, in which pecked lines indicate contours, arrows the direction of down hill and triangles the highest point; C, vegetation management with white areas indicating paths (rides, glades) through denser vegetation (e.g., wood); D, reserve margins in which a buffer zone (headlands to fields, set aside) is superior to a sharp transition; E, reserve margins in which a scalloped periphery to a reserve is a superior format to a straight edge; F, reserve margins, in which an open wind break (pecked line) is superior to a tall, hard wind break (continuous line) or none at all (no line).

of the butterfly communities) (see Warren and Stephens, 1989, for a full discussion of this topic), but there could well be inversion of a ruling for individual species and circumstances (e.g., Fig. B9.4c(D), depending on how shade-tolerant species are and the orientation of the rides). Once again, the reader should consider what these may be. With coniferous woodland, such solutions (rides and glades) are critical for butterflies; greatly expanded ride sections and box junctions can provide extensive glades, large enough to be sizeable habitats for butterflies like the heath fritillary *Melitaea athalia*. British butterflies found in woodlands fall into three broad categories (Warren and Stephens, 1989):

1 Truly woodland species which use tree hostplants and have all their other resources in trees or directly under them (e.g., purple hairstreak *Favonius quercus*).

2 Woodland fringe or margin species, whose resources are associated with plants and shade conditions intimately linked to woodland (e.g., wood white *Leptidea sinapis*).

3 Grassland species whose herb resources exist outside woodlands but which can also thrive in open spaces within woodlands (e.g., meadow brown *Maniola jurtina*).

Warren and Stephens (1989) identified three factors ensuring the suitability of woodland rides and glades for butterflies, all of which correspond to the ideas of a functional resource-based habitat:

1 Structure of the ride (i.e., vegetation gradation from bare ground/short herbs centrally to taller herbs and finally shrub conditions fronting the wood).

2 Composition of the vegetation (i.e., presence of consumer resources, nectar flowers and larval hostplants).

3 Variable shade and light conditions as species vary in the light conditions required; speckled wood *Pararge aegeria* (at high summer) prefers moderately shaded

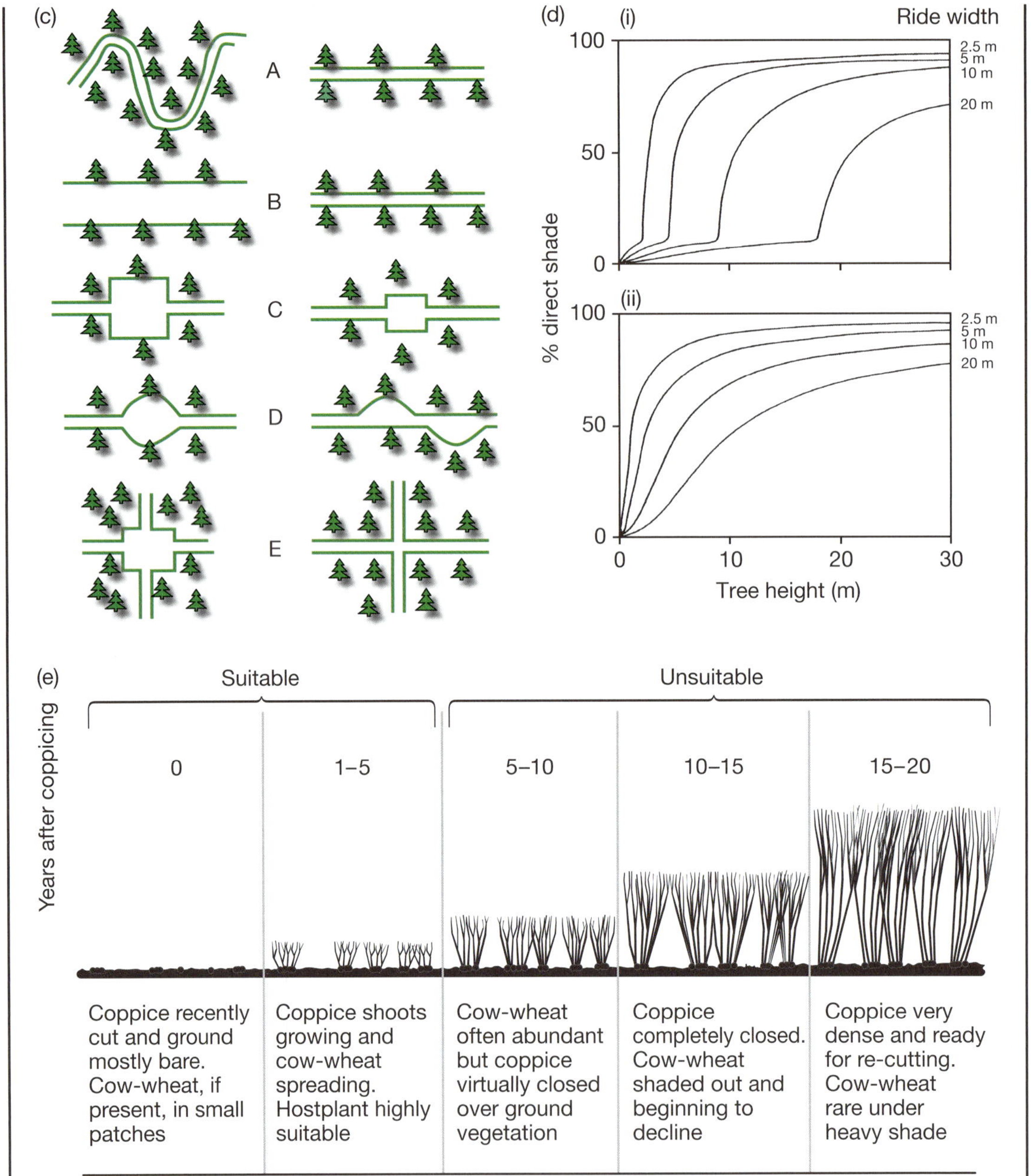

Fig. B9.4 (*continued*) (c) Alternative designs for open spaces (rides, glades) in woodlands. A, sinuous rides versus linear rides; B, wide rides versus narrow rides; C, expanded rides sections versus ones of limited extent; D, joint ride expansion versus alternate ride expansion; E, box junctions at intersections versus lack of them. (d) Predicted relationship between percentage direct shade during June and tree height in hypothetical model rides of different widths (2.5–20 m) and orientation: (i) east to west rides; (ii) north to south rides. (From Warren, 1985c, courtesy of Elsevier.) (e) The coppice cycle of a typical *Melitaea athalia* habitat. (From Warren, 1984a, courtesy of the author and Butterfly Conservation.)

rides, whereas white admiral *Limenitis camilla* (for breeding) and ringlet *Aphantopus hyperantus* prefer light shade, whilst many grassland species require bright sunshine. See Warren (1992b:266) for a full classification of the demands of British butterflies for light environment.

Warren and Stephens also recognized that rides and glades have at least three important components for butterfly conservation:

1 The height of the surrounding trees.
2 The width of the ride or glade.
3 The composition of the rides.

As the marginal woodland rises in height, so less light reaches the floor of rides and glades. The designs in Fig. B9.4c give the impression that all the woodlands are of the same composition, height and density, which, of course, may not be the case. In Fig. B9.4d, the amount of light reaching the ride floor (% shade) is shown to differ strongly for north–south compared with east–west rides for different combinations of tree height and ride width. Of course, the provision of open space in woodland does not depend entirely on the construction and maintenance of rides and glades. A crucial – the most important – aspect of woodland restoration is the management of entire woodland complexes. Much as with grasslands, a patchwork approach is required. For *Melitaea athalia*, this involves a regular coppice cycle carried out as a mosaic throughout the woodland (Warren, 1984a, 1985a, 1985b, 1987a, 1987b, 1987c). This work by Martin Warren is one of the most important pieces of research conducted on British butterflies. The reasons for the rapid decline of the butterfly in Blean Woods, Kent was clear:

• Conversion of deciduous woodland to conifers.
• Increase in the length of the coppice cycle.

Figure B9.4c illustrates in a simple way just how precarious the balance of hostplant cow-wheat *Melampyrum pratense* is for the butterfly for different stages in the woodland coppice cycle.

• **P9.45: Neighbouring patches for conservation are best served by mutual exposure of their longest edges to assist colonization.**
• **P9.46: A compact arrangement of patches is to be preferred for conservation purposes to a linear arrangement.**
• **P9.47: Patches connected by corridors are to be preferred for conservation than ones entirely isolated by matrix.**

These points should, by now, be familiar, as deriving from island biogeography (see MacArthur and Wilson, 1963, 1967). The patch size argument (**P9.41**) forms a cornerstone of comparative island biogeography; larger patches have the capacity for larger populations, with lower probability of extinction. Patches can be too small – insufficient area of hostplant (milk-parsley *Peucedanum palustre*) was a key reason why *Papilio machaon* failed to persist on Wicken Fen, Cambridgeshire (Dempster *et al.*, 1976). Even if a large patch and a small one are initiated from scratch, in time, greater biotope (habitat) variability will be generated in the larger patch and consequently it will house more species. The SLOSS argument (**P9.42**) (Game, 1980) is particularly vulnerable to exceptions if only because separate small patches may each achieve, stochastically, different vegetation dynamics, maintaining greater variability through greater individual vulnerability to external influences, and at any one point in time having more contrasting biotopes (habitats) than one large patch of equal area. The fact remains that rarely is everything equal and the principle schools caution. To adopt an axiom from a different science, an accountant will undoubtedly advise clients not to put all one's eggs, and in this case we can readily read this as butterfly eggs, in one basket. Almost certainly there will be exceptions to the benefits of circular patches to those that are elongated, particularly for what are referred to occasionally as 'edge' species (e.g., wall brown *Lasiommata megera*) (**P9.43**). The basic idea often propounded here is that elongated patches will experience more interference from surrounding influences, and if these are negative (e.g., herbicide spraying) then populations of the species being conserved will decline. Another argument is that losses from emigration will be greater from elongated patches as patch margins will be encountered more frequently by flying butterflies than in more compact patches. The correct patch shape is not separable from considerations of context, both patchwork and network, but it is sensible to consider that patch (habitat) shape, as the shape of individual resource items within patches, has consequences for conservation (see Chapter 3).

The remaining four 'rules' apply to multiple patches. Proximity to 'source' populations is a key and familiar island biogeography concept and metapopulation principle (**P9.44**). Patches nearer to potential source populations on other patches are likely to be colonized earlier, more frequently and by more species than ones distant from sources. In practice, restoration is best served by distances between patches no greater than would allow recolonization within a few years, but recall the point (**P9.25**) made earlier about intermediate distances being preferred (Hanski *et al.*, 2004). The more distant patches are planned to be from potential source populations, and greater reliance will necessarily be placed on artificial introductions. Maximizing the target frontage should also ensure that colonization from source areas is most rapid (**P9.45**). Patches where the longest side is orientated at right angles to circular sources (or parallel to linear sources) are likely to be colonized more rapidly and more frequently than those where the shortest side faces the potential source. The last two 'rules' (**P9.46** and **P9.47**) both work to minimize isolation and increase colonization and persistence. This can be achieved either by clustering patches rather than allowing them to be drawn out in a line, and by providing some means of effective mutual links (i.e., corridors). The reader should, by now, be aware of counter arguments; if not then it would be wise to re-read Chapters 6 and 7.

Restoring a landscape should be a task for biodiversity catering for a plethora of demands. Even among butterfly species, habitat patch shape can be species specific, and planning the restoration of landscapes for an assemblage of butterfly species will require integration of patches of different shapes and sizes. This is not as difficult as it sounds as edge conditions, providing key consumer resources for one species (e.g., blackthorn *Prunus spinosa* for brown hairstreak *Thecla betulae*; alder buckthorn *Frangula alnus* for brimstone *Gonepteryx rhamni*), typically provide essential windbreaks and other consumer resources (e.g., *Rubus fruticosus* nectar for meadow brown *Maniola jurtina* and gatekeeper *Pyronia tithonus*) for butterflies whose habitats are more typically open areas.

Internal habitat issues for single patches

Having assessed the geometry of the habitat patch, attention can be turned to patch composition, though it is wise to consider how this may direct basic geometry. What follows is a reminder of what has been discussed more fully beforehand (see Box 9.4):

- **P9.48: Patches with biotope variation and resource mosaics have more conservation potential than those lacking them.**
- **P9.49: Patches with topographic variation have more habitat potential (resources) than those patches lacking topographic variation.**
- **P9.50: Creation of 'breaks' (tracks, rides, glades) and barriers through patches increases habitat variation and protection of habitats from catastrophes.**
- **P9.51: Patches with 'soft' edges permit greater transfers of species between the patch and adjacent biotopes than those with 'hard' edges.**
- **P9.52: Patches benefit from the development of semi-natural vegetation biotopes (set aside) beyond the patch margins if surrounded by intensive monocultures.**
- **P9.53: Wind breaks at the margins of patches benefit resident species and are better when permeable than when impermeable.**

A key factor in biodiversity is 'habitat' (biotope) heterogeneity and thus patches with more biotopes will have a greater variety of species (**P9.48**). Part of habitat heterogeneity is the creation of resource mosaics – a patchwork quilt of resources and structure – which establish supplementary resource outlets in varying conditions for the same species. Owing to modern land use practices, biotopes and vegetation types for British butterflies often have limited resources, and enhancing patches for butterflies necessarily requires particular attention being given to appropriate resources linked to specific vegetation communities and biotopes. Those butterflies lost from a region, and in need of re-introduction, as in the case of *Carterocephalus palaemon* in Lincolnshire between 1994 and 1999 (Asher *et al.*, 2001) and *Lycaena dispar* in the East Anglian Fens (Pullin *et al.*, 1995, 1998), inevitably require the restoration of original biotopes (viz., pasture woodland, fenland) on substantial enough spatial scales to maintain, at least, metapopulations of the species – a formidable undertaking (see Box 9.3 and Fig. 9.3). British butterflies tend to be largely associated with herb-rich biotopes, even in woodlands. Thus, tall shrub and woodland biotopes benefit from the creation of open spaces (viz., rides, glades, clearings, coppiced areas, meadows) within them. In certain cases, highly specific, short-lived biotopes (e.g., coppicing of woods for the heath fritillary *Melitaea athalia*; Warren, 1987a, 1987b, 1987c, 1991; Kemp *et al.*, 2008) are required and need constant maintenance. Many woodland butterfly species are nectar specialists and open spaces provide these nectar sources (Tudor *et al.*, 2004).

An essential aspect of restoration is topographic variation (**P9.49**), as this provides slopes with varying aspects and therefore different heat and moisture conditions for butterflies and the resources needed by them (see Box 2.4). Steep slopes also have the property of slowing down vegetation succession; once done, less intensive management is needed to maintain bare ground and early herb seres. Not only does topographic variation deliver different habitats for insects, but it also generates supplementary resources for the same species (e.g., *Cardamine pratensis* on mires and *Alliaria petiolata* on drier slopes as larval hostplants for *Anthocharis cardamines*). Occasionally, the landscape is already moulded for future butterfly habitats, as in the case of abandoned quarries and sand and gravel pits. Where it is too expensive to create hummocks and hollow topography, planting trees at patch margins and as a mosaic inside patches, layered in width and height, provide not only suitable shelter belts but also hot spots (e.g., for *Lasiommata megera*; Dennis and Bramley, 1985) and shaded sections (e.g., for *Pararge aegeria*; Shreeve 1985, 1986, 1987) used by different species. The importance of ensuring this to be a mosaic is that it generates numerous territories for competing males and outlets for egg-laying females harassed by males (e.g., *Pyronia tithonus*; Dennis, 2004a). Clearly 'edges' need to be tailored for specific local climate contrasts in relation to requirements of different species and different developmental stages.

Contrasts in local climate are also generated by 'breaks' through patches, but these also serve to provide distinct biotopes (e.g., bare substrates where tracks cut through heath land, herb-rich grassland as rides and glades through woodland) and serve as barriers against catastrophic events such as fires (**P9.50**). Breaks sometimes have to be barriers (e.g., deer fencing), where protection is required from intensive grazing. Different species require different types of edge conditions for activities such as mate location, but for wider resource use, generally 'hard' edges tend to create contrasts in butterfly communities whereas 'soft' edges allow integration (**P9.51**) (Pryke and Samways, 2001). Hard edges include linear features dominated by abrupt changes in vegetation height, whereas soft edges are curvilinear with gradual transitions in vegetation structures. Such interdigitation of butterfly communities is likely to be enhanced where the surrounding crop land is fringed with 'set aside' seminatural biotopes as a filter to pernicious influences from intensive farming (**P9.52**). Other, economic opportunities in the future may also carry benefits for butterfly habitat structure (e.g., the development of perennial biomass crops such as short rotation willow *Salix* sp.; Haughton *et al.*, 2009). In Britain, dominated by prevailing, strong, westerly winds, butterflies on new patches are particularly vulnerable to wind chill, and wind breaks at patch margins become essential conservation tools (**P9.53**); a layered open structure, rather than a wall of trees, reduces lee vortex effects (see Box 2.4).

Principles for patchwork creation

It is well appreciated now that species cannot be sustained in a single habitat patch. For long-term persistence, a multipatch environment provides the minimum framework for conservation. Some principles can be established for creating patches:

- **P9.54: Species persist longer on extensive patchworks than limited patchworks.**
- **P9.55: A combination of differently sized patches enhances the conservation potential of landscapes for butterflies.**
- **P9.56: Species persist better in patchworks that comprise both neighbourhood patchworks (metapopulation structures) and isolated patchworks (separate metapopulations) than in situations where all patches are either close neighbours or isolated from one another.**
- **P9.57: Stepping stones enhance persistence of species in metapopulations only if fully resourced.**

The first principle is a key one from metapopulation studies; in a patchwork of stochastic extinction and colonization on patches, increasing the number of patches insures against metapopulation extinction (**P9.54**) (Hanski and Gilpin, 1997). **MVM** or **minimum viable metapopulation** size (Hanski *et al.*, 1996; see Hanski and Gilpin, 1997) is analogous to **MVP** (**minimum viable population**) size (Shaffer, 1981) except that MVM involves both the minimum viable number of populations, thus an adequate number of suitable habitat patches, as well as minimum overall population size. Although, occasionally, specific patch numbers have been quoted as providing a safeguard against loss of species, there is danger, of course, in prescribing a precise number of patches. 'In practice, use of these concepts may degenerate into specious "magic numbers"' (Hanski and Simberloff, 1997:22); these authors suggest more constructive approaches that direct attention to specific features rather than reliance on simplistic rules.

The same considerations apply to patch size; the immediate reflex action is to rely on systems of large patches rather than small ones. Recently, Tscharntke *et al.* (2002) have revisited the SLOSS issue of whether butterfly species' richness is best conserved by protecting many small fragments of grassland biotope or fewer large ones occupying the same area (**P9.55**). Unexpectedly, as regards species' incidence, they showed that a group of smaller sites could hold '*en masse*' more species. Data for endangered species was even more convincing with 100% of species contained within the 30 small fragments, compared with 40% in the equivalent large fragments. Small fragments represent a more heterogeneous collection of habitats over a wider range of environmental conditions than an equivalent contiguous block of land. They concluded that for effective landscape-scale conservation, fragments of different sizes offered different attributes and both needed conserving: for example, large fragments have fewer species but low extinction risk, while a collection of many small fragments represents higher biodiversity but higher extinction risks.

When broken down to hostplant specificity, it became clear that the number of monophagous (specialist) species in a fragment was positively related to increasing fragment area, oligophagous ones were neither positively nor negatively related to fragment area, but polyphagous (generalist) species' richness had a negative relationship with increasing fragment area. As polyphagous species tend to have greater migration capacity this relationship coincides with expectations from studies of life history strategies (Dennis *et al.*, 2004).

Isolated patches do not make for a metapopulation; connectivity among patches needs to be considered in relationship to the migration capacity of the least vagile species. However, there are better reasons to have isolated metapopulations in contrast to isolated local populations (**P9.56**); they insure against regional extinctions, for example, from invasions of parasites (Hanski *et al.*, 2004). Stepping stones (see Chapter 6), disconnected nodes, may have a part to play here but their value is difficult enough to determine in the case of single species. In a multispecies case a particular stepping stone may be a sink for one species, a true stepping stone for a second species and a long-term habitat for yet another. To be effective, such small patches need, at least, to house sufficient resources to be adequate fuelling stops (**P9.57**). There is an argument that making provision for breeding with stepping stones could result in greater mortality of progeny than if not available, as fertile females could otherwise continue their migration to a suitably large sized habitat patch. This argument, on its own even, is highly suspect: if the target species is highly mobile, fertile females would lay eggs and continue on to other patches; if it lacks mobility, it may never make it to other patches – it is a balance determined by circumstances. In a multispecies case, a variety of organisms are catered for and the landscape is enhanced.

Principles for creating networks, hierarchies and surfaces

From patchworks more complex landscapes can be constructed for species, graduating from networks, through hierarchies with patches contrasting in size and complexity, therefore in species' diversity, to surfaces – mosaics of suitable biotopes and vegetation classes, a 'field' of neighbouring patches, something beginning to simulate landscapes before agriculture. It is of huge importance that Butterfly Conservation now has some 67 landscape-scale projects, in which buffering key sites is a major aim (see Box 9.3 and Fig. 9.1). This involves taking over and/or managing land within dispersal distances (5–10 km) of sites already housing populations of threatened species (C. Bulman, personal communication). Some principles can be drawn up:

- **P9.58: Converting patchworks into networks by linking up patches using linear features (hedgerows, walls, verges) enhances conditions for butterflies that use the links as 'flyways'.**
- **P9.59: Functioning of networks is increased by ensuring a fully integrated system.**
- **P9.60: Connectivity is further improved by converting links from 'flyways' with limited resources for refuelling and recovery into fully resourced habitat corridors.**
- **P9.61: The wider the links in networks the more useful they are likely to become as habitats.**
- **P9.62: Valuable linkages for patchworks are provided by natural (riparian, ridge tops) and human-made (road verges, rail embankments and cuttings, canal sides) features.**

- **P9.63: Connected nodes will sustain more species than disconnected nodes, but disconnected nodes may receive advantages through their relative isolation.**
- **P9.64: As blind edges (terminator vertices) in links may heighten mortality they should be avoided.**
- **P9.65: An ideal landscape is one of resource surfaces (mosaics) where connectivity for resource types is maximized and boundaries (edges) between different resources in themselves provide for different resource types and lines of movement.**

Moving from a patchwork (e.g., East Anglian prairies) to a network (e.g., West Country dairy farmland) involves a quantum shift for butterfly conservation. The argument in favour of this is that links between patches (e.g., hedgerows, banks, drainage ditches, verges) are preferentially used by butterflies for movements than cross-matrix transfers (**P9.58**). Some of these can be created rapidly, such as ditches using giant diggers (e.g., the giant American big wheel rotary ditcher used by Natural England and the Royal Society for the Protection of Birds (RSPB) in Cheshire). This notion is strongly supported by studies in appropriate farmland contexts (Dover, 1996; Merckx *et al.* 2008). It is possible that even fence lines could bias movements (Dover and Fry, 2001). Just how effective linkages are depends on orientation in relation to sun angle and prevailing winds which change diurnally and seasonally. As such, a fully integrated system, in which all patches are linked up, maximizes opportunities for transfers for invariant linear structures (**P9.59**). To improve the situation further, it is necessary to increase the number of resources available along links (e.g., wild flower sowing along the Middlewood Way, a disused railway line in Cheshire, UK; see Fig. 6.6a), the limit determined by habitat requirements of the target species (**P9.60**). Such corridors become increasingly effective with increased width (**P9.61**) (Samways, 2007). Much as in the case of patch area, with increased corridor width potential is increased for resource abundance as well as heterogeneity along it. In creating corridors, it is clearly sensible to make use of already extant natural lines and edges (**P9.62**), particularly as these provide distinct resource types. Thus, different resources and habitats are provisioned by river banks, road verges, hedgerows, ridge tops, green lanes and so on. Sites of antiquity can be particularly valuable for butterfly habitats and corridors (Davis, 1979; Wells, 1985; Warren, 1993), protected as they are by Ancient Monument legislation (Ancient Monuments and Archaeological Areas Act 1979; see www.opsi.gov.uk/RevisedStatutes/Acts/ukpga/1979/).

In a networked landscape there is room for 'disconnected' nodes (**P9.63**). They may have fewer species than linked nodes, owing to lower immigration, but relative isolation may provide an advantage in protecting populations from hazards associated with the lines of movement, not least of all those conveyed by humans. Even so, blind vertices at the end of links, especially nodes insufficiently resourced as habitats (e.g., nectar flower concentrations), may result in aggregations of individuals which in turn attract predators leading to high mortality (**P9.64**).

The ultimate landscape for all wildlife is one in which vegetation and substrate patches vary in composition, structure and scale, are continuous and contiguous, a mosaic of resources and habitats for different organisms, and where a variety of boundary features exist for aggregations, movements and barriers (**P9.65**). It deserves emphasis that in this formula variation over time is not hazardous as long as it is not unidirectional, encompassing the entire site. But, to be successful, it is necessary that management practices (e.g., agri-environment schemes) are adopted at landscape scales and not just land-ownership (i.e., single farm) scales (Merckx *et al.*, 2008, 2009). From a single species' vantage, butterfly species are likely to do better with increasing amounts of habitat in the new landscape and with fewer barriers. Part of this habitat availability is variability in resources that would occur with topographic variability (i.e., slope and aspect) and with increasing altitude. These findings emerge in a study of three related fritillaries, the small pearl-bordered fritillary *Boloria selene*, Titania's fritillary *B. titania* and lesser marbled fritillary *Brenthis ino* in the northeast Swiss Alps (Cozzi *et al.*, 2007). There is still the basic framework for this kind of landscape in the Morecambe Bay area of Lancashire in Britain, a hot spot of butterfly diversity with many fritillary species, though this landscape too has detractors, for example, the intensively groomed and densely occupied caravan parks, improved pasture fields and unmanaged woodland. Fine-grained mosaic landscapes form a sound basis for supporting wildlife generally as well as being an excellent indicator for human well-being. Just what in essence such areas should contain is driven by geology, topography, soils and current wildlife; earlier chapters indicate how these factors influence the butterfly fauna.

Introductions

Occupancy of the restored landscape depends on two processes operating together:

1 Natural colonization.

2 Artificial introductions.

New sites will inevitably be colonized by species, plants and animals including butterfly species, from the adjoining countryside. An interesting experiment involving restoration of a landfill site near St Osyth, Essex – combining natural colonization and introduction of the marbled white *Melanargia galathea* and *Hipparchia semele* – illustrates this process very well (Box 9.5) (Davis, 1989). It also drives home that a crucial part of restoration is to provide adequate resources for all species targeted. But restoration is often species specific with definite targets in mind, as in the case of *Lycaena dispar* (Duffey, 1968) and *Maculinea arion* (Simcox and Bourn, 2006). A few principles are worth bearing in mind:

Box 9.5 Habitat creation for butterflies on Martin's Farm landfill site near St Osyth, Essex, UK

Habitat restoration or creation is not simply limited to creating physical structures, whether composed of rock, soil or vegetation. Crucially, it has to consider the key consumer resources for butterflies: larval hostplants and nectar flowers. The selection of hostplants is closely prescribed as the range used by butterfly species is limited (see Appendix 2). Butterflies use a wider range of nectar flowers than larval hostplants (Hardy *et al.*, 2007), but there are still important considerations as to which flowers are suitable for butterfly species differing in proboscis length and wing loading (see Chapter 2) (Corbet, 2000).

These decisions were paramount in butterfly habitat creation on the Martin's Farm landfill site near St Osyth, Essex (grid reference TM117175; Davis, 1989). The Institute of Terrestrial Ecology (ITE) was invited to undertake studies at the site before it was converted to amenity use by Essex County Council. The objective was to create grassland areas that would attract a wide range of butterfly species. Experimental plots were developed on the west slope to test two mixtures of grasses and wildflowers (Fig. B9.5a), one of tall herbs and the second of short herbs (Table B9.5). These were mixed randomly in a rectangular array with two other plot types, one cultivated but unsown (called 'mound' because they each had an elongated mound of fine limestone (*c.* 20 mm chippings) 20 m long and 1 m high) and the other a control area (not managed). All plots, except the controls, were cultivated in August 1983 and seed was sown in the experimental plots in September 1983. The mound plots were harrowed lightly in October 1987 to provide open conditions for the grayling *Hipparchia semele* and small amounts of fertilizer to promote flowering were spread on the experimental plots in April 1987. The only vegetation occurring on the site before the experiment was found in the southeast corner (dense couch grass *Elymus repens*). All but two butterfly species colonized the site from the surrounding countryside. Both *H. semele* and marbled white *Melanargia galathea*, unlikely to colonize the site naturally, were introduced in 1987.

Vegetation on the plots was surveyed using quadrats at least annually, and butterfly recording was carried out using Butterfly Monitoring Scheme (BMS) criteria (compared with 12 sites in the East Midlands and East Anglia; see Pollard and Yates, 1993a). The low nutrient status of the soil on the site (0.6 m subsoil of mainly sandy clay loam with a large amount of coarse gravel), and its compaction, led to slow development of the vegetation. Even so, 27 of the sown species established on the site and more than 140 other plant species colonized the site naturally. In all, 18 butterfly species were recorded on the site, seven of which established local breeding populations (i.e., small skipper *Thymelicus sylvestris*, Essex skipper *T. lineola*, common blue *Polyommatus icarus*, small tortoiseshell *Aglais urticae*, gatekeeper *Pyronia tithonus*, meadow brown *Maniola jurtina* and small heath *Coenonympha pamphilus*). These have shown consistent growth in their populations on the site compared with those of sites in the surrounding region (Fig. B9.5b). Some additional prominent features emerge from this study:

• Not all the hostplants sown survived on the site. Crucifers have not done well. Cuckoo flower in particular failed, and sown hedge garlic in 1984 (*c.* 18 000 seeds) produced very few flowers, essential for egg laying by orange tips *Anthocharis cardamines* and used by green-veined white *Pieris napi*. This impresses on site management the need to cater for alternative (supplementary) resources for species.

• Many useful plants were able to colonize the site from the surrounding countryside (e.g., important nectar sources, the thistles *Cirsium arvense* and *C. vulgare*, bristly ox-tongue *Picris echioides*, teasel *Dipsacus*

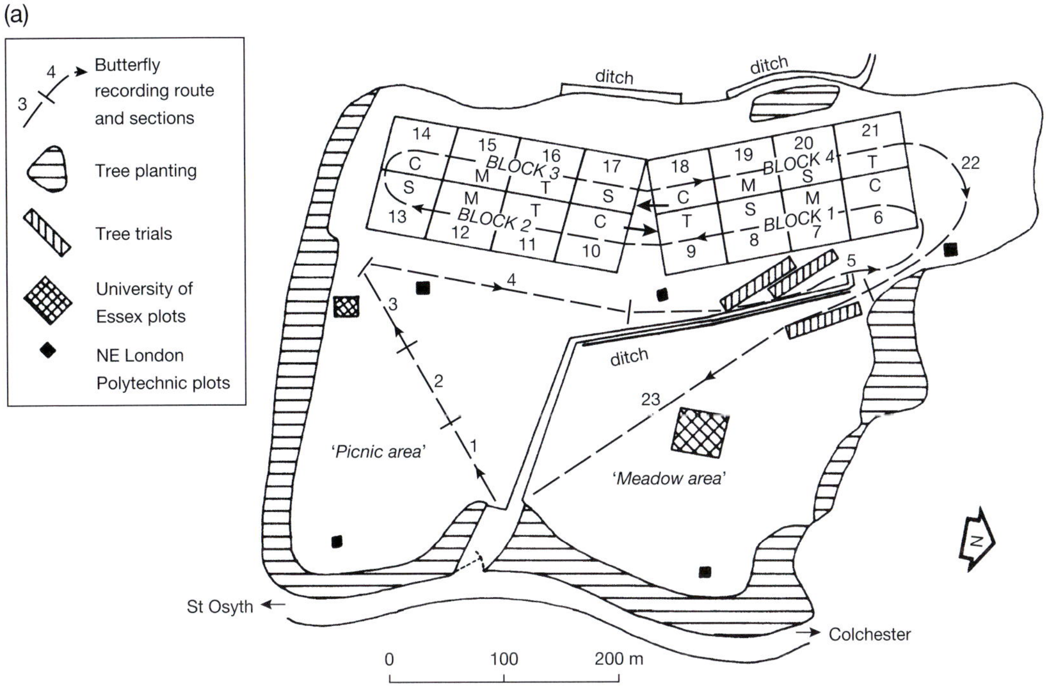

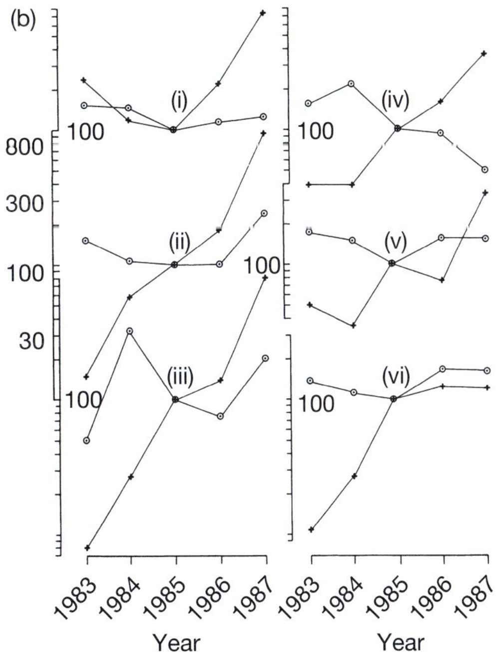

Fig. B9.5 (a) Site plan of the Martin's Farm landfill site showing the experimental plots, butterfly recording routes and sections. C, control; M, mound; S, short growing plants; T, tall growing plants. (b) Changes in annual index values for 1983–1987 for butterflies from Martin's Farm (+) compared with general trends in the East Midlands/East Anglian region (⊙), setting 1985 values at 100: (i) *Thymelicus sylvestris/lineola*; (ii) *Polyommatus icarus* first generation; (iii) small tortoiseshell *Aglais urticae*; (iv) *Pyronia tithonus*; (v) *Maniola jurtina*; (vi) *Coenonympha pamphilus*. (Courtesy of Brian N. K. Davis and the Royal Entomological Society.)

Table B9.5 Grasses and wildflowers sown at Martin's Farm landfill site in short growing (S) and tall growing (T) mixtures.

Grasses*	Wildflowers
Alopecurus pratensis ST	*Anthyllis vulneraria* * S
Anthoxanthum odoratum S	*Cardamine pratensis* *SX
Briza media S	*Centaurea nigra* T
Cynosurus cristatus T	*Centaurea scabiosa* TX
Dactylis glomerata T	*Daucus carota* T
Festuca rubra ST	*Galium verum* S
Holcus lanatus T	*Hypochaeris radicata* T
Phleum pratense T	*Knautia arvensis* TX
Poa pratensis T	*Leontodon hispidus* S
Trisetum flavescens S	*Leucanthemum vulgare* T
	Lotus corniculatus *T
	Medicago lupulina *S
	Plantago lanceolata T
	Plantago media S
	Primula veris S
	Prunella vulgaris S
	Ranunculus acris T
	Rhinanthus minor S
	Rumex acetosa *S
	Sanguisorba minor T
	Silene alba XT

* Intended larval hostplants.
X, failure.

fullonum). Similarly all but two butterflies colonized the site naturally.

- Accidental or unplanned events can benefit habitat creation. A clear example was in section 22 where a deposit of sewage sludge in 1982 encouraged a dense patch of nettles for nymphalids.
- Fertilizer application may not always be a bad thing. Some valuable grasses responded only after a spring application of fertilizer in 1987 (i.e., quaking grass *Briza media* and smooth-stalked meadow grass *Poa pratensis*, both hostplants of butterfly species (see Appendix 2)).
- Vegetation growth had structural benefits, especially the nettle patch which had prominent lee shelter for species.
- Finally, although soil compaction slowed plant growth, it also retained bare areas longer and reduced management (cutting or other disturbances).

This is just one example of what can be achieved with site restoration or creation. Here the focus has been on the provision of consumer resources rather than modification to topography. The team have had to use what was available on the site. In habitat creation, one must be prepared for surprises. For instance, *Polyommatus icarus* clearly thrived more in the tall-herb plots than the short-herb plots (Davis, 1989). At first this does not seem very surprising as *Lotus*, a key hostplant, was sown only in the tall-herb plots. However, no larvae were located on this tall growing cultivar of *Lotus*, but only on black medick *Medicago lupulina*, growing as prostrate mats. Another example is the failure of small copper *Lycaena phlaeas* to colonize the site, despite *Rumex acetosa* being made available. Perhaps this would have been successful if the seeds for *R. acetosa* had been applied to the short-herb plots. Whatever, it once again underlies the subtlety of habitat quality for butterfly populations to persist.

Unfortunately, Essex County Council re-opened the pit and re-covered some of the plots with fresh landfill waste (B. N. K. Davis, personal communication).

(Courtesy of Brian Davis and The Royal Entomological Society.)

- **P9.66: Most introduced populations survive for a short time only and are more successful if based on many individuals.**
- **P9.67: A requisite for successful introductions is the provision of all resource requirements, complete habitats, for species.**
- **P9.68: Long-term persistence of introduced populations requires either large tracts of habitat or the introduction of the species into a network of sites.**
- **P9.69: Restoration and introductions should always be followed up with monitoring progress and adjusting management to changing conditions.**
- **P9.70: Keys to management are the type of action to be taken, its intensity and timing.**

Most attempted introductions of species fail (**P9.66**) (Oates and Warren, 1990). In fact, the numbers failing are probably underestimated as most introductions are not reported in the literature and, if they fail, then evidence is lacking that they were ever attempted. Although introductions have been successfully based on few individuals, they are likely to be more successful if based on large samples, particularly of females. One of the classic examples was the introduction of 90 individuals of *Plebejus argus* from the Great Ormes Head, North Wales to the Dulas Valley, some 15 km away in 1942 (Merchant, 1956). Although the populations passed through periods of genetic drift, the butterfly was found to be thriving in subsequent years (Dennis, 1971, 1972b, 1977) and continues to thrive (Fig. 9.5) (Thomas and Harrison, 1992; Lewis *et al.*, 1997). In this case, no restoration of landscape was involved. But, as many introductions fail, this would suggest that one of two things were not in place for the introduction: either adequate resources (**P9.67**) and/or insufficient habitat patches (**P9.68**). In the latter case, the above sections concerning metapopulations and networks are germane (Hanski *et al.*, 2004). It should be pointed out that introductions of species in the UK

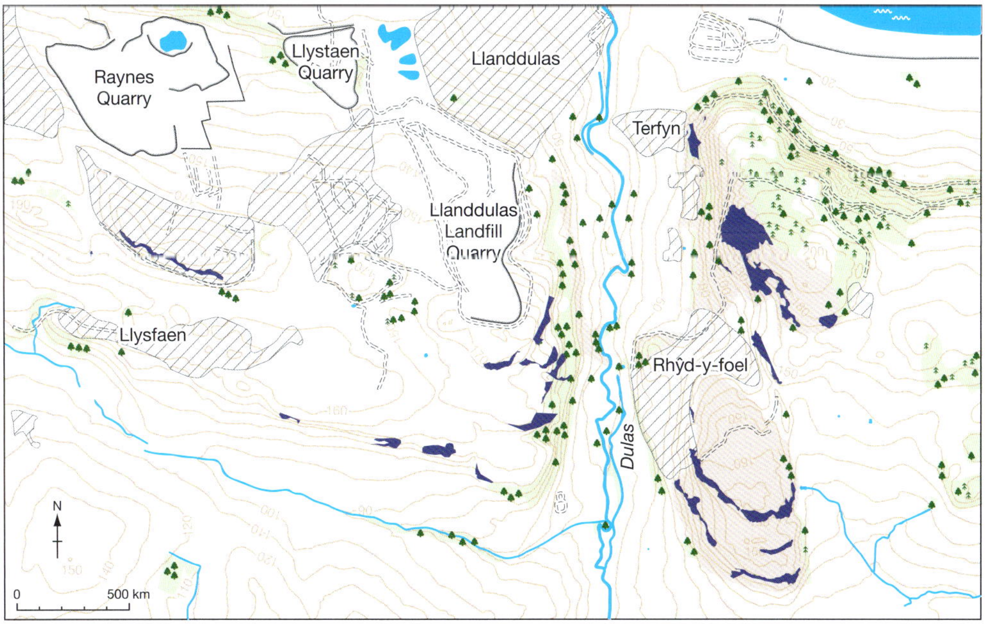

Fig. 9.5 A successful introduction of the silver-studded blue *Plebejus argus* (purple areas) to the Dulas Valley, North Wales. Ninety individuals of *P. argus* were introduced to the Dulas Valley in 1942 (Merchant, 1956). It was scarce there between 1943 and 1949, but thereafter was abundant. It has been mapped in the valley on several occasions, first by the author in the early 1970s (Dennis, 1977), thereafter by Professor C. D. Thomas and his research team, and is continuing to be mapped at 7-year intervals. This figure shows the most recent map of the butterfly by J. Hodgson for 2004. (Courtesy of Jenny Hodgson.)

without proper direction from Butterfly Conservation (www.butterfly-conservation.org) are not helpful to conservation programmes, particularly if the circumstances are not recorded. They also have the consequence of confounding science, for instance, understanding migration and colonization, which forms a central issue in conserving butterflies. In six cases handling of butterfly species anywhere in the country is illegal (see Appendix 1a, fully protected status); many more are subject to local bye-laws or other regulations (e.g., National Trust land).

Restoration of landscapes for butterflies can be a matter as simple as discarding a piece of land to invasions of plants and butterfly immigrants, or a process so complex that it expands across generations of management teams and hundreds of years. Classic examples of what may be required over the long term are forests frequented by the silver-washed fritillary *Argynnis paphia* and *Apatura iris*; planting trees is one thing, creating a woodland with all its complex plants, soil fauna, ancient timber, arthropods, bryophytes, birds, mammals and butterflies is another (Merryweather, 2007) – conservation should never be visualized as a unitary process in which what happens to a single species is all that matters. Restoration for target species requires knowledge and that too takes many years of research to acquire, as evident from work on *Maculinea arion* (Thomas, 1995; Thomas *et al.*, 1998a, 1998b; Settele *et al.*, 2005). For restoration, we need databanks of butterfly resources, such as those in progress (see Appendix 2a), sound knowledge of the implications for other organisms, and experience in landscape management. Finally, we learn nothing from restoration and introductions unless species and their resources are subsequently monitored (**P9.69**) (Spellerberg, 2005).

This book has not been about prescriptions for individual species; the reason is simple: for many species we simply do not know what works in any locality. It is not enough to prescribe an action (e.g., cattle grazing of pastures); much of success depends on the intensity of the action and its timing (e.g., how many of what kind of animals per unit area, for how long, during what period(s)? What turf height maintenance for the winter months?) (**P9.70**). Warren (1993) discussed this point with regard to *Euphydryas aurinia* and *Cupido minimus* in south central England; both species are adversely affected by sheep grazing, suggesting lighter grazing regimes and winter grazing as appropriate actions, but pointing to lack of firm guidelines for recommendations. Much remains to be understood. For instance, increased grazing intensity by farm animals may have as yet not fully understood consequences for butterflies; increased herbivory can induce silica defences in grasses which then may reduce digestibility and nitrogen uptake in herbivores such as caterpillars, slowing growth and development (Massey and Hartley, 2007). There is another vital element to management regimes (T. G. Shreeve, personal communication). Many prescribed grazing regimes dictate duration and stocking density. Historically, stocking density and timing was partly dictated by stock prices and weather; this introduced a pseudo-random element into the management of individual fields. There are instances where hay meadows have been purchased for conservation and their value has declined (i.e., species have been lost) under single management regimes, showing that single management regimes are not the solution (Stewart and Pullin, 2008). It may well be that management should consider not just spatial heterogeneity of resources but temporal heterogeneity of management of those resources as well.

BUTTERFLIES AS INDICATORS AND FLAGSHIP SPECIES

Many reasons are given for conserving organisms, but they largely fall into two groups that can be summarized as 'being of public interest' and 'in the public interest'. These are distinguished by what is of intrinsic value to people, and for that reason alone supported by funding and management, and what is in all humankind's interest, with fundamental implications for human survival. From the vantage of being of interest to the public, butterflies have established conservation in their own right; they are beautiful creatures and undoubtedly give immense pleasure. As a vital component 'in the public interest' they have potential value as bio-indicators. The concept here is a simple one, even if the machinations are complex: what humans do to natural ecosystems will eventually reverberate on human systems. It is no good putting conservation over in terms of what is good for nature; humans, as 'efficient Darwinian demons' are biased to short-term self-interest and appreciate what happens to their pockets and comfort ahead of what will happen to their descendants and ecosystems. Thankfully, there is increasing emphasis on quality of life targets which incorporate habitat banking and green space initiatives (see Natural England, 1995, 1997;

http://www.englishnature.org.uk/special/greenspace/), despite pressure for increased housing density, industry and communications which threaten these objectives.

Bio-indication as a serious science was launched by the 1992 Convention on Biological Diversity (Glowska *et al.*, 1994) and the 2010 target to reduce the rate of global biodiversity loss (United Nations, 2002; UNEP, 2003). The objectives of bio-indication have been economically summarized by McGeoch (2007): to develop an efficient approach to gather maximum information with minimum resources, bio-indicators are needed that reflect and represent the state of the environment. Three types of bio-indication have been identified (McGeoch, 1998):

1 Environmental: to detect and monitor change in some environmental state, usually distinguished as being abiotic (nutrients, soil water pH, water levels, toxins, etc.).
2 Ecological: to demonstrate and monitor stressors on changes in biota.
3 Biodiversity: to estimate biodiversity and monitor changes in biodiversity.

Insects have a number of potential advantages as bio-indicators, not least of which is that most animal organisms on the planet are insects (>50%; Gullan and Cranston, 2005). Butterflies share these advantages and have others:

- Their short life cycles ensure rapid response to environmental changes.
- Food specialization determines sensitivity to changes.
- Butterflies are apparent, easily observed and recorded.
- There is limited need for taxonomic expertise or for direct capture, and therefore disturbance or loss of individuals in a sampling process.

Key features as bio-indicators are high fidelity to sites and high specificity to conditions (Table 9.1); butterflies display a spectrum of these attributes that ensure their use for different purposes. The UK Butterfly Monitoring Scheme (BMS; www.ukbms.org) has built on these advantages. It has now been running since 1976 (Pollard and Yates, 1993a) and, in conjunction with Butterfly Conservation, is in the process of expanding imminently to many (>600) new sites. In the last 30 years, recorders have made over 140 000 weekly visits to 1228 separate sites covering some 350 000 km and counting 10.5 million butterflies (Greatorex-Davies *et al.*, 2007).

Table 9.1 Characteristics of species making for good bio-indicators. (Redrawn from McGeoch *et al.*, 2002, courtesy of the British Entomological Society and John Wiley and Sons Ltd.)

		'Fidelity' (occupancy or frequency of occurrence)		
		Low	Medium	High
Ecological specificity	Low	**Rural**		**Tramp**
	Medium	←	**Indicator** Detector species	→
	High	**Vulnerable**		**Indicator** Characteristic species

Species are classified as indicators on the basis of their degree of environmental specificity and 'fidelity' to sites (indicator value or IndVal of Dufrêne and Legendre, 1997). Those species with high specificity (habitat specialists) and fidelity (frequency of occurrence) form characteristic bio-indicators for the state of the biotope they occupy, whereas those species with medium specificity form useful indicators of environmental changes. Characteristic species are unlikely to be good indicators of environmental change as they are highly specific in terms of resource requirements and thus decline rapidly to a point of being 'unapparent' to observation. Detector species will be ubiquitous and abundant enough for changes in numbers over sites to be usefully monitored.

Reviews of bio-indication (McGeoch, 1998, 2007; Caro and O'Doherty, 1999; Hilty and Merenlender, 2000) recognize the importance of bio-indicators but also the need for critical evaluation, for objectivity and scientific method, and validation procedures. Lacking validation, some authors contest the value of insects as potential bio-indicators (Niemi and McDonald, 2004). To quote van Straalen (1998:162), a bio-indicator will never 'be a freeze-dried, talking bug on a stick, i.e., the simple standardised and easily applied measuring rod, asked for by regulatory authorities' as available for abiotic indications. Bockstaller and Girardin (2003) identify a hierarchy in the validation process of bio-indication; key components in this are establishing a relationship, the statistical significance and strength of relationship, between indicator and end product, and demonstrating the robustness of the relationship over space and time. Butterflies are regularly used as bio-indicators of habitat loss (Thomas, 1984, 2005), biodiversity erosion (J. A. Thomas *et al.*, 2004; Thomas, 2005) and the effects of climate change (J. K. Hill *et al.*, 2002; C. D. Thomas *et al.*, 2004; Franco *et al.*, 2006). For those that walk over the landscape and observe it closely, the evidence for losses and change seem overwhelming. But those used to laboratory science would argue that it is actually not proven; clear links between butterfly species, their composite resource use, loss of those resources, and loss of resources and habitats for other organisms, have yet to be firmly demonstrated – a critical opinion not helped by any biases that inevitably arise in data collection such as atlas mapping (Barbour, 2007; but see Dennis, 2009b). Just what do butterflies represent in resource terms? For what organisms do they provide suitable surrogates? A resource-based definition of habitat is the first step to testing their suitability as bio-indicators and for providing answers to these and related questions.

There is nothing new, perhaps, in the call for conserving wildlife – articles abound from the 19th century in the entomology literature alone. But, the urgency should at least match the exponential losses in populations and projected extinction of species. In this book, an attempt has been made to focus on what resources species need for their survival. To understand resource use, emphasis has been placed on linking observations of individual behaviour with resource use and on spatially explicit studies, so that it can be properly understood what species use, when and where. Butterflies and the British landscape have been used to illustrate the themes and points made. The approach, however, is relevant for all organisms. Thinking about habitats as resources immediately draws attention to the spatial scale of resources needed in relation to an organism's ability to access them. Some organisms are fortunate enough to have all their needs in a single patch, a field or wood, or perhaps over a slightly wider zone, within a parish. Others have seasonal resources scattered throughout countries and continents, as do so many birds. Conservation for them requires joined-up cooperation across states. Strong national conservation bodies form an essential kernel for conserving global species and their habitats. So, to finish on an upbeat note, one can do little better than highlight the dedicated work done by the many volunteers in the UK for Butterfly Conservation in its 40th year.

> We have calculated that over 10 000 people are helping in our fight to halt the widespread decline in butterflies and moths, who between them contribute over 80 000 days of their valuable time which is the equivalent of 380 full time staff. This is valued at over £5.4 million of voluntary effort every year. (Bulman *et al.*, 2008)

Appendix 1

BIODIVERSITY ACTION PLAN (BAP) STATUS, LEGAL PROTECTION AND TAXONOMIC RELATIONSHIPS FOR BRITISH BUTTERFLIES

APPENDIX 1a: RESIDENT AND RECENTLY EXTINCT SPECIES

Species*	BAP status
Hesperiidae	
Carterocephalus palaemon (Pallas, 1771) The chequered skipper	Priority; species action plan; protected for sale only
Thymelicus sylvestris (Poda, 1761) The small skipper	Not listed
Thymelicus lineola (Ochsenheimer, 1806) The Essex skipper	Not listed
Thymelicus acteon (Rottemburg, 1775) The Lulworth skipper	Priority; protected for sale only
Hesperia comma (Linnaeus, 1758) The silver-spotted skipper	Removed from priority list in 2008; protected for sale only
Ochlodes sylvanus (Esper, 1777) The large skipper	Not listed
Erynnis tages (Linnaeus, 1758) The dingy skipper	Priority
Pyrgus malvae (Linnaeus, 1758) The grizzled skipper	Priority
Papilionidae	
Papilio machaon Linnaeus, 1758 The swallowtail	Not listed; fully protected
Pieridae	
Leptidea sinapis (Linnaeus, 1758)/*L. reali* Reissinger, 1990 The wood white	Priority; protected for sale only
Colias crocea (Geoffroy, 1785) The clouded yellow	Not assessed (migrant)
Gonepteryx rhamni (Linnaeus, 1758) The brimstone	Not listed; protected in Northern Ireland

A Resource-Based Habitat View for Conservation, 1st edition. By Roger L. H. Dennis. Published 2010 by Blackwell Publishing.

Species*	BAP status
Pieris brassicae (Linnaeus, 1758) The large white	Not listed
Pieris rapae (Linnaeus, 1758) The small white	Not listed
Pieris napi (Linnaeus, 1758) The green-veined white	Not listed
Anthocharis cardamines (Linnaeus, 1758) The orange-tip	Not listed
Lycaenidae	
Callophrys rubi (Linnaeus, 1758) The green hairstreak	Not listed
Thecla betulae (Linnaeus, 1758) The brown hairstreak	Priority; protected for sale only
Favonius (Quercusia) quercus (Linnaeus, 1758) The purple hairstreak	Not listed
Satyrium w-album (Knoch, 1782) The white-letter Hairstreak	Priority; protected for sale only
Satyrium pruni (Linnaeus, 1758) The Black hairstreak	Not listed; protected for sale only
Lycaena phlaeas (Linnaeus, 1761) The small copper	Not listed
Lycaena dispar (Haworth, 1803) The large copper	Removed from priority list in 2008 (extinct); fully protected
Cupido minimus (Fuessly, 1775) The small blue	Priority; protected in GB for sale only; fully protected in Northern Ireland
Plebejus argus (Linnaeus, 1758) The silver-studded blue	Priority; protected for sale only
Aricia agestis ([Dennis and Schiffermüller], 1775) The brown argus	Not listed
Aricia artaxerxes (Fabricius, 1793) The northern brown argus	Priority; protected for sale only
Polyommatus icarus (Rottemburg, 1775) The common blue	Not listed
Polyommatus (*Lysandra*) *coridon* (Poda, 1761) The chalkhill blue	Not listed; protected for sale only
Polyommatus (*Lysandra*) *bellargus* (Rottemburg, 1775) The Adonis blue	Removed from priority list in 2008; protected for sale only
Celastrina argiolus (Linnaeus, 1758) The holly blue	Not listed
Glaucopsyche (*Maculinea*) *arion* (Linnaeus, 1758) The large blue†	Priority; fully protected
Hamearis lucina (Linnaeus, 1758) The Duke of Burgundy fritillary	Priority; protected for sale only
Nymphalidae	
Limenitis camilla (Linnaeus, 1764) The white admiral	Priority
Apatura iris (Linnaeus, 1758) The purple emperor	Not listed; protected for sale only
Vanessa atalanta (Linnaeus, 1758) The red admiral	Not assessed (migrant)
Vanessa cardui (Linnaeus, 1758) The painted lady	Not assessed (migrant)
Nymphalis (*Aglais*) *urticae* (Linnaeus, 1758) The small tortoiseshell	Not listed
Nymphalis polychloros (Linnaeus, 1758) The large tortoiseshell	Not assessed (migrant); protected for sale only
Nymphalis (*Inachis*) *io* (Linnaeus, 1758) The peacock	Not listed
Nymphalis (*Polygonia*) *c-album* (Linnaeus, 1758) The comma	Not listed
Boloria selene ([Dennis and Schiffermüller], 1775) The small pearl-bordered fritillary	Priority
Boloria euphrosyne (Linnaeus, 1758) The pearl-bordered fritillary	Priority; protected for sale only
Argynnis adippe ([Dennis and Schiffermüller], 1775) The high brown fritillary	Priority; fully protected
Argynnis aglaja (Linnaeus, 1758) The dark green fritillary	Not listed
Argynnis paphia (Linnaeus, 1758) The silver-washed fritillary	Not listed
Euphydryas aurinia (Rottemburg, 1775) The marsh fritillary	Priority; fully protected
Melitaea cinxia (Linnaeus, 1758) The Glanville fritillary	Priority; protected for sale only
Melitaea athalia (Rottemburg, 1775) The heath fritillary	Priority; fully protected
Pararge aegeria (Linnaeus, 1758) The speckled wood	Not listed
Pararge (*Lasiommata*) *megera* (Linnaeus, 1767) The wall	Priority
Erebia epiphron (Knoch, 1783) The small mountain ringlet	Priority; protected for sale only

Species*	BAP status
Erebia aethiops (Esper, 1777) The Scotch argus	Not listed
Melanargia galathea (Linnaeus, 1758) The marbled white	Not listed
Hipparchia semele (Linnaeus, 1758) The grayling	Priority
Maniola (Pyronia) tithonus (Linnaeus, 1771) The gatekeeper	Not listed
Maniola jurtina (Linnaeus, 1758) The meadow brown	Not listed
Coenonympha pamphilus (Linnaeus, 1758) The small heath	Priority
Coenonympha tullia (Müller, 1764) The large heath	Priority; protected for sale only
Aphantopus hyperantus (Linnaeus, 1758) The ringlet	Not listed

A red list report produced by Butterfly Conservation for the Joint Nature Conservation Committee (JNCC) is on the BC website at http://www.butterfly-conservation.org/downloads/78/Research_Reports.html. All 'priority' species have a Species Action Plan available from www.butterfly-conservation.org/downloads/73/Species_Action_Plans.html.

* See Bradley and Fletcher (1986) for a list of British Lepidoptera; butterfly names as in Kudrna (2002). For Denis and Schiffermüller species, see Sattler and Tremewan (2009).

† *Phengaris arion* (Linnaeus, 1758) in Settele *et al*. (2008).

APPENDIX 1b: RARE MIGRANTS, INTRODUCTIONS AND/OR LONG-EXTINCT SPECIES

Hesperiidae
Heteropterus morpheus (Pallas, 1771)

Papilionidae
Parnassius apollo (Linnaeus, 1758)
Iphiclides podalirius (Linnaeus, 1758)

Pieridae
Colias hyale (Linnaeus, 1758)
Colias alfacariensis Ribbe, 1905
Gonepteryx cleopatra (Linnaeus, 1767)
Aporia crataegi (Linnaeus, 1758)
Pieris (Pontia) daplidice (edusa) (Linnaeus, 1758)
Euchloe simplonia (Freyer, 1829)

Lycaenidae
Lycaena tityrus (Poda, 1761)
Lampides boeticus (Linnaeus, 1767)
Cupido (Everes) argiades (Pallas, 1771)
Polyommatus (Cyaniris) semiargus (Rottemburg, 1775)
Cacyreus marshalli (Butler, 1898)

Nymphalidae
Vanessa virginiensis (Drury, 1773)
Nymphalis xanthomelas (Esper, 1781)
Nymphalis antiopa (Linnaeus, 1758)
Araschnia levana (Linnaeus, 1758)
Issoria lathonia (Linnaeus, 1758)
Argynnis pandora ([Dennis and Schiffermüller], 1775)
Melitaea didyma (Esper, 1779)
Erebia ligea (Linnaeus, 1758)
Hipparchia arethusa ([Dennis and Schiffermüller], 1775)
Danaus plexippus (Linnaeus, 1758)

APPENDIX 1c: TAXONOMIC AFFINITIES OF BRITISH BUTTERFLIES

Distances in the dendrogram relate to shifts in taxonomic level (from base up: sibling species, species, subgenus, genus, tribe, subfamily, family and superfamily). (Courtesy of Nick J. B. Isaac, CEH.)

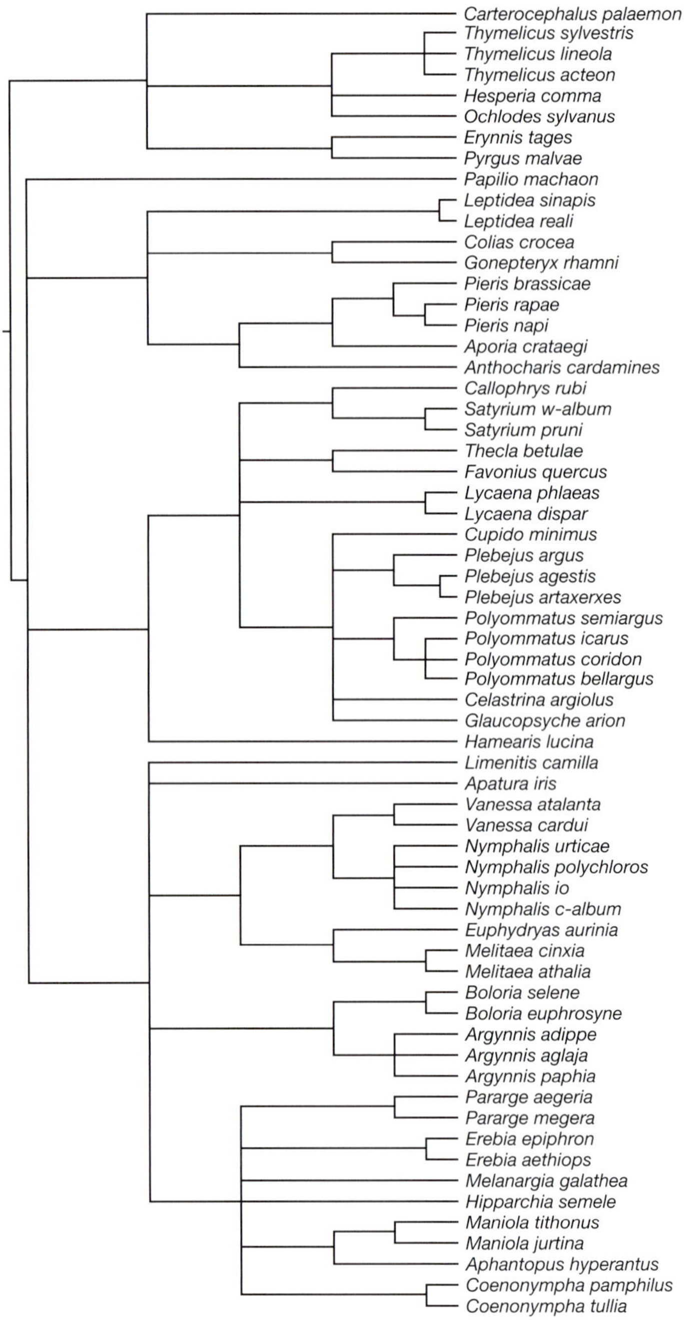

Appendix 2

LARVAL HOSTPLANTS FOR BRITISH BUTTERFLIES

APPENDIX 2a: STATUS OF HOSTPLANTS

Butterfly species*	Hostplants†
C. palaemon	***Brachypodium pinnatum* (L.) P. Beauv., Poaceae**, P, G (England only)
	***Brachypodium sylvaticum* (Huds.) P. Beauv., Poaceae**, P, G (England only)
	***Molinia caerulea* (L.) Moench, Poaceae**, P, G (Scotland only)
	[*Bromopsis ramosa* (Huds.) Holub, Poaceae, P, G (in captivity)]
T. sylvestris	***Holcus lanatus* L., Poaceae**, P, G
	Alopecurus pratensis L., Poaceae, P, G
	Arrhenatherum elatius P. Beauv. ex J. & C. Presl, Poaceae, P, G
	Brachypodium pinnatum (L.) P. Beauv., Poaceae, P, G
	Brachypodium sylvaticum (Huds.) P. Beauv., Poaceae, P, G
	Dactylis glomerata L., Poaceae, P, G
	Holcus mollis L., Poaceae, P, G
	Phleum bertolonii DC, Poaceae, P, G
	Phleum pratense L., Poaceae, P, G
T. lineola	***Dactylis glomerata* L., Poaceae**, P, G
	***Holcus mollis* L., Poaceae**, P, G
	Alopecurus pratensis L., Poaceae, P, G
	Arrhenatherum elatius P. Beauv. ex J. & C. Presl, Poaceae, P, G
	Brachypodium pinnatum (L.) P. Beauv., Poaceae, P, G
	Brachypodium sylvaticum (Huds.) P. Beauv., Poaceae, P, G
	Elytrigia repens (L.) Desv. ex Nevski, Poaceae, P, G
	Holcus lanatus L., Poaceae, P, G
	Phleum pratense L., Poaceae, P, G
T. acteon	***Brachypodium pinnatum* (L.) P. Beauv., Poaceae**, P, G
	Holcus lanatus L., Poaceae, P, G
	Phleum spp., Poaceae, P, G
	[*Brachypodium sylvaticum* (Huds.) P. Beauv., Poaceae, P, G (in captivity)]
	[*Elytrigia repens* (L.) Desv. ex Nevski, Poaceae, P, G (in captivity)]
	[*Holcus mollis* L., Poaceae, P, G (in captivity)]
	[*Phleum pratense* L., Poaceae, P, G (in captivity)]
	[*Poa annua* L., Poaceae, A/P, G (in captivity)]

A Resource-Based Habitat View for Conservation, 1st edition. By Roger L. H. Dennis. Published 2010 by Blackwell Publishing.

Butterfly species*	Hostplants†
H .comma	***Festuca ovina* L., Poaceae**, P, G
	Deschampsia cespitosa (L.) P. Beauv., Poaceae, P, G
O. sylvanus	***Brachypodium sylvaticum* (Huds.) P. Beauv., Poaceae**, P, G
	***Dactylis glomerata* L., Poaceae**, P, G
	***Molinia caerulea* (L.) Moench, Poaceae**, P, G
	Brachypodium pinnatum (L.) P. Beauv., Poaceae, P, G
	Calamagrostis epigejos (L.) Roth, Poaceae, P, G
	Elytrigia repens (L.) Desv. ex Nevski, Poaceae, P, G
	Holcus lanatus L., Poaceae, P, G
	Holcus mollis L., Poaceae, P, G
	Phleum pratense L., Poaceae, P, G
	Triticum sp., Poaceae, A, G
E. tages	***Lotus corniculatus* L., Fabaceae**, P, F
	Hippocrepis comosa L., Fabaceae, P, F
	Lotus pedunculatus Cav., Fabaceae, P, F
P. malvae	***Agrimonia eupatoria* L., Rosaceae**, P, F
	***Fragaria vesca* L., Rosaceae**, P, F
	***Potentilla reptans* L., Rosaceae**, P, F
	Geum urbanum L., Rosaceae, P, F
	Potentilla anserina L., Rosaceae, P, F
	Potentilla erecta (L.) Raeusch., Rosaceae, P, F
	Potentilla sterilis (L.) Garcke, Rosaceae, P, F
	Rosa canina L., Rosaceae, P, S
	Rubus fruticosus L. agg., Rosaceae, P, S
	Rubus idaeus L., Rosaceae, P, S
	Sanguisorba minor Scop., Rosaceae, P, F
	[*Fragaria x ananassa* (Duchesne) Duchesne, Rosaceae, P, F (non-native)]
P. machaon	***Peucedanum palustre* (L.) Moench, Apiaceae**, B/P, F
	Angelica sylvestris L., Apiaceae, P, F (migrants only)
	Daucus carota L., Apiaceae, A/B, F (migrants only)
	[*Aegopodium podagraria* L., Apiaceae, P, F (non-native)]
	[*Anethum graveolens* L., Apiaceae, A, F (non-native)]
	[*Foeniculum vulgare* Mill., Apiaceae, P, F (non-native)]
	[*Petroselinum crispum* (Mill.) Nyman ex A.W. Hill, Apiaceae, B, F (non-native)]
L. sinapis/L. reali	***Lathyrus linifolius* (Reichard) Bässler, Fabaceae**, P, F
	***Lathyrus pratensis* L., Fabaceae**, P, F
	***Lotus corniculatus* L., Fabaceae**, P, F
	***Lotus pedunculatus* Cav., Fabaceae**, P, F
	***Vicia cracca* L., Fabaceae**, P, F
	[*Lathyrus latifolius* L., Fabaceae, P, F (non-native)]
	[*Lathyrus odoratus* L., Fabaceae, A, C (non-native)]
	[*Pisum sativum* L., Fabaceae, A, F (non-native)]
	[*Lathyrus tuberosus* L., Fabaceae, P, F (in captivity)]
C. crocea	***Medicago sativa* L., Fabaceae**, P, F
	***Trifolium pratense* L., Fabaceae**, P, F
	***Trifolium repens* L., Fabaceae**, P, F
	Anthyllis vulneraria L., Fabaceae, P, F

Butterfly species*	Hostplants†
	Lotus corniculatus L., Fabaceae, P, F
	Lotus subbiflorus Lag., Fabaceae, A, F
	Medicago arabica (L.) Huds., Fabaceae, A, F
	Medicago lupulina L., Fabaceae, A/P, F
	Vicia spp., Fabaceae, P, F
	Hippocrepis comosa L., Fabaceae, P, F
	[*Melilotus officinalis* (L.) Pall., Fabaceae, B, F (non-native)]
	[*Securigera varia* (L.) Lassen, Fabaceae, P, F (non-native)]
G. rhamni	***Frangula alnus* Mill., Rhamnaceae**, P, S
	***Rhamnus cathartica* L., Rhamnaceae**, P, S
	[*Rhamnus alaternus* L., Rhamnaceae, P, S (non-native)]
	[*Rhamnus alpina* L., Rhamnaceae, P, S (non-native)]
P. brassicae	***Brassica oleracea* L., Brassicaceae**, B/P, F
	Alliaria petiolata (M. Bieb) Cavara & Grande, Brassicaceae, B, F
	Barbarea vulgaris W.T. Aiton, Brassicaceae, B/P, F
	Brassica nigra (L.) W.D.J. Koch, Brassicaceae, A, F
	Crambe maritima L., Brassicaceae, P, F
	Diplotaxis tenuifolia (L.) DC., Brassicaceae, P, F
	Raphanus raphanistrum L. ssp. maritimus (Sm.) Thell., Brassicaceae, A, F
	Reseda lutea L., Resedaceae, B/P, F
	Reseda luteola L., Resedaceae, B, F
	Rorippa nasturtium-aquaticum (L.) Hayek, Brassicaceae, P, F
	Sisymbrium officinale (L.) Scop., Brassicaceae, A/B, F
	[*Armoracia rusticana* P. Gaertn., B. Mey. & Scherb., Brassicaceae, P, F (non-native)]
	[*Brassica napus* L., Brassicaceae, A/B, F (non-native)]
	[*Brassica rapa* L., Brassicaceae, A/B, F (non-native)]
	[*Cleome sesquiorygalis* Nadin ex C. Huber, Capparaceae, A, F (non-native)]
	[*Erysimum cheiri* (L.) Crantz, Brassicaceae, P, F (non-native)]
	[*Hesperis matronalis* L., Brassicaceae, B/P, F (non-native)]
	[*Lepidium sativum* L., Brassicaceae, A, F (non-native)]
	[*Lunaria annua* L., Brassicaceae, B, F (non-native)]
	[*Malcolmia maritima* (L.) W.T. Aiton, Brassicaceae, A, F (non-native)]
	[*Raphanus raphanistrum* L., Brassicaceae, B/P, F (non-native)]
	[*Raphanus sativus* L., Brassicaceae, A/B, F (non-native)]
	[*Reseda odorara* L., Resedaceae, A/B, F (non-native)]
	[*Sinapis alba* L., Brassicaceae, A, F (non-native)]
	[*Tropaeolum majus* L., Tropaeolaceae, A, C (non-native)]
P. rapae	***Brassica oleracea* L., Brassicaceae**, B/P, F
	***Rorippa nasturtium-aquaticum* (L.) Hayek, Brassicaceae**, P, F
	Alliaria petiolata (M. Bieb) Cavara & Grande, Brassicaceae, B, F
	Barbarea vulgaris W.T. Aiton, Brassicaceae, B/P, F
	Brassica nigra (L.) W.D.J. Koch, Brassicaceae, A, F
	Cardamine hirsuta L., Brassicaceae, A/B, F
	Coronopus squamatus (Forssk.) Asch., Brassicaceae, A/B, F
	Capsella bursa-pastoris (L.) Medik., Brassicaceae, A/B, F
	Crambe maritima L., Brassicaceae, P, F
	Reseda lutea L., Resedaceae, B/P, F
	Reseda luteola L., Resedaceae, B, F

Butterfly species*	Hostplants†
	Sinapis arvensis L., Brassicaceae, A, F
	Sisymbrium officinale (L.) Scop., Brassicaceae, A/B, F
	[*Alyssum saxatile* L., Brassicaceae, P, F (non-native)]
	[*Aubrieta deltoidea* (L.) DC., Brassicaceae, P, F (non-native)]
	[*Brassica napus* L., Brassicaceae, A/B, F (non-native)]
	[*Brassica rapa* L., Brassicaceae, A/B, F (non-native)]
	[*Cleome sesquiorygalis* Nadin ex C. Huber, Capparaceae (non-native)]
	[*Coronopus didymus* (L.) Sm., Brassicaceae, A/B, F (non-native)]
	[*Erysimum cheiri* (L.) Crantz, Brassicaceae, P, F (non-native)]
	[*Lepidium draba* L., Brassicaceae, P, F (non-native)]
	[*Raphanus sativus* L., Brassicaceae, A/B, F (non-native)]
	[*Reseda odorara* L., Resedaceae, A/B, F (non-native)]
	[*Tropaeolum majus* L., Tropaeolaceae, A, C (non-native)]
	[*Tropaeolum peregrinum* L., Tropaeolaceae, A, C (non-native)]
P. napi	***Alliaria petiolata* (M. Bieb) Cavara & Grande, Brassicaceae**, B, F
	***Cardamine amara* L., Brassicaceae**, P, F
	***Cardamine hirsuta* L., Brassicaceae**, A/B, F
	***Cardamine pratensis* L., Brassicaceae**, P, F
	***Rorippa microphylla* (Boenn.) Hyl. ex Á. & D. Löve, Brassicaceae**, P, F
	***Rorippa nasturtium-aquaticum* (L.) Hayek, Brassicaceae**, P, F
	***Sisymbrium officinale* (L.) Scop., Brassicaceae**, A/B, F
	Arabis hirsuta (L.) Scop., Brassicaceae, B/P, F
	Barbarea vulgaris W.T. Aiton, Brassicaceae, B/P, F
	Brassica oleracea L., Brassicaceae, B/P, F
	Cakile maritima Scop., Brassicaceae, A, F
	Lepidium heterophyllum Benth., Brassicaceae, P, F
	Rorippa sylvestris (L.) Besser, Brassicaceae, P, F
	Sinapis arvensis L., Brassicaceae, A, F
	[*Armoracia rusticana* P. Gaertn., B. Mey. & Scherb., Brassicaceae, P, F (non-native)]
	[*Aubrieta deltoidea* (L.) DC., Brassicaceae, P, F (non-native)]
	[*Brassica napus* L., Brassicaceae, A/B, F (non-native)]
	[*Coronopus didymus* (L.) Sm., Brassicaceae, A/B, F (non-native)]
	[*Erysimum cheiri* (L.) Crantz, Brassicaceae, P, F (non-native)]
	[*Hesperis matronalis* L., Brassicaceae, B/P, F (non-native)]
	[*Raphanus raphanistrum* L., Brassicaceae, B/P, F (non-native)]
	[*Tropaeolum majus* L., Tropaeolaceae, A, C (non-native)]
	[*Reseda lutea* L., Resedaceae, B/P, F (in captivity)]
A. cardamines	***Alliaria petiolata* (M. Bieb) Cavara & Grande, Brassicaceae**, B, F
	***Cardamine pratensis* L., Brassicaceae**, P, F
	Arabidopsis thaliana (L.) Heynh., Brassicaceae, A, F
	Arabis glabra (L.) Bernh., Brassicaceae, B, F
	Arabis hirsuta (L.) Scop., Brassicaceae, B/P, F
	Armoracia rusticana P. Gaertn., B. Mey. & Scherb., Brassicaceae, P, F
	Barbarea vulgaris W.T. Aiton, Brassicaceae, B/P, F
	Brassica nigra (L.) W.D.J. Koch, Brassicaceae, A, F
	Capsella bursa-pastoris (L.) Medik., Brassicaceae, A/B, F
	Cardamine amara L., Brassicaceae, P, F
	Cardamine hirsuta L., Brassicaceae, A/B, F
	Cochlearia officinalis L., Brassicaceae, B/P, F

Butterfly species*	Hostplants†
	Hesperis matronalis L., Brassicaceae, B/P, F
	Lepidium campestre (L.) W.T. Aiton, Brassicaceae, A/B, F
	Reseda lutea L., Resedaceae, B/P, F
	Reseda luteola L., Resedaceae, B, F
	Rorippa nasturtium-aquaticum (L.) Hayek, Brassicaceae, P, F
	Rorippa sylvestris (L.) Besser, Brassicaceae, P, F
	Sinapis arvensis L., Brassicaceae, A, F
	Sisymbrium officinale (L.) Scop., Brassicaceae, A/B, F
	Thlaspi caerulescens J. & C. Presl, Brassicaceae, B/P, F
	Rapistrum rugosum (L.) J.P. Bergeret, Brassicaceae, A, F
	[*Arabis caucasica* Willd. ex Schltdl., Brassicaceae, P, F (non-native)]
	[*Arabis* spp., Brassicaceae, B/P, F (non-native)]
	[*Brassica napus* L., Brassicaceae, A/B, F (non-native)]
	[*Brassica rapa* L., Brassicaceae, A/B, F (non-native)]
	[*Lunaria annua* L., Brassicaceae, B, F (non-native)]
	[*Matthiola incana* (L.) W.T. Aiton, Brassicaceae, A/P, F (non-native)]
	[*Raphanus raphanistrum* L., Brassicaceae, A, F (non-native)]
	[*Sisymbrium* spp., Brassicaceae, A/B, F (non-native)]
	[*Tropaeolum majus* L., Tropaeolaceae, A, C (non-native)]
C. rubi	***Cytisus scoparius* (L.) Link, Fabaceae**, P, S
	***Genista tinctoria* L., Fabaceae**, P, S
	***Helianthemum nummularium* (L.) Mill., Cistaceae**, P, F
	***Lotus corniculatus* L., Fabaceae**, P, F
	***Ulex europaeus* L., Fabaceae**, P, S
	***Vaccinium myrtillus* L., Ericaceae**, P, S
	Arctostaphylos uva-ursi (L.) Spreng, Ericaceae, P, S
	Calluna vulgaris (L.) Hull, Ericaceae, P, S
	Cornus sanguinea L., Cornaceae, P, S
	Empetrum nigrum L., Empetraceae, P, S
	Erica cinerea L., Ericaceae, P, S
	Erica tetralix L., Ericaceae, P, S
	Frangula alnus Mill., Rhamnaceae, P, S
	Genista anglica L., Fabaceae, P, S
	Lotus pedunculatus Cav., Fabaceae, P, F
	Rhamnus cathartica L., Rhamnaceae, P, S
	Rubus fruticosus L. agg., Rosaceae, P, S
	Trifolium spp., Fabaceae, A/P, F
	Ulex gallii Planch., Fabaceae, P, S
	Ulex minor Roth., Fabaceae, P, S
	Vaccinium oxycoccus L., Ericaceae, P, S
	Vaccinium vitis-idaea L., Ericaceae, P, S
	Vicia spp., Fabaceae, P, F
	[*Laburnum anagyroides* Medic., Fabaceae, P, T (in captivity)]
	[*Phaseolus coccineus* L., Fabaceae, P, C (in captivity)]
	[*Pisum sativum* L., Fabaceae, A, F (in captivity)]
T. betulae	***Prunus spinosa* L., Rosaceae**, P, T
	[*Prunus domestica* L., Rosaceae, P, T (non-native)]
	[*Prunus domestica* L. ssp. *institia* (L.) Bonnier & Layens, Rosaceae, P, T (non-native)]
	[*Prunus domestica* L. ssp. *italica* (Borkh.) Garns ex Hegi, Rosaceae, P, T (non-native)]

Butterfly species*	Hostplants†
F. quercus	***Quercus petraea* (Matt.) Liebl., Fagaceae**, P, T ***Quercus robur* L., Fagaceae**, P, T [*Quercus cerris* L., Fagaceae, P, T (non-native)] [*Quercus ilex* L., Fagaceae, P, T (non-native)]
S. w-album	***Ulmus glabra* Huds., Ulmaceae**, P, T ***Ulmus procera* Salisb., Ulmaceae**, P, T *Ulmus minor* Mill., Ulmaceae, P, T [*Ulmus japonica* (Rehd.) Sarg., Ulmaceae, P, T (non-native)]
S. pruni	***Prunus spinosa* L., Rosaceae**, P, T [*Prunus domestica* L., Rosaceae, P, T (non-native)]
L. phlaeas	***Rumex acetosa* L., Polygonaceae**, P, F ***Rumex acetosella* L., Polygonaceae**, P, F *Rumex hydrolapathum* Huds., Polygonaceae, P, F *Rumex obtusifolius* L., Polygonaceae, P, F *Rumex pulcher* L., Polygonaceae, P, F
L. dispar	***Rumex hydrolapathum* Huds., Polygonaceae**, P, F [*Rumex obtusifolius* L., Polygonaceae, P, F (in captivity)]
C. minimus	***Anthyllis vulneraria* L., Fabaceae**, P, F [*Melilotus officinalis* (L.) Pall., Fabaceae, B, F (non-native)]
P. argus	***Calluna vulgaris* (L.) Hull, Ericaceae**, P, S ***Erica cinerea* L., Ericaceae**, P, S ***Helianthemum nummularium* (L.) Mill., Cistaceae**, P, F ***Helianthemum oelandicum* (L.) Dum. Cours., Cistaceae**, P, F ***Lotus corniculatus* L., Fabaceae**, P, F ***Ornithopus perpusillus* L., Fabaceae**, A, F ***Ulex europaeus* L., Fabaceae**, P, S *Anthyllis vulneraria* L., Fabaceae, P, F *Erica tetralix* L., Ericaceae, P, S *Genista anglica* L., Fabaceae, P, S *Hippocrepis comosa* L., Fabaceae, P, F *Ononis repens* L., Fabaceae, P, F *Ononis spinosa* L., Fabaceae, P, F *Thymus polytrichus* A. Kern. ex Borbás, Lamiaceae, P, F *Trifolium arvense* L., Fabaceae, A, F *Ulex gallii* Planch., Fabaceae, P, S [*Cytisus scoparius* (L.) Link, Fabaceae, P, S (in captivity)]
A. agestis	***Erodium cicutarium* (L.) L'Hér., Geraniaceae**, A, F ***Geranium molle* L., Geraniaceae**, A, F ***Helianthemum nummularium* (L.) Mill., Cistaceae**, P, F *Geranium dissectum* L., Geraniaceae, A, F *Geranium pratense* L., Geraniaceae, P, F *Geranium pyrenaicum* Burm., Geraniaceae, P, F [*Pelargonium x hybridum* Aiton, Geraniaceae, P, F (in captivity)]
A. artaxerxes	***Helianthemum nummularium* (L.) Mill., Cistaceae**, P, F *Erodium cicutarium* (L.) L'Hér., Geraniaceae, A, F *Geranium sanguineum* L., Geraniaceae, P, F

Butterfly species*	Hostplants†
P. icarus	***Lotus corniculatus* L., Fabaceae**, P, F
	Lotus pedunculatus Cav., Fabaceae, P, F
	Medicago lupulina L., Fabaceae, A/P, F
	Ononis repens L., Fabaceae, P, F
	Ornithopus perpusillus L., Fabaceae, A, F
	Trifolium campetsre Schreb., Fabaceae, A, F
	Trifolium dubium Sibth., Fabaceae, A, F
	Trifolium pratense L., Fabaceae, P, F
	Trifolium repens L., Fabaceae, P, F
	Ononis spinosa L., Fabaceae, P, F
	[*Melilotus altissimus* Thuill., Fabaceae, B/P, F (non-native)]
	[*Phaseolus* spp., Fabaceae, A/P, C (in captivity)]
	[*Pisum sativum* L., Fabaceae, A, F (in captivity)]
P. coridon	***Hippocrepis comosa* L., Fabaceae**, P, F
	Anthyllis vulneraria L., Fabaceae, P, F
	Astragalus glycyphyllos L., Fabaceae, P, F
	Lotus corniculatus L., Fabaceae, P, F
	Ornithopus perpusillus L., Fabaceae, A, F
	Trifolium pratense L., Fabaceae, P, F
	Trifolium repens L., Fabaceae, P, F
	[*Securigera varia* (L.) Lassen, Fabaceae, P, F (non-native)]
	[*Coronilla valentina* L., ssp. *glauca* (L.) Battand, Fabaceae, P, S (in captivity)]
	[*Pisum sativum* L., Fabaceae, A, F (in captivity)]
P. bellargus	***Hippocrepis comosa* L., Fabaceae**, P, F
	Trifolium repens L., Fabaceae, P, F
	[*Coronilla valentina* L., ssp. *glauca* (L.) Battand, Fabaceae, P, S (in captivity)]
	[*Pisum sativum* L., Fabaceae, A, F (in captivity)]
C. argiolus	***Ilex aquifolium* L., Aquifoliaceae**, P, T
	***Hedera helix* L., Araliaceae**, P, C
	Calluna vulgaris (L.) Hull, Ericaceae, P, S
	Cornus sanguinea L., Cornaceae, P, S
	Cytisus scoparius (L.) Link, Fabaceae, P, S
	Euonymus europaeus L., Celastraceae, P, S
	Frangula alnus Mill., Rhamnaceae, P, S
	Humulus lupulus L., Cannabaceae, P, C
	Lythrum salicaria L., Lythraceae, P, F
	Rhamnus spp., Rhamnaceae, P, S
	Rubus fruticosus L. agg., Rosaceae, P, S
	Rubus idaeus L., Rosaceae, P, S
	Salix cinerea L., Salicaceae, P, T
	Ulex europaeus L., Fabaceae, P, S
	Viburnum lantana L., Caprifoliaceae, P, S
	[*Buddleja davidii* Franch., Buddlejaceae, P, S (non-native)]
	[*Prunus laurocerasus* L., Rosaceae, P, T (non-native)]
	[*Prunus lusitanica* L., Rosaceae, P, T (non-native)]
	[*Pyracantha atalantoides* (Hance) Stapf, Rosaceae, P, S (non-native)]
	[*Pyracantha coccinea* M. Roem. 'Lalandei', Rosaceae, P, S (non-native)]
	[*Securigera varia* (L.) Lassen, Fabaceae, P, F (non-native)]

Butterfly species*	Hostplants†
	[*Spartium junceum* L., Fabaceae, P, F (non-native)] [*Symphoricarpos albus* (L.) S.F. Blake, Caprifoliaceae, P, S (non-native)] [*Wisteria sinensis* (Sims) DC, Fabaceae, P, C (non-native)] [*Escallonia macrantha* Hook. & Am., Grossulariaceae, P, S (in captivity)]
M. arion	***Thymus polytrichus* A. Kern. ex Borbás, Lamiaceae**, P, F *Origanum vulgare* L., Lamiaceae, P, F
H. lucina	***Primula veris* L., Primulaceae**, P, F ***Primula vulgaris* Huds., Primulaceae**, P, F *Primula elatior* (L.) Hill, Primulaceae, P, F *Primula veris* L. × *P. vulgaris* Huds., Primulaceae, P, F [*Primula* 'Pacific Series', Primulaceae, P, F (in captivity)]
L. camilla	***Lonicera periclymenum* L., Caprifoliaceae**, P, C [*Symphoricarpos albus* (L.) S.F. Blake, Caprifoliaceae, P, S (in captivity)]
A. iris	***Salix caprea* L., Salicaceae**, P, T *Salix cinerea* L., Salicaceae, P, T *Salix fragilis* L., Salicaceae, P, T [*Populus* spp., Salicaceae, P, T (in captivity)] [*Salix alba* L., Salicaceae, P, T (in captivity)]
V. atalanta	***Urtica dioica* L., Urticaceae**, P, F *Humulus lupulus* L., Cannabaceae, P, C *Parietaria judaica* L., Urticaceae, P, F *Urtica urens* L., Urticaceae, A, F [*Parietaria officinalis* L., Urticaceae, P, F (non-native)]
V. cardui	***Carduus crispus* L., Asteraceae, A/B, F** ***Carduus nutans* L., Asteraceae, A/B, F** ***Cirsium arvense* (L.) Scop., Asteraceae**, P, F ***Cirsium palustre* (L.) Scop., Asteraceae**, B, F ***Cirsium vulgare* (Savi) Ten., Asteraceae**, B, F *Althaea officinalis* L., Malvaceae, P, F *Arctium lappa* L., Asteraceae, B, F *Arctium minus* (Hill) Bernh., Asteraceae, B, F *Artemisia vulgaris* L, Asteraceae, P, F *Carduus tenuiflorus* Curtis, Asteraceae, A/B, F *Carlina vulgaris* L., Asteraceae, B, F *Centaurea nigra* L., Asteraceae, P, F *Centaurea scabiosa* L., Asteraceae, P, F *Cirsium acaule* (L.) Scop., Asteraceae, P, F *Echium vulgare* L., Boraginaceae, B, F *Filago vulgaris* Lam., Asteraceae, A, F *Malva sylvestris* L., Malvaceae, P, F *Parietaria judaica* L., Urticaceae, P, F *Plantago lanceolata* L., Plantaginaceae, P, F *Symphytum officinale* L., Boraginaceae, P, F *Urtica dioica* L., Urticaceae, P, F *Urtica urens* L., Urticaceae, A, F [*Artemisia stelleriana* Besser, Asteraceae, P, F (non-native)] [*Borago officinalis* L., Boraginaceae, A, F (non-native)]

Butterfly species*	Hostplants†
	[*Cynara cardunculus* L., Asteraceae, P, F (non-native)] [*Helichrysum bracteatum* (Vent.) Andrews, Asteraceae, A, F (non-native)] [*Onopordum acanthium* L., Asteraceae, B, F (non-native)] [*Plantago major* L., Plantaginaceae, P, F (in captivity)]
A. urticae	***Urtica dioica* L., Urticaceae**, P, F ***Urtica urens* L., Urticaceae**, A, F [*Humulus lupulus* L., Cannabaceae, P, C (in captivity)]
N. polychloros	***Ulmus glabra* Huds., Ulmaceae**, P, T ***Ulmus procera* Salisb., Ulmaceae**, P, T *Betula pubescens* Ehrh, Betulaceae, P, T *Populus nigra* L., Salicaceae, P, T *Populus tremula* L., Salicaceae, P, T *Prunus avium* (L.) L., Rosaceae, P, T *Prunus spinosa* L., Rosaceae, P, T *Salix alba* L., Salicaceae, P, T *Salix caprea* L., Salicaceae, P, T *Salix cinerea* L., Salicaceae, P, T *Salix fragilis* L., Salicaceae, P, T *Salix viminalis* L., Salicaceae, P, S *Sorbus aria* (L.) Crantz, Rosaceae, P, T *Ulmus minor* Mill., Ulmaceae, P, T [*Populus alba* L., Salicaceae, P, T (non-native)] [*Prunus cerasus* L., Rosaceae, P, T (non-native)] [*Pyrus communis* L., Rosaceae, P, T (non-native)] [*Pyrus pyraster* (L.) Burgsd., Rosaceae, P, T (non-native)]
I. io	***Urtica dioica* L., Urticaceae**, P, F *Humulus lupulus* L., Cannabaceae, P, C *Urtica urens* L., Urticaceae, A, F
P. c-album	***Ulmus glabra* Huds., Ulmaceae**, P, T ***Ulmus procera* Salisb., Ulmaceae**, P, T ***Urtica dioica* L., Urticaceae**, P, F *Humulus lupulus* L., Cannabaceae, P, C *Ribes uva-crispa* L., Grossulariaceae, P, S *Salix caprea* L., Salicaceae, P, T [*Ribes nigrum* L., Grossulariaceae, P, S (non-native)] [*Ribes rubrum* L., Grossulariaceae, P, S (non-native)] [*Corylus avellana* L., Betulaceae, P, T (in captivity)]
B. selene	***Viola palustris* L., Violaceae**, P, F ***Viola riviniana* Rchb., Violaceae**, P, F *Viola hirta* L., Violaceae, P, F *Viola lutea* Huds., Violaceae, P, F *Viola odorata* L., Violaceae, P, F *Viola tricolor* L., Violaceae, A/P, F [*Viola x wittrockiana* Gams ex Kappert, Violaceae, A/P, F (in captivity)]
B. euphrosyne	***Viola riviniana* Rchb., Violaceae**, P, F ***Viola palustris* L., Violaceae**, P, F *Viola canina* L., Violaceae, P, F

Butterfly species*	Hostplants†
	Viola hirta L., Violaceae, P, F
	Viola lutea Huds., Violaceae, P, F
	Viola odorata L., Violaceae, P, F
	Viola tricolor L., Violaceae, A/P, F
	[*Viola x wittrockiana* Gams ex Kappert, Violaceae, A/P, F (in captivity)]
A. adippe	***Viola riviniana* Rchb., Violaceae**, P, F
	Viola canina L., Violaceae, P, F
	Viola hirta L., Violaceae, P, F
	Viola lactea Sm., Violaceae, P, F
	Viola odorata L., Violaceae, P, F
	[*Viola tricolor* L., Violaceae, A/P, F (in captivity)]
	[*Viola x wittrockiana* Gams ex Kappert, Violaceae, A/P, F (in captivity)]
A. aglaja	***Viola riviniana* Rchb., Violaceae**, P, F
	***Viola hirta* L., Violaceae**, P, F
	***Viola palustris* L., Violaceae**, P, F
	Viola lutea Huds., Violaceae, P, F
	Viola reichenbachiana Jord. ex Boreau, Violaceae, P, F
	Viola tricolor L., Violaceae, A/P, F
	[*Viola odorata* L., Violaceae, P, F (in captivity)]
	[*Viola x wittrockiana* Gams ex Kappert, Violaceae, A/P, F (in captivity)]
A. paphia	***Viola riviniana* Rchb., Violaceae**, P, F
	Viola canina L., Violaceae, P, F
	Viola odorata L., Violaceae, P, F
	[*Viola tricolor* L., Violaceae, A/P, F (in captivity)]
	[*Viola x wittrockiana* Gams ex Kappert, Violaceae, A/P, F (in captivity)]
E. aurinia	***Succisa pratensis* Moench, Dipsaceae**, P, F
	Centaurea scabiosa L., Asteraceae, P, F
	Knautia arvensis (L.) Coult., Dipsaceae, P, F
	Leontodon spp., Asteraceae, P, F
	Lonicera periclymenum L., Caprifoliaceae, P, C
	Plantago lanceolata L., Plantaginaceae, P, F
	Plantago major L., Plantaginaceae, P, F
	Scabiosa columbaria L., Dipsaceae, P, F
	Teucrium scorodonia L., Lamiaceae, P, F
	[*Dipsacus fullonum* L., Dipsaceae, B, F (in captivity)]
	[*Symphoricarpos albus* (L.) S.F. Blake, Caprifoliaceae, P, S (in captivity)]
M. cinxia	***Plantago lanceolata* L., Plantaginaceae**, P, F
	***Plantago maritima* L., Plantaginaceae**, P, F
	Plantago coronopus L., Plantaginaceae, A/P, F
	Veronica chamaedrys L., Scrophulariaceae, P, F
	Veronica hederifolia L., Scrophulariaceae, A, F
	Veronica officinalis L., Scrophulariaceae, P, F
M. athalia	***Melampyrum pratense* L., Scrophulariaceae**, A, F
	***Plantago lanceolata* L., Plantaginaceae**, P, F
	***Veronica chamaedrys* L., Scrophulariaceae**, P, F
	Achillea millefolium L., Asteraceae, P, F
	Digitalis purpurea L., Scrophulariaceae, B/P, F

Butterfly species*	Hostplants†
	Melampyrum sylvaticum L., Scrophulariaceae, A, F *Plantago major* L., Plantaginaceae, P, F *Teucrium scorodonia* L., Lamiaceae, P, F *Veronica hederifolia* L., Scrophulariaceae, A, F *Veronica serpyllifolia* L., Scrophulariaceae, P, F [*Linaria* spp., Scrophulariaceae, A/P, F (in captivity)] [*Nepeta* spp., Lamiaceae, P, F (in captivity)]
P. aegeria	***Brachypodium sylvaticum* (Huds.) P. Beauv., Poaceae**, P, G ***Dactylis glomerata* L., Poaceae**, P, G ***Elytrigia repens* (L.) Desv. ex Nevski, Poaceae**, P, G *Holcus lanatus* L., Poaceae, P, G *Poa nemoralis* L., Poaceae, P, G *Agrostis canina* L., Poaceae, P, G *Agrostis capillaris* L., Poaceae, P, G *Agrostis stolonifera* L., Poaceae, P, G *Bromopsis ramosa* (Huds.) Holub, Poaceae, P, G *Festuca rubra* L., Poaceae, P, G *Lolium perenne* L., Poaceae, P, G *Phleum pratense* L., Poaceae, P, G *Poa trivialis* L., Poaceae, P, G [*Poa annua* L., Poaceae, A/P, G (in captivity)]
L. megera	***Brachypodium pinnatum* (L.) P. Beauv., Poaceae**, P, G ***Brachypodium sylvaticum* (Huds.) P. Beauv., Poaceae**, P, G ***Dactylis glomerata* L., Poaceae**, P, G ***Deschampsia flexuosa* (L.) Trin., Poaceae**, P, G ***Holcus lanatus* L., Poaceae**, P, G *Agrostis capillaris* L., Poaceae, P, G *Agrostis gigantea* Roth, Poaceae, P, G *Bromus* spp., Poaceae, P, G *Deschampsia cespitosa* (L.) P. Beauv., Poaceae, P, G *Elytrigia repens* (L.) Desv. ex Nevski, Poaceae, P, G *Festuca ovina* L., Poaceae, P, G *Lolium* spp., Poaceae, P, G [*Poa annua* L., Poaceae, A/P, G (in captivity)]
E. epiphron	***Nardus stricta* L., Poaceae**, P, G *Festuca ovina* L., Poaceae, P, G [*Aira praecox* L., Poaceae, A, G (in captivity)] [*Deschampsia cespitosa* (L.) P. Beauv., Poaceae, P, G (in captivity)] [*Deschampsia flexuosa* (L.) Trin., Poaceae, P, G (in captivity)] [*Poa annua* L., Poaceae, A/P, G (in captivity)]
E. aethiops	***Molinia caerulea* (L.) Moench, Poaceae**, P, G ***Sesleria caeulea* (L.) Ard., Poaceae**, P, G *Agrostis canina* L., Poaceae, P, G *Aira praecox* L., Poaceae, A, G *Deschampsia cespitosa* (L.) P. Beauv., Poaceae, P, G *Festuca* spp., Poaceae, P, G *Poa annua* L., Poaceae, A/P, G *Poa pratensis* L., Poaceae, P, G [*Elytrigia repens* (L.) Desv. ex Nevski, Poaceae, P, G (in captivity)]

Butterfly species*	Hostplants†
M. galathea	***Festuca rubra* L., Poaceae**, P, G
	Brachypodium pinnatum (L.) P. Beauv., Poaceae, P, G
	Bromus spp., Poaceae, P, G
	Dactylis glomerata L., Poaceae, P, G
	Festuca ovina L., Poaceae, P, G
	Holcus lanatus L., Poaceae, P, G
	Phleum pratense L., Poaceae, P, G
	Poa annua L., Poaceae, A/P, G
	[*Agrostis capillaris* L., Poaceae, P, G (in captivity)]
	[*Arrhenatherum elatius* P. Beauv. ex J. & C. Presl, Poaceae, P, G (in captivity)]
	[*Bromopsis erecta* (Huds.) Fourr., Poaceae, P, G (in captivity)]
	[*Festuca arundinacea* Schreb., Poaceae, P, G (in captivity)]
	[*Festuca pratensis* Huds., Poaceae, P, G (in captivity)]
	[*Lolium perenne* L., Poaceae, P, G (in captivity)]
H. semele	***Agrostis curtisii* Kerguélen, Poaceae**, P, G
	***Aira praecox* L., Poaceae**, A, G
	***Ammophila arenaria* (L.) Link, Poaceae**, P, G
	***Brachypodium pinnatum* (L.) P. Beauv., Poaceae**, P, G
	***Brachypodium sylvaticum* (Huds.) P. Beauv., Poaceae**, P, G
	***Deschampsia cespitosa* (L.) P. Beauv., Poaceae**, P, G
	***Vulpia myuros* (Kunth) Nees, Poaceae**, A, G
	Elytrigia repens (L.) Desv. ex Nevski, Poaceae, P, G
	Festuca ovina L., Poaceae, P, G
	Festuca rubra L., Poaceae, P, G
	Koeleria spp., Poaceae, P, G
	Poa annua L., Poaceae, A/P, G
P. tithonus	***Agrostis stolonifera* L., Poaceae**, P, G
	***Dactylis glomerata* L., Poaceae**, P, G
	***Festuca ovina* L., Poaceae**, P, G
	***Lolium perenne* L., Poaceae**, P, G
	***Poa annua* L., Poaceae**, A/P, G
	***Poa trivialis* L., Poaceae**, P, G
	Agrostis capillaris L., Poaceae, P, G
	Anthoxanthum odoratum L., Poaceae, P, G
	Elytrigia repens (L.) Desv. ex Nevski, Poaceae, P, G
	Festuca rubra L., Poaceae, P, G
	Poa pratensis L., Poaceae, P, G
M. jurtina	***Dactylis glomerata* L., Poaceae**, P, G
	***Elytrigia repens* (L.) Desv. ex Nevski, Poaceae**, P, G
	***Poa nemoralis* L., Poaceae**, P, G
	Agrostis stolonifera L., Poaceae, P, G
	Brachypodium sylvaticum (Huds.) P. Beauv., Poaceae, P, G
	Bromus spp., Poaceae, P, G
	Festuca ovina L., Poaceae, P, G
	Festuca rubra L., Poaceae, P, G
	Helictotrichon pubescens (Huds.) Pilg., Poaceae, P, G
	Holcus lanatus L., Poaceae, P, G
	Lolium spp., Poaceae, P, G
	Phleum spp., Poaceae, P, G

Butterfly species*	Hostplants†
	Poa annua L., Poaceae, A/P, G *Poa pratensis* L., Poaceae, P, G *Poa trivialis* L., Poaceae, P, G
A. hyperantus	***Brachypodium sylvaticum* (Huds.) P. Beauv., Poaceae**, P, G ***Dactylis glomerata* L., Poaceae**, P, G ***Deschampsia cespitosa* (L.) P. Beauv., Poaceae**, P, G ***Elytrigia repens* (L.) Desv. ex Nevski, Poaceae**, P, G *Agrostis stolonifera* L., Poaceae, P, G *Calamagrostis epigejos* (L.) Roth, Poaceae, P, G *Carex sylvatica* Huds., Cyperaceae, P, sedge *Festuca ovina* L., Poaceae, P, G *Holcus lanatus* L., Poaceae, P, G *Holcus mollis* L., Poaceae, P, G *Milium effusum* L., Poaceae, P, G *Poa pratensis* L., Poaceae, P, G [*Poa annua* L., Poaceae, A/P, G (in captivity)]
C. pamphilus	***Agrostis stolonifera* L., Poaceae**, P, G ***Festuca ovina* L., Poaceae**, P, G ***Festuca rubra* L., Poaceae**, P, G ***Poa pratensis* L., Poaceae**, P, G *Agrostis curtisii* Kerguélen, Poaceae, P, G *Brachypodium* spp., Poaceae, P, G *Cynosurus cristatus* L, Poaceae, P, G *Nardus stricta* L., Poaceae, P, G [*Festuca pratensis* Huds., Poaceae, P, G (in captivity)] [*Poa annua* L., Poaceae, A/P, G (in captivity)] [*Poa nemoralis* L., Poaceae, P, G (in captivity)]
C. tullia	***Eriophorum vaginatum* L., Cyperaceae**, P, sedge ***Molinia caerulea* (L.) Moench, Poaceae**, P, G *Deschampsia flexuosa* (L.) Trin., Poaceae, P, G *Eriophorum angustifolium* Honck., Cyperaceae, P, sedge *Festuca pratensis* Huds., Poaceae, P, G *Juncus articulatus* L., Juncaceae, P, Rush [*Festuca ovina* L., Poaceae, P, G (in captivity)] [*Poa annua* L., Poaceae, A/P, G (in captivity)] [*Rhyncospora alba* (L.) Vahl, Cyperaceae, P, sedge (often cited, but unconfirmed and unlikely)]

A, annual; A/B; annual/biennial, A/P, annual/perennial; B, biennial; B/P, biennial/perennial; C, climber; F, forb (herb); G, grass; P, perennial; S, shrub; T, tree.

* For full species names, see Appendix 1a.

† Plant names in **bold** type are main hostplants, and those in ordinary type are secondary hostplants. Square brackets indicate novel/alien hostplants, or plants only recorded as used in captivity. A full list of larval hostplant sources is given at: www.geocities.com/pgll@btopenworld.com/resources/resources.htm. See also http://www.staffs.ac.uk/schools/sciences/geography/links/IESR/staff_honfellows_rd_butterflydatabase.shtml and http://pbh-butterflies.yolasite.com.

APPENDIX 2b: HOSTPLANT FAMILIES, RANGE OF BUTTERFLY HERBIVORY AND HOSTPLANT PHENOLOGY

	Hostplant family																																					Hostplant range				Hostplant phenology				
Butterfly species*	Ulmaceae	Cannabaceae	Urticaceae	Fagaceae	Betulaceae	Polygonaceae	Malvaceae	Cistaceae	Violaceae	Salicaceae	Brassicaceae	Resedaceae	Empetraceae	Ericaceae	Primulaceae	Grossulariaceae	Rosaceae	Fabaceae	Lythraceae	Cornaceae	Celastraceae	Aquifoliaceaeae	Rhamnaceae	Geraniaceae	Tropaeolaceae	Araliaceae	Apiaceae	Boraginaceae	Lamiaceae	Plantaginaceae	Buddlejaceae	Scrophulariaceae	Caprifoliaceae	Dipsacaceae	Asteraceae	Juncaceae	Cyperaceae	Pocaeae	monophag	oligophag	polyphag	annual	annual/biannual	biennial	shortlivedper	longperennial
C. palaemon	0	0	0	0	0	0	0	0	0	0	0	0	0	0	0	0	0	0	0	0	0	0	0	0	0	0	0	0	0	0	0	0	0	0	0	0	0	1	0	0	1	0	0	0	1	0
T. sylvestris	0	0	0	0	0	0	0	0	0	0	0	0	0	0	0	0	0	0	0	0	0	0	0	0	0	0	0	0	0	0	0	0	0	0	0	0	0	1	0	0	1	0	0	0	1	0
T. lineola	0	0	0	0	0	0	0	0	0	0	0	0	0	0	0	0	0	0	0	0	0	0	0	0	0	0	0	0	0	0	0	0	0	0	0	0	0	1	0	0	1	0	0	0	1	0
T. acteon	0	0	0	0	0	0	0	0	0	0	0	0	0	0	0	0	0	0	0	0	0	0	0	0	0	0	0	0	0	0	0	0	0	0	0	0	0	1	1	0	0	0	0	0	1	0
H. comma	0	0	0	0	0	0	0	0	0	0	0	0	0	0	0	0	0	0	0	0	0	0	0	0	0	0	0	0	0	0	0	0	0	0	0	0	0	1	1	0	0	1	0	0	1	0
O. sylvanus	0	0	0	0	0	0	0	0	0	0	0	0	0	0	0	0	0	0	0	0	0	0	0	0	0	0	0	0	0	0	0	0	0	0	0	0	0	1	0	0	1	0	0	0	1	0
E. tages	0	0	0	0	0	0	0	0	0	0	0	0	0	0	0	0	0	1	0	0	0	0	0	0	0	0	0	0	0	0	0	0	0	0	0	0	0	0	0	0	1	0	0	0	1	1
P. malvae	0	0	0	0	0	0	0	0	0	0	0	0	0	0	0	0	1	0	0	0	0	0	0	0	0	0	0	0	0	0	0	0	0	0	0	0	0	0	0	0	1	0	1	0	1	0
P. machaon	0	0	0	0	0	0	0	0	0	0	0	0	0	0	0	0	0	0	0	0	0	0	0	0	0	0	1	0	0	0	0	0	0	0	0	0	0	0	0	1	0	0	1	0	1	0
L. sinapis	0	0	0	0	0	0	0	0	0	0	0	0	0	0	0	0	0	1	0	0	0	0	0	0	0	0	0	0	0	0	0	0	0	0	0	0	0	0	0	0	1	1	0	0	0	0
C. crocea	0	0	0	0	0	0	0	0	0	0	0	0	0	0	0	0	0	1	0	0	0	0	0	0	0	0	0	0	0	0	0	0	0	0	0	0	0	0	0	0	1	1	1	0	1	0
G. rhamni	0	0	0	0	0	0	0	0	0	0	0	0	0	0	0	0	0	0	0	0	0	0	1	0	0	0	0	0	0	0	0	0	0	0	0	0	0	0	0	0	1	0	0	0	0	1
P. brassicae	0	0	0	0	0	0	0	0	0	0	1	2	0	0	0	0	0	0	0	0	0	0	0	0	3	0	0	0	0	0	0	0	0	0	0	0	0	0	0	0	1	1	1	1	1	0
P. rapae	0	0	0	0	0	0	0	0	0	0	1	2	0	0	0	0	0	0	0	0	0	0	0	0	3	0	0	0	0	0	0	0	0	0	0	0	0	0	0	0	1	1	1	1	1	0
P. napi	0	0	0	0	0	0	0	0	0	0	1	0	0	0	0	0	0	0	0	0	0	0	0	0	3	0	0	0	0	0	0	0	0	0	0	0	0	0	0	0	1	1	1	1	1	0
A. cardamines	0	0	0	0	0	0	0	0	0	0	1	2	0	0	0	0	0	0	0	0	0	0	0	0	3	0	0	0	0	0	0	0	0	0	0	0	0	0	0	0	1	1	1	0	1	0
C. rubi	0	0	0	0	0	0	0	1	0	0	0	0	2	1	0	0	2	1	0	2	0	0	2	0	0	0	0	0	0	0	0	0	0	0	0	0	0	0	0	0	1	0	0	0	0	1
T. betulae	0	0	0	0	0	0	0	0	0	0	0	0	0	0	0	0	1	0	0	0	0	0	0	0	0	0	0	0	0	0	0	0	0	0	0	0	0	0	1	0	0	0	0	0	0	1
F. quercus	0	0	0	1	0	0	0	0	0	0	0	0	0	0	0	0	0	0	0	0	0	0	0	0	0	0	0	0	0	0	0	0	0	0	0	0	0	0	1	0	0	0	0	0	0	1
S. w-album	1	0	0	0	0	0	0	0	0	0	0	0	0	0	0	0	0	0	0	0	0	0	0	0	0	0	0	0	0	0	0	0	0	0	0	0	0	0	0	1	0	0	0	0	0	1
S. pruni	0	0	0	0	0	0	0	0	0	0	0	0	0	0	0	0	1	0	0	0	0	0	0	0	0	0	0	0	0	0	0	0	0	0	0	0	0	0	1	0	0	0	0	0	0	1
L. phlaeas	0	0	0	0	0	1	0	0	0	0	0	0	0	0	0	0	0	0	0	0	0	0	0	0	0	0	0	0	0	0	0	0	0	0	0	0	0	0	0	1	0	0	1	0	0	0
L. dispar	0	0	0	0	0	1	0	0	0	0	0	0	0	0	0	0	0	0	0	0	0	0	0	0	0	0	0	0	0	0	0	0	0	0	0	0	0	0	1	0	0	0	0	0	1	0
C. minimus	0	0	0	0	0	0	0	0	0	0	0	0	0	0	0	0	0	1	0	0	0	0	0	0	0	0	0	0	0	0	0	0	0	0	0	0	0	0	0	1	0	0	0	0	1	0
P. argus	0	0	0	0	0	0	0	1	0	0	0	0	0	1	0	0	0	1	0	0	0	0	0	0	0	0	0	0	1	0	0	0	0	0	0	0	0	0	0	1	0	1	0	0	1	0
A. agestis	0	0	0	0	0	0	0	1	0	0	0	0	0	0	0	0	0	0	0	0	0	0	0	1	0	0	0	0	0	0	0	0	0	0	0	0	0	0	0	0	1	1	0	0	1	1

A. artaxerxes	0	0	0	0	0	0	0	1	0	0	0	0	0	0	0	0	0	0	0	0	0	0	0	2	0	0	0	0	0	0	0	0	0	0	0	0	0	0	0	1	0	1	0	0	1	1
P. icarus	0	0	0	0	0	0	0	0	0	0	0	0	0	0	0	0	0	1	0	0	0	0	0	0	0	0	0	0	0	0	0	0	0	0	0	0	0	0	0	0	1	1	0	0	0	1
L. coridon	0	0	0	0	0	0	0	0	0	0	0	0	0	0	0	0	0	1	0	0	0	0	0	0	0	0	0	0	0	0	0	0	0	0	0	0	0	0	1	0	0	1	0	0	0	1
L. bellargus	0	0	0	0	0	0	0	0	0	0	0	0	0	0	0	0	0	1	0	0	0	0	0	0	0	0	0	0	0	0	0	0	0	0	0	0	0	0	1	0	0	0	0	0	0	1
C. argiolus	0	2	0	0	0	0	0	0	0	2	0	0	0	2	0	0	2	2	2	2	2	1	2	0	0	1	0	0	0	0	3	0	2	0	0	0	0	0	0	0	1	0	0	0	0	1
M. arion	0	0	0	0	0	0	0	0	0	0	0	0	0	0	0	0	0	0	0	0	0	0	0	0	0	0	0	0	1	0	0	0	0	0	0	0	0	0	1	0	0	0	0	0	0	1
H. lucina	0	0	0	0	0	0	0	0	0	0	0	0	0	0	1	0	0	0	0	0	0	0	0	0	0	0	0	0	0	0	0	0	0	0	0	0	0	0	0	1	0	0	0	0	0	1
L. camilla	0	0	0	0	0	0	0	0	0	0	0	0	0	0	0	0	0	0	0	0	0	0	0	0	0	0	0	0	0	0	0	0	1	0	0	0	0	0	1	0	0	0	0	0	0	1
A. iris	0	0	0	0	0	0	0	0	0	1	0	0	0	0	0	0	0	0	0	0	0	0	0	0	0	0	0	0	0	0	0	0	0	0	0	0	0	0	0	1	0	0	0	0	0	1
V. atalanta	0	2	1	0	0	0	0	0	0	0	0	0	0	0	0	0	0	0	0	0	0	0	0	0	0	0	0	0	0	0	0	0	0	0	0	0	0	0	1	0	0	1	0	0	1	0
V. cardui	0	0	2	0	0	0	2	0	0	0	0	0	0	0	0	0	0	0	0	0	0	0	0	0	0	0	0	2	0	2	0	0	0	0	1	0	0	0	0	0	1	1	0	0	1	0
A. urticae	0	0	1	0	0	0	0	0	0	0	0	0	0	0	0	0	0	0	0	0	0	0	0	0	0	0	0	0	0	0	0	0	0	0	0	0	0	0	1	0	0	1	0	0	1	0
N. polychloros	1	0	0	0	2	0	0	0	0	2	0	0	0	0	0	0	2	0	0	0	0	0	0	0	0	0	0	0	0	0	0	0	0	0	0	0	0	0	0	1	0	0	0	0	0	1
I. io	0	0	1	0	0	0	0	0	0	0	0	0	0	0	0	0	0	0	0	0	0	0	0	0	0	0	0	0	0	0	0	0	0	0	0	0	0	0	1	0	0	1	0	0	1	0
P. c-album	1	2	1	0	0	0	0	0	0	2	0	0	0	0	0	2	0	0	0	0	0	0	0	0	0	0	0	0	0	0	0	0	0	0	0	0	0	0	0	1	0	0	0	0	1	0
B. selene	0	0	0	0	0	0	0	0	1	0	0	0	0	0	0	0	0	0	0	0	0	0	0	0	0	0	0	0	0	0	0	0	0	0	0	0	0	0	0	1	0	0	0	0	1	0
B. euphrosyne	0	0	0	0	0	0	0	0	1	0	0	0	0	0	0	0	0	0	0	0	0	0	0	0	0	0	0	0	0	0	0	0	0	0	0	0	0	0	0	1	0	0	0	0	1	0
A. adippe	0	0	0	0	0	0	0	0	1	0	0	0	0	0	0	0	0	0	0	0	0	0	0	0	0	0	0	0	0	0	0	0	0	0	0	0	0	0	0	1	0	0	0	0	1	0
A. aglaja	0	0	0	0	0	0	0	0	1	0	0	0	0	0	0	0	0	0	0	0	0	0	0	0	0	0	0	0	0	0	0	0	0	0	0	0	0	0	0	1	0	0	0	0	1	0
A. paphia	0	0	0	0	0	0	0	0	1	0	0	0	0	0	0	0	0	0	0	0	0	0	0	0	0	0	0	0	0	0	0	0	0	0	0	0	0	0	0	1	0	0	0	0	1	0
E. aurinia	0	0	0	0	0	0	0	0	0	0	0	0	0	0	0	0	0	0	0	0	0	0	0	0	0	0	0	0	0	2	0	0	2	1	2	0	0	0	0	1	0	0	0	0	1	0
M. cinxia	0	0	0	0	0	0	0	0	0	0	0	0	0	0	0	0	0	0	0	0	0	0	0	0	0	0	0	0	0	1	0	2	0	0	0	0	0	0	0	1	0	1	0	0	1	0
M. athalia	0	0	0	0	0	0	0	0	0	0	0	0	0	0	0	0	0	0	0	0	0	0	0	0	0	0	0	0	2	1	0	1	0	0	2	0	0	0	0	1	0	1	0	0	1	0
P. aegeria	0	0	0	0	0	0	0	0	0	0	0	0	0	0	0	0	0	0	0	0	0	0	0	0	0	0	0	0	0	0	0	0	0	0	0	0	0	1	0	0	1	0	0	0	1	0
L. megera	0	0	0	0	0	0	0	0	0	0	0	0	0	0	0	0	0	0	0	0	0	0	0	0	0	0	0	0	0	0	0	0	0	0	0	0	0	1	0	0	1	0	0	0	1	0
E. epiphron	0	0	0	0	0	0	0	0	0	0	0	0	0	0	0	0	0	0	0	0	0	0	0	0	0	0	0	0	0	0	0	0	0	0	0	0	0	1	1	0	0	0	0	0	1	0
E. aethiops	0	0	0	0	0	0	0	0	0	0	0	0	0	0	0	0	0	0	0	0	0	0	0	0	0	0	0	0	0	0	0	0	0	0	0	0	0	1	0	1	0	1	0	0	1	0
M. galathea	0	0	0	0	0	0	0	0	0	0	0	0	0	0	0	0	0	0	0	0	0	0	0	0	0	0	0	0	0	0	0	0	0	0	0	0	0	1	0	0	1	0	0	0	1	0
H. semele	0	0	0	0	0	0	0	0	0	0	0	0	0	0	0	0	0	0	0	0	0	0	0	0	0	0	0	0	0	0	0	0	0	0	0	0	0	1	0	0	1	1	0	0	1	0
P. tithonus	0	0	0	0	0	0	0	0	0	0	0	0	0	0	0	0	0	0	0	0	0	0	0	0	0	0	0	0	0	0	0	0	0	0	0	0	0	1	0	0	1	0	0	0	1	0
M. jurtina	0	0	0	0	0	0	0	0	0	0	0	0	0	0	0	0	0	0	0	0	0	0	0	0	0	0	0	0	0	0	0	0	0	0	0	0	0	1	0	0	1	0	0	0	1	0
A. hyperantus	0	0	0	0	0	0	0	0	0	0	0	0	0	0	0	0	0	0	0	0	0	0	0	0	0	0	0	0	0	0	0	0	0	0	0	0	0	1	0	0	1	0	0	0	1	0
C. pamphilus	0	0	0	0	0	0	0	0	0	0	0	0	0	0	0	0	0	0	0	0	0	0	0	0	0	0	0	0	0	0	0	0	0	0	0	0	0	1	0	0	1	0	0	0	1	0
C. tullia	0	0	0	0	0	0	0	0	0	0	0	0	0	0	0	0	0	0	0	0	0	0	0	0	0	0	0	0	0	0	0	0	0	0	0	2	3	1	1	0	0	0	0	0	1	0

Codes for plant families: 1, main hostplants; 2, secondary hostplants; 3, novel and/or alien hostplants. Hcst range and phenology: 0, no; 1, yes.

* For full species names, see Appendix 1a.

APPENDIX 2c: LARVAL HOSTPLANT BIOTOPES, PHENOLOGY, GROWTH FORMS, ENVIRONMENTS AND LIFE HISTORY STRATEGIES*

		Primary biotopes												Ellenberg values					
Family	Hostplant	Aq	Mi	Sk	Ar	Pa	Sp	Wa	Wo	Cultivated	Garden	Phenology	Growth form	L	F	R	N	S	Strategy
Ulmaceae	*Ulmus glabra*	1	2	2	1	2	1	1	5	y	3	D, P	T	4	5	7	6	0	C/SC
	Ulmus japonica									?		D, P	T						C/SC
	Ulmus minor								+	y	3	D, P	T	5	5	7	7	0	C/SC
	Ulmus procera	1	1	1	1	2	1	1	5	Y	3	D, P	T	5	5	8	6	0	C/SC
Cannabaceae	*Humulus lupulus*	1	3	1	1	2	1	3	5	y	2	P	C	6	7	7	8	0	C
Urticaceae	*Parietaria judaica*	1	1	5	1	2	2	2	3	n	2	P	H	7	4	8	5	0	CR/CSR
	Parietaria officinalis									?		P	H						?
	Urtica dioica	1	3	3	3	2	3	3	3	n	3	P	H	6	6	7	8	0	C
	Urtica urens	1	1	1	5	2	3	1	1	n	2	A	H	8	5	6	8	0	R/CR
Fagaceae	*Quercus cerris*	1	1	1	1	1	1	5	+	Y	4	D, P	T	6	4	6	6	0	SC
	Quercus ilex								+	Y	4	E, P	T	6	3	7	4	0	SC
	Quercus petraea	1	1	1	1	2	1	3	5	y	4	D, P	T	6	6	3	4	0	SC
	Quercus robur	1	1	1	1	2	1	3	5	y	4	D, P	T	7	5	5	4	0	SC
Betulaceae	*Betula pubescens*	1	3	1	1	1	2	2	5	y	4	D, P	T	7	7	4	4	0	C/SC
Polygonaceae	*Rumex acetosa*	1	3	2	3	4	3	3	1	n	3	P	H	7	5	5	4	0	CSR
	Rumex acetosella	1	1	3	3	3	3	4	2	n	3	P	H	7	5	4	3	0	SR/CSR
	Rumex hydrolapathum	3	5	1	1	2	1	1	1	n	1	P	H	7	10	7	6	0	C/CSR
	Rumex obtusifolius	1	3	2	5	2	4	3	1	n	3	P	H	7	5	7	9	0	C/CSR
	Rumex pulcher					c		c		n		P	H	7	6	7	7	0	CR/CSR
Malvaceae	*Althaea officinalis*		c							n		P	H	7	7	8	4	2	C
	Malva sylvestris	1	1	4	3	2	4	4	1	y	3	P	H	8	4	8	7	0	CR
Cistaceae	*Helianthemum nummularium*	1	1	4	1	4	3	3	1	n	1	E, P	Ss	7	4	7	2	0	S
	Helianthemum oelandicum			+		+				n	1	E, P	Ss	8	3	8	1	0	S
Violaceae	*Viola arvensis*	1	1	1	5	2	3	3	1	n	1	A	H	8	4	6	6	0	R
	Viola canina	1	1	1	1	1	1	5	1	n	2	P	H	8	4	5	2	0	S/CSR
	Viola hirta	1	1	2	1	4	1	4	3	n	1	P	H	7	4	8	2	0	S/CSR
	Viola lactea					+		+		n	1	P	H	7	6	2	2	0	S/CSR
	Viola odorata	1	1	1	1	3	1	2	5	y		P	H	5	5	7	7	0	CSR
	Viola palustris	1	5	1	1	2	1	1	1	n	1	P	H	7	9	3	2	0	SR/CSR
	Viola reichenbachiana	1	1	3	1	1	1	1	5	n		P	H	4	6	7	5	0	S/CSR
	Viola riviniana	1	1	3	1	4	2	3	3	n	2	P	H	6	5	5	4	0	S/CSR
	Viola tricolor	1	1	1	5	2	4	3	1	n	2	A, P	H	8	4	6	4	0	R/SR
Salicaceae	*Populus alba*								+	Y		D, P	T	6	6	7	6	0	SC
	Populus nigra		+						+	y		D, P	T	6	8	7	7	0	SC
	Populus tremula	1	1	1	1	2	1	1	5	y	3	D, P	T	6	5	5	6	0	SC
	Salix caprea	1	2	1	1	1	3	3	3	n	3	D, P	S	7	7	7	7	0	C/SC
	Salix cinerea	1	4	2	1	1	3	3	2	n	3	D, P	S	7	8	6	5	0	C/SC
	Salix fragilis	3	4	1	1	2	1	1	3	y	3	D, P	T	6	8	7	7	0	C/SC
	Salix viminalis	1	5	1	1	2	1	1	3	y		D, P	S	7	8	6	6	0	C/SC

Capparaceae	*Cleome sesquiorygalis*									?		A	H						?
Brassicaceae	*Alliaria petiolata*	1	1	3	1	2	3	5	3	n	2	Be	H	5	6	7	8	0	CR
	Arabidopsis thaliana	1	1	4	1	3	3	3	1	n		A	H	8	3	6	2	0	SR
	Arabis caucasica	1	1	5	1	-	2	1	1	Y		P	H						CSR
	Arabis glabra							+		n		A, Be	H	7	3	8	5	0	R/SR
	Arabis hirsuta	1	1	5	1	3	3	3	1	n	2	Be, P	H	7	5	8	3	0	S/CSR
	Armoracia rusticana	1	1	1	3	2	4	4	1	Y	3	P	H	8	5	7	7	0	C/CSR
	Aubrieta deltoidea			+						Y	5	P	H						CSR
	Barbarea vulgaris	1	3	1	1	2	3	5	1	n	2	Be, P	H	7	6	7	8	0	R/CSR
	Brassica napus									Y	3	A, Be	H	7	4	7	7	0	CR
	Brassica nigra						+	+		n	3	A	H	8	5	7	6	0	CR
	Brassica oleracea			sc						Y	5	Be, P	H	8	4	7	8	3	CR
	Brassica rapa	1	1	2	4	2	3	4	1	Y	3	A, Be	H	7	5	7	6	0	CR
	Cakile maritima							sl		n		A	H	9	6	7	7	3	R/CR
	Capsella bursa-pastoris	1	1	1	5	2	3	3	1	n		A	H	7	5	7	7	0	R
	Cardamine amara	1	5	1	1	2	1	3	3	n	2	P	H	6	9	7	6	0	CR
	Cardamine hirsuta	1	1	3	1	3	4	3	1	n	4	A	H	8	5	6	6	0	SR
	Cardamine pratensis	1	5	1	1	4	1	1	1	n	2	P	H	7	8	5	4	0	R/CSR
	Cochlearia officinalis		sm		+					n		Be, P	H	8	6	7	5	3	SR/CSR
	Coronopus didymus	1	1	1	5	1	1	3	1	n	2	A	H	9	5	6	7	0	R
	Coronopus squamatus	1	1	1	4	4	3	3	1	n		A	H	7	5	7	7	0	R
	Crambe maritima							sh		y	1	P	H	9	5	8	7	3	C/CSR
	Diplotaxis tenuifolia	1	1	1	1	1	4	3	1	n	2	P	H	8	5	7	6	0	R/CSR
	Erysimum cheiri	1	1	5	1	2	1	1	1	Y		P	H	8	4	8	5	0	CSR
	Hesperis matronalis	1	2	2	1	2	1	3	5	Y	4	Be, P	H	7	7	7	7	0	CR/CSR
	Lepidium campestre	1	1	1	1	3	3	4	1	n		A	H	7	4	7	6	0	R/SR
	Lepidium draba	1	1	1	3	2	3	5	1	n	2	P	H	8	4	8	6	1	CR
	Lepidium heterophyllum	1	1	1	1	2	5	3	1	n		P	H	7	4	5	4	0	SR/CSR
	Lepidium sativum									Y		A	H						R
	Lunaria annua								+	Y	5	Be	H						R/CR
	Malcolmia maritima									Y		A	H						R/SR
	Matthiola incana			sc						Y		A, P	H	9	3	8	2	3	SR
	Raphanus raphanistrum	1	1	1	5	2	3	3	1	n	3	A	H	7	5	6	6	0	R/CR
	Raphanus sativus									Y		A	H						R
	Rapistrum rugosum	1	1	1	1	1	4	1	1	n		A	H						R/CR
	Rorippa microphylla	4	4	1	1	1	1	1	1	y		P	H	7	10	7	6	0	CR
	Rorippa nasturtium-aquaticum	4	4	1	1	2	1	1	1	y	2	P	H	7	10	7	7	0	CR
	Rorippa sylvestris	2	3	1	5	1	1	2	1	n	2	P	H	8	8	7	7	0	CR
	Sinapis alba				+					Y		A	H	7	4	7	6	0	R/CR
	Sinapis arvensis	1	1	1	5	2	3	2	1	n	2	A	H	8	5	7	7	0	R/CR
	Sisymbrium officinale	1	1	1	4	2	4	4	1	n	2	A	H	7	4	7	7	0	R/CR
	Thlaspi caerulescens	1	1	1	1	1	5	1	1	n		Be, P	H	8	4	6	1	0	S/SR
Resedaceae	*Reseda luteola*	1	1	2	2	2	5	3	1	n	2	Be	H	7	4	8	6	0	R/CR
	Reseda lutea	1	1	1	1	1	5	1	1	n	2	Be, P	H	7	4	7	5	0	SR/CSR
	Reseda odorata									Y		A, Be	H						?
Empetraceae	*Empetrum nigrum*	1	3	3	1	4	2	4	1	n		E, P	Ss	7	6	2	1	0	S/SC

Family	Hostplant	Primary biotopes								Cultivated	Garden	Phenology	Growth form	Ellenberg values					Strategy
		Aq	Mi	Sk	Ar	Pa	Sp	Wa	Wo					L	F	R	N	S	
Ericaceae	*Arctostaphylos uva-ursi*	1	1	5	1	5	1	1	1	n		E, P	Ss	7	5	2	2	0	S/SC
	Calluna vulgaris	1	3	3	1	4	3	3	1	n	3	E, P	Ss	7	6	2	2	0	S/SC
	Erica cinerea	1	1	3	1	4	4	3	1	n	3	E, P	Ss	7	5	2	2	0	S/SC
	Erica tetralix	1	5	1	1	3	1	2	1	n	2	E, P	Ss	8	8	2	1	0	S/SC
	Vaccinium oxycoccus									n	1	E, P	Ss	8	9	2	1	0	S
	Vaccinium vitis-idaea	1	3	1	1	4	3	3	3	n		E, P	Ss	6	5	2	2	0	S/SC
Primulaceae	*Primula vulgaris*	1	3	3	1	3	3	2	5	y	3	P	H	5	5	6	4	0	S/CSR
	Primula elatior								+	n		P	H	4	5	7	6	0	S/CSR
	Primula veris	1	1	3	1	4	2	4	1	y	3	P	H	7	4	7	3	0	S/CSR
Grossulariaceae	*Ribes nigrum*	1	4	1	1	1	1	3	4	Y	4	D, P	S	5	9	6	6	0	SC
	Ribes rubrum	1	3	2	1	1	1	1	5	Y	4	D, P	S	5	7	7	6	0	SC
	Ribes uva-crispa	1	1	1	1	2	1	1	5	Y		D, P	S	5	5	7	6	0	SC
Rosaceae	*Agrimonia eupatoria*	1	1	1	1	3	1	5	2	n	2	P	H	7	4	7	4	0	CSR
	Aphanes arvensis	1	1	3	5	3	3	3	1	n	2	A	H	8	4	6	4	0	R/SR
	Crataegus monogyna	1	1	2	1	2	1	1	5	y	5	D, P	S	6	5	7	6	0	SC
	Fragaria vesca	1	1	3	1	3	3	3	3	y	3	P	H	6	5	6	4	0	S/CSR
	Fragaria × ananassa	1	1	1	1	1	2	5	1	Y		P	H	6	5	8	7	0	CR/CSR
	Geum urbanum	1	1	3	1	2	1	2	5	n	3	P	H	4	6	7	7	0	CR/CSR
	Malus domestica								+	Y	4	D, P	T	7	5	6	7	0	SC
	Potentilla anserina	1	4	1	3	4	2	3	1	n		P	H	8	7	7	6	0	CR/CSR
	Potentilla erecta	1	3	1	1	4	1	4	1	n	2	P	H	7	7	3	2	0	S/CSR
	Potentilla reptans	1	3	2	1	3	3	4	2	n	2	P	H	7	5	7	5	0	CR/CSR
	Potentilla sterilis	1	1	3	1	4	3	3	3	n	2	P	H	5	5	5	5	0	SR/CSR
	Prunus avium	1	1	1	1	2	1	1	5	y		D, P	T	4	5	6	6	0	SC
	Prunus cerasus								+	y		D, P	S	6	5	6	5	0	SC
	Prunus domestica								+	Y	4	D, P	T	7	5	7	6	0	SC
	Prunus laurocerasus								+	Y		E, P	S	4	6	5	6	0	SC
	Prunus lusitanica								+	Y		E, P	S						SC
	Prunus spinosa	1	1	1	1	2	1	1	5	n	3	D, P	S	6	5	7	6	0	SC
	Pyracantha atalantoides									?		E, P	S						SC
	Pyracantha coccinea									Y	5	E, P	S						SC
	Pyrus communis								+	y	4	D, P	T	7	5	6	7	0	SC
	Pyrus pyraster								+	n	4	D, P	T	6	5	5	4	0	SC
	Rosa canina	1	1	3	1	1	3	3	3	n	3	D, P	S	6	5	7	6	0	SC
	Rubus fruticosus	1	2	2	2	2	3	3	4	y	4	D, P	S	6	6	6	6	0	SC
	Rubus idaeus	1	2	3	1	2	1	3	4	Y	4	D, P	S	6	5	5	5	0	SC
	Sanguisorba minor	1	1	3	1	4	3	3	1	y	2	P	H	7	4	8	3	0	S/CSR
	Sorbus aria			+					+	y		D, P	T	6	4	7	4	0	SC
Fabaceae	*Anthyllis vulneraria*	1	1	4	1	3	4	3	1	n	2	P	H	8	4	7	2	0	SR/CSR
	Astragalus glycyphyllos	1	1	1	1	1	1	4	2	n		P	H	6	4	7	3	0	C/CSR
	Colutea arborescens							+		y	4	D, P	S	7	4	8	3	0	SC
	Cytisus scoparius	1	1	1	1	2	4	4	1	y	4	D, P	S	8	5	4	4	0	SC
	Genista anglica	1	2	1	1	3	1	5	1	n		D, P	Ss	8	5	3	2	0	S/SC
	Genista tinctoria	1	1	2	1	4	3	3	1	n	4	D, P	Ss	8	6	7	2	0	SC

	Hippocrepis comosa			+		–				n	2	P	H	8	3	8	2	0	S/CSR
	Lathyrus latifolius	1	1	1	1	-	3	4	4	Y	2	P	H	7	4	8	3	0	C/CSR
	Lathyrus linifolius	1	1	1	1	4	2	5	1	n	2	P	H	6	5	4	3	0	SR/CSR
	Lathyrus odoratus									Y		A	C						CR
	Lathyrus pratensis	1	3	1	2	4	3	4	1	n	2	P	H	7	6	6	5	0	CSR
	Lathyrus sylvestris							+		n	2	P	H	7	4	8	2	0	C/CSR
	Lotus corniculatus	1	1	3	1	4	4	4	1	n	3	P	H	7	4	6	2	0	S/CSR
	Lotus pedunculatus	1	5	1	1	3	1	4	1	n	3	P	H	7	8	6	4	0	C/CSR
	Lotus subbiflorus							+		n		A	H	7	5	6	5	0	R/SR
	Medicago arabica							+		n		A	H	7	5	6	5	0	R/CSR
	Medicago lupulina	1	1	4	4	3	4	3	1	n	2	A, P	H	7	4	8	4	0	R/CSR
	Medicago sativa	1	1	2	1	2	3	5	1	Y	2	P	H	7	4	6	5	0	C/CSR
	Melilotus altissimus	1	1	1	1	1	4	5	1	n		Be, P	H	8	6	7	7	0	CR/CSR
	Melilotus officinalis	1	1	1	1	1	5	3	1	n		Be	H	8	5	7	5	0	CR/CSR
	Ononis repens	1	1	4	1	3	4	4	1	n	2	P	Hw	8	4	6	3	0	SC/CSR
	Ononis spinosa	1	1	4	1	1	4	3	1	n		P	Hw	8	4	8	3	0	SC/CSR
	Ornithopus perpusillus	1	1	1	3	3	1	5	1	n		A	H	7	4	4	3	0	SR/CSR
	Pisum sativum									Y	4	A	H	7	4	7	7	0	R/CR
	Securigera varia	1	1	1	1	1	1	5	1	y	2	P	H						C/CSR
	Spartium junceum							+		Y		D, P	S						SC
	Trifolium arvense	1	1	2	1	2	4	4	1	n		A	H	9	3	5	2	0	R/SR
	Trifolium dubium	1	1	4	4	3	4	3	1	n	3	A	H	7	4	6	5	0	R/SR
	Trifolium pratense	1	2	1	5	4	3	3	1	y	4	P	H	7	5	7	5	0	CSR
	Trifolium repens	1	3	1	4	4	3	3	1	y	4	P	H	7	5	6	6	0	CR/CSR
	Ulex europaeus	1	2	1	1	3	3	4	1	n	3	E, P	S	7	5	5	3	0	SC
	Ulex gallii	1	3	2	1	3	4	3	1	n		E, P	S	7	6	3	2	0	SC
	Ulex minor	1	1	1	1	1	1	5	1	n		E, P	Ss	8	6	1	2	0	SC
	Vicia cracca	1	3	2	3	3	3	5	1	n	2	P	H	7	6	7	5	0	C/CSR
	Wisteria sinensis									Y		D, P	C						SC
Lythraceae	*Lythrum salicaria*	2	4	1	1	2	1	2	1	y		P	H	7	9	7	5	0	C/CR
Cornaceae	*Cornus sanguinea*	1	1	3	1	1	1	2	5	y	4	D, P	S	7	5	7	6	0	SC
Celastraceae	*Euonymus europaeus*	1	1	1	1	2	1	3	5	n	4	D, P	T	5	5	8	5	0	SC
Aquifoliaceae	*Ilex aquifolium*	1	1	2	1	2	1	1	5	y	5	E, P	T	5	5	5	5	0	SC
Rhamnaceae	*Frangula alnus*		+						+	y	3	D, P	S	6	8	5	5	0	SC
	Rhamnus alaternus							+		Y	4	E, P	S						SC
	Rhamnus cathartica	1	1	5	1	2	3	3	3	n	3	D, P	S	7	5	7	6	0	SC
Geraniaceae	*Erodium cicutarium*	1	1	4	3	4	3	4	1	n	2	A	H	8	4	6	4	0	R/CSR
	Geranium dissectum	1	1	4	3	2	3	4	1	n	2	A	H	7	5	7	6	0	R/CR
	Geranium molle	1	1	4	4	3	3	3	1	n	2	A	H	7	5	6	5	0	R/CSR
	Geranium pratense	1	1	3	2	3	3	5	2	n	2	P	H	7	6	7	7	0	C/CSR
	Geranium pyrenaicum	1	1	1	1	2	4	4	1	y	2	P	H	8	4	7	6	0	CSR
	Geranium sanguineum									Y	2	P	H	7	4	7	3	0	S/CSR
	Pelargonium spp.									Y	4	P	Hw						?
Tropaeolaceae	*Tropaeolum majus*									Y	5	A	H						R/CR
	Tropaeolum peregrinum									Y		P	C						R/CR
Araliaceae	*Hedera helix*	1	1	3	1	2	1	1	4	y	5	E, P	C	4	5	7	6	0	SC

Family	Hostplant	Primary biotopes								Cultivated	Garden	Phenology	Growth form	Ellenberg values					Strategy
		Aq	Mi	Sk	Ar	Pa	Sp	Wa	Wo					L	F	R	N	S	
Apiaceae	*Aegopodium podagraria*	1	1	1	3	2	2	4	4	n		P	H	6	5	6	7	0	CR
	Anethum graveolens									Y		A	H						R/SR
	Angelica sylvestris	1	4	1	1	2	3	4	3	n	2	P	H	7	8	6	5	0	C/CR
	Daucus carota	1	1	3	1	2	3	4	1	y	2	A, Be	H	8	4	7	3	2	SR/CSR
	Foeniculum vulgare	1	1	1	1	2	3	5	1	y		P	H	9	5	8	5	3	SC/CSR
	Petroselinum crispum			+						Y		Be	H	8	4	7	5	1	R/CSR
	Peucedanum palustre		+							n	1	Be, P	H	7	9	7	5	0	CSR
Boraginaceae	*Borago officinalis*									Y		A	H						CR
	Echium vulgare	1	1	2	3	1	4	3	1	n	2	Be	H	8	4	7	4	1	R/CSR
	Symphytum officinale	1	3	1	1	2	1	4	2	y	2	P	H	7	7	7	8	0	C/CSR
Lamiaceae	*Origanum vulgare*	1	1	3	1	3	4	4	1	y	3	P	H	6	4	7	4	0	SC/CSR
	Teucrium scorodonia	1	1	4	1	3	3	3	3	n		P	H	6	4	4	3	0	SC/CSR
	Thymus polytrichus	1	1	4	1	4	3	1	1	n	2	E, P	Ss	8	4	6	2	0	S
Plantaginaceae	*Plantago coronopus*	1	1	3	1	2	1	5	1	n	2	A, P	H	8	6	6	4	2	SR/CSR
	Plantago lanceolata	1	1	3	3	4	3	3	1	n	3	P	H	7	5	6	4	0	CSR
	Plantago major	1	3	3	4	3	3	4	1	n		P	H	7	5	6	7	0	R/CSR
	Plantago maritima		sm	sc						n		P	H	8	7	6	4	3	CSR
Buddlejaceae	*Buddleja davidii*	1	1	1	1	1	5	4	1	y		D, P	S	7	5	7	5	0	C
Scrophulariaceae	*Digitalis purpurea*	1	2	3	3	2	3	2	4	y	3	Be, P	H	6	6	4	5	0	SR/CSR
	Melampyrum pratense	1	1	1	1	4	3	1	5	n	2	A	H	5	5	2	3	0	R/SR
	Melampyrum sylvaticum								+	n		A	H	4	5	2	2	0	R/SR
	Veronica chamaedrys	1	1	3	1	4	3	3	2	n	2	P	H	6	5	6	5	0	CSR
	Veronica hederifolia	1	1	3	5	3	1	2	4	n		A	H	6	5	7	6	0	R/SR
	Veronica officinalis	1	1	3	1	4	3	3	3	n		P	H	6	5	4	4	0	SR/CSR
	Veronica serpyllifolia	1	3	1	3	4	1	4	3	n		P	H	7	5	6	5	0	R/CSR
Caprifoliaceae	*Lonicera periclymenum*	1	2	1	1	2	1	1	5	n	5	D, P	C	5	6	5	5	0	SC
	Symphoricarpos albus	1	1	1	1	1	4	4	1	n	5	D, P	S	5	5	6	7	0	C/SC
	Viburnum lantana								+	y		D, P	S	7	5	7	5	0	SC
Dipsacaceae	*Knautia arvensis*	1	1	2	2	4	3	4	1	n	2	P	H	7	3	8	4	0	CSR
	Scabiosa columbaria	1	1	4	1	4	3	3	1	n	2	P	H	8	3	8	2	0	SR/CSR
	Succisa pratensis	1	3	1	1	4	1	4	1	n	2	P	H	7	7	5	2	0	S/CSR
Asteraceae	*Arctium lappa*	1	3	1	1	2	1	5	3	n		Be	H	9	5	7	9	0	C/CR
	Arctium minus	1	1	1	3	2	4	4	3	n	2	Be	H	6	4	7	5	0	C/CR
	Artemisia stelleriana									Y	2	P	H						C/CSR
	Artemisia vulgaris	1	1	2	4	2	4	3	1	n	3	P	H	7	4	8	7	0	C/CR
	Carduus crispus	1	1	2	2	2	3	4	4	n		A, Be	H	7	4	8	7	0	CR
	Carduus nutans	1	1	4	3	1	3	4	1	n	2	Be	H	7	4	8	5	0	R/CR
	Carduus tenuiflorus					c		c		n		A, Be	H	8	4	7	4	0	CR
	Carlina vulgaris	1	1	5	1	2	4	3	1	n		Be	H	8	4	7	2	0	SR
	Centaurea nigra	1	3	3	1	3	3	4	1	n	3	P	H	7	5	6	5	0	CSR
	Centaurea scabiosa	1	1	4	1	2	3	4	1	n	3	P	H	8	3	8	3	0	SC/CSR
	Cirsium acaule	1	1	2	1	5	3	3	1	n		P	H	9	4	8	3	0	SC/CSR
	Cirsium arvense	1	2	2	4	3	4	3	1	n	3	P	H	8	6	7	6	0	C
	Cirsium palustre	1	5	2	1	4	2	3	1	n		Be	H	7	8	5	4	0	CR/CSR
	Cirsium vulgare	1	1	3	3	3	4	3	1	n	3	Be	H	7	5	6	6	0	CR
	Cynara cardunculus									Y		P	H						CR
	Filago vulgaris	1	1	4	1	1	3	5	1	n	2	A	H	7	4	6	4	0	SR
	Helichrysum bracteatum									Y		P	Hw						?
	Onopordum acanthium							+		Y	2	Be	H	8	4	6	7	0	CR

Juncaceae	*Juncus articulatus*	2	5	1	1	3	1	3	1	n	2	P	H	8	9	6	3	0	CR/CSR
Cyperaceae	*Carex sylvatica*	1	2	1	1	3	1	3	5	n		P	H	4	5	6	5	0	S/CSR
	Eriophorum angustifolium	1	5	3	1	3	1	1	1	n	2	P	H	8	9	4	1	0	S/SC
	Eriophorum vaginatum	1	5	1	1	4	1	1	1	n	2	P	H	8	8	2	1	0	S/SC
	Rhyncospora alba		+							n		P	H	8	9	2	1	0	S/CSR
Poaceae	*Agrostis canina*	1	5	3	1	2	1	3	1	n	3	P	H	7	7	3	3	0	SR/CSR
	Agrostis capillaris	1	1	3	2	4	3	3	2	n	3	P	H	6	5	4	4	0	CSR
	Agrostis curtisii					–		+		n	2	P	H	7	6	2	1	0	S/SC
	Agrostis stolonifera	1	3	2	4	3	3	3	1	n	3	P	H	7	6	7	6	1	CR
	Aira praecox	1	1	3	1	3	4	3	1	n	2	A	H	8	2	4	2	0	SR
	Alopecurus pratensis	1	3	2	1	4	3	2	2	n	2	P	H	7	5	6	7	0	C/CSR
	Ammophila arenaria							sd		n	2	P	H	9	4	6	3	3	SC
	Anthoxanthum odoratum	1	3	3	1	4	3	3	2	n	2	P	H	7	6	4	3	0	SR/CSR
	Arrhenatherum elatius	1	1	4	2	3	3	3	3	n	2	P	H	7	5	7	7	0	C/CSR
	Brachypodium pinnatum	1	1	2	1	2	3	5	2	n	2	P	H	7	3	8	3	0	SC
	Brachypodium sylvaticum	1	1	3	1	3	2	1	4	n	3	P	H	6	5	6	5	0	SC/CSR
	Bromopsis ramosa	1	1	3	1	2	2	3	4	n	2	P	H	4	6	7	7	0	CSR
	Calamagrostis epigejos	1	2	1	1	1	3	3	3	n	2	P	H	7	7	7	6	0	C/SC
	Cynosurus cristatus	1	3	2	2	4	2	3	1	n		P	H	7	5	6	4	0	R/CSR
	Dactylis glomerata	1	1	3	3	3	3	3	2	n	3	P	H	7	5	7	6	0	C/CSR
	Deschampsia cespitosa	1	3	2	3	3	3	3	3	n	2	P	H	6	6	5	4	0	SC/CSR
	Deschampsia flexuosa	1	2	3	1	4	3	3	3	n	2	P	H	6	5	2	3	0	S/SC
	Elytrigia repens	1	1	1	5	3	3	3	2	n	3	P	H	7	5	7	7	0	C/CR
	Festuca ovina	1	1	3	1	4	3	3	1	n	3	P	H	7	5	4	2	0	S
	Festuca rubra	1	3	3	2	3	3	4	1	y	3	P	H	8	5	6	5	2	CSR
	Helictotrichon pubescens	1	1	3	1	4	2	4	1	n	2	P	H	7	4	7	3	0	S/CSR
	Holcus lanatus	1	3	3	3	4	4	3	2	n	3	P	H	7	6	6	5	0	CSR
	Holcus mollis	1	3	2	1	3	2	3	3	n	3	P	H	6	6	3	3	0	C/CSR
	Lolium perenne	1	1	3	4	4	3	3	1	Y	4	P	H	8	5	6	6	0	CR/CSR
	Milium effusum	1	1	1	1	2	1	1	5	n		P	H	4	5	6	5	0	S/CSR
	Molinia caerulea	1	5	2	1	4	1	3	1	n	2	P	H	7	8	3	2	0	SC
	Nardus stricta	1	3	1	1	4	2	3	1	n	2	P	H	7	7	3	2	0	S/SC
	Phleum bertolonii	1	1	4	2	4	3	3	1	n		P	H	8	4	7	4	0	SR/CSR
	Phleum pratense	1	3	1	5	4	3	3	1	y	3	P	H	8	5	7	6	0	CSR
	Poa annua	1	2	3	5	3	4	3	1	n		A, P	H	7	5	6	7	0	R
	Poa nemoralis	1	1	3	1	2	1	1	5	y		P	H	4	5	6	5	0	SR/CSR
	Poa pratensis	1	2	3	2	3	3	4	1	n	3	P	H	7	5	6	5	0	CSR
	Poa trivialis	1	3	3	3	3	3	3	3	n	3	P	H	7	6	6	6	0	R/CSR
	Sesleria caerulea					+		+		n	2	P	H	7	6	8	2	0	SC/CSR
	Vulpia myuros	1	1	1	1	1	5	3	1	n		A	H	8	3	6	3	0	R/SR

* Codes and abbreviations for the different features covered in this appendix:
Primary biotopes: Aq, aquatic; Mi mire; Sk, skeletal; Ar, arable; Pa, pasture; Sp, spoil; Wa, wasteland; Wo, woodland. Scores: 1, largely absent; 2, infrequent and uncharacteristic; 3, widespread; 4, common; 5, very common and characteristic; –, main habitat. Some species have a more maritime distribution (and are therefore not catered for by Grime *et al.*, 1988): c, typically associated with a coastal variant of the habitat; sc, sea cliff; sd, sand dune; sh, maritime shingle; sl, strand line; sm, salt marsh.
Cultivated: n, not or seldom cultivated; y sizeable populations in cultivation but also in wild; Y, nearly or always a cultivated plant.
Garden: coded as for primary biotopes; blanks, data absent.
Phenology: A, annual; Ba, biannual; Be, biennial; D, deciduous; E, evergreen; P, perennial.
Growth form: C, climber; H, herb; Hw, herb with woody stem; S, shrub; Ss, small shrub; T, tree.
Ellenberg values: L, light; F, moisture; R, reaction (soil pH); N, nitrogen (soil fertility); S, salinity. (From M. O. Hill *et al.*, 1999, courtesy of the Centre for Ecology and Hydrology.)
Strategy and above data: From Grime *et al.* (1988), amendments and additions courtesy of John G. Hodgson.

Appendix 3

NECTAR SOURCES OF BRITISH BUTTERFLIES

APPENDIX 3a: KEY FLOWERING NECTAR PLANTS USED BY BUTTERFLY SPECIES

Butterfly species*	Plant	Family	Rank
C. palaemon	*Ajuga reptans* L.	Lamiaceae	1
	Hyacinthoides non-scripta (L.) Chouard ex Rothm.	Liliaceae	2
	Cirsium palustre (L.) Scop.	Asteraceae	3
	Rubus fruticosus L. agg.	Rosaceae	4
	Glechoma hederacea L.	Lamiaceae	5
	Dactylorhiza maculata (L.) Soó	Orchidaceae	6
T. sylvestris	*Lotus corniculatus* L.	Fabaceae	1
	Centaurea nigra L.	Asteraceae	2
	Cirsium arvense (L.) Scop.	Asteraceae	3
	Trifolium pratense L.	Fabaceae	4
	Centaurea scabiosa L.	Asteraceae	5
	Vicia cracca L.	Fabaceae	6
	Buddleja davidii Franch.	Buddlejaceae	7
	Prunella vulgaris L.	Lamiaceae	8
	Lavandula x intermedia Loisel.	Lamiaceae	9
	Senecio jacobaea L.	Asteraceae	10
T. lineola†	*Anthriscus sylvestris* (L.) Hoffm.	Apiaceae	1
	Cirsium arvense (L.) Scop.	Asteraceae	2
	Medicago sativa L.	Fabaceae	3
	Centaurea nigra L.	Asteraceae	4
	Trifolium pratense L.	Fabaceae	5
	Hyssopus officinalis L.	Lamiaceae	6
	Buddleja davidii Franch.	Buddlejaceae	7
	Senecio jacobaea L.	Asteraceae	8
	Lotus corniculatus L.	Fabaceae	9
	Cirsium palustre (L.) Scop.	Asteraceae	10
T. acteon†	*Echium vulgare* L.	Boraginaceae	1
	Ononis repens L.	Fabaceae	2
	Origanum vulgare L.	Lamiaceae	3
	Lotus corniculatus L.	Fabaceae	4

A Resource-Based Habitat View for Conservation, 1st edition. By Roger L. H. Dennis. Published 2010 by Blackwell Publishing,

Butterfly species*	Plant	Family	Rank
	Rubus fruticosus L. agg.	Rosaceae	5
	Dipsacus fullonum L.	Dipsacaceae	6
	Knautia arvensis (L.) Coult.	Dipsacaceae	7
	Centaurea nigra L.	Asteraceae	8
	Senecio jacobaea L.	Asteraceae	9
	Scabiosa columbaria L.	Dipsacaceae	10
H. comma	*Cirsium acaule* (L.) Scop.	Asteraceae	1
	Centaurea nigra L.	Asteraceae	2
	Gentianella amarella (L.) Börner	Gentianaceae	3
	Leontodon saxatilis Lam.	Asteraceae	4
	Lotus corniculatus L.	Fabaceae	5
	Knautia arvensis (L.) Coult.	Dipsacaceae	6
	Achillea millefolium L.	Asteraceae	7
	Leontodon hispidus L.	Asteraceae	8
	Silybum marianum (L.) Gaertn.	Asteraceae	9
	Achillea millefolium L.	Asteraceae	10
O. sylvanus	*Lotus corniculatus* L.	Fabaceae	1
	Rubus fruticosus L. agg.	Rosaceae	2
	Trifolium pratense L.	Fabaceae	3
	Gladiolus illyricus W.D.J. Koch	Iridaceae	4
	Centranthus ruber (L.) DC	Valerianaceae	5
	Cirsium arvense (L.) Scop.	Asteraceae	6
	Vicia cracca L.	Fabaceae	7
	Buddleja davidii Franch.	Buddlejaceae	8
	Dianthus barbatus L.	Caryophyllaceae	9
	Lavandula × *intermedia* Loisel.	Lamiaceae	10
E. tages†	*Lotus corniculatus* L.	Fabaceae	1
	Calluna vulgaris (L.) Hull	Ericaceae	2
	Glechoma hederacea L.	Lamiaceae	3
	Armeria maritima Willd.	Plumbaginaceae	4
	Lychnis flos-cuculi L.	Caryophyllaceae	5
	Stellaria media (L.) Vill.	Caryophyllaceae	6
	Ajuga reptans L.	Lamiaceae	7
	Prunella vulgaris L.	Lamiaceae	8
	Leucanthemum vulgare Lam.	Asteraceae	9
	Pilosella officinarum F.W. Schultz & Sch. Bip.	Asteraceae	10
P. malvae†	*Lotus corniculatus* L.	Fabaceae	1
	Taraxacum 'officinale' agg.	Asteraceae	2
	Ranunculus acris L.	Ranunculaceae	3
	Ajuga reptans L.	Lamiaceae	4
	Bellis perennis L.	Asteraceae	5
	Calluna vulgaris (L.) Hull	Ericaceae	6
	Myosotis arvensis (L.) Hill	Boraginaceae	7
	Glechoma hederacea L.	Lamiaceae	8
	Trifolium pratense L.	Fabaceae	9
	Fragaria vesca L.	Rosaceae	10
P. machaon†	*Lychnis flos-cuculi* L.	Caryophyllaceae	1
	Dianthus barbatus L.	Caryophyllaceae	2

Butterfly species*	Plant	Family	Rank
	Cirsium dissectum (L.) Hill	Asteraceae	3
	Cirsium palustre (L.) Scop.	Asteraceae	4
	Silene dioica (L.) Clairv.	Caryophyllaceae	5
	Trifolium pratense L.	Fabaceae	6
	Buddleja davidii Franch.	Buddlejaceae	7
	Pedicularis palustris L.	Scrophulariaceae	8
	Centaurea nigra L.	Asteraceae	9
	Eupatorium cannabinum L.	Asteraceae	10
L. sinapis	*Lotus corniculatus* L.	Fabaceae	1
	Lathyrus linifolius (Reichard) Bässler	Fabaceae	2
	Geranium robertianum L.	Geraniaceae	3
	Centaurea nigra L.	Asteraceae	4
	Ajuga reptans L.	Lamiaceae	5
	Prunella vulgaris L.	Lamiaceae	6
	Vicia cracca L.	Fabaceae	7
	Lathyrus pratensis L.	Fabaceae	8
	Vicia sativa L.	Fabaceae	9
	Glechoma hederacea L.	Lamiaceae	10
C. crocea†	*Buddleja davidii* Franch.	Buddlejaceae	1
	Taraxacum 'officinale' agg.	Asteraceae	2
	Centaurea nigra L.	Asteraceae	3
	Cirsium vulgare (Savi) Ten.	Asteraceae	4
	Leontodon saxatilis Lam.	Asteraceae	5
	Lotus corniculatus L.	Fabaceae	6
	Medicago sativa L.	Fabaceae	7
	Aster x salignus Willd.	Asteraceae	8
	Pulicaria dysenterica (L.) Bernh.	Asteraceae	9
	Trifolium pratense L.	Fabaceae	10
G. rhamni	*Buddleja davidii* Franch.	Buddlejaceae	1
	Dipsacus fullonum L.	Dipsacaceae	2
	Phaseolus coccineus L.	Fabaceae	3
	Aubrieta deltoidea (L.) DC	Brassicaceae	4
	Lunaria annua L.	Brassicaceae	5
	Centranthus ruber (L.) DC	Valerianaceae	6
	Taraxacum 'officinale' agg.	Asteraceae	7
	Erysimum cheiri (L.) Crantz	Brassicaceae	8
	Glechoma hederacea L.	Lamiaceae	9
	Aster novi-belgii L.	Asteraceae	10
P. brassicae†	*Buddleja davidii* Franch.	Buddlejaceae	1
	Lavandula × *intermedia* Loisel.	Lamiaceae	2
	Centranthus ruber (L.) DC	Valerianaceae	3
	Phaseolus coccineus L.	Fabaceae	4
	Lunaria annua L.	Brassicaceae	5
	Aubrieta deltoidea (L.) DC	Brassicaceae	6
	Hesperis matronalis L.	Brassicaceae	7
	Aster novi-belgii L.	Asteraceae	8
	Tagetes erecta L.	Asteraceae	9
	Sedum spectabile Boreau	Crassulaceae	10

Butterfly species*	Plant	Family	Rank
P. rapae	*Buddleja davidii* Franch.	Buddlejaceae	1
	Lavandula x intermedia Loisel.	Lamiaceae	2
	Armoracia rusticana P. Gaertn., B. Mey. & Scherb.	Brassicaceae	3
	Cirsium arvense (L.) Scop.	Asteraceae	4
	Rubus fruticosus L. agg.	Rosaceae	5
	Centranthus ruber (L.) DC	Valerianaceae	6
	Hesperis matronalis L.	Brassicaceae	7
	Origanum vulgare L.	Lamiaceae	8
	Aster tripolium L.	Asteraceae	9
	Taraxacum 'officinale' agg.	Asteraceae	10
P. napi	*Buddleja davidii* Franch.	Buddlejaceae	1
	Lavandula x intermedia Loisel.	Lamiaceae	2
	Rubus fruticosus L. agg.	Rosaceae	3
	Cirsium arvense (L.) Scop.	Asteraceae	4
	Aubrieta deltoidea (L.) DC	Brassicaceae	5
	Taraxacum 'officinale' agg.	Asteraceae	6
	Hesperis matronalis L.	Brassicaceae	7
	Epilobium hirsutum L.	Onagraceae	8
	Lunaria annua L.	Brassicaceae	9
	Centranthus ruber (L.) DC	Valerianaceae	10
A. cardamines	*Lunaria annua* L.	Brassicaceae	1
	Aubrieta deltoidea (L.) DC	Brassicaceae	2
	Hesperis matronalis L.	Brassicaceae	3
	Cardamine pratensis L.	Brassicaceae	4
	Alliaria petiolata (M. Bieb) Cavara & Grande	Brassicaceae	5
	Taraxacum 'officinale' agg.	Asteraceae	6
	Erysimum cheiri (L.) Crantz	Brassicaceae	7
	Hyacinthoides hispanica (Mill.) Rothm.	Liliaceae	8
	Brassica oleracea L.	Brassicaceae	9
	Lobularia maritima (L.) Desv.	Brassicaceae	10
C. rubi†	*Salix cinerea* L.	Salicaceae	1
	Hyacinthoides non-scripta (L.) Chouard ex Rothm.	Liliaceae	2
	Lotus corniculatus L.	Fabaceae	3
	Vaccinium myrtillus L.	Ericaceae	4
	Ulex europaeus L.	Fabaceae	5
	Stellaria holostea L.	Caryophyllaceae	6
	Ranunculus acris L.	Ranunculaceae	7
	Ligustrum vulgare L.	Oleaceae	8
	Viburnum opulus L.	Caprifoliaceae	9
	Calluna vulgaris (L.) Hull	Ericaceae	10
T. betulae	*Rubus fruticosus* L. agg.	Rosaceae	1
	Pulicaria dysenterica (L.) Bernh.	Asteraceae	2
	Eupatorium cannabinum L.	Asteraceae	3
	Senecio jacobaea L.	Asteraceae	4
	Angelica sylvestris L.	Apiaceae	5
	Heracleum sphondylium L.	Apiaceae	6
	Argyranthemum frutescens (L.) Schultz Bip.	Asteraceae	7
	Cirsium arvense (L.) Scop.	Asteraceae	8
	Cirsium eriophorum (L.) Scop.	Asteraceae	9

Butterfly species*	Plant	Family	Rank
F. quercus†	*Rubus fruticosus* L. agg.	Rosaceae	1
	Eupatorium cannabinum L.	Asteraceae	2
	Buddleja davidii Franch.	Buddlejaceae	3
	Lychnis coronaria (L.) Murray	Caryophyllaceae	4
	Pulicaria dysenterica (L.) Bernh.	Asteraceae	5
	Lythrum salicaria L.	Lythraceae	6
	Senecio aquaticus Hill	Asteraceae	7
	Heracleum sphondylium L.	Apiaceae	8
	Origanum vulgare L.	Lamiaceae	9
	Frangula alnus Mill.	Rhamnaceae	10
S. w-album†	*Rubus fruticosus* L. agg.	Rosaceae	1
	Cirsium arvense (L.) Scop.	Asteraceae	2
	Ligustrum vulgare L.	Oleaceae	3
	Senecio jacobaea L.	Asteraceae	4
	Origanum vulgare L.	Lamiaceae	5
	Eupatorium cannabinum L.	Asteraceae	6
	Chamerion angustifolium (L.) Holub	Onagraceae	7
	Ligustrum ovalifolium Hassk.	Oleaceae	8
	Leucanthemum x superbum (Bergmans ex J.W. Ingram) D.H. Kent	Asteraceae	9
	Cirsium vulgare (Savi) Ten.	Asteraceae	10
S. pruni	*Viburnum lantana* L.	Caprifoliaceae	1
	Ligustrum vulgare L.	Oleaceae	2
	Rubus fruticosus L. agg.	Rosaceae	3
	Rosa canina L.	Rosaceae	4
	Viburnum opulus L.	Caprifoliaceae	5
	Ranunculus acris L.	Ranunculaceae	6
	Heracleum sphondylium L.	Apiaceae	7
L. phlaeas	*Senecio jacobaea* L.	Asteraceae	1
	Sedum spectabile Boreau	Crassulaceae	2
	Buddleja davidii Franch.	Buddlejaceae	3
	Cirsium arvense (L.) Scop.	Asteraceae	4
	Calluna vulgaris (L.) Hull	Ericaceae	5
	Origanum vulgare L.	Lamiaceae	6
	Pulicaria dysenterica (L.) Bernh.	Asteraceae	7
	Aster x salignus Willd.	Asteraceae	8
	Aster novi-belgii L.	Asteraceae	9
	Lavandula x intermedia Loisel.	Lamiaceae	10
L. dispar	*Lythrum salicaria* L.	Lythraceae	1
	Origanum vulgare L.	Lamiaceae	2
	Cirsium palustre (L.) Scop.	Asteraceae	3
C. minimus	*Anthyllis vulneraria* L.	Fabaceae	1
	Lotus corniculatus L.	Fabaceae	2
	Hippocrepis comosa L.	Fabaceae	3
	Ligustrum vulgare L.	Oleaceae	4
P. argus†	*Rubus fruticosus* L. agg.	Rosaceae	1
	Lotus corniculatus L.	Fabaceae	2
	Thymus polytrichus A. Kern. Ex Borbás	Lamiaceae	3

Butterfly species*	Plant	Family	Rank
	Brassica rapa L.	Brassicaceae	4
	Helianthemum nummularium (L.) Mill.	Cistaceae	5
	Scabiosa columbaria L.	Dipsacaceae	6
	Erica cinerea L.	Ericaceae	7
	Carduus tenuiflorus Curtis	Asteraceae	8
	Galium verum L.	Rubiaceae	9
	Ligustrum ovalifolium Hassk.	Oleaceae	10
A. agestis†	*Senecio jacobaea* L	Asteraceae	1
	Origanum vulgare L.	Lamiaceae	2
	Pulicaria dysenterica (L.) Bernh.	Asteraceae	3
	Stellaria nemorum L.	Caryophyllaceae	4
	Rubus fruticosus L. agg.	Rosaceae	5
	Anthyllis vulneraria L.	Fabaceae	6
	Cirsium arvense (L.) Scop.	Asteraceae	7
	Ulex europaeus L.	Fabaceae	8
	Eupatorium cannabinum L.	Asteraceae	9
	Trifolium repens L.	Fabaceae	10
A. artaxerxes	*Thymus polytrichus* A. Kern. Ex Borbás	Lamiaceae	1
	Lotus corniculatus L.	Fabaceae	2
	Geranium sanguineum L.	Geraniaceae	3
P. icarus	*Lotus corniculatus* L.	Fabaceae	1
	Lavandula x intermedia Loisel.	Lamiaceae	2
	Calluna vulgaris (L.) Hull	Ericaceae	3
	Succisa pratensis Moench	Dipsacaceae	4
	Buddleja davidii Franch.	Buddlejaceae	5
	Origanum vulgare L.	Lamiaceae	6
	Trifolium repens L.	Fabaceae	7
	Trifolium dubium Sibth.	Fabaceae	8
	Senecio jacobaea L.	Asteraceae	9
	Pulicaria dysenterica (L.) Bernh.	Asteraceae	10
L. coridon†	*Origanum vulgare* L.	Lamiaceae	1
	Knautia arvensis (L.) Coult.	Dipsacaceae	2
	Centaurea nigra L.	Asteraceae	3
	Lotus corniculatus L.	Fabaceae	4
	Scabiosa columbaria L.	Dipsacaceae	5
	Carlina vulgaris L.	Asteraceae	6
	Rubus fruticosus L. agg.	Rosaceae	7
	Senecio jacobaea L.	Asteraceae	8
	Succisa pratensis Moench	Dipsacaceae	9
	Centaurea scabiosa L.	Asteraceae	10
L. bellargus†	*Origanum vulgare* L.	Lamiaceae	1
	Hippocrepis comosa L.	Fabaceae	2
	Lotus corniculatus L.	Fabaceae	3
	Pilosella officinarum F.W. Schultz & Sch. Bip.	Asteraceae	4
	Centaurea nigra L.	Asteraceae	5
	Rubus fruticosus L. agg.	Rosaceae	6
	Thymus polytrichus A. Kern. Ex Borbás	Lamiaceae	7
	Eupatorium cannabinum L.	Asteraceae	8

Butterfly species*	Plant	Family	Rank
	Knautia arvensis (L.) Coult.	Dipsacaceae	9
	Senecio jacobaea L.	Asteraceae	10
C. argiolus†	*Buddleja davidii* Franch.	Buddlejaceae	1
	Rubus fruticosus L. agg.	Rosaceae	2
	Hebe x franciscana (Eastw.) Souster	Scrophulariaceae	3
	Ilex aquifolium L.	Aquifoliaceae	4
	Hedera helix L.	Araliaceae	5
	Origanum vulgare L.	Lamiaceae	6
	Phaseolus coccineus L.	Fabaceae	7
	Escallonia bifida Link & Otto	Grossulariaceae	8
	Lavandula x intermedia Loisel.	Lamiaceae	9
	Eupatorium cannabinum L.	Asteraceae	10
M. arion	*Thymus polytrichus* A. Kern. Ex Borbás	Lamiaceae	1
	Ajuga reptans L.	Lamiaceae	2
	Origanum vulgare L.	Lamiaceae	3
	Cirsium acaule (L.) Scop.	Asteraceae	4
H. lucina†	*Ajuga reptans* L.	Lamiaceae	1
	Ranunculus acris L.	Ranunculaceae	2
	Crataegus monogyna Jacq.	Rosaceae	3
	Euphorbia amygdaloides L.	Euphorbiaceae	4
	Ranunculus acris L.	Ranunculaceae	5
	Fragaria vesca L.	Rosaceae	6
	Potentilla anserina L.	Rosaceae	7
	Potentilla erecta (L.) Raeusch.	Rosaceae	8
	Polygala vulgaris L.	Polygalaceae	9
	Conopodium majus (Gouan) Loret	Apiaceae	10
L. camilla	*Rubus fruticosus* L. agg.	Rosaceae	1
	Buddleja davidii Franch.	Buddlejaceae	2
	Frangula alnus Mill.	Rhamnaceae	3
	Heracleum sphondylium L.	Apiaceae	4
	Torilis japonica (Houttt.) DC	Apiaceae	5
	Stachys officinalis (L.) Trevis.	Lamiaceae	6
	Ligustrum ovalifolium Hassk.	Oleaceae	7
	Ligustrum vulgare L.	Oleaceae	8
	Lonicera periclymenum L.	Caprifoliaceae	9
	Cirsium arvense (L.) Scop.	Asteraceae	10
A. iris	*Buddleja davidii* Franch.	Buddlejaceae	1
	Castanea sativa Mill.	Fagaceae	2
V. atalanta	*Buddleja davidii* Franch.	Buddlejaceae	1
	Hedera helix L.	Araliaceae	2
	Sedum spectabile Boreau	Crassulaceae	3
	Aster novi-belgii L.	Asteraceae	4
	Centranthus ruber (L.) DC	Valerianaceae	5
	Aster x salignus Willd.	Asteraceae	6
	Buddleja x weyeriana Weyer	Buddlejaceae	7
	Hebe x franciscana (Eastw.) Souster	Scrophulariaceae	8

Butterfly species*	Plant	Family	Rank
	Lavandula x intermedia Loisel.	Lamiaceae	9
	Eupatorium cannabinum L.	Asteraceae	10
V. cardui	*Buddleja davidii* Franch.	Buddlejaceae	1
	Centranthus ruber (L.) DC	Valerianaceae	2
	Sedum spectabile Boreau	Crassulaceae	3
	Oenanthe silaifolia M. Bieb.	Apiaceae	4
	Aster novi-belgii L.	Asteraceae	5
	Senecio aquaticus Hill	Asteraceae	6
	Cirsium arvense (L.) Scop.	Asteraceae	7
	Aster x salignus Willd.	Asteraceae	8
	Centaurea nigra L.	Asteraceae	9
	Rubus fruticosus L. agg.	Rosaceae	10
A. urticae	*Buddleja davidii* Franch.	Buddlejaceae	1
	Cirsium arvense (L.) Scop.	Asteraceae	2
	Eupatorium cannabinum L.	Asteraceae	3
	Sedum spectabile Boreau	Crassulaceae	4
	Taraxacum 'officinale' agg.	Asteraceae	5
	Aster novi-belgii L.	Asteraceae	6
	Medicago sativa L.	Fabaceae	7
	Senecio jacobaea L.	Asteraceae	8
	Aster x salignus Willd.	Asteraceae	9
	Aubrieta deltoidea (L.) DC	Brassicaceae	10
N. polychloros	*Buddleja davidii* Franch.	Buddlejaceae	1
	Prunus spinosa L.	Rosaceae	2
	Rubus fruticosus L. agg.	Rosaceae	3
	Salix caprea L.	Salicaceae	4
	Sedum spectabile Boreau	Crassulaceae	5
	Ligustrum ovalifolium Hassk.	Oleaceae	6
	Callistephus chinensis (L.) Nees	Asteraceae	7
	Tagetes erecta L.	Asteraceae	8
I. io	*Buddleja davidii* Franch.	Buddlejaceae	1
	Dipsacus fullonum L.	Dipsacaceae	2
	Pulicaria dysenterica (L.) Bernh.	Asteraceae	3
	Cirsium arvense (L.) Scop.	Asteraceae	4
	Taraxacum 'officinale' agg.	Asteraceae	5
	Sedum spectabile Boreau	Crassulaceae	6
	Eupatorium cannabinum L.	Asteraceae	7
	Aster novi-belgii L.	Asteraceae	8
	Aubrieta deltoidea (L.) DC	Brassicaceae	9
	Centranthus ruber (L.) DC	Valerianaceae	10
P. c-album	*Buddleja davidii* Franch.	Buddlejaceae	1
	Aster x salignus Willd.	Asteraceae	2
	Sedum spectabile Boreau	Crassulaceae	3
	Aster novi-belgii L.	Asteraceae	4
	Rubus fruticosus L. agg.	Rosaceae	5
	Cirsium arvense (L.) Scop.	Asteraceae	6
	Hedera helix L.	Araliaceae	7

Butterfly species*	Plant	Family	Rank
	Centranthus ruber (L.) DC	Valerianaceae	8
	Hebe x franciscana (Eastw.) Souster	Scrophulariaceae	9
	Origanum vulgare L.	Lamiaceae	10
B. selene†	*Lotus corniculatus* L.	Fabaceae	1
	Ranunculus acris L.	Ranunculaceae	2
	Cirsium palustre (L.) Scop.	Asteraceae	3
	Lychnis flos-cuculi L.	Caryophyllaceae	4
	Ajuga reptans L.	Lamiaceae	5
	Leucanthemum vulgare Lam.	Asteraceae	6
	Calluna vulgaris (L.) Hull	Ericaceae	7
	Thymus polytrichus A. Kern. Ex Borbás	Lamiaceae	8
	Dactylorhiza fuchsii (Druce) Soó	Orchidaceae	9
	Rubus fruticosus L. agg.	Rosaceae	10
B. euphrosyne†	*Lotus corniculatus* L.	Fabaceae	1
	Ajuga reptans L.	Lamiaceae	2
	Ranunculus acris L.	Ranunculaceae	3
	Taraxacum 'officinale' agg.	Asteraceae	4
	Glechoma hederacea L.	Lamiaceae	5
	Hyacinthoides non-scripta (L.) Chouard ex Rothm.	Liliaceae	6
	Trifolium pratense L.	Fabaceae	7
	Lychnis flos-cuculi L.	Caryophyllaceae	8
	Potentilla erecta (L.) Raeusch.	Rosaceae	9
	Prunella vulgaris L.	Lamiaceae	10
A. adippe†	*Cirsium palustre* (L.) Scop.	Asteraceae	1
	Rubus fruticosus L. agg.	Rosaceae	2
	Succisa pratensis Moench	Dipsacaceae	3
	Cirsium vulgare (Savi) Ten.	Asteraceae	4
	Buddleja davidii Franch.	Buddlejaceae	5
	Trifolium pratense L.	Fabaceae	6
	Senecio jacobaea L.	Asteraceae	7
	Carduus crispus L.	Asteraceae	8
	Stachys officinalis	Lamiaceae	9
	Centaurea scabiosa L.	Asteraceae	10
A. aglaja†	*Centranthus ruber* (L.) DC	Valerianaceae	1
	Buddleja davidii Franch.	Buddlejaceae	2
	Cirsium vulgare (Savi) Ten.	Asteraceae	3
	Cirsium palustre (L.) Scop.	Asteraceae	4
	Trifolium pratense L.	Fabaceae	5
	Eupatorium cannabinum L.	Asteraceae	6
	Centaurea nigra L.	Asteraceae	7
	Senecio jacobaea L.	Asteraceae	8
	Verbena x hybrida Voss	Verbenaceae	9
	Rubus fruticosus L. agg.	Rosaceae	10
A. paphia†	*Rubus fruticosus* L. agg.	Rosaceae	1
	Buddleja davidii Franch.	Buddlejaceae	2
	Calluna vulgaris (L.) Hull	Ericaceae	3
	Centaurea nigra L.	Asteraceae	4
	Eupatorium cannabinum L.	Asteraceae	5

Butterfly species*	Plant	Family	Rank
	Centaurea scabiosa L.	Asteraceae	6
	Cirsium arvense (L.) Scop.	Asteraceae	7
	Centranthus ruber (L.) DC	Valerianaceae	8
	Cirsium vulgare (Savi) Ten.	Asteraceae	9
	Senecio jacobaea L.	Asteraceae	10
E. aurinia†	*Leontodon saxatilis* Lam.	Asteraceae	1
	Ranunculus acris L.	Ranunculaceae	2
	Potentilla erecta (L.) Raeusch.	Rosaceae	3
	Stachys officinalis (L.) Trevis.	Lamiaceae	4
	Cirsium palustre (L.) Scop.	Asteraceae	5
	Polygala vulgaris L.	Polygalaceae	6
	Lychnis flos-cuculi L.	Caryophyllaceae	7
	Centaurea nigra L.	Asteraceae	8
	Ajuga reptans L.	Lamiaceae	9
	Taraxacum 'officinale' agg.	Asteraceae	10
M. cinxia	*Armeria maritima* Willd.	Plumbaginaceae	1
	Anthyllis vulneraria L.	Fabaceae	2
	Lotus corniculatus L.	Fabaceae	3
	Hippocrepis comosa L.	Fabaceae	4
	Ononis repens L.	Fabaceae	5
	Euphorbia amygdaloides L.	Euphorbiaceae	6
	Cruciata laevipes Opiz	Rubiaceae	7
	Leontodon saxatilis Lam.	Asteraceae	8
	Taraxacum 'officinale' agg.	Asteraceae	9
M. athalia	*Ranunculus acris* L.	Ranunculaceae	1
	Chamerion angustifolium (L.) Holub	Onagraceae	2
	Rubus fruticosus L. agg.	Rosaceae	3
	Ajuga reptans L.	Lamiaceae	4
	Leucanthemum vulgare Lam.	Asteraceae	5
	Ranunculus acris L.	Ranunculaceae	6
	Lychnis flos-cuculi L.	Caryophyllaceae	7
	Potentilla erecta (L.) Raeusch.	Rosaceae	8
	Melampyrum pratense L.	Scrophulariaceae	9
	Cirsium palustre (L.) Scop.	Asteraceae	10
P. aegeria	*Buddleja davidii* Franch.	Buddlejaceae	1
	Aster x salignus Willd.	Asteraceae	2
	Rubus fruticosus L. agg.	Rosaceae	3
	Sedum spectabile Boreau	Crassulaceae	4
	Senecio jacobaea L.	Asteraceae	5
	Hedera helix L.	Araliaceae	6
	Aster novi-belgii L.	Asteraceae	7
	Origanum vulgare L.	Lamiaceae	8
	Eupatorium cannabinum L.	Asteraceae	9
	Taraxacum 'officinale' agg.	Asteraceae	10
L. megera†	*Buddleja davidii* Franch.	Buddlejaceae	1
	Centranthus ruber (L.) DC	Valerianaceae	2
	Lavandula x intermedia Loisel.	Lamiaceae	3
	Origanum vulgare L.	Lamiaceae	4

Butterfly species*	Plant	Family	Rank
	Tagetes patula L.	Asteraceae	5
	Cirsium arvense (L.) Scop.	Asteraceae	6
	Senecio jacobaea L.	Asteraceae	7
	Ranunculus acris L.	Ranunculaceae	8
	Centaurea nigra L.	Asteraceae	9
	Lotus corniculatus L.	Fabaceae	10
E. epiphron	*Potentilla erecta* (L.) Raeusch.	Rosaceae	1
	Thymus polytrichus A. Kern. Ex Borbás	Lamiaceae	2
	Vaccinium myrtillus L.	Ericaceae	3
	Galium saxatile L.	Rubiaceae	4
	Taraxacum 'officinale' agg.	Asteraceae	5
E. aethiops†	*Succisa pratensis* Moench	Dipsacaceae	1
	Centaurea nigra L.	Asteraceae	2
	Cirsium arvense (L.) Scop.	Asteraceae	3
	Senecio jacobaea L.	Asteraceae	4
	Erica tetralix L.	Ericaceae	5
	Cirsium vulgare (Savi) Ten.	Asteraceae	6
	Achillea millefolium L.	Asteraceae	7
	Rubus fruticosus L. agg.	Rosaceae	8
	Valeriana officinalis L.	Valerianaceae	9
	Leontodon saxatilis Lam.	Asteraceae	10
M. galathea†	*Buddleja davidii* Franch.	Buddlejaceae	1
	Knautia arvensis (L.) Coult.	Dipsacaceae	2
	Centaurea nigra L.	Asteraceae	3
	Origanum vulgare L.	Lamiaceae	4
	Cirsium vulgare (Savi) Ten.	Asteraceae	5
	Centaurea scabiosa L.	Asteraceae	6
	Cirsium arvense (L.) Scop.	Asteraceae	7
	Rubus fruticosus L. agg.	Rosaceae	8
	Trifolium pratense L.	Fabaceae	9
	Lavandula x intermedia Loisel.	Lamiaceae	10
H. semele†	*Buddleja davidii* Franch.	Buddlejaceae	1
	Rubus fruticosus L. agg.	Rosaceae	2
	Calluna vulgaris (L.) Hull	Ericaceae	3
	Thymus polytrichus A. Kern. Ex Borbás	Lamiaceae	4
	Erica cinerea L.	Ericaceae	5
	Dipsacus fullonum L.	Dipsacaceae	6
	Centaurea nigra L.	Asteraceae	7
	Centranthus ruber (L.) DC	Valerianaceae	8
	Senecio jacobaea L.	Asteraceae	9
	Cirsium vulgare (Savi) Ten.	Asteraceae	10
P. tithonus	*Rubus fruticosus* L. agg.	Rosaceae	1
	Origanum vulgare L.	Lamiaceae	2
	Buddleja davidii Franch.	Buddlejaceae	3
	Senecio jacobaea L.	Asteraceae	4
	Calluna vulgaris (L.) Hull	Ericaceae	5
	Hebe x franciscana (Eastw.) Souster	Scrophulariaceae	6
	Lavandula x intermedia Loisel.	Lamiaceae	7
	Cirsium arvense (L.) Scop.	Asteraceae	8

Butterfly species*	Plant	Family	Rank
	Eupatorium cannabinum L.	Asteraceae	9
	Thymus polytrichus A. Kern. Ex Borbás	Lamiaceae	10
M. jurtina	*Buddleja davidii* Franch.	Buddlejaceae	1
	Cirsium arvense (L.) Scop.	Asteraceae	2
	Senecio jacobaea L.	Asteraceae	3
	Origanum vulgare L.	Lamiaceae	4
	Rubus fruticosus L. agg.	Rosaceae	5
	Succisa pratensis Moench	Dipsacaceae	6
	Centaurea nigra L.	Asteraceae	7
	Lavandula x intermedia Loisel.	Lamiaceae	8
	Eupatorium cannabinum L.	Asteraceae	9
	Centranthus ruber (L.) DC	Valerianaceae	10
C. pamphilus†	*Ranunculus acris* L.	Ranunculaceae	1
	Succisa pratensis Moench	Dipsacaceae	2
	Lotus corniculatus L.	Fabaceae	3
	Rubus fruticosus L. agg.	Rosaceae	4
	Stellaria media (L.) Vill.	Caryophyllaceae	5
	Trifolium repens L.	Fabaceae	6
	Calluna vulgaris (L.) Hull	Ericaceae	7
	Bellis perennis L.	Asteraceae	8
	Senecio jacobaea L.	Asteraceae	9
	Cirsium palustre (L.) Scop.	Asteraceae	10
C. tullia†	*Erica tetralix* L.	Ericaceae	1
	Calluna vulgaris (L.) Hull	Ericaceae	2
	Erica cinerea L.	Ericaceae	3
	Potentilla erecta (L.) Raeusch.	Rosaceae	4
	Ranunculus acris L.	Ranunculaceae	5
	Leontodon hispidus L.	Asteraceae	6
	Rubus fruticosus L. agg.	Rosaceae	7
	Trifolium repens L.	Fabaceae	8
	Thymus polytrichus A. Kern. Ex Borbás	Lamiaceae	9
	Centaurea nigra L.	Asteraceae	10
A. hyperantus†	*Rubus fruticosus* L. agg.	Rosaceae	1
	Calluna vulgaris (L.) Hull	Ericaceae	2
	Origanum vulgare L.	Lamiaceae	3
	Buddleja davidii Franch.	Buddlejaceae	4
	Pulicaria dysenterica (L.) Bernh.	Asteraceae	5
	Senecio jacobaea L.	Asteraceae	6
	Trifolium repens L.	Fabaceae	7
	Cirsium arvense (L.) Scop.	Asteraceae	8
	Lotus corniculatus L.	Fabaceae	9
	Eupatorium cannabinum L.	Asteraceae	10

Plant species are ranked in decreasing order of importance.
* For full species names, see Appendix 1a.
† Butterfly species with other plants (considered less abundant) that have the same rank as the last named plant. For the following butterfly species, plants identified to genus only are important nectar sources and would rank among the 10 most important if identified to species: *Gonepteryx rhamni*, *Favonius quercus*, *Aricia artaxerxes*, *Polyommatus icarus*, *Boloria euphrosyne* (*Cirsium* sp.); *Erebia epiphron*, *Pyronia tithonus* (*Mentha* sp.)

APPENDIX 3b: NECTAR PLANTS SUPPORTING 10 OR MORE BUTTERFLY SPECIES

Nectar plant	Number of butterfly species	Nectar plant	Number of butterfly species
Cirsium sp.	47	*Tagetes erecta*	18
Rubus fruticosus agg.	42	*Tanacetum parthenium*	18
Buddleja davidii	35	*Helichrysum bracteatum*	17
Origanum vulgare	35	*Medicago sativa*	17
Centaurea sp.	34	*Phlox drummondii*	17
Lotus corniculatus	33	*Prunella vulgaris*	17
Senecio jacobaea	33	*Scabiosa columbaria*	17
Calluna vulgaris	32	*Aster x salignus*	16
Centaurea nigra	32	*Dianthus barbatus*	16
Cirsium arvense	32	*Geranium* sp.	16
Centranthus ruber	30	*Hyssopus officinalis*	16
Taraxacum sp.	29	*Lobularia maritima*	16
Thymus polytrichus	29	*Myosotis* sp.	16
Pulicaria dysenterica	27	*Nepeta x faassenii*	16
Eupatorium cannabinum	26	*Verbena x hybrida*	16
Hieracium sp.	26	*Centaurea scabiosa*	15
Trifolium pratense	26	*Coreopsis tinctoria*	15
Knautia arvensis	25	*Dahlia pinnata*	15
Leontodon saxatilis	25	*Erysimum cheiri*	15
Scabiosa sp.	25	*Heliotropium arborescens*	15
Succisa pratensis	25	*Lychnis flos-cuculi*	15
Lavandula × intermedia	24	*Verbena bonariensis*	15
Ligustrum ovalifolium	24	*Argyranthemum frutescens*	14
Dipsacus fullonum	23	*Aster tripolium*	14
Bellis perennis	22	*Callistephus chinensis*	14
Cirsium palustre	22	*Carlina vulgaris*	14
Leucanthemum vulgare	22	*Heracleum sphondylium*	14
Other yellow Composite	22	*Leucanthemum x superbum*	14
Sedum spectabile	22	*Mentha aquatica*	14
Solidago canadensis	22	*Echinops* sp.	13
Trifolium repens	22	*Glechoma hederacea*	13
Iberis umbellata	21	*Hedera helix*	13
Mentha sp.	21	*Hesperis matronalis*	13
Ranunculus sp.	21	*Hyacinthoides non-scripta*	13
Tagetes patula	21	*Hydrangea macrophylla*	13
Ajuga reptans	20	*Hypochaeris radicata*	13
Cirsium vulgare	20	*Silene dioica*	13
Hebe × franciscana	20	*Vicia* sp.	13
Achillea millefolium	19	*Viola x wittrockiana*	13
Aster novi-belgii	19	*Cardamine pratensis*	12
Aubrieta deltoidea	19	*Chamerion angustifolium*	12
Lobelia sp.	19	*Ligustrum vulgare*	12
Potentilla sp.	19	*Rudbeckia* sp.	12
Ligularia dentata	18	*Stachys officinalis*	12
Lythrum salicaria	18	*Viola* sp.	12
Ranunculus acris	18	*Allium schoenoprasum*	11

Nectar plant	Number of butterfly species	Nectar plant	Number of butterfly species
Anaphalis margaritacea	11	*Chrysanthemum segetum*	10
Anthemis arvensis	11	*Crataegus monogyna*	10
Arctium lappa	11	*Dianthus* sp. (pink)	10
Calendula officinalis	11	*Echium vulgare*	10
Ceanothus x delileanus	11	*Epilobium hirsutum*	10
Centaurea cyanus	11	*Fragaria x ananassa*	10
Escallonia bifida	11	*Gilia capitata*	10
Lunaria annua	11	*Helenium autumnale*	10
Phaseolus coccineus	11	*Lavatera* sp.	10
Primula vulgaris	11	*Malcolmia maritima*	10
Rosa sp.	11	*Matthiola incana*	10
Carduus sp.	10	*Stellaria holostea*	10

APPENDIX 3c: NECTAR PLANT FAMILIES SUPPORTING SIX OR MORE BUTTERFLY SPECIES*

Nectar plant family	Number of butterfly species	Number of records	Nectar plant family	Number of butterfly species	Number of records
Asteraceae	57	3755	Lythraceae	17	32
Lamiaceae	52	630	Onagraceae	16	120
Rosaceae	50	816	Primulaceae	16	35
Fabaceae	44	663	Caprifoliaceae	15	45
Dipsacaceae	38	248	Rhamnaceae	14	26
Buddlejaceae	35	1495	Hydrangeaceae	13	22
Caryophyllaceae	34	161	Rubiaceae	13	16
Ranunculaceae	33	149	Grossulariaceae	12	32
Ericaceae	33	143	Orchidaceae	12	17
Valerianaceae	32	84	Araliaceae	11	85
Brassicaceae	31	482	Polygonaceae	11	21
Oleaceae	29	113	Malvaceae	11	14
Apiaceae	29	77	Balsaminaceae	10	20
Scrophulariaceae	27	111	Tropaeolaceae	9	12
Boraginaceae	25	98	Convolvulaceae	8	19
Crassulaceae	24	172	Clusiaceae	8	9
Liliaceae	23	118	Salicaceae	7	31
Geraniaceae	21	65	Solanaceae	7	16
Verbenaceae	20	101	Saxifragaceae	7	10
Violaceae	20	36	Rutaceae	7	10
Campanulaceae	19	36	Cistaceae	7	9
Polemoniaceae	19	33	Gentianaceae	7	8
Plumbaginaceae	18	42			

* A total of 75 flowering plant families were recorded, but this appendix is limited to those that seem to support six or more butterfly species.

APPENDIX 3d: NECTAR PLANTS USED BY BUTTERFLIES MORE OFTEN OR LESS OFTEN THAN EXPECTED

Nectar plants	Residual: more or less than expected*	Observed number of butterfly species	Predicted number of butterfly species	Nectaring records
Cirsium sp.	2.76	47	30.5	262
Other yellow Composite	2.37	22	12.7	22
Scabiosa sp.	1.94	25	16.5	35
Tanacetum parthenium	1.75	18	11.6	19
Ligularia dentata	1.75	18	11.6	19
Origanum vulgare	1.72	35	25.8	111
Thymus polytrichus	1.57	29	21.4	62
Red Labiates	1.52	15	9.9	15
Coreopsis tinctoria	1.52	15	9.9	15
Rubus fruticosus agg.	1.5	42	33	563
Verbena × hybrida	1.5	16	10.7	17
Lunaria annua	−1.55	10	15.8	32
Epilobium sp.	−1.56	1	3.5	4
Senecio squalidus	−1.58	6	10.7	17
Verbena bonariensis	−1.63	15	22.2	69
Erica tetralix	−1.63	3	6.8	9
Tussilago farfara	−1.7	4	8.4	12
Muscari sp.	−1.84	5	10.3	16
Stellaria sp.	−1.88	1	4.2	5
Chamerion angustifolium	−1.91	12	19.9	52
Hesperis matronalis	−1.98	13	21.5	63
Salix cinerea	−2.07	5	11.2	18
Buddleja × weyeriana	−2.18	9	17.2	38
Geranium robertianum	−2.18	6	13	23
Alliaria petiolata	−2.63	6	14.9	29
Cardamine pratensis	−2.67	12	23.8	84
Epilobium hirsutum	−2.71	9	19.7	51
Brassica oleracea	−2.89	7	17.7	40
Hedera helix	−2.9	11	23.6	82
Aster × salignus	−2.99	16	31.1	301
Vicia cracca	−3.17	9	22.1	68
Trifolium dubium	−3.57	2	12	20

* Residual in standard error units. Positive numbers indicate nectar plants used by butterflies more often than expected, negative numbers indicate use less often than expected.

APPENDIX 3e: ADULT FEEDING: NECTAR AND NON-NECTAR SOURCES

	Nectar†		Non-nectar‡						
Butterfly species*	**herb/flower**	**shrub/tree flower**	**ergot**	**honeydew**	**sap**	**decaying plants**	**animal**	**mineral**	**hostplant coincidence**
C. palaemon	1	2	0	0	0	0	0	0	0
T. sylvestris	1	1	0	0	0	0	0	2	0
T. lineola	1	2	0	0	0	0	0	0	0
T. acteon	1	2	0	0	0	0	0	0	0
H. comma	1	0	0	0	0	0	0	0	0
O. sylvanus	1	1	0	0	0	0	0	0	0
E. tages	1	0	0	3	0	0	0	0	1
P. malvae	1	2	0	0	0	0	0	0	1
P. machaon	1	2	0	0	0	0	0	0	0
L. sinapis	1	1	0	0	1	0	0	1	1
C. crocea	1	1	0	0	0	0	2	0	1
G. rhamni	1	1	0	0	0	0	3	0	3
P. brassicae	1	1	0	2	0	0	0	0	2
P. rapae	1	1	0	0	0	0	2	2	2
P. napi	1	1	0	0	0	3	2	1	1
A. cardamines	1	2	0	0	0	0	0	0	1
C. rubi	1	1	0	2	2	0	0	0	1
T. betulae	1	1	0	1	0	0	0	0	3§
F. quercus	1	1	0	1	0	0	3	2	3§
S. w-album	1	1	0	1	0	0	0	2	3§
S. pruni	2	1	0	1	0	0	0	0	3§
L. phlaeas	1	1	0	0	0	0	2	3	0
L. dispar	1	0	0	0	0	0	0	0	0
C. minimus	1	0	0	0	0	0	3	3	1
P. argus	1	1	0	0	0	0	3	3	1
A. agestis	1	1	0	0	0	0	3	3	2
A. artaxerxes	1	0	0	0	0	0	0	0	1
P. icarus	1	1	0	2	0	0	3	3	1
L. coridon	1	2	0	0	0	0	1	2	3
L. bellargus	1	2	0	0	0	0	2	2	3
C. argiolus	1	1	0	2	2	0	1	1	1
M. arion	1	0	0	2	0	0	0	0	1
H. lucina	1	0	0	0	0	0	0	0	0
L. camilla	1	1	1	1	0	1	1	1	0
A. iris	0	2	2	1	1	0	1	1	0
V. atalanta	1	1	0	2	1	1	2	2	0
V. cardui	1	1	0	3	2	0	0	0	1
A. urticae	1	1	0	0	0	3	3	2	0
N. polychloros	1	1	0	0	1	0	0	0	2
I. io	1	1	0	2	2	2	3	3	0
P. c-album	1	1	0	2	2	1	2	2	0
B. selene	1	2	3	0	0	0	0	0	2
B. euphrosyne	1	0	0	0	0	0	0	0	2
A. adippe	1	1	0	0	0	0	0	0	0
A. aglaja	1	1	0	0	0	0	0	0	0
A. paphia	1	1	0	2	0	0	0	2	0
E. aurinia	1	0	0	0	0	0	0	0	2

Butterfly species*	Nectar†		Non-nectar‡						
	herb/flower	shrub/tree flower	ergot	honeydew	sap	decaying plants	animal	mineral	hostplant coincidence
M. cinxia	1	0	0	0	0	0	0	0	0
M. athalia	1	2	0	0	0	0	2	0	2
P. aegeria	1	1	2	2	0	1	0	1	0
L. megera	1	1	0	0	0	0	0	0	0
E. epiphron	1	0	0	0	0	0	0	0	0
E. aethiops	1	2	0	0	0	0	0	2	0
M. galathea	1	1	0	0	0	0	0	0	0
H. semele	1	1	0	0	2	0	0	2	0
P. tithonus	1	1	0	0	0	0	0	0	0
M. jurtina	1	1	0	1	0	0	0	0	0
A. hyperantus	1	1	0	0	0	0	0	0	0
C. pamphilus	1	0	0	0	0	0	0	0	0
C. tullia	1	1	0	0	0	0	0	0	0

1, major source of nutrition; 2, minor source of nutrition; 3, source reported by T. G. Shreeve only.
* For full species names, see Appendix 1a.
† Shrub/tree flower: includes *Buddleja davidii* and *Rubus fruiticosus* agg.
‡ Decaying plants: fallen fruit and ripe fruit and berries. Animal: carrion and dung. Mineral: mud, soil and burnt wood.
§ Herb/flower: use of honey-dew on leaves of hostplant, not flowers; all other entries in this column refer to flowers.

Appendix 4

STATISTICS ON LARVAL HOST USE AND ADULT FEEDING IN BRITISH BUTTERFLIES

Butterfly species*	Nectar sources (number)†	Estimated nectar feed mean‡	Nectar feed records	Non-nectar feed records§	Nectar feeding specialism¶	Adult feed specialism**	All larval hostplants	Principal larval hostplants	Core larval hostplants	Annual/biennial hostplants (%)	% nectar source species that are hostplants	% nectar records that are hostplants	Introduced host plants (%)	Introduced nectar sources (%)
C. palaemon	7	6.9	29	0	−0.681	−0.609	3	3	3	0	0	0	0(0)	0(0)
T. sylvestris	110	36.5	447	2	−0.171	−0.085	9	9	1	0	0	0	0(0)	38(55)
T. lineola	29	23.4	50	0	0.192	0.268	9	9	2	0	0	0	0(0)	9(31)
T. acteon	13	12.9	23	0	0.124	0.195	2	2	1	0	0	0	0(0)	0(0)
H. comma	14	12.7	31	0	−0.116	−0.043	2	2	1	0	0	0	0(0)	2(14)
O. sylvanus	90	42.0	222	0	0.229	0.315	9	9	3	0	0	0	0(0)	34(38)
E. tages	19	20.4	33	0	0.313	0.387	4	4	1	0	4.7	21.2	0(0)	0(0)
P. malvae	20	19.5	33	0	0.213	0.287	11	10	3	0	10.5	12.1	1(8)	0(0)
P. machaon	19	17.5	37	0	0.067	0.141	7	5	3	43	0	0	4(57)	2(11)
L. sinapis/ L. realis	33	27.1	95	0	−0.117	−0.036	8	8	5	25	12.5	26.3	3(37)	2(6)
C. crocea	47	32.0	82	1	0.316	0.383	12	12	3	40	10.3	12.7	2(17)	16(34)
G. rhamni	85	42.6	226	0	0.123	0.208	4	2	2	0	0	0	2(50)	40(47)
P. brassicae	179	46.1	639	1	0.048	0.139	25	22	2	60	3.5	3.7	14(56)	107(60)
P. rapae	232	51.4	927	5	0.055	0.145	26	20	3	62	3.7	3.5	12(48)	129(56)
P. napi	168	47.5	694	11	−0.084	−0.003	21	20	9	43	4.5	10.9	8(36)	77(46)
A. cardamines	60	34.5	197	0	−0.041	0.043	29	24	2	57	10.2	40.0	10(32)	24(40)
C. rubi	33	28.7	46	2	0.418	0.461	21	21	7	0	9.7	6.4	0(0)	6(18)
T. betulae	10	9.7	38	10	−0.563	−0.654	1	1	1	0	0	0	1(50)	1(10)
F. quercus	15	12.9	26	26	0.124	−0.409	4	2	2	0	0	0	2(50)	6(40)
S. w-album	35	21.8	127	17	−0.307	−0.326	4	3	2	0	0	0	1(25)	11(31)
S. pruni	8	7.0	28	7	−0.653	−0.759	1	1	1	0	0	0	1(50)	0(0)
L. phlaeas	104	41.7	351	1	0.023	0.112	5	5	2	0	0	0	0(0)	52(50)
L. dispar	4	4.0	4	0	0.353	0.413	1	1	1	0	0	0	0(0)	0(0)
C. minimus	4	4.0	15	1	−0.710	−0.642	2	2	1	50	25.0	40.0	1(50)	0(0)

A Resource-Based Habitat View for Conservation, 1st edition. By Roger L. H. Dennis. Published 2010 by Blackwell Publishing.

Butterfly species*	Nectar sources (number)†	Estimated nectar feed mean‡	Nectar feed records	Non-nectar feed records§	Nectar feeding specialism¶	Adult feed specialism**	All larval hostplants	Principal larval hostplants	Core larval hostplants	Annual/biennial hostplants (%)	% nectar source species that are hostplants	% nectar records that are hostplants	Introduced host plants (%)	Introduced nectar sources (%)
P. argus	27	25.5	45	0	0.315	0.390	16	15	9	13	22.2	37.8	0(0)	11(41)
A. agestis	26	23.9	36	0	0.418	0.492	6	6	3	60	4.0	2.8	0(0)	7(27)
A. artaxerxes	4	4.0	5	0	0.174	0.235	3	3	1	33	25.0	20	0(0)	0(0)
P. icarus	110	46.0	344	1	0.117	0.203	11	10	1	36	5.6	27.3	1(9)	45(41)
P. coridon	22	17.6	33	4	0.263	0.232	8	8	1	13	5.6	6.9	1(12)	4(18)
P. bellargus	17	13.8	25	4	0.198	0.172	2	2	1	0	7.1	8.7	0(0)	1(6)
C. argiolus	143	58.3	269	24	0.511	0.531	20	14	2	0	5.9	22.1	9(37)	85(59)
M. arion	4	4.0	13	1	−0.595	−0.587	2	1	1	0	25.0	69.2	0(0)	0(0)
H. lucina	11	10.0	22	1	−0.065	−0.032	4	4	2	0	0	0	0(0)	0(0)
L. camilla	11	9.1	50	22	−0.747	−0.967	1	1	1	0	10.0	2.0	0(0)	3(27)
A. iris	2	2.0	4	70	−0.340	−2.609	3	3	1	0	0	0	0(0)	2(100)
V. atalanta	121	32.7	739	44	−0.403	−0.359	5	5	1	20	0.9	0.2	1(20)	84(69)
V. cardui	105	38.3	484	0	−0.260	−0.169	28	27	5	52	8.2	14.9	5(19)	58(55)
A. urticae	216	42.4	1185	5	−0.267	−0.174	2	2	2	50	0	0	0(0)	127(59)
N. polychloros	10	10.0	16	3	0.154	0.085	18	17	2	0	40	37.5	4(22)	6(60)
I. io	156	39.6	705	4	−0.150	−0.062	3	3	1	33	0	0	0(0)	92(59)
P. c-album	110	36.2	439	40	−0.073	−0.057	8	8	3	0	0.9	1.0	2(25)	66(60)
B. selene	29	24.2	54	0	0.146	0.207	6	6	2	17	3.8	1.9	0(0)	2(7)
B. euphrosyne	19	16.6	47	0	−0.108	−0.033	7	7	2	14	5.6	2.2	0(0)	0(0)
A. adippe	20	15.2	67	0	−0.418	−0.341	5	5	1	0	0	0	0(0)	2(10)
A. aglaja	27	19.8	65	0	−0.074	0.004	6	6	3	14	0	0	0(0)	5(19)
A. paphia	31	19.6	111	5	−0.341	−0.293	3	3	1	0	0	0	0(0)	9(29)
E. aurinia	19	16.7	31	0	0.152	0.225	7	7	3	0	11.8	9.7	0(0)	0(0)
M. cinxia	10	10.0	14	0	0.261	0.329	6	6	2	33	0	0	0(0)	0(0)
M. athalia	25	21.9	41	1	0.230	0.285	10	10	3	60	13.0	9.8	0(0)	0(0)
P. aegeria	92	43.2	225	43	0.275	0.224	13	6	3	0	0	0	0(0)	53(58)
L. megera	90	48.0	158	0	0.441	0.524	10	10	5	0	0	0	0(0)	43(48)
E. epiphron	6	6.0	13	0	−0.190	−0.122	2	2	1	0	0	0	0(0)	0(0)
E. aethiops	20	17.9	55	1	−0.183	−0.121	6	6	2	33	0	0	0(0)	0(0)
M. galathea	34	23.6	100	0	−0.190	−0.109	6	6	2	14	0	0	0(0)	10(29)
H. semele	30	26.3	52	7	0.318	0.287	11	11	7	27	0	0	0(0)	8(27)
P. tithonus	129	50.4	297	0	0.395	0.482	11	9	6	10	0	0	0(0)	76(59)
M. jurtina	111	37.7	472	1	−0.147	−0.059	12	11	3	8	0	0	0(0)	45(41)
A. hyperantus	52	35.1	99	0	0.256	0.336	11	10	4	0	0	0	0(0)	19(37)
C. pamphilus	63	43.8	93	0	0.529	0.609	8	8	4	0	0	0	0(0)	2(4)
C. tullia	12	12.0	18	0	0.241	0.311	6	6	2	0	0	0	0(0)	0(0)

From Hardy *et al*. (2007), Kemp *et al*. (2008) and Hardy and Dennis (2008).

* For full species names, see Appendix 1a.

† The number of plants used to obtain nectar.

‡ Mean number of nectar sources per 100 records estimated from 100 random samples with replacement (Monte Carlo).

§ Nectar feed records and non nectar feed records sum to make up total number of adult feed records.

¶ Residuals from a regression analysis of the number of nectar sources on log (nectar) records.

** Residuals from a regression analysis of the number of nectar sources on log (adult feeding) records.

Appendix 5

UTILITY RESOURCES AND LIFE HISTORY DATA ON BRITISH BUTTERFLIES

APPENDIX 5a: ADULT ENVIRONMENT

	Roost sites and rest sites (in adverse weather)												Mate location sites†												Basking sites‡						
Butterfly species*	rock/soil	trunk/fence	short turf	grasses	short herb	tall herb	shrub	tree canopy	liana	on hostplant	off hostplant	manmade structures	rock or cliff	bare earth	short herbs	grass	tall herbs	shrubs	tree canopy	nectar site	hostplant site	physical edge site	light edge site	hilltop	rock/bare earth	short herb	grasses	tall herbs	shrub	tree canopy	manmade surface
C. palaemon	0	0	0	1	0	0	0	0	0	1	1	0	0	0	0	0	1	0	0	0	0	1	0	0	0	1	1	0	0	0	0
T. sylvestris	0	0	0	1	0	1	0	0	0	1	1	0	0	0	1	0	1	0	0	1	0	1	0	0	0	1	1	1	1	0	0
T. lineola	0	0	0	1	0	1	0	0	0	1	1	0	0	0	1	0	1	0	0	1	0	1	0	0	1	0	1	1	0	0	0
T. acteon	0	0	0	1	0	1	0	0	0	1	1	0	0	0	1	0	1	0	0	1	0	1	0	0	0	0	1	1	0	0	0
H. comma	0	0	0	1	0	0	1	0	0	0	1	0	0	0	1	0	0	0	0	1	0	1	0	0	1	1	1	0	0	0	0
O. sylvanus	0	0	0	1	0	1	1	0	0	1	1	0	0	0	1	1	1	0	0	1	1	1	0	0	1	1	1	1	1	0	0
E. tages	0	0	1	1	0	1	0	0	0	0	1	0	0	0	1	0	1	0	0	0	0	1	0	0	1	1	1	1	0	0	0
P. malvae	0	0	0	1	1	1	1	0	0	0	1	0	0	0	1	1	0	1	0	0	0	1	0	0	1	1	1	0	0	0	0
P. machaon	0	0	0	0	0	1	1	0	0	0	1	0	0	0	0	0	1	1	0	1	1	0	0	0	0	0	0	1	1	0	0
L. sinapis	0	0	0	0	0	1	0	0	0	1	1	0	0	0	1	1	1	0	0	1	1	1	0	0	0	0	1	1	0	0	0
C. crocea	0	0	1	1	0	1	1	0	0	1	1	0	0	0	1	0	0	0	0	1	1	0	0	0	1	1	0	0	0	0	0
G. rhamni	0	1	0	0	0	1	1	0	1	1	1	0	0	0	1	0	0	1	0	1	1	1	0	0	1	1	1	1	1	1	0
P. brassicae	0	0	0	0	0	1	1	0	0	1	1	0	0	0	0	0	1	0	0	1	1	0	0	0	0	1	0	1	1	0	0
P. rapae	0	0	0	0	0	1	1	0	0	1	1	0	0	1	1	0	1	1	0	1	1	0	0	0	1	1	1	1	1	0	1
P. napi	0	0	0	1	1	1	0	0	0	1	1	0	0	0	1	1	1	1	0	1	1	1	0	0	0	1	1	1	1	0	0
A. cardamines	0	0	0	0	1	1	0	0	0	1	1	0	0	0	1	1	1	1	0	1	1	1	0	0	0	1	1	1	1	0	0
C. rubi	0	0	0	0	0	0	1	0	0	1	1	0	0	0	0	0	1	1	0	0	0	1	0	0	1	1	1	1	1	0	0
T. betulae	0	0	0	0	0	0	1	0	0	1	0	0	0	0	0	0	0	1	1	0	1	1	0	0	0	0	0	1	1	1	0
F. quercus	0	0	0	0	0	0	1	1	0	1	0	0	0	0	0	0	0	0	1	0	1	1	0	0	0	0	1	1	0	1	0
S. w-album	0	0	0	0	0	0	1	1	0	1	0	0	0	0	0	0	0	1	1	0	1	1	1	0	0	0	0	1	1	1	0
S. pruni	0	0	0	0	0	0	1	1	0	1	0	0	0	0	0	0	0	1	0	0	1	1	0	0	0	0	0	1	1	0	0
L. phlaeas	0	0	1	1	1	1	1	0	0	1	1	0	0	1	1	1	1	0	0	0	1	1	0	0	1	1	1	1	1	0	1
L. dispar	0	0	0	0	0	1	0	0	0	0	1	0	0	0	0	0	1	0	0	1	0	0	0	0	0	0	1	1	0	0	0

A Resource-Based Habitat View for Conservation, 1st edition. By Roger L. H. Dennis. Published 2010 by Blackwell Publishing.

Butterfly species*	Roost sites and rest sites (in adverse weather)												Mate location sites†												Basking sites‡						
	rock/soil	trunk/fence	short turf	grasses	short herb	tall herb	shrub	tree canopy	liana	on hostplant	off hostplant	manmade structures	rock or cliff	bare earth	short herbs	grass	tall herbs	shrubs	tree canopy	nectar site	hostplant site	physical edge site	light edge site	hilltop	rock/bare earth	short herb	grasses	tall herbs	shrub	tree canopy	man-made surface
C. minimus	0	0	1	1	0	1	0	0	0	0	1	0	0	0	0	1	1	1	0	0	0	1	0	0	1	1	1	1	1	0	0
P. argus	1	0	1	1	1	1	1	0	0	0	1	0	1	1	1	0	0	1	0	0	1	1	0	0	1	1	1	0	0	0	0
A. agestis	0	0	0	1	0	1	0	0	0	0	1	0	0	0	1	0	0	0	0	0	0	1	0	0	1	1	1	1	0	0	0
A. artaxerxes	0	0	1	1	1	1	0	0	0	0	1	0	0	0	1	0	0	0	0	0	0	1	0	0	1	1	1	0	1	0	0
P. icarus	0	0	0	1	1	0	1	0	0	0	1	0	0	0	1	1	0	0	0	1	1	1	0	0	1	1	1	1	1	0	0
P. coridon	0	0	1	1	1	1	0	0	0	0	1	0	0	0	1	0	0	0	0	1	1	0	0	0	1	1	1	0	0	0	0
P. bellargus	0	0	1	1	0	1	0	0	0	0	1	0	0	0	1	1	0	0	0	1	1	0	0	0	1	1	1	0	0	0	0
C. argiolus	0	0	0	0	0	0	1	1	0	1	1	0	0	0	0	0	0	1	1	0	1	1	0	0	0	0	0	1	1	1	0
M. arion	0	0	0	1	0	1	1	0	0	0	1	0	0	1	1	0	0	1	0	1	0	0	0	0	1	1	0	0	1	0	0
H. lucina	0	0	0	0	0	1	1	0	0	0	1	0	0	0	1	0	1	1	0	0	0	1	0	0	1	1	1	1	1	0	0
L. camilla	0	0	0	0	0	0	0	1	1	1	1	0	0	0	0	0	0	1	1	0	0	0	1	0	1	0	0	1	1	1	1
A. iris	0	0	0	0	0	0	1	1	0	1	1	0	0	0	0	0	1	0	1	1	0	1	1	0	1	0	0	0	1	1	0
V. atalanta	0	1	0	0	1	1	1	0	0	1	1	0	0	0	0	0	1	1	0	1	0	1	1	1	1	1	1	1	1	1	1
V. cardui	1	0	0	0	0	1	1	0	0	1	1	0	1	1	0	0	1	0	0	1	1	1	0	1	1	0	1	1	1	0	1
A. urticae	0	1	0	0	0	1	1	0	0	1	1	1	0	1	0	1	1	0	0	0	1	1	0	1	1	1	1	1	1	0	1
N. polychloros	0	0	0	0	0	0	0	1	0	1	1	1	0	0	0	0	0	0	1	0	1	1	1	0	0	0	0	1	1	1	0
I. io	0	1	0	0	1	0	1	0	1	0	1	1	0	0	0	0	1	0	0	0	1	1	0	0	1	1	1	1	1	0	1
P. c-album	0	1	0	0	0	1	1	1	0	0	1	0	0	0	0	0	1	1	0	0	1	1	0	0	1	1	1	1	1	0	0
B. selene	0	0	0	0	0	1	1	0	0	0	1	0	0	0	1	0	1	1	0	1	1	0	0	0	0	1	1	1	0	0	0
B. euphrosyne	0	0	0	0	0	1	1	0	0	0	1	0	0	0	1	0	1	1	1	1	1	0	0	0	1	0	1	1	1	0	0
A. adippe	0	0	0	0	0	0	0	1	0	0	1	0	0	1	0	0	0	1	1	0	1	1	0	0	1	1	0	1	1	1	0
A. aglaja	0	0	0	1	0	1	0	0	0	0	1	0	0	0	1	0	1	1	0	1	1	0	0	0	0	1	0	1	1	0	0
A. paphia	0	0	0	0	0	0	0	1	0	0	1	0	0	0	0	0	0	1	1	0	1	1	1	0	0	1	1	1	1	1	0
E. aurinia	0	0	0	0	0	1	0	0	0	0	1	0	0	0	1	1	1	0	0	0	1	1	0	0	0	1	1	1	0	0	0
M. cinxia	0	0	0	1	0	1	0	0	0	0	1	0	0	0	1	0	1	0	0	1	1	0	0	0	0	1	1	0	0	0	0
M. athalia	0	0	0	1	0	0	0	0	0	0	1	0	0	0	1	0	1	0	0	0	1	0	0	0	0	1	1	1	1	0	0
P. aegeria	0	0	0	0	0	1	1	1	0	0	1	0	0	1	1	0	0	1	1	0	0	1	1	0	1	1	1	1	1	1	1
L. megera	1	1	0	1	0	1	1	0	0	0	1	1	1	1	0	0	1	0	0	0	1	1	0	1	1	1	1	0	0	0	1
E. epiphron	0	0	0	1	0	0	0	0	0	1	1	0	0	0	0	1	1	0	0	0	1	0	0	0	0	0	1	0	0	0	0
E. aethiops	0	0	0	1	0	0	0	0	0	1	1	0	0	0	0	1	1	0	0	0	1	0	0	0	0	0	1	1	1	0	0
M. galathea	0	0	0	1	0	1	0	0	0	1	1	0	0	0	1	1	1	0	0	1	1	0	0	0	0	0	1	1	1	0	0
H. semele	1	0	1	0	0	0	0	0	0	0	1	0	1	1	0	0	0	0	0	0	0	1	0	1	1	1	0	0	0	0	0
P. tithonus	0	0	0	1	0	1	1	0	0	1	1	0	0	0	1	1	1	1	0	0	1	1	0	0	0	1	1	1	1	0	0
M. jurtina	1	0	1	1	0	1	1	1	0	1	1	0	0	0	1	1	1	1	0	1	1	0	0	0	1	1	1	1	1	0	0
A. hyperantus	0	0	0	1	0	1	1	0	0	1	1	0	0	0	0	1	1	0	0	1	1	0	0	0	0	0	1	1	1	0	0
C. pamphilus	0	0	1	1	0	1	1	0	0	0	1	0	1	1	1	1	0	0	0	0	1	1	0	0	1	1	1	0	0	0	0
C. tullia	0	0	0	1	0	1	0	0	0	1	1	0	0	0	0	0	1	0	0	0	1	0	0	0	0	1	1	1	1	0	0

* For full species names, see Appendix 1a. 0, not recorded as being used by butterfly; 1, used by butterfly.

† Other possible categories are: base of hill, gully or depression, or south-facing aspect.

‡ Includes adult feeding sites; another possible category is micro-landforms.

APPENDIX 5b: EGG ENVIRONMENT

Butterfly species*	Egg-laying locality							Egg environment								Egg substrate																		
	bare earth or ground artifact	short turf/herbs	tall/mature herbs	shrub	tree trunk	canopy	liana	light	partial shade	shade	shelter/cover important	edge of hostplant patch	edge of non-hostplant	recess used	south-facing oviposition site	soil, rock or ground architecture	non-hostplant parts	roots	dead leaves of host	flower buds and flowers	flower pedicel	leaf bud or shoots	leaf petiole	leaf upperside	leaf underside	bract underside	flower sheath	suckers	stem/twig (nodes)	pod/berry	base of spine or stem/junction of twigs	tree trunk (bark)	young growth on old plants	dead plant or plant part
C. palaemon	0	1	0	0	0	0	0	1	1	0	0	0	0	0	0	0	0	0	0	0	0	0	0	1	1	0	0	0	0	0	0	0	0	0
T. sylvestris	0	0	1	0	0	0	0	1	0	0	0	0	0	0	0	0	0	0	0	0	0	0	0	0	0	0	1	0	0	0	0	0	0	0
T. lineola	0	0	1	0	0	0	0	1	0	0	0	0	0	0	0	0	0	0	0	0	0	0	0	0	0	0	1	0	0	0	0	0	0	0
T. acteon	0	0	1	0	0	0	0	1	0	0	0	0	0	0	0	0	0	0	0	0	0	0	0	0	0	0	1	0	0	0	0	0	0	1
H. comma	1	1	0	0	0	0	0	1	0	0	0	0	0	1	1	0	1	0	0	0	0	0	0	1	1	0	0	0	0	0	0	0	0	0
O. sylvanus	0	0	1	0	0	0	0	1	0	0	1	1	1	0	0	0	0	0	0	0	0	0	0	0	1	0	0	0	0	0	0	0	0	0
E. tages	0	1	0	0	0	0	0	1	0	0	0	0	0	0	0	0	0	0	0	0	0	0	1	1	0	0	0	0	0	0	0	0	0	0
P. malvae	0	1	0	0	0	0	0	1	0	0	0	0	0	0	0	0	0	0	0	0	0	0	0	0	1	0	0	1	0	0	0	0	0	0
P. machaon	0	0	1	0	0	0	0	1	0	0	1	?	1	0	0	0	0	0	0	0	0	0	0	1	1	0	0	0	0	0	0	0	1	0
L. sinapis	0	0	1	0	0	0	0	1	1	0	0	0	0	0	0	0	0	0	0	0	0	0	0	0	1	1	0	0	0	0	0	0	0	0
C. croceus	0	1	0	0	0	0	0	1	0	0	0	0	0	0	0	0	0	0	0	0	0	0	0	1	0	0	0	0	0	0	0	0	0	0
G. rhamni	0	0	0	1	0	0	0	1	1	0	0	0	1	0	1	0	0	0	0	0	0	1	0	0	1	0	0	0	1	0	0	0	0	0
P. brassicae	0	1	1	0	0	0	0	1	1	0	0	0	0	0	0	0	0	0	0	0	0	0	0	1	1	0	0	0	0	0	0	0	0	0
P. rapae	0	1	1	0	0	0	0	1	1	0	1	0	1	0	0	0	0	0	0	0	0	0	0	0	1	0	0	0	0	0	0	0	0	0
P. napi	0	1	1	0	0	0	0	1	1	1	1	0	0	0	0	0	0	0	0	0	0	0	0	0	1	0	0	0	0	0	0	0	0	0
A. cardamines	0	1	1	0	0	0	0	1	1	0	0	1	1	0	0	0	0	0	0	1	1	0	1	0	0	0	0	0	0	0	0	0	1	0
C. rubi	0	1	1	1	0	0	0	1	0	0	0	0	0	0	0	0	0	0	0	1	0	1	1	0	0	0	0	0	0	0	0	0	1	0
T. betulae	0	0	0	1	0	0	0	1	0	0	0	1	1	0	1	0	0	0	0	0	0	0	0	0	0	0	0	0	1	0	1	0	0	0
F. quercus	0	0	0	0	0	1	0	1	0	0	1	0	0	0	0	0	0	0	0	0	0	1	0	0	0	0	0	0	1	0	0	0	0	0
S. w-album	0	0	0	1	0	1	0	1	1	0	0	0	0	0	0	0	0	0	0	0	0	1	0	0	0	0	0	0	1	0	1	0	0	0
S. pruni	0	0	0	1	0	0	0	1	1	0	1	1	1	0	1	0	0	0	0	0	0	0	0	0	0	0	0	0	1	0	0	0	0	0
L. phlaeas	0	1	1	0	0	0	0	1	1	0	0	0	0	0	0	0	0	0	0	0	0	0	0	0	1	0	0	0	0	0	0	0	0	0
L. dispar	0	0	1	0	0	0	0	1	0	0	0	0	0	0	0	0	0	0	0	0	0	0	0	0	0	0	0	0	0	0	0	0	0	0
C. minimus	0	1	1	0	0	0	0	1	0	0	0	0	0	0	0	0	0	0	0	1	0	0	0	0	0	0	0	0	0	0	0	0	0	0
P. argus	1	1	0	0	0	0	0	1	0	0	1	1	1	1	0	1	1	0	0	0	0	0	0	0	1	0	0	0	0	0	0	0	0	0
A. agestis	0	1	0	0	0	0	0	1	0	0	0	0	0	0	0	0	0	0	0	0	0	0	0	0	1	0	0	0	0	0	0	0	0	0
A. artaxerxes	0	1	0	0	0	0	0	1	0	0	0	0	0	0	1	0	0	0	0	0	0	0	0	1	0	0	0	0	0	0	0	0	0	0
P. icarus	0	1	0	0	0	0	0	1	0	0	1	1	1	0	0	1	1	0	0	0	0	0	1	1	0	0	0	0	0	0	0	0	0	0
P. coridon	1	1	0	0	0	0	0	1	0	0	0	0	0	0	0	1	1	0	0	0	0	0	0	0	0	0	0	0	1	0	0	0	0	0
P. bellargus	1	1	0	0	0	0	0	1	0	0	0	0	0	1	1	0	0	0	0	0	0	0	0	0	1	0	0	0	0	0	0	0	0	0
C. argiolus	0	0	0	1	0	0	1	0	1	0	0	0	0	0	0	0	0	0	0	1	1	0	0	0	0	0	0	0	0	0	0	0	0	0
M. arion	0	1	0	0	0	0	0	1	0	0	0	0	0	0	0	0	0	0	0	1	0	0	0	0	0	0	0	0	0	0	0	0	0	0
H. lucina	0	1	0	0	0	0	0	1	0	0	0	0	0	0	0	0	1	0	0	0	0	0	0	0	1	0	0	0	0	0	0	0	0	0
L. camilla	0	0	0	0	0	0	1	0	1	1	0	0	1	0	0	0	0	0	0	0	0	0	0	1	0	0	0	0	0	0	0	0	0	0
A. iris	0	0	0	1	0	0	0	0	1	1	0	0	0	0	0	0	0	0	0	0	0	0	0	1	0	0	0	0	0	0	0	0	0	0
V. atalanta	0	0	1	0	0	0	0	1	0	0	0	0	0	0	0	0	0	0	0	0	0	0	0	1	0	0	0	0	0	0	0	0	0	0
V. cardui	0	0	1	0	0	0	0	1	0	0	0	0	0	0	0	0	0	0	0	0	0	0	0	1	0	0	0	0	0	0	0	0	0	0
A. urticae	0	1	1	0	0	0	0	1	0	0	0	1	0	1	1	0	0	0	0	0	0	0	0	0	1	0	0	0	0	0	0	0	0	0
N. polychloros	0	0	1	0	0	1	0	0	1	0	0	0	0	0	0	0	0	0	0	0	0	0	0	0	0	0	0	0	1	0	0	0	0	0
I. io	0	0	1	0	0	0	0	1	0	0	1	1	1	1	0	0	0	0	0	0	0	0	0	0	1	0	0	0	0	0	0	0	0	0
P. c-album	0	0	1	0	0	0	1	0	1	0	?	0	1	0	0	0	0	0	0	0	0	0	0	1	0	0	0	0	0	0	0	0	0	0
B. selene	1	1	0	0	0	0	0	1	0	0	0	0	0	0	0	1	1	0	0	0	0	0	0	1	0	0	0	0	0	0	0	0	0	0
B. euphrosyne	1	1	0	0	0	0	0	1	1	0	0	0	0	0	0	1	1	0	0	0	0	0	0	1	1	0	0	0	0	0	0	0	0	0
A. adippe	1	1	0	0	0	0	0	1	1	0	0	0	0	0	0	1	1	0	0	0	0	0	0	0	0	0	0	0	0	0	0	0	0	0

	Egg-laying locality							Egg environment								Egg substrate																		
Butterfly species*	bare earth or ground artifact	short turf/herbs	tall/mature herbs	shrub	tree trunk	canopy	liana	light	partial shade	shade	shelter/cover important	edge of hostplant patch	edge of non-hostplant	recess used	south-facing oviposition site	soil, rock or ground architecture	non-hostplant parts	roots	dead leaves of host	flower buds and flowers	flower pedicel	leaf bud or shoots	leaf petiole	leaf upperside	leaf underside	bract underside	flower sheath	suckers	stem/twig (nodes)	pod/berry	base of spine or stem/junction of twigs	tree trunk (bark)	young growth on old plants	dead plant or plant part
A. aglaja	1	1	0	0	0	0	0	1	0	0	0	0	0	0	0	1	1	0	0	0	0	0	0	0	0	0	0	0	0	0	0	0	0	0
A. paphia	0	0	0	0	1	0	0	0	1	1	0	0	0	0	0	1	0	0	0	0	0	0	0	0	0	0	0	0	0	0	0	1	0	0
E. aurinia	0	0	1	0	0	0	0	1	0	0	0	0	0	0	0	0	0	0	0	0	0	0	0	0	1	0	0	0	0	0	0	0	0	0
M. cinxia	0	1	0	0	0	0	0	1	0	0	0	0	0	0	0	0	0	0	0	0	0	0	0	0	1	0	0	0	0	0	0	0	0	0
M. athalia	0	1	0	0	0	0	0	1	1	0	0	0	0	0	0	0	1	0	0	0	0	0	0	0	1	0	0	0	0	0	0	0	0	0
P. aegeria	0	1	0	0	0	0	0	1	1	1	0	0	0	0	0	0	0	0	0	0	0	0	0	0	1	0	0	0	0	0	0	0	0	0
L. megera	0	1	0	0	0	0	0	1	1	0	1	1	1	1	0	1	0	1	0	0	0	0	0	1	1	0	0	0	0	0	0	0	0	0
E. epiphron	0	1	0	0	0	0	0	1	0	0	0	0	0	0	0	0	0	0	0	0	0	0	0	1	1	0	0	0	0	0	0	0	0	0
E. aethiops	0	1	0	0	0	0	0	1	0	0	1	0	0	0	0	1	1	0	0	0	0	0	0	1	1	0	0	0	0	0	0	0	0	0
M. galathea	1	1	1	0	0	0	0	1	0	0	0	0	0	0	0	1	1	0	0	0	0	0	0	0	0	0	0	0	0	0	0	0	0	0
H. semele	1	1	0	0	0	0	0	1	0	0	0	1	0	1	0	1	0	0	0	0	0	0	0	1	1	0	0	0	0	0	0	0	0	0
P. tithonus	0	1	0	0	0	0	0	0	1	0	0	0	1	0	0	1	1	0	0	0	0	0	0	1	1	0	0	0	0	0	0	0	0	0
M. jurtina	0	1	0	0	1†	0	0	1	0	0	0	0	0	0	0	1	1	0	0	0	0	0	0	1	1	0	0	0	0	0	0	1	0	0
A. hyperantus	0	1	0	0	0	0	0	0	1	0	0	0	0	0	0	1	1	0	0	0	0	0	0	0	0	0	0	0	0	0	0	0	0	0
C. pamphilus	0	1	0	0	0	0	0	1	0	0	0	0	0	0	0	1	1	0	0	0	0	0	0	1	1	0	0	0	0	0	0	0	0	0
C. tullia	0	1	0	0	0	0	0	0	1	0	0	1	0	?	0	0	0	0	1	0	0	0	0	0	0	0	0	0	0	0	0	0	0	0

* For full species names, see Appendix 1a. 0, not recorded as being used by butterfly; 1, used by butterfly.
† From Dennis (2003).

APPENDIX 5c: LARVAL ENVIRONMENT

	Hostplant growth form occupied†						Hostplant part used				Plant maturity		Hostplant patchiness			Larval zone occupied‡					
Butterfly species*	**short herb/grass**	**tall herb**	**shrub**	**tree**	**non-plant**	**liana**	**flowers/pod**	**leaf**	**bud**	**stem**	**small/immature**	**old**	**large patch**	**medium patch**	**small/single**	**buried**	**groundlayer**	**fieldlayer**	**shrub layer**	**canopy layer**	**attended**
C. palaemon	1	0	0	0	0	0	0	1	0	0	1	1	1	1	0	0	0	1	0	0	0
T. sylvestris	1	0	0	0	0	0	0	1	0	0	0	1	1	1	0	0	0	1	0	0	0
T. lineola	1	0	0	0	0	0	0	1	0	0	0	1	1	1	0	0	0	1	0	0	0
T. acteon	1	0	0	0	0	0	0	1	0	0	0	1	1	1	0	0	0	1	0	0	0
H. comma	1	0	0	0	0	0	0	1	0	0	1	1	1	1	0	0	1	0	0	0	0
O. sylvanus	1	0	0	0	0	0	0	1	0	0	0	1	1	1	0	0	0	1	0	0	0
E. tages	1	0	0	0	0	0	0	1	0	0	1	1	0	1	1	0	1	1	0	0	0
P. malvae	1	0	0	0	0	0	0	1	0	0	1	1	0	1	1	0	1	1	0	0	0
P. machaon	0	1	0	0	0	0	1	1	1	1	0	1	0	1	1	0	0	1	0	0	0
L. sinapis	0	1	0	0	0	1	0	1	0	0	1	1	0	1	1	0	0	1	0	0	0
C. crocea	0	1	0	0	0	0	0	1	0	0	1	1	1	1	1	0	0	1	0	0	0
G. rhamni	0	0	1	0	0	0	0	1	1	0	0	1	0	1	1	0	0	0	1	0	0
P. brassicae	0	1	0	0	0	0	0	1	0	0	0	1	1	1	0	0	1	1	0	0	0
P. rapae	1	1	0	0	0	0	0	1	0	0	1	1	1	1	1	0	1	1	0	0	0
P. napi	1	1	0	0	0	0	0	1	0	0	1	1	0	1	1	0	1	1	0	0	0
A. cardamines	1	1	0	0	0	0	1	0	0	0	0	1	1	1	1	0	0	1	0	0	0
C. rubi	1	1	1	0	0	0	1	1	1	0	1	1	1	1	1	0	1	1	1	0	0
T. betulae	0	0	1	0	0	0	0	1	1	0	0	1	1	1	0	0	0	0	1	0	1
F. quercus	0	0	0	1	0	0	1	1	1	0	0	1	1	1	0	0	0	0	0	1	0
S. w-album	0	0	1	1	0	0	1	1	1	0	0	1	1	1	0	0	0	0	1	1	1
S. pruni	0	0	1	0	0	0	1	1	1	0	0	1	1	1	0	0	0	0	1	0	0
L. phlaeas	1	1	0	0	0	0	0	1	0	0	1	1	0	1	1	0	1	0	0	0	0
L. dispar	0	1	0	0	0	0	0	1	0	0	0	1	0	1	1	0	1	1	0	0	1
C. minimus	0	1	0	0	0	0	1	0	0	0	0	1	0	1	1	0	0	1	0	0	1
P. argus	1	0	1	0	0	0	1	1	1	0	1	1	1	1	1	0	1	1	0	0	1
A. agestis	1	0	0	0	0	0	0	1	0	0	1	1	0	1	1	0	1	1	0	0	1
A. artaxerxes	1	0	0	0	0	0	0	1	0	0	1	1	0	1	1	0	1	1	0	0	1
P. icarus	1	0	0	0	0	0	0	1	0	0	1	1	1	1	1	0	1	1	0	0	1
P. coridon	1	0	0	0	0	0	0	1	0	0	1	1	1	1	1	0	1	0	0	0	1
P. bellargus	1	0	0	0	0	0	0	1	0	0	1	1	1	1	1	0	1	0	0	0	1
C. argiolus	0	1	1	1	0	1	1	1	0	0	0	1	1	1	1	0	0	0	1	1	1
M. arion	1	0	0	0	1	0	1	0	0	0	0	1	1	1	0	1	1	1	0	0	1
H. lucina	1	0	0	0	0	0	0	1	0	0	0	1	1	1	0	0	1	1	0	0	0
L. camilla	0	0	0	0	0	1	0	1	0	0	0	1	0	0	1	0	0	0	1	1	0
A. iris	0	0	1	1	0	0	0	1	0	0	0	1	0	1	1	0	0	0	1	0	0
V. atalanta	0	1	0	0	0	0	0	1	0	0	1	1	1	1	0	0	0	1	0	0	0
V. cardui	0	1	0	0	0	0	0	1	0	0	0	1	1	1	1	0	0	1	0	0	0
A. urticae	0	1	0	0	0	0	0	1	0	0	1	1	1	1	0	0	0	1	0	0	0

Butterfly species*	Hostplant growth form occupied†						Hostplant part used				Plant maturity		Hostplant patchiness			Larval zone occupied‡					
	short herb/grass	tall herb	shrub	tree	non-plant	liana	flowers/pod	leaf	bud	stem	small/immature	old	large patch	medium patch	small/single	buried	groundlayer	fieldlayer	shrub layer	canopy layer	attended
N. polychloros	0	0	1	1	0	0	0	1	0	0	0	1	1	1	0	0	0	0	1	1	0
I. io	0	1	0	0	0	0	0	1	0	0	1	1	1	1	0	0	0	1	0	0	0
P. c-album	0	1	1	1	0	0	0	1	0	0	0	1	0	1	1	0	0	1	1	0	0
B. selene	1	0	0	0	0	0	0	1	0	0	1	1	1	1	0	0	1	1	0	0	0
B. euphrosyne	1	0	0	0	0	0	0	1	0	0	1	1	1	1	0	0	1	1	0	0	0
A. adippe	1	0	0	0	0	0	0	1	0	0	1	1	0	1	1	0	1	1	0	0	0
A. aglaja	1	0	0	0	0	0	0	1	0	0	1	1	1	1	0	0	1	1	0	0	0
A. paphia	1	0	0	0	0	0	0	1	0	0	1	1	0	1	1	0	1	1	0	0	0
E. aurinia	1	1	0	0	0	0	0	1	0	0	0	1	1	1	0	0	1	1	0	0	0
M. cinxia	1	1	0	0	0	0	0	1	0	0	0	1	1	1	0	0	1	0	0	0	0
M. athalia	1	1	0	0	0	0	0	1	0	0	0	1	1	1	0	0	1	0	0	0	0
P. aegeria	1	0	0	0	0	0	0	1	0	0	1	1	0	1	1	0	0	1	0	0	0
L. megera	1	0	0	0	0	0	0	1	0	0	1	1	1	1	1	0	0	1	0	0	0
E. epiphron	1	0	0	0	0	0	0	1	0	0	1	1	1	1	0	0	0	1	0	0	0
E. aethiops	1	0	0	0	0	0	0	1	0	0	1	1	1	1	0	0	0	1	0	0	0
M. galathea	1	0	0	0	0	0	0	1	0	0	1	1	1	1	0	0	0	1	0	0	0
H. semele	1	0	0	0	0	0	0	1	0	0	1	1	1	1	1	0	1	1	0	0	0
P. tithonus	1	0	0	0	0	0	0	1	0	0	1	1	1	1	0	0	0	1	0	0	0
M. jurtina	1	0	0	0	0	0	0	1	0	0	1	1	1	1	0	0	1	1	0	0	0
A. hyperantus	1	0	0	0	0	0	0	1	0	0	1	1	1	1	0	0	0	1	0	0	0
C. pamphilus	1	0	0	0	0	0	0	1	0	0	1	1	1	1	1	0	1	0	0	0	0
C. tullia	1	0	0	0	0	0	0	1	0	0	1	1	1	1	1	0	0	1	0	0	0

* For full species names, see Appendix 1a. 0, not recorded as being used by butterfly; 1, used by butterfly.
† Other possible category is grass tussock/tuft.
‡ Other possible categorgies: shelter and south-facing; attended refers to ant attendance.

APPENDIX 5d: PUPAL ENVIRONMENT

Butterfly species*	Pupal zone occupied†					
	buried	ground layer	field layer	shrub layer	canopy layer	attended
C. palaemon	0	1	1	0	0	0
T. sylvestris	0	1	0	0	0	0
T. lineola	0	1	0	0	0	0
T. acteon	0	1	0	0	0	0
H. comma	0	1	0	0	0	0
O. sylvanus	0	0	1	0	0	0
E. tages	0	1	1	0	0	0
P. malvae	0	1	1	0	0	0
P. machaon	0	0	1	0	0	0
L. sinapis	0	0	1	0	0	0
C. crocea	0	0	1	0	0	0
G. rhamni	0	0	1	1	0	0
P. brassicae	0	0	1	1	0	0
P. rapae	0	0	1	0	0	0
P. napi	0	0	1	0	0	0
A. cardamines	0	0	1	0	0	0
C. rubi	1	1	0	0	0	1
T. betulae	1	1	0	0	0	1
F. quercus	1	1	0	0	0	1
S. w-album	0	0	0	1	1	0
S. pruni	0	0	0	1	0	0
L. phlaeas	0	1	1	0	0	0
L. dispar	0	0	1	0	0	0
C. minimus	1	1	0	0	0	0
P. argus	1	0	0	0	0	1
A. agestis	1	1	0	0	0	1
A. artaxerxes	1	1	0	0	0	0
P. icarus	1	1	0	0	0	1
P. coridon	1	1	0	0	0	1
P. bellargus	1	1	0	0	0	1

Butterfly species*	Pupal zone occupied†					
	buried	ground layer	field layer	shrub layer	canopy layer	attended
C. argiolus	0	1	0	1	0	0
M. arion	1	0	0	0	0	1
H. lucina	0	1	0	0	0	0
L. camilla	0	0	0	1	0	0
A. iris	0	0	0	1	0	0
V. atalanta	0	0	1	0	0	0
V. cardui	0	0	1	0	0	0
A. urticae	0	0	1	0	0	0
N. polychloros	0	0	0	1	0	0
I. io	0	0	1	0	0	0
P. c-album	0	0	1	0	0	0
B. selene	0	1	0	0	0	0
B. euphrosyne	0	1	0	0	0	0
A. adippe	0	1	0	0	0	0
A. aglaja	0	1	0	0	0	0
A. paphia	0	1	0	0	0	0
E. aurinia	0	1	1	0	0	0
M. cinxia	0	1	0	0	0	0
M. athalia	0	1	0	0	0	0
P. aegeria	0	0	1	0	0	0
L. megera	0	0	1	0	0	0
E. epiphron	0	1	1	0	0	0
E. aethiops	1	1	0	0	0	0
M. galathea	0	1	0	0	0	0
H. semele	1	0	0	0	0	0
P. tithonus	0	0	1	0	0	0
M. jurtina	0	1	1	0	0	0
A. hyperantus	0	1	1	0	0	0
C. pamphilus	0	1	1	0	0	0
C. tullia	0	1	1	0	0	0

* For full species names, see Appendix 1a. 0, not recorded as being used by butterfly; 1, used by butterfly.
† Other possible category is on hostplant; attended refers to ant attendance.

APPENDIX 5e: LIFE HISTORY

Butterfly species*	Hibernation site							Overwintering phase				Symbionts (ants)			Voltinism (generations per year)†						Adult appearance‡	Mobility§
	buried	surface	short sward	tall sward	shrub	tree	liana	egg	larva	pupa	adult	mono	oligo	poly	biennial	univoltine	uni+partial	bivoltine	multivoltine	maximum voltinism score	flight period (days)	movement score
C. palaemon	0	0	0	1	0	0	0	0	1	0	0	0	0	0	0	1	0	0	0	1	21.2	1
T. sylvestris	0	0	0	1	0	0	0	1	0	0	0	0	0	0	0	1	0	0	0	1	35.3	5
T. lineola	0	0	0	1	0	0	0	0	1	0	0	0	0	0	0	1	0	0	0	1	18.8	4
T. acteon	0	0	0	1	0	0	0	0	1	0	0	0	0	0	0	1	0	0	0	1	37.7	1
H. comma	0	1	1	0	0	0	0	1	0	0	0	0	0	0	0	1	0	0	0	1	21.2	2
O. sylvanus	0	0	0	1	0	0	0	0	1	0	0	0	0	0	0	1	0	0	0	1	35.1	5
E. tages	0	0	1	0	0	0	0	0	1	0	0	0	0	0	0	0	1	0	0	2	25.9	2
P. malvae	0	0	1	0	0	0	0	0	0	1	0	0	0	0	0	0	1	0	0	2	24.5	2
P. machaon	0	0	0	1	0	0	0	0	0	1	0	0	0	0	0	0	1	0	0	2	52.5	5
L. sinapis	0	0	0	1	0	0	0	0	0	1	0	0	0	0	0	0	1	0	0	2	29.6	2
C. crocea	0	0	0	1	0	0	0	0	0	1	0	0	0	0	0	0	0	0	1	4	26.6	8
G. rhamni	0	0	0	0	1	1	1	0	0	0	1	0	0	0	0	1	0	0	0	1	80.1	6
P. brassicae	0	0	0	1	1	0	0	0	0	1	0	0	0	0	0	1	0	1	1	4	69.3	8
P. rapae	0	0	0	1	1	0	0	0	0	1	0	0	0	0	0	0	0	1	1	4	79.7	8
P. napi	0	0	0	1	0	0	0	0	0	1	0	0	0	0	0	0	1	1	1	4	92.2	7
A. cardamines	0	0	1	1	0	0	0	0	0	1	0	0	0	0	0	1	1	0	0	2	25.1	5
C. rubi	1	1	0	0	0	0	0	0	0	1	0	0	0	1	0	1	0	0	0	1	19.8	3
T. betulae	0	0	0	0	1	0	0	1	0	0	0	0	0	1	0	1	0	0	0	1	9.9	1
F. quercus	0	0	0	0	0	1	0	1	0	0	0	0	0	1	0	1	0	0	0	1	12.7	3
S. w-album	0	0	0	0	1	1	0	1	0	0	0	0	0	1	0	1	0	0	0	1	10.6	2
S. pruni	0	0	0	0	1	0	0	1	0	0	0	0	0	1	0	1	0	0	0	1	4.9	1
L. phlaeas	0	0	1	1	0	0	0	0	1	0	0	0	0	0	0	1	1	1	1	4	83.3	6
L. dispar	0	1	1	0	0	0	0	0	1	0	0	0	0	0	0	1	0	0	0	1	12.4	1
C. minimus	1	1	0	0	0	0	0	0	1	0	0	0	0	1	0	1	0	1	0	3	36.2	2
P. argus	0	1	1	0	0	0	0	1	0	0	0	0	1	0	0	1	0	0	0	1	38.4	2
A. agestis	0	1	1	0	0	0	0	0	1	0	0	0	0	1	0	1	0	1	0	3	46.6	2
A. artaxerxes	0	1	1	0	0	0	0	0	1	0	0	0	0	1	0	1	0	0	0	1	37.9	2
P. icarus	0	1	0	0	0	0	0	0	1	0	0	0	0	1	0	1	1	1	1	4	55.0	6
P. coridon	0	1	0	0	0	0	0	1	0	0	0	0	0	1	0	1	0	0	0	1	40.2	2
P. bellargus	0	1	1	0	0	0	0	0	1	0	0	0	0	1	0	0	0	1	0	2	48.8	1
C. argiolus	1	0	0	0	0	0	0	0	0	1	0	0	0	1	0	1	0	1	0	3	33.4	6
M. arion	1	0	0	0	0	0	0	0	1	0	0	1	0	0	0	1	0	0	0	1	35.5	1
H. lucina	0	0	0	1	0	0	0	0	0	1	0	0	0	1	0	1	0	0	0	1	12.7	1
L. camilla	0	0	0	0	0	0	1	0	1	0	0	0	0	0	0	1	0	0	0	1	19.0	5
A. iris	0	0	0	0	1	0	0	0	1	0	0	0	0	0	0	1	0	0	0	1	4.9	3
V. atalanta	0	0	0	0	1	1	1	0	0	0	1	0	0	0	0	0	1	1	0	3	64.9	8
V. cardui	0	0	0	1	1	1	0	0	0	0	1	0	0	0	0	0	0	0	1	4	39.0	9
A. urticae	0	0	0	1	1	1	1	0	0	0	1	0	0	0	0	1	1	1	1	4	108.8	8
N. polychloros	0	0	0	0	1	1	1	0	0	0	1	0	0	0	0	1	0	0	0	1	?	6
I. io	0	0	0	1	1	1	1	0	0	0	1	0	0	0	0	1	1	0	0	2	68.7	8

Butterfly species*	Hibernation site							Overwintering phase				Symbionts (ants)			Voltinism (generations per year)†						Adult appearance‡	Mobility§
	buried	surface	short sward	tall sward	shrub	tree	liana	egg	larva	pupa	adult	mono	oligo	poly	biennial	univoltine	uni+partial	bivoltine	multivoltine	maximum voltinism score	flight period (days)	movement score
P. c-album	0	0	0	1	1	1	1	0	0	0	1	0	0	0	0	1	1	1	0	3	88.6	6
B. selene	0	1	1	0	0	0	0	0	1	0	0	0	0	0	0	0	1	0	0	2	23.7	3
B. euphrosyne	0	1	0	0	0	0	0	0	1	0	0	0	0	0	0	0	1	0	0	2	20.4	2
A. adippe	0	1	0	0	0	0	0	1	0	0	0	0	0	0	0	1	0	0	0	1	26.6	3
A. aglaja	0	1	0	0	0	0	0	0	1	0	0	0	0	0	0	1	0	0	0	1	27.0	5
A. paphia	0	0	0	0	1	0	0	0	1	0	0	0	0	0	0	1	0	0	0	1	30.4	6
E. aurinia	0	0	0	1	0	0	0	0	1	0	0	0	0	0	0	1	0	0	0	1	13.5	2
M. cinxia	0	1	0	1	0	0	0	0	1	0	0	0	0	0	0	1	0	0	0	1	29.6	1
M. athalia	0	1	1	0	0	0	0	0	1	0	0	0	0	0	0	0	1	0	0	2	33.4	1
P. aegeria	0	0	0	1	0	0	0	0	1	1	0	0	0	0	0	1	1	1	1	4	121.3	5
L. megera	0	0	0	1	0	0	0	0	1	0	0	0	0	0	0	0	0	1	1	4	33.9	5
E. epiphron	0	0	0	1	0	0	0	0	1	0	0	0	0	0	1	1	0	0	0	1	17.7	1
E. aethiops	0	0	0	1	0	0	0	0	1	0	0	0	0	0	0	1	0	0	0	1	30.7	2
M. galathea	0	0	0	1	0	0	0	0	1	0	0	0	0	0	0	1	0	0	0	1	28.7	2
H. semele	0	1	1	0	0	0	0	0	1	0	0	0	0	0	0	1	0	0	0	1	33.7	3
P. tithonus	0	0	1	1	0	0	0	0	1	0	0	0	0	0	0	1	0	0	0	1	40.8	4
M. jurtina	0	0	1	1	0	0	0	0	1	0	0	0	0	0	0	1	0	0	0	1	63.2	6
A. hyperantus	0	0	0	1	0	0	0	0	1	0	0	0	0	0	0	1	0	0	0	1	32.1	3
C. pamphilus	0	1	1	0	0	0	0	0	1	0	0	0	0	0	0	1	1	1	1	4	74.1	4
C. tullia	0	0	0	1	0	0	0	0	1	0	0	0	0	0	0	1	0	0	0	1	20.1	1

* For full species names, see Appendix 1a. See Emmet and Heath (1991) for life history data on all British Lepidoptera. 0, not recorded as being used by butterfly; 1, used by butterfly.

† Voltinism varies over the British Isles.

‡ Last record minus first record but with correction for multibrooded species. (Butterfly Monitoring Scheme data, courtesy of David Roy and the Centre for Ecology and Hydrology. *Maculinea arion* data courtesy of Jeremy A. Thomas.)

§ Mobility score (Dennis, 1993a) is the sum of ranks for eight attributes, with 1 the least movement and 8 the most movement. See Cook *et al*. (2001) for a test of the movement scores, and Dennis *et al*. (2005) for an alternative 9 point score system.

Appendix 6

ADULT AND LARVAL BEHAVIOUR IN BRITISH BUTTERFLIES

APPENDIX 6a: ADULT BEHAVIOUR*

	Mate location method			Roosting mode		Basking method				Egg-laying mode			Egg batch details	
Butterfly species†	**perching**	**territorial perching**	**lecking**	**exposed**	**communal**	**dorsalabsorb**	**doesalreflect**	**lateral**	**wing and body appression**	**single**	**small batch**	**large batch**	**egg batch size (max)**	**egg aggregation occurs**
C. palaemon	1	1	0	1	0	1	0	0	0	1	0	0	1	?
T. sylvestris	1	1	0	1	0	1	0	0	0	0	1	0	3 to 8	33
T. lineola	1	1	0	0	0	1	0	0	0	0	1	0	4 to 5	14
T. acteon	1	1	0	1	0	1	0	0	0	0	1	0	5 to 15	15
H. comma	1	0	0	1	0	1	0	0	0	1	0	0	1	Yes
O. sylvanus	1	1	0	1	0	1	0	0	0	1	0	0	1	Probably
E. tages	1	1	0	1	0	1	0	0	1‡	1	0	0	1	
P. malvae	1	1	0	1	0	1	0	0	0	1	0	0	1	
P. machaon	0	0	0	1	0	1	0	0	0	1	0	0	1	Probably
L. sinapis	0	0	0	1	0	0	0	1	0	1	0	0	1	
C. crocea	0	0	0	1	0	0	0	1	0	1	0	0	1	
G. rhamni	0	0	0	1	0	0	0	1	0	1	0	0	1	Yes
P. brassicae	0	0	0	1	0	0	1	0	0	0	0	1	40	Yes
P. rapae	0	0	0	1	0	0	1	0	0	1	0	0	1	Probably
P. napi	0	0	0	1	0	0	1	0	0	1	0	0	1	
A. cardamines	0	0	0	1	0	0	1	0	0	1	0	0	1	Yes
C. rubi	1	1	0	1	0	0	0	1	0	1	0	0	1	Yes
T. betulae	0	1	1	1	0	0	1	0	0	1	0	0	1	Yes
F. quercus	0	1	1	1	0	0	1	0	0	1	0	0	1 to 2	Probably
S. w-album	0	1	1	1	0	0	0	1	0	1	0	0	1	Yes
S. pruni	0	1	1	1	0	0	0	1	0	1	0	0	1	Probably
L. phlaeas	1	1	0	1	0	0	1	0	0	1	0	0	1	20
L. dispar	1	0	0	1	0	0	1	0	0	1	1	0	1	
C. minimus	1	0	0	1	1	0	1	0	0	1	0	0	1	Yes

A Resource-Based Habitat View for Conservation, 1st edition. By Roger L. H. Dennis. Published 2010 by Blackwell Publishing,

Butterfly species†	Mate location method			Roosting mode		Basking method				Egg-laying mode			Egg batch details	
	perching	territorial perching	lecking	exposed	communal	dorsalabsorb	doesalreflect	lateral	wing and body appression	single	small batch	large batch	egg batch size (max)	egg aggregation occurs
P. argus	1	0	0	1	1	0	1	0	0	1	0	0	1	Yes
A. agestis	1	0	0	1	1	1	0	0	0	1	0	0	1	
A. artaxerxes	0	0	0	1	1	1	0	0	0	1	0	0	1	
P. icarus	1	0	0	1	1	1	1	0	0	1	0	0	1	Yes
P. coridon	0	0	0	1	1	0	1	0	0	1	0	0	1	
P. bellargus	0	0	0	1	1	0	1	0	0	1	0	0	1	
C. argiolus	1	0	0	1	0	0	0	0	0	1	0	0	1	
M. arion	0	0	0	0	0	0	1	0	0	1	0	0	1	Yes
H. lucina	0	1	0	1	0	1	0	0	0	0	1	0	1 to 8	Yes num?
L. camilla	0	0	0	1	0	1	0	0	0	1	0	0	1	Yes
A. iris	0	1	0	1	0	1	0	0	1	1	0	0	1	Yes
V. atalanta	0	1	0	0	0	1	0	0	1	1	0	0	1	
V. cardui	1	0	0	1	0	1	0	0	1	1	0	0	1	Yes
A. urticae	1	1	0	0	0	1	0	0	1	0	0	1	80+	Yes
N. polychloros	0	1	0	1	0	1	0	0	1	0	0	1	?	212
I. io	1	1	0	0	0	1	0	0	1	0	0	1	300 to 400	Yes
P. c-album	1	0	0	0	0	1	0	0	1	1	0	0	1	
B. selene	0	0	0	1	0	1	0	0	0	1	0	0	1	
B. euphrosyne	0	0	0	1	0	1	0	0	0	1	0	0	1 or 2	
A. adippe	0	0	0	1	0	1	0	0	0	1	0	0	1	
A. aglaja	0	0	0	1	0	1	0	0	0	1	0	0	1	
A. paphia	1	0	0	1	0	1	0	0	0	1	0	0	1	Yes
E. aurinia	1	1	0	1	0	1	0	0	0	0	0	1	45 to 600	Yes
M. cinxia	0	0	0	1	1	1	0	0	0	0	0	1	50 to 200	Yes
M. athalia	0	0	0	1	0	1	0	0	0	0	0	1	15 to 150	Yes
P. aegeria	1	1	0	1	0	1	0	0	1	1	0	0	1	Yes
L. megera	1	1	0	1	0	1	0	0	0	1	0	0	1 to 3	Yes
E. epiphron	0	0	0	1	0	1	0	0	0	1	0	0	1	
E. aethiops	1	0	0	0	0	1	0	0	0	1	0	0	1	
M. galathea	0	0	0	1	0	1	0	0	0	1	0	0	1	
H. semele	1	1	0	0	0	0	0	1	0	1	0	0	1	
P. tithonus	1	0	0	1	0	1	0	0	0	1	0	0	1	
M. jurtina	1	0	0	1	0	1	0	0	0	1	0	0	1	Yes
A. hyperantus	0	0	0	1	0	1	0	0	0	1	0	0	1	
C. pamphilus	1	1	0	1	0	0	0	1	0	1	0	0	1	
C. tullia	0	0	0	0	0	0	0	1	0	1	0	0	1	Yes

* Data in Appendix 6 are deficient for several variables (not entered): polyandry, female mate-refusal posture (e.g., active, passive or possum; see Shreeve *et al*., 2006) or roosting position (e.g., head upmost, down or sideways aligned; wings depressed and clasping substrate).

† For full species names, see Appendix 1a. 0, not recorded as being used by butterfly; 1, used by butterfly. As for other appendices, a zero in a cell may simply mean that the resource or behaviour has not been witnessed; in time it is expected that quantitative data will be collected on utility resources and behaviour.

‡ From Dennis (2010).

APPENDIX 6b: LARVAL BEHAVIOUR

	Larval shelter construction						Larval feeding time†					Larval resting position											Pupal attachment			
Butterfly species*	none	larval platforms	no-cut	hibernaculum/cocoon	pupal shelter/pad/cocoon	pupal cremastral pad and/or girdle	dawn	evening	night	daytime	feeding architecture	pier extension	petiole	leaf edge	stem	flower	leaf surface (upper)	leaf surface (under)	leaf midrib	seed pod	bud	surface of soil or ground debris	suspended	attached	loose (in litter)	loose (buried)
C. palaemon	0	0	1	1	1	1	0	0	1	0	1	0	0	0	0	0	1	1	0	0	0	0 tube	1	0	0	0 low in grass tent
T. sylvestris	0	0	1	1	1	1	0	0	1	0	0	0	0	0	0	0	1	0	0	0	0	0 tube	1	0	0	0 low in cocoon in grass tent
T. lineola	0	0	1	0	1	1	0	1	1	0	0	0	0	0	0	0	1	0	0	0	0	0 tube	1	0	0	0 low in cocoon in grass tent
T. acteon	0	0	1	1	1	1	0	0	1	0	0	0	0	0	0	0	1	0	0	0	0	0 tube?	1	0	0	0 low in cocoon in grass tent
H. comma	0	0	1	0	1	1	0	0	1	0	0	0	0	0	0	0	0	0	0	0	0	1 tunnel‡	1	0	0	0 low in cocoon in grass tent
O. sylvanus	0	0	1	1	1	1	0	0	1	0	1	0	0	0	0	0	1	0	0	0	0	0 tube	1	0	0	0 low in cocoon in grass tent
E. tages	0	0	1	1	1	1	0	0	1	0	0	0	0	0	0	0	0	0	0	0	0	0 tent	0	0	1	0 in silk web
P. malvae	0	0	1	0	1	1	1	1	0	0	0	0	0	0	0	0	1	0	1	0	0	0	0	1	0	0 low in cocoon of foodplant
P. machaon	0	0	0	0	0	1	0	0	0	1	0	0	0	0	1	1	1	0	0	0	0	0	0	1	0	0 low
L. sinapis	0	0	0	0	0	1	0	0	1	1	0	0	1	1	1	0	0	0	0	0	0	0	0	1	0	0 low
C. crocea	0	0	0	0	0	1	0	0	1	1	0	0	0	1	1	0	0	0	1	0	0	0	0	1	0	0 low
G. rhamni	0	0	0	0	0	1	0	0	1	1	0	0	1	1	1	0	1	0	1	0	0	0	1	0	0	0 low
P. brassicae	0	0	0	0	0	1	0	0	1	1	0	0	0	0	0	0	1	1	0	0	0	0	0	1	0	0 low
P. rapae	0	0	0	0	0	1	0	0	1	1	0	0	0	0	0	0	1	0	1	0	0	0	0	1	0	0 low
P. napi	0	0	0	0	0	1	0	0	1	1	0	0	0	0	1	0	1	0	0	0	0	0	0	1	0	0 low
A. cardamines	0	0	0	0	0	1	0	0	1	1	0	0	0	0	1	0	0	0	0	1	0	0	0	1	0	0 low
C. rubi	0	1	0	0	0	0	0	0	1	1	0	0	0	0	1	0	0	0	0	0	1	0	0	1	1	1 low
T. betulae	0	1	0	0	0	1	0	0	1	0	0	0	0	0	0	0	0	1	0	0	1	0	0	1	1	1
F. quercus	0	1	0	0	0	1	0	0	1	0	0	0	1	0	1	0	0	0	0	0	1	0	0	0	1	0
S. w-album	0	0	0	0	0	1	0	0	1	0	0	0	1	0	1	0	0	1?	0	0	1	0	0	1	0	0 high
S. pruni	0	0	0	0	0	1	0	0	1	1	0	0	0	0	0	0	1	0	0	0	1	0	0	1	0	0 high
L. phlaeas	0	1	0	0	0	1	0	0	1	1	0	0	0	0	1	0	0	1	0	0	0	0	0	1	0	0 low
L. dispar	0	0	0	0	0	1	0	0	1	1	0	0	0	0	0	0	0	1	0	0	0	0	1	0	0	0 low
C. minimus	0	0	0	1	0	1	0	0	1	1	0	0	0	0	0	1	0	0	0	0	0	0	1	0	1	1 low
P. argus	0	0	0	0	0	1?	0	0	1	1	0	0	0	0	0	1	0	0	0	0	1	0	0	0	0	1

A. agestis	0	0	0	0	0	1?	0	0	0	1	0	0	0	0	0	0	0	1	0	0	0	0		0	0	1	1	
A. artaxerxes	0	0	0	0	0	1?	0	0	1	1	0	0	0	0	0	0	0	1	0	0	0	0		0	0	1	1	
P. icarus	0	0	0	0	0	1?	0	0	1	1	0	0	0	0	1	0	1	1	0	0	0	0		0	0	1	1	
P. coridon	0	0	0	0	0	1?	0	1	1	0	0	0	0	0	1	0	0	1	0	0	0	0		0	0	1	1	
P. bellargus	0	0	0	0	1	1?	0	0	0	1	0	0	0	0	0	0	0	1	0	0	0	1		0	0	1	1	
C. argiolus	0	0	0	0	0	1	0	0	0	1	0	0	0	0	1	1	1	0	0	1	1	0		0	0	1	0	low
M. arion	0	1	0	0	0	0	0	0	0	1	0	0	0	0	0	1	0	0	0	0	0	0		0	0	0	1	
H. lucina	0	0	0	0	0	1	0	0	1	0	0	0	0	0	1	0	1	1	0	0	0	1		0	0	1	0	
L. camilla	0	1	0	1	0	1	0	0	1	1	1	1	0	0	1	0	0	0	1	0	0	0		1	0	0	0	high
A. iris	0	1	0	0	0	1	0	0	1	1?	0	0	0	0	1	0	0	0	1	0	0	0		1	0	0	0	high
V. atalanta	0	0	1	0	1	1	0	0	1	1	0	0	1	1	1	0	1	1	0	0	0	0	tent	1	0	0	0	low
V. cardui	0	1	1	0	1	1	0	0	1	1	0	0	0	0	0	0	0	1	0	0	0	0		1	0	0	0	low
A. urticae	0	1	1	0	0	1	0	0	1	1	0	0	0	0	1	0	1	1	0	0	0	0	web	1	0	0	0	low
N. polychloros	0	1	1	0	0	1	0	0	1	1	0	0	0	0	1	0	1	1	0	0	0	0	web	1	0	0	0	high
I. io	0	1	1	0	0	1	0	0	1	1	0	0	0	0	1	0	1	1	0	0	0	0	web	1	0	0	0	low
P. c-album	0	1	0	0	0	1	0	0	1	1	0	0	0	0	1	0	1	1	0	0	0	0		1	0	0	0	low
B. selene	0	0	0	0	0	1	0	0	0	1	0	0	0	0	1	0	0	1	0	0	0	1		1	0	1	0	low
B. euphrosyne	0	0	0	0	0	1	0	0	0	1	0	0	0	0	0	0	0	1	0	0	0	1		1	0	1	0	low
A. adippe	0	0	0	0	1	1	0	0	0	1	0	0	0	0	0	0	0	0	0	0	0	1		0	1	0	0	low in loose tent of vegetation
A. aglaja	0	0	0	0	1	1	0	0	0	1	0	0	0	0	1	0	0	1	0	0	0	1		0	1	0	0	low in loose tent of vegetation
A. paphia	0	0	0	0	0	1	0	0	0	1	0	0	0	0	1	0	0	0	0	0	0	1		1	0	0	0	low
E. aurinia	0	1	1	0	0	1	0	0	0	1	0	0	0	0	1	0	1	0	1	0	0	0	web§	0	1	0	0	low
M. cinxia	0	1	1	0	0	1	0	0	0	1	0	0	0	0	1	0	1	0	1	0	0	0	web§	1	0	0	0	low in loose tent of vegetation
M. athalia	0	1	1	1	0	1	0	0	0	1	0	0	0	0	0	0	0	0	0	0	0	1		1	0	0	0	low
P. aegeria	0	0	0	0	0	1	0	0	1	1	0	0	0	0	1	0	0	1	0	0	0	0		1	0	0	0	low
L. megera	0	0	0	0	0	1	0	0	1	1	0	0	0	0	0	0	0	1	0	0	0	1?		1	0	0	0	low
E. epiphron	0	0	0	0	1	1	0	0	1	1	0	0	0	0	0	0	0	0	0	0	0	1		1	0	0	0	low in loose tent of vegetation
E. aethiops	0	0	0	0	1	0	0	0	1	1	0	0	0	0	0	0	0	0	0	0	0	1		0	0	1	1	
M. galathea	0	0	0	0	0	0	0	0	1	1	0	0	0	0	1	0	0	0	0	0	0	1		0	0	1	0	
H. semele	0	0	0	1	1	0	0	0	1	1	0	0	0	0	1	0	0	0	0	0	0	1	head down	0	0	0	1	
P. tithonus	0	0	0	0	0	1	0	0	1	1	0	0	0	0	1	0	0	0	0	0	0	1	head down	1	0	0	0	low
M. jurtina	0	0	0	0	0	1	0	0	1	1	0	0	0	0	1	0	0	0	0	0	0	1		1	0	0	0	low
A. hyperantus	0	0	0	0	1	0	0	0	1	0	0	0	0	0	1	0	0	0	0	0	0	1		0	0	1	0	in loose cocoon
C. pamphilus	0	0	0	0	0	1	0	0	1	0	0	0	0	0	1	0	0	0	0	0	0	0		1	0	0	0	low
C. tullia	0	0	0	0	0	1	0	0	1	1	0	0	0	0	1	0	0	0	0	0	0	0		1	0	0	0	low

For details on larval and pupal crypsis, see Dennis (1992a). Data courtesy of E. John, P. B. Hardy, J. E. Pateman, P. J. Russell and T. G. Shreeve.

* For full species names, see Appendix 1a. 0, not recorded as being used by butterfly; 1, used by butterfly. As for other appendices, a zero in a cell may simply mean that the resource or behaviour has not been witnessed; in time it is expected that quantitative data will be collected on utility resources and behaviour as well as on larval shelter construction of varying sophistication (see Greeney and Jones (2003) for shelter construction of different 'cut' numbers).

† Insufficient data exist on the creation of feeding architecture as for *Limenitis camilla* (Fox, 1996), though some Hesperiids may use simple structures (leaf cuts) to improve food nutrient quality (see Ravenscroft, 1994c).

‡ Makes a silk tunnel at the base of the small grass clump upon which it is feeding.

§ Bask on surface to thermoregulate.

BIOTOPES FOR BRITISH BUTTERFLIES

	Biotopes																							
Butterfly species*	**broadleaved woodland: dependence on climax component**	**broadleaved woodland: understorey component; coppice**	**broadleaved woodland: grass rides; permanent clearings**	**conifer plantations: rides; early growth stages**	**shrubs and scrub: scrub, grasses and herbs; bracken**	**acid & neutral grassland: long grass & tall herb; dense cover**	**acid & neutral grassland: short turf; sparse taller herbs/grasses**	**calcareous grassland: long grass & tall herb; dense cover**	**calcareous grassland: short turf; sparse taller herbs/grasses**	**dry lowland heath**	**wet lowland heath**	**moorland: upland heath & coarse grass**	**blanket bog and raised bogs**	**marshland: bog & valley mires; ditches**	**basic flush and fen**	**river bank and lakesides, levees; cutoffs**	**cliffs, scree & rock outcrops, walls, quarries & derelict land**	**salt marsh**	**sand-dune and dune slacks**	**arable and improved pasture**	**gardens, allotments, market gardens, fields of *Brassica* crops**	**hedgerow**	**road verge, field edge and banks**	**embankment & cuttings: railway; motorway**
C. palaemon	0	1	2	1	2	1	0	0	0	0	0	0	0	2	1	0	0	0	0	0	0	0	0	0
T. sylvestris	0	1	2	2	2	2	1	2	1	0	0	0	0	2	0	1	1	0	1	0	1	2	2	1
T. lineola	0	0	2	2	2	2	1	2	1	1	0	0	0	0	2	2	1	2	0	0	0	2	2	1
T. acteon	0	0	0	0	0	0	0	2	0	0	0	0	0	0	0	0	1	0	0	0	0	0	0	1
H. comma	0	0	0	0	0	0	0	0	2	0	0	0	0	0	0	0	0	0	0	0	0	0	0	0
O. sylvanus	0	1	2	2	2	2	0	2	0	1	2	0	0	2	0	2	1	0	1	0	1	2	2	1
E. tages	0	0	2	1	0	0	1	0	2	1	0	0	0	0	0	0	2	0	2	0	0	0	0	1
P. malvae	0	1	2	1	1	0	1	0	2	1	0	0	0	0	0	0	1	0	1	0	0	0	0	1
P. machaon	0	0	0	0	0	0	0	0	0	0	0	0	0	0	2	0	0	0	0	0	0	0	0	0
L. sinapis/L. realis	0	2	2	2	2	0	0	0	0	0	0	0	0	0	0	0	1	0	1	0	0	0	0	2
C. crocea	0	0	0	0	0	1	0	2	0	0	0	0	0	0	0	0	1	0	0	2	0	0	1	0
G. rhamni	0	2	2	1	2	0	0	0	0	0	0	0	0	0	2	0	0	0	0	0	1	2	0	1
P. brassicae	0	0	0	0	0	0	0	0	0	0	0	0	0	0	0	1	2	0	1	2	2	1	1	1
P. rapae	0	0	1	1	0	0	0	0	0	0	0	0	0	1	1	1	1	0	1	2	2	0	1	0
P. napi	0	1	2	1	0	1	1	0	0	0	1	0	0	2	2	2	2	0	1	1	1	2	2	1
A. cardamines	0	2	2	1	0	1	1	0	0	0	1	0	0	2	2	2	1	0	0	0	1	2	2	1
C. rubi	0	0	2	1	2	1	0	2	2	2	0	2	2	0	0	0	1	0	0	0	0	0	0	2
T. betulae	1	0	0	0	2	0	0	0	0	0	0	0	0	0	0	0	0	0	0	0	0	2	0	1
F. quercus	2	0	0	0	0	0	0	0	0	0	0	0	0	0	0	0	0	0	0	0	1	1	0	1
S. w-album	2	1	0	0	0	0	0	0	0	0	0	0	0	0	0	0	0	0	0	0	0	2	0	1
S. pruni	1	2	0	0	2	0	0	0	0	0	0	0	0	0	0	0	0	0	0	0	0	2	0	1
L. phlaeas	0	1	1	1	2	2	2	2	2	2	0	2	0	2	0	2	2	0	2	1	1	1	1	2
L. dispar	0	0	0	0	0	0	0	0	0	0	0	0	0	0	2	0	0	0	0	0	0	0	0	0
C. minimus	0	0	0	0	0	0	0	1	2	0	0	0	0	0	0	0	2	0	1	0	0	0	0	2
P. argus	0	0	1	0	0	0	0	0	2	2	2	0	1	0	0	0	2	0	2	0	0	0	0	0
A. agestis	0	1	1	1	1	0	1	0	2	0	0	0	0	0	0	1	2	0	2	0	1	0	1	2
A. artaxerxes	0	0	0	0	0	0	0	2	2	0	0	0	0	0	0	0	2	0	0	0	0	0	0	0
P. icarus	0	1	1	1	1	2	2	1	2	1	0	1	0	2	0	1	1	0	2	0	0	1	1	2
P. coridon	0	0	0	0	0	0	0	0	2	0	0	0	0	0	0	0	1	0	0	0	0	0	0	1
P. bellargus	0	0	0	0	0	0	0	0	2	0	0	0	0	0	0	0	1	0	0	0	0	0	0	1
C. argiolus	2	2	0	0	2	0	0	0	0	0	0	0	0	0	0	0	1	0	0	0	2	1	0	1
M. arion	0	0	0	0	0	0	1	0	1	1	0	0	0	0	0	0	1	0	0	0	0	0	0	0

A Resource-Based Habitat View for Conservation, 1st edition. By Roger L. H. Dennis. Published 2010 by Blackwell Publishing,

	Biotopes																							
Butterfly species*	broadleaved woodland: dependence on climax component	broadleaved woodland: understorey component; coppice	broadleaved woodland: grass rides; permanent clearings	conifer plantations: rides; early growth stages	shrubs and scrub: scrub, grasses and herbs; bracken	acid & neutral grassland: long grass & tall herb; dense cover	acid & neutral grassland: short turf; sparse taller herbs/grasses	calcareous grassland: long grass & tall herb; dense cover	calcareous grassland: short turf; sparse taller herbs/grasses	dry lowland heath	wet lowland heath	moorland: upland heath & coarse grass	blanket bog and raised bogs	marshland: bog & valley mires; ditches	basic flush and fen	river bank and lakesides, levees cutoffs	cliffs, scree & rock outcrops, walls, quarries & derelict land	salt marsh	sand-dune and dune slacks	arable and improved pasture	gardens, allotments, market gardens, fields of *Brassica* crops	hedgerow	road verge, field edge and banks	embankment & cuttings: railway; motorway
H. lucina	0	2	2	1	2	1	1	2	2	0	0	0	0	0	0	0	0	0	0	0	0	0	0	0
L. camilla	2	2	2	1	0	0	0	0	0	0	0	0	0	0	0	0	0	0	0	0	0	0	0	0
A. iris	2	2	2	1	0	0	0	0	0	0	0	0	0	0	0	0	0	0	0	0	0	0	0	0
V. atalanta	0	1	2	2	1	1	0	1	0	0	0	0	0	1	0	1	1	0	0	1	2	1	1	1
V. cardui	0	1	2	2	1	1	0	2	2	0	0	0	0	2	0	1	1	0	2	1	2	2	1	1
A. urticae	0	1	2	2	2	2	0	2	0	1	0	1	0	2	0	2	1	0	2	2	2	2	2	2
N. polychloros	2	0	0	0	0	0	0	0	0	0	0	0	0	0	0	0	0	0	0	0	0	1	1	0
I. io	0	2	2	2	1	0	0	0	0	0	0	0	0	1	0	1	0	0	1	1	1	1	1	1
P. c-album	0	2	2	2	2	0	0	0	0	0	0	0	0	0	0	0	0	0	0	0	2	2	1	1
B. selene	0	2	2	2	2	0	2	0	1	0	2	2	0	2	1	0	2	0	0	0	0	0	0	0
B. euphrosyne	0	2	2	2	2	0	0	1	0	0	0	0	0	0	0	0	1	0	0	0	0	0	0	0
A. adippe	0	2	1	1	2	0	1	0	0	0	0	0	0	0	0	0	0	0	0	0	0	0	0	0
A. aglaja	0	1	1	1	2	1	1	1	2	1	0	2	0	0	0	0	2	0	2	0	0	0	0	0
A. paphia	2	2	2	0	0	0	0	0	0	0	0	0	0	0	0	0	0	0	0	0	0	1	1	0
E. aurinia	0	0	1	1	0	2	2	1	2	0	1	0	0	2	1	1	0	0	1	0	0	0	0	1
M. cinxia	0	0	0	0	0	0	0	0	0	0	0	0	0	0	0	0	2	0	0	0	0	0	0	0
M. athalia	0	2	1	2	0	0	1	0	0	2	0	1	0	0	0	0	0	0	0	0	0	0	0	0
P. aegeria	0	2	2	2	2	0	0	0	0	0	0	0	0	0	0	0	1	0	0	0	1	1	0	1
L. megera	0	1	1	1	0	0	2	0	2	2	0	0	0	1	0	2	2	0	2	1	1	2	2	2
E. epiphron	0	0	0	0	0	0	0	0	0	0	0	2	0	0	0	0	0	0	0	0	0	0	0	0
E. aethiops	0	0	2	1	2	2	0	2	0	0	0	2	0	2	0	0	0	0	0	0	0	0	0	1
M. galathea	0	0	2	1	2	2	0	2	0	0	0	0	0	0	0	0	1	0	1	0	0	1	1	1
H. semele	0	0	0	1	0	0	2	0	2	2	0	0	0	0	0	0	2	0	2	0	0	0	0	1
P. tithonus	0	1	2	1	2	2	0	1	0	2	0	0	0	0	0	0	1	1	1	0	2	2	1	2
M. jurtina	0	1	2	2	2	2	0	2	0	1	0	0	0	2	0	2	1	1	2	0	1	2	2	2
A. hyperantus	0	2	2	2	2	2	0	2	0	0	0	0	0	2	0	0	0	1	1	0	0	2	2	2
C. pamphilus	0	1	2	0	2	0	2	0	2	2	2	2	0	0	0	0	1	0	2	0	1	1	1	2
C. tullia	0	0	0	0	0	0	0	0	0	0	0	2	2	1	0	0	0	0	0	0	0	0	0	0

* For full species names, see Appendix 1a. 0, not recorded as being used by butterfly; 1, used by butterfly; 2, biotopes in which species are most frequently found.

REFERENCES

Aandahl, A. R. (1948). The characterisation of slope positions and their influence on the total N content of a few virgin soils in Western Iowa. *Soil Science Society of America, Proceedings* 13: 449–454.

Abrams, P. A. (1992). Why don't predators have positive effects on prey populations? *Ecological Entomology* 6: 449–457.

Ackery, P. R. (1984). Systematic and faunistic studies on butterflies. In *The Biology of Butterflies*, ed. R. I. Vane-Wright and P. R. Ackery. London: Academic Press, pp. 9–21.

Ackery, P. R. (1988). Hostplants and classification: nymphalid butterflies. *Biological Journal of the Linnean Society* 33: 95–203.

Ackery, P. R. and Vane-Wright, R. I. (1984). *Milkweed Butterflies. Their Cladistics and Biology*. London: British Museum (Natural History).

Allee, W. C., Emerson, A. E., Park, O., Park, T. and Schmidt, K. P. (1949). *Principles of Animal Ecology*. Philadelphia: W. B. Saunders Company.

Allen, J. R. L. (1992). Trees and their response to wind – Mid Flandrian strong winds, Severn Estuary and Inner Bristol Channel, Southwest Britain. *Philosophical Transactions of the Royal Society London B* 338: 335–364.

Alpert, P., Bone, E. and Holzapfel, C. (2000). Invasiveness, invasibility and the role of environmental stress in the spread of non-native plants. *Perspectives in Plant Ecology, Evolution and Systematics* 3: 52–66.

Andel, J. Van, Bakker, J. P. and Grootjans, A. P. (1993). Mechanisms of vegetation succession: a review of concepts and perspectives. *Acta Botanica Neerlandica* 42: 413–433.

Anderson, S. H. (2003). The relative importance of birds and insects as pollinators of the New Zealand flora. *New Zealand Journal of Ecology* 27: 83–94.

Andrewartha, H. G. and Birch, L. C. (1954). *The Distribution and Abundance of Animals*. Chicago: University of Chicago Press.

Andrews, E. J. (2008). *The Effect of Sward Height on the Distribution of Boloria selene in Shropshire*. MSc thesis, University of Birmingham.

Angold, P. G., Sadler, J. P., Hill, M. O. *et al.* (2006). Biodiversity in urban habitat patches. *Science of the Total Environment* 360: 196–204.

Anon (1990). *Handbook for Phase 1 Habitat Survey – a Technique for Environmental Audit*. Peterborough: Nature Conservancy Council (Natural England).

Anon (1991). *CORINE Biotopes Manual*. Luxembourg: Office for Official Publications of the European Communities.

Anon (1992). CORINE land cover: a European Community project presented in the framework of the International Space Year. 1992 European Conference of the International Space Year. Brussels: Commission of the European Communities.

Anon (2007). Warwickshire site works for butterflies. *Butterfly* 94: 33.

Anton, C., Musche, M. and Settele, J. (2007). Spatial patterns of host exploitation in a larval parasitoid of the predatory dusky large blue *Maculinea nausithous*. *Basic and Applied Ecology* 8: 66–74.

Arponen, A., Heikkinen, R. K., Thomas, C. D. and Moilanen, A. (2005). The value of biodiversity in reserve selection: representation, species weighting, and benefit functions. *Conservation Biology* 19: 2009–2014.

Asher, J., Warren, M., Fox, R., Harding, P., Jeffcoate, G. and Jeffcoate, S. (2001). *The Millennium Atlas of Butterflies in Britain and Ireland*. Oxford: Oxford University Press.

Askew, R. R. and Shaw, M. R. (1986). Parasitoid communities: their size, structure and development. In *Insect Parasitoids*, ed. J. Waage and D. Greathead. Symposium of the Royal Entomological Society 13. London: Academic Press, pp. 225–264.

Aspinall, R. J. (1987). The geography of the small copper butterfly *Lycaena phlaeas* (L.) in Northumberland. *Entomologist's Gazette* 38: 159–165.

Auckland, J. N., Debinski, D. M. and Clark, W. R. (2004). Survival, movement, and resource use of the butterfly *Parnassius clodius*. *Ecological Entomology* 29: 139–149.

Ausden, M. (2007). *Habitat Management for Conservation. A Handbook of Techniques*. Oxford: Oxford University Press.

Avery, B. W. (1973). Soil classification in the Soil Survey of England and Wales. *Journal of Soil Science* 24: 324–338.

Baguette, M. and Van Dyck, H. (2007). Landscape connectivity and animal behavior: functional grain as a key determinant for dispersal. *Landscape Ecology* 22: 1117–1129.

Baguette, M., Petit, S. and Quéva, F. (2000). Population spatial structure and migration of three butterfly species within the same habitat network: consequences for conservation. *Journal of Applied Ecology* 37: 100–108.

Bailey, E. B. (1952). *Geological Survey of Great Britain*. London: HMSO.

Bailey, N. (2007). An early red admiral *Vanessa atalanta* at high altitude. *Atropos* 31: 74.

Baker, R. R. (1969). The evolution of the migratory habitat in butterflies. *Journal of Animal Ecology* 38: 703–746.

Baker, R. R. (1972). Territorial behaviour of the nymphalid butterflies, *Aglais urticae* (L.) and *Inachis io* (L.). *Journal of Animal Ecology* 41: 453–469.

Baker, R. R. (1978). *The Evolutionary Ecology of Animal Migration*. London: Hodder and Stoughton.

Baker, R. R. (1984). The dilemma: when and how to go or stay. In *The Biology of Butterflies*, ed. R. I. Vane-Wright and P. R. Ackery. London: Academic Press, pp. 279–296.

Ball, S. and Henshall, M. (2006). Using data to interpret changes in the UK's biodiversity. *Bulletin of the British Ecological Society* December: 51–54.

Banks, J. E. (1998). The scale of landscape fragmentation affects herbivore response to vegetation heterogeneity. *Oecologia* 117: 239–246.

Barbour, D. A. (2007). Measuring changes in range-size of British butterflies: a critique of sub-sampling methods. *Entomologist's Gazette* 58: 259–265.

Barker, T. (2008). The economics of avoiding dangerous climate change. An editorial essay on The Stern Review. *Climate Change* 89: 173–194.

Barr, C., Bunce, R. G. H., Clarke, R. T. *et al.* (1993). *Countryside Survey 1990*. Main Report. London: Department of the Environment.

Barry, R. G. and Chorley, R. J. (1982). *Atmosphere, Weather and Climate*. London: Methuen.

Baughman, J. F., Murphy, D. D. and Ehrlich, P. R. (1988). Population structure of a hill-topping butterfly. *Oecologia* 75: 593–600.

Baughman, J. F., Murphy, D. D. and Ehrlich, P. R. (1990). A re-examination of hill-topping in *Euphydryas editha*. *Oecologia* 83: 259–260.

Beaumont, S. (1995). *The Ecology of Larvae of the Scotch Argus butterfly Erebia aethops at Arnside Knott SSSI, Cumbria*. MSc thesis, University of London (Wye College).

Beck, J. and Fiedler, K. (2009). Adult life spans of butterflies (Lepidoptera: Papilionoidea and Hesperioidea): broadscale contingencies with adult and larval traits in multi-species comparisons. *Biological Journal of the Linnean Society* 96: 166–184.

Beirne, B. P. (1943). The distribution and origin of the British Lepidoptera. *Proceedings of the Royal Irish Academy* 49: 27–59.

Beirne, B. P. (1947). The origin and history of the British Macrolepidoptera. *Transactions of the Royal Entomological Society* 98: 275–372.

Beirne, B. P. (1955). Natural fluctuations in abundance of British Lepidoptera. *Entomologist's Gazette* 6: 21–52.

Belfrage, K., Bjorklund, J. and Salomonsson, L. (2006). The effects of farm size and organic farming on diversity of birds, pollinators, and plants in a Swedish landscape. *Ambio* 34: 582–588.

Bennett, D. P. and Humphries, D. A. (1974). *Introduction to Field Biology*. London: Edward Arnold.

Bennett, K. D. (1989). A provisional map of forest types for the British Isles 5000 year ago. *Journal of Quaternary Science* 4: 141–144.

Bergman, K. O., Askling, J., Ekberg, O., Ignell, H., Wahlman, H. and Milberg, P. (2004). Landscape effects on butterfly assemblages in an agricultural region. *Ecography* 27: 619–628.

Bergstrom, J. and Wiklund, C. (2002). Effects of size and nuptial gifts on butterfly reproduction: can females compensate for a smaller size through male-derived nutrients? *Behavioral Ecology and Sociobiology* 52: 296–302.

Berney, F. (2002). *The Status, Mobility and Habitat Requirements of the Small Pearl-bordered Fritillary Boloria selene in County Durham*. MSc thesis, University of Sunderland.

Berwaerts, K., Aerts, P. and Van Dyck, H. (2006). On the sex-specific mechanisms of butterfly flight: flight performance relative to flight morphology, wing kinematics, and sex in *Pararge aegeria*. *Biological Journal of the Linnean Society* 89: 675–687.

Bettinger, P., Bradshaw, G. A. and Weaver, G. W. (1996). Effects of geographic information system vector-raster-vector data conversion on landscape indices. *Canadian Journal of Forest Research* 26: 1416–1425.

Beveridge, W. I. B. (1950). *The Art of Scientific Investigation*. London: Heinemann.

Biggs, J., Williams, P., Whitfield, M., Nicolet, P. and Weatherby, A. (2005). 15 years of pond assessment in Britain: results and lessons learned from the work of Pond Conservation. *Aquatic Conservation: Marine and Freshwater Ecosystems* 15: 693–714.

Bink, F. A. (1985). Hostplant preference of some grass feeding butterflies. In *Proceedings of the 3rd Congress of European Lepidopterology, Cambridge 1982*. pp. 23–29. Karlsruhe: Societas Europaea Lepidopterologica.

Bink, F. A. (1986). Acid stress in *Rumex hydrolapathum* (Polygonaceae) and its influence on the phytophage *Lycaena dispar* (Lepidoptera: Lycaenidae). *Oecologia* 70: 447–451.

Bink, F. A. (1992). *Ecologische Atlas van de Dagvlinders van Noordwest-Europa*. Haarlem: Schuyt and Co.

Bink, F. A. and Siepel, H. (1996). Nitrogen and phosphorous in *Molinia caerulea* (Gramineae) and its impact on the larval development in the butterfly-species *Lasiommata megera* (Lepidoptera: Satyridae). *Entomologia Generalis* 20: 271–280.

Binzenhofer, B., Schroder, B., Strauss, B., Biedermann, R. and Settele, J. (2005). Habitat models and habitat connectivity analysis for butterflies and burnet moths – the example of *Zygaena carniolica* and *Coenonympha arcania*. *Biological Conservation* 126: 247–259.

Bland, K. P. (1999). Further observations on predation of hibernating *Aglais urticae* (Lep.: Nymphalidae) by a wren. *Entomologist's Record and Journal of Variation* 111: 137.

Bland, K. P. (2009). An unusual hibernation site for *Aglais urticae* (Linnaeus, 1758) (Lepidoptera: Nymphalidae). *Entomologist's Gazette* 60: 36.

Blossey, B. and Nötzold, R. (1995). Evolution of increased competitive ability in invasive nonindigenous plants: a hypothesis. *Journal of Ecology* 83: 887–889.

Boardman, P. (2005). *A Landscape Approach to the Conservation of the Pearl-bordered Fritillary Boloria euphrosyne (and Other Priority Species of Lepidoptera) in the Sites of Special Scientific Interest of the Oswestry Hills Joint Character Area and Adjacent Areas of Shropshire*. Wareham, UK: Butterfly Conservation.

Bockstaller, C. and Girardin, P. (2003). How to validate environmental indicators. *Agricultural Systems* 76: 639–653.

Bogaert, J., Van Hecke, P., Savador-Van Eysenrode, D. and Impens, I. (2000). Landscape fragmentation assessment using a single measure. *Wildlife Society Bulletin* 28: 875–881.

Boggs, C. L. (1987). Ecology of nectar and pollen feeding in Lepidoptera. In *Nutritional Ecology of Insects, Mites and Spiders*, ed. F. Slansky and J. G. Rodriguez. New York: John Wiley and Sons, pp. 369–391.

Boggs, C. L. (1990). A general model of the role of male-donated nutrients in female insects' reproduction. *American Naturalist* 136: 598–617.

Boggs, C. L. and Gilbert, L. E. (1979). Male contribution to egg production in butterflies: evidence for transfer of nutrients at mating. *Science* 206: 83–84.

Boggs, C. L. and Ross, C. L. (1993). The effect of adult food limitation on life-history traits in *Speyeria mormonia* (Lepidoptera: Nymphalidae). *Ecology* 74: 433–441.

Boiteau, G. (2001). Electronic tags for the tracking of insects in flight: effect of weight on flight performance of adult Colorado beetles. *Entomologia Experimentalis et Applicata* 100: 187–193.

Boiteau, G. and Colpitts, B. (2004). The potential of portable harmonic radar technology for the tracking of beneficial insects. *International Journal of Pest Management* 50: 233–242.

Boorman, S. A. and Levitt, P. R. (1973). Group selection on the boundary of a stable population. *Theoretical Population Biology* 4: 85–128.

Boothby, J. and Hull, A. P. (1994). A census of ponds in Cheshire, northwest England. *Aquatic Conservation: Marine and Freshwater Ecosystems* 7: 75–79.

Boppré, M. (1984). Chemically mediated interactions between butterflies. In *The Biology of Butterflies*, ed. R. I. Vane-Wright and P. R. Ackery. London: Academic Press, pp. 259–275.

Bosschieter, L. and Goedhart, P. W. (2005). Gap crossing decisions by reed warblers (*Acrocephalus scirpaceus*) in agricultural landscapes. *Landscape Ecology* 20: 455–468.

Boughton, D. A. (2000). The dispersal of a butterfly: a test of source-sink theory suggests the intermediate-scale hypothesis. *American Naturalist* 156: 131–144.

Bourn, N. A. D. (2007). New priorities list presents challenges for everyone. *Butterfly* 96: 19–21.

Bourn, N. A. D. and Thomas, J. A. (1993). The ecology and conservation of the brown argus butterfly *Aricia agestis* in Britain. *Biological Conservation* 63: 67–74.

Bourn, N. A. D. and Thomas, J. A. (2002). The challenge of conserving grassland insects at the margins of their range in Europe. *Biological Conservation* 104: 285–292.

Bourn, N. A. D., Whitfield, K. E. J., Pearman, G. S. and Roberts, E. (1999). *Site Dossier and Status of the Adonis Blue Polyommatus (Lysandra) bellargus in Dorset between 1997 and 1999*. Wareham, UK: Butterfly Conservation.

Bourn, N. A. D., Joy, J. and McCracken, M. (2005). *The Status and Habitat Requirements of the Small Pearl-bordered Fritillary (Boloria selene) in Clocaenog Forest*. Wareham, UK: Butterfly Conservation.

Bowen, G. W. and Burgess, R. L. (1981). *A Quantitative Analysis of Forest Island Pattern in Selected Ohio Landscapes*. ORNL Environmental Sciences Division, Publication No. 1719, ORNL/TM-7759. Oak Ridge, TN: Oak Ridge National Laboratory, 111 pp.

Bowersox, M. A. and Brown, D. G. (2001). Measuring the abruptness of patchy ecotones – a simulation-based comparison of landscape pattern statistics. *Plant Ecology* 156: 89–103.

Bowley, A. (2007). The Great Fen – a waterland for the future. *British Wildlife* 18: 415–423.

Boyce, R. R. and Clark, W. A. V. (1964). The concept of shape in geography. *Geographical Review* 54: 561–572.

Boyd-Wallis, W. (1994). *Survey of Creag Meagaidh NNR for Colonies of the Small Mountain Ringlet Butterfly*. Unpublished report to Scottish Natural Heritage, Edinburgh.

Bradley, J. D. and Fletcher, D. S. (1986). *An Indexed List of British Butterflies and Moths*. Orpington: Kedleston Press.

Bradshaw, A. D. and Chadwick, M. J. (1980). *The Restoration of Land: the Ecology and Reclamation of Derelict and Degraded Land*. Oxford: Blackwell Scientific Publications.

Brakefield, P. M. (1979). *An Experimental Study of the Maintenance of Variation in Spot Pattern in Maniola jurtina*. PhD thesis, University of Liverpool.

Brakefield, P. M. (1982a). Ecological studies on the butterfly *Maniola jurtina* in Britain. I. Adult behaviour, microdistribution and dispersal. *Journal of Animal Ecology* 51: 713–726.

Brakefield, P. M. (1982b). Ecological studies on the butterfly *Maniola jurtina* in Britain. II. Population dynamics: the present position. *Journal of Animal Ecology* 51: 727–738.

Brakefield, P. M. and Larsen, T. B. (1984). The evolutionary significance of dry and wet seasonal forms in some tropical butterflies. *Biological Journal of the Linnean Society* 22: 1–12.

Brakefield, P. M., Shreeve, T. G. and Thomas, J. A. (1992). Avoidance, concealment, and defence. In *Ecology of Butterflies in Britain*, ed. R. L. H. Dennis. Oxford: Oxford University Press, pp. 93–119.

Braschler, B. and Hill, J. K. (2007). Role of larval host plants in the climate-driven range expansion of the butterfly *Polygonia c-album*. *Journal of Animal Ecology* 76: 415–423.

Braun-Blanquet, J. and Tuxen, R. (1952). Irische Pflanzengesellschaften. In *Die Pflanzenwelt Irlands*, ed. W. Ludi. *Veroffentlichungen des Geobotanischen Institutes Rubel in Zurich* 25: 224–415.

Brereton, T. M. (1997). *Ecology and Conservation of the Butterfly Pyrgus malvae (Grizzled Skipper) in South-East England*. PhD thesis, University of East London.

Brereton, T. M. (2004). Farming and butterflies in Britain. *Biologist* 51: 1–5.

Brereton, T. M. and Warren, M. S. (1999). Ecology of the pearl-bordered fritillary butterfly in Scotland and possible threats from bracken eradication measures in woodland grant schemes. In *Bracken Perceptions and Bracken Control in the British Uplands*, ed. J. Taylor. Special Publication No. 3. Lancaster: International Bracken Group, pp. 62–73.

Brereton, T. M., Warren, M. S. and Roberts, E. (1998). *Action for the Heath Fritillary: Status, Monitoring and Conservation Progress 1996 and 1997*. Confidential Report. Wareham, UK: Butterfly Conservation.

Brereton, T. M., Wigglesworth, T., Warren, M. S. and Stewart, K. (2005). *Agri-environment Schemes and Butterflies: Reassessing the Impacts and Improving Delivery of BAP Targets*. Butterfly Conservation Final Project Report No. BD1446. London: DEFRA.

Brereton, T. M., Warren, M. S., Roy, D. B. and Stewart, K. (2008). The changing status of the chalkhill blue butterfly *Polyommatus coridon* in the UK: the impacts of conservation policies and environmental factors. *Journal of Insect Conservation* 12: 629–638.

Briggs, K., Riley, D. and Tolley, H. (1979). *Data Response Exercises in Physical and Human Geography*. Oxford: Oxford University Press.

Brown, J. H. (1984). On the relationship between abundance and distribution of species. *American Naturalist* 124: 255–279.

Brown, J. H. (1995). *Macroecology*. Chicago: University of Chicago Press.

Brown, J. H. and Kodric-Brown, A. (1977). Turnover rates in insular biogeography: effects of immigration on extinction. *Ecology* 58: 445–449.

Brown, N. J., Wright, S. M. and Fuller, R. M. (1996). A technique for the removal of outliers during a computerized map generalization process. *British Cartographic Journal* 33: 11–16.

Brown, W. D. and Alcock, J. (1990). Hill-topping by the red qdmiral butterfly: mate searching alongside congeners. *Journal of Research on the Lepidoptera* 29: 1–10.

Brunzel, S., Elligsen, H. and Frankl, R. (2004). Distribution of the cinnabar moth *Tyria jacobaeae* L. at landscape scale: use of linear landscape structures in egg laying on larval hostplant exposures. *Landscape Ecology* 19: 21–27.

Bryant, S. R. and Shreeve, T. G. (2002). The use of artificial neural networks in ecological analysis: estimating microhabitat temperature. *Ecological Entomology* 27: 424–432.

Bryant, S. R., Thomas, C. D. and Bale, J. S. (1997). Nettle-feeding nymphalid butterflies: temperature, development and distribution. *Ecological Entomology* 22: 390–398.

Bryant, S. R., Thomas, C. D. and Bale, J. S. (2000). Thermal ecology of gregarious and solitary nettle-feeding nymphalid butterfly larvae. *Oecologia* 122: 1–10.

Bulman, C. R. (2001). *Conservation Biology of the Marsh Fritillary Butterfly Eurodryas aurinia*. PhD thesis, University of Leeds.

Bulman, C. R. (2004). The heath fritillary. *Butterfly* 87: 21–23.

Bulman, C. R., Wilson, R. J., Holt, A. R., Gálvez Bravo, L., Early, R. I., Warren, M. S. and Thomas, C. D. (2007). Minimum viable metapopulation size, extinction debt, and the conservation of a declining species. *Ecological Applications* 17: 1460–1473.

Bulman, C. R., Bourn, N. A. D. and Warren, M. S. (2008). *Conservation Review, 2000–2008*. Wareham, UK: Butterfly Conservation. Also at ww.butterfly-conservation.org/conservationreview/.

Bunce, R. G. H., Barr, C. J., Clarke, R. T., Howard, D. C. and Lane, A. M. J. (1996). ITE Merlewood land classification of Great Britain. *Journal of Biogeography* 23: 625–634.

Bunge, W. (1966). *Theoretical Geography*, 2nd edn. Lund Studies in Geography, Series C, No. 1. Lund, Sweden.

Burden, J. P., Possee, R. D., Sait, S. M., King, L. A. and Hails, R. S. (2006). Phenotypic and genotypic characterisation of persistent baculovirus infections in populations of the cabbage moth (*Mamestra brassicae*) within the British Isles. *Archives of Virology* 151: 635–649.

Burghardt, F. and Fiedler, K. (1996). The influence of diet on growth and secretion behaviour of myrmecophilous *Polyommatus icarus* caterpillars (Lepidoptera: Lycaenidae). *Ecological Entomology* 21: 1–8.

Burke, S., Pullin, A. S., Wilson, R. J. and Thomas, C. D. (2005). Selection for discontinuous life-history traits along a continuous thermal gradient in the butterfly *Aricia agestis*. *Ecological Entomolgy* 30: 613–619.

Burnham, C. P. (1980). The soils of England and Wales. *Field Studies* 5: 349–363.

Burrough, P. A. (1986). *Principles of Geographic Information Systems for Land Resource Assessment*. Oxford: Oxford University Press.

Burrough, P. A. and McDonnell, R. A. (1998). *Principles of Geographic Information Systems*. New York: Oxford University Press.

Burrows, C. J. (1990). *Processes of Vegetation Change*. London: Unwin Hyman.

Butler, B. E. (1980). *Soil Classification for Soil Survey*. Oxford: Clarendon Press.

BUTT (Butterflies Under Threat Team) (1986). *The Management of Chalk Grassland for Butterflies*. Focus on Nature Conservation, No. 17. Peterborough: Nature Conservancy Council.

Cabeza, M., Araújo, M. B., Wilson, R. J., Thomas, C. D., Cowley, M. R. J. and Moilanen, A. (2004). Combining probabilities of occurrence with spatial reserve design. *Journal of Applied Ecology* 41: 252–262.

Calow, P. (1999). *Blackwell's Concise Encyclopedia of Ecology*. Oxford: Blackwell Sciences.

Campbell, A. L., Naik, R. R., Sowards, L. and Stone, M. O. (2002). Biological infrared imaging and sensing. *Micron* 33: 211–225.

Cant, E. T., Smith, A. D., Reynolds, D. R. and Osborne, J. L. (2005). Tracking butterfly flight paths across the landscape with harmonic radar. *Proceedings of the Royal Society of London B* 272: 785–790.

Caporn, S., Lee, J., Studholme, C. and Carroll, J. (2005). Recovery of ombrotrophic sphagnum moss in relation to air pollution in the Southern Pennines. Moors of the Future Conference Presentation. http://www.moorsforthefuture.org.uk/mftf/main/Publications.htm.

Carey, D. B. (1994). Patch dynamics of *Glaucopsyche lygdamus* (Lycaenidae). Correlations between butterfly density and host species-diversity. *Oecologia* 99: 337–342.

Carey, P. D., Preston, C. D., Hill, M. O., Usher, M. B. and Wright, S. M. (1995). An environmentally defined biogeographical zonation of Scotland designed to reflect species distributions. *Journal of Ecology* 83: 833–845.

Caro, T. M. and O'Doherty, G. (1999). On the use of surrogate species in conservation biology. *Conservation Biology* 13: 805–814.

Caton, P. G. F. (1976). *Maps of Hourly Mean Wind Speed over the United Kingdom 1965–1973*. Climatological Memorandum No. 79. Bracknell, UK: Meteorological Office.

Caughley, G. and Sinclair, A. R. E. (1994). *Wildlife Ecology and Management*, 4. Oxford: Blackwell Science.

Chambers, F. M. (ed.) (1993). *Climate Change and Human Impact on the Landscape*. London: Chapman and Hall.

Cherrill, A. (1994). A comparison of three landscape classifications and investigation of the potential for using remotely sensed land cover data for landscape classification. *Journal of Rural Studies* 10: 275–289.

Cherrill, A. J., Lane, A. and Fuller, R. M. (1994). The use of classified Landsat-5 thematic mapper imagery in the characterization of landscape composition: a case study in northern England. *Journal of Environmental Management* 40: 357–377.

Cherrill, A. J., McClean, C., Lane, A. and Fuller, R. M. (1995). A comparison of land cover types in an ecological field survey in northern England and a remotely sensed land cover map of Great Britain. *Biological Conservation* 71: 313–323.

Chew, F. S. (1977). Coevolution of pierid butterflies and their cruciferous foodplants. II. The distribution of eggs of potential foodplants. *Evolution* 31: 568–579.

Chuvieco, E. (1999). Measuring changes in landscape pattern from satellite images: short-term effects of fire on spatial diversity. *International Journal of Remote Sensing* 20: 2331–2346.

Cizek, L., Fric, Z. and Konvicka, M. (2006). Host plant defences and voltinism in European butterflies. *Ecological Entomology* 31: 337–344.

Clark, C. D., Evans, D. J. A., Khatwa, A. *et al.* (2004). Map and GIS database of landforms and features related to the last British ice sheet. *Boreas* 33(4): 359–375.

Clark, P. J. and Evans, F. C. (1954). Distance to nearest neighbour as a measure of spatial relationships in populations. *Ecology* 35: 445–453.

Clarke, S. A. (2005). *High Brown Fritillary Breeding Habitat Survey 2004. Malvern Hills*. Wareham, UK: Butterfly Conservation.

Clausen, H. D., Holbeck, H. B. and Reddersen, J. (2001). Factors influencing abundance of butterflies and burnet moths in the uncultivated habitats of an organic farm in Denmark. *Biological Conservation* 98: 167–178.

Clement, E. J., Smith, D. P. J. and Thirlwell, I. R. (2005). *Illustrations of Alien Plants of the British Isles*. London: Botanical Society of the British Isles.

Clements, F. E. (1904). *Development and Structure of Vegetation*. Report of the Botanical Survey of Nebraska No. 7. Loncoln, NB: University of Nebraska.

Clements, F. E. (1916). *Plant Succession. An Analysis of the Development of Vegetation*. No. 242. Washington: Carnegie Institute.

Clench, H. K. (1966). Behavioral thermoregulation in butterflies. *Ecology* 47: 1021–1034.

Clowes, A. and Comfort, P. (1983). *Process and Landform: an Outline of Contemporary Geomorphology*. Edinburgh: Oliver and Boyd.

Clunas, A. (1986). *The Biology and Habitat Requirements of Aricia artaxerxes Fabricius (Lepidoptera: Lycaenidae)*. MSc thesis, University of Edinburgh.

Cole, J. P. and King, C. A. M. (1968). *Quantitative Geography*. London: John Wiley and Sons.

Coleman, A., Isbell, J. E. and Sinclair, G. (1974). The comparative statics approach to British land-use trends. *Cartographic Journal* 11: 34–41.

Collenette, C. L. (1935). Notes concerning attacks by British birds on butterflies. *Proceedings of the Zoological Society of London* 1935: 201–217.

Collier, R. V. (1972). *Chequered Skipper (Carterocephalus palaemon)*. Unpublished report to the Nature Conservancy Council.

Collier, R. V. (1984). *Chequered Skipper (Carterocephalus palaemon) survey 1984*. Unpublished report to the Nature Conservancy Council.

Collier, R. V. (1986). *The Conservation of the Chequered Skipper in Britain*. Focus on Nature Conservation No. 16. Peterborough: Nature Conservancy Council, 16 pp.

Collin, P. H. (1988). *Dictionary of Ecology and the Environment*. Teddington, UK: Peter Collin Publishing.

Collins, G. A. (1995). *Butterflies of Surrey*. Woking, UK: Surrey Wildlife Trust.

Condry, W. (1973). *The Snowdonia National Park*. London: Fontana.

Conradt, L., Bodsworth, E. J., Roper, T. J. and Thomas, C. D. (2000). Non-random dispersal in the butterfly *Maniola jurtina*: implication for metapopulation models. *Proceedings of the Royal Society of London B* 267: 1505–1570.

Cook, L. M., Dennis, R. L. H. and Hardy, P. B. (2001). Butterfly–hostplant fidelity, vagrancy and a measurement of mobility from distribution maps. *Ecography* 24: 497–504.

Cooke, R. U. and Doornkamp, J. C. (1974). *Geomorphology in Environmental Management*. Oxford: Oxford University Press.

Coope, G. R. (1987). The response of Late Quaternary insect communities to sudden climatic changes. In *Organization of Communities Past and Present*, ed. J. H. R. Gee and P. S. Giller. Oxford: Blackwell Scientific Publications, pp. 421–438.

Coppolillo, P., Gomez, H., Maisels, F. and Wallace, R. (2004). Selection criteria for suites of landscape species as a basis for site-based conservation. *Biological Conservation* 115: 419–430.

Corbet, S. A. (2000). Butterfly nectaring flowers: butterfly morphology and flower form. *Entomologia Experimentalis et Applicata* 96: 289–298.

Corbett, A. and Plant, R. E. (1993). Role of movement in the response of natural enemies to agroecosystem diversification: a theoretical evaluation. *Environmental Entomology* 22: 519–531.

Coulthard, N. (1982). *An Investigation of the Habitat Requirements and Behavioural Ecology of the Small Blue Butterfly Cupido minimus, in Relation to its Distribution and Abundance*. MSc thesis, University of Aberdeen.

Coulthard, T. J., Macklin, M. G. and Kirkby, M. J. (2002). A cellular model of Holocene upland river basin and alluvial fan evolution. *Earth Surface Processes and Landforms* 27: 269–288.

Courtney, F. M. and Trudgill, S. T. (1976). *The Soil. An Introduction to Soil Study in Britain*. London: Edward Arnold.

Courtney, S. P. (1980). *Studies on the Biology of the Butterflies Anthocharis cardamines L. and Pieris napi L. in Relation to Speciation in the Pierinae*. PhD thesis, University of Durham.

Courtney, S. P. (1986). The ecology of pierid butterflies: dynamics and interactions. *Advances in Ecological Research* 15: 51–116.

Courtney S. P. (1988). Oviposition on peripheral hosts by dispersing *Pieris napi* (L.) (Pieridae). *Journal of Research on the Lepidoptera* 26: 58–63.

Courtney, S. P. and Courtney, S. (1982). The edge effect in butterfly oviposition: causality in *Anthocharis cardamines* and related species. *Ecological Entomology* 7: 131–137.

Courtney, S. P. and Duggan, A. E. (1983). The population biology of the orange tip butterfly *Anthocharis cardamines* in Britain. *Ecological Entomology* 8: 271–281.

Cowley, M. J. R., Thomas, C. D., Thomas, J. A. and Warren, M. S. (1999). Flight areas of British butterflies: assessing species status and decline. *Proceedings of the Royal Society of London B* 266: 1587–1592.

Cowley, M. J. R., Wilson, R. J., León-Cortés, J. L., Gutiérrez, D., Bulman, C. R. and Thomas, C. D. (2000). Habitat-based statistical models for predicting the spatial distribution of butterflies and day-flying moths in a fragmented landscape. *Journal of Applied Ecology* 37: 60–72.

Cowley, M. J. R., Thomas, C. D., Roy, D. B. *et al.* (2001a). Density–distribution relationships in British butterflies. I. The effect of mobility and spatial scale. *Journal of Animal Ecology* 70: 410–425.

Cowley, M. J. R., Thomas, C. D., Wilson, R. J., León-Cortés, J. L., Gutiérrez, D. and Bulman, C. R. (2001b). Density–distribution relationships in British butterflies. II. An assessment of mechanisms. *Journal of Animal Ecology* 70: 426–441.

Cozzi, G., Müller, C. B. and Krauss, J. (2007). How do local habitat management and landscape structure at different spatial scales affect fritillary butterfly distribution on fragmented wetlands? *Landscape Ecology* 23: 269–283.

Crawford, A. (1996). Perching to advantage? The purple hairstreak (*Quercusia quercus*). Observations during July/August 1996 on a common by the M25. *Bulletin of the Amateur Entomologists' Society* 55: 261–262.

Creaser, A., Pateman, J. A. and Dennis, R. L. H. (2008). A surfeit of female *Pieris rapae* (L.) (Lepidoptera, Pieridae) in New Zealand: a suite of coincidences or something more sinister? *Antenna* 32: 199–204.

Creed, E. R., Dowdeswell, W. H., Ford, E. B. and McWhirter, K. G. (1970). Evolutionary studies on *Maniola jurtina* (Lepidoptera: Satyridae): the 'boundary phenomenon' in southern England. In *Essays in Evolution and Genetics*, ed. M. K. Hecht and W. C. Steere. New York: Appleton-Century-Crofts, pp. 263–287.

Crocker, R. L. and Major, J. (1955). Soil development in relation to vegetation and surface age at Glacier Bay, Alaska. *Journal of Ecology* 43: 427–448.

Crowe, P. R. (1971). *Concepts in Climatology*. Harlow, UK: Longman.

Crowle, A. (2007). Letting our carbon go free. The sustainable management of carbon and blanket peat in the English uplands. *British Wildlife* 19: 28–34.

Croxton, P. J., Hann, J. P., Greatorex-Davies, J. N. and Sparks, T. H. (2005). Linear hotspots? The floral and butterfly diversity of green lanes. *Biological Conservation* 121: 579–584.

Cruickshank, M. M. and Tomlinson, R. W. (1996). Application of CORINE land cover methodology to the UK – some issues raised from Northern Ireland. *Global Ecology and Biogeography Letters* 5: 235–248.

Crutzen, P. J. and Streimer, E. F. (2000). The 'Anthropocene'. *Global Change Newsletter* 41: 12–13.

Cullinan, V. I. and Thomas, J. M. (1992). A comparison of quantitative methods for examining landscape pattern and scale. *Landscape Ecology* 7: 211–227.

Curtis, L. F., Doornkamp, J. C. and Gregory, K. J. (1965). The description of relief in field studies of soils. *Journal of Soil Science* 16: 16–30.

Dansereau, P. (1957). *Biogeography: an Ecological Perspective*. New York: Ronald Press.

Darwin, F. (ed.) (1888). *Life and Letters of Charles Darwin*. London: John Murray.

Davies, G. H. N., Frazer, J. F. D. and Tynam, A. M. (1958). Population numbers in a colony of *Lysandra bellargus* Rott. (Lep., Lycaenidae). *Proceedings of the Royal Entomological Society of London A* 33: 31–36.

Davies, G. L. H. (1983). *Sheets of Many Colours: the Mapping of Ireland's Rocks, 1750–1890*. Dublin: Royal Dublin Society.

Davies, M. (1992). *The White-Letter Hairstreak Butterfly*. Colchester: Butterfly Conservation.

Davies, N. B. (1978). Territorial defence in the speckled wood butterfly (*Pararge aegeria*): the resident always wins. *Animal Behaviour* 26: 138–147.

Davies, Z. G. and Pullin, A. S. (2007). Are hedgerows effective corridors between fragments of woodland habitat? An evidence-based approach. *Landscape Ecology* 22: 333–351.

Davies, Z. G., Wilson, R. J., Brereton, T. M. and Thomas, C. D. (2005). The re-expansion and improving status of the silver-spotted skipper butterfly (*Hesperia comma*) in Britain: a metapopulation success story. *Biological Conservation* 124: 189–198.

Davies, Z. G., Wilson, R. J., Coles, S. and Thomas, C. D. (2006). Changing habitat associations of a thermally constrained species, the silver-spotted skipper butterfly, in response to climate warming. *Journal of Animal Ecology* 75: 247–256.

Davis, A. J., Jenkinson, L. S., Lawton, J. H., Shorrocks, B. and Wood S. (1998). Making mistakes when predicting shifts in species range in response to global warming. *Nature* 391: 783–786.

Davis, B. N. K. (1979). Chalk and limestone quarries as wildlife habitats. *Minerals and Environment* 1: 48–56.

Davis, B. N. K. (1989). Habitat creation for butterflies on a landfill site. *The Entomologist* 108: 109–122.

Davis, B. N. K., Lakhani, K. H. and Yates, T. J. (1991). The hazards of insecticides to butterflies of field margins. *Agriculture, Ecosystems and Environment* 36: 151–161.

Davis, B. N. K., Brown, M. J. and Frost, A. J. (1993). Selection of receptors for measuring spray drift deposition and comparison with bioassyas with special reference to the shelter effect of hedges. *Brighton Crop Protection Conference: Weeds* 3: 139–144.

Davis, M. A., Grime, J. P. and Thompson, K. (2000). Fluctuating resources in plant communities: a general theory of invasibility. *Journal of Ecology* 88: 528–534.

Davis, W. M. (1909). *Geographical Essays*. Boston: Ginn. (Reprinted, 1954. Dover.)

D'Eon, R. G. and Glenn, S. M. (2000). Perceptions of landscape patterns: do the numbers count? *Forestry Chronicle* 76: 475–480.

De Whalley, L., de Whalley, B., Green, P., Gammon, N. and Shreeves, W. (2006). Digging scrapes to enhance silver-studded blue *Plebejus argus* habitat at Broadcroft Quarry, Isle of Portland, Dorset, England. *Conservation Evidence* 3: 39–43.

Delibes, M., Ferreras, P. and Gaona, P. (2001). Attractive sinks, or how individual behaviour decisions determine source–sink dynamics. *Ecological Letters* 4: 401–403.

Dell, D., Sparks, T. H. and Dennis R. L. H. (2005). Climate change and the effect of increasing spring temperatures on emergence dates of the flagship butterfly *Apatura iris* (Lepidoptera: Nymphalidae). *European Journal of Entomology* 102: 161–167.

Dempster, J. P. (1967). The control of *Pieris rapae* with DDT. I. The natural mortality of the young stages of *Pieris*. *Journal of Applied Ecology* 4: 485–500.

Dempster, J. P. (1968). The control of *Pieris rapae* with DDT. II. Survival of the young stages of *Pieris* after spraying. *Journal of Applied Ecology* 5: 451–462.

Dempster, J. P. (1971). Some observations on a population of the small copper butterfly *Lycaena phlaeas* (Linnaeus) (Lep.: Lycaenidae). *Entomologist's Gazette* 22: 199–204.

Dempster, J. P. (1984). The natural enemies of butterflies. In *The Biology of Butterflies*, ed. R. I. Vane-Wright and P. R. Ackery. Symposium of the Royal Entomological Society No. 11. London: Academic Press, pp. 97–104.

Dempster, J. P. (1989). Insect introductions: natural dispersal and population persistence in insects. *The Entomologist* 108: 5–13.

Dempster, J. P. (1991). Fragmentation, isolation, and mobility of insect populations. In *The Conservation of Insects and their Habitats*, ed. N. M. Collins and J. A. Thomas. London: Academic Press, pp. 143–154.

Dempster, J. P. (1995). The ecology and conservation of *Papilio machaon* in Britain. In *The Ecology and Conservation of Butterflies*, ed. A. S. Pullin. London: Chapman and Hall, pp. 137–149.

Dempster, J. P. and Pollard, E. (1981). Fluctuations in resource availability and insect populations. *Oecologia* 50: 412–416.

Dempster, J. P., King, M. L. and Lakhani, K. H. (1976). The status of the swallowtail butterfly in Britain. *Ecological Entomology* 1: 71–84.

Dempster, J. P., Atkinson, D. A. and Cheesman, O. D. (1995). The spatial population dynamics of insects exploiting a patchy food resource. *Oecologia* 104: 340–353.

Den Boer, P. J. (1968). Spreading the risk and the stabilization of animal numbers. *Acta Biotheoretica (Leiden)* 18: 165–194.

Den Boer, P. J. (1981). On the survival of populations in a heterogeneous and variable environment. *Oecologia* 50: 39–53.

Den Boer, P. J. and Reddingius, J. (1996). *Regulation and Stabilization Paradigms in Population Ecology*. London: Chapman and Hall.

Dennis, B. and Taper, M. L. (1994). Density dependence in time series observations of natural populations: estimation and testing. *Ecological Monographs* 64: 205–224.

Dennis, R. L. H. (1971). *The Aspect of Weichselian and Flandrian Geomorphological and Climatic Evidence in the Establishment of the North Wales Lepidoptera Rhopalocera*. BA dissertation, University of Durham.

Dennis, R. L. H. (1972a). *Eumenis semele* (L.) *thyone* Thompson (Lep., Satyridae). A microgeographical race. *Entomologist's Record and Journal of Variation* 84: 1–11, 38–44.

Dennis, R. L. H. (1972b). *Plebejus argus* (L.) *caernensis* Thompson (Lep., Lycaenidae). A stenoecious geotype. *Entomologist's Record and Journal of Variation* 84: 100–108.

Dennis, R. L. H. (1977). *The British Butterflies. Their Origin and Establishment.* Faringdon, UK: E. W. Classey.

Dennis, R. L. H. (1982a). Observations on habitats and dispersion made from oviposition markers in north Cheshire *Anthocharis cardamines* (L.) (Lepidoptera: Pieridae). *Entomologist's Gazette* 33: 151–159.

Dennis, R. L. H. (1982b). Patrolling behaviour in orange tip butterflies within the Bollin valley in north Cheshire, and a comparison with other pierids. *Vasculum* 67: 17–25.

Dennis, R. L. H. (1982c). *Erebia aethiops (Esper) (Satyridae) on Arnside Knott.* Confidential Report for the NCC. Peterborough: Nature Conservancy Council.

Dennis, R. L. H. (1982–1983). Mate-location strategies in the wall brown butterfly, *Lasiommata megera* L. (Lep., Satyridae). Wait or seek? *Entomologist's Record and Journal of Variation* 94: 209–214 (1982); 95: 7–10 (1983).

Dennis, R. L. H. (1983a). Hierarchy and pattern in the spatial responses of ovipositing *Anthocharis cardamines* (Lep.). *Vasculum* 68: 27–43.

Dennis, R. L. H. (1983b). Spatial patterning in butterfly oviposition: density influences and the 'edge effect' in the orange-tip butterfly (*Anthocharis cardamines* (L.) Lep., Pieridae). *Bulletin of the Amateur Entomologists' Society* 42: 55–60.

Dennis, R. L. H. (1983c). Egg-laying cues in the wall brown butterfly *Lasiommata megera* (L.) (Lepidoptera: Satyridae). *Entomologist's Gazette* 34: 89–95.

Dennis, R. L. H. (1984a). Egg-laying sites of the common blue butterfly, *Polyommatus icarus* (Rottemburg) (Lepidoptera: Lycaenidae): the edge effect and beyond the edge. *Entomologist's Gazette* 35: 85–93.

Dennis, R. L. H. (1984b). The edge effect in butterfly oviposition: batch siting in *Aglais urticae* (L.) (Lepidoptera: Nymphalidae). *Entomologist's Gazette* 35: 157–173.

Dennis, R. L. H. (1985a). *Polyommatus icarus* (Rottemburg) (Lepidoptera: Lycaenidae) on Brereton Heath in Cheshire: voltinism and switches in resource exploitation. *Entomologist's Gazette* 36: 175–179.

Dennis, R. L. H. (1985b). The 'edge effect' in butterfly oviposition. Hostplant condition, edge effect breakdown and opportunism. *Entomologist's Gazette* 36: 285–291.

Dennis, R. L. H. (1985c). Choice of egg-laying sites in the green-veined white butterfly (*Artogeia napi* (L.)) (Lep., Pieridae). *Bulletin of the Amateur Entomologists' Society* 44: 199–219.

Dennis, R. L. H. (1985d). Voltinism in British *Aglais urticae* (L.) (Lep., Nymphalidae). Variation in space and time. *Proceedings and Transactions of the British Entomological and Natural History Society* 18: 51–61.

Dennis, R. L. H. (1986a). Selection of roost sites by *Lasiommata megera* (L.) (Lep., Satyridae) on fencing at Brereton Heath country park, Cheshire. *Nota Lepidopterologica* 9: 39–46.

Dennis, R. L. H. (1986b). Motorways and cross-movements. An insect's 'mental map' of the M56 in Cheshire. *Bulletin of the Amateur Entomologists' Society* 45: 228–243.

Dennis, R. L. H. (1987). Hill-topping as a mate-location strategy in a Mediterranean population of *Lasiommata megera* (L.) (Lep., Satyridae). *Nota Lepidopterologica* 10: 65–70.

Dennis, R. L. H. (ed.) (1992a). *The Ecology of Butterflies in Britain.* Oxford: Oxford University Press.

Dennis, R. L. H. (1992b). Expanded wing posture in lateral-basking *Hipparchia semele* (L.) (Lep., Nymphalidae, Satyrinae). *Entomologist's Gazette* 43: 45–46.

Dennis, R. L. H. (1993a). *Butterflies and Climate Change.* Manchester: Manchester University Press.

Dennis, R. L. H. (1993b). Predation in a northern population of *Pieris napi* (L.) (Lep., Pieridae). Evidence from wing fragments. *Entomologist's Gazette* 44: 157–159.

Dennis, R. L. H. (2000). Contrasts in status of butterfly species among open and woodland biotopes of a northern English SSSI. *Entomologist's Gazette* 51: 257–273.

Dennis, R. L. H. (2001). Progressive bias in species status is symptomatic of fine-grained mapping units subject to repeated sampling. *Biodiversity and Conservation* 10: 483–494.

Dennis, R. L. H. (2003). Arboreal substrate for an egglaying meadow brown. *Entomologist's Record and Journal of Variation* 115: 241–242.

Dennis, R. L. H. (2004a). Butterfly habitats, broad-scale biotope affiliations, and structural exploitation of vegetation at finer scales: the matrix revisited. *Ecological Entomology* 29: 744–752.

Dennis, R. L. H. (2004b). Just how important are structural elements as habitat components? Indications from a declining lycaenid butterfly with conservation status. *Journal of Insect Conservation* 8: 37–45.

Dennis, R. L. H. (2004c). Landform resources for territorial nettle-feeding nymphalid butterflies: biases at different spatial scales. *Animal Biodiversity and Conservation* 27: 37–45.

Dennis, R. L. H. (2005a). Alternative to a nectar source for a thirsty *Pararge aegeria* Linnaeus (Satyrinae). *Entomologist's Record* 117: 150.

Dennis, R. L. H. (2005b). Pupal mortality in a metapopulation of *Zygaena filipendulae* (Linnaeus, 1758) (Lepidoptera: Zygaenidae): heterogeneity for season, between and within local populations. *Entomologist's Gazette* 56: 25–31.

Dennis, R. L. H. (2008a). How abundant are key resources for common butterflies? Insights from nymphalid butterflies in a conservation area. *Entomologist's Gazette* 59: 79–84.

Dennis, R. L. H. (2008b). Roadways – a burial ground for dead butterflies, or the flip-side to using tarmac as a substrate for thermoregulation and territorial perching? *Entomologist's Gazette* 59: 172–173.

Dennis, R. L. H. (2009a). Where are we now and where to go? In *Ecology of Butterflies in Europe*, ed. J. Settele, H. van Dyck, M. Konvicka and T. G. Shreeve. Cambridge: Cambridge University Press, pp. 1–6.

Dennis, R. L. H. (2009b). Changes in butterfly distributions: a simple correction for bias caused by sub-sampling of atlas records makes no difference to BAP status. *Entomologist's Gazette*, in press.

Dennis, R. L. H. (2009c). Butterflies on Alderley Edge: resources, habitats and changes. In: *Living with the Edge: Alderley's Story*, ed. A. J. N. W. Prag. Manchester: Manchester University Press, in press.

Dennis, R. L. H. (2010). Wing and body substrate appression in thermoregulating *Erynnis tages* (Linnaeus, 1758) (Rhopalocera: Hesperiidae). *Entomologists's Gazette* 60: 220.

Dennis, R. L. H. and Asher, J. (2009). Head-down basking – a postural trait common to territorial nymphalids perching on vertical surfaces. *Entomologist's Gazette* 60: 237–242.

Dennis, R. L. H. and Bardell, P. (1996). The impact of extreme weather on Great Orme populations of *Hipparchia semele* (Linnaeus, 1758) and *Plebejus argus* (Linnaeus, 1758) (Papilionoidea: Satyrinae and Lycaenidae): hindsight, inference and lost opportunities. *Entomologist's Gazette* 47: 211–225.

Dennis, R. L. H. and Bramley, M. J. (1985). The influence of man and climate on dispersion patterns within a population of adult *Lasiommata megera* (L.) (Satyridae) at Brereton Heath, Cheshire. *Nota Lepidopterologica* 8: 309–324.

Dennis, R. L. H. and Dennis, M. P. (2006). Hill-topping in British butterflies: incidence and cues in a cool, windy climate? *Entomologist's Gazette* 57: 17–20.

Dennis, R. L. H. and Dennis, M. P. (2007). Hill-topping in British butterflies: a test of prevalence, place and purpose. *Entomologist's Gazette* 58: 17–22.

Dennis, R. L. H. and Dennis, M. P. (2008). Territorial hill topping in British butterflies: high summits present a special case. *Entomologist's Gazette* 59: 227–232.

Dennis, R. L. H. and Eales, H. T. (1997). Patch occupancy in *Coenonympha tullia* (Lepidoptera: Satyridae): habitat quality matters as much as patch size and isolation. *Journal of Insect Conservation* 1: 167–176.

Dennis, R. L. H. and Eales, H. T. (1999). Probability of patch site occupancy in *Coenonympha tullia* (Müller) (Lepidoptera: Satryinae) determined from geographical and ecological data. *Biological Conservation* 87: 295–301.

Dennis, R. L. H. and Hardy, P. B. (1999). Targeting squares for recording: predicting species richness and incidence of species for a butterfly atlas. *Global Ecology and Biogeography Letters* 8: 443–454.

Dennis, R. L. H. and Hardy, P. B. (2001). Loss rates of butterfly species with urban development. A test of atlas data and sampling artefacts at a fine scale. *Biodiversity and Conservation* 10: 1831–1837.

Dennis, R. L. H. and Hardy, P. B. (2006). Excessive *Anthocharis cardamines* (L.) (Pieridae) egg load on cuckoo flower host-plants. *Entomologist's Gazette*, 57: 13–15.

Dennis, R. L. H. and Hardy, P. B. (2007). Support for mending the matrix: resource seeking by butterflies in apparent non-resource zones. *Journal of Insect Conservation* 11: 157–168.

Dennis, R. L. H. and Schmitt, T. (2009). Faunal structures, palaeogeography and historical inference. In *Ecology of Butterflies in Europe*, ed. J. Settele, H. van Dyck, M. Konvicka and T. G. Shreeve. Publisher, pp. 250–280.

Dennis, R. L. H. and Shreeve, T. G. (1988). Hostplant-habitat structure and the evolution of butterfly mate-location behaviour. *Zoological Journal of the Linnean Society* 94: 301–318.

Dennis, R. L. H. and Shreeve, T. G. (1989). Butterfly wing morphology variation in the British Isles. The influence of climate, behavioural posture and the hostplant-habitat. *Biological Journal of the Linnean Society* 38: 323–348.

Dennis, R. L. H. and Shreeve, T. G. (1991). Climatic change and the British butterfly fauna. Opportunities and constraints. *Biological Conservation* 55: 1–16.

Dennis, R. L. H. and Shreeve, T. G. (1996). *Butterflies on British and Irish Offshore Islands*. Wallingford, UK: Gem Publishing Company.

Dennis, R. L. H. and Shreeve, T. G. (1997). Diversity of butterfly species on British islands: ecological influences underlying the roles of area, isolation and faunal source. *Biological Journal of the Linnaean Society* 60: 257–275.

Dennis, R. L. H. and Shreeve, T. G. (2003). Gains and losses of French butterflies: tests of predictions, under-recording and regional extinction from data in a new atlas. *Biological Conservation* 110: 131–139.

Dennis, R. L. H. and Sparks, T. H. (2005). Landscape resources for the territorial nymphalid butterfly *Inachis io*: micro-site landform selection and behavioural responses to environmental conditions. *Journal of Insect Behavior* 18: 725–742.

Dennis, R. L. H. and Sparks, T. H. (2006). When is a habitat not a habitat? Dramatic resource use changes under differing weather conditions for the butterfly *Plebejus argus*. *Biological Conservation* 129: 291–301.

Dennis, R. L. H. and Sparks, T. H. (2007). Climate signals are reflected in an 89 year series of British Lepidoptera records. *European Journal of Entomology* 104: 763–767.

Dennis, R. L. H. and Thomas, C. D. (2000). Bias in butterfly distribution maps: the influence of hot spots and access. *Journal of Insect Conservation* 4: 73–77.

Dennis, R. L. H. and Williams, W. R. (1986). Butterfly 'diversity'. Regressing and a little latitude. *Antenna* 10: 108–112.

Dennis, R. L. H. and Williams, W. R. (1987). Mate-location behaviour in the butterfly *Ochlodes venata* (Br. and Grey) (Hesperiidae). Flexible strategies and spatial components. *Journal of the Lepidopterists' Society* 41: 45–64.

Dennis, R. L. H., Porter, K. and Williams, W. R. (1984). Ocellation in *Coenonympha tullia* (Mueller) (Lep., Satyridae). I. Structures in correlation matrices. *Nota Lepidopterologica* 7: 199–219.

Dennis, R. L. H., Porter, K. and Williams, W. R. (1986). Ocellation in *Coenonympha tullia* (Mueller) (Lep., Satyridae). II. Population differentiation and clinal variation in the context of climatically-induced anti-predator defence strategies. *Entomologist's Gazette* 37: 133–172.

Dennis, R. L. H., Williams, W. R. and Shreeve, T. G. (1991). A multivariate approach to the determination of faunal units among European butterfly species (Lepidoptera: Papilionoidea, Hesperioidea). *Zoological Journal of the Linnean Society* 101: 1–49.

Dennis, R. L. H., Shreeve, T. G. and Williams, W. R. (1995). Taxonomic differentiation in species diversity gradients among European butterflies: contribution of macroevolutionary dynamics. *Ecography* 18: 27–40.

Dennis, R. L. H., Shreeve, T. G. and Sparks, T. H. (1998a). The effects of island area, isolation and source population size on the presence of *Hipparchia semele* (L.) (Lepidoptera: Satyrinae) on British and Irish offshore islands. *Biodiversity and Conservation* 7: 765–776.

Dennis, R. L. H., Sparks, T. H. and Shreeve, T. G. (1998b). Geographical factors influencing the probability of *Hipparchia semele* (L.) (Lepidoptera: Satyrinae) occurring on British and Irish offshore islands. *Global Ecology and Biogeography Letters* 7: 205–214.

Dennis, R. L. H., Williams, W. R. and Shreeve, T. G. (1998c). Faunal structures among European butterflies: evolutionary implications of bias for geography, endemism and taxonomic affiliations. *Ecography* 21: 181–203.

Dennis, R. L. H., Sparks, T. H. and Hardy, P. B. (1999). Bias in butterfly distribution maps: the effects of sampling effort. *Journal of Insect Conservation* 3: 33–42.

Dennis, R. L. H., Donato, B., Sparks, T. H. and Pollard, E. (2000a). Ecological correlates of island incidence and geographical range among British butterflies. *Biodiversity and Conservation* 9: 343–359.

Dennis, R. L. H., Shreeve, T. G., Olivier, A. and Coutsis, J. G. (2000b). Contemporary geography dominates butterfly diversity gradients within the Aegean archipelago (Lepidoptera: Papilionoidea, Hesperioidea). *Journal of Biogeography* 27: 1365–1384.

Dennis, R. L. H., Shreeve, T. G. and L'honore, J. (2002). A comparison of geographical and neighbourhood models for improving atlas databases. The case of the French butterfly atlas. *Biological Conservation* 108: 143–159.

Dennis, R. L. H., Shreeve, T. G. and Van Dyck, H. (2003). Towards a functional resource-based concept for habitat: a butterfly biology viewpoint. *Oikos* 102: 417–426.

Dennis, R. L. H., Hodgson, J. G., Grenyer, R., Shreeve, T. G. and Roy, D. B. (2004). Host plants and butterfly biology. Do host-plant strategies drive butterfly status? *Ecological Entomology* 29: 12–26.

Dennis, R. L. H., Shreeve, T. G., Arnold, H. R. and Roy, D. B. (2005). Does diet breadth control herbivorous insect range size? Predictions and tests using butterflies. *Journal of Insect Conservation* 9: 187–200.

Dennis, R. L. H., Shreeve, T. G., Isaac, N. J. B., Roy, D. B., Hardy, P. B., Fox, R. and Asher, J. (2006a). The effects of visual apparency on bias in butterfly recording and monitoring. *Biological Conservation* 128: 486–492.

Dennis, R. L. H., Shreeve, T. G. and Van Dyck, H. (2006b). Habitats and resources: the need for a resource-based definition to conserve butterflies. *Biodiversity and Conservation* 15: 1943–1966.

Dennis, R. L. H., Stefanescu, C. and Tremewan, W. G. (2006c). Why does *Vanessa atalanta* (Linnaeus) (Lepidoptera: Nymphalidae) engage in late summer territorial disputes when close relatives are feeding up for winter? *Entomologist's Gazette* 57: 83–89.

Dennis, R. L. H., Shreeve, T. G. and Sheppard, D. A. (2007). Species conservation and landscape management: a habitat perspective. In *Insect Conservation Biology*, ed. A. Stewart, O. Lewis and T. New. Proceedings of the Royal Entomological Society's 23rd Symposium. Wallingford, UK: CABI, pp. 92–126.

Dennis, R. L. H., Hardy, P. B. and Shreeve, T. G. (2008). The importance of resource databanks for conserving insects: a butterfly biology perspective. *Journal of Insect Conservation* 12: 711–719.

Dennis, R. L. H., Hardy, P. B. and Kinder, P. M. (2009). High-level territorial perching behaviour by *Pararge aegeria* (Linnaeus, 1758) (Lepidoptera: Nymphalidae, Satyrinae) in a British woodland. *Entomologist's Gazette* 60: 84.

Dethier, V. G. (1954). Evolution of feeding preferences in phytophagous insects. *Evolution* 8: 33–54.

Diamond, J. M. (1973). Distributional ecology of New Guinea birds. *Science* 179: 759–769.

Diamond, J. M. (1975). The island dilemma: lessons of modern biogeographic studies for the design of nature reserves. *Biological Conservation* 7: 129–145.

Dockery, M. A. (2005). Grooming behaviour in two noctuid moths. *Journal of the Lancashire and Cheshire Entomological Society* 129: 26–33.

Douglas, G. (2003). *Carterocephalus palaemon Pallas, chequered skipper (Lepidoptera: Hesperiidae)*. Invertebrate Species Dossier. Edinburgh: Scottish Natural Heritage.

Dover, J. W. (1988). Butterflies and game cover. *Game Conservancy Review of 1987* 19: 84–87.

Dover, J. W. (1989). The use of flowers by butterflies foraging in cereal field margins. *Entomologist's Gazette* 40: 283–291.

Dover, J. W. (1990). Butterflies and wildlife corridors. *Game Conservancy Review of 1989* 21: 62–64.

Dover, J. W. (1991). The conservation of insects on arable farmland. In *The Conservation of Insects and their Habitats*, ed. N. M. Collins and J. A. Thomas. London: Academic Press, pp. 293–318.

Dover, J. W. (1994). Arable field margins: factors affecting butterfly distribution and abundance. In *Field Margins: Integrating Agriculture and Conservation*, ed. N. D. Boatman. Farnham: British Crop Protection Council, pp. 59–66.

Dover, J. W. (1996). Factors affecting the distribution of butterflies on arable farmland. *Journal of Applied Ecology* 33: 723–734.

Dover, J. W. (1997). Conservation headlands: effects on butterfly distribution and behaviour. *Agriculture Ecosystems and Environment* 63: 31–49.

Dover, J. W. and Fry, G. L. A. (2001). Experimental simulation of some visual and physical components of a hedge and the effects on butterfly behaviour in an agricultural landscape. *Entomologia Experimentalis et Applicata* 100: 221–233.

Dover, J. and Settele, J. (2009). The influences of landscape structure on butterfly distribution and movement: a review. *Journal of Insect Conservation* 13: 3–27.

Dover, J. and Sparks, T. (2000). A review of the ecology of butterflies in British hedgerows. *Journal of Environmental Management* 60: 51–63.

Dover, J., Sotherton, N. and Gobbett, K. (1990). Reduced pesticide inputs on cereal field margins – the effects on butterfly abundance. *Ecological Entomology* 15: 17–24.

Dover, J. W., Clarke, S. A. and Rew, L. (1992). Habitats and movement patterns of satyrid butterflies (Lepidoptera: Satyridae) on arable farmland. *Entomologist's Gazette* 43: 29–44.

Dover, J. W., Sparks, T. H. and Greatorex-Davies, J. N. (1997). The importance of shelter for butterflies in open landscapes. *Journal of Insect Conservation* 1: 89–97.

Dover, J. W., Sparks, T., Clarke, S., Gobbett, K. and Glossop, S. (2000). Linear features and butterflies: the importance of green lanes. *Agriculture Ecosystems and Environment* 80: 227–242.

Dowdeswell, W. H. (1961). Experimental studies on natural selection in the butterfly, *Maniola jurtina*. *Heredity* 16: 39–52.

Dowdeswell, W. H. (1981). *The Life of the Meadow Brown*. London: Heinnemann.

Dramstad, W. E., Olson, J. D. and Forman, R. T. T. (1996). *Landscape Ecology, Principles in Landscape Architecture and Land-Use Planning*. Cambridge, MA: Harvard University.

Driscoll, D. A. (2005). Is the matrix a sea? Habitat specificity in a naturally fragmented landscape. *Ecological Entomology* 30: 8–16.

Duffey, E. (1968). Ecological studies on the large copper butterfly *Lycaena dispar* Haw *batavus* Obth at Woodwalton Fen national nature reserve, Huntingdonshire. *Journal of Applied Ecology* 5: 69–96.

Dufrêne, M. and Legendre, P. (1997). Species assemblages and indicator species: the need for a flexible asymmetrical approach. *Ecological Monographs* 67: 345–366.

Dunn, T. C. and Parrack, J. D. (1986). The moths and butterflies of Northumberland and Durham. *Vasculum* Suppl. 2: iv + 284 pp.

Dunning, J. B., Danielson, B. J. and Pulliam, H. R. (1992). Ecological processes that affect populations in complex landscapes. *Oikos* 65: 169–175.

Dwyer, G. (1991). The roles of density, stage, and patchiness in the transmission of an insect virus. *Ecology* 72: 559–574.

Dwyer, G. and Elkinton, J. S. (1995). Host dispersal and the spatial spread of insect pathogens. *Ecology* 76: 1262–1275.

Eales, H. T. and Dennis, R. L. H. (1998). Predicting site occupancy for *Coenonympha tullia* (Müller, 1764) (Lepidoptera: Satyrinae) using habitat parameters. *Entomologist's Gazette* 49: 3–16.

Ebdon, D. (1977). *Statistics in Geography. A Practical Approach*. Oxford: Basil Blackwell.

Ehrlich, P. R. and Hanski, I. (2004). *On the Wings of Checkerspots. A Model System for Population Biology*. Oxford: Oxford University Press.

Ehrlich, P. R. and Raven, P. H. (1965). Butterflies and plants: a study in coevolution. *Evolution* 18: 586–608.

Ehrlich, P. R. and Wheye, D. (1986). 'Nonadaptive' hilltopping behavior in male checkerspot butterflies (*Euphydryas editha*). *American Naturalist* 127: 477–483.

Ellenberg, H. (1988). *Vegetation Ecology of Central Europe*. Cambridge: Cambridge University Press.

Ellenberg, H., Weber, W., Düll, R., Wirth, V., Werner, W. and Paulissen, D. (1992). *Zeigerwerte von Pflanzen in Mitteleuropa*, 2nd edn. Scripta Geobotanica No. 18. Göttingen: Erich Goltze Verlag, 258 pp.

Ellis, S. (1995). *Ecological Studies of the Butterflies of Magnesian Limestone Grassland*. PhD thesis, University of Sunderland.

Ellis, S. (1999). *Metapopulation Structure and Grazing Management of the Durham Argus Butterfly Aricia artaxerxes ssp. Salmacis in North-East England*. Unpublished report to English Nature, Peterborough.

Ellis, S. (2000). *The Small Pearl-Bordered Fritillary Boloria selene in County Durham*. Contract report to English Nature, Durham County Council and Northumbrian Water.

Ellis, S. (2001). *Population Ecology and Conservation Management of the Small Pearl-Bordered Fritillary Boloria selene in County Durham*. Contract report to English Nature, Peterborough.

Ellis, S. (2003). Habitat quality and management for the northern brown Argus butterfly *Aricia artaxerxes* in North-East England. *Biological Conservation* 113: 285–294.

Elton, C. S. (1958). *The Ecology of Invasions by Animals and Plants*. London: Methuen.

Elton, C. S. (1966). *The Pattern of Animal Communities*. London: Methuen.

Elton, C. S. and Miller, R. S. (1954). The ecological survey of animal communities: with a practical system of classifying habitats by structural characters. *Journal of Ecology* 42: 460–496.

Emden, H. F., van, Storek, A. P., Douloumpaka, S., Eleftherianos, I., Poppy, G. M. and Powell, W. (2008). Plant chemistry and aphid parasitoids (Hymenoptera: Braconidae): imprinting and memory. *European Journal of Entomology* 105: 477–483.

Emmet, A. M. and Heath, J. (1990). *The Butterflies of Great Britain and Ireland*. Colchester: Harley Books.

Emmet, A. M. and Heath, J. (eds) (1991). *The Moths and Butterflies of Great Britain and Ireland*, Vol. 7, Part 2. Colchester: Harley Books.

Endler, J. A. (1984). Progressive background matching in moths, and a quantitative measure of crypsis. *Biological Journal of the Linnean Society* 22: 187–231.

Erhardt, A. and Mevi-Schütz, J. (2009). Adult food resources. In *Ecology of Butterflies in Europe*, ed. J. Settele, T. G. Shreeve, M. Konvicka and H. Van Dyck. Cambridge: Cambridge University Press, pp. 9–16.

Evans, D. J. A., Clark, C. D. and Mitchell, W. A. (2005). The last British ice sheet: a review of the evidence utilised in the compilation of the Glacial Map of Britain. *Earth Science Reviews* 70(3–4): 253–312.

Evans, I. S. (1972). General geomorphology, derivatives of altitude, and descriptive statistics. In *Spatial Analysis in Geomorphology*, ed. R. J. Chorley. London: Methuen, pp. 17–90.

Evans, I. S. (1977). World-wide variations in the direction and concentration of cirque and glacier aspects. *Geografiska Annaler* 59: 151–175.

Evans, I. S. (1979). *An Integrated System of Terrain Analysis and Slope Mapping*. Final report on Grant DA-ERO-591-G0040. Department of Geography, University of Durham, 192 pp.

Evans, I. S. (1987). The morphometry of specific landforms. In *International Geomorphology*, Part II, ed. V. Gardiner. New York: John Wiley and Sons, pp. 105–124.

Evans, I. S. (1998). What do terrain statistics really mean? In *Landform Monitoring, Modelling and Analysis*, ed. S. N. Lane, K. S. Richards and J. H. Chandler. New York: John Wiley and Sons, pp. 119–137.

Evans, I. S. (2001). Some geomorphometric characteristics of real land surfaces. In *GIS Research in the UK. Proceedings of the GIS Research UK 9th Annual Conference GISRUK 2001*, ed. D. B. Kidner and G. Higgs. Glamorgan, Wales: University of Glamorgan, pp. 408–410.

Evans, I. S. and Cox, N. J. (1999). Relations between land surface properties: altitude, slope and curvature. In *Process Modelling and Landform Evolution*, ed. S. Hergarten and H. J. Neugebauer. Berlin: Springer, pp. 13–45.

Evans, I. S. and McClean, C. J. (1995). The land surface is not unifractal: variograms, cirque scale and allometry. *Zeitschrift für Geomorphologie NF* 101: 127–147.

Evans, J. G. (1975). *The Environment of Early Man in the British Isles*. London: Paul Elek.

Eyre, M. D. (2006). A strategic interpretation of beetle (Coleoptera) assemblages, biotopes, habitats and distribution, and the conservation implications. *Journal of Insect Conservation* 10: 151–160.

Fahrig, L. and Merriam, G. (1985). Habitat patch connectivity and population survival. *Ecology* 66: 1762–1768.

Fauth, J. E., Bernard, J., Camara, M., Resetarits, W. J. Jr, Van Buskirk, J. and McCollum, S. A. (1996). Simplifying the jargon of community ecology: a conceptual approach. *American Naturalist* 147: 282–286.

Fayt, P., Dufrêne, M., Branquart, E., Hastir, P., Pontégnie, C., Henin, J-N. and Versteirt, V. (2006). Contrasting responses of saproxylic insects to focal habitat resources: the example of longhorn beetles and hoverflies in Belgian deciduous forests. *Journal of Insect Conservation* 10: 129–150.

Feber, R. E., Smith, H. and Macdonald, D. W. (1994). The effects of field margin restoration on the meadow brown butterfly (*Maniola jurtina*). In *Field Margins: Integrating Agriculture and Conservation*, ed. N. D. Boatman. Farnham: British Crop Protection Council, pp. 295–300.

Feber, R. E., Brereton, T. M., Warren, M. S. and Oates, M. (2001). The impacts of deer on woodland butterflies: the good, the bad and the complex. *Forestry* 74: 271–276.

Feeny, P. (1975). Biochemical evolution between plants and their insect herbivores. In *Coevolution of Animals and Plants*, ed. L. E. Raven and P. H. Raven. Austin, TX: University of Texas Press, pp. 3–19.

Feltwell, J. (1982). *Large White Butterfly – the Biology, Biochemistry and Physiology of Pieris brassicae (Linnaeus)*. Hague: Dr W. Junk Publishers.

Field, R. G., Gardiner, T., Mason, C. F. and Hill, J. (2005). Agri-environment schemes and butterflies: the utilization of 6m grass margins. *Biodiversity and Conservation* 14: 1969–1976.

Field, R. G., Gardiner, T. and Watkins, G. (2007). The use of farmland by butterflies: a study on mixed farmland and field margins. *Entomologist's Gazette* 58: 3–15.

Fielding, A. H. and Haworth P. E. (1995). Testing the generality of bird-habitat models. *Conservation Biology* 9: 1466–1481.

Findlay, R., Young, M. R. and Findlay, J. A. (1983). Orientation behaviour in the grayling butterfly: thermoregulation or crypsis? *Ecological Entomology* 8: 145–153.

Fischer, K. and Fiedler, K. (2000). Response of the copper butterfly *Lycaena tityrus* to increased leaf nitrogen in natural foods: evidence against the nitrogen limitation hypothesis. *Oecologia* 124: 235–241.

Fischer, K. and Fiedler, K. (2001). Spatio-temporal dynamics in a population of the copper butterfly *Lycaena hippothoe*. *Nota lepidopterologica* 24: 77–86.

Floater, G. J. (2001). Habitat complexity, spatial interference, and 'minimum risk distribution': a framework for population stability. *Ecological Monographs* 71: 447–468.

Ford, E. B. (1945). *Butterflies*. London: Collins. (Third edition, 1957.)

Ford, E. B. (1964). *Ecological Genetics*. London: Methuen. (Second edition, 1975.)

Ford, H. (2008). *High Browns versus Dark Greens, a Fritillary Response to Management in Morecambe Bay and Exmoor*. MSc dissertation, University of York.

Ford, T. H., Shaw, M. R. and Robertson, D. M. (2000). Further host records of some West Palaearctic Tachinidae (Diptera). *Entomologist's Record and Journal of Variation* 112: 25–36.

Forman, R. T. T. (1991). Landscape corridors: from theoretical foundation to public policy. In *Nature Conservation: the Role of Corridors*, ed. D. A. Saunders and R. J. Hobbs. Chipping Norton, Australia: Surrey Beatty & Sons, pp. 71–84.

Forman, R. T. T. (1995a). *Land Mosaics: the Ecology of Landscapes and Regions*. Cambridge: Cambridge University Press.

Forman, R. T. T. (1995b). Some general principles of landscape and regional ecology. *Landscape Ecology* 10: 133–142.

Forman, R. T. T. and Godron, M. (1986). *Landscape Ecology*. New York: Wiley.

Fortin, M-J. and Dale, M. (2005). *Spatial Analysis. A Guide for Ecologists*. Cambridge: Cambridge University Press.

Fowler, C. W. and Baker, J. D. (1991). A review of animal population dynamics at extremely reduced population levels. *Report of the International Whaling Commisssion* 41: 545–554.

Fowles, A. P. and Smith, R. G. (2006). Mapping the habitat quality of patch networks for the marsh fritillary *Euphydryas aurinia* (Rottemburg, 1775) (Lepidoptera, Nymphalidae) in Wales. *Journal of Insect* Conservation 10: 161–177.

Fox, B. W. (1996). *Alternative Feeding Strategies of the White Admiral (Ladoga camilla L.) and the Broad-bordered Bee-Hawk Moth (Hemaris fuciformis L.) on Honeysuckle (Lonicera periclymenum L.)*. PhD thesis, University of Southampton.

Fox, B. W. (2005). The larva of the white admiral butterfly, *Limenitis camilla* (Linnaeus, 1764) – a master builder. *Entomologist's Gazette* 56: 225–236.

Fox, R. (2007). Red admiral *Vanessa atalanta* (L.) (Lep.: Nmyphalidae) courtship in winter. *Entomologist's Record and Journal of Variation* 119: 75–76.

Fox, R. and Sleep, T. (2005). Red admirals: under-sexed and over here? *Atropos* 25: 66–67.

Fox, R. and Williams, L. (2006). A commentary on recent changes to butterfly distributions in the London area. *Entomologist's Record and Journal of Variation* 118: 69–84.

Fox, R., Asher, J., Brereton, T., Roy, D. and Warren, M. (2006). *The State of Butterflies in Britain and Ireland*. Newbury, UK: Pisces Publications.

Franco, A. M. A., Hill, J. K., Kitschke, C. *et al.* (2006). Impacts of climate warming and habitat loss on extinctions at species' low-latitude range boundaries. *Global Change Biology* 12: 1545–1553.

Freeman, C., Ostle, N. and Kang, H. (2001). An enzymic 'latch' on a global carbon store. *Nature* 409: 149.

Freitas, A. V. L. and Oliveira, P. S. (1996). Ants as selective agents on herbivore biology: effects on the behaviour of a non-myrmecophilous butterfly. *Journal of Animal Ecology* 65: 205–210.

Fretwell, S. D. and Lucas, H. L. Jr (1970). On territorial behaviour and other factors influencing habitat distribution in birds. *Acta biotheoretica* 19: 16–36.

Frohawk, F. W. (1934). *The Complete Book of British Butterflies*. London: Ward Locke.

Fry, G. (1995). Landscape ecology of insect movement in arable ecosystems. In *Ecology and Integrated Farming Systems*, ed. D. M. Glen, M. P. Greaves and H. M. Anderson. Chichester: John Wiley and Sons, pp. 177–202.

Fry, G. L. A. and Main, A. R. (1993). Restoring seemingly natural communities on agricultural land. In *Nature Conservation 3: Reconstruction of Fragmented Ecosystems*, ed. D. A. Saunders, R. J. Hobbs, and P. R. Ehrlich. Chipping Norton, UK: Surry Beatty and Sons, pp. 225–241.

Fry, G. L. A. and Robson, W. J. (1994). The effects of field margins on butterfly movement. In *Field Margins: Integrating Agriculture and Conservation*, ed. N. D. Boatman. Farnham: British Crop Protection Council, pp. 111–116.

Fuller, R. M. (1993). The land cover map of Great Britain. *Earth Space Review* 2: 13–18.

Fuller, R. M. and Brown, N. J. (1994). *A CORINE Map of Great Britain by Automated Means: a Feasibility Study*. ITE report to the Department of the Environment, London.

Fuller, R. M. and Brown, N. J. (1996). A CORINE map of Great Britain by automated means. Techniques for automated generalization of the Land Cover Map of Great Britain. *International Journal of Geographical Information Systems* 8: 937–953.

Fuller, R. M., Groom, G. B. and Jones, A. R. (1994a). The land cover map of Great Britain: an automated classification of Landsat Thematic Mapper data. *Photogrammetric Engineering and Remote Sensing* 60: 553–562.

Fuller, R. M., Groom, G. B. and Wallis, S. M. (1994b). The availability of Landsat TM images for Great Britain. *International Journal of Remote Sensing* 15: 1357–1362.

Futuyama, D. J. (1976). Foodplant specialisation and environmental predictability in Lepidoptera. *American Naturalist* 110: 285–292.

Futuyama, D. J. (1983). Selective factors in the evolution of host choice by phytophagous insects. In *Insects. Host-Seeking Behaviours and Mechanisms*, ed. S. Ahmas. New York: Academic Press, pp. 227–244.

Game, M. (1980). Best shape for nature reserves. *Nature* 287: 630–632.

Garcia-Barros, E. (2000). Comparative data on the adult biology, ecology and behaviour of species belonging to the genera *Hipparchia*, *Chazara* and *Kanetisa* in Central Spain. *Nota Lepidopterologica* 23: 119–140.

Gardiner, M. J. and Radford, T. (1980). *Ireland: General Soil Map*, 2nd edn. Dublin: Teagasc (formerly An Foras Taluntais).

Gardiner, T. and Gardiner, M. (2008). Altitudinal limits of insects on Snowdon. *Bulletin of the Amateur Entomologist's Society* 67: 204–206.

Garland, M. S. and Davis, A. K. (2002). An examination of monarch butterfly (*Danaus plexippus*) autumn migration in coastal Virginia. *American Midland Naturalist* 147: 170–174.

Garland, S. P. (1981). *Butterflies of the Sheffield Area*. Sheffield: Sorby Natural History Society and Sheffield City Museums.

Gaston, K. J. (1994). *Rarity*. London: Chapman and Hall.

Gaston, K. J. and McArdle, B. H. (1993). All else is not equal: temporal population variability in insect conservation. In *Perspectives on Insect Conservation*, ed. K. J. Gaston, T. R. New and M. J. Samways. Andover, UK: Intercept Ltd, pp. 171–184.

Gaston, K. J., Blackburn, T. M. and Lawton, J. H. (1997). Interspecific abundance–range size relationships: an appraisal of mechanisms. *Journal of Animal Ecology* 66: 579–601.

Gaston, K. J., Cush, P., Ferguson, S. *et al.* (2007). Improving the contribution of urban gardens for wildlife: some guiding propositions. *British Wildlife* 18: 171–177.

Geiger, R. (1965). *The Climate near the Ground*. Cambridge, MA: Harvard University Press.

Gilbert, L. E. (1980). Food web organisation and the conservation of neotropical diversity. In *Conservation Biology: an Evolutionary-Ecological Perspective*, ed. M. Soulé and B. A. Wilcox. Sunderland, MA: Sinauer Associates, Inc., pp. 11–33.

Gilbert, L. E. and Singer, M. C. (1973). Dispersal and gene flow in a butterfly species. *American Naturalist* 107: 58–72.

Gilbertson, D. D., Kent, M. and Pyatt, F. B. (1985). *Practical Ecology*. London: Chapman and Hall.

Giles, R. H. and Trani, M. K. (1999). Key elements of landscape pattern measures. *Environmental Management* 23: 477–481.

Gill, W. J. (1987). *The Conservation of Frangula alnus on Brereton Heath Country Park, Cheshire, UK*. BA dissertation, University of Durham.

Gillam, B. (1998). White-letter hairstreak ovipositing sites. *Entomologist's Record and Journal of Variation* 110: 133.

Gilpin, M. E. (1991). The genetic effective size of a metapopulation. In *Metapopulation Biology*, ed. I. A. Hanski and M. E. Gilpin. London: Academic Press, pp. 165–175.

Gkaraveli, A., Williams, J. H. and Good, J. E. G. (2001). Fragmented native woodlands in Snowdonia (UK): assessment and amelioration. *Forestry* 74: 89–103.

Glasser, N. F. (2002). Scottish landform example 27: the large roches moutonnées of upper Deeside. *Scottish Geographical Journal* 118(2): 129–138.

Glenn-Lewin, D. C. and Van der Maarel, E. (1992). Patterns and processes of vegetation dynamics. In *Plant Succession*, ed. D. C. Glenn-Lewin, R. K. Peet and Th. T. Veblen. London: Chapman and Hall, pp. 11–59.

Glowska, L., Burnhenne-Guilmin, F., Synge, H., McNeely, J. A. and Gundling, L. (1994). *A Guide to the Convention on Biological Diversity*. Gland, Switzerland: IUCN.

Godwin, H. and Conway, V. M. (1939). The ecology of a raised bog near Tregaron, Cardiganshire. *Journal of Ecology* 27: 313–359.

Golden, D. M. and Crist, T. O. (1999). Experimental effects of habitat fragmentation on old-field canopy insects: community, guild and species responses. *Oecologia* 118: 371–380.

Gotelli, N. J. (1991). Metapopulation models: the rescue effect, the propagule rain, and the core–satellite hypothesis. *American Naturalist* 138: 768–776.

Gotelli, N. J. and Kelly, W. G. (1993). A general model of metapopulation dynamics. *Oikos* 68: 36–44.

Goudie, A. (1993). *The Human Impact on the Natural Environment*. Oxford: Blackwell.

Goulson, D. (1991). *Maintenance of Phenotypic Variation in the Butterfly, Maniola jurtina*. PhD thesis, Oxford Polytechnic.

Goulson, D. (1993). The evolutionary significance of bimodal emergence in the butterfly, *Maniola jurtina* (Lepidoptera: Satyrinae) (L.). *Biological Journal of the Linnean Society* 49: 127–139.

Goulson, D. (2000). Why do pollinators visit proportionally fewer flowers in large patches? *Oikos* 91: 485–492.

Goulson, D. and Cory, J. S. (1993). Flower constancy and learning in foraging preferences of the green-veined white butterfly *Pieris napi*. *Ecological Entomology* 18: 315–320.

Goulson, D., Ollerton, J. and Schuman, C. (1997a). Foraging strategies in the small skipper butterfly, *Thymelicus flavus*: when to switch? *Animal Behaviour* 53: 1009–1016.

Goulson, D., Stout, J. C. and Hawson, S. A. (1997b). Can flower constancy in nectaring butterflies be explained by Darwin's interference hypothesis? *Oecologia* 112: 225–231.

Graves, S. D. and Shapiro, A. M. (2003). Exotics as host plants of the California butterfly fauna. *Biological Conservation* 110: 413–433.

Gray, J. M. (2001). Geomorphological conservation and public policy in England: a geomorphological critique of English Nature's 'Natural Areas' approach. *Earth Surface Processes and Landforms* 26: 1009–1023.

Greatorex-Davies, J. N., Sparks, T. H., Hall, M. L. and Marrs, R. H. (1993). The influence of shade on butterflies in rides of coniferised lowland woods in southern England and implications for conservation management. *Biological Conservation* 63: 31–41.

Greatorex-Davies, J. N., Brereton, T. B., Roy, D. B., Middlebrook, I. and Cruickshanks, K. L. (2007). *United Kingdom Butterfly Monitoring Scheme report for 2006*. Huntingdon, UK: CEH Monks Wood.

Greeney, H. F. and Jones, M. T. (2003). Shelter building in the Hesperiidae: a classification scheme for larval shelters. *Journal of Research on the Lepidoptera* 37: 27–36.

Greig-Smith, P. (1964). *Quantitative Plant Ecology*. London: Butterworths.

Grime, J. P. (1974). Vegetation classification by reference to strategies. *Nature* 250: 26–31.

Grime, J. P. (1979). *Plant Strategies and Vegetation Processes*. Chichester: Wiley.

Grime, J. P., Hodgson, J. G. and Hunt, R. (1988). *Comparative Plant Ecology. A Functional Approach to Common British Species*. London: Unwin Hyman. (2nd edition, 2007. Dalbeattie: Castlepoint Press.)

Grossi, L., Zurlini, G. and Rossi, O. (2001). Statistical selection of multiscale landscale patterns. *Environmental and Ecological Statistics* 8: 253–267.

Guisan, A. and Thuiller, W. (2005). Predicting species distribution: offering more than simple habitat models. *Ecology Letters* 8: 993–1009.

Guisan, A. and Zimmerman, N. E. (2000). Predictive habitat distribution models in ecology. *Ecological Modelling* 135: 147–186.

Gullan, P. J. and Cranston, P. S. (2005). *The Insects. An Outline of Entomology*. Oxford: Blackwell Publishing.

Gutiérrez, D. (2005). Effectiveness of existing reserves in the long-term protection of a regionally rare butterfly. *Conservation Biology* 19: 1586–1597.

Gutiérrez, D. and Thomas, C. D. (2000). Marginal range expansion in a host-limited butterfly species *Gonepteryx rhamni*. *Ecological Entomology* 25: 165–170.

Gutiérrez, D., Thomas, C. D. and Léon-Cortés, J. L. (1999). Dispersal, distribution, patch-network and metapopulation dynamics of the dingy skipper butterfly (*Erynnis tages*). *Oecolgia* 121: 506–517.

Gutiérrez, D., León-Cortés, J. L., Menéndez, R., Wilson, R. J., Cowley, M. J. R. and Thomas, C. D. (2001). Metapopulations of four Lepidopteran herbivores on a single host plant, *Lotus corniculatus*. *Ecology* 82: 1371–1386.

Gyllenberg, M., Hanski, I. A. and Hastings, A. (1997). Structured metapopulation models. In *Metapopulation Biology. Ecology, Genetics, and Evolution*, ed. I. A. Hanski and M. E. Gilpin. London: Academic Press, pp. 93–122.

Haggett, P. (1965). *Locational Analysis in Human Geography*. London: Edward Arnold.

Haikola, S., Fortelius, W., O'Hara, R. B. *et al.* (2001). Inbreeding depression and the maintenance of genetic load in *Melitaea cinxia* metapopulations. *Conservation Genetics* 2: 325–335.

Haines-Young, R. and Chopping, M. (1996). Quantifying landscape structure: a review of landscape indices and their application to forested landscapes. *Progress in Physical Geography* 20: 418–445.

Haining, R. (1990). *Spatial Data Analysis in the Social and Environmental Sciences*. Cambridge: Cambridge University Press.

Hainsworth, F. R. and Hamill, T. (1993). Foraging rules for nectar – food choice by painted ladies. *American Naturalist* 142: 857–867.

Hall, J. P. W. and Willmott, K. R. (2000). Patterns of feeding behaviour in adult riodinid butterflies and their relationship to morphology and ecology. *Biological Journal of the Linnean Society* 69: 1–23.

Hall, L. S., Krausman, P. R. and Morrison, M. L. (1997). The habitat concept and a plea for standard terminology. *Wildlife Society Bulletin* 25: 173–182.

Hanski, I. A. (1982). Dynamics of regional distribution: the core and satellite species hypothesis. *Oikos* 38: 210–221.

Hanski, I. A. (1991). Single-species metapopulation dynamics. In *Metapopulation Dynamics: Empirical and Theoretical Investigations*, ed. M. Gilpin and I. Hanski. London: Academic Press, pp. 17–38.

Hanski, I. A. (1994a). A practical model of metapopulation dynamics. *Journal of Animal Ecology* 63: 151–162.

Hanski, I. A. (1994b). Patch-occupancy dynamics in fragmented landscapes. *Trends in Ecology and Evolution* 9: 131–135.

Hanski, I. A. (1997). Metapopulation dynamics. From concepts and observations to predictive models. In *Metapopulation Biology. Ecology, Genetics, and Evolution*, ed. I. A. Hanski and M. E. Gilpin. London: Academic Press, pp. 69–91.

Hanski, I. A. and Gilpin, M. E. (1997). *Metapopulation Biology. Ecology, Genetics, and Evolution*. London: Academic Press.

Hanski, I. A. and Kussaari, M. (1995). Butterfly metapopulation dynamics. In *Population Dynamics: New Approaches and Synthesis*, ed. N. Cappuccino and P. W. Price. San Diego, CA: Academic Press, pp. 149–172.

Hanski, I. A. and Pöyry, J. (2007). Insect populations in fragmented habitats. In *Insect Conservation Biology*, ed. A. Stewart, O. Lewis and T. New. Proceedings of the Royal Entomological Society's 23rd Symposium. Wallingford, UK: CABI, pp. 175–202.

Hanski, I. A. and Simberloff, D. (1997). The metapopulation approach, its history, conceptual domain, and application to conservation. In *Metapopulation Biology. Ecology, Genetics, and Evolution*, ed. I. A. Hanski and M. E. Gilpin. London: Academic Press, pp. 5–26.

Hanski, I. A. and Singer, M. C. (2001). Extinction-colonization dynamics and host-plant choice in butterfly metapopulations. *American Naturalist* 158: 341–353.

Hanski, I. A. and Thomas, C. D. (1994). Metapopulation dynamics. A spatially explicit model applied to butterflies. *Biological Conservation* 68: 167–180.

Hanski, I., Kouki, J. and Halkka, A. (1993a). Three explanations of the positive relationship between distribution and abundance of species. In *Historical and Geographical Determinants of Community Diversity*, ed. R. Ricklefs and D. Schluter. Chicago: University of Chicago Press, pp. 108–116.

Hanski, I., Woiwod, I. P. and Perry, J. N. (1993b). Density-dependence, population persistence and largely futile arguments. *Oikos* 95: 595–598.

Hanski, I., Moilanen, A. and Gyllenberg, M. (1996). Minimum viable metapopulation size. *American Naturalist* 147: 527–541.

Hanski, I., Ehrlich, P. R., Nieminen, M., Murphy, D. D., Hellmann, J. J., Boggs, C. L. and McLaughlin, J. F. (2004). Checkerspots and conservation biology. In *On the Wings of Checkerspots. A Model System for Population Biology*, ed. P. R. Ehrlich and I. Hanski. Oxford: Oxford University Press, pp. 264–287.

Hardy, P. B. (1998). *Butterflies of Greater Manchester*. Sale, UK: PGL Enterprises.

Hardy, P. B. and Dennis, R. L. H. (1999). The impact of urban development on butterflies within a city region. *Biodiversity and Conservation* 8: 1261–1279.

Hardy, P. B. and Dennis, R. L. H. (2007). Seasonal and daily shifts in substrate use by settling butterflies: conserving resources for invertebrates has a behavioral dimension. *Journal of Insect Behavior* 20: 181–199.

Hardy, P. B. and Dennis, R. L. H. (2008). Resources for British butterflies (Lepidoptera: Hesperioidea, Papilionoidea). The alien consumer component and its significance for butterfly habitats. *European Journal of Entomology*, 105: 649–657.

Hardy, P. B. and Kinder, P. M. (2000). Hill-topping by red admirals *Vanessa atalanta* L. (Lep.: Nymphalidae). *Entomologist's Record and Journal of Variation* 112: 22–23.

Hardy, P. B. and Kinder, P. M. (2007). Not quite hill-topping in butterflies? *Entomologist's Gazette* 58: 169–170.

Hardy, P. B., Hind, S. H. and Dennis, R. L. H. (1993). Range extension and distribution-infilling among selected butterfly species in north-west England: evidence for inter-habitat movements. *Entomologist's Gazette* 44: 247–255.

Hardy, P. B., Sparks, T. H., Isaac, N. J. B. and Dennis, R. L. H. (2007) Specialism for larval and adult consumer resources among British butterflies: implications for conservation. *Biological Conservation* 138: 440–452.

Harker, R. J. and Shreeve, T. G. (2008). Recording the abundance of butterflies: how reliable is transect recording? *Journal of Insect Conservation* 12: 125–133.

Harmer, R. (1994a). Natural regeneration of broadleaved trees in Britain: I. Historical aspects. *Forestry* 67: 179–188.

Harmer, R. (1994b). Natural regeneration of broadleaved trees in Britain: II. Seed production and predation. *Forestry* 67: 287–286.

Harmer, R. (1995). Natural regeneration of broadleaved trees in Britain: III. Germination and establishment. *Forestry* 68: 1–9.

Harmer, R. (1999). Survival and new shoot production by artifically browsed seedlings of ash, beech, oak and sycamore grown under different levels of shade. *Forest Ecology and Management* 116: 39–50.

Harmer, R. (2001). The effect of plant competition and simulated summer browsing by deer on tree regeneration. *Journal of Applied Ecology* 38: 1094–1103.

Harmer, R., Peterken, G., Kerr, G. and Poulton, P. (2001). Vegetation changes during 100 years of development of two secondary woodlands on abandoned arable land. *Biological Conservation* 101: 291–304.

Harrison, S. (1991). Local extinction in a metapopulation context: an empirical evaluation. In *Metapopulation Dynamics: Empirical and Theoretical Investigations*, ed. M. E. Gilpin and I. A. Hanski. London: Academic Press, pp. 73–88.

Harrison, S. (1994). Metapopulations and conservation. In *Large-Scale Ecology and Conservation Biology*, ed. P. J. Edwards, N. R. Webb and R. M. May. Oxford: Blackwell, pp. 111–128.

Harrison, S. and Taylor, A. D. (1997). Empirical evidence for metapopulation dynamics. In *Metapopulation Biology*, ed. I. A. Hanski and M. E. Gilpin. London: Academic Press, pp. 27–42.

Harrison, S., Murphy, D. D. and Ehrlich, P. R. (1988). Distribution of the bay checkerspot butterfly *Euphydryas editha bayensis*. Evidence for a metapopulation model. *American Naturalist* 132: 360–382.

Hart, M. G. (1986). *Geomorphology Pure and Applied*. London: George Allen and Unwin.

Harvey, P. H. and Pagel, M. D. (1991). *The Comparative Method in Evolutionary Ecology*. Oxford: Oxford University Press.

Haslett, J. R. (2007). *European Strategy for the Conservation of Invertebrates*. Nature and Environment No. 145. Strasbourg Cedex, France: Council of Europe Publishing, 91 pp.

Haughton, A. J., Bond, A. J., Lovett, A. A. *et al.* (2009). A novel, integrated approach to assessing social, economic implications of changing rural land-use: a case study of perennial biomass crops. *Journal of Applied Ecology* 46: 315–322.

Hawkins, B. A. and Porter, E. E. (2003). Water–energy balance and the geographical pattern of species richness of western Palearctic butterflies. *Ecological Entomology* 28: 678–686.

Heath, J. (ed.) (1976). *The Moths and Butterflies of Great Britain and Ireland*, Vol. 1. London: Curwen.

Heath, J., Pollard, E. and Thomas, J. A. (1983). *Atlas of Butterflies in Britain and Ireland*. Middlesex: Viking.

Hedelin, H. and Rydell, J. (2007). Daily habitat shifts by the Neotropical butterfly *Manataria maculata* (Nymphalidae: Satyrinae) is driven by predation. *Journal of the Lepidopterists' Society* 61: 67–71.

Heikkinen, R. K., Luoto, M., Kuussaari, M. and Pöyry, J. (2005). New insights into butterfly–environment relationships using partitioning methods. *Proceedings of the Royal Society B* 272: 2203–2210.

Heinz, S. K., Wissel, C. and Frank, K. (2006). The viability of metapopulations: individual dispersal behaviour matters. *Landscape Ecology* 21: 77–89.

Hellmann, J. J. (2002). The effect of an environmental change on mobile butterfly larvae and the nutritional quality of their hosts. *Journal of Animal Ecology* 71: 925–936.

Hellmann, J. J., Weiss, S. B., McLaughlin, J. F., Ehrlich, P. R., Murphy, D. D. and Launer, A. E. (2004). Structure and dynamics of *Euphydryas editha* populations. In *On the Wings of Checkerspots. A Model System for Population Biology*, ed. P. R. Ehrlich and I. Hanski. Oxford: Oxford University Press, pp. 34–62.

Helps, M. B. (1994). Field margins – an agricultural perspective. In *Field Margins: Integrating Agriculture and Conservation*, ed. N. D. Boatman. Farnham: British Crop Protection Council, pp. 21–30.

Henderson, P. A. (2003). *Practical Methods in Ecology*. Oxford: Blackwell Publishing.

Herzog, F. and Lausch, A. (2001). Supplementing land-use statistics with landscape metrics: some methodological considerations. *Environmental Monitoring and Assessment* 72: 37–50.

Heslop Harrison, J. W. (1950). A dozen year's biogeography researches in the Inner and Outer Hebrides. *Proceedings of the University of Durham Philosophical Society* 10: 516–524.

Hickling, R., Roy, D. B., Hill, J. K., Fox, R. and Thomas, C. D. (2006). The distributions of a wide range of taxonomic groups are expanding polewards. *Global Change Biology* 1: 450–455.

Hill, J. K., Thomas, C. D. and Lewis, O. T. (1996). Effects of habitat patch size and isolation on dispersal by *Hesperia comma* butterflies: implications for metapopulation structure. *Journal of Animal Ecology* 65: 725–735.

Hill, J. K., Thomas, C. D. and Huntley, B. (1999a). Climate and habitat availability determine 20th century changes in a butterfly's range margin. *Proceedings of the Royal Society of London B* 266: 1197–1206.

Hill, J. K., Thomas, C. D. and Lewis, O. T. (1999b). Flight morphology in fragmented populations of a rare British butterfly, *Hesperia comma*. *Biological Conservation* 87: 277–283.

Hill, J. K., Collingham, Y. C., Thomas, C. D., Balkeley, D. S., Fox, R., Moss, D. and Huntley, B. (2001). Impacts of landscape structure on butterfly range expansion. *Ecology Letters* 4: 313–321.

Hill, J. K., Thomas, C. D., Fox, R., Telfer, M. G., Willis, S. G., Asher, J. and Huntley, B. (2002). Responses of butterflies to twentieth century climate warming: implications for future ranges. *Proceedings of the Royal Society of London B* 269: 2163–2171.

Hill, M. O., Mountford, J. O., Roy, D. B. and Bunce, R. G. H. (1999). *Ellenberg's Indicator Values for British Plants*. ECOFACT Vol. 2 Technical Annex. Huntingdon: Centre for Ecology and Hydrology.

Hill, M. O., Roy, D. B. and Thompson, K. (2002). Hemeroby, urbanity and ruderality: bioindicators of disturbance and human impact. *Journal of Applied Ecology* 39: 708–720.

Hill, M. O., Baker, R., Broad, G. *et al.* (2005). Audit of non-native species in England. *English Nature Research Reports* 662: 1–81.

Hilty, J. and Merenlender, A. (2000). Faunal indicator taxa selection for monitoring ecosystem health. *Biological Conservation* 92: 185–197.

Hinton, H. E. (1951). *Myrmecophilous lycaenidae* – a summary. *Proceedings of the South London Entomological and Natural History Society* 1949–1950: 111–175.

Hobbs, R. J. (1992). The role of corridors in conservation: solution or bandwagon? *Trends in Ecology and Evolution* 7: 389–392.

Hobson, R., Bourn, N. A. D. and Warren, M. S. (2002). Conserving the marsh fritillary in Britain. *British Wildlife* 13: 404–411.

Hochberg, M. E. (1991). Extra-host interactions between a braconid endoparasite, *Apanteles glomeratus*, and a baculovirus for larvae of *Pieris brassicae*. *Journal of Animal Ecology* 60: 65–77.

Hochberg, M. E., Thomas, J. A. and Elmes, G. W. (1992). A modelling study of the population dynamics of a large blue butterfly, *Maculinea rebeli*, a parasite of red ant nests. *Journal of Animal Ecology* 61: 397–409.

Hodges, J. (2005). Cheap food and feeding the world sustainably. *Livestock Production Science* 92: 1–16.

Hodgson, J. (2007). *Butterfly Metapopulations in Dynamic Habitats*. PhD thesis, University of York.

Hodgson, J. G. (1993). Commonness and rarity in British butterflies. *Journal of Applied Ecology* 30: 407–427.

Hodgson, J. G., Wilson, P. J., Hunt, R., Grime, J. P. and Thompson, K. (1999). Allocating C-S-R plant functional types: a soft approach to a hard problem. *Oikos* 85: 282–294.

Hodgson, M. and Thompson, R. (1985). Uncovering the secrets of soil. *New Scientist* 14 Nov: 44–47.

Holloway, G. J., Griffiths, G. H. and Richardson, P. (2003). Conservation strategy maps: a tool to facilitate biodiversity action planning illustrated using the heath fritillary butterfly. *Journal of Applied Ecology* 40: 413–421.

Holt, R. D. (1985). Population dynamics in two-patch environments: some anomalous consequences of an optimal habitat distribution. *Theoretical Population Biology* 28: 181–208.

Holt, R. D. (1990). The microevolutionary consequences of climate change. *Trends in Ecology and Evolution* 5: 311–315.

Holt, R. D. (1997). From metapopulation dynamics to community structure. Some consequences of spatial heterogeneity. In *Metapopulation Biology. Ecology, Genetics, and Evolution*, ed. I. A. Hanski and M. E. Gilpin. London: Academic Press, pp. 149–164.

Holyoak, M. (1993). New insights into testing for density-dependence. *Oecologia* 93: 435–444.

Holyoak, M. and Lawton, J. H. (1993). Comment arising from a paper by Wolda and Dennis: using and interpreting the results of tests for density-dependence. *Oecologia* 95: 592–594.

Hoole, J. C., Joyce, D. A. and Pullin, A. S. (1999). Estimates of gene flow between populations of the swallowtail butterfly, *Papilio machaon* in Broadland, UK and implications for conservation. *Biological Conservation* 89: 293–299.

Hoskins, W. G. (1969). *The Making of the English Landscape*. London: Hodder and Stoughton.

Howe, M., Whitehouse, A. and Knight, G. (2008). Life on the edge – key coastal soft cliffs for invertebrates in England and Wales. *British Wildlife* 19: 172–181.

Howe, P. D. (2004). *The Ecological Consequences of Morphological Variation in the Common Blue Butterfly Polyommatus icarus (Rott.) in the United Kingdom*. PhD thesis, Oxford Brookes University.

Howe, P. D., Bryant, S. R. and Shreeve, T. G. (2007). Predicting body temperature and activity of adult *Polyommatus icarus* using neural network models under current and projected climate scenarios. *Oecologia* 153: 857–869.

Hubbes, T. (1999). The American elm and Dutch elm disease. *Forestry Chronicle* 75: 265–273.

Huby, M. (2001). The sustainable use of resources on a global scale. *Social Policy and Administration* 35: 521–537.

Hugie, D. M. and Grand, T. C. (2003). Movements between habitats by unequal competitors: effects of finite population size on ideal free distributions. *Evolutionary Ecology Research* 5: 131–153.

Huisman, J., Olff, H. and Fresco, L. F. M. (1993). A hierarchical set of models for species response analysis. *Journal of Vegetation Science* 4: 37–46.

Hulshoff, R. M. (1995). Landscape indexes describing a Dutch landscape. *Landscape Ecology* 10: 101–111.

Hunsaker, C. T., O'Neill, R. V., Jackson, B. L., Timmins, S. P., Levine, D. A. and Norton, D. J. (1994). Sampling to characterize landscape pattern. *Landscape Ecology* 9: 207–226.

Hunter, M. D. (2003). Effects of plant quality on the population ecology of parasitoids. *Agricultural and Forest Entomology* 5: 1–8.

Huntley, B. (1991). How plants respond to climatic change: migration rates, individualism and the consequences for plant communities. *Annals of Botany* 67: 15–22.

Hutchinson, G. E. (1957). Concluding remarks. *Cold Spring Harbor Symposia on Quantitative Biology* 22: 415–427.

Hutchinson, G. E. (1978). *An Introduction to Population Ecology*. New Haven, CT: Yale University Press.

Idle, E. T. (1995). Conflicting priorities in site management in England. *Biodiversity and Conservation* 4: 929–937.

IUCN (International Union for the Conservation of Nature and Natural Resources) (1980). *World Conservation Strategy*. Gland, Switzerland: IUCN.

Isingrini, M., Lenoir, A. and Jaisson, P. (1985). Preimaginal learning as a basis of colony-brood recognition in the ant *Cataglyphis* cursor. *Proceedings of the National Academy of Sciences* 82: 8545–8547.

Jaeger, J. A. G. (2000). Landscape division, splitting index, and effective mesh size: new measures of landscape fragmentation. *Landscape Ecology* 15: 115–130.

Jaenike, J. (1978). On optimal oviposition behavior in phytophagous insects. *Theoretetical Population Biology* 14: 350–356.

Janz, N. (2003). The cost of polyphagy: oviposition decision time *vs* error rate in a butterfly. *Oikos* 100: 493–496.

Janz, N., Nylin, S. and Wedell, N. (1994). Host plant utilization in the comma butterfly: sources of variation and evolutionary implications. *Oecologia* 99: 132–140.

Jeffcoate, S. (2006). Seasonal variation in the use of vegetation resources by *Leptidea sinapis* (Linnaeus, 1758) (Lepidoptera: Pieridae), a multivoltine species in southern Britain: implications for its conservation at the edge of its range and in the context of climate change. *Entomologist's Gazette* 57: 69–82.

Jiménez-Valverde, A., Lobo, J. M. and Hortal, J. (2008). Not as good as they seem: the importance of concepts in species distribution modelling. *Diversity and Distributions*, 14: 885–890.

John, E. (2001). Migration of *Vanessa cardui* (L.) (Lep.: Nymphalidae) through Cyprus, March 2001. *Entomologist's Record and Journal of Variation* 113: 269–281.

John, E. (2004). *Sturmia bella* (Meigen) (Dipt.: Tachinidae) new to Wales. *Entomologist's Record and Journal of Variation* 116: 75–77.

John, E., Gascoigne-Pees, M. and Larsen, T. B. (2010). *Ypthima asterope* (Klug, 1832) (Lepidoptera: Nymphalidae, Satyrinae): its biogeography, lifecycle, ecology and present status in Cyprus, with additional notes from Rhodes and the eastern Mediterranean. *Entomologist's Gazette* 61, in press.

Johnson, A. R., Milne, B. T. and Wiens, J. A. (1992). Diffusion in fractal landscapes; simulations and experimental studies of tenebrionid beetle movements. *Ecology* 73: 1968–1983.

Johnson, C. G. (1969). *Migration and Dispersal of Insects by Flight*. London: Methuen.

Johnson, N., Payton, P. and Spalding, A. (1996). *The Conservation of Metalliferous Mine Sites in Cornwall*. Truro: Cornwall County Council, Cornwall Archaeological Unit and Institute of Cornish Studies.

Johst, K., Drechsler, M., Thomas, J. and Settele, J. (2006). Influence of mowing on the persistence of two endangered large blue butterfly species. *Journal of Applied Ecology* 43: 333–342.

Jones, R. E. (1977). Movement patterns and egg distribution in cabbage butterflies. *Journal of Animal Ecology* 46: 195–212.

Jonson, A. C. J., Land, M. F., Osorio, D. C. and Nilsson, D. E. (1998). Relationships between pupil working range and habitat luminance in flies and butterflies. *Journal of Comparative Physiology A – Sensory Neural and Behavioral Physiology* 182: 1–9.

Jordan, A. and Bourn, N. A. D. (2007). *The Status, Mobility and Habitat Requirements of the Small Pearl-bordered Fritillary (Boloria selene) in Clocaenog Forest: a Summary of the 2006 Status*. Wareham, UK: Butterfly Conservation.

Joy, J. (1991). The ecology and life history of the large heath butterfly (*Coenonympha tullia*) on the Shropshire/Clwyd mosses. *British Ecological Society Bulletin* 22: 114–118.

Joy, J. (1996a). The larval habits of the grayling butterfly *Hipparchia semele* (L.) (Lep: Satyridae) at inland lead sites in Shropshire. *Entomologist's Gazette* 47: 139–142.

Joy, J. (1996b). How common is hill-topping in the painted lady, *Cynthia cardui*. *Butterfly Conservation News* 63: 17.

Joy, J. (2002a). *The Status and Habitat Requirements of the Small Pearl-Bordered Fritillary (Boloria selene) on Shropshire and Staffordshire Sites in 2002*. Unpublished report to English Nature, Peterborough.

Joy, J. (2002b). The Pearl-bordered Fritillary in the Wyre Forest 2002. Unpublished report to English Nature and Forest Enterprise.

Joy, J. (2005). *Small Pearl-bordered Fritillary (Boloria selene) in the Wyre Forest 2005. Survey Report*. Butterfly Conservation Report No. S05-48. Wareham, UK: Butterfly Conservation.

Joy, J. and Pullin, A. S. (1997). The effects of flooding on the survival and behaviour of overwintering large heath butterfly *Coenonympha tullia* larvae. *Biological Conservation* 82: 61–66.

Joy, J. and Pullin, A. S. (1999). Field studies on flooding and survival of overwintering large heath butterfly *Coenonympha tullia* larvae on Fenn's and Whixall mosses in Shropshire and Wrexham, UK. *Ecological Entomology* 24: 426–431.

Joy, J., Dennis, R. L. H., Miles, A. and Hinde, J. (1999). Atypical habitat choice by white admiral butterflies, *Ladoga camilla* (L.) (Lepidoptera: Nymphalidae), at the edge of their range. *Entomologist's Gazette* 50: 169–179.

Joyce, D. A. and Pullin, A. S. (2003). Conservation implications of the distribution of genetic diversity at different scales: a case study using the marsh fritillary butterfly (*Euphydryas aurinia*). *Biological Conservation* 114: 453–461.

Kadlec, K., Benes, J., Jarosik, V. and Konvicka, M. (2008). Revisiting urban refuges: changes of butterfly and burnet fauna in Prague reserves over three decades. *Landscape and Urban Planning* 85: 1–11.

Kandori, I. and Ohsaki, N. (1998). Effect of experience on foraging behaviour towards artificial nectar guide in the cabbage butterfly, *Pieris rapae crucivora* (Lepidoptera: Pieridae). *Applied Entomology and Zoology* 33: 35–42.

Kant, E. (1946). Den inre omflyttningen I Estland I samband med de estniska städernas omland. *Svensk Gegrafiska Arsbok* 22: 83–124.

Karieva, P. M. and Shigesada, N. (1983). Analysing insect movement as a correlated random walk. *Oecologia* 56: 234–238.

Karlsson, B. (1995). Resource allocation and mating systems in butterflies. *Evolution* 49: 955–961.

Karlsson, B. (1996). Male reproductive reserves in relation to mating system in butterflies: a comparative approach. *Proceedings of the Royal Society of London B* 263: 187–192.

Karlsson, B. and Wiklund, C. (2005). Butterfly life history and temperature adaptations; dry open habitats select for increased fecundity and longevity. *Journal of Animal Ecology* 74: 99–104.

Keane, R. M. and Crawley, M. J. (2002). Exotic plant invasions and the enemy release hypothesis. *Trends in Ecology and Evolution* 17: 164–170.

Keese, M. C. (1997). Does escape to enemy-free space explain host specialization in two closely related leaf-feeding beetles (Coleoptera: Chrysomelidae)? *Oecologia* 112: 81–86.

Kelber, A., Thunell, C. and Arikawa, K. (2001). Polarised-dependent colour vision in *Papilio* butterflies. *Journal of Experimental Biology* 204: 2469–2480.

Kellman, M. C. (1975). *Plant Geography*. London: Methuen and Co. Ltd.

Kelly, C. J. (2008). *Spatial Pattern and Association of Host Plant and Oviposition Choice of the Heath Fritillary Butterfly Melitaea athalia on Unimproved Grassland*. BSc dissertation, University of Plymouth.

Kemp, D. J. and Wiklund, C. (2001). Fighting without weaponry: a review of male–male contest competition in butterflies. *Behavioral Ecology and Sociobiology* 49: 429–442.

Kemp, R. (1998). Butterflies – importance of larval foodplant's life cycle with particular reference to the brown argus (*Aricia agestis*). *Bulletin of the Amateur Entomologists' Society* 57: 225–227.

Kemp, R., Hardy, P. B., Roy, D. and Dennis, R. L. H. (2008). The relative exploitation of annuals as larval host plants by phytophagous Lepidoptera. *Journal of Natural History* 42: 1079–1093.

Kent, D. H. (1992). *List of Vascular Plants of the British Isles*. London: Botanical Society of the British Isles.

Kershaw, K. A. (1973). *Quantitative and Dynamic Plant Ecology*, 2nd edn. London: Edward Arnold.

Kinder, P. M. (2009). *Pieris brassicae* (Linnaeus, 1758) (Lepidoptera: Pieridae) breeding on rape plants in a small suburban garden in Stockport, Cheshire. *Entomologist's Gazette* 60: 28.

Kindlmann, P., Aviron, S., Burel, F. and Ouin, A. (2004). Can the assumption of a non-random search improve our prediction of butterfly fluxes between resource patches? *Ecological Entomology* 29: 447–456.

Kindlmann, P., Aviron, S. and Burel, F. (2005). When is landscape matrix important for determining animal fluxes between resource patches? *Ecological Complexity* 2: 150–158.

King, C. A. M. (1976). *The Geomorphology of the British Isles. Northern England*. London: Methuen and Co. Ltd.

Kingsolver, J. G. (1985). Butterfly thermoregulation: organismic mechanisms and population consequences. *Journal of Research on the Lepidoptera* 24: 1–20.

Kinoshita, M. and Arikawa, K. (2000). Colour constancy of the swallowtail butterfly *Papilio xuthus*. *Journal of Experimental Biology* 203: 3521–3530.

Kinoshita, M., Shimada, N. and Arikawa, K. (1999). Colour vision of the foraging swallowtail butterfly *Papilio xuthus*. *Journal of Experimental Biology* 202: 95–102.

Kirby, K. J., Reed, C. M., Thomas, R. C. and Goldsmith, F. B. (1998). Preliminary estimates of fallen dead wood and standing dead trees in managed and unmanaged forests in Britain. *Journal of Applied Ecology* 35: 148–155.

Kirkland, P. (1996). *A Review of the Distribution, Ecology and Behaviour of the Scotch Argus (Erebia aethiops Esper, 1777)*. Unpublished report to the British Ecological Society, London.

Kitahara, M. and Watanabe, M. (2003). Diversity and rarity hotspots and conservation of butterfly communities in and around the Aokigahara woodland of Mount Fuji, central Japan. *Ecological Research* 18: 503–522.

Kitahara, M., Yumoto, M. and Kobayashi, T. (2008). The relationship of butterfly diversity with nectar plant species richness in and around the Aokigahara primary woodland of Mount Fiji, central Japan. *Biodiversity and Conservation* 17: 2713–2734.

Kivinen, S., Luoto, M., Kuussaari, M. and Helenius, J. (2006). Multi-species richness of boreal agricultural landscapes: effects of climate, biotope, soil and geographical location. *Journal of Biogeography* 33: 862–875.

Kleijn, D. and van Langevelde, F. (2006). Interacting effects of landscape context and habitat quality on flower visiting insects in agricultural landscapes. *Basic and Applied Ecology* 7: 201–214.

Kluge, A. G. (1990). Species as historical individuals. *Biology and Philosophy* 5: 417–431.

Knight, T. W. and Morris, D. W. (1996). How many habitats do landscapes contain? *Ecology* 77: 1756–1764.

Kolev, Z. (1998). *Maculinea arion* (L.) in Finland – distribution, state of knowledge and conservation. *Journal of Insect Conservation* 2: 91–93.

Komonen, A., Grapputo, A., Kaitala, V., Kotiaho, J. S. and Paininen, J. (2004). The role of niche breadth, resource availability and range position on the history of butterflies. *Oikos* 105: 41–54.

Krack-Roberg, E., Riege-Wcislo, W. and Wirthmann, A. (1995). *UGR Materials. Concept of an Accounting System for*

Land Use and Land Cover. Final report to the working group 'Physical Environmental Accounting'. Wiesbaden: German Federal Statistical Office, 175 pp.

Krcho, J. (1973). Morphometric analysis of relief on the basis of geometric aspects of field theory. *Acta Geographica Universitatis Comenianae, Geographico-physica* 1: 7–233. (Slovak Pedagogical Publishers, Bratislava.)

Krebs, C. J. (1989). *Ecological Methodology*. New York: Harper Collins Publishers.

Krebs, J. R. and Davies, N. B. (1993). *An Introduction to Behavioural Ecology*, 3rd edn. Oxford: Blackwell Scientific Publications.

Krenn, H. W., Zulka, K. P. and Gatschnegg, T. (2001). Proboscis morphology and food preferences in nymphalid butterflies (Lepidoptera: Nymphalidae). *Journal of Zoology* 254: 17–26.

Kristensen, C. O. (1994). Investigations on the natural mortality of eggs and larvae of the large white *Pieris brassicae* (L.) (Lep. Pieridae). *Journal of Applied Entomology – Zeitschriftfur Angeeandte Entomologie* 117: 92–98.

Krummel, J. R., Gardner, R. H., Sugihara, G., O'Neill, R. V. and Coleman, P. R. (1987). Landscape patterns in a disturbed environment. *Oikos* 48: 321–324.

Kudrna, O. (2002). The distribution atlas of European butterflies. *Oedippus* 20: 1–342.

Kühn, I. and Klotz, S. (2007). From ecosystem invasibility to local, regional and global patterns of invasive species. In *Biological Invasions, Ecological Studies 139*, ed. W. Nentwig. Berlin, Heidelberg: Springer-Verlag, pp. 181–196.

Kunte, K. (2008). Competition and species diversity: removal of dominant species increases diversity in Costa Rican butterfly communities. *Oikos* 117: 69–76.

Kuussaari, M., Saccheri, I., Camara, M. and Hanski, I. (1998). Allee effect and population dynamics in the Glanville fritillary butterfly. *Oikos* 82: 384–392.

Kuussaari, M., Singer, M. C. and Hanski, I. (2000). Local specialisation and landscape-level influence on host use in an herbivorous insect. *Ecology* 81: 2177–2187.

Lanyon, L. E. and Hall, G. F. (1983). Land surface morphology. *Soil Science* 136: 291–299, 382–386.

Lavery, T. A. (1993). A review of the distribution, ecology and status of the marsh fritillary *Euphydryas aurinia* Rottemburg, 1775 (Lepidoptera: Nymphalidae) in Ireland. *Irish Naturalists' Journal* 24: 192–199.

Lees, E. and Archer, D. M. (1974). Ecology of *Pieris napi* (L.) (Lep., Pieridae) in Britain. *Entomologist's Gazette* 25: 231–237.

Legendre, P. and Legendre, L. (2000). *Numerical Ecology*. Amsterdam: Elsevier.

Legendre, P., Dale, M. R. T., Fortin, M-J., Gurevitch, J., Hohn, M. and Myers, D. (2002). The consequences of spatial structure for the design and analysis of ecological field surveys. *Ecography* 25: 601–615.

Léon-Cortés, J. L., Cowley, M. J. R. and Thomas, C. D. (1999). Detecting decline in a formerly widespread species: how common is the common blue butterfly *Polyommatus icarus*? *Ecography* 22: 643–650.

León-Cortés, J. L., Cowley, M. J. R. and Thomas, C. D. (2000). The distribution and decline of a widespread butterfly *Lycaena phlaeas* in a pastoral landscape. *Ecological Entomology* 25: 285–294.

León-Cortés, J. L., Cowley, M. J. R. and Thomas, C. D. (2003a). Ecological dynamics of extinct species in empty habitat networks. 1. The role of habitat pattern and quantity, stochasticity and dispersal. *Oikos* 102: 449–464.

León-Cortés, J. L., Lennon, J. L. and Thomas, C. D. (2003b). Ecological dynamics of extinct species in empty habitat networks. 2. The role of host plant dynamics. *Oikos* 102: 465–477.

Leverton, R. (1998). Mass predation of hibernating *Aglais urticae* (L.) (Lep.: Nymphalidae) by wren. *Entomologist's Record and Journal of Variation* 110: 294.

Leverton, R. (2007). Where are our new butterflies? *Atropos* 31: 67.

Levins, R. (1968). *Evolution in Changing Environments: Some Theoretical Explorations*. Princeton, NJ: Princeton University Press.

Levins, R. (1969). Some demographic and genetic consequences of environmental heterogeneity for biological control. *Bulletin of the Entomological Society of America* 15: 237–240.

Levins, R. (1970). Extinction. In *Some Mathematical Problems in Biology*, ed. M. Gerstenhaber. Providence, RI: American Mathematical Society, pp. 75–107.

Lewis, O. (2008). Why is the small tortoiseshell declining? In *United Kingdom Butterfly Monitoring Scheme Report for 2007*, ed. M. S. Botham, T. M. Brereton, I. Middlebrook, K. L. Cruickshanks and B. B. Roy. Wallingford, UK: Centre for Ecology and Hydrology, pp. 20–21.

Lewis, O. T., Thomas, C. D., Hill, J. K. *et al.* (1997). Three ways of assessing metapopulation structure in the butterfly *Plebejus argus*. *Ecological Entomology* 22: 283–293.

Lewis, T. (1969). The distribution of insects near a low hedgerow. *Journal of Applied Ecology* 6: 443–452.

Liebhold, A. M., Rossi, R. E. and Kemp, W. P. (1993). Geostatistics and geographical information systems in applied insect ecology. *Annual Review of Entomology* 38: 303–327.

Lincoln, R. J., Boxshall, G. A. and Clark, P. F. (1982). *A Dictionary of Ecology, Evolution and Systematics*. Cambridge: Cambridge University Press.

Loader, C. and Dalman, H. (1991). Nitrogen content of foodplants and vulnerability of *Pieris rapae* to natural enemies. *Ecology* 72: 1586–1590.

Lockwood, J. A. and Latchininsky, A. V. (2008). Philosophical justifications for the extirpation of non-indigenous species: the case of the grasshopper *Schistocerca nitens* (Orthoptera) on the Island of Nihoa, Hawaii. *Journal of Insect Conservation* 12: 429–445.

Longley, P. A., Goodchild, M. F., Maguire, D. J. and Rhind, D. W. (2001). *Geographical Information Systems and Science*. New York: John Wiley and Sons.

Loram, A. (2004). *The Ecological Factors Governing Butterflies in Urban Areas*. PhD thesis, University of Birmingham.

Loram, A., Joy, J. and Pullin A. S. (2003). The habitat requirements of the grayling, *Hipparchia semele* (Linnaeus, 1758) (Lepidoptera: Nymphalidae, Satyrinae) in a semi-natural inland landscape in Shropshire, England. *Entomologist's Gazette* 54: 153–165.

Loram, A., Tratalos, J., Warren, P. H. and Gaston, K. J. (2007). Urban domestic gardens (X): the extent and structure of the resource in five major cities. *Landscape Ecology* 22: 601–615.

Loram, A., Thompson, K., Warren, P. H. and Gaston, K. J. (2008a). Urban domestic gardens (XII): the richness and composition of the flora in five UK cities. *Journal of Vegetation Science* 19: 321–330.

Loram, A., Warren, P. H. and Gaston, K. J. (2008b). Urban domestic gardens (XIV): the characteristics of gardens in five cities. *Environmental management* 42: 361–376.

Ludwig, J. A. and Reynolds, J. F. (1988). *Statistical Ecology. A Primer on Methods and Computing*. New York: John Wiley and Sons.

MacArthur, R. (1958). Population ecology of some warblers of northeast coniferous forests. *Ecology* 39: 599–619.

MacArthur, R. H. and Wilson, E. O. (1963). An equilibrium theory of insular zoogeography. *Evolution (Lawrence, Kansas)* 17: 373–387.

MacArthur, R. H. and Wilson, E. O. (1967). *The Theory of Island Biogeography*. Princeton, NJ: Princeton University Press.

MacNally, R., Fleishman, E., Bulluck, L. P. and Betrus, C. J. (2004). Comparative influence of spatial scale on beta diversity within regional assemblages of birds and butterflies. *Journal of Biogeography* 31: 917–929.

Maes, D. and Van Dyck, H. (2001). Butterfly diversity loss in Flanders (north Belgium): Europe's worst case scenario? *Biological Conservation* 99: 263–276.

Maes, D., Gilbert, M., Titeux, N., Goffart, P. and Dennis, R. L. H. (2003). Prediction of butterfly diversity hotspots in Belgium: a comparison of statistically focused and land use-focused models. *Journal of Biogeography* 30: 1907–1920.

Maes, D., Vanreusel, W., Talloen, W. and Van Dyck, H. (2004). Functional conservation units for the endangered butterfly *Maculinea alcon* in Belgium (Lepidoptera: Lycaenidae). *Biological Conservation* 120: 229–241.

Maier, C. (1998). *The Behaviour and Wing Morphology of the Meadow Brown Butterfly (Maniola jurtina L.) in Britain: the Influence of Weather and Location*. PhD thesis, Oxford Brookes University.

Mallon, D. (2004). Current status of butterflies in the Peak District. *News, Cheshire and Peak District Branch of Butterfly Conservation* 58: 8–18.

Maltman, A. (1998). *Geological Maps. An Introduction*. Chichester: John Wiley and Son.

Manel, S., Williams, H. C. and Ormerod, S. J. (2001). Evaluating presence–absence models in ecology: the need to account for prevalence. *Journal of Applied Ecology* 38: 921–931.

Massey, F. and Hartley, S. (2007). Grasses bite back. http://www.nerc.ac.uk/publications/planetearth/2007/winter/win07-grasses.pdf.

Matter, S. F. and Roland, J. (2002). An experimental examination of the effects of habitat quality on the dispersal and local abundance of the butterfly *Parnassius smintheus*. *Ecological Entomology* 27: 308–316.

Matter, S. F., Roland, J., Keyghobadi, N. and Sabourin, K. (2003). The effects of isolation, habitat area and resources on the abundance, density and movement of the butterfly *Parnassius smintheus*. *American Midland Naturalist* 150: 26–36.

Matter, S. F., Roland, J., Moilanen, A. and Hanski, I. (2004). Migration and survival of *Parnassius smintheus*: detecting effects of habitat for individual butterflies. *Ecological Applications* 14: 1526–1534.

Mattock, G., Gange, A. C. and Gange, E. G. (2007). Spring fungi are fruiting earlier. *British Wildlife* 18: 267–272.

May, R. M. (ed.) (1976). *Theoretical Ecology, Principles and Applications*. Oxford: Blackwell Scientific Publications.

May, R. M. (1994). The effects of spatial scale on ecological questions and answers. In *Large-Scale Ecology and Conservation Biology*, ed. P. J. Edwards, R. M. May and N. R. Webb. Oxford: Blackwell Science, pp. 1–17.

Mayhew, P. J. (1997). Adaptive patterns of host-plant selection by phytophagous insects. *Oikos* 79: 417–428.

McCoy, E. D. and Bell, S. S. (1991). Habitat structure: the evolution and diversification of a complex topic. In *Habitat Structure: the Physical Arrangement of Objects in Space*, ed. S. S. Bell, E. D. McCoy and H. R. Mushinsky. London: Chapman and Hall, pp. 3–27.

McCullagh, P. (1978). *Modern Concepts in Geomorphology*. Oxford: Oxford University Press.

McFarland, N. (1976). Hill-topping and defensive behaviour in a diurnal agaristid moth. *Australian Entomology Magazine* 3: 25–29.

McGeoch, M. A. (1998). The selection, testing and application of terrestrial insects as bioindicators. *Biological Review* 73: 181–201.

McGeoch, M. A. (2007). Insects and bioindication. In *Insect Conservation Biology*, ed. A. Stewart, O. Lewis and T. New. Proceedings of the Royal Entomological Society's 23rd Symposium. Wallingford, UK: CABI, pp. 144–174.

McGeoch, M. A., Van Rensburg, B. J. and Botes, A. (2002). The verification and application of bioindicators: a case study of dung beetles in a savanna ecosystem. *Journal of Applied Ecology* 39: 661–672.

McIntyre, N. E. and Wiens, J. A. (2000). A novel use of the lacunarity index to discern landscape pattern. *Landscape Ecology* 15: 313–321.

McKay, H. V. (1991). Egg-laying requirements of woodland butterflies; brimstones (*Gonepteryx rhamni*) and alder buckthorn (*Frangula alnus*). *Journal of Applied Ecology* 28: 731–43.

McNeely, C. and Singer, M. C. (2001). Contrasting roles of learning in butterflies foraging for nectar and oviposition sites. *Animal Behaviour* 61: 847–852.

Melling, T. M. (1987). *The Ecology and Population Structure of a Butterfly Cline*. PhD thesis, University of Newcastle.

Mendel, H. and Parsons, E. (1987). Observations on the life history of the silver-studded blue, *Plebejus argus* L. *Transactions of Suffolk Natualists' Society* 23: 2–8.

Menéndez, R., Megías, A. G., Hill, J. K. *et al.* (2006). Species richness changes lag behind climate change. *Proceedings of the Royal Society B* 273: 1465–1470.

Merchant, A. J. (1956). *Plebejus argus* L. (Lep. Lycaenidae) *caernensis* Thompson at Llandulas and Rhyd-y-Foel. *Entomologist* 89: 234.

Merckx, T. and Van Dyck, H. (2002). Interrelations among habitat use, behavior, and flight related morphology in two co-occurring satyrine butterflies, *Maniola jurtina* and *Pyronia tithonus*. *Journal of Insect Behavior* 15: 541–561.

Merckx, T. and Van Dyck, H. (2005). Mate location behaviour of the butterfly *Pararge aegeria* in woodland and fragmented landscapes. *Animal Behaviour* 70: 411–416.

Merckx, T., Van Dyck, H., Karlsson, B. and Leimar, O. (2003). The evolution of movements and behaviour at boundaries in different landscapes: a common arena experiment with butterflies. *Proceedings of the Royal Society of London B* 270: 1815–1821.

Merckx, T., Karlsson, B. and Van Dyck, H. (2006). Sex – and landscape-related differences in flight ability under suboptimal temperatures in a woodland butterfly. *Functional Ecology* 20: 436–411.

Merckx, T., Feber, R. E., Dulieu, L. R. *et al.* (2008). Effect of field margins on moths depends on species mobility: field-based evidence for landscape-scale conservation. *Agriculture, Ecosystems and Environment* 129: 302–309.

Merckx, T., Feber, R. E., Riordan, P., Townsend, M. C., Bourn, N. A. D., Parsons, M. S. and Macdonald, D. W. (2009). Optimizing the biodiversity gain from agri-environment schemes. *Agriculture, Ecosystems and Environment* 130: 177–182.

Merryweather, J. (2007). Planting trees or woodlands? An ecologist's perspective. *British Wildlife* 18: 250–258.

Mevi-Schütz, J. and Erhardt, A. (2005). Amino acids in nectar enhance butterfly fecundity: a long-awaited link. *American Naturalist* 165: 411–419.

Mevi-Schütz, J., Goverde, M. and Erhardt, A. (2004). Effects of fertilization and elevated CO_2 on larval food and butterfly nectar amino acid preference in *Coenonympha pamphilus* L. *Behavioral Ecology and Sociobiology* 54: 36–43.

Middleton, A. and Goodyear, L. (2008). Some notes on the territorial activity of brown hairstreak *Thecla betulae* (L.) (Lep.: Lycaenidae) in particular, its timing. *Entomologist's Record and Journal of Variation* 120: 183–186.

Miles, J. (1979). *Vegetation Dynamics*. London: Chapman and Hall.

Miles, J. (1987). Vegetation succession: past and present perceptions. In *Colonization, Succession and Stability*, ed. A. J. Gray, M. J. Crawley and P. J. Edwards. Oxford: Blackwell Scientific Publications, pp. 1–29.

Miller, J. N., Brooks, R. P. and Croonquist, M. J. (1997). Effects of landscape patterns on biotic communities. *Landscape Ecology* 12: 137–153.

Milne, A. (1957). The natural control of insect populations. *Canadian Entomologist* 89: 193–213.

Mitchell, J. B. (1965). *Historical Geography*. London: English Universities Press.

MLURI (Macaulay Land Use Research Institute) (1993). *The Land Cover of Scotland 1988*. Aberdeen: MLURI.

Moilanen, A. and Hanski, I. (1998). Metapopulation dynamics: effect of habitat quality and landscape structure. *Ecology* 79: 2503 2515.

Moilanen, A. and Nieninen, M. (2002). Simple connectivity measures in spatial ecology. *Ecology* 83: 1131–1145.

Moilanen, A. and Wintle, B. A. (2007). The boundary quality penalty a quantitative method for approximating species responses to fragmentation in reserve selection. *Conservation Biology* 21: 355–364.

Moilanen, A., Franco, A. M. A., Early, R. I., Fox, R., Wintle, B. and Thomas, C. D. (2005). Prioritizing multiple-use landscapes for conservation: methods for large multi-species planning problems. *Proceedings of the Royal Society B* 272: 1885–1891.

Morris, D. W. (1992). Scales and costs of habitat selection in heterogeneous landscapes. *Evolutionary Ecology* 6: 412–432.

Morris, M. G., Thomas, J. A., Ward, L. K., Snazell, R. G., Pywell, R. F., Stevenson, M. J. and Webb, N. R. (1994). Recreation of early successional stages for threatened butterflies – an ecological engineering approach. *Journal of Environmental Management* 42: 119–135.

Morton, A. C. G. (1985). *The Population Biology of an Insect with a Restricted Distribution: Cupido minimus Fuessly (Lepidoptera: Lycaenidae)*. PhD thesis, University of Southampton.

Moser, B., Jaeger, J. A. G., Tappeiner, U., Tasser, E. and Eiselt, B. (2007). Modification of the effective mesh size for measuring landscape fragmentation to solve the boundary problem. *Landscape Ecology* 22: 447–459.

Mountford, M. D. (1988). Population regulation, density dependence, and heterogeneity. *Journal of Animal Ecology* 57: 845–858.

Munguira, M. L. and Thomas, J. A. (1992). Use of road verges by butterfly and burnet populations, and the effects of roads on adult dispersal and mortality. *Journal of Applied Ecology* 29: 316–329.

Munguira, M. L., Martin, J., Garcia-Barros, E. and Viejo, J. L. (1997). The use of space and resources in a Mediterranean population of the butterfly *Euphydryas aurinia*. *Acta Oecologica – International Journal of Ecology* 18: 597–612.

Murdoch, W. W. (1994). Population regulation in theory and practice. *Ecology* 75: 271–287.

Murphy, D. D. (1984). Butterflies and their nectar plants: the role of the checkerspot butterfly *Euphydryas editha* as a pollen vector. *Oikos* 43: 113–116.
Murphy, D. D., Launer, A. E. and Ehrlich, P. R. (1983). The role of adult feeding in egg production and population dynamics of the checkerspot butterfly *Euphydryas editha*. *Oecologia* 56: 257–263.
Murphy, D. D., Menninger, M. S. and Ehrlich, P. R. (1984). Nectar source as a determinant of oviposition host species in *Euphydryas chalcedona*. *Oecologia* 62: 269–271.
Murray, B. G. (1994). On density dependence. *Oikos* 69: 520–523.
Musk, L. F. (1988). *Weather Systems*. Cambridge: Cambridge University Press.
Natural England (1995). *Accessible Natural Greenspace in Towns and Cities: a Review of Appropriate Size and Distance Criteria*. Research Report No. 153. Sheffield: Natural England.
Natural England (1997). *A Framework for the Future: Green Networks with Multiple Uses in and around Towns and Cities*. Research Report No. 256. Sheffield: Natural England.
Nee, S., May, R. M. and Hassell, M. P. (1997). Two-species metapopulation models. In *Metapopulation Biology. Ecology, Genetics, and Evolution*, ed. I. A. Hanski and M. E. Gilpin. London: Academic Press, pp. 123–147.
Nelson, B., Hughes, M., Nash, R. and Warren, M. (2001). *Leptidea reali* Reissinger 1989 (Lep.:Pieridae): a butterfly new to Britain and Ireland. *Entomologist's Record and Journal of Variation* 113: 97–101.
New, T. R. (2005). Recreation and reserves: values of golf courses for insect conservation. *Journal of Insect Conservation* 9: 1–2.
New, T. R. (2006). *Conservation Biology in Australia. An Introduction*, 2nd edn. Oxford: Oxford University Press.
New, T. R. (2007). Recovery plans for insects: needs, design, and implementation. *Newsletter of the Richmond Birdwing Recovery Network Inc*. 9: 26–36.
Newland, D. E. (2009) Observations of purple hairstreaks at canopy level. *British Wildlife* 20: 317–320.
Nicholson, A. J. (1933). The balance of animal populations. *Journal of Animal Populations* 2: 132–178.
Nicholson, A. J. (1954). An outline of the dynamics of animal populations. *Australian Journal of Zoology* 2: 9–65.
Niemi, G. J. and McDonald, M. E. (2004). Application of ecological indicators. *Annual Review of Ecology Evolution and Systematics* 35: 89–111.
Nieminen, M., Siljander, M. and Hanski, I. (2004). Structure and dynamics of *Melitaea cinxia* populations. In *On the Wings of Checkerspots. A Model System for Population Biology*, ed. P. R. Ehrlich and I. Hanski. Oxford: Oxford University Press, pp. 63–91.
Noss, R. F. and Harris, L. (1986). Nodes, networks and MUMs: preserving diversity at all scales. *Environmental Management* 10: 299–309.
Novak, J. and Konvicka, M. (2006). Proximity of valuable habitats affects succession patterns in abandoned quarries. *Ecological Engineering* 26: 113–122.
Nylin, S. (1988). Host plant specialization and seasonality in a polyphagous butterfly, *Polygonia c-album* (Nymphalidae). *Oikos* 53: 381–386.
Nylin, S. (1992). Seasonal plasticity in life history traits: growth and development in *Polygonia c-album*. *Biological Journal of the Linnean Society* 47: 301–323.
Nylin, S., Janz, N. and Wedell, N. (1996). Oviposition plant preferences and offspring performance in the comma butterfly: correlations and conflicts. *Entomologia Experimentalis et Applicata* 80: 141–144.
O'Neill, R. V., Hunsaker, C. T., Timmins, S. P., Jackson, B. L., Jones, K. B., Tiitters, K. H. and Wickham, J. D. (1996). Scale problems in reporting landscape pattern at the regional scale. *Landscape Ecology* 11: 169–180.
O'Neill, R. V., Ritters, K. H., Wickham, J. D. and Jones, K. B. (1999). Landscape pattern metrics and regional assessment. *Ecosystem Health* 5: 225–233.
Oates, M. R. (1986). *The Ecology and Conservation of the Duke of Burgundy Fritillary*. Unpublished report to the World Wide Fund for Nature, Godalming.
Oates, M. R. (2000). The Duke of Burgundy – conserving the intractable. *British Wildlife* 11: 250–257.
Oates, M. R. (2008). The myth of the master tree: mate-location strategies of the purple emperor butterfly. *British Wildlife* 19:330–337.
Oates, M. R. and Warren, M. S. (1990). *A Review of Butterfly Introductions in Britain and Ireland*. Godalming, UK: World Wildlife Fund for Nature.
Oates, M. R., Shreeves, W., Steel, C., Toynton, P. and Wilmott, K. J. (1986). Duke of Burgundy. In *The Management of Chalk Grassland for Butterflies*, ed. A. Stubbs, J. Bacon and P. Oswald. Focus on Nature Conservation No. 17. Peterborough: Nature Conservancy Council, pp. 32–34.
Oates, M. R., Goodyear, L., Middleton, A. and Willmott, K. J. (2005). *Purple Emperor Master Tree Project 2004*. Unpublished report to Butterfly Conservation, Wareham.
Odum, E. P. (1963). *Ecology*. New York: Holt, Rinehart and Wilston.
Offer, D., Edwards, M. and Edgar, P. (2003). *Grazing Heathland: a Guide to Impact Assessment for Insects and Reptiles*. English Nature Research Reports No. 497. Peterborough: English Nature.
Ohsaki, N. and Sato, Y. (1999). The role of parasitoids in evolution of habitat and larval food plant preference by three *Pieris* butterflies. *Researches on Population Ecology* 41: 107–119.
Okubo, A. (1980). *Diffusion and Ecological Problems: Mathematical Models*. New York: Springer-Verlag.
Olson, J. S. (1958). Rates of succession and soil changes on Southen Lake, Michigan, sand dunes. *Botanical Gazette* 119: 125–170.
Omura, H. and Honda, K. (2005). Priority of color over scent during flower visitation by adult *Vanessa indica* butterflies. *Oecologia* 142: 588–596.

Omura, H., Honda, K. and Hayaski, N. (1999). Chemical and chromatic bases for preferential visiting by the cabbage butterfly *Pieris rapae* to rape flowers. *Journal of Chemical Ecology* 25: 1897–1906.

Omura, H., Honda, K. and Hayashi, N. (2000). Identification of feeding attractants in oak sap of two nymphalid butterflies, *Kaniska canace* and *Vanessa indica*. *Physiological Entomology* 25: 281–287.

Oostermeijer, J. G. B. and van Swaay, C. A. M. (1998). The relationship between butterflies and environmental indicator values: a tool for conservation in a changing landscape. *Biological Conservation* 86: 271–280.

Ouin, A., Aviron, S., Dover, J. and Burel, F. (2004). Complementation/supplementation of resources for butterflies in agricultural landscapes. *Agriculture Ecosystems and Environment* 103: 473–479.

Ovaskainen, O. (2004). Habitat-specific movement parameters estimated using mark–recapture data and a diffusion model. *Ecology* 85: 42–257.

Owen, D. F. (1949). The Macrolepidoptera of Moorgate, London, bombed sites. *Entomologist* 82: 59–62.

Owen, D. F. (1951). Bombed site Lepidoptera. *Entomologist* 84: 265–272.

Parker, R. (2003). Suffolk grassland as a butterfly habitat. *Transactions of the Suffolk Natural History Society* 39: 4–21.

Parmesan, C., Ryrholm, N., Stefanescu, C. *et al.* (1999). Polewards shifts in geographical ranges of butterfly species associated with regional warming. *Nature* 399: 579–583.

Patterson, D. and Wells, L. (2006) A colony of ringlet *Aphantopus hyperantus* new to Sutherland. *Atropos* 30: 40–41.

Pears, N. (1977). *Basic Biogeography*. London: Longman.

Pe'er, G., Saltz, D., Thulke, H. H. and Motro, U. (2004). Response to topography in a hilltopping butterfly and implications for modelling nonrandom dispersal. *Animal Behaviour* 68: 825–839.

Pe'er, G., Saltz, D. and Frank, K. (2005). Virtual corridors for conservation management. *Conservation Biology* 19: 1997–2003.

Pe'er, G., Heinz, S. K. and Frank, K. (2006). Connectivity in heterogeneous landscapes: analyzing the effect of topography. *Landscape Ecology* 21: 47–61.

Pennock, D. J., Zebarth, B. J. and de Jong, E. (1987). Landform classification and soil distribution in hummocky terrain, Saskatchewan, Canada. *Geoderma* 40: 297–315.

Peretti, J. H. (1998). Nativism and nature: rethinking biological invasion. *Environmental Values* 7: 183–192.

Peterken, G. F. (1967). *Guide to Checksheet for IBP Areas*. IBP Handbook No. 4. Oxford: Blackwell Scientific Publications.

Peterken, G. F. (2009). Woodland origins and meadows. *British Wildlife* 20: 161–170.

Pierce, N. E. and Easteal, S. (1986). The selective advantage of attendant ants for the larvae of a lycaenid butterfly, *Glaucopsyche lygdamus*. *Journal of Animal Ecology* 55: 451–462.

Pierce, N. E., Braby, M. F., Heath, A., Lohman, D. J., Mathew, J., Rand, D. B. and Travassos, M. A. (2002). The ecology and evolution of ant association in the Lycaenidae (Lepidopetra). *Annual Review of Entomology* 47: 733–771.

Pinheiro, C. E. G. (1990). Territorial hill-topping of three swallowtail butterflies (Lep.: Papilionidae) in western Brazil. *Journal of Research on the Lepidoptera* 29: 134–142.

Plant, C. (1987). *The Butterflies of the London Area*. London: Passmore Edwards Museum.

Pollard, E. (1979). Population ecology and changes in range of the white admiral butterfly *Ladoga camilla* L. in England. *Ecological Entomology* 4: 61–74

Pollard, E. (1981). Aspects of the ecology of the meadow brown butterfly, *Maniola jurtina* (L.). *Entomologist's Gazette* 32: 67–74.

Pollard, E. (1982). Monitoring butterfly abundance in relation to the management of a nature reserve. *Biological Conservation* 24: 317–328.

Pollard, E. (1988). Temperature, rainfall and butterfly numbers. *Journal of Applied Ecology* 25: 819–828.

Pollard, E. and Cooke, A. S. (1994). Impact of muntjac deer *Muntiacus reevesi* on egg-laying sites of the white admiral butterfly *Ladoga camilla* in a Cambridgeshire wood. *Biological Conservation* 70: 189–191.

Pollard, E. and Greatorex-Davies, J. N. (1997). Flight-periods of the small heath butterfly, *Coenonympha pamphilus* (Linnaeus) (Lepidoptera: Nymphalidae, Satyrinae) on chalk downs and in woodland in southern Engalnd. *Entomologist's Gazette* 48: 3–7.

Pollard, E. and Greatorex-Davies, J. N. (1998). Increased abundance of the red admiral butterfly *Vanessa atalanta* in Britain: the roles of immigration, overwintering and breeding within the country. *Ecology Letters* 1: 77–81.

Pollard, E. and Hall, M. L. (1980). Possible movement of *Gonepteryx rhamni* (L.) (Lepidoptera: Pieridae) between hibernating and breeding areas. *Entomologist's Gazette* 31: 217–220.

Pollard, E. and Moss, D. (1995). Historical records of the occurrence of butterflies in Britain: examples showing associations between annual numbers of records and weather. *Global Change Biology* 1: 107–113.

Pollard, E. and Yates, T. J. (1993a). *Monitoring Butterflies for Ecology and Conservation*. London: Chapman and Hall.

Pollard, E. and Yates, T. J. (1993b). Population fluctuations of the holly blue butterfly, *Celastrina argiolus* (L.) (Lepidoptera: Lycaenidae). *Entomologist's Gazette* 44: 3–9.

Pollard, E., Rothery, P. and Yates, T. J. (1996). Annual growth rates in newly established populations of the butterfly *Pararge aegeria*. *Ecological Entomology* 21: 365–369.

Pollard, E., Greatorex-Davies, J. N. and Thomas, J. A. (1997). Drought reduces breeding success of the butterfly *Aglais urticae*. *Ecological Entomology* 22: 315–318.

Pollard, E., Van Swaay, C. A. M., Stefanescu, C., Lundsten, K. E., Maes, D. and Greatorex-Davies, J. N. (1998). Migration of the painted lady butterfly *Cynthia cardui* in Europe: evidence from monitoring. *Diversity and Distributions* 4: 243–253.

Porter, K. (1982). Basking behaviour in larvae of the butterfly *Euphydryas aurinia*. *Oikos* 38: 308–312.

Porter, K. (1992). Eggs and egg-laying. In *Ecology of Butterflies in Britain*, ed. R. L. H. Dennis. Oxford: Oxford University Press, pp. 46–72.

Porter, K., Steel, C. A. and Thomas, J. A. (1992). Butterflies and communities. In *Ecology of Butterflies in Britain*, ed. R. L. H. Dennis. Oxford: Oxford University Press, pp. 139–177.

Pounds, N. J. (1963). *Political Geography*. New York: McGraw-Hill.

Poysa, H., Elmberg, J., Sjoberg, K. and Nummi, P. (1998). Habitat selection rules in breeding mallards (*Anas platyrhynchos*): a test of two competing hypotheses. *Oecologia* 114: 283–287.

Prater, B. (2007). Progress with conservation of the dingy skipper *Erynnis tages* (L.) and the grizzled skipper *Pyrgus malvae* (L.) in the East Midlands. *Atropos* 30: 18–23.

Price, P. W. (2002). Resource-driven terrestrial interaction webs. *Ecological Research* 17: 241–247.

Prudic, K. L., Oliver, J. C. and Bowers, M. D. (2005). Soil nutrient effects on oviposition preference, larval performance, and chemical defense of a specialist insect herbivore. *Oecologia* 143: 578–587.

Pryke, S. R. and Samways, M. J. (2001). Width of grassland linkages for the conservation of butterflies in South African afforested areas. *Biological Conservation* 101: 85–96.

Pulliam, H. R. (1988). Sources, sinks and population regulation. *American Naturalist* 132: 652–661.

Pullin, A. S. (1986). Influence of the food plant, *Urtica dioica*, on larval development, feeding efficiency, and voltinism of a specialist insect, *Inachis io*. *Holarctic Ecology* 9: 72–78.

Pullin, A. S. (1987). Changes in leaf quality following clipping and regrowth in *Urtica dioica*, and consequences for a specialist insect herbivore *Aglais urticae*. *Oikos* 49: 39–45.

Pullin, A. S. (1988). Environmental cues and variable voltinism patterns in *Aglais urticae*. *Entomologist's Gazette* 39: 101–112.

Pullin, A. S. (1997). Habitat requirements of *Lycaena dispar batavus* and implications for re-establishment in England. *Journal of Insect Conservation* 1: 177–185.

Pullin, A. S. and Knight, T. M. (2001). Effectiveness in conservation practice. Pointers from medicine and public health. *Conservation Biology* 15: 50–54.

Pullin, A. S., McLean, I. F. G. and Webb, M. R. (1995). Ecology and conservation of *Lycaena dispar*: British and European perspectives. In *The Ecology and Conservation of Butterflies*, A. S. Pullin. London: Chapman and Hall, pp. 150–164.

Pullin, A. S., Balint, Z., Balletto, E. *et al.* (1998). The status, ecology and conservation of *Lycaena dispar* (Lycaenidae: Lycaenini) in Europe. *Nota lepidopterologica* 21: 94–100.

Purvis, A. and Rambaut, A. (1995). Comparative-Analysis by Independent Contrasts (CAIC) – an Apple-Macintosh application for analyzing comparative data. *Computer Applications Biosciences* 11: 247–251.

Pyörnilä, M. (1976–1977). Parasitism in *Aglais urticae* (L.) (Lep., Nymphalidae). *Annales Entomologici Fennici* 42: 26–33, 133–139, 151–161; 43: 21–27.

Pywell, R. F., Warman, E. A., Sparks, T. H. *et al.* (2004). Assessing habitat quality for butterflies in intensively managed arable farmland. *Biological Conservation* 118: 313–325.

Quinn, R. M., Gaston, K. J., Blackburn, T. M. and Eversham, B. C. (1997). Abundance–range size relationships of macrolepidoptera in Britain: the effects of taxonomy and life history variables. *Ecological Entomology* 22: 453–461.

Quinn, R. M., Gaston, K. J. and Roy, D. B. (1998). Coincidence in the distributions of butterflies and their foodplants. *Ecography* 21: 279–288.

Rackham, O. (1986). *The History of the Countryside*. London: J. M. Dent and Sons Ltd.

Rands, M. R. W. and Sotherton, N. W. (1986). Pesticide use on cereal crops and changes in the abundance of butterflies on arable farmland in England. *Biological Conservation* 36: 71–82.

Rathcke, B. J. and Poole, R. W. (1975). Coevolutionary race continues butterfly larval adaptation to plant trichomes. *Science* 187: 175–176.

Raunkiaer, C. (1934). *The Life Forms of Plants and Statistical Plant Geography*. Oxford: Clarendon Press.

Ravenscroft, N. O. M. (1990). The ecology and conservation of the silver-studded blue (*Plebejus argus* L.) on the Sandlings of East Anglia. *Biological Conservation* 53: 21–36.

Ravenscroft, N. O. M. (1994a). The ecology of the chequered skipper butterfly *Carterocephalus palaemon* Pallas in Scotland. I. Microhabitat. *Journal of Applied Ecology* 31: 613–622.

Ravenscroft, N. O. M. (1994b). The ecology of the chequered skipper butterfly *Carterocephalus palaemon* in Scotland. II. Foodplant quality and population range. *Journal of Applied Ecology* 31: 623–630.

Ravenscroft, N. O. M. (1994c). The feeding behaviour of *Carterocephalus palaemon* (Lepidoptera: Hesperiidae) caterpillars: does it avoid host defences or maximize nutrient intake? *Ecological Entomology* 19: 26–30.

Reavey, D. and Lawton, J. H. (1991). Larval contribution to fitness in leaf-eating insects. In *Reproductive Behaviour of Insects: Individuals and Populations*, ed. W. J. Bailey and J. Ridsdill-Smith. London: Chapman and Hall, pp. 293–329.

Reilly, W. J. (1929). *Methods for the Study of Retail Relationships*. University of Texas Bulletin No. 2944. University of Texas.

Renkonen, O. (1938). Statisch-okologische Untersuchungen uber die terrestiche kaferwelt der finnischen bruchmoore. *Annales Botanici Societatis Vanamo* 6: 1–231.

Resetarits, W. J. (1996). Oviposition site choice and life history evolution. *American Zoologist* 36: 205–215.

Richards, O. W. (1940). The biology of the small white butterfly (*Pieris rapae*) with special reference to the factors controlling its abundance. *Journal of Animal Ecology* 9: 143–88.

Ries, L. and Debinski, D. M. (2001). Butterfly responses to habitat edges in the highly fragmented prairies of Central Iowa. *Journal of Animal Ecology* 70: 840–852.

Rietdorf, K. and Steidle, J. L. M. (2002). Was Hopkins right? Influence of larval and early adult experience on the olfactory response in the granary weevil *Sitophilus granarius* (Coleoptera, Curculionidae). *Physiological Entomology* 27: 223–227.

Riley, J. R. (2002). Design considerations for harmonic radar to investigate the flights of insects at low altitude. *Computers and Electronics in Agriculture* 35: 151–169.

Ritters, K. H., O'Neill, R. V., Hunsaker, C. T. *et al.* (1995). A factor-analysis of landscape pattern and structure metrics. *Landscape Ecology* 10: 23–39.

Roberts, B. K. (1977). *Rural Settlement in Britain*. London: Hutchinson.

Roberts, N. (1989). *An Environmental History*. Oxford: Blackwell.

Robinson, A. (2008). *Habitat Requirements of the Grayling Butterfly, Hipparchia semele*. MSc dissertation, University of East Anglia.

Rodwell, J. S. (1991–2000). *British Plant Communities*, Vols 1–5. Cambridge: Cambridge University Press.

Rohde, K. (2006). *Nonequilibrium Ecology*. Cambridge: Cambridge University Press.

Roland, J., Keyghobadi, N. and Fownes, S. (2000). Alpine *Parnassius* butterfly dispersal. Effects of landscape and population size. *Ecology* 81: 1642–1653.

Ronce, O. (2001). When source become sinks: migration meltdown in heterogeneous habitats. *Evolution* 55: 1520–1531.

Root, R. B. (1967). The niche exploitation pattern of the blue-gray gnatcatcher. *Ecological Monographs* 37: 317–350.

Roper, P. (1992). Buddleia butchery. *Butterfly Conservation News* 52: 35–36.

Roper, P. (1993). Buddleia butchery – the response from members. *Butterfly Conservation News* 54: 28–30.

Rosenzweig, M. L. (1991). Habitat selection and population interactions – the search for mechanism. *American Naturalist* 137: S5–S28.

Ross, J. A., Matter, S. F. and Roland, J. (2005). Edge avoidance and movement of the butterfly *Parnassius smintheus* in matrix and non-matrix habitat. *Landscape Ecology* 20: 127–135.

Rothschild, M., Valadon, G. and Mummery, R. (1977). Carotenoids of the pupae of the large white butterfly (*Pieris brassicae*) and the small white butterfly (*Pieris rapae*). *Journal of Zoology* 181: 323–339.

Roy, D. B. and Thomas, J. A. (2003). Seasonal variation in the niche, habitat availability and population fluctuations of a bivoltine thermophilous insect near its range margin. *Oecologia* 134: 439–444.

Roy, D. B., Rothery, P., Moss, D., Pollard, E. and Thomas, J. A. (2001). Butterfly numbers and weather: predicting historical trends in abundance and the future effects of climate change. *Journal of Animal Ecology* 70: 201–217.

Rusterholz, H. P. and Erhardt, A. (1998). Effects of elevated CO_2 on flowering phenology and nectar production of nectar plants important for butterflies of calcareous grasslands. *Oecologia* 113: 341–349.

Rusterholz, H. P. and Erhardt, A. (2000). Can nectar properties explain sex-specific flower preferences in the Adonis blue *Lysandra bellargus*. *Ecological Entomology* 25: 81–90.

Rutowski, R. L. (1991). The evolution of mate-locating behaviour in butterflies. *American Naturalist* 138: 1121–1139.

Rydin, H. and Borgegård, S-O. (1991). Plant characteristics over a century of primary succession on islands: Lake Hjälmaren. *Ecology* 72: 1089–1101.

Saarinen, K. and Jantunen, J. (2005). Grassland butterfly fauna under traditional animal husbandry: contrasts in diversity in mown meadows and grazed pastures. *Biodiversity and Conservation* 14: 3201–3213.

Saarinen, K., Marttila, O. and Jantunen, J. (1998). Species richness and distribution of butterflies (Lepidoptera, Hesperiidae, Papilioniddae) in an agricultural environment in SE Finland. *Entomologica Fennica* 9: 9–18.

Saccheri, I. J., Kuussaari, M., Kankare, M., Vikman, P., Fortelius, W. and Hanski, I. (1998). Inbreeding and extinction in a butterfly metapopulation. *Nature* 392: 491–494.

Salmon, M. A. (2000). *The Aurelian Legacy – British Butterflies and their Collectors*. Colchester: Harley Books.

Samways, M. J. (2007). Implementing ecological networks for conserving insect and other biodiversity. In *Insect Conservation Biology*, ed. A. J. A. Stewart, T. R. New and O. T. Lewis. Wallingford, UK: CABI, pp. 127–143.

Sanderson, E. W., Redford, K. H., Vedder, A., Coppolillo, P. B. and Ward, S. E. (2002). A conceptual model for conservation planning based on landscape species requirements. *Landscape Urban Planning* 58: 41–56.

Sanz, J. J. (1998). Effect of habitat and latitude on nestling diet of pied flycatchers *Ficedula hypoleuca*. *Ardea* 86: 81–88.

Sattler, K. and Tremewan, W. G. (2009). The authorship of the so-called 'Wiener Verzeichnis'. *Nota Lepidopterologica* 32: 3–10.

Saura, S. and Martinez-Millan, J. (2001). Sensitivity of landscape pattern metrics to map spatial extent. *Photogrammetric Engineering and Remote Sensing* 67: 1027–1036.

Scali, V. (1971). Imaginal diapause and gonadal maturation of *Maniola jurtina* (Lepidopetera, Satyrinae) from Tuscany. *Journal of Animal Ecology* 40: 467–472.

Schoener, T. W. (1989). The ecological niche. In *Ecological Concepts. The Contribution of Ecology to an Understanding of the Natural World*, ed. J. M. Cherret. Oxford: Blackwell Scientific Publications, pp. 79–113.

Schoonhoven, L. M. (1990). Host-marking pheromones in Lepidoptera, with special reference to two *Pieris* species. *Journal of Chemical Ecology* 16: 3043–3052.

Schops, K. and Hanski, I. (2001). Population level correlation between pre-alighting and post-alighting host plant preference in the Glanville fritillary butterfly. *Ecological Entomology* 26: 517–524.

Schröder, B., Strauss, B., Binzenhöfer, B., Biedermann, R. and Settele, J. (2009). Predictive species distribution modelling in butterflies. In *Ecology of Butterflies in Europe*, ed. J. Settele, T. G. Shreeve, M. Konvicka and H. Van Dyck. Cambridge: Cambridge University Press, pp. 62–78.

Schtickzelle, N., Joiris, A., Van Dyck, H. and Baguette, M. (2007). Quantitative analysis of changes in movement behaviour within and outside habitat in a specialist butterfly. *BMC Evolutionary Biology* 7: 4 (DOI:10.1186/1471-2148-7-4).

Schultz, C. B. and Crone, E. E. (2001). Edge-mediated dispersal behaviour in a prairie butterfly. *Ecology* 82: 1879–1892.

Schumaker, N. H. (1996). Using landscape indices to predict habitat connectivity. *Ecology* 77: 1210–1225.

Scott, J. A. (1968). Hill-topping as a mating mechanism to aid the survival of low density species. *Journal of Research on the Lepidoptera* 7: 191–204.

Scott, J. A. (1974). Mate-locating behavior of butterflies. *American Midland Naturalist* 91: 103–117.

Scott, J. M., Csuti, B., Smith, K., Estes, J. E. and Caicco, S. (1991). Gap analysis of species richness and vegetation cover: an integrated biodiversity conservation strategy. In *Balancing on the Brink of Extinction: the Endangered Species Act and Lessons for the Future*, ed. K. A. Kohm. Washington, DC: Island Press, pp. 282–297.

Scriber, J. M. (1973). Latitudinal gradients in larval feeding specialisation of the world Papilionidae (Lepidopetra). *Psyche* 80: 355–373.

Scriber, J. M. and Lederhouse, R. C. (1992). The thermal environment as a resource dictating geographic patterns of feeding specialization of insect herbivores. In *Effects of Resource Distribution on Animal–Plant Interactions*, ed. M. R. Hunter, T. Ohgushi and P. W. Price. New York: Academic Press, pp. 429–466.

Sculley, C. E. and Boggs, C. L. (1996). Mating systems and sexual division of foraging effort affect puddling behaviour by butterflies. *Ecological Entomology* 21: 193–197.

Seto, K. C., Fleishman, E., Fay, J. P. and Betrus, C. J. (2004). Linking spatial patterns of bird and butterfly species richness with Landsat TM derived NDVI. *International Journal of Remote Sensing* 25: 4309–4324.

Settele, J., Feldmann, R. and Reinhardt, R. (1999). *Die tagfalter Deutschlands*. Stuttgart: Ulmer.

Settele, J., Kühn, E. and Thomas J. A. (2005). *Studies on the Ecology and Conservation of Butterflies in Europe*, Vol. 2: *Species Ecology along a European Gradient: Maculinea Butterflies as a Model*. Moscow: Pensoft.

Settele, J., Kudrna, O., Harpke, A. *et al.* (2008). *Climate Risk Atlas of European Butterflies*. Sofia-Moscow: Pensoft.

Severns, P. M. (2008). Exotic grass invasion impacts fitness of an endangered prairie butterfly, *Icaricia icarioides fenderi*. *Journal of Insect Conservation* 12: 651–661.

Shaffer, M. L. (1981). Minimum population sizes for species conservation. *BioScience* 31: 131–134.

Shannon, S. (1995). *A Study of the Small Mountain Ringlet Erebia epiphron Knoch. in South West Cumbria*. MSc thesis, University of Leicester.

Shaw, B. T. (2005). The most widespread butterflies in Cheshire (1980–2004). *Cheshire and Peak District Branch News of Butterfly Conservation* 60: 12–14.

Shaw, M. R. (1997). *Rearing Parastic Hymenoptera*. Orpington, UK: Amateur Entomologists's Society.

Shaw, M. R. (2006). Habitat considerations for parasitic wasps (Hymenoptera). *Journal of Insect Conservation* 10: 117–127.

Shaw, M. R. and Askew, R. R. (1976). Parasites. In: *The Moths and Butterflies of Great Britain and Ireland*. Vol. 1, ed. J. Heath. Colchester: Harley Books, pp. 240–256.

Shaw, M. R. and Bland, K. P. (1994). Foodplant and parasitoids (Hymenoptera: Braconidae and Diptera: Tachinidae) of *Callophrys rubi* (L.) (Lepidoptera: Lycaenidae) on two wet heaths in Argyll, Scotland. *Entomologist's Gazette* 45: 27–28.

Shaw, M. R., Stefanescu, C. and Van Nouhuys, S. (2009). Parasitoids of European butterflies. In *Ecology of Butterflies in Europe*, ed. J. Settele, T. G. Shreeve, M. Konvicka and H. Van Dyck. Cambridge: Cambridge University Press, pp. 130–156.

Shea, K. and Chesson, P. (2002). Community ecology theory as a framework for biological invasions. *Trends in Ecology and Evolution* 17: 170–176.

Sheppard, D. A. (2002). The importance of scrub and bare ground. In *The Species Recovery Programme Proceedings of the 10th Anniversary Conference*, ed. D. Stone, J. Tither and P. Lacey. Peterborough: English Nature, pp. 67–71.

Shields, O. (1967). Hill-topping. An ecological study of summit congregation behavior of butterflies on a southern California hill. *Journal of Research on the Lepidoptera* 6: 69–178.

Shreeve, T. G. (1984). Habitat selection, mate-location, and microclimatic constraints on the activity of the speckled wood butterfly *Pararge aegeria*. *Oikos* 42: 371–377.

Shreeve, T. G. (1985). *The Population Biology of the Speckled Wood Butterfly Pararge aegeria (L.) (Lepidoptera: Satyridae)*. PhD thesis (CNAA), Oxford Polytechnic.

Shreeve, T. G. (1986). Egg-laying by the speckled wood butterfly (*Pararge aegeria*) – the role of female behaviour, host plant abundance and temperature. *Ecological Entomology* 11: 229–236.

Shreeve, T. G. (1987). The mate location behaviour of the speckled wood butterfly, *Pararge aegeria*, and the effect of phenotypic differences in hind-wing spotting. *Animal Behaviour* 35: 682–690.

Shreeve, T. G. (1989). The extended flight period of *Maniola jurtina* (Lepidoptera: Satyridae) on chalk downland: seasonal changes of the adult phenotype and evidence for a population of mixed origins. *The Entomologist* 108: 202–215.

Shreeve, T. G. (1990). Microhabitat use and hindwing phenotype in *Hipparchia semele* (L.) (Lepidoptera: Satyrinae):

thermoregulation and background matching. *Ecological Entomology* 15: 201–213.

Shreeve, T. G. (1992a). Adult behaviour. In *The Ecology of Butterflies in Britain*, ed. R. L. H. Dennis. Oxford: Oxford University Press, pp. 22–45.

Shreeve, T. G. (1992b). Monitoring butterfly movements. In *The Ecology of Butterflies in Britain*, ed. R. L. H. Dennis. Oxford: Oxford University Press, pp. 120–138.

Shreeve T. G. (1995). Butterfly mobility. In *Ecology and Conservation of Butterflies*, ed. A. S. Pullin. London: Chapman and Hall, pp. 37–45.

Shreeve, T. G. and Dennis, R. L. H. (2009). Cross-water transfer by common butterflies. Clear examples of ignoring the highway. *Entomologist's Gazette* 60: 107–109.

Shreeve, T. G., Dennis, R. L. H. and Pullin, A. S. (1996a). How marginal is the British butterfly fauna and what are the implications for research opportunities and conservation? *Biodiversity and Conservation* 5: 1131–1141.

Shreeve, T. G., Dennis, R. L. H. and Williams, W. R. (1996b). Uniformity of wing spotting of *Maniola jurtina* (L.) (Lep. Satyrinae) in relation to environmental heterogeneity. *Nota Lepidopterologica* 18: 77–92.

Shreeve, T. G., Dennis, R. L. H., Roy, D. B. and Moss, D. (2001). An ecological classification of British butterflies: ecological attributes and biotope occupancy. *Journal of Insect Conservation* 5: 145–161.

Shreeve, T. G., Dennis, R. L. H. and Van Dyck, H. (2004). Resources, habitats and metapopulations – whither reality? *Oikos* 106: 404–408.

Shreeve, T. G., Dennis, R. L. H. and Wakeham-Dawson, A. (2006). Phylogenetic, habitat and behavioural aspects of possum behaviour in European Lepidoptera. *Journal of Research on the Lepidoptera* 39: 80–85.

Silberglied, R. E. (1984). Visual communication and sexual selection among butterflies. In *The Biology of Butterflies*, ed. R. I. Vane-Wright and P. R. Ackery. London: Academic Press, pp. 207–224.

Simberloff, D. (1974). Equilibrium theory of island biogeography and ecology. *Annual Review of Ecology and Systematics* 5: 161–182.

Simberloff, D., Farr, J. A., Cox, J. and Mehlman, D. W. (1992). Movement corridors: conservation bargains or poor investments? *Conservation Biology* 6: 493–504.

Simcox, D. and Bourn, N. (2006). The large blue flies again. *Butterfly* 93: 11–13.

Singer, M. C. (1971). Evolution of food-plant preferences in the butterfly *Euphydryas editha*. *Evolution* 25: 383–389.

Singer, M. C. (1972). Complex components of habitat suitability within a butterfly colony. *Science* 176: 75–77.

Singer, M. C. (2004). Measurement, correlates, and importance of oviposition preference in the life of checkerspots. In *On the Wings of Checkerspots. A Model System for Population Biology*, ed. P. R. Ehrlich and I. Hanski. Oxford: Oxford University Press, pp. 112–137.

Singer, M. C. and Lee, J. R. (2000). Discrimination within and between host species by a butterfly: implications for design of preference experiment. *Ecology Letters* 3: 101–105.

Singer, M. C. and Thomas, C. D. (1992). The difficulty of deducing behavior from resource use: an example from hill-topping in checkerspot butterflies. *American Naturalist* 40: 654–664.

Singer, M. C., Ng, D. and Thomas, C. D. (1988). Heritability of oviposition preference and its relationship to offspring performance within a single insect population. *Evolution* 42: 977–985.

Slater, M. (2007). Creation of a drystone wall to create egglaying habitat for grizzled skipper *Pyrgus malvae* at Ryton Wood Meadows Butterfly Conservation Reserve, Warwickshire, England. *Conservation Evidence* 4: 35–40.

Slobodkin, L. B. and Sanders, H. L. (1969). On the contribution of environmental predictability to species diversity. In *Diversity and Stability in Ecological Systems*, eds. G. M. Woodwell and H. H. Smith. Brookhaven Symposia in Biology No. 22, pp. 82–95.

Smith, A. G. (1978). Environmental factors influencing pupal colour determination in Lepidoptera. I. Experiments with *Papilio polytes*, *Papilio demoleus* and *Papilio polyxenes*. *Proceedings of the Royal Society of London B* 200: 295–329.

Smith, A. G. (1980). Environmental factors influencing pupal colour determination in Lepidoptera. II. Experiments with *Pieris rapae*, *Pieris napi* and *Pieris brassicae*. *Proceedings of the Royal society of London B* 207: 163–186.

Smith, R. G. and Clark, M. J. (2005). *Brown Hairstreak (Thecla betulae) in Carmarthenshire: Egg Surveys and Habitat Assessment 2002 to 2004*. Unpublished report to Butterfly Conservation Wales and the Carmarthenshire LBAP. Wareham, UK: Butterfly Conservation.

Snep, R. P. H., Opdam, P. F. M., Baveco, J. M., WallisDeVries, M. F., Timmermans, W., Kwak, R. G. M. and Kuypers, V. (2006). How peri-urban areas can strengthen animal populations within cities: a modeling approach. *Biological Conservation* 127: 345–355.

Southwood, T. R. E. (1961). The number of species of insect associated with various trees. *Journal of Animal Ecology* 30: 1–8.

Southwood, T. R. E. (1973). The insect/plant relationship – an evolutionary perspective. *Symposium of the Royal Entomological Society of London* 6: 3–30.

Southwood, T. R. E. (1977). Habitat, the templet for ecological strategies. *Journal of Animal Ecology* 46: 337–365.

Southwood, T. R. E. (1988). Tactics, strategies and templets. *Oikos* 52: 3–18.

Sparks, B. W. (1972). *Geomorphology*. London: Longman.

Sparks, T. H. (2006). Phenology changes in the UK (and further afield). British Naturalist's Association. *Counrtyside Magazine* Winter 2005/2006: 31–34.

Sparks, T. H. and Parish, T. (1995). Factors affecting the abundance of butterflies in field boundaries in Swavesey fens, Cambridgeshire, UK. *Biological Conservation* 73: 221–227.

Sparks, T. H., Porter, K., Greatorex-Davies, J. N., Hall, M. L. and Marrs, R. H. (1994). The choice of oviposition sites in woodland by the Duke of Burgundy butterfly *Hamearis lucina* in England. *Biological Conservation* 70: 257–264.

Sparks, T. H., Greatorex-Davies, J. N., Mountford, J. O., Hall, M. L. and Marrs, R. H. (1996). The effects of shade on the plant communities of rides in plantation woodland and implication for butterfly conservation. *Forest Ecology and Management* 80: 197–207.

Sparks, T. H., Dover, J. W., Warren, M. S. and Cox, R. (1997). How well can we model the distribution of butterflies at the landscape scale? In *Landscape Ecology: Theory and Application*, ed. G. H. Griffiths. Aberdeen: International Association for Landscape Ecology, pp. 24–31.

Sparks, T. H., Hann, J. P. and Greatorex-Davies, J. N. (1999). The influence of field margin structure on butterflies. *Aspects of Applied Biology* 54: 235–240.

Sparks, T. H., Roy, D. B. and Dennis, R. L. H. (2005). The influence of temperature on migration of Lepidoptera into Britain. *Global Change Biology* 11: 507–514.

Sparks, T. H., Huber, K. and Dennis, R. L. H. (2006). Complex phenological responses to climate warming trends? Lessons from history. *European Journal of Entomology* 103(2): 379–386.

Sparks, T. H., Dennis, R. L. H., Croxton, P. J. and Cade, M. (2007). Increased migration of Lepidoptera linked to climate change. *European Journal of Entomology* 104: 139–143.

Speight, J. G. (1968). Parametric description of land form. In *Land Evaluation*, ed. G. A. Stewart. Melbourne: Macmillan, pp. 239–250.

Spellerberg, I. F. (2005). *Monitoring Ecological Change*. Cambridge: Cambridge University Press.

Stace, C. (1997). *New Flora of the British Isles*, 2nd edn. Cambridge: Cambridge University Press.

Stefanescu, C., Pintureau, B., Tschorsnig, H-P. and Pujade-Villar, J. (2003). The parasitoid complex of the butterfly *Iphiclides podalirius feisthamelii* (Lepidoptera: Papilionidae) in north-east Spain. *Journal of Natural History* 37: 379–396.

Stefanescu, C., Herrando, S. and Paramo, F. (2004). Butterfly species richness in the north-west Mediterranean Basin: the role of natural and human-induced factors. *Journal of Biogeography* 31: 905–915.

Steffan-Dewenter, I. and Tscharntke, T. (1997). Early succession of butterfly and plant communities on set-aside fields. *Oecologia* 109: 294–302.

Steffan-Dewenter, I. and Tscharntke, T. (2000). Butterfly community structure in fragmented habitats. *Ecology Letters* 3: 449–456.

Stern, V. M. and Smith, R. F. (1960). Factors affecting egg production and oviposition in populations of *Colias philodice eurytheme Boisduval* (Lepidoptera: Pieridae). *Hilgardia* 29: 411–454.

Stevens, P. A., Bell, S. A., Brittain, S. A., Hughes, S. and Lowe, J. A. H. (1995). *Soil/plant Interactions in Lowland Grasslands – Great Orme study, Final Report*. CCW Contract Science Report No. 142. Bangor, Wales: Countryside Council for Wales. www.ccw.gov.uk.

Stewart, A. J. A. (2001). The impact of deer on lowland woodland invertebrates: a review of the evidence and priorities for future research. *Forestry* 74: 259–270.

Stewart, G. B. and Pullin, A. S. (2008). The relative importance of grazing stock type and grazing intensity for conservation of mesotrophic 'old meadow' pasture. *Journal of Nature Conservation* 16: 175–185.

Stewart, K. E. J. and Bourn, N. A. D. (2004). The Status, Mobility and Habitat *Requirements of the Small Pearl-bordered Fritillary (Boloria selene) in Clocaenog Forest: an Overview of Three Years Work*. Butterfly Conservation Report No. S04-11. Wareham, UK: Butterfly Conservation.

Stewart, K. E. J., Bourn, N. A. D., Warren, M. S. and Brereton, T. M. (2001). *The Heath Fritillary on Exmoor: Changing Status 1980–2000 and Conservation Recommendations*. Butterfly Conservation Report No. S01-30. Wareham, UK: Butterfly Conservation.

Stouffer, S. A. (1940). Intervening opportunities: a theory relating mobility and distance. *American Sociological Review* 5: 845–867.

Sugden, D. E., Glasser, N. and Clapperton, C. M. (1992). Evolution of large roches moutonnees. *Geografiska Annaler, Series A, Physical Geography* 74: 253–264.

Summerville, K. S. and Crist, T. O. (2001). Effects of experimental habitat fragmentation on patch use by butterflies and skippers (Lepidoptera). *Ecology* 82: 1360–1370.

Summerville, K. S., Veech, J. A. and Crist, T. O. (2002). Does variation in patch use among butterfly species contribute to nestedness at fine spatial scales? *Oikos* 97: 195–204.

Sutcliffe, O. L. and Thomas, C. D. (1996). Open corridors appear to facilitate dispersal by ringlet butterflies (*Aphantopus hyperantus*) between woodland clearings. *Conservation Biology* 10: 1359–1365.

Sutcliffe, O. L., Thomas, C. D. and Peggie, D. (1997a). Area-dependent migration by ringlet butterflies generates a mixture of patchy population and metapopulation attributes. *Oecologia* 109: 229–234.

Sutcliffe, O. L., Thomas, C. D., Yates, T. J. and Greatorex-Davies, J. N. (1997b). Correlated extinctions, colonizations and population fluctuations in a highly connected ringlet butterfly metapopulation. *Oecologia* 109: 235–241.

Sutherland, W. J., Pullin, A. S., Dolman, P. M. and Knight, T. M. (2004). The need for evidence-based conservation. *Trends in Ecology and Evolution* 19: 305–308.

Tansley, A. G. (1916). The development of vegetation. *Journal of Ecology* 4: 198–204.

Tansley, A. G. (1935). The use and abuse of vegetation concepts and terms. *Ecology* 16: 284–307.

Tansley, A. G. (1939). *The British Islands and their Vegetation*. Cambridge: Cambridge University Press, 2 vols.

Taylor, P. D., Fahrig, L., Henein, K. and Merriam, G. (1993). Connectivity is a vital element of landscape structure. *Oikos* 68: 571–573.

Tennent, W. J. (1995). Field observations of the 'hill-topping' phenomenon in northwest Africa – and an introduction to 'ravining' (Lep: Rhopalocera). *Entomologist's Record and Journal of Variation* 107: 57–67.

Ter Braak, C. F. J. and Smilauer, P. (2002). *CANOCO Reference Manual and CanoDraw for Windows User's Guide: Software for Canonical Community Ordination*, Version 4.5. Ithaca, NY: Microcomputer Power.

Tews, J., Brose, U., Grimm, V., Tielborger, K., Wichmann, M. C., Schwager, M. and Jeltsch, F. (2004). Animal species diversity driven by habitat heterogeneity/diversity: the importance of keystone structures. *Journal of Biogeography* 31: 79–92.

Thomas, C. D. (1983). *The Ecology and Status of Plebejus argus L. in North-West Britain*. MSc thesis, University College of North Wales, Bangor.

Thomas, C. D. (1985a). Specializations and polyphagy of *Plebejus argus* (Lepidoptera: Lycaenidae) in North Wales. *Ecological Entomology* 10: 325–340.

Thomas, C. D. (1985b). The status and conservation of the butterfly *Plebejus argus* L. (Lep: Lycaenidae) in North Wales. *Biological Conservation* 33: 29–51.

Thomas, C. D. and Abery, J. C. G. (1995). Estimating rates of butterfly decline from distribution maps: the effect of scale. *Biological Conservation* 73: 59–65.

Thomas, C. D. and Hanski, I. A. (1997). Butterfly metapopulations. In *Metapopulation Biology. Ecology, Genetics, and Evolution*, ed. I. A. Hanski and M. E. Gilpin. London: Academic Press, pp. 359–386.

Thomas, C. D. and Harrison, S. (1992). Spatial dynamics of a patchily distributed butterfly species. *Journal of Animal Ecology* 61: 437–446.

Thomas, C. D. and Jones, C. D. (1993). Partial recovery of skipper butterfly (*Hesperia comma*) from population refuges: lessons for conservation in a fragmented landscape. *Journal of Animal Ecology* 62: 472–481.

Thomas, C. D. and Kunin, W. E. (1999). The spatial structure of populations. *Journal of Animal Ecology* 68: 647–657.

Thomas, C. D., Simcox, D. J. and Clarke, R. T. (1986). Ecology and declining status of the silver-spotted skipper butterfly (*Hesperia comma*) in Britain. *Journal of Applied Ecology* 23: 365–380.

Thomas, C. D., Warren, M. S. and Thomas, J. A. (1992). Distributions of occupied and vacant butterfly habitats in fragmented landscapes. *Oecologia* 92: 563–567.

Thomas, C. D., Hill, J. K. and Lewis, O. T. (1998). Evolutionary consequences of habitat fragmentation in a localized butterfly. *Journal of Animal Ecology* 67: 485–97.

Thomas, C. D., Glen, S. W. T., Lewis, O. T., Hill, J. K. and Blakeley, D. S. (1999). Population differentiation and conservation of endemic races: the butterfly *Plebejus argus*. *Animal Conservation* 2: 15–21.

Thomas, C. D., Bodsworth, E. J., Wilson, R. J., Simmons, A. D., Davies, Z. G., Musche, M. and Conradt, L. (2001). Ecological and evolutionary processes at expanding range margins. *Nature* 411: 577–581.

Thomas, C. D., Cameron, A., Green, R. E. *et al.* (2004). Extinction risk from climate change. *Nature* 427: 145–148.

Thomas, J. A. (1974). *Factors Influencing the Numbers and Distribution of the Brown Hairstreak, Thecla betulae, and Black Hairstreak, Styrymonidia pruni*. PhD thesis, University of Leicester.

Thomas, J. A. (1975a). The ecology of the brown wairstreak butterfly. *Report of Insitute of Terrestrial Ecology* 1974: 24–25.

Thomas, J. A. (1975b). Some observations of the early stages of the purple hairstreak butterfly, *Quercusia quercus* (Linnaeus) (Lep., Lycaenidae). *Entomologist's Gazette* 26: 224–226.

Thomas, J. A. (1975c). *The Black Hairstreak: Conservation Report*. Unpublished report by the Institute of Terrestrial Ecology and the Nature Conservancy Council.

Thomas, J. A. (1976). *The Ecology and Conservation of the Large Blue Butterfly Maculinea arion L.* Unpublished report to the Institute of Terrestrial Ecology.

Thomas, J. A. (1977a). The ecology of the large blue butterfly. In *Institute of Terrestrial Ecology Annual Report 1976*. London: HMSO, pp. 25–28.

Thomas, J. A. (1977b). *Second Report on the Large Blue Butterfly*. Unpublished report to the Institute of Terrestrial Ecology.

Thomas, J. A. (1978). Insect conservation in Britain: some case histories. *Atala* 6: 31–36.

Thomas, J. A. (1980a) Why did the large blue become extinct in Britain. *Oryx, Journal of the Fauna Preservation Society* XV: 243–247.

Thomas, J. A. (1980b). The extinction of the large blue and the conservation of the black hairstreak butterflies (a contrast of failure and success). In *Institute of Terrestrial Ecology Annual Report 1979*. London: HMSO, pp. 19–23.

Thomas, J. A. (1983a). The ecology and conservation of *Lysandra bellargus* (Lepidoptera; Lycaenidae) in Britain. *Journal of Applied Ecology* 20: 59–83.

Thomas, J. A. (1983b). A 'WATCH' census of common British butterflies. *Journal of Biological Education* 17: 333–338.

Thomas, J. A. (1983c). The ecology and status of *Thymelicus acteon* (Lepidoptera: Hesperiidae) in Britain. *Ecological Entomology* 8: 427–435.

Thomas, J. A. (1984). The conservation of butterflies in temperate countries: past efforts and lessons for the future. In *The Biology of Butterflies*, ed. R. I. Vane-Wright and P. R. Ackery. London: Academic Press, pp. 333–353.

Thomas, J. A. (1993). Holocene climate changes and warm man-made refugia may explain why a sixth of British butterflies possess unnatural early-successional habitats. *Ecography* 16: 278–284.

Thomas, J. A. (1994). Why small cold-blooded insects pose different conservation problems to birds in modern landscapes. *Ibis* 137: 112–119.

Thomas, J. A. (1995). The ecology and conservation of *Maculinea arion* and other European species of butterfly. In *The Ecology and Conservation of Butterflies*, ed. A. S. Pullin. London: Chapman and Hall, pp. 180–97.

Thomas, J. A. (2005). Monitoring change in the abundance and distribution of insects using butterflies and other indicator groups. *Philosophical Transactions of the Royal Society B* 360: 339–357.

Thomas, J. A. and Elmes, G. W. (1993). Specialised searching and the hostile use of allomones by a parasitoid whose host, the butterfly *Maculinea rebeli*, inhabits ant nests. *Animal Behaviour* 45: 593–602.

Thomas, J. A. and Elmes, G. W. (1998). Higher productivity at the cost of increased host-specificity when *Maculinea* butterfly larvae exploit ant colonies through trophallaxis rather than by predation. *Ecological Entomology* 23: 457–464.

Thomas, J. A. and Lewington, R. (1991). *The Butterflies of Britain and Ireland.* London: Dorling Kindersley.

Thomas, J. A. and Simcox, D. (1982). A quick method for estimating larval populations of *Melitaea cinxia* L. during surveys. *Biological Conservation* 22: 315–322.

Thomas, J. A. and Wardlaw, J. C. (1990). The effect of queen ants on the survival of *Maculinea arion* larvae in *Myrmica* ant nests. *Oecologia* 85: 87–91.

Thomas, J. A. and Wardlaw, J. C. (1992). The capacity of a *Myrmica* ant nest to support a predacious species of *Maculinea* butterfly. *Oecologia* 91: 101–109.

Thomas, J. A., Elmes, G. W., Wardlaw, J. C. and Woyciechowski, M. (1989). Host specificity among *Maculinea* butterflies in *Myrmica* ant nests. *Oecologia* 79: 452–457.

Thomas, J. A., Moss, D. and Pollard, E. (1994). Increased fluctuations of butterfly populations towards the northern edges of species' ranges. *Ecography* 17: 215–220.

Thomas, J. A., Clarke, R. T., Elmes, G. W. and Hochberg, M. E. (1998a). Population dynamics in the genus *Maculinea* (Lepidoptera: Lycaenidae). In *Insect Populations: In Theory and Practice*, ed. J. P. Dempster and I. F. McLean. Symposium of the Royal Entomological Society 19. London: Chapman and Hall, pp. 261–290.

Thomas, J. A., Simcox, D. J., Wardlaw, J. C., Elmes, G. W., Hochberg, M. E. and Clarke, R. T. (1998b). Effects of latitude, altitude and climate on the habitat and conservation of the endangered butterfly *Maculinea arion* and its *Myrmica* ant hosts. *Journal of Insect Conservation* 2: 39–46.

Thomas, J. A., Rose, R. J., Clarke, R. T., Thomas, C. D. and Webb, N. R. (1999). Intraspecific variation in habitat availability among ectotherm animals near their climatic limits and their centres of range. *Functional Ecology* 13: 55–64.

Thomas, J. A., Bourn, N. A. D., Clarke, R. T. *et al.* (2001). The quality and isolation of habitat patches both determine where butterflies persist in fragmented landscapes. *Proceedings of the Royal Society of London Series B* 268: 1791–1796.

Thomas, J. A., Telfer, M. G., Roy, D. B. *et al.* (2004). Comparative losses of British butterflies, birds, and plants and the global extinction crisis. *Science* 303: 1879–1881.

Thomas, J. A., Simcox, D. J. and Clarke, R. T. (2009). Successful conservation of a threatened *Maculinea* butterfly. *Science* 325: 80–83.

Thompson, D. (ed.) (1995). *The Concise Oxford Dictionary*, 9th edn. Oxford: Clarendon Press.

Thompson, J. N. (1988). Evolutionary ecology of the relationship between oviposition preference and performance of offspring in phytophagous insects. *Entomologia Experimentalis Applicata* 47: 3–14.

Thomson, G. (1987). *Enzyme Variation at Morphological Boundaries in Maniola and related Genera.* PhD thesis, University of Stirling.

Tilman, D. (1982). *Resource Competition and Community Structure.* Monographs in Population Biology No. 17. Princeton, NJ: Princeton University Press.

Tinbergen, N. (1972). The courtship of the grayling *Eumenis* (*Satyrus*) *semele* L. (Lep., Satyridae). In *The Animal and its World*, Vol. 1, ed. Tinbergen, N. Field Studies 1932–1972. London: Allen and Unwin, pp. 197–249.

Tiple, A. D., Deshmukh, V. P. and Dennis, R. L. H. (2005). Factors influencing nectar plant resource visits by butterflies on a university campus: implications for conservation. *Nota Lepidopterologica* 28: 213–224.

Tiple, A. D., Khurada, A. M. and Dennis, R. L. H. (2009). Adult butterfly feeding–nectar flower associations: constraints of taxonomic affiliation, butterfly, and nectar flower morphology. *Journal of Natural History* 43: 855–884.

Tischendorf, L. (2001). Can landscape indices predict ecological processes consistently? *Landscape Ecology* 16: 235–254.

Townsend, P. A. and Levey, D. J. (2005). An experimental test of whether habitat corridors affect pollen transfer. *Ecology* 86: 466–475.

Traub, B. and Kleinn, C. (1999). Measuring fragmentation and structural diversity. *Forstwissenschaftliches Centralblatt* 118: 39–50.

Troels-Smith, J. (1960). Ivy, mistletoe and elm decline climatic indicators – fodder plants. *Danmarks Geologiske Undersogelse* 4: 1–32.

Troll, C. (1939). Luftbildplan und okologische Bodenforschung. *Zeitschrift der Gesellschaft für Erdkunde zu Berlin* 74: 241–298.

Trudgill, S. T. (1977). *Soil and Vegetation Systems.* Oxford: Clarendon Press.

Tscharntke, T., Steffan-Dewenter, I., Kruess, A. and Thies, C. (2002). Contribution of small habitat fragments to conservation of insect communities of grassland–cropland landscapes. *Ecological Applications* 12: 354–363.

Tucker, J. M. (1991). *Problems Posed by the Hibernation and Migration Habits of the Red Admiral Butterfly (Vanessa atalanta).* Butterfly Conservation Occasional Paper No. 1. Colchester: Butterfly Conservation.

Tudor, O., Dennis, R. L. H., Greatorex-Davies, J. N. and Sparks, T. H. (2004). Flower preferences of woodland butterflies in the UK: nectaring specialists are species of conservation concern. *Biological Conservation* 119: 397–403.

Turchin, P. (1989). Beyond simple diffusion: models of not-so-simple movement of animals and cells. *Comments on Theoretical Biology* 1: 65–83.

Turlure, C., Van Dyck, H., Schtickzelle, N. and Baguette, M. (2009). Resource-based habitat definition, niche overlap and conservation of two sympatric glacial relict butterflies. Oikos 118: 950–960.

Turner, E. C., Granroth, H. M. V., Johnson, H. R. *et al.* (2009). Habitat preference and dispersal of the Duke of Burgundy butterfly (*Hamearis lucina*) on an abandoned chalk quarry in Bedfordshire, UK. *Journal of Insect Conservation* 13: 475–486.

Turner, J. R. G. (1963). A quantitative study of a Welsh colony of the large heath butterfly, *Coenonympha tullia Müller* (Lepidoptera). *Proceedings of the Royal Entomological Society of London A* 38: 101–112.

Turner, J. R. G. (1986). Why are there so few butterflies in Liverpool? Homage to Alfred Russel Wallace. *Antenna* 10: 18–24.

Turner, J. R. G., Gatehouse, C. M. and Corey, C. A. (1987). Does solar energy control organic diversity? Butterflies, moths and the British climate. *Oikos* 48: 195–205.

Turner, M. G. (ed.) (1987). *Landscape Heterogeneity and Disturbance*. New York: Springer-Verlag.

Turner, M. G. (1989). Landscape ecology: the effect of pattern on process. *Annual Review of Ecology and Systematics* 20: 171–197.

Underhill-Day, J. C. (2005). *A Literature Review of Urban Effects on Lowland Heaths and their Wildlife*. English Nature Research Report No. 623. Peterborough: English Nature.

UNEP (2003). *Proposed Biodiversity Indicators Relevant to the 2010 Target*. UNEP/CBD/SBSTTA/9/INF/26. Montreal: United Nations Environment Programme.

United Nations (2002). *Report of the World Summit on Sustainable Development*. A/CONF.199/20. Johannesburg: United Nations.

Valimaki, P. and Itamies, J. (2003). Migration of the clouded Apollo butterfly *Parnassius mnemosyne* in a network of suitable habitats – effects of patch characteristics. *Ecography* 26: 679–691.

Van Dyck, H. (2003). Mate location: a matter of design? Adaptive morphological variation in the speckled wood butterfly. In *Butterflies. Ecology and Evolution Taking Flight*, ed. C. L. Boggs, W. B. Watt and P. R. Ehrlich. Chicago: Chicago University Press, pp. 353–366.

Van Dyck, H. and Baguette, M. (2005). Dispersal behaviour in fragmented landscapes: routine or special movements? *Basic and Applied Ecology* 6: 535–545.

Van Dyck, H., Matthysen, E. and Dhondt, A. A. (1997). Mate-locating strategies are related to relative body length and wing colour in the speckled wood butterfly *Pararge aegeria*. *Ecological Entomology* 22: 116–120.

Van Dyck, H., Oostermeijer, J. G. B., Talloen, W., Feenstra, V., van der Hidde, A. and Wynhoff, I. (2000). Does the presence of ant nests matter for oviposition to a specialised myrmecophilous *Maculinea* butterfly. *Proceedings of the Royal Society of London B* 267: 861–866.

Van Kleunen, M. and Johnson, S. D. (2007). Effects of self-compatibility on the distribution range of invasive European plants in North America. *Conservation Biology* 21: 1537–1544.

Van Nouhuys, S. and Hanski, I. (2004). Natural enemies of checkerspots. In *On the Wings of Checkerspots. A Model System for Population Biology*, ed. P. R. Ehrlich and I. Hanski. Oxford: Oxford University Press, pp. 161–180.

Van Straalen, N. M. (1998). Evaluation of bio-indicator systems derived from soil arthropod communities. *Applied Soil Ecology* 9: 429–437.

Van Swaay, C. A. M. and Warren, M. S. (1999). *Red Data Book of European Butterflies (Rhopalocera)*. Nature and Environment No. 99. Strasbourg: Council of Europe Publishing.

Vandermeer, J. H. (1972). Niche theory. *Annual Review of Ecology and Systematics* 3: 107–132.

Vane-Wright, R. I. and Boppré, M. (1993). Visual and chemical signaling in butterflies – functional and phylogenetic perspectives. *Philosophical Transactions of the Royal Society of London B* 340: 197–205.

Vanreusel, W. and Van Dyck, H. (2007). When functional habitat does not match vegetation types: a resource-based approach to map butterfly habitat. *Biological Conservation* 135: 202–211.

Vanreusel, W., Maes, D. and Van Dyck, H. (2006). Transferability of species distribution models: a functional habitat approach for two regionally threatened butterflies. *Conservation Biology* 21: 201–212.

Varley, G. C., Gradwell, G. R. and Hassell, M. P. (1973). *Insect Population Ecology, an Analytic Approach*. Oxford: Blackwell Science.

Vera, F. W. M. (2000). *Grazing Ecology and Forest History*. Wallingford, UK: CABI Publishing.

Vincent, P. (1990). *The Biogeography of the British Isles. An Introduction*. London: Routledge.

Von Bertalanffy, L. (1951). An outline of general systems theory. *British Journal for the Philosophy of Science* 1: 134–165.

Von Bertalanffy, L. (1962). General systems theory: a critical review. *General Systems* 7: 1–20.

Vos, C. C., Verboom, J., Opdam, P. F. M. and Ter Braak, C. J. F. (2001). Toward ecologically scaled landscape indices. *American Naturalist* 157: 24–41.

Vulliamy, B. and Bulman, C. R. (2009). *Monitoring the Population and Dispersal of the Heath Fritillary (Melitaea athalia) in Greenscoombe Wood SSSI in 2008, Two Years after Reintroduction*. Butterfly Conservation Report No. S08-28. Wareham, UK: Butterfly Conservation.

Wacher, J. (1998). Successful overwintering of painted lady *Cynthia cardui* in the UK. *Atropos* 5: 19–20.

Wainwright, D. (2005a). *Conservation and Habitat Requirements of the Large Heath Butterfly (Coenonympha tullia)*. PhD thesis, University of Sunderland.

Wainwright, D. (2005b). *Conserving the Dingy Skipper Erynnis tages Butterfly in North East England*. Butterfly Conservation Report No. S05-34. Wareham, UK: Butterfly Conservation.

Wainwright, D. and Ellis, S. (2004). *Status of the Large Heath Coenonympha tullia on the North York Moors*. Confidential

Butterfly Conservation Report No. S04-34. Lulworth, UK: Butterfly Conservation.

WallisDeVries, M. F. and Van Swaay, C. A. M. (2006). Global warming and excess nitrogen may induce butterfly decline by microclimatic cooling. *Global Change Biology* 12: 1620–1626.

Waring, P. (2006). Caterpillars of magpie moth *Abraxas grossulariata* (L.) abundant at the Ring of Brogar, Orkney Isles, 1st June 2005. *Bulletin of the Amateur Entomologists' Society* 65: 229–231.

Warren, M. S. (1984a). The future of the heath fritillary in Britain. *Butterfly Conservation News* 32: 19–30.

Warren, M. S. (1984b). The biology and status of the wood white butterfly, *Leptidea sinapis* (L.) (Lepidoptera: Pieridae), in the British Isles. *Entomologist's Gazette* 35: 207–223.

Warren, M. S. (1985a). The ecology and conservation of the heath fritillary butterfly, *Mellicta athalia*. Unpublished report for the Nature Conservancy Council.

Warren, M. S. (1985b). The status of the heath fritillary butterfly, *Mellicta athalia* Rott in relation to changing woodland management in the Blean Woods, Kent. *Quarterly Journal of Forestry* 19: 174–182.

Warren, M. S. (1985c). The influence of shade on butterfly numbers in woodland rides, with special reference to the wood white *Leptidea sinapis*. *Biological Conservation* 33: 147–164.

Warren, M. S. (1987a). The ecology and conservation of the heath fritillary butterfly, *Mellicta athalia*. I. Host selection and phenology. *Journal of Applied Ecology* 24: 467–482.

Warren, M. S. (1987b). The ecology and conservation of the heath fritillary butterfly, *Mellicta athalia*. II. Adult population structure and mobility. *Journal of Applied Ecology* 24: 483–498.

Warren, M. S. (1987c). The ecology and conservation of the heath fritillary butterfly, *Mellicta athalia*. III. Population dynamics and the effect of habitat management. *Journal of Applied Ecology* 24: 499–513.

Warren, M. S. (1991). The successful conservation of an endangered species, the heath fritillary butterfly *Mellicta athalia*, in Britain. *Biological Conservation* 55: 37–56.

Warren, M. S. (1992a). Butterfly populations. In *Ecology of Butterflies in Britain*, ed. R. L. H. Dennis. Oxford: Oxford University Press, pp. 73–92.

Warren, M. S. (1992b). The conservation of British Butterflies. In *Ecology of Butterflies in Britain*, ed. R. L. H. Dennis. Oxford: Oxford University Press, pp. 246–274.

Warren, M. S. (1993). A review of butterfly conservation in central southern Britain: II. Site management and habitat selection of key species. *Biological Conservation* 64: 37–49.

Warren, M. S. (1994). The UK status and suspected metapopulation structure of a threatened European butterfly, the marsh fritillary, *Eurodryas aurinia*. *Biological Conservation* 67: 239–249.

Warren, M. S. (1995). Managing local micro-climates for the high brown fritillary. In *The Ecology and Conservation of Butterflies*, ed. A. S. Pullin. London: Chapman and Hall, pp. 198–210.

Warren, M. S. and Oates, M. R. (1995). The importance of bracken habitats to fritillary butterflies and their management for conservation. In *Bracken: an Environmental Issue*, ed. R. T. Smith and J. A. Taylor. Leeds: International Bracken Group, pp. 178–181.

Warren, M. S. and Stephens, D. E. A. (1989). Habitat design and management for butterflies. *The Entomologist* 108: 123–134.

Warren, M. S., Thomas, C. D. and Thomas, J. A. (1984). The status of heath fritillary butterfly, *Mellicta athalia* Rott., in Britain. *Biological Conservation* 29: 287–305.

Warren, M. S., Pollard, E. and Bibby, T. J. (1986). Annual and long-term changes in a population of the wood white butterfly *Leptidea sinapis*. *Journal of Animal Ecology* 55: 707–720.

Warren, M. S., Thomas, J. A. and Wilson, R. J. (1999). *Management Options for the Silver-spotted Skipper Butterfly: a Study of the Timing of Grazing at Beacon Hill NNR, Hampshire, 1986–98*. Wareham, UK: Butterfly Conservation.

Warren, M. S., Clarke, S. and Currie, F. (2001a). The coppice for butterflies challenge: a targeted grant scheme for threatened species. *British Wildlife* 13: 21–28.

Warren, M. S., Hill, J. K., Thomas, J. A. *et al.* (2001b). Rapid responses of British butterflies to opposing forces of climate and habitat change. *Nature* 414: 65–69.

Warrington, S. and Brayford, J. P. (1995). Some aspects of the population ecology and dispersal of the small skipper butterfly *Thymelicus sylvestris* in a series of linked grasslands. *The Entomologist* 114: 201–209.

Watkinson, A. R. and Sutherland, W. J. (1995). Sources, sinks and pseudo-sinks. *Journal of Animal Ecology* 64: 126–130.

Watt, A. S. (1947). Pattern and process in the plant community. *Journal of Ecology* 35: 1–22.

Waugh, D. (1990). *Geography. An Integrated Approach*. Edinburgh: Thomas Nelson and Sons Ltd.

Webb, J. R. and Lott, D. A. (2006). The development of ISIS: a habitat-based invertebrate assemblage classification system for assessing conservation interest in England. *Journal of Insect Conservation* 10: 179–188.

Webb, M. R. and Pullin, A. S. (2000). Egg distribution in the large copper butterfly *Lycaena dispar batavus* (Lepidoptera: Lycaenidae): host plant versus habitat mediated effects. *European Journal of Entomology* 97: 363–367.

Webb, N. R. (1993). Heathland fragmentation and the potential for expansion. In *Landscape Ecology in Britain*, ed. R. Haines-Young. Nottingham: Nottingham University, pp. 49–54.

Weddell, B. J. (2002). *Conserving Living Natural Resources*. Cambridge: Cambridge University Press.

Wedell, N., Nylin, S. and Janz, N. (1997). Effects of larval host plant and sex on the propensity to enter diapause in the comma butterfly. *Oikos* 78: 569–575.

Weiss, M. R. (2001). Vision and learning in some neglected pollinators: beetles, flies, moths, and butterflies. In *Cognitive*

Ecology of Pollination – Animal Behaviour and Floral Evolution, ed. L. Chittka, J. D. Thomson. Cambridge: Cambridge University Press, pp. 171–190.

Weiss, M. R. and Papaj, D. R. (2003). Colour learning in two behavioural contexts: how much can a butterfly keep in mind? *Animal Behaviour* 65: 425–434.

Weiss, S. B., Murphy, D. D. and White, R. R. (1988). Sun, slope, and butterflies: topographic determinants of habitat quality for *Euphydryas editha*. *Ecology* 69: 1486–1496.

Wells, T. C. E. (1985). The botanical and ecological interest of ancient monuments. In *Archaeology and Nature Conservation*, ed. G. Lambrick. Oxford: Oxford University Department of External Studies, pp. 1–9.

West, R. G. (1969). *Pleistocene Geology and Biology with Especial Reference to the British Isles*. London: Longmans.

White, A., Watt, A. D., Hails, R. S. and Hartley, S. E. (2000). Patterns of spread in insect–pathogen systems: the importance of pathogen dispersal. *Oikos* 89: 137–145.

White, E. J. and Perry, A. H. (1989). Classification of the climate of England and Wales based on agricultural data. *International Journal of Climatology* 9: 271–291.

White, E. J. and Smith, R. I. (1982). *Climatological Maps of Great Britain*. Cambridge: ITE, Natural Environment Research Council.

Whitfield, K. E. J. (1999). *An Investigation of the Status and Recolonisation Patterns of Adonis Blue Butterfly Colonies in the South Wessex Downs*. MSc thesis, University of Warwick.

Whittaker, R. H. (1975). *Communities and Ecosystems*. New York: Macmillan.

Whittaker, R. H., Levin, S. A. and Root, R. B. (1973). Niche, habitat, and ecotope. *American Naturalist* 107: 321–338.

Whittow, J. (1984). *Dictionary of Physical Geography*. London: Penguin.

Wickham, J. D., O' Neill, R. V., Ritters, K. H., Wade, T. G. and Jones, K. B. (1997). Sensitivity of selected landscape pattern metrics to land-cover misclassification and differences in land-cover composition. *Photogrammetric Engineering and Remote Sensing* 63: 397–402.

Wickman, P-O. (1988). Dynamics of mate searching behaviour in a hilltopping butterfly *Lasiommata megera* (L.). The effects of weather and male density. *Zoological Journal of the Linnean Society* 93: 357–377.

Wickman, P-O. (1992). Sexual selection and butterfly design – a comparative study. *Evolution* 46: 1525–1536.

Wickman, P-O. (2009). Mating behaviour in butterflies. In *Ecology of Butterflies in Europe*, ed. J. Settele, T. G. Shreeve, M. Konvicka and H. Van Dyck. Cambridge: Cambridge University Press, pp. 17–28.

Wiens, J. A. (1989). Spatial scaling in ecology. *Functional Ecology* 3: 385–397.

Wiens, J. A. (1996). Wildlife in patchy environments: metapopulations, mosaics, and management. In *Metapopulations and Wildlife Conservation*, ed. D. R. McCullough. Washington DC: Island Press, pp. 53–84.

Wiens, J. A. (1997). Metapopulation dynamics and landscape ecology. In *Metapopulation Biology. Ecology, Genetics, and Evolution*, ed. I. A. Hanski and M. E. Gilpin. London: Academic Press, pp. 43–62.

Wiklund, C. (1973). Host plant suitability and the mechanism of host selection in larvae of *Papilio machaon*. *Entomologia Experimentalis et Applicata* 16: 232–242.

Wiklund, C. (1974a). Oviposition preferences in *Papilio machaon* in relation to the host plants of the larvae. *Entomologia Experimentalis et Applicata* 17: 189–198.

Wiklund, C. (1974b). The concept of oligophagy and the natural habitats and host plants of *Papilio machaon* L. in Fennoscandia. *Entomologica Scandinavica* 5: 151–160.

Wiklund, C. (1975). The evolutionary relationship between adult oviposition preferences and larval host plant range in *Papilio machaon* L. *Oecologia* 18: 185 197.

Wiklund, C. (1977). Oviposition, feeding and spatial separation of breeding and foraging habitats in a population of *Leptidea sinapis* (Lepidoptera). *Oikos* 28: 56–68.

Wiklund, C. (1978). Host plants, nectar source plants, and habitat selection of males and females of *Anthocharis cardamines* (Lepidoptera). *Oikos* 31: 169–183.

Wiklund, C. (1981). Generalist vs. specialist oviposition behaviour in *Papilio machaon* (Lepidoptera) and functional aspects on the hierarchy of oviposition preferences. *Oikos* 36: 163–170.

Wiklund, C. (1984). Egg laying patterns in butterflies in relation to their phenology and the visual apparency and abundance of their host plants. *Oecologia* 63: 23–29.

Wiklund, C. (1985). Male determined mating duration in butterflies? *Journal of the Lepidopterists' Society* 39: 341–342.

Wiklund, C. and Åhrberg, C. (1978). Hostplants, nectar source plants, and habitat selection of males and females of *Anthocharis cardamines* (Lepidoptera). *Oikos* 31: 169–183.

Wiklund, C. and Karlsson, B. (1984). Egg size variation in satyrid butterflies: adaptive versus non historical 'Bauplan', and mechanistic explanations. *Oikos* 43: 391–400.

Wilcockson, A. (2002). *The Functional Significance of Wing Morphology Variation in the Green-Veined White Butterfly (Pieris napi (L.))*. PhD thesis, Oxford Brookes University.

Williams, C. B. (1958). *Insect Migration*. London: Collins. (Second edition, 1965.)

Williams, C. L. and Bourn, N. A. D. (2008). *The Status, Mobility and Habitat Requirements of the Small Pearl-bordered Fritillary (Boloria selene) in Clocaenog Forest*. Wareham, UK: Butterfly Conservation.

Williamson, M. H. (1983). *Island Populations*. Oxford, Oxford University Press.

Willmott, K. J. (1985). A survey of Glanville fritillary roosting sites. *News of the British Butterfly Conservation Society* 35: 35–36.

Willmott, K. J. (1987). *The Ecology and Conservation of the Purple Emperor butterfly* (*Apatura iris*). Confidential Report on Project BSP/2 for the World Wildlife Fund 1984–1986. v + 140 pp, 5 appendices.

Willmott, K. J. (1990). *The Purple Emperor Butterfly*. Colchester: Butterfly Conservation.

Willmott, K. J. (1994). Locating and conserving the elusive purple emperor. *British Wildlife* 5: 288–295.

Willmott, K. J. (1999). *The Holly Blue Butterfly*. Colchester: Butterfly Conservation.

Wilson, A. (1985). Flavonoid pigments in the marbled white butterfly (*Melanargia galathea*) are dependent on flavonoid content of larval diet. *Journal of Chemical Ecology* 11: 1161–1179.

Wilson, J. D., Morris, A. J., Arroyo, B. E., Clark, S. C. and Bradbury, R. B. (1999). A review of the abundance and diversity of invertebrate and plant foods of granivorous birds in Northern Europe in relation to agricultural change. *Agriculture, Ecosystems and Environment* 75: 13–30.

Wilson, J. P. and Gallant, J. C. (eds) (2000). *Terrain Analysis – Principles and Applications*. New York: Wiley.

Wilson, R. J. (1999). *The Spatiotemporal Dynamics of Three Lepidopteran Herbivores of Helianthemum chamaecistus*. PhD thesis, University of Leeds.

Wilson, R. J., Ellis, S., Baker, J. S., Lineham, M. E., Whitehead, R. W. and Thomas, C. D. (2002). Large-scale patterns of distribution and persistence at the range margins of a butterfly. *Ecology* 83: 3357–3368.

Wilson, R. J., Thomas, C. D., Fox, R., Roy, D. B. and Kunin, W. E. (2004). Spatial patterns in species distributions reveal biodiversity change. *Nature* 432: 393–396.

Wolda, H. and Dennis, B. (1993). Density-dependence tests, are they? *Oecologia* 95: 581–591.

Wolda, H., Dennis, B. and Taper, M. L. (1994). Density-dependence tests and largely futile comments: answers to Holyoak and Lawton (1993) and Hanski, Woiwod and Perry (1993). *Oecologia* 98: 229–234.

Wood, B. C. and Pullin, A. S. (2002). Persistence of species in a fragmented landscape: the importance of dispersal ability and habitat availability for grassland butterflies. *Biodiversity and Conservation* 11: 1451–1468.

Woodcock, N. (1994). *Geology and Environment in Britain and Ireland*. London: UCL Press.

Wooldridge, S. W. and Linton, D. L. (1955). *Structure, Surface and Drainage in South-East England*. London: George Philip.

Wright, S. (1940). Breeding structure of populations in relation to speciation. *American Naturalist* 74: 232–248.

Wyatt, B. K., Greatorex-Davies, N. G., Bunce, R. G. H., Fuller, R. M. and Hill, M. O. (1994). *Comparison of Land Cover Definitions*. Countryside 1990 Series, Vol. 3. London: Department of the Environment.

Wynhoff, I., Oostermeijer, J. G. B., Scheper, M. and Van der Made, J. G. (1996). Effects of habitat fragmentation on the butterfly *Maculinea alcon* in the Netherlands. In *Species Survival in Fragmented Landscapes*, ed. J. Settele, C. R. Margules, R. Poschlod and K. Henle. Dordretch: Kluwer Academic Publishers, pp. 15–23.

Yip, J. Y., Corlett, R. T. and Dudgeon, D. (2004). A fine-scale gap analysis of the existing protected area system in Hong Kong, China. *Biodiversity and Conservation* 13: 943–957.

Yip, J. Y., Corlett, R. T. and Dudgeon, D. (2006). Selecting small reserves in a human-dominated landscape: a case study of Hong Kong, China. *Journal of Environmental Management* 78: 86–96.

Zalucki, M. P. (1993). Sex around the milkweed patch – the significance of patches of host plants in monarch reproduction. In *Biology and Conservation of the Monarch Butterfly*, ed. S. B. Malcolm and M. P. Zalu. Los Angeles: Natural History Museum of Los Angeles County, pp. 69–76.

Zonneveld, I. S. (2005). The land unit as a black box: a Pandora's box? In *Issues and Perspectives in Landscape Ecology*, ed. J. Wiens and M. Moss. Cambridge: Cambridge University Press, pp. 331–345.

INDEX

Note: page numbers in *italics* refer to figures, those in **bold** refer to tables and boxes.